Peanut Genomics and Biotechnology

In a world where food security and sustainable agriculture face unprecedented challenges, the cultivated peanut emerges as an important food crop. Representing an essential quality oil and food legume crop grown across over 100 countries, peanuts provide vital nutrition and economic stability. *Peanut Genomics and Biotechnology: Status and Prospects for Sustainability* delves into the fascinating journey of peanuts, from their evolutionary origins as a natural tetraploid hybrid to their critical roles in addressing global food and oil demands amidst changing climates.

Advances in peanut genomics and biotechnology have transformed the possibilities for sustainable production. This book is a comprehensive guide to these breakthroughs, offering insights into peanut genome sequencing, breeding programs, and the innovative biotechnological tools that are driving progress. From decoding the complexities of the peanut genome to addressing abiotic and biotic stresses, this resource provides actionable solutions to develop resilient cultivars with superior yields and nutritional content.

Peanut Genomics and Biotechnology: Status and Prospects for Sustainability is useful to students, researchers, and practitioners, bringing together knowledge on genome editing technologies, omics, and bioinformatics applications. By providing a single, consolidated source of systemic information, it empowers scientists and agricultural professionals to revolutionize peanut breeding programs and contribute to sustainable agriculture.

Peanut Genomics and Biotechnology

Status and Prospects for Sustainability

Edited by Weijian Zhuang, Rajeev K. Varshney,
Xingjun Wang, and Xinyou Zhang

CRC Press
Taylor & Francis Group
Boca Raton London New York

CRC Press is an imprint of the
Taylor & Francis Group, an **informa** business

Designed cover image: Shutterstock

First edition published 2026
by CRC Press
2385 NW Executive Center Drive, Suite 320, Boca Raton FL 33431

and by CRC Press
4 Park Square, Milton Park, Abingdon, Oxon, OX14 4RN

CRC Press is an imprint of Taylor & Francis Group, LLC

ISBN: 978-1-032-42049-3 (hbk)
ISBN: 978-1-032-42054-7 (pbk)
ISBN: 978-1-003-36098-8 (ebk)

DOI: 10.1201/9781003360988

Typeset in Times
by Deanta Global Publishing Services, Chennai, India

Contents

Editors

Prof. Weijian Zhuang is a senior scientist specializing in genomics, molecular and cell biology, genetics, and breeding, with over 38 years of service at Fujian Agriculture and Forestry University (FAFU, China), UC Berkeley, UCLA, UC Davis, and Tuskegee University as a research fellow, postdoctoral appointee, and visiting professor. He currently serves at FAFU as the Director of the Oil Crop Research Institute and the Center for Legume Plant Genetics and Systems Biology. He has served as the head of the Fujian Province Key Laboratory of Plant Molecular and Cell Biology and is currently a guest professor and principal investigator at several universities, as well as provincial, ministerial, and state key laboratories in Fujian and other provinces. He also serves as Vice-Chairman and executive member of the Plant Biotechnology and Industrialization Professional Committee of the Chinese Society for Plant Biology as well as holding positions in other national and provincial crop and academic societies. Additionally, he has organized and chaired several national conferences on plant biotechnology and oil crops.

Prof. Zhuang has led more than 40 national and provincial research projects. His research focuses primarily on peanuts, through which he has published over 200 papers in national and international journals, authored one book and several book chapters, released 18 peanut varieties, obtained 28 patents, and received 7 provincial and ministerial awards for contributions to science and technology. In 2019, by leading an international research team, his group was the first to decode the tetraploid cultivated peanut genome, a milestone that significantly advanced peanut scientific research. He has been recognized as Distinguished Expert of Fujian Province and was awarded the State Council Allowance of China.

Prof. Rajeev K. Varshney FRS is an agricultural research scientist specializing in genomics and molecular breeding with more than 20 years of service in international agriculture while working in India, Germany, Australia, Mexico, and several countries in Africa. At present, he is serving Murdoch University (Australia) as a director of the Centre for Crop & Food Innovation; director of the State Agricultural Biotechnology Centre; and International Chair in Agriculture & Food Security with the Food Futures Institute. He is an Honorary or Adjunct Professor with more than ten universities/organizations in China, Australia, Africa, and India.

Prof. Varshney is a globally recognized leader for his work on genome sequencing, cataloguing and utilizing genetic diversity, genomics-assisted breeding, seed systems, and capacity building in developing countries. He has made centrally important contributions in the area of assembling genomes, developing genomic resources, and integrating genomic technologies in crop improvement programs in many tropical crops. His research has contributed towards enhancing farmers' profitability and environmental sustainability in Australia and improving food and nutrition security in India and several countries in Africa and Asia.

Prof. Xingjun Wang is a senior scientist at the Institute of Crop Germplasm Resources, Shandong Academy of Agricultural Sciences (SAAS), Jinan, China. He is the former director general of Biotechnology Research Center, SAAS. He is the Master and PhD student supervisor of Shandong University and Shandong Normal University. He is the vice chairman of the Shandong Genetics Society. He is an executive member of the Oil Crop Branch of the Crop Science Society of China and an executive member of the High Technology Branch of the China Association of Agricultural Science Society. He is a Taishan Scholar Expert and a Distinguished Expert of Shandong Province. He received his Ph.D. from the Life Science College of the National University of Singapore (NUS) and completed his postdoctoral research at the University of California, Los Angeles (UCLA). Professor Wang has focused on peanut functional genomics, genomics, gene identification, DNA marker development, germplasm innovation, and molecular breeding since 2006 in SAAS. He is the principal investigator on more than 20 national and provincial projects, including the regular NSFC project and NSFC-CGIAR project. He has published more than 180 papers in national and international journals, edited the book *Studies on Peanut Biotechnology*, and wrote chapters for several books. He has been awarded six prizes for Progress in Science and Technology from Shandong Province and the Ministry of Agriculture and Rural Affairs of the People's Republic of China. He won the Outstanding Contribution Award for peanut genomics studies. He has given talks at different national and international congresses and hosted the 11th International Conference of the Peanut Research Community on Advances in *Arachis* through Genomics and Biotechnology (AAGB), held in 2019 in Jinan, China.

Prof. Xinyou Zhang is a member of Chinese Academy of Engineering (CAE), the president of Henan Academy of Agricultural Sciences (HAAS), and the chief scientist of China's National Peanut Industry Technology System affiliated to the Ministry of Agriculture and Rural Affairs. He has been serving as a researcher specializing in the genetic improvement of peanut at HAAS for over 40 years, where he initiated a distant hybridization program and made great technological progress in deriving valuable germplasm lines integrating elite traits from wild *Arachis* species. The variety "Yuanza 9102," developed and released by his group, has become the most widely grown peanut variety developed through interspecific hybridization. He has also actively participated in organizing and implementing the International Peanut Genome Initiative, contributing to the completion of genome sequencing for peanut cultivars and their diploid ancestors.

Prof. Xinyou Zhang is one of the leading scientists in peanut breeding in China. He has made significant contributions in variety improvement and breeding technology innovations. The "Yuhua" and "Yuanza" series of peanut varieties released by his group have been grown over a total area of over 10 million hectares, substantially facilitating the development of China's peanut industry, assisting Chinese farmers in increasing their income, and ensuring a stable supply of edible vegetable oil in China.

Contributors

Sheidu Abdullaziz
Center of Legume Plant Genetics and System Biology, Oil Crops Research Institute, College of Agriculture, Fujian Agriculture and Forestry University (FAFU), Fuzhou, China
and
Department of Agronomy
Faculty of Agriculture
Nasarawa State University, Keffi (Lafia Campus)
Keffi, Nasarawa State
Nigeria

Sandaru Malaweera Arachchige
Agronomy Department, University of Florida, Gainesville, Florida, USA

Laura Chalup
Instituto de Botánica del Nordeste (CONICET-UNNE, Fac. Cs. Agrarias), Argentina

Wen-Chi Chang
Institute of Tropical Plant Sciences and Microbiology, College of Biosciences and Biotechnology, National Cheng Kung University, Taiwan, China

Charles Chen
Crop, Soil and Environmental Sciences, Auburn University, Auburn, Alabama, USA

Hua Chen
Center of Legume Plant Genetics and System Biology, Oil Crops Research Institute, College of Agriculture, Fujian Agriculture and Forestry University (FAFU), Fuzhou, China

Na Chen
Shandong Peanut Research Institute, China

Xiangyu Chen
Fujian Academy of Agricultural Sciences, Fuzhou, China

Xiaoping Chen
Crops Research Institute, Guangdong Academy of Agricultural Sciences, Guangdong Provincial Key Laboratory of Crop Genetic Improvement, South China Peanut Sub-Center of National Center of Oilseed Crops Improvement, Guangzhou, Guangdong, China

Xiaoyuan Chi
Shandong Peanut Research Institute, China

Yu-hsuan Chiu
Institute of Tropical Plant Sciences and Microbiology, College of Biosciences and Biotechnology, National Cheng Kung University, Taiwan, China

Chi-Nga Chow
Institute of Tropical Plant Sciences and Microbiology, College of Biosciences and Biotechnology, National Cheng Kung University, Taiwan, China

D. Deshmukh
International Crop Research Institute for Semi-Arid Tropics (ICRISAT), India

Pei Du
The Shennong Laboratory, Henan Academy of Crop Molecular Breeding, Henan Academy of Agricultural Sciences, Henan, China

Jake C. Fountain
Department of Plant Pathology, University of Georgia, Griffin, Georgia, USA

Sunil S. Gangurde
CIMMYT-China Wheat and Maize Research Center, Taian, Shandong, China

Alejandra V. García
Instituto de Botánica del Nordeste (CONICET-UNNE, Fac. Cs. Agrarias), Argentina

A. S. Gerrano
ARC-Vegetable, Industrial and Medicinal Plants, Private Bag X293, Pretoria, South Africa; Food Security and Safety, Focus Area, Faculty of Natural and Agricultural Sciences, North-West University, Mafikeng, South Africa
and
Department of Food Security and Safety, Faculty of Natural and Agricultural Sciences, North-West University, Mafikeng, South Africa

Baozhu Guo
Crop Genetics and Breeding Research Unit, United States Department of Agriculture-Agricultural Research Service (USDA-ARS), Tifton, Georgia, USA

S. Han
Henan Academy of Agricultural Sciences, Zhengzhou, China

Guohao He
Tuskegee University, Tuskegee, Alabama, USA

Lu Huang
Crops Research Institute, Guangdong Academy of Agricultural Sciences, Guangdong Provincial Key Laboratory of Crop Genetic Improvement, South China Peanut Sub-Center of National Center of Oilseed Crops Improvement, Guangzhou, Guangdong, China

Ruihua Huang
Guangdong Key Laboratory of Biotechnology for Plant Development, College of Life Sciences, South China Normal University, China

Xiao Jiang
Shandong Peanut Research Institute, China

Shahid Ali Khan
Pakistan Agricultural Research Council-National Tea and High Value Crops Research Institute, Shinkiari, Pakistan

M. T. Labuschagne
Department of Plant Sciences, University of the Free State, Bloemfontein, South Africa

Graciela I. Lavia
Instituto de Botánica del Nordeste (CONICET-UNNE, Fac. Cs. Agrarias), Argentina

Yong Lei
Oil Crops Research Institute of Chinese Academy of Agricultural Sciences (OCRI-CAAS), Wuhan, China

Hongqing Li
Guangdong Key Laboratory of Biotechnology for Plant Development, College of Life Sciences, South China Normal University, China

Li Li
Hebei University of Engineering, Handan, China

Boshou Liao
Oil Crops Research Institute of Chinese Academy of Agricultural Sciences (OCRI-CAAS), Wuhan, China

Qing Lu
Crops Research Institute, Guangdong Academy of Agricultural Sciences, Guangdong Provincial Key Laboratory of Crop Genetic Improvement, South China Peanut Sub-Center of National Center of Oilseed Crops Improvement, Guangzhou, Guangdong, China

Huaiyong Luo
Oil Crops Research Institute of Chinese Academy of Agricultural Sciences (OCRI-CAAS), Wuhan, China

Xingli Ma
Henan Agricultural University, China

Mahesh K. Mahatma
ICAR-National Research Centre on Seed Spices (ICAR-NRCSS), India

T. W. Mekonnen
Department of Plant Sciences, University of the Free State, Bloemfontein 9301, South Africa

Anjanasree K. Neelakandan
Iowa State University, Ames, Iowa, USA

Alejandra M. Ortiz
Instituto de Botánica del Nordeste (CONICET-UNNE, Fac. Cs. Agrarias), Argentina

Kamal Krishna Pal
ICAR-Directorate of Groundnut Research (ICAR-DGR), India

Lijuan Pan
Shandong Peanut Research Institute, China

Manish K. Pandey
International Crops Research Institute for the Semi-Arid Tropics (ICRISAT), Hyderabad, India

Jinesh Patel
Crop, Soil and Environmental Sciences, Auburn University, Auburn, Alabama, USA

Laura Perez
Instituto de Botánica del Nordeste (CONICET-UNNE, Fac. Cs. Agrarias), Argentina

Lixian Qiao
College of Agronomy, Qingdao Agricultural University

Ali Raza
Center of Legume Plant Genetics and System Biology, Oil Crops Research Institute, College of Agriculture, Fujian Agriculture and Forestry University (FAFU), Fuzhou, China

Ali Razzaq
Agronomy Department, University of Florida, Gainesville, Florida, USA

Germán Robledo
Instituto de Botánica del Nordeste (CONICET-UNNE, Fac. Cs. Agrarias), Argentina

Sergio S. Samoluk
Instituto de Botánica del Nordeste (CONICET-UNNE, Fac. Cs. Agrarias), Argentina

Guillermo Seijo
Instituto de Botánica del Nordeste (CONICET-UNNE, Fac. Cs. Agrarias), Argentina

Priya Shah
International Crops Research Institute for the Semi-Arid Tropics (ICRISAT), Hyderabad, India

Yasir Sharif
Center of Legume Plant Genetics and System Biology, Oil Crops Research Institute, College of Agriculture, Fujian Agriculture and Forestry University (FAFU), Fuzhou, China

Xiaolong Shi
Peanut Research Institute, College of Agronomy, Shenyang Agricultural University, Shenyang, Liaoning, China

Celeste Silvestri
Instituto de Botánica del Nordeste (CONICET-UNNE, Fac. Cs. Agrarias), Argentina

Pooja Soni
International Crops Research Institute for the Semi-Arid Tropics (ICRISAT), India

Binita Subedi
Tuskegee University, Tuskegee, Alabama, USA

S. P. Tallury
USDA-ARS, 1109 Experiment Street, Griffin, Georgia, USA

Yanyan Tang
College of Agronomy, Qingdao Agricultural University, China

Radhakrishnan Thankappan
ICAR-Directorate of Groundnut Research (ICAR-DGR), India

Sy M. Traore
Tuskegee University, Tuskegee, Alabama, USA

Muhammad Jawad Umer
Crops Research Institute, Guangdong Academy of Agricultural Sciences, Guangdong Provincial Key Laboratory of Crop Genetic Improvement, South China Peanut Sub-Center of National Center of Oilseed Crops Improvement, Guangzhou, Guangdong, China

Papa Rao Vaikuntappu
ICAR-Directorate of Groundnut Research (ICAR-DGR), India

Rajeev K. Varshney
Centre for Crop and Food Innovation, State Agricultural Biotechnology Centre, Murdoch University, Perth, Western Australia, Australia

Suman Veeramasu
Tuskegee University, Tuskegee, Alabama, USA

Jianping Wang
Agronomy Department, University of Florida, Gainesville, Florida, USA

Jing Wang
Peanut Research Institute, College of Agronomy, Shenyang Agricultural University, Shenyang, Liaoning, China

Mingli Wang
USDA-ARS Plant Genetic Resources Conservation, Griffin, Georgia, USA

Xingjun Wang
Institute of Crop Germplasm Resources, Shandong Academy of Agricultural Sciences, PR China

Xiyin Wang
North China University of Science and Technology, Tangshan, China

David A. Wright
Iowa State University, Ames, Iowa, USA

Nai-Yun Wu
Institute of Tropical Plant Sciences and Microbiology, College of Biosciences and Biotechnology, National Cheng Kung University, Taiwan, China

Jing Xu
Shandong Peanut Research Institute, China

Liying Yan
Oil Crops Research Institute of Chinese Academy of Agricultural Sciences (OCRI-CAAS), Wuhan, China

Chien-Wen Yang
Institute of Tropical Plant Sciences and Microbiology, College of Biosciences and Biotechnology, National Cheng Kung University, Taiwan, China

Xinlei Yang
Hebei Agricultural University, Baoding, China

Dongmei Yin
College of Agronomy & Center for Peanut Functional Genomics and Molecular Breeding, Henan Agricultural University, Zhengzhou, Henan, China

Xiangzhen Yin
Shandong Peanut Research Institute, China

Haiqiu Yu
Peanut Research Institute, College of Agronomy, Shenyang Agricultural University, Shenyang, Liaoning, China

Weichang Yu
China Good Crop Company (Shenzhen) Limited, Shenzhen, China

Xiaona Yu
College of Agronomy, Qingdao Agricultural University, China

Mei Yuan
Shandong Peanut Research Institute, China

Chong Zhang
Center of Legume Plant Genetics and System Biology, Oil Crops Research Institute, College of Agriculture, Fujian Agriculture and Forestry University (FAFU), Fuzhou, China

He Zhang
Peanut Research Institute, College of Agronomy, Shenyang Agricultural University, Shenyang, Liaoning, China

Hui Zhang
South China Agricultural University, Guangzhou, China

Shengchun Zhang
Guangdong Key Laboratory of Biotechnology for Plant Development, College of Life Sciences, South China Normal University, China

Xingguo Zhang
College of Agronomy & Center for Peanut Functional Genomics and Molecular Breeding, Henan Agricultural University, Zhengzhou, Henan, China

Xinyou Zhang
Henan Academy of Agricultural Science, Zhengzhou, Henan, China

Chuanzhi Zhao
Institute of Crop Germplasm Resources, Shandong Academy of Agricultural Sciences, Jinan, China

Xinhua Zhao
Peanut Research Institute, College of Agronomy, Shenyang Agricultural University, Shenyang, Liaoning, China

Xuhong Zhao
Shandong Peanut Research Institute, China

Hong Zhu
College of Agronomy, Qingdao Agricultural University, Qingdao, China

Weijian Zhuang
Center of Legume Plant Genetics and System Biology, Oil Crops Research Institute, College of Agriculture, Fujian Agriculture and Forestry University (FAFU), Fuzhou, Fujian, China

Yuhui Zhuang
College of Life Sciences, Fujian Agriculture and Forestry University (FAFU), Fuzhou, China

Preface

The peanut, a legume of vast economic and nutritional importance, has played a pivotal role in sustaining communities across the globe. In Africa and Asia, it serves not only as a vital source of oil and protein but also as a cornerstone of local economies. Despite its importance, peanut has remained an orphan crop in the scientific world—largely neglected in the realm of advanced agricultural research and biotechnology until recent years. A quiet revolution of peanut cultivation is underway, one that holds transformative potential for global agriculture, food security, and livelihoods.

This book, *Peanut Genomics and Biotechnology: Status and Prospects for Sustainability*, seeks to shed light on this revolution. It captures the journey of the peanut, from its domestication through natural hybridization to the breakthroughs of modern genomic research. As we face the threats posed by climate change, which exacerbate both abiotic and biotic stresses on crops, the peanut is emerging as a symbol of resilience. Through genetic and biotechnological advancements, this crop holds the potential to redefine sustainable agriculture, paving the way for enhanced food security and climate adaptation.

With advancements in next-generation sequencing technologies, the peanut genome has been decoded, unlocking new avenues for peanut improvement. Scientists are now empowered with tools to develop stress-resilient peanut cultivars that can withstand a broad spectrum of stresses—biotic factors like pests and diseases, and abiotic challenges such as extreme temperatures, salinity, drought, and other climate-related pressures. These stress-resilient varieties promise not only superior yields but also greater stability across diverse and changing environments. This volume compiles cutting-edge research in peanut genomics and evolution, gene discovery, and marker development, uncovering molecular mechanisms underlying life activities and trait formation and the biotechnological advancements that are poised to shape the future of peanut farming.

WHAT LIES AHEAD IN THESE PAGES

As you turn the pages of this book, you will delve into a wealth of knowledge that spans a wide range of topics critical to understanding the future of peanut cultivation, both from a biological and technological standpoint. The opening chapters introduce you to the global significance of peanuts, not just as an agricultural commodity but also as a vital component of human nutrition and economic stability. Readers will discover the biological and genetic journey of the peanut, beginning with its hybrid origins from two diploid species, *Arachis duranensis* and *Arachis ipaensis*. This deep dive into the plant's evolutionary history will set the stage for the cutting-edge research that follows.

A substantial portion of the book focuses on the advancements in peanut genetic resources and the latest developments in the peanut genome. Early chapters (Chapters 1 and 2) provide a general overview of peanut production, the current state of genetic resources, and the ongoing research in the field. The book delves into key topics such as species diversity, molecular markers, genetic maps, and interspecific hybridization—tools essential for improving traits like yield, quality, disease resistance, and climate adaptability.

As the narrative progresses, the book shifts focus to advances in cytogenetics (Chapter 3), exploring the role of sophisticated techniques like fluorescence in situ hybridization (FISH) and genome in situ hybridization (GISH) in unraveling the complex structure of peanut chromosomes. Following that, Chapter 4 details the extensive research efforts and breakthroughs in decoding the peanut genome, highlighting the sequencing of both cultivated and wild progenitor genomes, three-dimensional chromatin organization, and the implications of these findings for advancing peanut breeding and productivity.

Chapter 5 provides an in-depth review of genomic advancements in peanut research, examining the complete genome sequences of cultivated peanuts and their wild ancestors, revealing chromosomal organization, evolutionary history, domestication, and the genetic foundations of key agronomic traits to inform breeding strategies. Chapter 6 covers the application of genome-wide association studies (GWAS), QTL mapping, QTLseg, BSRseq, and so on, techniques that have revolutionized the identification of key agronomic traits. Chapter 7 delves into molecular marker development, including the use of SSR, AFLP, SNPs, and other markers used in peanut genomic research. Chapter 8 explores the identification and functional analysis of key peanut genes involved in quality traits, stress responses, and metabolic pathways, utilizing genomic and proteomic tools to enhance peanut breeding and resilience. Chapter 9 explores the molecular mechanisms underlying peanut growth, development, and stress responses, focusing on hormone regulation, signaling pathways, epigenetic modifications, and disease resistance, while also highlighting advanced breeding tools like MAS and CRISPR/Cas9 for developing resilient peanut varieties. Chapter 10 looks at the promise of genomic-assisted breeding, discussing modern breeding techniques such as marker-assisted selection (MAS) and high-throughput genomic tools. It highlights how these innovations have propelled peanut breeding programs forward, improving the crop's resilience to both abiotic and biotic stresses.

With this understanding, readers are equipped to explore the exciting realm of genomics-assisted breeding, which has already shown tremendous promise in crop improvement. The focus on modern breeding techniques, such as MAS and high-throughput genomic tools, illustrates how far peanut breeding programs have come in a relatively short time.

In an era defined by increasing environmental volatility, the book's chapters on stress tolerance take on heightened importance (Chapters 11–17). Climate change brings forth numerous challenges, including drought, salinity, extreme temperatures, and pests, each posing a significant threat to peanut production globally. Chapter 11 explores the role of multi-omics techniques in understanding and improving peanut tolerance to biotic stresses, highlighting recent advancements, key biotic stressors, and the potential for sustainable peanut production. Chapter 12 explores advancements in multi-omics research on peanuts under abiotic stresses, focusing on genes, physiological responses, and regulatory mechanisms that enhance peanut resilience in semi-arid regions. Chapter 13 explores how modern biotechnological advancements, such as gene editing, high-throughput genotyping, and marker-assisted selection, have deepened our understanding of molecular networks that drive peanut adaptation to abiotic stresses, providing valuable insights for resilience-focused breeding applications. Chapter 14 discusses the challenges biotic stresses pose to peanut production, highlighting progress in developing resistant cultivars using molecular tools like MAS, genomics, and gene editing, while emphasizing the need for further research to overcome resistance complexities and identify elite germplasm. Another critical issue addressed in this book is aflatoxin contamination (Chapters 15 and 16), a problem that plagues peanut production and compromises food safety. Chapter 17 explores the molecular mechanisms of peanut resistance to bacterial wilt caused by *Ralstonia solanacearum*, focusing on key genes, signaling pathways, and biochemical responses, and discusses how integrating these insights into breeding strategies like MAS and genomic selection can lead to the development of disease-resistant peanut cultivars, supporting sustainable agriculture and improved productivity. The chapters dedicated to abiotic and biotic stress provide insights into how modern biotechnological tools, such as CRISPR/Cas-based genome editing and multi-omics approaches, are being harnessed to create cultivars that can withstand these challenges. By developing peanuts with enhanced resilience, scientists aim to secure food-supply chains for both developed and underdeveloped regions of the world.

Beyond conventional breeding, the integration of omics technologies—genomics, transcriptomics, proteomics, metabolomics, and miRNAomics—represents a frontier of innovation that has the potential to transform peanut production. The book discusses how these tools are enabling a more precise understanding of the biological pathways involved in stress responses, growth, and development, leading to more effective strategies for peanut improvement. The advent of genome editing technologies, particularly CRISPR/Cas, represents another monumental leap forward (Chapter 18). The book explores how these technologies can be applied to peanut breeding, offering the potential to introduce traits of yield, quality, and stress tolerance at an unprecedented pace. By marrying genome editing with fast-breeding concepts, we are witnessing the dawn of a new era where crop improvement timelines are being dramatically shortened.

This book does more than simply document the progress that has been made; it serves as a roadmap for the future of peanut genomics and biotechnology. The closing chapters discuss the integration of modern bioinformatics tools in peanut breeding programs and the immense potential these tools hold for accelerating breeding progress (Chapters 19 and 20). Chapter 19 highlights recent advancements in bioinformatic tools and genomic resources for peanut research, focusing on how next-generation sequencing, genome assembly, and data platforms accelerate breeding for disease resistance and stress resilience, offering new opportunities for genomics-led genetic studies and peanut cultivar development. Chapter 20 explores the importance of bio-databases in plant genomics, focusing on peanut-specific platforms like PeanutBase, PGR, and LIS-ArachisMine and large databases such as NCBI and Phytozome, which facilitate genome sequencing, gene annotation, and trait identification, while discussing the integration of multi-omics data, machine learning, and AI to advance peanut research and breeding. In addition, the authors explore the prospects of fast breeding and 5G breeding concepts to revolutionize peanut production further (Chapter 21). The chapter reviews the evolution of peanut breeding from traditional to molecular and smart breeding, highlighting the role of biotechnology, including genomics, transcriptomics, and Big Data technologies, in enhancing breeding efficiency, speed, and accuracy, and discusses the future potential of intelligent breeding to improve yield and quality and reduce production costs.

By consolidating research from multiple disciplines ranging from genomics and bioinformatics to molecular breeding, this book offers a holistic view of what it takes to cultivate a crop that can meet the demands of a growing global population while withstanding the pressures of a changing climate.

AN INVITATION TO INNOVATE

We offer this comprehensive body of knowledge not only to researchers and students but also to every curious mind invested in the quest for a sustainable future. The insights and innovations detailed within these pages are more than scientific milestones; they are stepping stones toward addressing global food security challenges. By delving into the intricate genetics of the peanut, we illuminate a path forward, where biotechnological advancements hold the key to nourishing a growing population under increasingly uncertain conditions.

This book promises to be an indispensable resource for those invested in agricultural sciences, plant biotechnology, and sustainable development. Beyond its technical depth, it is also a testament to the power of scientific discovery in driving solutions to some of the world's most pressing challenges. We invite you to embark on this journey of exploration and innovation, to witness how a simple legume could reshape the future of food production.

In the chapters that follow, you will find the building blocks of the next agricultural revolution, a revolution where sustainability, resilience, and innovation converge to meet the challenges of tomorrow's world.

1 An Introduction to Peanut Importance, Genomics, and Biotechnology

Past, Present, and Future Outlooks

Ali Raza, Sheidu Abdullaziz, Xingjun Wang, Xinyou Zhang, Rajeev K. Varshney, and Weijian Zhuang

1.1 INTRODUCTION

The cultivated peanut (*Arachis hypogaea* L.), a tetraploid with 2.54 Gb genome, is a vital oil- and food-legume crop grown worldwide. Peanut is largely grown in tropical and subtropical zones (Zhuang et al., 2019; Bertioli et al., 2019). During 2023–2024, peanut production was 51.314 million tons worldwide (FAS, USDA, 2024; https://ipad.fas.usda.gov/cropexplorer/cropview/commodityView.aspx?cropid=2221000). Peanut is a member of the Leguminosae family, tribe *Aeschynomeneae*, and subtribe *Stylosanthenae*. Many centuries ago, domesticating plants enhanced food supplies and paved the way for the development of advanced cultures. Only a small number of the thousands of wild varieties were domesticated into crops, which currently supply the majority of the food used by human beings (Salman-Minkov et al., 2016; Zhuang et al., 2019; Bohra et al., 2022). Around 9,400 years ago, humans might have played a role in the domestication of the cultivated peanut (*Arachis hypogaea* L.), as projected by collected morphological differences (Bertioli et al., 2016). The cultivated peanut resulted from a hybridization event between two diploid species, *Arachis duranensis* and *Arachis ipaensis*, which contributed the A and B genomes, respectively (Bertioli et al., 2016). During domestication, the cultivated peanut developed several key agronomic traits for human use, including high oil content, larger seed size, and disease resistance (Zhuang et al., 2019; Bertioli et al., 2019). *Arachis hypogaea* L. has two sets of chromosome pairs, one set from each of the diploid species *Arachis duranensis* and *Arachis ipaensis*, which are two different species. Because it contains these two distinct sets of chromosomes, it is classified as an allotetraploid crop (Zhuang et al., 2019; Bertioli et al., 2019). In global markets, there is significant demand for peanuts and peanut-based products, particularly in the sweets industry. Peanuts are an excellent source of protein and oil, as well as a range of essential minerals, vitamins, antioxidants, and bioactive compounds that promote health. Due to these nutritional benefits, peanuts are widely recognized and promoted as a nutritious food (Variath and Janila, 2017). Peanuts are consumed worldwide in diverse forms, including cooked seeds, oil, protein, candy, and peanut cake, with peanut oil being the most widely used product, followed by peanut butter. Additionally, peanuts hold different cultural and culinary significance across various regions of the globe (Variath and Janila, 2017). However, peanut production in many developing nations remains relatively low due to several constraints, including abiotic and biotic stressors, limited access to efficient seed distribution systems, insufficient scientific knowledge, and low-input use (Desmae et al., 2019). Among these restrictions, harmful insects, bacteria, viruses, drought, soil salinity, extreme temperature, nutrient imbalance, etc., are the most important biotic and abiotic stresses. Small-scale farmers in underdeveloped nations often grow peanuts, but often lack assets and productivity-boosting technology. Thus, peanut scientists worldwide must leverage genetic knowledge and contemporary biotechnology methods in crop advancement initiatives to produce stress-smart peanut cultivars to benefit farmers across the globe.

Over the previous decade, a substantial number of ground-breaking innovations have been achieved using high-throughput next-generation sequencing (NGS), genomics, biotechnology, and fast-track breeding technologies. Genomics (mainly genomics-assisted breeding, GAB) and a set of biotechnological tools have become a vital component of breeding operations to hasten the production of better varieties (Varshney et al., 2005; Pandey et al., 2012; Liu et al., 2013; Varshney et al., 2018, 2019, 2021, 2021aPandey et al., 2020; Varshney, Bohra, Yu, et al., 2021; Varshney, Bohra, Roorkiwal, et al., 2021; Liu et al., 2022). With the breakthrough of genome sequencing of cultivated peanut (Bertioli et al., 2019; Zhuang et al., 2019; Chen et al., 2019) and its two diploid progenitors (Yin et al., 2018; Bertioli et al., 2016), rapid discoveries were made on peanut genomics and functional genomics, benefiting crop improvement in recent years. During the last 10 years, peanut has also observed noteworthy advances, including huge transcriptome and proteome resources, genomics resources, improvement in genetic maps and markers, marker-assisted breeding (MAB), and molecular breeding outcomes in advancing peanut cultivars (Pandey et al., 2014; Ozias-Akins et al., 2017; Liao, 2017; Vishwakarma et

DOI: 10.1201/9781003360988-1

al., 2017; Wang et al., 2017; Chen et al., 2019; Bertioli et al., 2019; Zhuang et al., 2019; Aravind et al., 2022; Bhat et al., 2022; Radhakrishnan et al., 2023).

Climate change gives rise to various environmental stresses, e.g., abiotic (such as drought, salinity, extreme temperature, waterlogging, etc.) and biotic (such as bacteria, viruses, fungi, pathogens, and harmful insects) (Suzuki et al., 2014; Signorelli et al., 2019; Raza et al., 2021; Thudi et al., 2021; Varshney, Barmukh, et al., 2021; Zandalinas et al., 2021; Raza, Mubarik, et al., 2022; Raza, Tabassum, Fakhar, et al., 2022; Raza, Charagh, et al., 2022; Raza, Tabassum, Zahid, et al., 2022; Raza, Salehi, et al., 2022; Aravind et al., 2022; Rivero et al., 2022; Gonzalez Guzman et al., 2022; Saeed et al., 2023). These stresses significantly impair peanut quality and productivity worldwide and create challenges for global peanut supply and food security. It is vital to recognize the biological characteristics and stress tolerance mechanisms of peanuts and to develop stress-smart peanut cultivars that can raise yields under shifting climate conditions to feed the growing population. Therefore, it is highly necessary to integrate genomics and modern biotechnology tools to breed advanced peanut varieties which can endure harsh conditions and deliver high yield and quality. In light of the aforementioned, the editors intended for this book to collect the information pertaining to the use of peanut genomics and biotechnology for crop advancement.

1.2 IMPORTANCE AND PRODUCTION STATUS OF PEANUT

The global peanut production landscape is vast, with over 100 countries involved in cultivation under various agro-ecological conditions. In Asia and Africa, peanut production surpasses soybean, making it a major crop in high demand (Zhuang et al., 2019; Chen et al., 2019; Figure 1.1). Due to its exceptionally high oil and protein content—up to 80% in kernels—peanut ranks as the second-largest source of plant-based protein globally, with demand steadily rising (Zhuang et al., 2019; Chen et al., 2019). According to data from the Foreign Agricultural Service (FAS) of the USDA, global peanut production in the 2023–2024 season reached 51.14 million tons, up from approximately 46 million tons in 2019, with China as the leading producer.

China accounted for 37% of global peanut production in 2024, producing 19.0 million tons, with a 5-year average of 17.65 million tons from 2017 to 2021. The primary peanut-growing regions in China include Henan, Shandong, Hebei, Hubei, and Guangdong. India is the second-largest producer, contributing 14% of global peanut production. India produced 7.1 million tons in 2024, with a 5-year average of 6.22 million tons. Key peanut-producing states include Gujarat, Rajasthan, Tamil Nadu, Andhra Pradesh, and Karnataka. Nigeria ranks as the third-largest peanut producer worldwide, responsible for 8% of global production. Nigeria produced 4.3 million tons in 2024, with a 5-year average of 4.37 million tons. Most production comes from the dry regions of North-West, North-East, and North-Central Nigeria. The United States, another significant peanut producer, contributed 6% of global production, yielding 2.922 million tons in 2024, with a 5-year average of 2.776 million tons. Major producing states include Georgia, Alabama, Texas, and Florida.

Other notable peanut-producing countries include Sudan (2.0 million tons), Senegal (1.7 million tons), Burma (1.75 million tons), Argentina (1.35 million tons), Guinea (1.0 million tons), and Tanzania (1.1 million tons). Collectively, China, India, and Nigeria account for over 50% of the world's total peanut production (Figure 1.2 and 1.3).

Peanut is widely used for oil production, for food consumption, and as animal feed. It plays a crucial role in human nutrition, especially in Africa and Asia, with China being a major consumer. Over the years peanut has witness significant improvement in yield due to concerted efforts

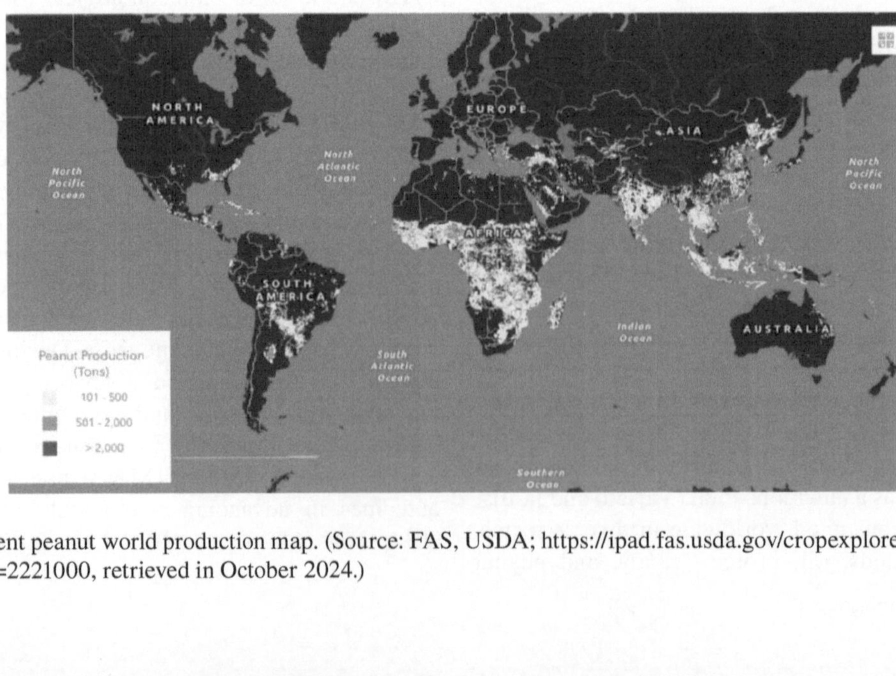

FIGURE 1.1 Current peanut world production map. (Source: FAS, USDA; https://ipad.fas.usda.gov/cropexplorer/cropview/commodityView.aspx?cropid=2221000, retrieved in October 2024.)

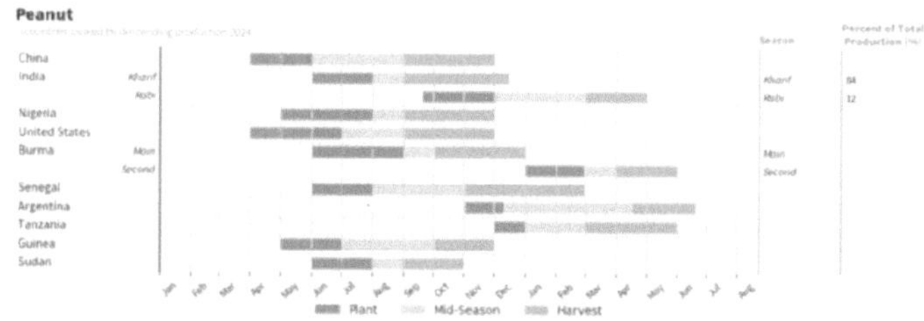

FIGURE 1.2 Calendar of peanut-producing countries ranked by descending production order. (Source: FAS, USDA; https://ipad.fas.usda.gov/cropexplorer/cropview/commodityView.aspx?cropid=2221000, retrieved in October 2024).

by scientist who have taken advantage of peanut genomics to fast-track improvements of various aspects of the peanut plant. Peanuts offer nutritional value through various forms, including kernels, foliage, and processed food products such as oil and candies. The crop is also utilized in food supplements due to its rich content of bioactive compounds such as tocopherols, resveratrol, phytosterols, arginine, phenolic acids, and flavonoids (Variath and Janila, 2017).

These compounds contribute to the health benefits of peanuts, enhancing their value beyond basic nutrition. Variath and Janila (2017) explore these aspects in detail, highlighting that although peanut pods and kernels are the most valuable part of the crop, other components of the peanut plant have significant uses as well. These include their application in various industrial, nutritional, and agricultural contexts. A brief overview below emphasizes the multifaceted nutritional and functional benefits of peanuts, underscoring their importance in global diets.

Plants: After harvesting, the crisped haulms (the over-ground portion of the plant) serve as a nutrient-dense feed for livestock or as compost. Peanut haulms are rich in proteins (8%–15%), minerals (9%–17%), lipids (1%–3%), crude fiber (22%–38%) and carbohydrates (38%–45%). Haulms are able to produce approximately 4000 kilocalories (kcal) of energy per kilogram of dry mass (Variath and Janila, 2017; Nagaraj, 1988).

Seeds (kernels): Crude, roasted, and boiled peanut seeds are eaten straight away. They are also turned into oil, sweets, and peanut flour to add taste. In order to fight malnutrition in underdeveloped nations, peanut kernels are also commonly utilized in the production of fully prepared medicinal and supplemental meals. Peanut kernels contain oil (44%–56%), protein (22%–30%), carbohydrates (10%–25%), vitamins (E, K, and B complex), minerals (Ca, P, Mg, Zn and Fe), and fiber (Variath and Janila, 2017). Peanut oil contains a

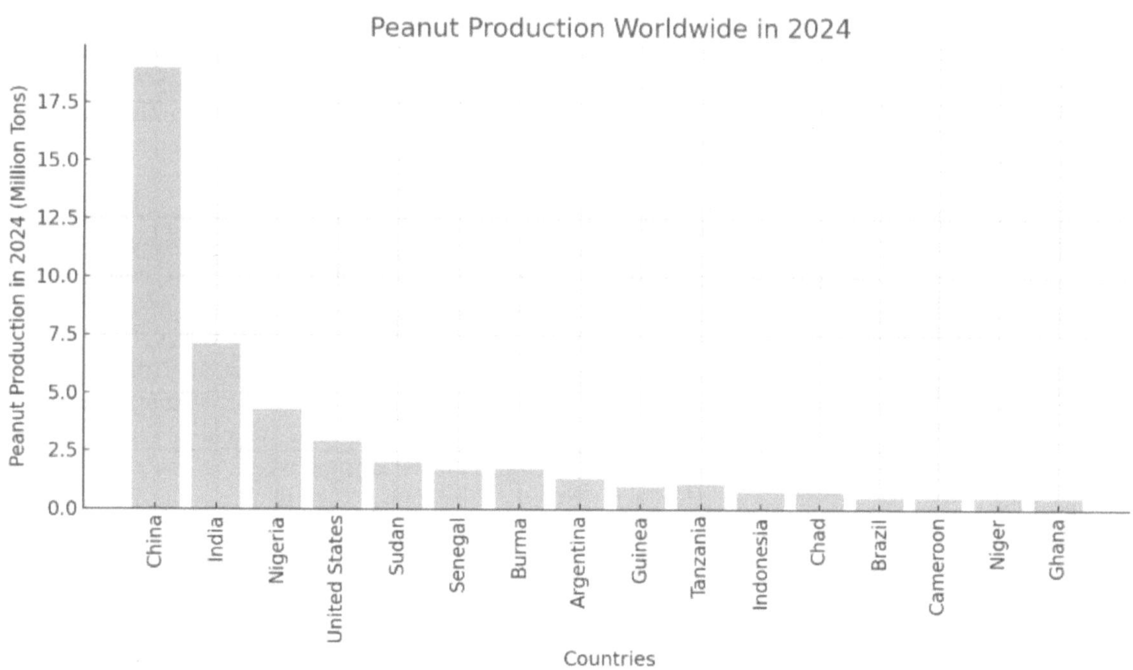

FIGURE 1.3 Peanut production in leading peanut-growing countries.

variety of fatty acids, with palmitic, oleic, and linoleic acids making up approximately 90% of its total fatty acid content. Among these, oleic acid is particularly notable; in high-oleic peanut varieties, it can comprise more than 75% of the total fatty acids. This high oleic acid content enhances the oil's stability and shelf life, making it particularly valuable for cooking and food preservation (Bera et al., 2018, 2019; Shasidhar et al., 2020; Jadhav et al., 2021). Despite being classified as a lower-grade oil in some markets, peanut oil is widely used as an edible oil, particularly in countries like China, where it is valued for its mild flavor and high smoke point, making it ideal for cooking. Beyond its culinary applications, peanut oil also has industrial uses, where it serves as a base for biodiesel production, cosmetics, and pharmaceuticals, leveraging its stability and chemical properties (Nagaraj, 1988; Aydin, 2007; Bell, 2008; Musa, 2010; Variath and Janila, 2017).

Pods: Peanut pods serve as the protective shell for the seeds and are often repurposed after shelling. In some regions, the shells are used as a source of fiber in animal feed, as a biofuel, or as a component in compost for soil improvement.

Shoots: The shoot system includes the stem and leaves. Peanut plants are low-growing with stems that spread along the ground, and their pinnate, alternate leaves play a crucial role in photosynthesis, supplying energy needed for growth and pod development.

Peanut coats: Peanut seed coats have been used for centuries in ancient Chinese medicine to cure chronic bleeding and pneumonia because they are rich in phenolics and other phytochemical compounds. Furthermore, they are used as additives in treated foods like peanut paste and butter to boost their taste, durability, and antioxidant potential (Variath and Janila, 2017; Hathorn and Sanders, 2012; Toomer, 2020).

Shells: Peanut shells are versatile byproducts with a wide range of applications. They are used as biofuel, as livestock feed, and as an additive in animal bedding. Additionally, they are incorporated into manure for soil enrichment and used in the production of building materials like hollow blocks. Peanut shells are also processed to extract cellulose, which can be used to produce rayon and paper; and mucilage, which is used as a natural adhesive. Some experimental processes ferment peanut shells to produce acetone and other chemicals, though this is less common (Bell, 2008; Variath and Janila, 2017; Perea-Moreno et al., 2018; Ge et al., 2020).

Roots: The roots of this legume crop are rich sources of nitrogen (100–152 kilogram per hectare nitrogen) and organic material in the soil (Variath and Janila, 2017; Nigam, 2014).

1.3 POTENTIAL OF GENOMICS TOOLS FOR PEANUT IMPROVEMENT

1.3.1 Advancements in Pre-Genomics Peanut Breeding

Before the advent of genomics, peanut breeding primarily relied on traditional methods, which posed significant challenges. Early breeding efforts aimed to improve key traits such as yield, stress tolerance, and disease resistance, but progress was often constrained by the narrow genetic base of cultivated peanuts. Traditional breeding methods involved crossing cultivated peanuts with wild *Arachis* species to introduce genetic variation. However, the complexity of peanut's tetraploid genome composed of two distinct diploid ancestors made introducing and stabilizing desirable traits challenging. Breeding efforts were often limited by the difficulty of achieving effective cross-species compatibility and transferring beneficial traits without compromising other agronomic qualities (Seijo et al., 2018). To increase genetic diversity, peanut breeders began utilizing core and mini-core germplasm collections, which provided access to a wider range of genetic material with desirable traits. These collections served as reservoirs of genetic variation, offering potential sources of beneficial traits like disease resistance, drought tolerance, and improved nutritional content. Although valuable, the use of germplasm collections was still limited in scope and efficiency due to the lack of molecular tools to identify and track these traits precisely. The development of molecular cytogenetics marked a turning point, allowing breeders to study the chromosomal structure of peanuts and understand chromosome behavior during breeding. Techniques like karyotyping and chromosome banding enabled researchers to observe structural variations, improving crossbreeding outcomes and trait stability. Additionally, early genetic marker systems, such as restriction fragment length polymorphism (RFLP) and amplified fragment length polymorphism (AFLP), and later the limited SSR markers, were introduced, providing breeders with preliminary tools to assess genetic variation (Seijo et al., 2018; Bertioli et al., 2020; Seijo et al., 2014; Seijo et al., 2018; Zhuang et al., 2019). However, these marker systems were low throughput, limiting the speed and accuracy of peanut breeding efforts. Despite these advancements, peanut breeding remained slow and limited. The complexity of the peanut genome, coupled with a lack of efficient genetic tools, created bottlenecks in developing high-yielding, resilient cultivars. Amphidiploids—hybrid plants derived from crossing different wild *Arachis* species—were introduced as a strategy to access previously untapped genetic diversity, but the lack of molecular precision restricted their impact.

1.3.2 Advances in Peanut Genomics

1.3.2.1 Sequencing of Peanut Genomes

The development of genomics technologies has facilitated advancements in peanut breeding, though practical applications remain in the early stages. Whole-genome sequencing projects launched in 2010 marked a turning point, leading to the complete sequencing of the two diploid peanut ancestors by 2016 (Bertioli et al., 2016), followed by the whole-genome sequencing of wild peanut (*A. monticola* L.) (Yin et al., 2018). Subsequent efforts have led to the sequencing of cultivated varieties such as Shitouqi, Fuhuasheng, and Tifrunner (Zhuang et al., 2019; Chen et al., 2019), providing a detailed structure of all coding genes and functional components. These genomic insights have expanded our understanding of peanut biology through systems biology approaches. Advances in genomics have also enabled the development of numerous peanut genetic markers, particularly single nucleotide polymorphism (SNP), insertion-deletion (Indel), structural variation (SV), and genome-wide simple sequence repeat (SSR) markers, all of which significantly enhance breeding programs by supporting marker-assisted selection.

1.3.2.2 Molecular Marker-Assisted Selection

Molecular markers are the most vital players in defining and exploiting accessible genetic differences between all genomic information (Pandey et al., 2014; Salgotra and Stewart, 2020; Varshney et al., 2021). With the advancement of peanut genomics, DNA markers have been gradually adopted to improve the productivity and accuracy of conventional peanut breeding by marker-assisted breeding (MAB) (Kumar et al., 2020; Pandey et al., 2014; Peng et al., 2016; Pandey et al., 2012). The massive number of quantitative trait loci (QTLs) mappings for peanut have provided plenty of connections between molecular markers and genes. These markers have been applied in numerous genetic investigations, including germplasm classification and trait mapping for agronomically significant attributes such as stress tolerance. They have also played a critical role in MAB, supporting the precise selection of desired traits in peanut breeding programs (Kumar et al., 2020; Pandey et al., 2014; Peng et al., 2016; Pandey et al., 2012; Mishra et al., 2015; Li et al., 2017; Luo, et al., 2019; Ghosh et al., 2022; Ding et al., 2022; Gangurde et al., 2022). Various marker systems have been employed for breeding programs. Among these markers, SSR and SNP markers are, at present, the ideal marker techniques for genetic investigations and breeding practice. Despite SSR markers being widely used in present plant breeding activities, SNPs appear to offer more opportunities owing to their high-throughput screening applicability.

1.3.2.3 Functional Genomics in Peanut Breeding

Functional genomics, a newer field in peanut breeding, involves using genome-wide methodologies to investigate gene functions, focusing on the molecular pathways that regulate important traits (Wang et al., 2011). This approach leverages high-throughput technologies, such as transcriptomics, proteomics, and metabolomics, and so on, to study large numbers of genes and their interactions. By analyzing data from these technologies, researchers gain a comprehensive understanding of the roles and interactions of genes, leading to improved crop traits (Bouchez and Hőfte, 1998). Recent advances in functional genomics have significantly enhanced peanut breeding efforts. The development of genomic resources, including molecular markers, genetic maps, and expressed sequence tags, has facilitated the identification of QTLs and genes associated with important traits (Pandey et al., 2012). Wild relatives of peanut have emerged as valuable sources of genetic diversity and resistance alleles, with transcriptome studies aiding in the discovery of beneficial genomic segments (Guimaraes et al., 2017). The increasing availability of Arachis transcriptomic resources has improved the understanding of genotype–phenotype correlations, potentially enabling more precise trait improvement (Guimaraes et al., 2017). Genetic transformation methods have been developed to facilitate the integration of wild alleles into cultivated peanut varieties (Brasileiro et al., 2014). The recent publication of reference genome sequences for peanut's diploid progenitors has further expanded genotyping tools for QTL mapping and gene-based selection, providing a foundation for improved disease resistance and reproductive traits (Ozias-Akins et al., 2017). Although functional genomics breeding is still emerging, it holds promise for creating peanut varieties with improved yield, resilience, and nutritional content in the future.

1.3.2.4 Unlocking Traits for Improved Resilience, Yield, and Quality

In recent years, the field of peanut genomics has experienced transformative advancements, largely enabled by high-throughput genomic technologies. The accumulation of extensive genomic data for both wild and cultivated Arachis species has led to the identification of genes and pathways responsible for crucial traits, including disease resistance, drought tolerance, and oil content. Through the analysis of complex trait loci and polygenic traits, researchers have made strides in understanding the genetic architecture of these traits, providing essential insights into peanut adaptation and productivity under varying stress conditions (Vishwakarma et al., 2017; Soni et al., 2020; Sun et al., 2022). Comparative genomic studies between wild and cultivated peanut species have yielded vital information about the evolutionary history and domestication process of peanuts. This enhanced understanding of genetic variation within wild progenitors has supported efforts to increase the genetic diversity of cultivated peanut varieties, a crucial step toward improving crop resilience and adaptability in the face of climate change. Notably, genomic tools have expanded the molecular marker repertoire, enabling breeders to pinpoint markers associated with desirable traits more efficiently. While high-throughput genotyping technologies such as genotyping-by-sequencing (GBS) have facilitated these advancements, the focus in peanut genomics remains on understanding the functional basis of these traits rather than merely on the application of these tools. By identifying candidate genes and characterizing their functions, researchers are unlocking potential for modern biotechnological interventions which can enhance yield and resilience in future peanut varieties. Moving forward, a deeper exploration of candidate genes and their functional roles will be essential. This research is critical to overcoming genetic bottlenecks, such as genome variation and polyploidization, and addressing longstanding challenges related to cross-compatibility between wild progenitors and cultivated peanuts. By advancing the genomics of peanuts, we are establishing a foundation for more efficient breeding strategies that will contribute to a more sustainable and resilient global peanut industry.

1.3.2.5 Application of Genomic-Assisted Breeding

Genomic-assisted breeding (GAB) provides hope for quicker peanut development since genomic technologies have been more widely available in recent times (Varshney et al., 2018, 2021; Pandey et al., 2012; Liu et al., 2022; Pandey et al., 2020; Varshney et al., 2005, 2019). Over the past decades, GAB has been widely used in peanut breeding programs, including trait improvement and stress/disease tolerance (Bhat et al., 2022; Varshney et al., 2013; Varshney, 2016; Guo et al., 2011; Wang et al., 2017; Ozias-Akins et al., 2017; Luo, Pandey, Khan, Guo, et al., 2019; Chu et al., 2011; Sun et al., 2022; Qi et al., 2022; Shanthala et al., 2022; Wang et al., 2022; Chen et al., 2019; Liu et al., 2022; Choudhary et al., 2019; Janila et al., 2013). By enhancing genetic gains in each selection phase, GAB may greatly boost breeding effectiveness in comparison with conventional breeding methods. The GAB technologies, such as marker-assisted backcrossing, marker-assisted recurrent selection, genomic selection, advanced backcross QTL analysis-based breeding, and haplotype breeding, are becoming more popular in crop breeding programs. Recently, a 5G (genome assembly, germplasm characterization, gene function identification, genomic breeding, and genome editing) breeding idea was introduced by Varshney et al. (2020), and 5G breeding approaches are very necessary for fast-track crop improvement in the era of climate change. As the genome assemblies of wild and cultivated peanuts are available, along with a huge collection of characterized germplasm around the world, this data can be used to identify marker-trait links and exceptional haplotypes. Systems biology techniques and sequencing-based mapping methodologies could be employed to discover genes implicated in systems that contribute to trait expression, giving diagnostic markers for objective attributes. These genes, markers, haplotypes, and whole-genome sequencing information could be applied in genomic breeding and genome editing approaches in arrangement with a fast-track breeding scheme (called speed breeding) (Luo et., 2019). Some success stories are available from diverse plant species about genome editing and speed breeding. Also, GAB-driven data is also available for peanut, advancing peanut improvement. However, in the near future, more efforts are required to integrate different 5G tools, such as speed breeding, and determine how its integration with 5G and genome editing can help to develop future peanut cultivars with high yield, better quality, and stress-tolerance traits. Therefore, the combination of genomics technologies could help in understanding the diversity of the tetraploid genome for genetic improvement and in extending the peanut genetic pool with desirable alleles.

1.3.2.6 The Global Impact of Genomic Resources

The advancement of genomic resources in peanut breeding has substantial implications across agricultural, economic, and nutritional sectors, particularly in regions where peanuts are critical for income and food security. One major breakthrough is the development of disease- and pest-resistant peanut varieties, which can significantly mitigate global crop losses. This reduction in crop vulnerability minimizes farmers' reliance on chemical pesticides, supporting both environmental sustainability and the health of farming communities. Nutritionally, genomics has facilitated the creation of high-oleic peanut varieties, which offer extended shelf life and lower saturated-fat content. These varieties, identified initially through phenotyping and now consistently selected through genomic tools (Jadhav et al., 2021), are appealing to health-conscious consumers, creating new market opportunities and fostering economic growth. Genomic advances also enable the development of biofortified peanuts, enriched with micronutrients like iron and zinc, which are essential for combating malnutrition in regions heavily reliant on peanuts as a staple. Climate resilience is another critical area enhanced by genomics, where researchers can identify genes that bolster drought and heat tolerance. Such varieties empower farmers to maintain productivity in increasingly harsh climates, supporting food security in affected areas. Furthermore, genomics fosters economic diversification by enabling breeders to develop peanut varieties with reduced allergens or unique flavors, catering to niche markets and allowing for premium product lines. This diversification broadens both local and international market access, especially beneficial in regions with limited distribution infrastructure. On a broader scale, genomic resources promote sustainable agriculture by reducing the need for pesticides and fertilizers, as resistant and nitrogen-efficient varieties lower both environmental impacts and input costs. Collectively, these genomic innovations have the potential to revolutionize peanut production globally, strengthening local food systems and enhancing the economic resilience of peanut-dependent regions.

1.4 GENETIC RESOURCE COLLECTION AND CHARACTERIZATION

1.4.1 Germplasm Collections and Their Importance

Global germplasm collections, such as those at the International Crops Research Institute for the Semi-Arid Tropics (ICRISAT) and the United States Department of Agriculture (USDA), play a pivotal role in conserving the genetic diversity of peanuts. These collections maintain thousands of accessions of wild and cultivated peanut species, providing a foundational resource for breeding programs aimed at addressing agricultural challenges and advancing crop improvement (Bramel, 2021; Desmae et al., 2019; Dulloo et al., 2010). ICRISAT's genebank, housing over 15,000 Arachis accessions, is among the world's largest peanut collections. It preserves a wide genetic base, comprising numerous accessions of both *Arachis hypogaea* and wild Arachis species. Similarly, the USDA manages a significant collection of about 9,000 *A. hypogaea* accessions and 500 wild species, primarily for distribution and research, ensuring preservation through storage at optimal temperatures for long-term viability (Dulloo et al., 2010; Mascher et al., 2019). Other major centers include the EMBRAPA Genetic Resources and Biotechnology center in Brazil and the National Bureau of Plant Genetic Resources in India, where efforts support peanut diversity preservation on a global scale (Desmae et al., 2019). These collections are not only essential for the conservation of genetic diversity but also for supporting sustainable agricultural advancements. By characterizing accessions with molecular markers, such as SSRs and SNPs, researchers gain precision in evaluating genetic diversity, enabling the identification of optimal parental lines for breeding. This level of genetic insight is fundamental for creating high-yield, disease-resistant, and climate-resilient cultivars, which are critical for sustaining peanut production amid global challenges like climate change and disease pressure (Dar et al., 2015; Upadhyaya et al., 2003; Desmae et al., 2019). In essence, the preservation and utilization of these genetic resources enable continued progress in peanut improvement, contributing to food security, enhanced agricultural productivity, and environmental resilience. By safeguarding these genetic resources and applying advanced molecular techniques, germplasm collections underpin future breeding strategies, ensuring adaptability and nutritional value for generations to come (Upadhyaya et al., 2003; Wang et al., 2011).

1.4.2 The Role of Wild Relatives in Expanding Genetic Diversity

The effective utilization and characterization of genetic resources are vital for the advancement of any crop improvement program. In peanuts, the narrow genetic base of cultivated varieties makes it essential to incorporate diverse germplasm from both wild relatives and cultivated accessions. This genetic diversity is crucial for developing peanut varieties with enhanced yields, improved nutritional quality, and greater resilience against biotic and abiotic stresses (Gepts, 2006; Barkley et al., 2016; Abeli et al., 2020). Wild peanut species, conserved in global germplasm collections, offer a vast pool of untapped genetic variation that is invaluable for breeding efforts and long-term crop sustainability. Technological advancements in cytogenetics, including fluorescent in situ hybridization (FISH) and genome in situ hybridization (GISH), have greatly facilitated the exploration of genetic relationships among wild and cultivated peanut species. These techniques help elucidate the organization of the secondary gene pool and the evolutionary processes that shaped the origin of cultivated peanuts. Such studies are fundamental to understanding the mechanisms of chromosomal differentiation and genome organization in peanuts, thus guiding breeders in effectively harnessing traits from wild species (Volis and Blecher, 2010; Barkley et al., 2016). Incorporating genetic material from wild relatives into cultivated peanuts has led to some of the most impactful improvements in peanut breeding. Wild species such as *Arachis duranensis*, *Arachis ipaensis*, *Arachis stenosperma*, and *Arachis magna* have contributed valuable genes for traits often lacking in cultivated peanuts. These traits include resistance to root-knot nematodes, leaf spots, and rust and tolerance to drought, which significantly enhance the adaptability and resilience of peanut crops under diverse environmental stresses (Barkley et al., 2016; Meilleur and Hodgkin, 2004). In particular, *Arachis ipaensis* and *Arachis duranensis* are recognized as the wild progenitors of cultivated peanut, and their genomic contributions have been instrumental in identifying genes linked to yield improvement and stress resistance. In situ conservation of wild peanut species has multiple advantages, including adaptation to local environmental conditions, natural gene flow, and the preservation of genetic diversity within native habitats. This approach ensures that the genetic potential of these wild species is safeguarded for future breeding efforts, particularly as climate change introduces new challenges to agriculture (Meilleur and Hodgkin, 2004; Abeli et al., 2020). By leveraging the genetic resilience found in wild relatives, breeders can continue to expand the genetic diversity of cultivated peanuts, supporting the development of varieties that are more productive, sustainable, and suited to meet global agricultural demands.

1.4.3 Peanut Germplasm Creation and Utilization

Historically, peanut breeding efforts focused on selecting desirable traits from the limited genetic material found within cultivated varieties. This approach restricted genetic improvements due to the narrow diversity within the gene pool, limiting the introduction of novel traits and advancements. Challenges such as aflatoxin contamination, a major global threat to peanut production, remained difficult to address with traditional breeding methods alone (Torres et al., 2014; Meneely et al., 2022; Jallow et al., 2021; Norlia et al., 2019; Krska et al., 2022). In recent decades, global breeding programs—particularly in China, India, and the United States—have made significant strides in identifying the genetic basis of resistance to aflatoxin contamination. Through extensive screening, researchers have discovered germplasm lines with potential aflatoxin tolerance, marking a critical advancement for future breeding initiatives aimed at reducing this toxin in peanut crops (Nigam et al., 2009; Wang et al., 2016, 2016; Faustinelli et al., 2017; Yu et al., 2019; Dieme et al., 2018; Sobolev et al., 2018; Pandey et al., 2019; Khan et al., 2020; Yu et al., 2020; Soni et al., 2020; Tengey et al., 2022; Ding et al., 2022; Zhou et al., 2022). Despite these advances, the full genetic potential of existing germplasm remains underutilized due to the limited availability of molecular and genomic tools that can facilitate effective trait transfer. To overcome these limitations, modern peanut breeding has incorporated a diverse range of populations designed to capture broader genetic variation. These include recombinant MAGIC and NAM populations, recombinant inbred lines (RILs), mutant populations, and backcross populations, all of which are rigorously characterized to assess traits such as yield, drought tolerance, and disease resistance. Molecular markers like SNPs and SSRs are increasingly used in conjunction with phenotypic data, providing breeders with a precise framework for selecting desirable traits. Additionally, collections from various peanut genebanks have been evaluated for morphological and agronomic characteristics, offering a critical foundation for enhancing peanut germplasm with targeted attributes (Liang et al., 2009; Korani et al., 2018; Arias et al., 2018). These advancements are not just improving individual varieties but are also contributing to a broader genetic base within peanut breeding programs. By integrating new genetic material and focusing on traits related to stress resistance and nutritional quality, breeders are establishing a pipeline for sustained improvement in peanut crop resilience and productivity.

Looking forward, the continued characterization and effective utilization of peanut germplasm will be essential for addressing emerging challenges in peanut production, especially as climate change brings new environmental stresses. Access to diverse genetic resources, particularly those conserved in global germplasm collections and wild species, will be fundamental for developing peanut varieties capable of withstanding unpredictable conditions. By combining traditional breeding with molecular techniques, such as marker-assisted selection (MAS) and quantitative trait locus (QTL) mapping, and using advanced genotyping tools, breeders are progressively developing peanut varieties that are more productive, resilient, and adaptable to the demands of a changing climate. This integrated approach ensures that the genetic diversity within peanut germplasm can be fully harnessed to meet future agricultural needs.

1.5 BIOTECHNOLOGICAL TOOLS IN PEANUT IMPROVEMENT

1.5.1 Modern Biotechnology Tools in Peanut Improvement

The landscape of peanut breeding has been transformed through the integration of cutting-edge biotechnology tools, including transcriptomics, proteomics, metabolomics, and genomics. This shift has expanded the scope of peanut research beyond conventional methods, enabling scientists to delve deeply into molecular mechanisms that influence essential traits like disease resistance, drought tolerance, and oil quality.

A major research focus has been the peanut plant's interaction with *Aspergillus flavus*, the fungus responsible for producing aflatoxin—a harmful contaminant that poses health risks and economic losses. Recent advancements, such as the development of a pan-genome for *A. flavus* by Gangurde et al. (2024), allow researchers to explore the genetic basis of aflatoxin contamination and identify potential pathways for engineering aflatoxin-resistant peanut varieties. By harnessing omics technologies, scientists can now pinpoint stress-related alleles, genes, proteins, and metabolites that contribute to peanut resilience under challenging environmental conditions.

Plant growth and productivity in peanuts are often hindered by abiotic stresses (e.g., drought, salinity) and biotic stresses (e.g., diseases, pests). These stresses impact peanut plants across various levels, from morphological and physiological responses to biochemical, cellular, and molecular adaptations. By analyzing these responses through a multi-omics approach, researchers gain a comprehensive view of how different genes, proteins, and metabolites respond to stress, paving the way for breeding more robust, high-yielding cultivars under biotic (Zhao et al., 2021; Gong et al., 2020; Guimaraes et al., 2015; Rathod et al., 2020; Mishra et al., 2015; Soni et al., 2020; Kumar and Kirti, 2015; Cui et al., 2022; Zhang et al., 2020; Mota et al., 2018; Sharma et al., 2021; Bhatnagar-Mathur et al., 2021; Wang et al., 2012) and abiotic stresses (Reddy et al., 2012; Aravind et al., 2022; Zhao et al., 2021; Jiang et al., 2021; Brasileiro et al., 2015; Gundaraniya et al., 2020; Patel et al., 2022; Wu et al., 2022; Katam et al., 2016; Zhang et al., 2020; Chen et al., 2016; Jiang et al., 2020; Sui et al., 2018; Thoppurathu et al., 2022; Cui et al., 2018). Advances in peanut genomics, coupled with available reference genome sequences, have accelerated the application of omics in peanut breeding. This integration has facilitated the identification of stress-associated genes, pathways, and phenotypes that inform breeding strategies. A wealth of omics-based studies has already contributed to peanut improvement, enhancing both biotic and abiotic stress tolerance through the identification of specific alleles and molecular markers associated with resilience (Raza et al., 2021, 2022a, 2022b, 2022c, 2022d, 2022e; Zhao et al., 2021; Jiang et al., 2021; Saeed et al., 2023).

Biotechnology has expanded the potential for peanut yield enhancement, quality improvement, and stress management, especially when the existing germplasm lacks certain advantageous traits. In cases where traditional breeding methods fall short, biotechnological interventions open up new avenues (Liu et al., 2013; Raza et al., 2022a; Krishna et al., 2015; Zhu et al., 2020). For instance, gene-editing technologies, transformation systems, and CRISPR/Cas-based techniques enable targeted modifications, allowing breeders to integrate desirable traits like drought tolerance or disease resistance with unprecedented precision and efficiency. Peanut breeders and researchers are at the frontier of using advanced genomics, phenomics, and machine learning to address the crop's limitations, thus securing its place in the future of sustainable agriculture. With each new development in peanut biotechnology, breeders can achieve improvements that are faster, more precise, and increasingly resilient to the unpredictable challenges of modern agriculture. The future of peanut breeding lies in these integrated, data-driven approaches that promise to transform the crop's role in both global food security and agricultural innovation.

This convergence of technology and traditional breeding is not only making peanut breeding smarter but also setting a benchmark for efficiency and sustainability in crop improvement—a compelling reason for breeders to delve deeper into the potential of modern biotechnology in agriculture.

1.5.2 Potentials and Application of Bioinformatics in Peanut Breeding

In the past decade, the integration of high-throughput sequencing technologies—such as next-generation sequencing (NGS), genotyping-by-sequencing (GBS), and RNA sequencing (RNA-seq)—has revolutionized peanut breeding by generating extensive genetic and genomic data. These advancements have empowered researchers to sequence entire genomes, identify single nucleotide polymorphisms (SNPs), and elucidate gene expression patterns, thereby establishing a robust genetic and molecular foundation for the enhancement of peanut traits. The emergence of advanced computational models and bioinformatics tools has further facilitated global collaboration across various disciplines, including genetics, molecular biology, and environmental science, enabling deeper insights into the complex interactions that influence peanut traits.

Bioinformatics has become indispensable in advancing peanut breeding by integrating statistical methodologies with sophisticated computational tools to manage, analyze, and store large volumes of complex molecular data. This encompasses genomic sequences, protein structures, and gene expression profiles—crucial elements for understanding the genetic architecture of important agronomic traits (Esposito et al., 2016; Iquebal et al., 2015; Kumar et al., 2022). By effectively handling and interpreting these datasets, bioinformatics enables the seamless integration of genomics and phenomics, driving the development of high-yielding and climate-resilient peanut cultivars (Marsh et al., 2021). The development of molecular markers, expressed sequence tags (ESTs), and comprehensive genomic libraries has significantly bolstered the characterization of cultivated peanut species. These resources are fundamental for pinpointing and isolating traits associated with productivity and resistance to biotic and abiotic stressors (Srikanth et al., 2025). Additionally, bioinformatics supports the construction of high-resolution genetic maps, aiding in the precise identification of quantitative trait loci (QTLs) linked to traits such as disease resistance and stress tolerance. Tools like GATK, Samtools, and specialized pipelines have proven effective for SNP calling and QTL mapping in peanuts (Han et al., 2018; Yu et al., 2020). The application of transcriptome analysis through RNA-seq has offered deeper insights into gene expression and SNP discovery. Bioinformatics pipelines that incorporate tools like HISAT2, StringTie, and EdgeR facilitate the alignment, quantification, and analysis of gene expression, helping identify differentially expressed genes (DEGs) related to oil biosynthesis, stress response, and other agronomic traits (Clevenger et al., 2016; Peng et al., 2020). This approach supports breeders in identifying key genes for targeted trait enhancement.

Bioinformatics also plays a critical role in the processing, storage, and distribution of large-scale data generated by genome programs, encompassing details about sequence variations, gene mutations, and functional annotations that inform peanut physiology and adaptation mechanisms (Vassilev et al., 2005). Centralized databases like PeanutBase and the Legume Information System (LIS) provide essential platforms for researchers and breeders to access, analyze, and share omics data, thus fostering collaborative research and efficient breeding strategies (Dash et al., 2016; Berendzen et al., 2021). Furthermore, the integration of bioinformatics tools into breeding programs has streamlined the development of new cultivars with improved traits. By incorporating genomic tools alongside traditional breeding practices, researchers can address the challenges posed by climate change and global food demand, enhancing traits such as yield, oil content, and stress resilience while minimizing environmental and economic impacts (Mallikarjuna and Varshney, 2014). High-throughput phenotyping technologies complement these efforts by providing real-time data on plant traits, enabling breeders to make informed decisions for introducing desirable characteristics into cultivated varieties (Marsh et al., 2021).

Looking forward, the future of bioinformatics in peanut breeding includes leveraging machine learning (ML) and artificial intelligence (AI) to decode gene regulatory networks and predict gene functions. This advancement could accelerate the discovery of novel resistance genes, optimize breeding strategies, and promote the development of peanut cultivars that are better adapted to environmental stressors, supporting sustainability and food security (Yang et al., 2021; Xu et al., 2023).

In conclusion, the strategic application of bioinformatics in peanut genomics has paved the way for the more effective utilization of plant genetic resources. This integration supports the breeding of elite cultivars that meet market needs while promoting sustainable agricultural practices, positioning peanut breeding at the forefront of agricultural innovation and contributing to global food security (Vassilev et al., 2005).

1.5.3 From Genomics to Genome Editing

The understanding of peanut genomics has paved the way for the application of advanced genome-editing technologies, such as CRISPR/Cas9, to precisely modify genes associated with desirable traits. CRISPR/Cas9 technology is already being utilized to enhance disease resistance, drought tolerance, and nutritional quality in peanuts (Shi et al., 2023; Huai et al., 2024). The ability to directly edit genes holds immense potential for accelerating the breeding process and addressing challenges that traditional breeding methods cannot overcome. Recent advances in peanut genomics and genome editing have opened new avenues for crop improvement. Protoplast-based systems have been optimized for transient CRISPR-Cas9 editing, enabling rapid validation of single guide RNAs (sgRNAs) targeting allergen genes like *Ara h 2* (Biswas et al., 2022). Additionally, base editing techniques using nCas9 fused with deaminases have shown promise in modifying *FAD2* genes to increase oleic acid content in peanut oil (Neelakandan et al., 2022). Various genome editing technologies, including ZFNs (zinc finger nucleases), TALENs (transcription activator-like effector nucleases), and CRISPR-Cas9, are being applied to improve traits such as yield, disease resistance, and drought tolerance in peanuts (Lamboro et al., 2021). These advancements have transformed peanuts from a "resource-poor" crop to a "genomic resource-rich" crop, enabling the use of genomics information and tools in breeding programs (Varshney et al., 2017). The integration of these genomic approaches is expected to accelerate peanut improvement, addressing critical challenges in food security and crop productivity. By leveraging these cutting-edge technologies, researchers can develop new peanut cultivars with enhanced traits, contributing to global food security and sustainable agricultural practices.

1.6 FUTURE PERSPECTIVES

Peanut genomics and biotechnology hold vast potential to transform and address critical challenges posed by biotic and abiotic factors. Precision breeding, particularly through genome editing, will enable more targeted genetic modifications, introducing traits such as improved drought tolerance, pest resistance, and enhanced nutritional value. The integration of multi-omics will help to understand the molecular mechanisms under the key agronomic traits. This multi-omics, in combination with advanced bioinformatics, will accelerate the discovery of novel genes and pathways critical for crop improvement. Developing climate-resilient peanut varieties will be an essential focus as abiotic and biotic stresses continue to escalate. Genomic tools will allow breeders to select traits that enable peanuts to thrive in increasingly extreme environmental conditions. Advances in molecular cytogenetics and genomic technologies will help overcome the challenges of utilizing wild germplasm to incorporate valuable traits such as disease resistance and environmental adaptation. The incorporation of AI and ML in breeding programs will revolutionize data-driven breeding efforts. These technologies will enhance the precision of trait prediction, streamline the identification of genetic markers, and optimize breeding strategies, leading to faster and more efficient development cycles for new peanut varieties. Additionally, as global demand for nutrient-dense, sustainable food increases, future breeding efforts will focus on developing peanuts with improved nutritional profiles, such as higher levels of healthy fats and bioactive compounds. Finally, global collaboration and data sharing are crucial for advancing peanut research. Platforms such as PeanutBase and the Peanut Genome Resource will facilitate international cooperation, enabling the dissemination of genomic data, germplasm, and breeding knowledge to accelerate peanut research. Together, these innovations will contribute to the creation of high-yield, stress-resistant, and nutritionally rich peanut varieties that support global food security.

1.7 ACKNOWLEDGEMENTS

Weijian Zhuang, Ali Raza, and Sheidu Abdullaziz are grateful to the National Natural Science Foundation of China, the Ministry of Science and Technology of China, the Department of Science and Technology of Fujian, and Fujian Agriculture and Forestry University, China for supporting our peanut genomics and biotechnology work at Key Laboratory of Ministry of Education for Genetics, Breeding and Multiple Utilization of Crops, Center of Legume Plant Genetics and Systems Biology, College of Agriculture, Oil Crops Research Institute, Fujian Agriculture and Forestry University (FAFU), China. Rajeev K. Varshney is thankful to the Bill & Melinda Gates Foundation, United States; and the Australia–India Strategic Research Fund from the Department of Biotechnology, Government of India for supporting the work at Center of Excellence in Genomics & Systems Biology, International Crops Research Institute for the Semi-Arid Tropics (ICRISAT), Hyderabad, India; and also the Food Futures Institute of Murdoch University, Australia for supporting this work in part. Xingjun Wang is thankful to the National Natural Science Foundation of China, Taishan Scholar Project of Shandong Province, Innovation Project of Shandong Academy of Agricultural Sciences. Lastly, the editors are also thankful to all the authors for their significant contribution to developing this up-to-date one-stop peanut book for the peanut community.

REFERENCES

Abeli, T., S. Dalrymple, S. Godefroid, A. Mondoni, J. V. Müller, G. Rossi, and S. Orsenigo. "Ex situ collections and their potential for the restoration of extinct plants." *Conservation Biology* 34 (2020): 303–313.

Aravind, B., S. N. Nayak, R. S. Choudhary, S. S. Gandhadmath, P. Prasad, M. K. Pandey, R. S. Bhat, N. Puppala, P. Latha, and P. Sudhakar. "Integration of genomics approaches in abiotic stress tolerance in groundnut (Arachis hypogaea L.): an overview." *Genomic Designing for Abiotic Stress Resistant Oilseed Crops* (2022): 149–197.

Arias, Renee S., Victor S. Sobolev, Alicia N. Massa, Valerie A. Orner, Travis E. Walk, Linda L. Ballard, Sheron A. Simpson et al. "New tools to screen wild peanut species for aflatoxin accumulation and genetic fingerprinting." *BMC Plant Biology* 18 (2018): 1–13.

Aydin, C. "Some engineering properties of peanut and kernel." *Journal of Food Engineering* 79 (3) (2007): 810–816.

Barkley, Noelle A., Hari D. Upadhyaya, Boshou Liao, and C. Corley Holbrook. "Global resources of genetic diversity in peanut." In *Peanuts*, 67–109. AOCS Press, 2016.

Bell, S. "Peanuts and their classification under the HTSUS." *An Informed Compliance Publication* 22 (2008): 1–21.

Bera, S. K., J. H. Kamdar, S. V. Kasundra, P. Dash, A. K. Maurya, M. D. Jasani, A. B. Chandrashekar, N. Manivannan, R. Vasanthi, and K. Dobariya. "Improving oil quality by altering levels of fatty acids through marker-assisted selection of ahfad2 alleles in peanut (*Arachis hypogaea* L.)." *Euphytica* 214 (2018): 1–15.

Bera, S. K., J. H. Kamdar, S. V. Kasundra, S. V. Patel, M. D. Jasani, A. Maurya, P. Dash, A. B. Chandrashekar, K. Rani, and N. Manivannan. "Steady expression of high oleic acid in peanut bred by marker-assisted backcrossing for fatty acid desaturase mutant alleles and its effect on seed germination along with other seedling traits." *PLoS One* 14 (12) (2019): e0226252.

Berendzen, Joel, Anne V. Brown, Connor T. Cameron, Jacqueline D. Campbell, Alan M. Cleary, Sudhansu Dash, Samuel Hokin et al. "The legume information system and associated online genomic resources." *Legume Science* 3 (3) (2021): e74.

Bertioli, D. J., S. B. Cannon, L. Froenicke, G. Huang, A. D. Farmer, E. K. Cannon, X. Liu, D. Gao, J. Clevenger, and S. Dash. "The genome sequences of Arachis duranensis and Arachis ipaensis, the diploid ancestors of cultivated peanut." *Nature Genetics* 48 (4) (2016): 438–446. https://org/10.1038/ng.3517

Bertioli, D. J., J. Jenkins, J. Clevenger, O. Dudchenko, D. Gao, G. Seijo, S. Leal-Bertioli, et al. "The genome sequence of segmental allotetraploid peanut *Arachis hypogaea*." *Nature Genetics* 51 (5) (2019): 877–884.

Bertioli, David J., Brian Abernathy, Guillermo Seijo, Josh Clevenger, and Steven B. Cannon. "Evaluating Two Different Models of Peanut's Origin." Nature Genetics 52 (6) (2020): 557–559.

Bhat, R., M. Jadhav, P. Patil, and K. Shirasawa. "Genomics-assisted breeding for resistance to leaf spots and rust diseases in peanut." In *Accelerated Plant Breeding, Volume 4: Oil Crops*, edited by Satbir Singh Gosal and Shabir Hussain Wani, 239–278. Cham: Springer, 2022.

Bhatnagar-Mathur, P., K. Yogendra, S. Parankusam, H. Sanivarapu, K. Prasad, S. B. Lingampali, and K. K. Sharma. "Comparative proteomics provide insights on the basis of resistance to Aspergillus flavus infection and aflatoxin production in peanut (Arachis hypogea L.)." *Journal of Plant Interactions* 16(1) (2021): 494–509.

Biswas, Sudip, Nancy J. Wahl, Michael J. Thomson, John M. Cason, Bill F. McCutchen, and Endang M. Septiningsih. "Optimization of protoplast isolation and transformation for a pilot study of genome editing in peanut by targeting the allergen gene Ara h 2." *International Journal of Molecular Sciences* 23 (2) (2022): 837.

Bohra, A., A. Tiwari, P. Kaur, S. A. Ganie, A. Raza, M. Roorkiwal, R. R. Mir, A. R. Fernie, P. Smýkal, and R. K. Varshney. "The key to the future lies in the past: insights from grain legume domestication and improvement should inform future breeding strategies." *Plant and Cell Physiology* 63 (11) (2022): 1554–1572.

Bouchez, D., and H. Hőfte. "Functional genomics in plants." *Plant physiology* 118 (3) (1998): 725–732.

Bramel P. Key issues facing genebanks in preserving crop genetic diversity ex situ: overview of the range of challenges. Plant genetic resources 15 (2021):139-153.

Brasileiro, A. C., C. V. Morgante, A. C. Araujo, S. C. Leal-Bertioli, A. K. Silva, A. C. Martins, C. C. Vinson, C. M. Santos, O. Bonfim, and R. C. Togawa. "Transcriptome profiling of wild Arachis from water-limited environments uncovers drought tolerance candidate genes." *Plant Molecular Biology Reporter* 33 (2015): 1876–1892.

Brasileiro Miranda A. C., A. Guerra Araujo, S. C. Leal-Bertioli, and P. M. Guimarães. "Genomics and Genetic Transformation in Arachis." International Journal of Plant Biology Research 2 (3) (2014): 1017.

Chen, N., M. Su, X. Chi, Z. Zhang, L. Pan, M. Chen, T. Wang, M. Wang, Z. Yang, and S. Yu. "Transcriptome analysis reveals salt-stress-regulated biological processes and key pathways in roots of peanut (Arachis hypogaea L.)." *Genes & Genomics* 38 (2016): 493–507.

Chen, X., Q. Lu, H. Liu, J. Zhang, Y. Hong, H. Lan, H. Li, J. Wang, H. Liu, and S. Li. "Sequencing of cultivated peanut, Arachis hypogaea, yields insights into genome evolution and oil improvement." *Molecular Plant* 12 (7) (2019): 920–934.

Choudhary, D., G. Agarwal, H. Wang, M. K. Pandey, A. K. Culbreath, R. K. Varshney, and B. Guo. "Molecular markers and genomic resources for disease resistance in peanut-A review." *Legume Research-An International Journal* 42 (2) (2019): 137–144.

Chu, Y., C. Wu, C. Holbrook, B. Tillman, G. Person, and P. Ozias-Akins. "Marker-assisted selection to pyramid nematode resistance and the high oleic trait in peanut." *The Plant Genome* 4 (2) (2011): 110–117.

Clevenger, Josh, Ye Chu, Brian Scheffler, and Peggy Ozias-Akins. "A developmental transcriptome map for allotetraploid Arachis hypogaea." *Frontiers in Plant Science* 7 (2016): 1446.

Cui, F., N. Sui, G. Duan, Y. Liu, Y. Han, S. Liu, S. Wan, and G. Li. "Identification of metabolites and transcripts involved in salt stress and recovery in peanut." *Frontiers in Plant Science* 9 (2018): 217.

Cui, M., S. Han, D. Wang, M. S. Haider, J. Guo, Q. Zhao, P. Du, Z. Sun, F. Qi, and Z. Zheng. "Gene co-expression network analysis of the comparative transcriptome identifies hub genes associated with resistance to Aspergillus flavus L. in cultivated peanut (Arachis hypogaea L.)." *Frontiers in Plant Science* 13 (2022): 899177.

Dar, Rafiq A., Manzoor Ahmad, Sunil Kumar, and M. Reshi. "Agricultural germplasm resources: A tool of conserving diversity." Scientific Research and Essays 10 (2015): 326–338.

Dash, Sudhansu, Ethalinda KS Cannon, Scott R. Kalberer, Andrew D. Farmer, and Steven B. Cannon. "PeanutBase and other bioinformatic resources for peanut." In *Peanuts*, 241–252. AOCS Press, 2016.

Desmae, Haile, P. Janila, P. Okori, M. K. Pandey, B. N. Motagi, E. Monyo, O. Mponda, D. Okello, D. Sako, C. Echeckwu, R. Oteng-Frimpong, A. Miningou, C. Ojiewo, and Rajeev K. Varshney. "Genetics, genomics and breeding of groundnut (Arachis hypogaea L.)." Plant Breeding 138 (4) (2019): 425–444.

Dieme, Richard Moise Alansou, Issa Faye, Yedomon Ange Bovys Zoclanclounon, Daniel Fonceka, Ousmane Ndoye, and Papa Madiallacke Diedhiou. "Identification of sources of resistance for peanut Aspergillus flavus colonization and aflatoxin contamination." *International Journal of Agronomy* 2018 (1) (2018): 5468602.

Ding, Y., X. Qiu, H. Luo, L. Huang, J. Guo, B. Yu, H. Sudini, M. Pandey, Y. Kang, and N. Liu. "Comprehensive evaluation of Chinese peanut mini-mini core collection and QTL mapping for aflatoxin resistance." *BMC Plant Biology* 22 (1) (2022): 207.

Dulloo, Ehsan M., Danny Hunter, and Teresa Borelli. "Ex situ and in situ conservation of agricultural biodiversity: Major advances and research needs." Notulae Botanicae Horti Agrobotanici Cluj-Napoca 38 (2010): 123–135.

Esposito, Alfonso, Chiara Colantuono, Valentino Ruggieri, and Maria Luisa Chiusano. "Bioinformatics for agriculture in the next-generation sequencing era." *Chemical and Biological Technologies in Agriculture* 3 (2016): 1–12.

Faustinelli, Paola C., Edwin R. Palencia, Victor S. Sobolev, Bruce W. Horn, Hank T. Sheppard, Marshall C. Lamb, Xinye M. Wang, Brian E. Scheffler, Jaime Martinez Castillo, and Renée S. Arias. "Study of the genetic diversity of the aflatoxin biosynthesis cluster in Aspergillus section Flavi using insertion/deletion markers in peanut seeds from Georgia, USA." *Mycologia* 109 (2) (2017): 200–209.

Gangurde, S. S., A. W. Khan, P. Janila, M. T. Variath, S. S. Manohar, P. Singam, A. Chitikineni, R. K. Varshney, and M. K. Pandey. "Whole-genome sequencing based discovery of candidate genes and diagnostic markers for seed weight in groundnut." *The Plant Genome* (2022): e20265.

Gangurde, Sunil S., Walid Korani, Prasad Bajaj, Hui Wang, Jake C. Fountain, Gaurav Agarwal, Manish K. Pandey et al. "Aspergillus flavus pangenome (AflaPan) uncovers novel aflatoxin and secondary metabolite associated gene clusters." *BMC Plant Biology* 24 (1) (2024): 354.

Ge, S., Y. Wu, W. Peng, C. Xia, C. Mei, L. Cai, S. Q. Shi, C. Sonne, S. S. Lam, and Y. F. Tsang. "High-pressure CO2 hydrothermal pretreatment of peanut shells for enzymatic hydrolysis conversion into glucose." *Chemical Engineering Journal* 385 (2020): 123949.

Gepts, Paul. "Plant genetic resources conservation and utilization: The accomplishments and future of a societal insurance policy." *Crop Science* 46 (5) (2006): 2278–2292.

Ghosh, S., S. S. Mahadevaiah, S. A. Gowda, S. S. Gangurde, M. P. Jadhav, A. A. Hake, P. Latha, T. Anitha, V. Chimmad, and K. K. Mirajkar. "Genetic mapping of drought tolerance traits phenotyped under varying drought stress environments in peanut (Arachis hypogaea L.)." *Euphytica* 218 (12) (2022): 168.

Gong, L., S. Han, M. Yuan, X. Ma, A. Hagan, and G. He. "Transcriptomic analyses reveal the expression and regulation of genes associated with resistance to early leaf spot in peanut." *BMC Research Notes* 13 (1) (2020): 1–6.

Gonzalez Guzman, M., F. Cellini, V. Fotopoulos, R. Balestrini, and V. Arbona. "New approaches to improve crop tolerance to biotic and abiotic stresses." *Physiologia Plantarum* 174 (1) (2022): e13547.

Guimaraes, P. M., L. A. Guimaraes, C. V. Morgante, O. B. Silva Jr, A. C. G. Araujo, A. C. Martins, M. A. Saraiva, T. N. Oliveira, R. C. Togawa, and S. C. Leal-Bertioli. "Root transcriptome analysis of wild peanut reveals candidate genes for nematode resistance." *PLoS One* 10 (10) (2015): e0140937.

Guimarães, Patricia M., A. C. M. Brasileiro, A. Mehta, and A. C. G. Araujo. "Functional genomics in peanut wild relatives." *The Peanut Genome* (2017): 149–164.

Gundaraniya, S. A., P. S. Ambalam, and R. S. Tomar. "Metabolomic profiling of drought-tolerant and susceptible peanut (Arachis hypogaea L.) genotypes in response to drought stress." *ACS Omega* 5 (48) (2020): 31209–31219.

Guo, B., C. Chen, Y. Chu, C. Holbrook, P. Ozias-Akins, and H. Stalker. "Advances in genetics and genomics for sustainable peanut production." *Sustainable Agriculture and New Biotechnologies* (2011): 341–367.

Han, Suoyi, Mei Yuan, Josh P. Clevenger, Chun Li, Austin Hagan, Xinyou Zhang, Charles Chen, and Guohao He. ""A SNP-based linkage map revealed QTLs for resistance to early and late leaf spot diseases in peanut (Arachis hypogaea L.)." *Frontiers in Plant Science* 9 (2018): 1012.

Hathorn, C. S., and T. H. Sanders. "Flavor and antioxidant capacity of peanut paste and peanut butter supplemented with peanut skins." *Journal of Food Science* 77 (11) (2012): S407–S411.

Huai, Dongxin, Xiaomeng Xue, Jie Wu, Manish K. Pandey, Nian Liu, Li Huang, Liying Yan et al. "Enhancing peanut nutritional quality by editing AhKCS genes lacking natural variation." *Plant Biotechnology Journal* 22 (11) (2024): 3015.

Iquebal, M. A., Sarika Jaiswal, C. S. Mukhopadhyay, Chiranjib Sarkar, Anil Rai, and Dinesh Kumar. "Applications of bioinformatics in plant and agriculture." *PlantOmics: The Omics of Plant Science* (2015): 755–789.

Jadhav, M. P., M. D. Patil, M. Hampannavar, P. Dattatreya, K. Shirasawa, J. Pasupuleti, M. K. Pandey, R. K. Varshney, and R. S. Bhat. "Enhancing oleic acid content in two commercially released peanut varieties through marker-assisted backcross breeding." *Crop Science* 61 (4) (2021): 2435–2443.

Jallow, Abdoulie, Huali Xie, Xiaoqian Tang, Zhang Qi, and Peiwu Li. "Worldwide aflatoxin contamination of agricultural products and foods: From occurrence to control." *Comprehensive Reviews in Food Science and Food Safety* 20 (3) (2021): 2332–2381.

Janila, P., S. Nigam, M. K. Pandey, P. Nagesh, and R. K. Varshney. "Groundnut improvement: use of genetic and genomic tools." *Frontiers in Plant Science* 4 (2013): 23.

Jiang, C., H. Zhang, J. Ren, J. Dong, X. Zhao, X. Wang, J. Wang, C. Zhong, S. Zhao, and X. Liu. "Comparative transcriptome-based mining and expression profiling of transcription factors related to cold tolerance in peanut." *International Journal of Molecular Sciences* 21 (6) (2020): 1921.

Jiang, C., X. Li, J. Zou, J. Ren, C. Jin, H. Zhang, H. Yu, and H. Jin. "Comparative transcriptome analysis of genes involved in the drought stress response of two peanut (Arachis hypogaea L.) varieties." *BMC Plant Biology* 21 (1) (2021): 1–14.

Katam, R., K. Sakata, P. Suravajhala, T. Pechan, D. M. Kambiranda, K. S. Naik, B. Guo, and S. M. Basha. "Comparative leaf proteomics of drought-tolerant and-susceptible peanut in response to water stress." *Journal of Proteomics* 143 (2016): 209–226.

Khan, Shahid Ali, Hua Chen, Ye Deng, Yuhua Chen, Chong Zhang, Tiecheng Cai, Niaz Ali et al. "High-density SNP map facilitates fine mapping of QTLs and candidate genes discovery for Aspergillus flavus resistance in peanut (Arachis hypogaea)." *Theoretical and Applied Genetics* 133 (2020): 2239–2257.

Krishna, G., B. K. Singh, E. K. Kim, V. K. Morya, and P. W. Ramteke. "Progress in genetic engineering of peanut (Arachis hypogaea L.)—A review." *Plant Biotechnology Journal* 13 (2) (2015): 147–162.

Kumar, D., and P. B. Kirti. "Transcriptomic and proteomic analyses of resistant host responses in Arachis diogoi challenged with late leaf spot pathogen, Phaeoisariopsis personata." *PLoS One* 10 (2) (2015): e0117559.

Kumar, Pradeep, Abhishek Singh, Vishnu D. Rajput, Ajit Kumar Singh Yadav, Pravin Kumar, Anil Kumar Singh, and Tatiana Minkina. "Role of artificial intelligence, sensor technology, big data in agriculture: next-generation farming." In *Bioinformatics in Agriculture*, 625–639. Academic Press, 2022.

Kumar, R., P. Janila, M. K. Vishwakarma, A. W. Khan, S. S. Manohar, S. S. Gangurde, M. T. Variath, Y. Shasidhar, M. K. Pandey, and R. K. Varshney. "Whole-genome resequencing-based QTL-seq identified candidate genes and molecular markers for fresh seed dormancy in groundnut." *Plant Biotechnology Journal* 18 (4) (2020): 992–1003.

Korani, Walid, Ye Chu, C. Corley Holbrook, and Peggy Ozias-Akins. "Insight into genes regulating postharvest aflatoxin contamination of tetraploid peanut from transcriptional profiling." *Genetics* 209 (1) (2018): 143–156.

Krska, Rudolf, John F. Leslie, Simon Haughey, Moira Dean, Yoneal Bless, Oonagh McNerney, Michelle Spence, and Chris Elliott. "Effective approaches for early identification and proactive mitigation of aflatoxins in peanuts: An EU–China perspective." *Comprehensive Reviews in Food Science and Food Safety* 21 (4) (2022): 3227–3243.

Lamboro, Abraham, Baixing Song, Yang Songnan, Xiao Han, Hao Mingguo, Xueying Li, Dan Yao, and Jun Zhang. "Genetic engineering and genome editing techniques in peanut plants." *Plant Science Today* 8 (3) (2021): 528–534.

Li, Y., L. Li, X. Zhang, K. Zhang, D. Ma, J. Liu, X. Wang, F. Liu, and Y. Wan. "QTL mapping and marker analysis of main stem height and the first lateral branch length in peanut (Arachis hypogaea L.)." *Euphytica* 213 (2) (2017): 57.

Liang, Xuanqiang, Guiyuan Zhou, Yanbin Hong, Xiaopin Chen, Haiyan Liu, and Shaoxiong Li. "Overview of research progress on peanut (Arachis hypogaea L.) host resistance to aflatoxin contamination and genomics at the Guangdong Academy of Agricultural Sciences." *Peanut Science* 36 (1) (2009): 29–34.

Liao, B. "Germplasm characterization and trait discovery in peanut." *The Peanut Genome* (2017): 53–68.

Liu, W., J. S. Yuan, and C. N. Stewart Jr. "Advanced genetic tools for plant biotechnology." *Nature Reviews Genetics* 14 (11) (2013): 781–793.

Liu, Y., L. Shao, J. Zhou, R. Li, M. K. Pandey, Y. Han, F. Cui, J. Zhang, F. Guo, and J. Chen. "Genomic insights into the genetic signatures of selection and seed trait loci in cultivated peanut." *Journal of Advanced Research* 42 (2022): 237–248.

Luo, H., M. K. Pandey, A. W. Khan, J. Guo, B. Wu, Y. Cai, L. Huang, X. Zhou, Y. Chen, and W. Chen. "Discovery of genomic regions and candidate genes controlling shelling percentage using QTL-seq approach in cultivated peanut (Arachis hypogaea L.)." *Plant Biotechnology Journal* 17 (7) (2019): 1248–1260.

Luo, H., M. K. Pandey, A. W. Khan, B. Wu, J. Guo, X. Ren, X. Zhou, Y. Chen, W. Chen, and L. Huang. "Next-generation sequencing identified genomic region and diagnostic markers for resistance to bacterial wilt on chromosome B02 in peanut (Arachis hypogaea L.)." *Plant Biotechnology Journal* 17 (12) (2019): 2356–2369.

Mallikarjuna, Nalini, and Rajeev K. Varshney, eds. *Genetics, Genomics and Breeding of Peanuts*. CRC Press, 2014.

Mascher, Martin, Matthias Schreiber, Uwe Scholz, Andreas Graner, Jochen C. Reif, and Nils Stein. "Genebank genomics bridges the gap between the conservation of crop diversity and plant breeding." Nature Genetics 51 (2019): 1076–1081.

Marsh, Jacob I., Haifei Hu, Mitchell Gill, Jacqueline Batley, and David Edwards. "Crop breeding for a changing climate: Integrating phenomics and genomics with bioinformatics." *Theoretical and Applied Genetics* 134 (2021): 1677–1690.

Meilleur, Brien A., and Toby Hodgkin. "In situ conservation of crop wild relatives: Status and trends." *Biodiversity & Conservation* 13 (2004): 663–684.

Meneely, Julie P., Oluwatobi Kolawole, Simon A. Haughey, Sarah J. Miller, Rudolf Krska, and Christopher T. Elliott. "The challenge of global aflatoxins legislation with a focus on peanuts and peanut products: A systematic review." *Exposure and Health* 15 (2) (2023): 467–487.

Mishra, G. P., T. Radhakrishnan, A. Kumar, P. Thirumalaisamy, N. Kumar, T. C. Bosamia, B. Nawade, and J. R. Dobaria. "Advancements in molecular marker development and their applications in the management of biotic stresses in peanuts." *Crop Protection* 77 (2015): 74–86.

Mota, A. P. Z., B. Vidigal, E. G. Danchin, R. C. Togawa, S. Leal-Bertioli, D. J. Bertioli, A. C. G. Araujo, A. C. M. Brasileiro, and P. M. Guimaraes. "Comparative root transcriptome of wild Arachis reveals NBS-LRR genes related to nematode resistance." *BMC Plant Biology* 18 (1) (2018): 1–16.

Musa, Ö. M. "Some nutritional characteristics of kernel and oil of peanut (Arachis hypogaea L.)." *Journal of Oleo Science* 59 (1) (2010): 1–5.

Nagaraj, G. "Chemistry and utilization." *Groundnut: Indian Council of Agricultural Research, New Delhi, India* (1988): 554–565.

Neelakandan, Anjanasree K., David A. Wright, Sy M. Traore, Xiangyu Chen, Martin H. Spalding, and Guohao He. "CRISPR/Cas9 based site-specific modification of FAD2 cis-regulatory motifs in peanut (Arachis hypogaea L)." *Frontiers in Genetics* 13 (2022): 849961.

Nigam, S. (2014). *Groundnut at a Glance*, 121. ICRISAT.

Nigam, S. N., Farid Waliyar, R. Aruna, S. V. Reddy, P. Lava Kumar, Peter Q. Craufurd, A. T. Diallo, B. R. Ntare, and Hari D. Upadhyaya. "Breeding peanut for resistance to aflatoxin contamination at ICRISAT." *Peanut Science* 36 (1) (2009): 42–49.

Norlia, Mahror, Selamat Jinap, Mahmud Ab Rashid Nor-Khaizura, Son Radu, Nik Iskandar Putra Samsudin, and Farah Asilah Azri. "Aspergillus section Flavi and aflatoxins: Occurrence, detection, and identification in raw peanuts and peanut-based products along the supply chain." *Frontiers in Microbiology* 10 (2019): 2602.

Ozias-Akins, Peggy, Ethalinda K. S. Cannon, and Steven B. Cannon. "Genomics Resources for Peanut Improvement." In The Peanut Genome, edited by Rajeev K. Varshney, Manish K. Pandey, and Naveen Puppala, 69–91. Compendium of Plant Genomes. Cham: Springer International Publishing, 2017.

Pandey, M. K., B. Guo, C. C. Holbrook, P. Janila, X. Zhang, D. J. Bertioli, S. Isobe, X. Liang, and R. K. Varshney. "Molecular markers, genetic maps and QTLs for molecular breeding in peanut." *Genetics, Genomics and Breeding of Peanuts*, edited by Rajeev K. Varshney, Manish K. Pandey, and Naveen Puppala, 61–113. Cham: Springer International Publishing, 2014.

Pandey, M. K., E. Monyo, P. Ozias-Akins, X. Liang, P. Guimarães, S. N. Nigam, H. D. Upadhyaya, P. Janila, X. Zhang, and B. Guo. "Advances in Arachis genomics for peanut improvement." *Biotechnology Advances* 30 (3) (2012): 639–651.

Pandey, M. K., A. K. Pandey, R. Kumar, C. V. Nwosu, B. Guo, G. C. Wright, R. S. Bhat, X. Chen, S. K. Bera, and M. Yuan. "Translational genomics for achieving higher genetic gains in groundnut." *Theoretical and Applied Genetics* 133 (2020): 1679–1702.

Patel, J., D. Khandwal, B. Choudhary, D. Ardeshana, R. K. Jha, B. Tanna, S. Yadav, A. Mishra, R. K. Varshney, and K. H. Siddique. "Differential physio-biochemical and metabolic responses of peanut (Arachis hypogaea L.) under multiple abiotic stress conditions." *International Journal of Molecular Sciences* 23 (2) (2022): 660.

Peng, Z., M. Gallo, B. L. Tillman, D. Rowland, and J. Wang. "Molecular marker development from transcript sequences and germplasm evaluation for cultivated peanut (Arachis hypogaea L.)." *Molecular Genetics and Genomics* 291 (2016): 363–381.

Peng, Ze, Zifan Zhao, Josh Paul Clevenger, Ye Chu, Dev Paudel, Peggy Ozias-Akins, and Jianping Wang. "Comparison of SNP calling pipelines and NGS platforms to predict the genomic regions harboring candidate genes for nodulation in cultivated peanut." *Frontiers in Genetics* 11 (2020): 222.

Perea-Moreno, M.-A., F. Manzano-Agugliaro, Q. Hernandez-Escobedo, and A.-J. Perea-Moreno. "Peanut shell for energy: properties and its potential to respect the environment." *Sustainability* 10 (9) (2018): 3254.

Qi, F., Z. Sun, H. Liu, Z. Zheng, L. Qin, L. Shi, Q. Chen, H. Liu, X. Lin, and L. Miao. "QTL identification, fine mapping, and marker development for breeding peanut (Arachis hypogaea L.) resistant to bacterial wilt." *Theoretical and Applied Genetics* 135 (4) (2022): 1319–1330.

Radhakrishnan, T., K. Rani, and T. C. Bosamia. "Molecular breeding for groundnut (peanut) improvement: present status and future prospects." In *QTL Mapping in Crop Improvement*, edited by Rajeev K. Varshney and Manoj K. Pandey, 325–354. Cham: Springer International Publishing, 2023.

Rathod, V., R. Hamid, R. S. Tomar, S. Padhiyar, J. Kheni, P. Thirumalaisamy, and N. S. Munshi. "Peanut (Arachis hypogaea) transcriptome revealed the molecular interactions of the defense mechanism in response to early leaf spot fungi (Cercospora arachidicola)." *Plant Gene* 23 (2020): 100243.

Raza, A., S. Charagh, P. García-Caparrós, M. A. Rahman, V. H. Ogwugwa, F. Saeed, and W. Jin. "Melatonin-mediated temperature stress tolerance in plants." *GM Crops & Food* 13 (1) (2022): 196–217.

Raza, A., J. Tabassum, H. Kudapa, and R. K. Varshney. "Can omics deliver temperature resilient ready-to-grow crops?" *Critical Reviews in Biotechnology* 41 (2021): 1209–1232.

Reddy, D. S., P. Bhatnagar-Mathur, V. Vadez, and K. K. Sharma. "Grain legumes (soybean, chickpea, and peanut): omics approaches to enhance abiotic stress tolerance." *Improving Crop Resistance to Abiotic Stress*, edited by N. Tuteja, S. S. Gill, A. F. Tiburcio, and R. Tuteja, 995–1032. Weinheim: Wiley-VCH,2012.

Rivero, R. M., R. Mittler, E. Blumwald, and S. I. Zandalinas. "Developing climate-resilient crops: improving plant tolerance to stress combination." *The Plant Journal* 109 (2) (2022): 373–389.

Saeed, F., U. K. Chaudhry, A. Raza, S. Charagh, A. Bakhsh, A. Bohra, S. Ali, A. Chitikineni, Y. Saeed, and R. G. Visser. "Developing future heat-resilient vegetable crops." *Functional & Integrative Genomics* 23 (1) (2023): 47.

Salgotra, R. K., and C. N. Stewart. "Functional markers for precision plant breeding." *International Journal of Molecular Sciences* 21 (13) (2020): 4792.

Salman-Minkov, A., N. Sabath, and I. Mayrose. "Whole-genome duplication as a key factor in crop domestication." *Nature Plants* 2 (8) (2016): 1–4.

Seijo, José Guillermo, Evelin Ivana Kovalsky, Laura María Chalup, Sergio Sebastián Samoluk, Alessandra Fávero, Germán Ariel Robledo, and others. "Karyotype stability and genome-specific nucleolar dominance in peanut, its wild 4× ancestor, and a synthetic AABB polyploid." Crop Science 58 (4) (2018): 1671–1683.

Shanthala, J., S. Gazala Parveen, and B. K. P. Jambagi. "Genomic-assisted breeding for enhanced harvestable (Pod) and consumable (seed) product, yield productivity in groundnut (Arachis hypogaea L.)." *Accelerated Plant Breeding, Volume 4: Oil Crops* (2022): 181–237.

Sharma, S., B. Choudhary, S. Yadav, A. Mishra, V. K. Mishra, R. Chand, C. Chen, and S. P. Pandey. "Metabolite profiling identified pipecolic acid as an important component of peanut seed resistance against Aspergillus flavus infection." *Journal of Hazardous Materials* 404 (2021): 124155.

Shasidhar, Y., M. T. Variath, M. K. Vishwakarma, S. S. Manohar, S. S. Gangurde, M. Sriswathi, H. K. Sudini, K. L. Dobariya, S. K. Bera, and T. Radhakrishnan. "Improvement of three popular Indian groundnut varieties for foliar disease resistance and high oleic acid using SSR markers and SNP array in marker-assisted backcrossing." *The Crop Journal* 8 (1) (2020): 1–15.

Shi, Lei, Xiaona Li, Lulu Xue, Jin Zhang, Bingyan Huang, Ziqi Sun, Zhongxin Zhang et al. "Creation of herbicide-resistance in allotetraploid peanut using CRISPR/Cas9-meditated cytosine base-editing." *Plant Biotechnology Journal* 21 (10) (2023): 1923.

Signorelli, S., Ł. P. Tarkowski, W. Van den Ende, and D. C. Bassham. "Linking autophagy to abiotic and biotic stress responses." *Trends in Plant Science* 24 (5) (2019): 413–430.

Sobolev, Victor, Renee Arias, Kerestin Goodman, Travis Walk, Valerie Orner, Paola Faustinelli, and Alicia Massa. "Suppression of aflatoxin production in Aspergillus species by selected peanut (Arachis hypogaea) stilbenoids." *Journal of Agricultural and Food Chemistry* 66 (1) (2018): 118–126.

Soni, P., S. N. Nayak, R. Kumar, M. K. Pandey, N. Singh, H. K. Sudini, P. Bajaj, J. C. Fountain, P. Singam, and Y. Hong. "Transcriptome analysis identified coordinated control of key pathways regulating cellular physiology and metabolism upon Aspergillus flavus infection resulting in reduced aflatoxin production in groundnut." *Journal of Fungi* 6 (4) (2020): 370.

Sui, J., P. Jiang, G. Qin, S. Gai, D. Zhu, L. Qiao, and J. Wang. "Transcriptome profiling and digital gene expression analysis of genes associated with salinity resistance in peanut." *Electronic Journal of Biotechnology* 32 (2018): 19–25.

Sun, Z., F. Qi, H. Liu, L. Qin, J. Xu, L. Shi, Z. Zhang, L. Miao, B. Huang, and W. Dong. "QTL mapping of quality traits in peanut using whole-genome resequencing." *The Crop Journal* 10 (1) (2022): 177–184.

Suzuki, N., R. M. Rivero, V. Shulaev, E. Blumwald, and R. Mittler. "Abiotic and biotic stress combinations." *New Phytologist* 203 (1) (2014): 32–43.

Srikanth, P., D. Sivakumar, A. Sharma, and N. Kaushik. "Recent developments in omics techniques for improving plant abiotic stress using microbes." *International Journal of Environmental Science and Technology* 22 (5) (2025): 3787–3810.

Tengey, Theophilus Kwabla, Frederick Kankam, Dominic Ngagmayan Ndela, Daniel Frempong, and William Ofori Appaw. "Growth and toxigenicity of A. flavus on resistant and susceptible peanut genotypes." *Toxins* 14 (8) (2022): 536.

Thoppurathu, F. J., Z. Ghorbanzadeh, A. K. Vala, R. Hamid, and M. Joshi. "Unravelling the treasure trove of drought-responsive genes in wild-type peanut through transcriptomics and physiological analyses of root." *Functional & Integrative Genomics* 22 (2) (2022): 215–233.

Thudi, M., R. Palakurthi, J. C. Schnable, A. Chitikineni, S. Dreisigacker, E. Mace, R. K. Srivastava, C. T. Satyavathi, D. Odeny, and V. K. Tiwari. "Genomic resources in plant breeding for sustainable agriculture." *Journal of Plant Physiology* 257 (2021): 153351.

Toomer, O. T. "A comprehensive review of the value-added uses of peanut (Arachis hypogaea) skins and by-products." *Critical Reviews in Food Science and Nutrition* 60 (2) (2020): 341–350.

Torres, Adriana Mabel, Germán Gustavo Barros, Sofia Alejendra Palacios, Sofia Noemi Chulze, and Paola Battilani. "Review on pre-and post-harvest management of peanuts to minimize aflatoxin contamination." *Food Research International* 62 (2014): 11–19.

Upadhyaya, Hari D., Rodomiro Ortiz, Paula J. Bramel, and Sube Singh. "Development of a groundnut core collection using taxonomical, geographical and morphological descriptors." *Genetic Resources and Crop Evolution* 50 (2003): 139–148.

U.S. Department of Agriculture. "Crop Explorer – Peanuts." International Production Assessment Division (IPAD), accessed September 17, 2024. https://ipad.fas.usda.gov/cropexplorer/cropview/commodityView.aspx?cropid=2221000.

Variath, M. T., and P. Janila. "Economic and academic importance of peanut." In *The Peanut Genome,* edited by Rajeev K. Varshney, Manish K. Pandey, and Naveen Puppala, 7–26. Cham: Springer International Publishing, 2017.

Varshney, R. K. "Exciting journey of 10 years from genomes to fields and markets: some success stories of genomics-assisted breeding in chickpea, pigeonpea and groundnut." *Plant Science* 242 (2016): 98–107.

Varshney, R. K., R. Barmukh, M. Roorkiwal, Y. Qi, J. Kholova, R. Tuberosa, M. P. Reynolds, F. Tardieu, and K. H. Siddique. "Breeding custom-designed crops for improved drought adaptation." *Advanced Genetics* 2 (3) (2021): e202100017.

Varshney, R. K., A. Bohra, M. Roorkiwal, R. Barmukh, W. Cowling, A. Chitikineni, H.-M. Lam, L. T. Hickey, J. Croser, and D. Edwards. "Rapid delivery systems for future food security." *Nature biotechnology* 39 (10) (2021a): 1179–1181.

Varshney, R. K., A. Bohra, J. Yu, A. Graner, Q. Zhang, and M. E. Sorrells. "Designing future crops: genomics-assisted breeding comes of age." *Trends in Plant Science* 26 (2021): 631–649.

Varshney, R. K., A. Graner, and M. E. Sorrells. "Genomics-assisted breeding for crop improvement." *Trends in Plant Science* 10 (12) (2005): 621–630.

Varshney, R. K., S. M. Mohan, P. M. Gaur, N. Gangarao, M. K. Pandey, A. Bohra, S. L. Sawargaonkar, A. Chitikineni, P. K. Kimurto, and P. Janila. "Achievements and prospects of genomics-assisted breeding in three legume crops of the semi-arid tropics." *Biotechnology Advances* 31 (8) (2013): 1120–1134.

Varshney, R. K., M. K. Pandey, A. Bohra, V. K. Singh, M. Thudi, and R. K. Saxena. "Toward the sequence-based breeding in legumes in the post-genome sequencing era." *Theoretical and Applied Genetics* 132 (3) (2019): 797–816.

Varshney, Rajeev K., Manish K. Pandey, and Naveen Puppala. *The Peanut Genome: An Introduction.* Springer International Publishing, 2017.

Varshney, R. K., V. K. Singh, A. Kumar, W. Powell, and M. E. Sorrells. "Can genomics deliver climate-change ready crops?" *Current Opinion in Plant Biology* 45 (2018): 205–211.

Varshney, R. K., P. Sinha, V. K. Singh, A. Kumar, Q. Zhang, and J. L. Bennetzen. "5Gs for crop genetic improvement." *Current Opinion in Plant Biology* 56 (2020): 190–196.

Vassilev, D., J. Leunissen, A. Atanassov, A. Nenov, and G. Dimov. "Application of Bioinformatics in Plant Breeding." *Biotechnology & Biotechnological Equipment* 19 (sup3) (2005): 139–152.

Vishwakarma, M. K., S. N. Nayak, B. Guo, L. Wan, B. Liao, R. K. Varshney, and M. K. Pandey. "Classical and molecular approaches for mapping of genes and quantitative trait loci in peanut." *The Peanut Genome,* edited by Rajeev K. Varshney, Manish K. Pandey, and Naveen Puppala, 93–116. Cham: Springer International Publishing, 2017.

Volis, Sergei, and Michael Blecher. "Quasi in situ: a bridge between ex situ and in situ conservation of plants." *Biodiversity and Conservation* 19 (2010): 2441–2454.

Wang, H., X. Guo, M. K. Pandey, X. Ji, R. K. Varshney, V. Nwosu, and B. Guo. "History and impact of the International Peanut Genome Initiative: the exciting journey toward peanut whole-genome sequencing." In *The Peanut Genome,* edited by Rajeev K. Varshney, Manish K. Pandey, and Naveen Puppala, 117–133. Cham: Springer International Publishing, 2017.

Wang, Houmiao, Yong Lei, Liying Yan, Liyun Wan, Xiaoping Ren, Silong Chen, Xiaofeng Dai, Wei Guo, Huifang Jiang, and Boshou Liao. "Functional genomic analysis of Aspergillus flavus interacting with resistant and susceptible peanut." *Toxins* 8 (2) (2016): 46.

Wang, M. L., S. Sukumaran, N. A. Barkley, Z. Chen, C. Y. Chen, B. Guo, H. T. Stalker, C. C. Holbrook, P. G. Pederson, and R. N. Pittman. "Population structure and marker–trait association analysis of the US peanut (Arachis hypogaea L.) mini-core collection." *Theoretical and Applied Genetics* 123 (6) (2011): 1307–1317.

Wang, M. L., H. Wang, C. Zhao, B. Tonnis, S. Tallury, X. Wang, J. Clevenger, and B. Guo. "Identification of QTLs for seed dormancy in cultivated peanut using a recombinant inbred line mapping population." *Plant Molecular Biology Reporter* 40 (1) (2022): 208–217.

Wang, Z., S. Yan, C. Liu, F. Chen, and T. Wang. "Proteomic analysis reveals an aflatoxin-triggered immune response in cotyledons of Arachis hypogaea infected with Aspergillus flavus." *Journal of Proteome Research* 11 (5) (2012): 2739–2753.

Wu, Q., L. Yang, H. Liang, L. Yin, D. Chen, and P. Shen. "Integrated analyses reveal the response of peanut to phosphorus deficiency on phenotype, transcriptome and metabolome." *BMC Plant Biology* 22 (1) (2022): 1–26.

Xu, M., C. Xu, M. Chen, Z. Xiao, Y. Wang, Y. Xu, and D. Xu. "Comparative analysis of commonly used bioinformatics software based on omics." *Gene Reports* (2023): 101800.

Yang, Y., M. A. Saand, L. Huang, W. B. Abdelaal, and M. H. Sirohi. "Applications of multi-omics technologies for crop improvement." *Frontiers in Plant Science* 12 (2021): 563953.

Yang, Yaodong, Mumtaz Ali Saand, Liyun Huang, Walid Badawy Abdelaal, Jun Zhang, Yi Wu, Jing Li, Muzafar Hussain Sirohi, and Fuyou Wang. "Applications of multi-omics technologies for crop improvement." *Frontiers in Plant Science* 12 (2021): 563953.

Yin, D., C. Ji, X. Ma, H. Li, W. Zhang, S. Li, F. Liu, K. Zhao, F. Li, and K. Li. "Genome of an allotetraploid wild peanut Arachis monticola: a de novo assembly." *Gigascience* 7 (6) (2018): giy066.

Yu, Bolun, Dongxin Huai, Li Huang, Yanping Kang, Xiaoping Ren, Yuning Chen, Xiaojing Zhou et al. "Identification of genomic regions and diagnostic markers for resistance to aflatoxin contamination in peanut (Arachis hypogaea L.)." *BMC Genetics* 20 (2019): 1–13.

Yu, Bolun, Huifang Jiang, Manish K. Pandey, Li Huang, Dongxin Huai, Xiaojing Zhou, Yanping Kang et al. "Identification of two novel peanut genotypes resistant to aflatoxin production and their SNP markers associated with resistance." *Toxins* 12 (3) (2020): 156.

Zandalinas, S. I., F. B. Fritschi, and R. Mittler. "Global warming, climate change, and environmental pollution: recipe for a multifactorial stress combination disaster." *Trends in Plant Science* 26 (6) (2021): 588–599.

Zhang, H., X. Zhao, Q. Sun, C. Yan, J. Wang, C. Yuan, C. Li, S. Shan, and F. Liu. "Comparative transcriptome analysis reveals molecular defensive mechanism of Arachis hypogaea in response to salt stress." *International Journal of Genomics* 2020 (2020). https://doi.org/10.1155/2020/6524093.

Zhao, C., L. He, H. Xia, X. Zhou, Y. Geng, L. Hou, P. Li, G. Li, S. Zhao, and C. Ma. "De novo full length transcriptome analysis of Arachis glabrata provides insights into gene expression dynamics in response to biotic and abiotic stresses." *Genomics* 113 (3) (2021): 1579–1588.

Zhu, H., C. Li, and C. Gao. "Applications of CRISPR–Cas in agriculture and plant biotechnology." *Nature Reviews Molecular Cell Biology* 21 (11) (2020): 661–677.

Zhuang, W., H. Chen, M. Yang, J. Wang, M. K. Pandey, C. Zhang, W.-C. Chang, et al. "The genome of cultivated peanut provides insight into legume karyotypes, polyploid evolution and crop domestication." *Nature Genetics* 51 (5) (2019): 865–876.

Zhou, Xiaojing, Xiaoping Ren, Huaiyong Luo, Li Huang, Nian Liu, Weigang Chen, Yong Lei, Boshou Liao, and Huifang Jiang. "Safe conservation and utilization of peanut germplasm resources in the Oil Crops Middle-term Genebank of China." *Oil Crop Science* 7 (1) (2022): 9–13.

2 Peanut Genetic Resources
Status, Challenges, and Use in Peanut Genetic Improvement

S. P. Tallury, A. S. Gerrano, D. Deshmukh, S. Han, T. W. Mekonnen, and M. T. Labuschagne

2.1 INTRODUCTION

Genetic resources are the foundation for peanut genetic enhancement and sustainable production. They serve as a reservoir to fight unforeseen and/or future threats to peanut production. Continuous progress in peanut improvement depends on the discovery and characterization of new sources of genetic variation and their judicious use for the development of new cultivars. Genetic resources conserved in genebanks offer huge potential for utilization in breeding programs worldwide as sources of desirable traits. Globally, the most important traits for sustainable peanut production include high yield, early maturity, resistance to biotic and abiotic stresses, high shelling percentage, high oil, increased oleic acid content, and dual-purpose types (Desmae et al., 2019). However, in the United States and other developed countries, under high-input, mechanized production systems, the breeding focus has been on maximizing yield, but in recent years, improving nutritional quality and flavor and biotic and abiotic resistances have become important priorities.

Peanut, also called groundnut, belongs to the genus *Arachis*, and the botanical name is *Arachis hypogaea* L., which in Greek means a weed with an underground chamber. It has a unique feature of reproduction whereby flowers are produced above ground, but pods/seeds are developed inside the soil. All *Arachis* species produce underground pods, a phenomenon known as geocarpy. Geocarpic reproduction is likely an adaptive survival mechanism against adverse environmental stresses (Tan et al., 2010). The wild *Arachis* species have accumulated tolerance genes for biotic and abiotic stresses. These traits make them attractive to use in breeding programs to develop superior cultivars with stress resistances.

The cultivated peanut, *A. hypogaea*, is an allotetraploid ($2n = 4x = 40$; AABB) and is the only domesticated species in the genus, although several other species were used for food, for forage, and as ground cover (Stalker and Simpson, 1995; Stalker et al., 2016). *Arachis hypogaea* is believed to have originated in northwestern Argentina and southern Bolivia as a happenstance hybrid of two diploid species, *A. duranensis* Krapov. & W.C. Greg. and *A. ipaënsis* Krapov. & W.C. Greg. (Kochert et al., 1996; Seijo et al., 2004, 2007). Following the hybridization event, polyploidization of the sterile diploid hybrid led to a fertile allotetraploid, which was likely selected by natives for larger plants, pods, and seeds, leading to the present-day *A. hypogaea*. The estimated genome size of *A. hypogaea* is 2.8 Gb, with each of two closely related subgenomes of 1.3 Gb (*A. duranensis*; A subgenome) and 1.5 Gb (*A. ipaënsis*; B subgenome) (Bertioli et al., 2016). Further, the diploid genome sequence comparisons to the cultivated peanut showed collinearity of chromosomal pseudomolecules, where the *A. ipaënsis* B-genome pseudomolecules were identical to those of *A. hypogaea* B-subgenome, whereas the A-subgenome pseudomolecules were less similar (Bertioli et al., 2016). Additionally, the genome sequence of tetraploid peanut (Bertioli et al., 2019; Zhuang et al., 2019) provided a broader understanding of functionally important gene sequences for economically important traits and for the development of genetic markers for continued peanut productivity and sustainability.

2.1 BOTANICAL CHARACTERISTICS FROM SEED TO MATURE POD

2.1.1 Seed and Seedling Development

Peanut seed is covered by a seed coat or testa, which varies in color from white to black and different shades of red or pink. The seeds vary in size from the large Virginia market types to the small, rounded Spanish types. The seed is composed of two cotyledons which contain stored food reserves for the young seedling during germination and early establishment. The cotyledons enclose the dormant shoot (plumule/leaf primordia) and the root initials (radicle). When planted under optimum soil moisture and temperature conditions, the process of seed germination is initiated in about 3–5 days with the emergence of lower radicle and upper hypocotyl. This is accompanied by the rapid elongation of the hypocotyl, which pushes the cotyledons above ground. The cotyledons split open to expose the shoot primordia, which extends to form the epicotyl. The epicotyl forms the main stem, whereas the radicle elongates to form the

DOI: 10.1201/9781003360988-2

primary or taproot (Gregory et al., 1973). From the taproot, lateral roots emerge and are extensively developed about 10 days after germination. Occasionally, adventitious roots are formed in mature plants where branches are in contact with soil.

2.1.2 Plant Growth and Morphology

Peanut plants are usually about 30–45 cm tall and display a wide range of growth habits, from an erect plant type with branches extending straight up from the base of the plant to a semi-spreading or prostrate type where lateral branches spread to about 30 cm wide from the base. Plants have compound leaves with four leaflets (tetrafoliate) located alternately on the main stem and lateral branches. The leaves are connected to the stems by an adnate stipule, and leaflets vary in size and shape. Further, the *A. hypogaea* subsp. *hypogaea* has dark green leaves compared to the lighter green leaves in subsp. *fastigiata*. The stems are angular and mostly green except for Valencia and aequatoriana types, which are reddish purple. Glabrous leaves and stems are seen in some varieties. Stem pigmentation and hairiness on stems and leaves have been shown to deter leaf-feeding insect pests (Campbell et al., 1976; Stalker and Campbell, 1983; Stalker et al., 1984; Sharma et al., 2003). Contrastingly, many wild species have a short main stem with lateral branches extending a few meters long from the base of the plant with usually oblong or lanceolate tetrafoliate leaves. However, three species from the section *Trierectoides*, namely *A. guaranitica*, *A. tuberosa*, and *A. sesquijuga*, have leaves with three leaflets (trifoliate).

2.1.3 Flowering, Fertilization, and Pod Development

Usually about 30 days after planting, the plants start producing flowers. All peanut species are self-pollinators. The inflorescence is a raceme containing at least three flower buds. At any given time, usually only one flower opens, and the interval between the openings of flowers in the same inflorescence varies up to several days. Peanut is an indeterminate plant, and as a result, flowers are produced throughout the growing season leading up to harvest. Flowers are usually present in leaf axils on the branches, but in the subspecies *fastigiata* types, they are also present on the main stem. Since the flowers contain both male and female reproductive tissues, natural self-pollination leads to the development of pods. The flower is papilionaceous with five brightly colored petals (corolla) consisting of a large standard (banner), two wing petals and two fused keel petals. The calyx is green with five lobes, one lobe opposite the keel, whereas the other four are fused and cover the back side of the standard. The standard is usually yellow or orange, with red veins on the inner face. The wings are usually yellow surrounding the keel. The keel is almost colorless and encloses the stamens and style. The androecium (male tissue) is monadelphous with filaments fused for two-thirds of their length and contains eight functional stamens and two small sterile ones. The stamens produce pollen for fertilization of the egg cell. The flower is attached to the stem, at the leaf axil, by a long tube-like structure called a hypanthium or "calyx tube." The flowers are subtended by a bract and are sessile, although they appear pedicellate because of the tubular hypanthium. The style is enclosed within the hypanthium and is connected to the ovary located at the base of the hypanthium. The tip of the style, called stigma, is at the same level or slightly above the anthers so pollen grains can reach it. Differences in stigma morphology were observed between *A. hypogaea* and the wild species. In *A. hypogaea*, the stigma is of a dry papillate type (Lakshmi and Shivanna, 1986) with no surrounding hairs and probably accommodates about 15 pollen grains (Moss and Rao, 1995). On the other hand, the annual *Arachis* species have a large stigmatic surface, whereas the perennial species have smaller, cuticularized stigmas with unicellular hairs accommodating a maximum of only three pollen grains (Lu et al., 1990). However, in the wild species, *A. lignosa*, the truncated shape of the stigma and its elevated position relative to the anthers restricts natural self-pollination (Banks, 1990). In this case, manual tripping of flowers is needed for pollen to reach stigma for fertilization and later pod development. Outcrossing is possible with bees or other insects; however, it is limited to less than 10% under natural field conditions (Hammons, 1973; Knauft et al., 1992).

The process of fertilization begins with anthesis, which occurs within a few hours after sunrise with the opening of the flower. The mature pollen grain is two-celled with two generative nuclei (Xi, 1991). The ovary usually has two ovules and up to three or more in some *ssp. fastigiata* types. Each ovule contains a mature embryo sac with a well-differentiated egg cell at the micropylar end and a central cell surrounded by starch grains. When pollen germinates on a receptive stigma, the pollen tube containing the mature pollen grain with the two generative nuclei travels through the style and eventually enters the embryo sac through the micropyle. One of the two generative nuclei fuses with the egg cell (syngamy or fertilization) to form the embryo and the other with the central cell to form the endosperm. After fertilization, the starch grains break down to provide initial nutrition for the proembryo to grow which eventually develops into a mature seed. The entire process of fertilization usually takes about 18–24 hours from anthesis to fertilization (Pattee et al., 1991). Each ovule develops into a peanut seed, and the ovary becomes the pod.

Following fertilization, pod development is initiated as a pointed, stalk-like structure that is called the "peg" (Smith, 1950). During the early embryo growth (24–72 h after fertilization), an intercalary meristem at the base of the ovary actively divides forming the peg with the ovules at its tip. Pegs are usually observed between 4 and 7 days after self-pollination, are positively geotropic (Zamski and Ziv, 1976), and require darkness for pod formation (Ziv, 1981). Before the peg enters the soil (aerial phase), the embryo remains in a quiescent stage, usually, as an 8-celled proembryo (Pattee and Mohapatra, 1987). It is only after soil penetration that the elongation of the peg is arrested, to initiate pod formation. The first sign of pod development involves the swelling of the peg tip along with the horizontal turning of the peg. The peg becomes diageotropic after soil penetration such that the ovules are always located on the upper wall of the pod, with the pod tip pointing away from the plant (Moss and Rao, 1995). Pod enlargement occurs from the base towards the tip with simultaneous faster development of the basal ovule (Smith, 1950). A mature peanut pod is developed in about 60–80 days after pollination. However, due to the indeterminate nature of plants, flowering continues throughout the growing season until harvest. Consequently, pods at different maturities are seen on plants even at harvest. Detailed descriptions of peanut embryology including the growth and development of pegs, pods, and seeds are well documented in literature (Smith, 1950; Gregory et al., 1973; Periasamy and Sampoornam, 1984; Pattee and Mohapatra, 1987; Xi, 1991; Moss and Rao, 1995; Tallury et al., 1995).

2.2 ORIGIN AND BIOGEOGRAPHICAL DISTRIBUTION OF *ARACHIS*

Genus *Arachis* is native to South America with natural populations growing in the highlands in Argentina, Bolivia, Brazil, Paraguay, and Uruguay (Valls et al., 1985). It is believed that the genus originated in the highlands of southwestern Brazil or northeast Paraguay (Hammons, 1973; Gregory et al., 1980; Simpson et al., 2001; Williams, 2022), where the most primitive, trifoliate species, *A. guaranitica* Chodat. and Hassl., and *A. tuberosa* Bong. Ex Benth. and *A. sesquijuga* were found and are still growing. Subsequently, with water movement, *Arachis* species spread to drier lowlands in all directions and evolved and adapted into various river valleys and drainage systems (Gregory and Gregory, 1979; Smartt and Stalker, 1982; Stalker and Simpson, 1995; Simpson et al., 2001), with species growing in widely varied ecogeographical regions (Simpson et al., 2001). One of the species, *A. burkartii* Handro, was collected in southern Brazil in black gummy clay mixed with small stones with a soil pH of 3.2 (Stalker and Simpson, 1995), indicating the wide adaptation of *Arachis* species to extremely diverse geographical environments (Smartt and Stalker, 1982; Stalker and Simpson, 1995; Simpson et al., 2001; Tallury, 2017; Williams, 2022).

2.2.1 Taxonomic Classification of Genus *Arachis*

To date, 84 *Arachis* species have been described (Cason et al., 2022; Williams, 2022; Seijo et al., 2025), and another 10–15 species are likely to be described soon (Charles Simpson and Guillermo Seijo, personal communication, 2022). Krapovickas and Gregory (1994, 2007) indicated that the genus *Arachis* is defined by its morphological features of the underground structures, including the pods, rhizomatous stems, root systems, and hypocotyls. They showed that these defining characters grouped the *Arachis* collections into different geographic areas and ecological features. Based on these observations coupled with species karyotypes/genomes and cross-compatibilities, they partitioned the genus into nine different taxonomic sections (Table 2.1). Further, it was suggested that different species originated sympatrically in different sections where isolation barriers developed leading to the geographical separation of the species (Stalker and Simpson, 1995; Seijo et al., 2014) and the evolution of different genomes occurred independently in different sections (Smartt and Stalker, 1982; Seijo et al., 2007; Robledo and Seijo, 2010; Seijo et al., 2014). Section *Trierectoides* was suggested as the oldest and most primitive of the nine sections with the trifoliate (3-leaflet) species, *A. tuberosa* and *A. guaranitica*. It was believed that the sections *Erectoides*, *Extranervosae*, *Triseminatae*, and *Heteranthae* evolved from these primitive species. The species of these four sections have varying affinities to the primitive section, *Trierectoides*, as reported by Gregory and Gregory (1979) and Krapovickas and Gregory (1994). The more advanced sections include *Arachis*, *Caulorrhizae*, *Procumbentes*, and *Rhizomatosae*. The affinities of the species among these sections are also varied, with very limited successes reported in crosses with section *Arachis* species (Gregory and Gregory, 1979; Krapovickas and Gregory, 1994; Mallikarjuna and Sastri, 2002). The geographic distribution of species in the most advanced section, *Arachis*, has overlapped with that of the other sections in many areas of South America. It is known that the most advanced species are more adaptable to varied environments and are able to rapidly spread to areas where ancient species have existed for many millennia (Stalker et al., 2013). Brazil contains the greatest number of *Arachis* species with at least 45 endemic species. Bolivia has the second largest number of species followed by Paraguay, Argentina, and Uruguay (Singh and Nigam, 2016). Unfortunately, global climate change and human activities are leading to the rapid loss of wild species habitats in South America (Guillermo Seijo, personal communication).

Section *Arachis* is the largest of the nine sections and contains about 40% of the species within the genus, currently 34 described species (Table 2.1). This section includes the tetraploid ($2n = 4x = 40$) cultivated species *A. hypogaea* and another tetraploid species, *A. monticola*. Three species, *A. decora*, *A. palustris*, and *A. praecox* are aneuploids ($2n = 2x = 18$) and the rest of the 29 species are diploids ($2n = 2x = 20$) (Krapovickas and Gregory, 1994, 2007; Stalker et al., 2013, 2016; Seijo et al., 2025). The remaining 50 described species have been assigned to the corresponding taxonomic sections as shown in Table 2.1 except for two species, *A. chiquitana* and *A. sesquijuga* with uncertain sectional designations at this time. Section *Heteranthae* contains all annual species. Both annual and perennial species are present in section *Arachis*, with the remaining seven sections containing only perennial species. Although most species have been assigned to

TABLE 2.1
Arachis Sections and Species with Their Ploidy and Genome Types (Modified from Stalker et al., 2016)

Section and Species	2n	Genome Type
Section *Arachis* (33 species)		
Arachis batizocoi Krapov. & Gregory	20	K
Arachis benensis Krapov., Gregory, & Simpson	20	F
Arachis cardenasii Krapov. & Gregory	20	A
Arachis correntina (Burkart) Krapov. & Gregory	20	A
Arachis cruziana Krapov., Gregory, & Simpson	20	K
Arachis decora Krapov., Gregory, & Valls	18	G
Arachis diogoi Hoehne	20	A
Arachis duranensis Krapov. & Gregory	20	A
Arachis glandulifera Stalker	20	D
Arachis gregoryi Simpson, Krapov., & Valls	20	B
Arachis helodes Martius ex Krapov. & Rigoni	20	A
Arachis herzogii Krapov., Gregory, & Simpson	20	A
Arachis hoehnei Krapov. & Gregory	20	A
Arachis hypogaea Linnaeus	40	AB
Arachis inflata Seijo, Atahuachi, Simpson, & Krapov.	20	B
Arachis ipaënsis Krapov. & Gregory	20	B
Arachis kempff-mercadoi Krapov., Gregory & Simpson	20	A
Arachis krapovickasii Simpson, Williams, Valls, & Vargas	20	K
Arachis kuhlmannii Krapov. & Gregory	20	A
Arachis linearifolia Valls, Krapov., & Simpson	20	A
Arachis magna Krapov., Gregory, & Simpson	20	B
Arachis microsperma Krapov., Gregory, & Valls	20	A
Arachis monticola Krapov. & Rigoni	40	AB
Arachis palustris Krapov., Gregory, & Valls	18	G
Arachis praecox Krapov., Gregory, & Valls	18	G
Arachis schininii Valls & Simpson	20	A
Arachis simpsonii Krapov. & Gregory	20	A
Arachis stenosperma Krapov. & Gregory	20	A
Arachis trinitensis Krapov. & Gregory	20	F
Arachis valida Krapov. & Gregory	20	B
Arachis vallsii Krapov. & Gregory	20	B
Arachis villosa Benth.	20	A
Arachis williamsii Krapov. & Gregory	20	B
Arachis woodii Seijo, Atahuachi & Krapov	20	unassigned
Section *Caulorrhizae* (2 species)		
Arachis pintoi Krapov. & Gregory	20	C
Arachis repens Handro	20	C
Section *Erectoides* (14 species)		
Arachis archeri Krapov. & Gregory	20	E
Arachis benthamii Handro	20	E
Arachis brevipetiolata Krapov. & Gregory	20	E
Arachis cryptopotamica Krapov. & Gregory	20	E
Arachis douradiana Krapov. & Gregory	20	E
Arachis gracilis Krapov. & Gregory	20	E
Arachis hatschbachii Krapov. & Gregory	20	E
Arachis hermannii Krapov. & Gregory	20	E
Arachis major Krapov. & Gregory	20	E
Arachis martii Handro	20	E
Arachis oteroi Krapov. & Gregory	20	E
Arachis paraguariensis	20	
ssp. *paraguariensis* Chodat & Hassl.		E
ssp. *capibarensis* Krapov. & Gregory		E
Arachis porphyrocalyx Valls & Simpson	18	E
Arachis stenophylla Krapov. & Gregory	20	E

(*Continued*)

TABLE 2.1 (CONTINUED)
Arachis Sections and Species with Their Ploidy and Genome Types (Modified from Stalker et al., 2016)

Section and Species	2*n*	Genome Type
Section *Extranervosae* (10 species)		
Arachis burchellii Krapov. & Gregory	20	EX
Arachis lutescens Krapov. & Rigoni	20	EX
Arachis macedoi Krapov. & Gregory	20	EX
Arachis marginata Gardner	20	EX
Arachis pietrarellii Krapov. & Gregory	20	EX
Arachis prostrata Benth.	20	EX
Arachis retusa Krapov., Gregory, & Valls	20	EX
Arachis setinervosa Krapov. & Gregory	20	EX
Arachis submarginata Valls, Krapov., & Simpson	20	EX
Arachis villosulicarpa Hoehne	20	EX
Section *Heteranthae* (6 species)		
Arachis dardani Krapov. & Gregory	20	H
Arachis giacomettii Krapov., Gregory, Valls, & Simpson	20	H
Arachis interrupta Valls & Simpson	20	H
Arachis pusilla Benth.	20	H
Arachis seridoensis Valls, Simpson, Krapov., & Veiga	20	H
Arachis veigae Santana & Valls	20	H
Section *Procumbentes* (9 species)		
Arachis appressipila Krapov. & Gregory	20	P
Arachis hassleri Valls & Simpson	20	P
Arachis jacobinensis Valls & Simpson	20	P
Arachis kretschmeri Krapov. & Gregory	20	P
Arachis lignosa (Chodat and Hassl.) Krapov. & Gregory	20	P
Arachis matiensis Krapov., Gregory, & Simpson	20	P
Arachis pflugeae Simpson, Krapov., & Valls	20	P
Arachis rigonii Krapov. & Gregory	20	P
Arachis subcoriacea Krapov. & Gregory	20	P
Section *Rhizomatosae* (4 species)	20	P
Ser. *Prorhizomatosae*		
Arachis burkartii Handro	20	R_1
Arachis glabrata	40	
var. *glabrata* Benth.		R_2
var. *hagenbeckii* Benth. (Harms ex. Kuntze) Herm.		R_2
Arachis nitida Valls, Krapov., & Simpson	40	R_2
Arachis pseudovillosa (Chodat & Hassl.) Krapov. & Gregory	40	R_2
Section *Trierectoides* (2 species)		
Arachis guaranitica Chodat & Hassl.	20	TE
Arachis tuberosa Bong. ex Benth.	20	TE
Section *Triseminatae* (1 species)		
Arachis Triseminata Krapov. & Gregory	20	T
Section Unassigned (2 species)		
Arachis chiquitana Krapov., Gregory, & Simpson	20	–
Arachis sesquijuga Valls, Costa, & Custodio	20	–

different taxonomic sections, there is still some ambiguity with the placement of some, and some others may have to be assigned to new sections (Leal-Bertioli et al., 2024). As additional genomic tools are becoming available, it is anticipated that such discrepancies will be resolved soon for further clarification of genus *Arachis*.

2.2.2 Taxonomic Classification of *A. hypogaea*

Arachis hypogaea L. was divided into two subspecies, subsp. *hypogaea* and subsp. *fastigiata*, based on the absence versus presence of flowers on the main stem and the sequence of floral and leaf nodes on the lateral branches (Krapovickas and Rigoni, 1960). The subsp. *hypogaea* is characterized by the absence of flowers on the main stem with alternate vegetative and floral nodes (two vegetative nodes alternate with two floral nodes) on the lateral branches and a long life cycle. The subsp. *fastigiata* is characterized by the presence of flowers on the main stem with a sequential order of vegetative and floral nodes on the lateral branches and a shorter life cycle. Further, based on pod features, they proposed two botanical varieties of subsp. *fastigiata*, var. *fastigiata* and var. *vulgaris*. Later, Krapovickas (1968) proposed that subsp. *hypogaea* should also be divided into two botanical varieties, var. *hypogaea* and var. *hirsuta*. With additional collections of *A. hypogaea*, Krapovickas and Gregory (1994, 2007) not only confirmed the two subspecies of *A. hypogaea* (subsp. *hypogaea* and subsp. *fastigiata*) but also expanded botanical varietal groups to six (vars. *hypogaea, hirsuta, fastigiata, peruviana, aequatoriana, and vulgaris*) based on plant growth habit, leaf color, branching patterns, and pod morphologies as described below (Tallury, 2017), including the four major market types grown in the United States.

A. ***A. hypogaea* subsp. *hypogaea* L.**

No flowers on main stem
Alternate pairs of floral and vegetative nodes on branches
Branches short and less hairy
Dark green leaves
Prostrate or spreading growth habit
Late maturing
Seed dormancy present
Origin in southeastern Bolivia

var. *hypogaea*:

Leaflets with underside glabrous and a few hairs on the midvein
Prostrate/spreading (runner) or bunch type growth habit

- **Market type: Virginia**
  - Less hairy main stem and leaves
  - Large pods, two seeded, slight constriction and reticulation
- **Market type: Runner**
  - Less hairy main stem and leaves
  - Small pods, two seeded, slight constriction and reticulation

var. *hirsuta*:

Leaflets with 1–2 mm long hairs on the underside
Long main stem and very hairy
Origin in Peru

- **Market type: Peruvian runner**
  - More hairy leaves
  - Late maturing
  - Long pods, 2–3 seeded, deep constriction and prominent reticulation

B. ***A. hypogaea* subsp. *fastigiata***

Flowers on main stem
Sequential floral and vegetative branches
Branches less hairy
Light green leaves
Bunch or erect growth habit
Early maturing
Seed dormancy absent
Origin in Peru and northwestern Bolivia

var. *fastigiata*

Leaflets with underside glabrous or hairs only on the midvein
Few branches, short and slender
Pods with smooth or little reticulation, more than two seeds
Origin in Paraguay

- **Market type: Valencia**
  - Sparsely branched; curved branches
  - Erect growth habit
  - Usually 3–4 seeded, long, thin pods
  - Red seed coat

var. *aequatoriana*

Erect plants with large leaves
Leaflets 1–2 mm long, hairy underside on entire surface
Long reproductive lateral branches
Long pods, 4–7 cm long, 3–4 seeded
Prominent longitudinal ribs on pods with deep pod reticulation
Seed coat is commonly violet/purple
Origin in Ecuador

var. *peruviana*

Thick, large leaves; leaflets glabrous on both sides
Hairy on the margins
Long, robust reproductive branches
Long 3–4-seeded pods
Prominent longitudinal ribs on pods with deep reticulation
Seed coat colors vary from black, violet, cream to variegated
Origin in Peru, also in Ecuador, northern Bolivia, and Brazil

var. *vulgaris*

Erect growth habit with many upright branches
Medium-sized leaves with glabrous surface, long hairs on margins
Mostly 2-seeded, small pods bunched at the base of the plant
Slight pod constriction and reticulation
Origin in Argentina, southern Brazil, and Paraguay

- **Market type: Spanish**
  - More branched; upright branches
  - Light green leaves

2.3 ORIGIN AND GEOGRAPHICAL DISTRIBUTION OF *A. HYPOGAEA*

Krapovickas and Gregory (1994, 2007) suggested that cultivated peanut likely originated in the region of southern Bolivia and northwestern Argentina, based on the presence of the parental diploid wild species of *A. hypogaea* in this region and the wide range of variations observed in pod and seed morphologies. They also noticed that the *A. hypogaea* types found in this region exhibited many primitive traits such as the runner growth habit; small, two-seeded pods with marked constriction and slight reticulation like the wild *Arachis* species; and seed dormancy (Tallury, 2017). Further, they concluded that southeastern Bolivia was the center of origin and diversity for subsp. *hypogaea*, whereas subsp. *fastigiata* differentiated in northwestern Bolivia and possibly in Peru, along with vars. *fastigiata*, *peruviana*, and *aequatoriana*. Interestingly, based on archeological evidence and a favorable environment for extended plant growth, Simpson et al. (2001) suggested the west coast of Peru and/or the eastern slopes of Cordillera in the Andes as possible alternate regions for the origin of *A. hypogaea*. Further investigations are needed to support this observation.

Although the taxonomic classification of *A. hypogaea* described in the previous section (Section 2.2) is valid and routinely used to assign accessions to the different subspecies and botanical varieties, molecular diversity studies revealed some discrepancies about the botanical varietal groupings. Ferguson et al. (2004) observed similarities between three botanical varieties of subsp. *fastigiata*, namely *fastigiata*, *vulgaris*, and *aequatoriana*, but did not support the inclusion of var. *peruviana* in subsp. *fastigiata*. Further, they also found that the botanical varieties *hypogaea* and *hirsuta* were not closely related and thus should not be grouped under subsp. *hypogaea*. Contrarily, He and Prakash (2001) demonstrated that vars. *aequatoriana* and *peruviana* were closer to subsp. *hypogaea* than to subspecies *fastigiata*. Interestingly, Grabiele et al. (2012) suggested that the six botanical varieties originated from a single genetic origin and that *A. monticola* is the immediate ancestor of *A. hypogaea*. Thus, there still exists considerable confusion about the taxonomic classification of the cultivated species. Among the market types, Gregory et al. (1980) and Hammons (1982) suggested that the Bolivian and Amazonian geographic regions are the possible sites for the origin of the large-seeded Virginia types. Hammons (1982) indicated that the Guarani area of northeastern Argentina, Paraguay, and southern Brazil is the center of variation for the Spanish (var. *vulgaris*) market type, whereas the Valencia type (var. *fastigiata*) probably spread from Paraguay and central Brazil (Hammons, 1982; Krapovickas, 1968).

Also, Krapovickas (1968) and Gregory and Gregory (1976) recognized six other regions in South America as the secondary centers of diversity for the cultivated species based on morphological variability of the landraces. Additionally, Africa, China, and India are considered as tertiary centers of diversity for *A. hypogaea* as many of the local landraces exhibit a large amount of variability in plant, pod, and seed traits resulting from centuries of growing peanuts under diverse environments and the selection of varied types by the locals (Gibbons et al., 1972).

2.4 GEOGRAPHICAL SPREAD OF *A. HYPOGAEA*

Arachis hypogaea spread from the centers of origin and diversity in South America to Europe following the Spanish and Portuguese explorations to South America. Likely, the Portuguese traders later spread it to Africa and Asia. There is no evidence that peanut was present in North America during this time. Historical evidence suggests that peanut was introduced into the United States and North America on slave trade ships from Africa via the coast of northeastern Brazil, where peanut was gathered as food source to complete the journey, strongly suggesting that the first peanut introductions into the United States were from Brazil rather than from Africa (Stalker and Simpson, 1995; Tallury, 2017). A detailed history and dissemination of peanut from its centers of origin and diversity in South America to Europe, Africa, and Asia was summarized by Williams (2022).

2.5 *ARACHIS* GERMPLASM RESOURCES AND THEIR MANAGEMENT

2.5.1 Germplasm Explorations and Conservation

Peanut germplasm exploration and collection efforts were initiated in the 1930s in South America to preserve native genetic resources in local facilities. The initial explorations were sponsored by the national/local agencies with later systematic collection trips starting in the 1960s through the 1980s with funding support from the Food and Agriculture Organization (FAO) of the United Nations, the International Board for Plant Genetic Resources (IBPGR), and the International Crops Research Institute for the Semi-Arid Tropics (ICRISAT). As a result of these efforts, the seeds collected were shared among the participating genebanks/countries to safeguard them at more than one location. The main goal of these genebanks is to conserve the genetic diversity available among the germplasm resources for peanut improvement research. Many genebanks maintain locally or regionally available landraces of *A. hypogaea* as well as *Arachis* wild species, especially the genebanks in South America.

Conservation and maintenance of genetic resources are critical for their effective utilization in peanut improvement. When properly preserved under optimum storage conditions, seeds of *A. hypogaea* will remain viable for 20 years or more. In the USDA genebank, *A. hypogaea* seeds are tightly packed in heat-sealed aluminum foil bags and stored at −18°C. The freezers are connected to backup generators in case of power outages and alarm alerts in case of mechanical malfunctions. Most accessions in the genebank were originally collected in local markets in South America as seed mixtures with varied pod/seed sizes, shapes, colors, and other features. Many others were local landraces with different plant growth habits, maturities, pod/seed sizes, colors, and shapes. Thus, in seed increase or regeneration nurseries, a representative sample with enough seeds must be planted to avoid genetic drift and maintain the purity/genetic integrity of the original accession. Routine management practices are followed to grow plants and harvest seeds to replenish fresh seeds in the genebank collection. Contrarily, the wild species accessions are stored as intact pods in heat-sealed foil bags at −18°C, and many remain viable for 10 or more years. The wild species are significantly more difficult to regenerate than cultivated peanut, especially in field plots. In the USDA collection, all wild species are propagated in greenhouses to prevent outcrossing and ease of maintaining plants to harvest pods. Wild species produce long, trailing branches, and thus care must be taken to keep the branches confined to the growing container to maintain genetic purity. Species in section *Rhizomatosae* do not or rarely produce pods, and thus are maintained as live plants. The world's largest wild species collection, containing 1,200 accessions, is housed at EMBRAPA in Brazil under the direction of Dr. Jose Valls. In the United States, Dr. Charles Simpson at the Texas A&M University AgriLife Research Center in Stephenville, TX, also maintains a wild species collection containing about 800 accessions.

The cultivated species, *A. hypogaea*, was described by Linnaeus in 1753. It was the first and the only known species of genus *Arachis* until Bentham in 1841 described the first wild species in the genus, *A. glabrata*. In the following 100 years, several individual as well as collective germplasm explorations in South America led to an increased number of *Arachis* materials of both cultivated as well as wild species being collected. However, the first systematic germplasm exploration and collection of living wild species plant materials was started in 1959 by W. C. Gregory, J. R. Pietrarelli, and A. Krapovickas, with funding support from their respective national agencies. During the next three decades, several national and international agencies supported additional germplasm collection trips in South America, leading to the description of 69 *Arachis* species and the publication of the first *Arachis* monograph in 1994 (Krapovickas and Gregory, 1994, 2007).

2.5.2 *Arachis* Genepools and Global Germplasm Collections

Since genus *Arachis* contains both tetraploid and diploid species which hybridize with *A. hypogaea* with varying degrees of seed fertility in the resulting hybrids, *Arachis* genetic resources were classified into four different genepools. The primary genepool includes the genetic resources that intercross readily with *A. hypogaea*, resulting in fertile hybrids. They include the native landraces and other botanical varieties from the centers of origin and diversity, farmer grown cultivars, advanced breeding lines, and other related genetic stocks (Singh and Nigam, 2016). Thus, the primary genepool provides useful genetic diversity that is readily available for peanut improvement. It also includes the tetraploid wild species *A. monticola* because of its genetic similarity and cross-compatibility with *A. hypogaea*. The secondary genepool contains the *Arachis* species that cross-hybridize relatively easily with *A. hypogaea*. They include the section *Arachis* diploid ($2n = 2x = 20$) species consisting of A, B, D, F, G, and K genomes (Table 2.1). Although these species intercross among themselves and with *A. hypogaea*, because of their ploidy and genomic differences, the resulting interspecific hybrids are usually semi-fertile. This genepool provided many species that were hybridized with *A. hypogaea* to transfer important disease resistances into *A. hypogaea* (Bertioli et al., 2021; Chu et al., 2021; Dutra et al., 2018; Mallikarjuna et al., 2004, 2011). The tertiary genepool represents *Arachis* species of sections *Erectoides*, *Heteranthae*, *Procumbentes*, and *Rhizomatosae*, which intercross with section *Arachis* species resulting in partially fertile interspecific hybrids which may be used as bridge crosses to introgress desirable genes into *A. hypogaea* (Simpson et al., 2018; Cason et al., 2019). Some of these species have also been shown to intercross with *A. hypogaea* but require embryo rescue techniques to obtain interspecific hybrids (Singh, 1998; Mallikarjuna and Sastri, 2002). Lastly, the quaternary genepool consists of *Arachis* species that are essentially cross-incompatible with *A. hypogaea*. The species in the sections *Caulorrhizae*, *Extranervosae*, *Trierectoides*, and *Triseminatae* comprise this genepool (Singh and Nigam, 2016). Overall, these genepools provide an understanding of the genetic structure and relatedness of the *Arachis* wild species to *A. hypogaea* and options for interspecific hybridization for peanut improvement.

Globally, *Arachis* genetic resources are being conserved at several different national, regional, and international genebanks (Pandey et al., 2012; Nigam, 2015). The genebank at the International Crops Research Institute for the Semi-Arid Tropics (ICRISAT) in India maintains the largest peanut germplasm collection in the world, comprising about 15,000 accessions of both *Arachis hypogaea* and different wild species. This germplasm collection was assembled in the 1970s with seeds donated by 95 countries and support from the Food and Agriculture Organization (FAO), the International Board for Plant Genetic Resources (IBPGR), and several national governmental agencies. Among the national genebanks, the USDA-ARS–managed National Plant Germplasm System (NPGS) maintains the US national peanut germplasm collection of about 9,000 accessions of *A. hypogaea* and 500 accessions of *Arachis* wild species in the genebank located on the University of Georgia-Griffin campus. This germplasm collection is freely distributed for research and educational uses worldwide. The germplasm collection is also stored as a safety backup collection at the National Laboratory for Genetic Resources Preservation (NLGRP) in Ft. Collins, CO, the United States. Other important national genebanks with large germplasm collections include the Botanical Institute of the Northeast (IBONE) and the National Institute of Agricultural Technology (INTA) in Argentina; Embrapa Genetic Resources and Biotechnology, CENARGEN in Brazil; the Oil Crops Research Institute of the Chinese Academy of Agricultural Sciences (OCRI-CAAS, China); and the National Bureau of Plant Genetic Resources (NBPGR) in India. Also, the International Center for Tropical Agriculture (CIAT) in Colombia maintains vegetatively propagated peanut germplasm of forage potential. ICRISAT and USDA have developed different descriptors for characterizing the genetic variability observed within the cultivated peanut (IBPGR-ICRISAT, 1992; Pittman, 1995). They include a standard set of several plant, pod, and seed traits and occasional data on resistances or other quality traits. The characterization data provides valuable information to researchers to select suitable germplasm for effective and efficient use in peanut improvement (Otyama et al., 2019, 2020; Wang et al., 2022). Further, a core and mini-core set of lines have been assembled for ease of working with a smaller, representative sample of lines representing the entire genetic diversity from the original collections of ICRISAT and USDA. The ICRISAT collections contains 1,704 and 184 lines in the core and mini-core sets, respectively (Upadhyaya et al., 2002, 2003). Similarly, the USDA collection has 831 lines as part of the core set (Holbrook et al., 1993; Otyama et al., 2020), and the mini core contains 112 lines (Holbrook and Dong, 2005; Otyama et al., 2019).

Additionally, Africa, and Asia have been viewed as secondary centers of peanut genetic diversity because of their varied climates, soil types, and agronomic production practices. Below is a summary of the genetic resources available across these different regions.

2.6 PEANUT GENETIC DIVERSITY AND GENETIC RESOURCE CREATION

2.6.1 Peanut Genetic Resources Status in India

The ICRISAT genebank maintains the world's largest collection of peanut genetic resources of over 15,000 accessions. Peanut germplasm is also conserved and maintained at the National Bureau of Plant Genetic Resources (NBPGR) and the Directorate of Peanut Research (DGR). Over the years, several advanced breeding and germplasm lines have been identified and developed worldwide for drought, leaf spots, rust, groundnut rosette virus, aflatoxin, rust, and quality traits (Isleib et al., 2001). From 1976 to 2014, 31 improved varieties of peanut have been released by Indian partners using breeding material from ICRISAT. However, studies have shown scanty use of germplasm in crop improvement programs globally. For example, peanut scientists at ICRISAT used 986 unique parents (1986–2002) to develop 8,279 advanced varieties in peanut but included only 132 unique germplasm accessions and ten wild *Arachis* species from more than 15,000 germplasm accessions available in its genebank. The two most often used cultivars were Robut 33-1 (ICG 799) and Chico (ICG 476) (Upadhyaya et al., 2006). Progress on cultivar and breeding line development using peanut genetic resources with major emphasis on diseases and insect pests has been the priority in breeding programs worldwide (Table 2.2).

2.6.2 Peanut Genetic Resources Status in Africa

Peanut is one of the most important cash crops in Africa, particularly for its quality, edible oil and higher protein content (Abady et al., 2019). Because of the varied ecogeographical conditions of peanut production, the African continent displays a large amount of peanut genetic diversity. As a result, Africa is considered as a secondary center of peanut genetic diversity. Peanut germplasm is maintained in several national genebank collections in Africa as well as by individual research programs due to its socioeconomic importance and for future utilization in peanut improvement. Additionally, ICRISAT maintains a regional genebank in Niamey (Niger) with about 11,000 peanut accessions (http://iscgbank.icrisatne.ne/). The ICRISAT-Niger genebank was established in 1991 and is an active distribution site. Its primary goal is to assist the National Agricultural Research Systems (NARS) in the West and Central Africa (WCA) region with access to peanut genetic diversity or germplasm resources and conservation of the NARS peanut germplasm in the genebank. All these genebanks not only preserve the local or regional germplasm but also provide it to researchers to develop improved cultivars for future food and nutrition security in Africa. Peanut germplasm resources contain diverse genotypes, which enables breeders and farmers to select among these genotypes with desired traits, especially if they are high yielding and adapted to specific environments. In those cases, the selected germplasm resources can be used directly as a variety in countries with similar environments (Abady et al., 2019; Halewood et al., 2018; Salgotra and Gupta, 2015). Collections of peanut germplasm provide the primary source of genetic diversity for peanut improvement for sustainable production and food security in the continent. Diverse genetic resources for different traits are essential for peanut breeding and various genebank establishments (Desmae et al., 2019). The production of this crop is mainly through traditional cropping practices in the continent. In Africa, germplasm resources are lost at an alarming rate due to anthropogenic effects such as climate change, pollution, genetic erosion, gross mismanagement, and population growth. Many peanut genetic resources are treated as endangered, and some have even gone extinct due to genetic erosion and environmental changes in north Africa (Williams, 2022). The peanut genetic resources status and distributions in African countries are quite different from

TABLE 2.2
Cultivar and Breeding Line Development Using Peanut Genetic Resources with Major Emphasis on Diseases and Pests

Major Trait	Target Trait/Features	Variety/ Breeding Line Developed using Germplasm	Source Germplasm	References
Diseases	Foliar fungal diseases	GPDB 4 (India), ICGV 87165 (Myanmar, Philippines), Sutherland and Kairi (Australia), Sempre Verde (Brazil)	*A. cardenasii* GKP10017	Bertioli et al. (2021)
	Early leaf spot, late leaf spot, and Spodoptera	Interspecific derivatives (BC_2F_2)	*A. kempff-mercadoi*	Mallikarjuna et al. (2004)
	Peanut bud necrosis virus (PBNV)	ICGV 86031 and JCGV 86388	NC Ac 2214 and NC Ac 2232 resp.	Dwivedi et al. (1995)
	Late leaf spot disease and tomato spotted wilt	PI 695391, PI 695392, and PI 695393.	(*A. ipaënsis* KG 30076 × *A. correntina* GKP 9530), (*A. ipaënsis* KG 30076 × *A. duranensis* KGBSPSc 30060), and (*A. valida* KG30011 × A. stenosperma V 10309)	Chu et al. (2021)
Pest (Nematode)	Peanut root-knot nematode (*Meloidogyne arenaria*)	Tifguard, TifGP-2	COAN (PI 610452)	Holbrook et al. (2012)

country to country due to the climate, ecologies, government policies, funding agencies, and research institute activities.

2.6.2.1 Cultivated Genetic Resources

Currently, peanut genetic resources are maintained at various genebanks and research institutions and programs in Africa (Desmae et al., 2019). Table 2.3 lists the number of landraces from the different regions of Africa. The most landraces were collected from Tanzania (773 landraces), Nigeria (661 landraces), Sudan (394 landraces), Uganda (336 landraces), and Zimbabwe (330 landraces) (Williams, 2022). In addition, several genebanks in African countries, such as the Ethiopian Biodiversity Institute (EBI), the Ethiopian Institute of Agricultural Research (EIAR), and the Agricultural Research Council, South Africa (Cilliers and Swanevelder, 2003) conduct research with collecting, characterizing, and conserving of local peanut genetic materials.

Climate change is having a significant negative impact on the environment (Ogwu et al., 2014). Climate varies from region to region; in this perspective, the availability, distribution, collection, and conservation systems are also quite different in African regions. Table 2.2 displays that the countries in eastern Africa provide about 37.11% of the total genetic resources, followed by western (27.76%) and southern Africa (21.78%). Effective characterization and evaluation of genetic resources is necessary for an efficient utilization of these resources for peanut improvement. The research work on peanut is based on introductions and local collections (landraces) in most of the African countries. Based on important agronomic traits and genetic diversity, the groundnut improvement network for Africa (GINA) assembled a core collection of 300 diverse accessions originating from nine different countries as a rich source of allele mining for peanut improvement across Africa (Conde et al., 2024).

2.6.2.2 Exotic or Wild Species Genetic Resources

The national peanut crop improvement efforts in the continent utilize exotic materials that have come through various international institutions and other research collaborations. A pre-breeding program of exotic peanut germplasm is actively progressing, especially in Senegal, which aims to incorporate genes from *Arachis* wild species in peanut improvement programs in the continent (Fonceka et al., 2012). These species play an important role in sustainable food security and the alleviation of malnutrition in Africa. It is important to genetically evaluate and select superior parental lines that have high yield potential and nutritional composition for small-scale farmers as well as to develop new peanut cultivars for the larger communities.

2.6.3 Peanut Genetic Resources Status in China

Cultivated Species Resources

China began to establish peanut germplasm collections in the 1950s. Several institutes across the country maintain large quantities of germplasm resources. At present, 8,173

TABLE 2.3
Number of Peanut Landraces and Country of Origin in Africa

East Africa			Southern Africa			Western Africa		
Countries	**Accessions**	**References**	**Countries**	**Accessions**	**References**	**Countries**	**Accessions**	**References**
Burundi	1	(ICRISAT, 2023)	Angola	93	(David, 2022)	Benin	7	(ICRISAT, 2023)
Comoros	1	(ICRISAT, 2023)	Botswana	274	(David, 2022)	Burkina Faso	95	(David, 2022)
Ethiopia	63	(David, 2022)	Malawi	34	(ICRISAT, 2023)	Côte d'Ivoire	2	(ICRISAT, 2023)
Kenya	150	(David, 2022)	Mozambique	279	(ICRISAT, 2023)	Gambia	23	(ICRISAT, 2023)
Madagascar	78	(David, 2022)	Namibia	54	(David, 2022)	Ghana	98	(David, 2022)
Mauritius	22	(ICRISAT, 2023)	South Africa	254	(David, 2022)	Guinea	184	(David, 2022)
Rwanda	2	(David, 2022)	Swaziland	7	(David, 2022)	Guinea-Bissau	179	(ICRISAT, 2023)
Somalia	9	(ICRISAT, 2023)	Zambia	284	(ICRISAT, 2023)	Liberia	2	(ICRISAT, 2023)
Uganda	336	(David, 2022)	Zimbabwe	330	(ICRISAT, 2023)	Mali	367	(David, 2022)
Tanzania	773	(David, 2022)	**Total**	**1,609**		Niger	95	(David, 2022)
Zambia	1307	(David, 2022)				Nigeria	661	(David, 2022)
Total	**2,742**					Senegal	206	(David, 2022)
						Sierra Leone	32	(David, 2022)
Middle Africa			North Africa			Togo	100	(David, 2022)
Countries	Acc	References	Countries	Acc	References	**Total**	**2,051**	
Cameroon	109	(David, 2022)	Algeria	2	(David, 2022)			
Central African Republic	119	(David, 2022)	Egypt	28	(David, 2022)			
Chad	163	(David, 2022)	Morocco	11	(David, 2022)			
Congo	161	(David, 2022)	Sudan	394	(David, 2022)			
Total	**552**		**Total**	**435**				

germplasm resources have been preserved in the National Crop Germplasm Resources long term storage conditions at −18°C. The National Oil Crop Medium Term Storage (−4 °C.) has 8,957 cultivar resources, including 3,025 peanut cultivars introduced from 33 countries. During 2011–2018, 1,425 peanut germplasm resources were newly collected and preserved in medium-term storage. "Zhonghua" series varieties are the main varieties released by the Chinese Academy of Agricultural Sciences. There were more than 3,000 germplasm resources of peanut cultivars preserved by Shandong Academy of Agricultural Sciences (SAAS). SAAS has the largest team of peanut research scientists and released the greatest number of cultivars. The cultivars series "Luhua," "Fenghua," "Huayu," "Shanhua," and "Weihua" are very popular. Moreover, the peanut team of Henan Academy of Agricultural Sciences (HAAS) has preserved more than 3,500 peanut germplasm resources. HAAS is also a leading institute in peanut breeding in terms of early maturity, high yield, high oil content, high oleic content, and multiple resistance. More than 60 peanut varieties of the "Yuhua" and "Yuanza" series have been released successively. The cultivated varieties account for 40%–50% of the peanut planting area in Henan Province. The Guangxi Academy of Agricultural Sciences (GAAS) has preserved nearly 2,000 cultivated peanut genetic resources. The scientific research institutions that have long been engaged in the breeding of peanut varieties in Guangxi include the Guangxi Zhuang Autonomous Region Academy of Agricultural Sciences and Hezhou Academy of Agricultural Sciences. In recent years, the Guilin Academy of Agricultural Sciences has also started some breeding research, and correspondingly bred "Guihua", "Heyou," and "Liyou" series of peanut varieties. Among them, the "Guihua" series represents the main varieties in Guangxi peanut production areas, accounting for about 70% of the current area of high-quality varieties in Guangxi peanut production. Ren et al. (2014) analyzed 196 peanut cultivars from different regions in China with simple sequence repeat (SSR) markers to assess genetic diversity and population structure. They observed significant genetic differentiation between cultivars from the south and the north and concluded that the cultivars represent valuable genetic diversity. Per the decree of the Ministry of Agriculture of the People's Republic of China (No. 1 in 2017) that the registration of non-major crop varieties will be implemented, 1,058 peanut cultivars were registered and released as of 2022 (http://www.a-seed.cn/).

2.6.3.1 Wild Species Resources

The National Germplasm Wuchang Wild Peanut Nursery is in Wuhan, Hubei Province, China, and is affiliated with the Institute of Oil Crops, Chinese Academy of Agricultural Sciences. The research direction of the nursery is to collect and preserve wild peanut resources, evaluate various important traits, and provide resources to relevant units for research and utilization. The National Germplasm Wuchang Wild Peanut Nursery (Wuhan) and the Nanning branch of theursery have preserved 338 wild peanut resources imported from India and the United States, covering 31 species (Zhou et al., 2022).

The Henan Academy of Agricultural Sciences has preserved 60 wild peanut resources. The Guangxi Academy of Agricultural Sciences has preserved 55 wild peanut resources. As the undertaking unit of the National Germplasm Wild Peanut Nursery in Nanning, most of the preserved wild peanut germplasm resources come from the National Germplasm Wild Peanut Nursery in Wuhan managed by the Institute of Oil Crops, Chinese Academy of Agricultural Sciences. Shandong Academy of Agricultural Sciences has preserved 25 wild peanut resources.

2.7 WILD SPECIES GENETIC RESOURCES AND INTERSPECIFIC HYBRIDIZATION FOR THE GENETIC ENHANCEMENT OF *A. HYPOGAEA*

2.7.1 Interspecific Hybridization for Broadening the Genetic Base for Biotic and Abiotic Stresses

Various biotic stresses such as foliar fungal diseases, early and late leaf spots, and other fungal diseases such as *Aspergillus flavus* Link ex Fries producing aflatoxin; diseases caused by viruses, such as peanut bud and stem necrosis, groundnut rosette, and tomato spotted wilt; and abiotic stresses such as drought, heat, and salinity affect the productivity of peanut worldwide. Adequate sources of resistance/tolerance to these stresses are lacking in the cultivated species. Such a lack of resistant sources in cultivated germplasm is attributed to its narrow genetic base, as *A. hypogaea* ($2n = 4x = 40$; AABB) was derived from a single, recent hybridization event between two wild diploid progenitor species followed by polyploidization, leading to a genetic bottleneck.

Many *Arachis* species have very high levels of resistances to common peanut insect-pests and pathogens (Cason, 2022; Stalker et al., 2013, 2016; Dwivedi et al., 2007; Stalker and Campbell, 1983). Consequently, researchers attempted to introgress these resistances into *A. hypogaea*. Hull and Carver (1938) were the first to attempt interspecific hybridization in the genus between two tetraploid species, *A. hypogaea* (section *Arachis*) and *A. glabrata* (section *Rhizomatosae*), albeit unsuccessfully. The first successful cross resulting in a vigorous but sterile interspecific hybrid was reported by Krapovickas and Rigoni (1951) between *A. hypogaea* and *A. villosa* var. *correntina*. Stalker et al. (2013) suggested that crosses are usually more successful when the higher ploidy level species, *A. hypogaea*, is used as the female parent. Moreover, considering the importance of allelic richness for broad spectrum resistances in wild species, research has been carried out to transfer multiple disease resistance from the wild species into the cultivated genetic background. As indicated in Table 2.1, *A. hypogaea* is an allotetraploid with two subgenomes (A and B), whereas most of the species are diploids, especially those from section *Arachis*, which are cross-compatible with *A. hypogaea*. Because of the ploidy level and genomic differences, often the progeny of these interspecific crosses remains sterile. However, as more knowledge of the genetic relationships among the species became available (Gregory and Gregory, 1979; Singh, 1985; Stalker, 1985; Valls et al., 1985; Moretzsohn et al., 2004, 2013; Tallury et al., 2005; Friend et al., 2010), many different combinations of interspecific hybrids were produced. Mallikarjuna et al. (2011) generated 17 amphiploids and autotetraploids of different genome combinations. Of these, 13 different interspecific AB-genome and three AA-genome hybrids were obtained. These sterile hybrids were polyploidized, and five combinations produced tetraploid progeny. Also, they were crossed with *A. hypogaea* resulting in 11 different hybrid combinations. This study confirms the successful transfer of resistance genes from wild species into cultivated peanut using artificial hybridization, thus enhancing the genetic variability in peanut genetic improvement programs. In another attempt, 376 interspecific lines from three different populations derived from crosses with the wild species *A. duranensis*, *A. ipaënsis*, *A. batizocoi*, and *A. valida* were phenotyped for two seasons and across two locations, Serere and Nakabango, in Uganda. Several genotypes showed good agronomic traits and stacking disease resistance to improve the peanut crop in sub-Saharan Africa (Essandoh et al., 2022).

In the United States, several introgression pathways have been utilized by peanut researchers (Simpson, 2001; Stalker, 1992; Stalker et al., 2013; Stalker, 2017) to transfer desirable genes and develop improved germplasm lines or cultivars (Bertioli et al., 2021; Chu et al., 2021; Holbrook et al., 2021; Simpson et al., 1993; Simpson and Starr, 2001; Simpson et al. 2003; Stalker and Beute, 1993; Stalker et al., 2002a, 2002b; Isleib et al., 2006, 2011; Tallury et al., 2014). The introgression of wild species germplasm resulted in the broadening of the genetic base of the cultivated peanut.

2.7.2 Introgression Pathways

Mainly, three different introgression pathways have been utilized to create populations of fertile tetraploid interspecific hybrids. The pathways have been utilized by different researchers for introgression of mostly disease and/or pest resistance from the diploid species into *A. hypogaea*.

2.7.2.1 Direct Cross

This approach is also known as the triploid–hexaploid pathway. It involves making direct cross between *A. hypogaea* ($2n = 4x = 40$) and the diploid ($2n = 2x = 20$) species, resulting in a sterile, triploid ($2n = 3x = 30$) interspecific hybrid. Once triploid hybrid plants are established, stem cuttings can be treated with colchicine to duplicate the chromosomes and restore fertility at the hexaploid ($2n = 6x = 60$) level. The next step involves ploidy level manipulations to bring down the chromosome number to tetraploid in these hybrids. This is achieved by backcrossing hexaploids to *A. hypogaea* and selecting tetraploids in the resulting progenies. Stalker et al. (2016) suggested that no useful germplasm resulted from backcrossing hexaploids with *A. hypogaea*. Further, it was also suggested that backcrossing hexaploids with diploids, though, will theoretically drop the chromosome number to the tetraploid level in one generation, the progenies resulting from such crosses were not seed fertile (Stalker et al., 2016). Contrarily, hexaploidy hybrids are allowed to self-pollinate for several cycles (5–8), leading to the spontaneous loss of chromosomes during this process with stable, tetraploid interspecific hybrids recovered. Although the loss of chromosomes may be infrequent and random, this process provides opportunities for forced genetic recombination between the different species chromosomes at a higher ploidy level. This pathway was successfully utilized at North Carolina State University (NCSU) in Raleigh, NC, the United States, to develop several disease-resistant germplasm lines from *A. hypogaea* × *A. cardenasii* hexaploids (Stalker and Beute, 1993; Stalker et al., 2002a, 2002b; Tallury et al., 2014). Additionally, ICRISAT scientists also utilized this approach to release leaf spot and rust resistant germplasm lines (Moss et al., 1998; Singh et al., 2003).

Occasionally, triploid hybrids produce a few pods via the formation of restitution nuclei and unreduced gametes, especially when the plants are maintained in the field or grown in the greenhouse for longer periods of time (Singh and Moss, 1984; Singsit and Ozias-Akins, 1992). The resulting progenies provide fertile segregants to select for further evaluations, overcoming some of the initial fertility barriers described above.

2.7.2.2 Autotetraploid Pathway

In this pathway, the chromosome number of the diploid species (AA or BB) is first doubled to tetraploid level using colchicine to produce an autotetraploid (AAAA or BBBB). Consequently, crosses can be made with *A. hypogaea* (AABB) at the tetraploid level, thus avoiding several sterility barriers as discussed in the triploid–hexaploid pathway. However, autotetraploids are less vigorous and exhibit infertility. Scientists at ICRISAT developed germplasm lines using this pathway (Singh, 1986a).

2.7.2.3 Allotetraploid Pathway

This is the most logical pathway to introgress wild species genes into *A. hypogaea*. Synthetic allotetraploids (AABB) with genome composition like *A. hypogaea* (AABB) are produced to make crosses at the tetraploid level. This is achieved by first producing a diploid interspecific hybrid combining different genomes (AB) and then duplicating the genomes using colchicine to produce a synthetic allotetraploid (AABB). This provides ease of crossing at the tetraploid level to produce fertile interspecific hybrids as well as avoiding the fertility barriers/incompatibilities discussed above. Additionally, this pathway provides the advantage of combining desirable traits from different diploid wild species to introgress into *A. hypogaea* (Bertioli et al.,

2021; Chu et al., 2021; Dutra et al., 2018). At Texas A&M University, TX, the United States, Simpson (1991) and Simpson (2001) proposed a modified allotetraploid pathway of using a tri-species diploid hybrid (diploid/tetraploid pathway) to produce a synthetic allotetraploid (AABB) to make crosses with *A. hypogaea* (AABB). Using this pathway, the first root-knot nematode (RKN)-resistant germplasm lines, TxAG-6 and TxAG-7, and the world's first RKN-resistant cultivar, COAN, were released (Simpson et al., 1993; Simpson and Starr, 2001). COAN has been the source of RKN resistance in many cultivars in the United States, especially in the southeastern peanut production regions (Holbrook et al., 2008; Holbrook et al., 2017). This pathway was also utilized by ICRISAT scientists to release several leaf spot and rust-resistant germplasm lines (Singh, 1986b; Moss et al., 1998; Singh et al., 2003). Conventionally, traits of interest from *Arachis* species have been difficult to follow in progenies of interspecific hybrids, and disease and insect resistance in 40-chromosome progenies has been lost because of low population sizes and inadequate methods for selection within very small populations (Stalker, 2017). However, the recent peanut genomic advances provided molecular markers and other genomic tools to efficiently select progenies with the desired traits early in the breeding programs, thus speeding up the development of new cultivars (Holbrook et al., 2008; Holbrook et al., 2017).

2.8 PEANUT GENETIC RESOURCES DATABASES

Globally, there are three main databases for access to peanut germplasm and related information/data. Below is a summary of these databases. The reader can access the individual database websites for information/data and register a user account to request seeds of the germplasm. Additionally, PeanutBase (Dash et al., 2016), discussed in Chapter 4, also provides many search tools to access the latest genetic and genomic information of *Arachis* (www.peanutbase.org).

2.8.1 Genesys (https://www.genesys-pgr.org/)

Genesys is the largest online platform for information/data about plant genetic resources conserved in genebanks worldwide and is managed by the Crop Trust. It contains more than 4 million accessions located in over 500 genebanks around the globe. The data can be accessed for individual crops including peanut/groundnut. Genesys contains data from about 28,000 groundnut accessions. Seeds can be requested online, and Genesys relays requests for material to the corresponding genebank. Most genebanks distribute a small quantity of seeds for research and breeding purposes, and access to seeds is subject to legal and phytosanitary regulations. Genesys is a completely free resource for users of genebank data and is free to join as a registered user.

2.8.2 ICRISAT Genebank (http://genebank.icrisat.org/)

As discussed earlier, ICIRSAT maintains about 15,000 *Arachis* accessions. The active collection of germplasm is stored at 4°C and ≤20% relative humidity and is the basic source for distribution and utilization. The base collection is maintained at −20°C in vacuum-packed laminated four-layer aluminum foil pouches at 3%–7% seed moisture content. The conserved germplasm has been characterized for important morpho-agronomic characteristics including 21 different plant, pod, and seed traits and agronomic evaluation data of late leaf spot, early leaf spot, rust, and groundnut rosette virus. Additionally, as mentioned earlier, the core and mini core collections have been assembled, displaying the entire genetic diversity of the total germplasm collection (Upadhyaya et al., 2002, 2003) and can be accessed on the genebank website. Based on the performance of mini core accessions against various biotic and abiotic stresses and other published information, a subset of 56 accessions from different countries and different traits was assembled to help breeders in developing elite cultivars (Upadhyaya et al., 2014). Further, a reference set of the most genetically diverse 300 accessions was also developed (Upadhyaya et al., 2014). This reference set captured 95% of the unique alleles, representing diversity from the entire spectrum of composite collection. Information on these subsets is also available on the genebank website. The subsets can be utilized for genetic and genomic investigations to understand the genetic basis of economically desirable traits for peanut improvement.

The genebank link (http://genebank.icrisat.org/) allows one to select the crop—for example, peanut (groundnut)—from the list of crops. The data includes passport information, characterization, evaluation, core and mini core sets, reference sets, and subsets for useful traits. Seeds can be requested online and are provided following the Plant Treaty guidelines.

2.8.3 Germplasm Resources Information Network-Global (GRIN-Global (https://www.grin-global.org/))

GRIN-Global is the United States National Plant Germplasm System (NPGS) database. It was jointly developed by the Agricultural Research Service of the USDA, the Global Crop Diversity Trust, and Bioversity International and is being used by many international genebanks. The USDA GRIN Global database contains information of about 9,000 *Arachis hypogaea* accessions and 500 *Arachis* wild species accessions. It is free to register as a user to access data and request seeds. The search tool allows one to search by genus or by the accession or plant introduction (PI) number. The data includes accession passport information, taxonomy, intellectual property rights, and characterization. The database is user-friendly with a shopping cart to submit the order. For each accession that is requested, 25 seeds of *A. hypogaea* and 10 pods of the wild species are sent. As with the other genebanks, distribution of seeds follows the Plant Treaty guidelines and other international regulations.

2.9 PERSPECTIVES ON PEANUT GENETIC RESOURCES CONSERVATION AND USE

Peanut is a native, New World taxon and has a wide range of adaptation to many different ecological conditions. It exhibits large morphological variation with different subspecies and botanical varieties. Despite the large morphological diversity, cultivated peanut exhibits little molecular genetic diversity. Additionally, there are many wild *Arachis* species which have demonstrated great morphological and molecular genetic diversity. These wild species have also been documented as sources of biotic and abiotic resistances. The cultivated as well as the wild-species genetic resources are essential for the genetic improvement and broadening of the genetic base of cultivated peanut. Many genebanks around the world have been preserving these genetic resources for continued research and educational uses. However, international regulations with germplasm access and benefit sharing have hampered the addition of new collections to the global genebanks. Further, the loss of native habitats to urban development and global climate change have been contributing to the rapid loss of native *Arachis* species. Preservation and maintenance of this genetic diversity is crucial to future genetic improvement of peanut and its sustained production. Additionally, characterization and evaluation of the genetic resources is essential to identify potential sources of desirable traits for the genetic enhancement of peanut. Once such sources are identified, introgression programs must be initiated to transfer the desirable genes for the development of populations and future cultivars. In this respect, the recent advances in peanut genome sequence analysis are beginning to provide an understanding of the molecular basis of complex multigenic traits such as yield and stress resistances and offer an exciting opportunity for genome manipulations and gene-linked markers to rapidly develop superior cultivars for sustained global peanut production.

REFERENCES

Abady S, Shimelis H, Janila P, Mashilo J. Groundnut (*Arachis hypogaea* L.) improvement in sub-Saharan Africa: a review. Acta agric scand sect B — Soil and Plant Science 69 (2019): 528–545.

Banks DJ. Hand-tripped flowers promote seed production in *Arachis lignosa*, a wild peanut. *Peanut Science* 17 (1990): 23–24.

Bertioli DJ, Cannon SB, Froenicke L, Huang G, Farmer AD, et al. The genome sequence of *Arachis duranensis* and *Arachis ipaënsis*, the diploid ancestors of cultivated peanut. *Nature Genetics* 48 (2016): 438–446.

Bertioli DJ, Gao D, Ballen-Taborda C, et al. Registration of GA-BatSten1 and GA-MagSten1, two induced allotetraploids derived from peanut wild relatives with superior resistance to leaf spots, rust, and root-knot nematode. *Journal of Plant Registrations* 15 (2021): 372–378.

Bertioli DJ, Jenkins J, Clevenger J, et al. The genome sequence of segmental allotetraploid peanut *Arachis hypogaea. Nature Genetics* 51 (2019): 877–884.

Campbell WV, Emery DA, Wynne JC. Resistance of peanuts to the potato leafhopper. *Peanut Science* 3 (1976): 40–43.

Cason JM, Simpson CE, Burow MD, Tallury SP, Pham H, Ravelombola W. Use of wild and exotic germplasm for resistance in peanut. *Journal of Plant Registrations* 17 (2022): 1–25.

Cason JM, Simpson CE, Rooney WL, Brady JA. First report of successful hybridization between *Arachis vallsii* and *Arachis dardani. Agrosystems, Geosciences & Environment* 2 (2019): 190013.

Chu Y, Stalker HT, Marasigan K, et al. Registration of three peanut allotetraploid interspecific hybrids resistant to late leaf spot disease and tomato spotted wilt. *Journal of Plant Registrations* 15 (2021): 562–572.

Cilliers AJ, Swanevelder CJ. The South African germplasm collection of peanut, (*Arachis hypogaea* L.), and its utility. *South African Journal of Plant and Soil* 20 (2003): 93–96.

Conde S, Rami JF, Okello DK, Sambou A, Muitia A, Oteng-Frimpong R, Makweti L, et al. The groundnut improvement network for Africa (GINA) germplasm collection: a unique genetic resource for breeding and gene discovery. G3(2024): *Genes, Genomes, Genetics* 14 (2024): 1–12. https://doi.org/10.1093/g3journal/jkad244.

Dash S, Cannon EKS, Kalberer SR, Farmer AD, Cannon SB. PeanutBase and other bioinformatic resources for peanut (Chapter 8). In Stalker HT, Wilson RF (Eds.), *Peanuts: genetics, processing, and utilization*. AOCS Press, 2016: 241–252.

Desmae H, Janila P, Okori P, Pandey MK, et al. Genetics, genomics and breeding of peanut (*Arachis hypogaea* L.). *Plant Breeding* 138 (2019): 425–444.

Dutra WF, Guerra YL, Ramos JPC, Fernandes PD, et al. Introgression of wild alleles into the tetraploid peanut crop to improve water use efficiency, earliness and yield. *PloS One*. 2018;13(6):e0198776 Epub 2018/06/12. 10.1371/journal.pone.0198776

Dwivedi SL, Bertioli DJ, Crouch JH, Valls JF, et al. Peanut. In Kole C (Ed.), *Genome mapping and molecular breeding in plants. Oilseeds*, vol. 2. Springer-Verlag, Berlin Heidelberg, 2007: 115–151.

Dwivedi SL, Nigam SN, Reddy DVR, Reddy AS, Ranga Rao GV. Progress in breeding peanut varieties resistant to peanut bud necrosis virus and its vector. In Buiel AAM, Parlevliet JE, Lenne JM (Ed.), Recent studies on peanut bud necrosis disease: proceedings of a Meeting, 20 Mar 1995. ICRISAT Asia Center, India, 1995: 35–40.

Essandoh DA, Odong T, Okello DK, Fonceka D, et al. Quantitative trait analysis shows the potential for alleles from the wild species *Arachis batizocoi* and *A. duranensis* to improve peanut disease resistance and yield in East Africa. *Agronomy* 12, no. 9 (2022). https://doi.org/10.3390/agronomy12092202

Ferguson ME, Bramel PJ, Chandra S. Gene diversity among botanical varieties in peanut (*Arachis hypogaea* L.). *Crop Science* 44 (2004): 1847–1854.

Fonceka D, Tossim HA, Rivallan R, Vignes H, et al. Construction of chromosome segment substitution lines in peanut (*Arachis hypogaea* L.) using a wild synthetic and QTL mapping for plant morphology. *PLoS One.* 7 (2012): doi: 10.1371/journal.pone.0048642.

Friend SA, Quandt D, Tallury SP, Stalker HT, Hilu KW. Species, genomes and section relationships in genus *Arachis* (Fabaceae): A molecular phylogeny. *Plant Systematics and Evolution* 290 (2010): 185–199.

Gibbons RW, Bunting AH, Smartt J. The classification of varieties of groundnut (*Arachis hypogaea* L.). *Euphytica* 21 (1972): 78–85.

Grabiele M, Chalup L, Robledo G, Seijo G. Genetic and geographic origin of domesticated peanut as evidenced by 5S rDNA and chloroplast DNA sequences. *Plant Systematics and Evolution* 298 (2012): 1151–1165.

Gregory MP, Gregory WC. Exotic germplasm of *Arachis* L. interspecific hybrids. *Journal of Heredity* 70 (1979): 185–193.

Gregory WC, Gregory MP. 1976. Groundnut. In Simmonds NW (Ed.), *Evolution of crop plants.* Longman Group Ltd., London, 1976: 151–154.

Gregory WC, Gregory MP, Krapovickas A, Smith BW, Yarbrough JA. *Structure and genetic resources of peanuts. Peanuts—Culture and uses.* American Peanut Research and Education Association, Stillwater, OK, 1973: 47–133.

Gregory WC, Krapovickas A, Gregory MP. Structure, variation, evolution and classification in *Arachis.* In Summerfield RJ, Bunting AH (Eds.), *Advances in legume science.* Royal Botanic Gardens, Kew, London, 1980: 469–481.

Halewood M, Chiurugwi T, Sackville Hamilton R, Kurtz B, Marden E, Welch E, Michiels F, et al. Plant genetic resources for food and agriculture: opportunities and challenges emerging from the science and information technology revolution. *New Phytologist* 217, no. 4 (2018): 1407–1419.

Hammons RO. *Genetics of Arachis hypogaea: Peanuts-culture and uses.* American Peanut Research and Education Association, Stillwater, OK, 1973: 135–173.

Hammons RO. Origin and early history of the peanut. In Pattee HE, Young CT (Eds.), *Peanut science and technology.* American Peanut Research and Education Society, Yoakum, TX, 1982: 1–20.

He G, Prakash C. Evaluation of genetic relationships among botanical varieties of cultivated peanut (*Arachis hypogaea* L.) using AFLP markers. *Genetic Resources of Crop Evolution* 48 (2001): 347–352.

Holbrook CC, Anderson WF, Pittman RN. Selection of a core collection from the U. S. germplasm collection of peanut. *Crop Science* 33 (1993): 859–861.

Holbrook CC, Dong W. Development and evaluation of a mini core collection for the U.S. peanut germplasm collection. *Crop Science* 45 (2005): 1540–1544.

Holbrook CC, Dong W, Timper P, Culbreath AK, Kvien CK. Registration of peanut germplasm line TifGP-2, a nematode-susceptible sister line of 'Tifguard'. *Journal of Plant Registrations* 6 (2012): 208–211.

Holbrook CC, Ozias-Akins P, Chu Y, Culbreath AK, Kvien CK, Brenneman TB. Registration of 'TifNV-high O/L'peanut. *Journal of Plant Registrations* 11, no. 3 (2017): 228–230.

Holbrook CC, Ozias-Akins P, Chu Y, Lamon S, Bertioli DJ, Leal-Bertioli SCM, Culbreath AK, Godoy I. Registration of TifGP-3 and TifGP-4 peanut germplasm lines. *Journal of Plant Registrations* 16, no. 1 (2021): 120–123.

Holbrook CC, Timper P, Culbreath AK, Kvien CK. Registration of Tifguard peanut. *Journal of Plant Registrations* 2 (2008): 92–94.

Hull FH, Carver WA. Peanut improvement. In *Annual report, Florida agricultural experiment station*, 1938: 39–40. University of Florida-Gainesville, USA.

IBPGR and ICRISAT. *Descriptors for peanut* (second revision). IBPGR, Rome and ICRISAT, Patancheru, India. 1992.

ICRISAT. The International Crops Research Institute for the Semi-Arid Tropics [WWWDocument]. https://doi.org/http://genebank.icrisat.org/IND/Passport?Crop=Groundnut. 2023.

Isleib TG, Holbrook CC, Gorbet DW. Use of plant introductions in peanut cultivar development. *Peanut Science* 28 (2001): 96–113.

Isleib TG, Milla-Lewis SR, Pattee HE, Copeland SC, Zuleta MC, Shew BB, Hollowell JE, et al. Registration of 'Bailey'peanut. *Journal of Plant Registrations* 5, no. 1 (2011): 27–39.

Isleib TG, Rice PW, Mozingo RW II, Copeland SC. Registration of N96076L peanut germplasm line. *Crop Science* 46, no. 5 (2006): 2329.

Knauft DA, Chiyembekeza AJ, Gorbet DW. Possible reproductive factors contributing to outcrossing in Peanut (*Arachis hypogaea* L.). Peanut Science 19 (1992): 29–31.

Kochert G, Stalker HT, Gimenes M, Galgaro L, Lopes CR, Moore K. RFLP and cytogenetic evidence on the origin and evolution of allotetraploid domesticated peanut, *Arachis hypogaea* (Leguminosae). *American Journal of Botany* 83, no. 10 (1996): 1282–1291.

Krapovickas A. The origin, variability and spread of the peanut (*Arachis hypogaea*). In Ucko RJ, Dimbleby CW (Eds.), *The domestication and exploitation of plants and animals.* Duckworth, London, 1968: 427–441.

Krapovickas A, Gregory WC. Taxonomia del genero *Arachis* (Leguminosae). *Bonplandia* 8, no 1–4 (1994): 1–186.

Krapovickas A, Gregory WC, Williams DE, Simpson CE. Taxonomy of the genus *Arachis* (Leguminosae). *Bonplandia* 16 (2007): 7–205.

Krapovickas A, Rigoni VA. Estudios Citologicas en el Genero *Arachis. Revista de Investigaciones Agricoles* 5 (1951): 289–293.

Krapovickas A, Rigoni VA. La nominclatura de las subspecies y variedades de *Arachis hypogaea* L. *Revista de Investigaciones Agricoles* 14 (1960): 197–228.

Krapovickas A, Gregory WC. Taxonomy of the genus Arachis (Leguminosae). Translated by Williams DE and Simpson CE. *Bonplandia* 16 (2007): 1–205.

Lakshmi VK, Shivanna KR. Structure and cytochemistry of the pistil in *Arachis hypogaea. Proceedings: Plant Sciences* 95 (1986): 357–363.

Leal-Bertioli SCM, de Blas FJ, Chavarro MC, Simpson CE, Valls JFM, Tallury SP, Moretzsohn MC, et al. Relationships of the wild peanut species, section Arachis: a resource for botanical classification, crop improvement, and germplasm management. *American Journal of Botany* 111, no. 6 (2024): e16357. https://doi.org/10.1002/ajb2.16357.

Lu J, Mayer A, Pickersgill B. Stigma morphology and pollination in *Arachis* L. (Leguminosae). *Annals of Botany* 66 (1990): 73–82.

Mallikarjuna N, Pande S, Jadhav DR, Sastri DC, Rao JN. Introgression of disease resistance genes from *Arachis kempff-mercadoi* into cultivated peanut. *Plant Breeding* 123 (2004): 573–576.

Mallikarjuna N, Sastri DC. Morphological, cytological and disease resistance studies of the intersectional hybrid between *Arachis hypogaea* L. and *A. glabrata* Benth. *Euphytica* 126 (2002):161–167.

Mallikarjuna N, Senthilvel S, Hoisington D. Development of new sources of tetraploid *Arachis* to broaden the genetic base of cultivated peanut (*Arachis hypogaea* L.). *Genetic Resources of Crop Evolution* 58 (2011): 889–907.

Moretzsohn MC, Gouvea EG, Inglis PW, Leal-Bertioli SCM, Valls JFM, Bertioli DJ. A study of the relationships of cultivated peanut (*Arachis hypogaea*) and its most closely related wild species using intron sequences and microsatellite markers. *Annals of Botany* 111, no. 1 (2013): 113–126.

Moretzsohn MC, Hopkins MS, Mitchell SE, Kresovich S, Valls JFM, Ferreira ME. Genetic diversity of peanut (*Arachis hypogaea* L.) and its wild relatives based on the analysis of hypervariable regions of the genome. *BMC Plant Biology* 4 (2004): 1–10.

Moss JP, Rao VR. The peanut reproductive development to plant maturity. In Pattee HE, Stalker HT (Eds.), *Advances in peanut science*. American Peanut Research and Education Society, Stillwater, OK, 1995: 1–13.

Moss JP, Singh AK, Nigam SN, Hildebrand GL, Govinden N, Ismael FM, Subrahmanyam P, Reddy LJ. Registration of ICGV-SM 86715 peanut germplasm. *Crop Science* 38, no. 2 (1998): 572–572.

Nigam SN. *Peanut at a glance. Feed the future innovation lab for peanut*. USAID, Athens, GA. 2015.

Ogwu MC, Osawaru ME, Ahana CM. Challenges in conserving and utilizing plant genetic resources (PGR). *International Journal of Genetics and Molecular Biology* 6 (2014): 16–23.

Otyama PI, Kulkarni R, Chamberlin K, Ozias-Akins P, Chu Y, Lincoln LM, MacDonald GE, et al. Genotypic characterization of the US peanut core collection. *G3: Genes, Genomes, Genetics* 10, no. 11 (2020): 4013–4026.

Otyama PI, Wilkey A, Kulkarni R, Assefa T, Chu Y, Clevenger J, O'Connor DJ, et al. Evaluation of linkage disequilibrium, population structure, and genetic diversity in the US peanut mini core collection. *BMC Genomics* 20 (2019): 1–17.

Pandey MK, Monyo E, Ozias-Akins P, Liang X, Guimarães P, Nigam SN, Upadhyaya HD, et al. Advances in Arachis genomics for peanut improvement. *Biotechnology Advances* 30, no. 3 (2012): 639–651.

Pattee HE, Mohapatra SC. Anatomical changes during ontogeny of the peanut (*Arachis hypogaea* L.) fruit: mature megagametophyte through heart shaped embryo. *Botanical Gazette* 148 (1987): 156–164.

Pattee HE, Stalker HT, Giesbrecht FG. Comparative peg, ovary, and ovule ontogeny of selected cultivated and wild-type Arachis species. *Botanical Gazette* 152, no. 1 (1991): 64–71.

Periasamy K, Sampoornam C. The morphology and anatomy of ovule and fruit development in *Arachis hypogaea* L. *Annals of Botany* 53 (1984): 399–411.

Pittman RN. *United States peanut descriptors*. USDA-ARS-132, US Government Printing Office, Washington, DC, 1995.

Ren X, Jiang H, Yan Z, Chen Y, Zhou X, Huang L, Lei Y, et al. Genetic diversity and population structure of the major peanut (*Arachis hypogaea* L.) cultivars grown in China by SSR markers. *PLoS One* 9, no. 2 (2014): e88091.

Robledo G, Seijo G. Species relationships among the wild B genome of *Arachis* species (section *Arachis*) based on FISH mapping of rDNA loci and heterochromatin detection: A new proposal for genome arrangement. *Theoretical Applied Genetics* 121 (2010): 1033–1046.

Salgotra RK, Gupta BB. *Plant genetic resources and traditional knowledge for food security*, 1st ed. Springer Singapore, Singapore. 2015.

Seijo G, Lavia GI, Fernández A, Krapovickas A, Ducasse DA, Bertioli DJ, et al. Genomic relationships between the cultivated peanut (*Arachis hypogaea*, Leguminosae) and its close relatives revealed by double GISH. *American Journal of Botany* 94, no. 12 (2007): 1963–1971.

Seijo G, Samoluk S, Chalup L, Grabiele M, Robledo G. Species, genomes and diversification in section *Arachis*. 8th International Conference of the Peanut Research Community on Advances in *Arachis* through Genomics and Biotechnology (AAGB). Savannah, GA, USA. http://www.peanutbioscience.com. 2014.

Seijo JG, Lavia GI, Fernández A, Krapovickas A, Ducasse D, Moscone EA. Physical mapping of the 5S and 18S–25S rRNA genes by FISH as evidence that Arachis duranensis and A. ipaensis are the wild diploid progenitors of A. hypogaea (Leguminosae). *American Journal of Botany* 91, no. 9 (2004): 1294–1303.

Sharma HC, Pampapathy G, Dwivedi SL, Reddy LJ. Mechanisms and diversity of resistance to insect pests in wild relatives of peanut. *Journal of Economic Entomology* 96 (2003): 1886–1897.

Simpson CE. Pathways for introgression of pest resistance in *Arachis hypogaea* L. *Peanut Science* 18 (1991): 22–26.

Simpson CE. Use of wild *Arachis* species introgression of genes into *A. hypogaea* L. *Peanut Science* 28 (2001): 114–116.

Simpson CE, Custodio AR, Rodriques LS, Penaloza AP, et al. Using *Arachis vallsii* Krapov. & W.C. Greg. as a bridge species for introgression in *Arachis*. Abstract. Proceedings. *American Peanut Resource and Education Society* 50 (2018): 124.

Simpson CE, Krapovickas A, Valls JFM. History of *Arachis* including evidence of *A. hypogaea* L. progenitors. *Peanut Science* 28 (2001): 78–80.

Simpson CE, Starr J. Registration of 'COAN' peanut. *Crop Science* 41 (2001): 918.

Simpson CE, Starr JL, Church GT, Burow MD, Paterson AH. Registration of "NemaTAM" peanut. *Crop Science* 43 (2003): 1561.

Simpson CE, Starr JL, Nelson SC, Woodard KE, Smith OD. Registration of TxAG-6 and TxAG-7 peanut germplasm. *Crop Science* 33 (1993): 1418.

Singh AK. Genetic introgression form compatible *Arachis* species into peanut. In J. P. Moss (Ed.), *Proceedings of an international workshop on cytogenetics of arachis*. ICRISAT, Hyderabad, India, 1985: 107–118.

Singh AK. Hybridization barriers among species of *Arachis* L., namely of the section *Arachis* (including the peanut) and *Erectoides*. *Genetic Resources Crop Evolution* 45 (1998): 41–45.

Singh AK. Utilization of wild relatives in the genetic improvement of *Arachis hypogaea* L. 7. Autotetraploid production prospects on interspecific breeding. *Theoretical Applied Genetics* 72 (1986a.): 164–169.

Singh AK. Utilization of wild relatives in the genetic improvement of *Arachis hypogaea* L. 8. Synthetic amphidiploids and their importance in interspecific breeding. *Theoretical Applied Genetics* 72 (1986b.): 433–439.

Singh AK, Dwivedi SL, Pande S, Moss JP, Nigam SN, Sastri DC. Registration of rust and late leafspot resistant peanut germplasm lines. *Crop Science* 43 no. 1 (2003): 440–441.

Singh AK, Moss JP. Utilization of wild relatives in genetic improvement of *Arachis hypogaea* L. VI. Fertility in triploids: Cytological basis and breeding implications. *Peanut Science* 11 (1984): 17–21.

Singh AK, Nigam SN. Arachis gene pools and genetic improvement in peanut. Chapter 2. In V. R. Rajpal et al. (eds.), *Gene pool diversity and crop improvement, sustainable development and biodiversity 10*. Springer International Publishing Switzerland. 2016: 17–75.

Singsit C, Ozias-Akins P. Rapid estimation of ploidy levels in *in vitro*-regenerated interspecific *Arachis* hybrids and fertile triploids. *Euphytica* 64 (1992): 183–188.

Smartt J, Stalker HT, 1982. Speciation and cytogenetics in Arachis. In Pattee HE, Young CT (Eds.), *Peanut science and technology*. American Peanut Research and Education Society, Yoakum, TX,1982: 21–49.

Smith BW. *Arachis hypogaea*. Aerial flower and subterranean fruit. *American Journal of Botany* 37 (1950): 802–815.

Stalker HT. Cytotaxonomy of *Arachis*. In Moss JP (Ed.), *International workshop on cytogenetics of arachis*. International Crops Research Institute for the Semi-Arid Tropics, Hyderabad, Andhra Pradesh, India, 1985: 65–79.

Stalker HT. Utilizing *Arachis* germplasm resources. In Nigam SN (Ed.), *Peanut, a global perspective: proceedings of an international workshop*, 25–29 November 1991. ICRISAT Center, Patancheru, Andhra Pradesh, India, 1992: 281–295.

Stalker HT. Utilizing wild species for peanut improvement. *Crop Science* 57 (2017): 1102–1120.

Stalker HT, Campbell WV. Resistance of wild species of peanut to an insect complex. *Peanut Science* 10 (1983): 30–33.

Stalker HT, Campbell WV, Wynne JC. Evaluation of cultivated and wild peanut species for resistance to the lesser corn stalk borer, *Elasmopalpus lignosellus* (Lepidoptera: Pyralidae). *Journal Economy Entomology* 77 (1984): 53–57.

Stalker HT, Beute MK. Registration of four leafspot resistant peanut germplasm lines. *Crop Science* 33 (1993): 1117.

Stalker HT, Beute MK, Shew BB, Barker KR. Registration of two root-knot nematode resistant peanut germplasm lines. *Crop Science* 42 (2002a): 312–313.

Stalker HT, Beute MK, Shew BB, Isleib TG. Registration of five leaf spot-resistant peanut germplasm lines. *Crop Science* 42 (2002b): 314–316.

Stalker HT, Simpson CE. Germplasm resources in *Arachis*. In Pattee HE, Stalker HT (Eds.), *Advances in peanut science*. Amer. Peanut Resources and Educational Society, Stillwater, OK, 1995: 14–53.

Stalker HT, Tallury SP, Ozias-Akins P, Bertioli D, Leal Bertioli SC. The value of diploid peanut relatives for breeding and genomics. *Peanut Science* 40 (2013): 70–88.

Stalker HT, Tallury SP, Seijo GR, Leal-Bertioli SC. Biology, speciation, and utilization of peanut species. In Stalker HT, Wilson RF (Eds.), *Peanuts: genetics, processing, and utilization*. Academic Press and AOCS Press, 2016: 27–66.

Tallury SP. Peanut origin and botanical descriptions. Chapter 3. In Varshney RK, Pandey MK, Puppala N (Eds.), *The peanut genome, Compendium of plant genomes*. Springer International Publishing, 2017: 27–36.

Tallury SP, Hilu KW, Milla SR, Friend SA, et al. Genomic affinities in *Arachis* section *Arachis* (Fabaceae): Molecular and cytogenetic evidence. *Theoretical Applied Genetics* 111 (2005.): 1229–1237.

Tallury SP, Isleib TG, Copeland SC, Rosas-Anderson P, Balota M, Singh D, et al. Registration of two multiple disease-resistant peanut germplasm lines derived from Arachis cardenasii Krapov. & WC Gregory, GKP 10017. *Journal of Plant Registrations* 8, no. 1 (2014): 86–89.

Tallury SP, Stalker HT, Pattee HE. Early reproductive ontogeny in interspecific crosses of *Arachis hypogaea* and Section *Arachis* Species. *Annals of Botany* 76 (1995): 397–404.

Tan DY, Zhang Y, Wang A. A review of geocarpy and amphicarpy in angiosperms, with special reference to their ecological adaptive significance. *Chinese Journal of Plant Ecology* 34 (2010): 72–88.

Upadhyaya HD, Bramel PJ, Ortiz R, Singh S. Developing a mini core of peanut for utilization of genetic resources. *Crop Science* 42 (2002): 2150–2156.

Upadhyaya HD, Ortiz,R. Bramel PJ, Singh S. Development of a groundnut core collection using taxonomical, geographical and morphological descriptors. *Genetic Resources and Crop Evolution* 50 (2003): 139–148.

Upadhyaya HD, Dwivedi S, Vadez V, Falalou H, Singh S, Varshney R, Liao B. Multiple resistant and nutritionally dense germplasm identified from mini core collection in groundnut. *Crop Science* 54 (2014): 679–693.

Upadhyaya HD, Reddy LJ, Gowda CLL, Singh S. Identification of diverse peanut germplasm: sources of early maturity in a core collection. *Field Crops Resources* 97 (2006): 261–271.

Valls JFM, Rao VR, Simpson CE, Krapovickas A. Current status of collection and conservation of South American peanut germplasm with emphasis on wild species of *Arachis*. In Moss JP (Ed.), *Proceedings of an international workshop on cytogenetics of Arachis*. October 31–November 2, 1983. ICRISAT, Patancheru, A. P. India, 1985: 15–35.

Wang ML, Tonnis BD, Chen CY, Li X, Pinnow DL, Tallury S, et al. Evaluation of variability in seed coat color, weight, oil content, and fatty acid composition within the entire USDA-cultivated peanut germplasm collection. *Crop Science* 62, no. 6 (2022): 2332–2346.

Williams DE. 2022. *Global strategy for the conservation and use of peanut genetic resources*. Global Crop Diversity Trust. Bonn, Germany.

Xi XY. Development and structure of pollen and embryo sac in peanut (*Arachis hypogaea* L.). *Botanical Gazette* 152 (1991): 164–172.

Zamski E, Ziv M. Pod formation and its geotropic orientation in the peanut, *Arachis hypogaea* L., in relation to light and mechanical stimulus. *Annals of Botany* 40 (1976): 631–636.

Zhou X, Ren X, Luo H, Huang L, Liu N, Chen W, et al. Safe conservation and utilization of peanut germplasm resources in the oil crops middle-term genebank of China. *Oil Crop Science* 7, no. 1 (2022): 9–13.

Zhuang W, Chen H, Yang M, Wang J, Pandey MK, Zhang C, et al. The genome of cultivated peanut provides insight into legume karyotypes, polyploid evolution and crop domestication. *Nature Genetics* 51, no. 5 (2019): 865–876.

Ziv M. Photomorphogenesis of the gynophore, pod and embryo in peanut, *Arachis hypogaea* L. *Annals of Botany* 48 (1981): 353–359.

3 Recent Advances in Conventional and Molecular Cytogenetics in Peanut

Guillermo Seijo, Pei Du, Alejandra V. García, Sergio S. Samoluk, Laura Chalup, Laura Perez, Celeste Silvestri, Alejandra M. Ortiz, Graciela I. Lavia, and Germán Robledo

ABBREVIATIONS

DAPI 4′, 6-Diamidino-2-phenylindole
FISH Fluorescent in situ hybridization
GISH Genome in situ hybridization

3.1 INTRODUCTION

The genus *Arachis* is composed of 83 species (Krapovickas and Gregory, 1994; Valls and Simpson, 2005; Valls et al., 2013; Santana and Valls, 2015; Valls and Simpson, 2017; Seijo et al., 2021), including the cultivated peanut and the tropical forage species *A. pintoi* and *A. glabrata*. Wild species are spontaneously distributed over a wide region of South America, extending from the eastern foothills of the Andes in Bolivia and northern Argentina to the Atlantic coast in Brazil, and from the southern limit of the Amazon region to the northern coast of the Rio de la Plata in Uruguay. All the species are herbs and characteristically present geocarpic fruits. According to morphology, cross-compatibilities, viability of hybrids, geographic distribution and cytogenetic characteristics, *Arachis* species have been arranged into nine taxonomic sections: *Trierectoides*, *Erectoides*, *Procumbentes*, *Rhizomatosae*, *Heteranthae*, *Caulorrhizae*, *Extranervosae*, *Triseminatae*, and *Arachis* (Krapovickas and Gregory, 1994). Of these sections, the most diverse and derived is section *Arachis*, which covers much of the geographic distribution of the genus, except in the extreme northeast of the genus range. Section *Arachis* includes 32 species (Table 3.1), 27 of which are diploid with $2n = 2x = 20$, three diploid with $2n = 2x = 18$, and two allotetraploid with $2n = 4x = 40$ (Fernández and Krapovickas, 1994; Lavia, 1996, 1998, 1994; Lavia et al., 2009; Peñaloza and Valls, 2005; Seijo et al., 2021). The latter includes the cultivated peanut (*A. hypogaea*) and the wild *A. monticola*.

All diploid species in this section are included within the secondary germplasm pool of peanut, although they have different life cycles, chromosome features, genomes, and degrees of reproductive compatibility among each other (Krapovickas and Gregory, 1994). Here we present a review of the current state of cytogenetic information on the species of section *Arachis*, the integration of which has allowed researchers to organize the secondary pool of peanut germplasm with a genomic criteria, to make inferences about the origin of cultivated peanuts, and to identify some of the most important mechanisms in the karyotype and genomic differentiation of the group.

3.2 GENERAL CHROMOSOMAL CHARACTERISTICS

Classical cytogenetic studies revealed that the genus *Arachis* is dibasic, with a large predominance of species with $x = 10$, and only four species with $x = 9$ (Fernández and Krapovickas, 1994; Lavia, 1996, 1998; Peñaloza and Valls, 2005; Seijo et al., 2021). Most species are diploid, except cultivated peanut and *A. monticola* of section *Arachis*, three tetraploid species of section *Rhizomatosae*, and one triploid accession of *A. pintoi* of section *Caulorrhizae*.

Pioneering cytogenetic works on peanuts (Husted, 1933, 1936) identified one pair of chromosomes that had an extended secondary constriction (defined as SAT chromosomes according to Fernández and Krapovickas, 1994) and one pair of small chromosomes (termed the A chromosome pair). This fact led to the proposal of the allopolyploid composition (Stebbins, 1947) of peanut with an AABB genomic constitution (Smartt and Gregory, 1967). The chromosomes of the A and B genomes of peanut are indistinguishable by classical techniques (except for the A pair, which is small and shows condensation allochrony). The karyotype formula is 38 m + 2 sm, with very little variation between the two subspecies *hypogaea* and *fastigiata* and the six taxonomic varieties (Fernández and Krapovickas, 1994; Lavia and Fernández, 2004). All varieties and *A. monticola* have a single (very rarely two) and usually distinguishable pair of SAT chromosomes (Seijo et al., 2018). The major karyotype difference among peanut varieties was observed in the type of SAT chromosomes. According to the SAT chromosome classification of Fernández and Krapovickas (1994), SAT type 6 was reported for var. *hirsuta* and SAT types 5 and 6 for var. *hypogaea*. Within subspecies *fastigiata*, type 3 was reported for var. *vulgaris* and var. *fastigiata*, and SAT type 5 for var. *aequatoriana* and var. *peruviana* (Lavia and Fernández, 2004).

DOI: 10.1201/9781003360988-3

TABLE 3.1
Species of Section Arachis, Their Chromosome Number, Genomic Constitution, and Life Cycle. Species Are Ordered By Chromosome Number, Then by Genome and Within Each Genome Alphabetically

Section *Arachis* Species	2n	Genome	Life Cycle
A. hypogaea L.	40	AB	Annual
A. monticola Krapov. & Rigoni	40	AB	Annual
A. cardenasii Krapov. & W. C. Greg.	20	A	Perennial
A. correntina (Burkart) Krapov. & W. C. Greg.	20	A	Perennial
A. diogoi Hoehne	20	A	Perennial
A. duranensis Krapov. & W.C. Greg.	20	A	Annual
A. helodes Mart. ex Krapov. & Rigoni	20	A	Perennial
A. herzogii Krapov., W. C. Greg. & C. E. Simpson	20	A	Perennial
A. hoehnei Krapov. & W. C. Greg.	20	A	Perennial
A. kempff-mercadoi Krapov., W. C. Greg., & C. E. Simpson	20	A	Perennial
A. kuhlmannii Krapov. & W. C. Greg.	20	A	Perennial
A. linearifolia Valls, Krapov. & C. E. Simpson	20	A	Perennial
A. microsperma Krapov., W.C. Greg., & Valls	20	A	Perennial
A. schininii Krapov., Valls, & C. E. Simpson	20	A	Perennial
A. simpsonii Krapov. & W. C. Greg.	20	A	Perennial
A. stenosperma Krapov. & W. C. Greg.	20	A	Annual/biennial
A. villosa Benth.	20	A	Perennial
A. gregoryi C. E. Simpson, Krapov., & Valls	20	B	Annual
A. ipaënsis Krapov. & W. C. Greg.	20	B	Annual
A. inflata Seijo, Atahuachi, & Krapov.	20	B	Annual
A. magna Krapov., W. C. Greg., & C. E. Simpson	20	B	Annual
A. valida Krapov. & W. C. Greg.	20	B	Annual
A. williamsii Krapov. & W. C. Greg.	20	B	Annual
A. glandulifera Stalker	20	D	Annual
A. benensis Krapov., W. C. Greg., & C.E. Simpson	20	F	Annual
A. trinitensis Krapov. & W. C. Greg.	20	F	Annual
A. batizocoi Krapov. & W. C. Greg.	20	K	Annual
A. cruziana Krapov., W. C. Greg. & C. E. Simpson	20	K	Annual
A. krapovickasii C. E. Simpson, D. E. Williams, Valls, & I. G. Vargas	20	K	Annual
A. decora Krapov., W. C. Greg., & Valls	18	G	Annual
A. palustris Krapov., W. C. Greg., & Valls	18	G	Annual
A. praecox Krapov., W. C. Greg., & Valls	18	G	Annual

Diploid species with $2n = 20$ of section *Arachis* have symmetric karyotypes composed mostly of metacentric chromosomes (Figure 3.1A), except *A. glandulifera* (D genome), which has an asymmetric karyotype with a formula of 8 m + 4 sm + 8 st (Stalker, 1991; Robledo and Seijo, 2008). Among the species with symmetric karyotypes, 15 are characterized by the presence of A chromosomes (Fernández and Krapovickas, 1994; Lavia, 1996, 1999; Peñaloza and Valls, 2005; Robledo et al., 2009) and a karyotype formula of 18 m + 2 sm (A genome) (Smartt et al., 1978; Robledo et al., 2009). The remaining species with 2n = 20 (B, F, and K genomes) do not have the small A chromosome (Smartt et al., 1978; Smartt and Stalker, 1982; Fernández and Krapovickas, 1994; Lavia, 1996; Peñaloza and Valls, 2005; Robledo and Seijo, 2010), and the most frequent karyotype formula is 20 m, although some species have between one and four pairs of submetacentric chromosomes (Fernández and Krapovickas, 1994; Robledo and Seijo, 2010; Seijo et al., 2021). Diploid species with 2n = 18 (G genome) have symmetrical karyotypes without the A chromosome pair (Lavia, 1996, 1998; Peñaloza and Valls, 1997; Silvestri et al., 2015). Almost all diploid species within section *Arachis* have only one pair of secondary constrictions on the long arms of chromosome pair 10, except *A. krapovickasii* and *A. gregoryi*, which have two pairs (Robledo and Seijo, 2008, 2010; Robledo et al., 2009; Custódio et al., 2013; Silvestri et al., 2015).

3.2.1 Meiosis

The 40 chromosomes of peanut generally pair to form 20 bivalents in meiosis (Figure 3.1B), although univalents, trivalents, and tetravalents are observed in low frequency in some materials (Husted, 1936; Raman, 1981). The high frequency of autosyndesis (association of chromosomes from the same genome in bivalents), together with the presence of

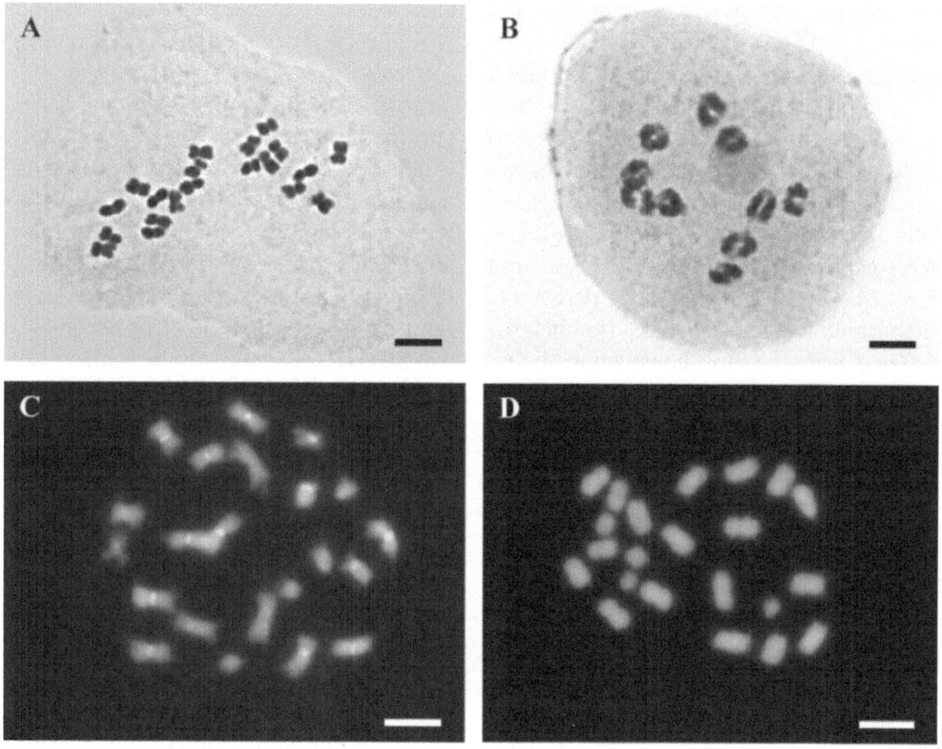

FIGURE 3.1 Mitotic and meiotic chromosomes of wild *Arachis* species. A and B stained with Feulgen reagent, C and D stained with DAPI. A: mitotic chromosomes of *A. benensis* $2n = 2x = 20$. B: Meiotic chromosomes of *A. ipaënsis*. All the chromosomes are paired in ten bivalents. C: Mitotic chromosomes of *A. duranensis*; centromeric heterochromatic bands are highlighted after being stained with DAPI. Note the characteristic A chromosome as the smallest one with a large heterochromatic band. D: Mitotic chromosomes of *A. ipaënsis*. Note the absence of conspicuous centromeric bands of heterochromatin. Scale bar = 5 μm.

a single pair of A and SAT chromosomes, were the main evidence for the allopolyploid nature of *A. hypogaea* (Husted, 1936). Open or closed bivalents present only one or two terminal chiasmata, respectively. The occasional occurrence of complex configurations in peanut, such as tetravalent associations, was initially interpreted as the product of A and B intergenomic translocations (Singh and Moss, 1982; Wynne and Halward, 1989). More recently, it was proposed that the occurrence of tetravalents is compatible with tetrasomal parings because of the existence of ancestral homology between the A and B genomes (Leal Bertioli et al., 2015a; Clevenger et al., 2017). This idea is compatible with the fact that in diploid hybrids between *A. ipaënsis* and *A. duranensis* up to eight bivalents were observed, demonstrating the existence of intergenomic homeology (García, 2015).

Diploid species ($2n = 20$) of section *Arachis* form ten bivalents in diakinesis-metaphase I (Smartt, 1964; Smartt et al., 1978; Resslar and Gregory, 1979; Stalker and Wynne, 1979; Singh and Moss, 1982; Lavia et al., 2001; Rodríguez et al., 2004). Most bivalents are closed and have chiasmata in distal and terminal position (Lavia et al., 2001; Rodríguez et al., 2004). In some wild species such as *A. benensis*, *A. duranensis*, *A. stenosperma*, and *A. villosa*, the occasional observation of one tetravalent was interpreted as the existence of reciprocal translocations in heterozygosis (Singh and Moss, 1982; Lavia et al., 2001; Rodríguez et al., 2004). Based on the observation of tetravalents in metaphases I and bridges with fragments in anaphases I of interspecific or interpopulation hybrids, it was suggested that structural re-arrangements would have been an important mechanism of karyotype diversification (Stalker, 1991; Fernández and Krapovickas, 1994). This was further confirmed comparing the genome sequences of the *A. duranensis* and *A. ipaënsis* (Bertioli et al., 2016).

3.2.2 Genomic Classification

Based on classical cytogenetics and cross-ability data, diploid species of section *Arachis* with $2n = 20$ have traditionally been arranged into two distinct genomic groups (A and non-A genomes) on the basis of the presence or absence, respectively, of the small chromosome pair (A chromosomes). These two groups of species show strong reproductive isolation evidenced by low hybrid production, low chromosome pairing, and low pollen viability in interspecific hybrids (Gregory and Gregory, 1979; Stalker et al., 1991; Krapovickas and Gregory, 1994; Tallury et al., 2005; Singh and Moss, 1984; Burow et al., 2009; Fávero et al., 2015; Paula et al., 2019; García et al., 2020, 2021; Custódio et al., 2021). The first species to segregate from these groups into a distinct genome, the D genome, was *A. glandulifera* because of its asymmetric karyotype and high reproductive isolation from the rest of the known diploid species (Stalker, 1991).

Heterochromatin, analyzed by C-DAPI banding; and ribosomal loci, analyzed by in situ hybridization, revealed different patterns among *Arachis* species and provided useful markers to distinguish chromosomes and genomes (Raina and Mukai, 1999; Seijo et al., 2004; Robledo and Seijo, 2008, 2010; Robledo et al., 2009; Custódio et al., 2013; Silvestri et al., 2015). In *A. hypogaea* and *A. monticola*, half of the chromosomes (those belonging to the A genome) show DAPI$^+$ centromeric bands (AT-rich), while the remaining (those of the B genome) lack detectable centromeric heterochromatin (Figure 3.2C). The heterochromatic bands differ in size, the most conspicuous being those of the A chromosomes (A9 pair). All varieties of *A. hypogaea* and *A. monticola* have a similar distribution and amount of heterochromatin, accounting for about 7% of the karyotype length (Seijo et al., 2004). Physical mapping of rDNA loci in the six botanical varieties of *A. hypogaea* and in *A. monticola* revealed two pairs of 5S rDNA sites and five pairs of 18S–26S rDNA (Figure 3.2C). In both species, the 5S rDNA loci are located proximally on short arms (pairs A3 and B3), whereas the 18S–26S rDNA loci are located proximally (pairs A2, A10, B3, and B10) or subterminally on B7 chromosomes. A 5S rDNA locus is syntenic with an 18S–26S site on pair B3. With the mapping of rDNA loci and heterochromatin, the first reference chromosome map for peanut was constructed, upon which chromosome markers based on other repetitive sequences have been positioned and tentative homeologies have been identified among diploid species of section *Arachis* (Seijo et al., 2004; Zhang et al., 2012; Chen et al., 2016; Samoluk et al., 2017, 2019; Du et al., 2018; Gao et al., 2022).

The pattern of heterochromatin is similar among diploid species with A chromosomes. Most species have conspicuous centromeric bands in all chromosomes, the biggest one being in chromosome A9 (Figure 3.1C). The number of ribosomal genes ranges from two (Figure 3.2A) to four pairs; the largest loci are located on the A10 pair, those of intermediate size are in the A2 pair, while the smallest and faintest signals (in cases where species have more than two loci) are in chromosomes A7 and A4. All these species were grouped under the A genome, but according to the variation in the number and positions of DAPI$^+$ bands and 18S–26S rDNA sites, three karyotypic subgroups were established. Each one was designated based on the preponderant geographical distribution of the species that compose them, that is, Chiquitano (*A. cardenasii*, *A. herzogii*, and *A. kempff-mercadoi*); Pantanal (*A. diogoi*, *A. helodes*, *A. kuhlmannii*, *A. linearifolia*, *A. simpsonii*, and *A. stenosperma*); and La Plata River Basin (*A. correntina*, *A. duranensis*, *A. microsperma*, *A. schininii*, and *A. villosa*) (Robledo et al., 2009). This grouping is partially supported by distance and phylogenetic analyses based on molecular markers (Raina et al., 2001; Milla et al., 2005; Tallury et al., 2005; Grabiele et al., 2012; Moretzsohn et al., 2013).

Species with $2n = 20$ but without the A chromosomes were segregated into four genomes: B *sensu stricto*, D, F, and K, based on the differential patterns of heterochromatin and ribosomal loci (Robledo and Seijo, 2010). The B *sensu stricto* (here after named as B) genome species (*A. inflata*, *A. ipaënsis*, *A. gregoryi*, *A. magna*, *A. valida*, and *A. williamsii*) have karyotypes without pericentromeric heterochromatin detectable by conventional banding methods (Figure 3.1D), between two and four 18S-26S rDNA loci, and one 5S rDNA locus (Figure 3.2B), and they present the highest homology with peanut B chromosomes (Robledo and Seijo, 2010; Seijo et al., 2021). The F genome (*A. benensis* and *A. trinitensis*) presents karyotypes with small, faint bands on seven or eight chromosome pairs and only one 5S rDNA locus on the F3 pair and one or two 18S–26S rDNA loci. The K genome (*A. batizocoi*, *A. cruziana*, and *A. krapovickasii*) is characterized by conspicuous heterochromatic bands on all chromosome pairs except K7, and three 5S rDNA loci and two or three 18S–26S rDNA loci. *Arachis glandulifera* has an asymmetric karyotype and a unique pattern of heterochromatin distribution (with multiple interstitial bands) and ten pairs of 18S–26S rDNA loci, making it the only species representing the D genome (Stalker, 1991; Fernández and Krapovickas, 1994; Robledo and Seijo, 2008).

This genomic assignment was supported by available data on species crossing ability, pollen viability of interspecific hybrids, morphological characters, geographic distribution (Smartt et al., 1978; Stalker, 1991; Krapovickas and Gregory, 1994; Tallury et al., 2005; Burow et al., 2009), and phylogenetic analyses based on chloroplast and nuclear sequences (Tallury et al., 2005; Grabiele et al., 2012; Moretzsohn et al., 2013; Leal Bertioli et al., 2015b). More recently, the development of new chromosomal markers provided additional karyotype markers that support the genomic groups established for section *Arachis* (Du et al., 2018).

Geographically, species included in each of the segregating genomes of the non-A group tend to be co-distributed. Species with the F genome are restricted to the lowland savannas of the Beni department in Bolivia, while those with the K genome are distributed in the northwest of the Boreal Chaco region. Species with the B genome are distributed mainly in the semi-deciduous forests and cerrado savannas associated with the Chiquitano highlands and west of the Pantanal. *Arachis ipaënsis*, known from a single population, was collected on the sandbanks of a stream in an ecotone between tucumano-oranence deciduous forest and Chaco xerophytic forest (Krapovickas and Gregory, 1994).

The three species (*A. decora*, *A. palustris*, and *A. praecox*) with basic chromosome number x = 9 of section *Arachis* were assigned to the G genome based on the sum of karyotype markers, the reproductive isolation and molecular marker analyses (Bechara et al., 2010; Friend et al., 2010; Moretzsohn et al., 2013; Silvestri et al., 2015). These species are characterized by small-sized DAPI$^+$ centromeric bands on all chromosomal pairs, except for *A. palustris*, which lacks these bands on the pair G5 (Silvestri et al., 2015). All three species have only one 18S–26S rDNA locus on the SAT pair (G9) and one 5S rDNA locus on the G6 pair (Silvestri et al., 2015).

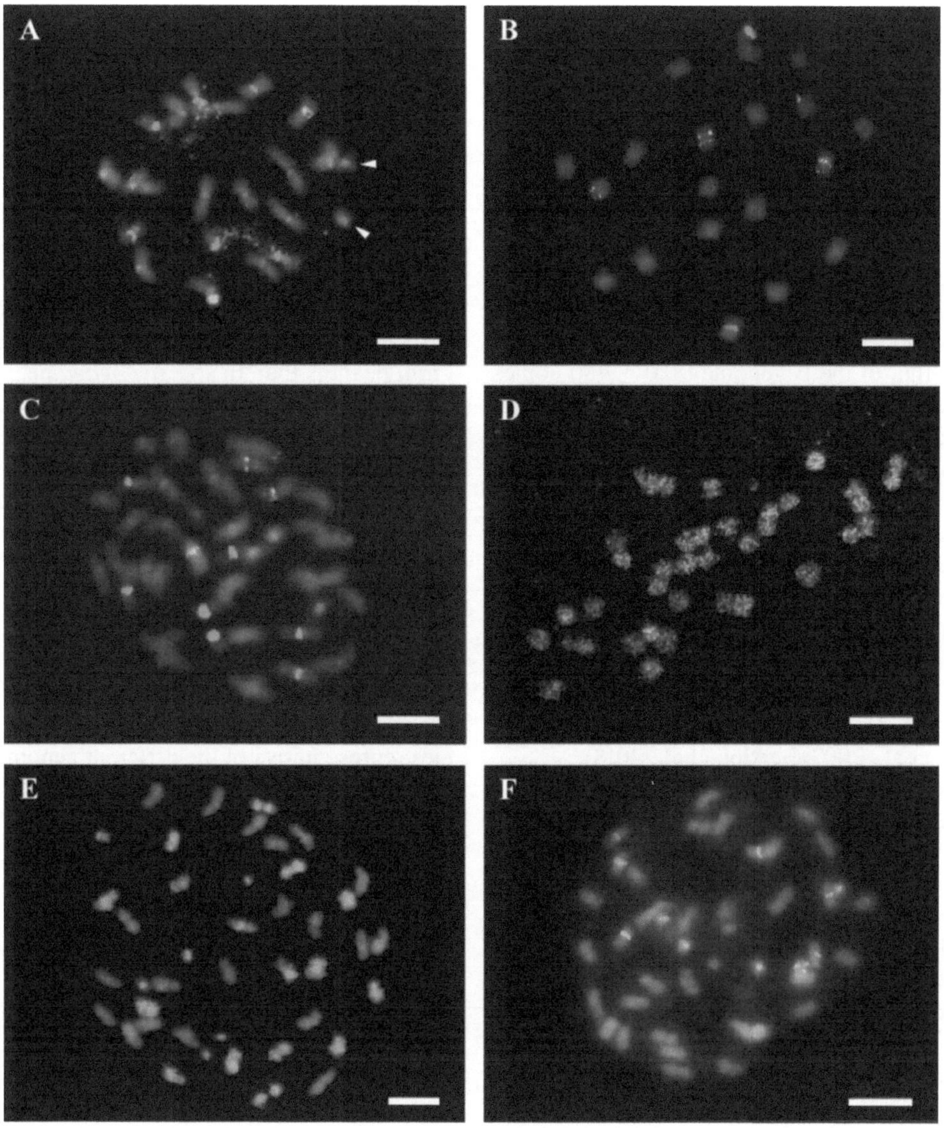

FIGURE 3.2 Somatic metaphases of *Arachis hypogaea* and its diploids genome donors. A–C: Mitotic chromosomes after double fluorescent in situ hybridization (FISH), showing yellow-green fluorescein isothiocyanate (FITC) signals from the 5S rDNA probe (pXV1) and red tetramethyl-rodamine isothiocyanate (TRITC) signals from the 18S–25S rDNA probe (R2). The 4′, 6-diamidino-2-phenylindole (DAPI) counterstaining (light blue) subsequent to the FISH procedure was used to highlight the heterochromatin bands and to stain euchromatin. A. *A. duranensis.* Note the A chromosomes indicated by arrowheads. The extended secondary constrictions of chromosomes A10 are in red dot strips connecting the proximal segment of the chromosome arm to the satellite. B. *A. ipaënsis.* Note chromosomes without conspicuous heterochromatic bands. C. *A. hypogaea.* Note half of the chromosomes with conspicuous heterochromatic bands (A genome) and half without (B genome). The FISH signals for rDNAs are the sum of those observed in *A. duranensis* and *A. ipaënsis.* D. Double genomic in situ hybridization (GISH) with a combination of total DNA probes from diploid wild *Arachis* species labeled with green fluorescein isothiocyanate (FITC) *Arachis duranensis* (Green) or red tetramethyl-rodamine isothiocyanate (TRITC) *A. ipaënsis* (red). E. *In situ* hybridization showing red (TRITC) signals from a probe derived from the FIDEL retroelement retrotanscriptase hybridized onto *A. hypogaea* chromosomes. Note the differential hybridization onto chromosomes with heterochromatic bands (A genome). F. *In situ* hybridization showing red (TRITC) signals corresponding to the TR- 2 satDNA hybridized to *A. hypogaea* chromosomes. Scale bar = 5 μm.

The arrangement proposed based on cytogenetic features allowed a better understanding of the genomic relationships within section *Arachis* and also the selection of diploid parental species for the development of fertile amphidiploids with compatible genomes for recombination with those of the cultivated peanut (Fávero et al., 2015, 2020, García et al., 2020, 2021, Gao et al., 2021).

3.3 CYTOGENETIC MARKERS DEVELOPED FROM THE MOBILOME

Further cytogenetic characterization involved the development of chromosome markers based on other repetitive sequences. This approach began in the last 15 years, but has intensified since the development of genomic tools and the genome sequencing projects of peanut and its wild diploid progenitors (*A. ipaënsis* and *A. duranensis*). Pioneering works performed before the genome sequences were available included the identification, cloning, characterization, quantification, and cytological mapping of the repetitive sequences. The early studies focused on the LTR retrotransposons Ty3-gypsy and Ty1-copia in these three species (Nielen et al., 2010, 2012). The Ty3-gypsy element, named FIDEL (fairly long inter-dispersed euchromatic LTR retrotransposon) shows a three times higher representation in the A genome (3,000 ± 950 copies per haploid genome, *A. duranensis*) than that in the B genome (820 ± 480 copies, *A. ipaënsis*) (Nielen et al., 2010) (Figure 3.2E). The copy number observed in peanut was almost the sum of those of the diploid parents (3,900 ± 1,500). The distribution of these elements studied by FISH showed a dispersed pattern in euchromatin with little or no hybridization at centromeres and distal portions of the chromosomes. The differential representation among the genomes suggests that their amplification/deletion constituted one of the most important mechanisms of genomic differentiation within section *Arachis*. In the case of FIDEL, the distribution in peanut chromosomes largely reflects the patterns observed with GISH (Seijo et al., 2007; Bertioli et al., 2013).

In contrast, similar studies performed with other retrotransposons such as Matita, a Ty1-copia from the Bianca lineage, showed that it is represented at moderate copy number and in an equivalent manner on the A and B genomes (Nielen et al., 2012). FISH experiments showed that Matita localizes mainly in the distal regions of chromosome arms, with a tendency to present more abundance in gene-rich regions. More recent studies on homeologous chromosomes of the A and B genomes evidenced the existence of a diverse group of complete and truncated copies of the LTR retrotransposon fraction covering more than 40% of the sequences analyzed (Bertioli et al., 2013, 2016, 2019). The results revealed that a substantial proportion of the highly repetitive component of the A genome is represented by relatively few LTR retrotransposons.

Non-LTR retroelements are generally less abundant than LTR retroelements in plant genomes, and *Arachis* genomes are no exception (Alix and Heslop-Harrison, 2004; Wang, Li, Fu et al., 2021). Reverse transcriptase-based phylogenetic analysis of a family belonging to the L1 clade of LINE elements showed that lineages were distributed independently of genome groups or karyotypes (Samoluk, Chalup et al., 2015). FISH experiments revealed a dispersed pattern with many hybridization signals located mainly in the interstitial and distal euchromatin of most chromosome arms, regardless of the genome type analyzed. Despite the fact that next-generation sequencing (NGS) analysis revealed that LINES cover less than 1% of the genome of diploid *Arachis* species, the genome sequences indicate that these elements account for 7%–12% of the genome sizes (Bertioli et al., 2016; Chen et al., 2016; Samoluk et al., 2022). These are the highest coverages for LINE reported so far for plant genomes (Samoluk, Chalup et al., 2015) and would merit further exploration as cytogenetic markers.

In contrast to the high abundance of retroelements, DNA transposons constitute about 5%–10% of the genome of *Arachis* species (Bertioli et al., 2016; Chen et al., 2016). Among the few elements studied, it was demonstrated that the MITE1-like elements are differentially represented in the *Arachis* species and subspecies. They are in high copy numbers in the genomes of *A. hypogaea*, *A. magna*, and *A. monticola*, but not in *A. duranensis* (Shirasawa et al., 2012). In addition, MITE1 representation is higher in spp. *fastigiata* peanuts than in spp. *hypogaea* peanuts (Gowda et al., 2011). However, these elements have been little explored in *Arachis* as markers at the chromosomal level.

The sequencing of the genomes of the two parent species and peanut cultivars as well as the advent of NGS techniques were crucial to the development of new chromosomal markers (Bertioli et al., 2016, 2019; Zhuang et al., 2019). Transposable elements contribute 61.7% and 68.5% of the genomes in *A. duranensis* and *A. ipaënsis*, respectively. These values are compatible with the 64% repeated DNA content estimated for cultivated peanut using renaturation kinetics and with values obtained by NGS for species representative of the different genomes (Dhillon, 1980; Samoluk et al., 2022).

Most of the transposon families are present in all genomes, and the macroscale positioning of the abundant families is similar in the two assembled diploid genomes. However, the microscale positioning and relative abundance differ significantly. Several of the most abundant Ty3-gypsy and non-autonomous retrotransposon families (autonomous/non-autonomous pairs FIDEL/Feral, Pipoka/Pipa, Apollo/Polo and the non-autonomous Gordo) were used for FISH assays, detecting dense accumulations in the proximal and interstitial regions of the chromosomes (Bertioli et al., 2013; Figure 3.2E). This chromosome distribution agrees with what was observed in the genome sequence of the species (Bertioli et al., 2016, 2019).

The different LTR elements tested as chromosomal markers so far evidenced that they are better indicatives of genomes than specific chromosomes or chromosomal regions, and are thus ideal for genome-specific recognition in intergenomic hybridizations and, potentially, for detection of large-scale translocations.

3.4 CYTOGENETIC MARKERS DEVELOPED FROM THE SATELLITOME

Among the diploid species of section *Arachis*, one of the most variable characters is the amount and distribution of heterochromatin (Seijo et al., 2004). This fraction of eukaryotic genomes is preponderantly composed of satellite DNA. The repetitive units of this type of DNA are of variable length arranged in blocks of up to 100 Mb and usually show discrete chromosomal locations, particularly at centromeres (Charlesworth et al., 1994; Schmidt and

Heslop-Harrison, 1998; Plohl et al., 2008), telomeres, and, less frequently, interstitially (Macas et al., 2000; Hudakova et al., 2001; Gindullis et al., 2001; Urdampilleta et al., 2009; Mukai et al., 1992). Thus, they have become a useful tool to study karyotype evolution in different groups of species (Lanfredi et al., 2001; Slamovits et al., 2001).

The cloning and characterization of various chromosomal markers from this fraction has resulted in a major step for FISH-based karyotyping of *Arachis* species. The first identified satellite sequence ATR-2 in seven diploid species (x = 10) was conserved among representatives of different genomes and karyotype groups of section *Arachis*, with a low spread of new monomeric variants (Robledo, 2010; Samoluk et al., 2017). However, quantitative analysis revealed interspecific variability in the abundance of different sat-DNAs from a set "library" of ancestral variants. FISH analysis with ATR-2 motifs as probe showed signals exclusively distributed in centromeric (Figure 3.2F) and interstitial DAPI⁺ heterochromatin, although it is not the only sequence that composes these bands (Zhang et al., 2016; Samoluk et al., 2017). Despite the sequence conservation of ATR-2, variation in its representation was proposed as a major mechanism that actively participated in the remodeling of heterochromatic patterns in diploid *Arachis* species (Samoluk et al., 2017, 2022).

Simultaneously, other genome and chromosome-specific FISH markers were developed through the cloning of de novo repetitive DNA from Cot-1 libraries of cultivated peanut and *A. duranensis* as well as from the screening of peanut BAC clones (Zhang et al., 2012). By applying these probes and using previously reported markers, most chromosomes of cultivated peanut were labeled with one or more markers, with the exception of chromosome B2. This approach generated a better resolved FISH-based karyotype (Zhang et al., 2016).

Further research on this fraction with the implementation of low coverage NGS and the analysis of individual datasets from eight different *Arachis* species allowed the characterization of 44 satDNA sequences, each one having at least 0.01% of genome representation (Samoluk et al., 2022). Almost all monomeric sequences clustered into four groups of satDNA sequences, named as AraSat1–AraSat4. The monomers of these groups ranged from 300 bp and ca. 80 % of AT nucleotides to 20 bp and 60%–75% AT rich. Seven clusters (Clusters 4, 36, 48, 96, 114, 118, and 141) were shared by at least four *Arachis* species, and five clusters (Clusters 4, 36, 48, 96, and 141) were found in all species analyzed. The variation in the proportions of the satDNA fraction in *Arachis* genomes ranged from 0.29% (*A. magna*) to 7.56% (*A. duranensis*), which is directly proportional to the DAPI⁺ heterochromatin. Among them, the most abundant ones (C4, C36, and C48) from the AraSat1 group accounted for more than 90% of the fraction. These clusters include sequences with homology to the previously characterized sequences ATR-2, H-b, and Agla-CL8sat (Samoluk et al., 2017; Zhang et al., 2016; Samoluk et al., 2019). These satDNAs are the main component of the DAPI⁺ heterochromatin in the A, F, K, and D genomes. However, they are barely represented in the species of the B genome, which are deprived of heterochromatic bands. Based on these results, it was proposed that high rates of turnover occurred in this superfamily of satDNAs, which led to the different amounts of DAPI⁺ heterochromatin, one of the main features that modeled the genomes in section *Arachis*.

Among the other satDNAs identified by NGS, C96 (group AraSat2) with homology to the previously characterized satDNA "clone 119" of *A. hypogaea* hybridized in the centromeres of 16 B genome chromosomes and six A genome chromosomes in peanut, being an excellent marker for different genome centromeres (Zhang et al., 2012; Samoluk et al., 2022). The remaining SatDNA were considered as members of other families, several of which would be good markers for cytogenetic studies.

The *Arachis* telomeres, the physical ends of chromosomes, are formed by the canonical plant telomeric repeat TTTAGGG (CPTR) (Bertioli et al., 2019). Variations among interstitial telomeric motives were reported in cultivated peanut and *A. duranensis*, on the basis of which doubts as to the single origin of the cultivated peanut were raised. However, from the analysis of the peanut genome sequences, a new type of tandem repeat with 10-bp basic motif TTTT(C/T)TAGGG named TAndem Repeat (TAR) 30 was identified (Gao et al., 2022). Fluorescence *in situ* hybridization of this satellite revealed interstitial locations in peanut chromosomes, but visible signals were not detected in the terminal ends of chromosomes, as expected for telomeric repeat. Moreover, the canonical telomeric repeat TTTAGGG is also present closely adjacent to TAR30 from both cultivated peanut and its wild relatives. The presence of this satellite in interstitial position, and the fact that its positions can vary after polyploidization as demonstrated in the artificial amphidiploid ValSten, suggest that evolutionary conclusions based on LTR should be considered with caution.

3.5 RECENT ADVANCES IN OLIGO-FISH IN PEANUT

Single-stranded oligonucleotides, short DNA or RNA fragments, can be modified with fluorophores and directly hybridized with chromosomal DNA, thereby potentially replacing traditional plasmid clones and genome DNA probes. Oligonucleotide fluorescence *in situ* hybridization (Oligo-FISH), using repetitive and single-copy sequences as probes, is a simple and efficient method that is being increasingly used in chromosome research of plants (Du et al., 2017; Jiang, 2019; De Oliveira Bustamante et al., 2021).

Two types of oligonucleotide probes are currently available: repetitive and single-copy. Repetitive oligonucleotide probes, generally clustered together on one or several genomes or chromosomes, are frequently used to develop high-resolution karyotypes and identify chromosomes. Repetitive oligonucleotides can be derived from simple sequence repeats, genomic sequence repeats, and multi-copy genetic sequences. The first repetitive oligonucleotide probes (eight) developed in peanut were used to establish high-resolution karyotypes of 16 *Arachis* species and provided useful tools for the differentiation of chromosomes and identification of structural chromosomal rearrangement (Du et al., 2018). Subsequently, 114 new oligonucleotide probes and two cocktails (Multiplex #3 and Multiplex #4) were developed based on the genome-wide tandem repeats (TRs) identified from the reference sequences of the peanut variety Tifrunner and the diploid species *A. ipaënsis* (Fu et al., 2021). A genome map-based karyotype was established after sequential FISH/GISH and in silico mapping that was used for the identification of chromosome variations in peanut following comparisons with their reference sequence positions.

Repetitive oligonucleotides can bind to DNA strands through triplex DNA or in some other unknown way in non-denaturing chromosomes. So, Oligo-FISH using repetitive oligonucleotide probes has produced signals not only in denaturing chromosomes but also in non-denaturing chromosomes (Cuadrado and Jouve, 2010). It was found that lower levels of complementarities without the polymerization of repeated sequence oligonucleotides facilitate their invasion of the chromosomes in peanut (Du et al., 2019). Using this approach, a new, low-cost, and simple oligonucleotide dye technique was developed to paint 21 accessions of the cultivated peanut and 30 wild *Arachis* species, providing additional information to address the genome relationships.

Genomic *in situ* hybridization (GISH) using genomic DNA from the diploid species *A. duranensis* and *A. ipaënsis* as probes can clearly distinguish A- and B-genome chromosomes (Raina and Mukai, 1999; Seijo et al., 2007) (Figure 3.2D). Based on clustering and *in silico* localization of tandem repeat sequences in the reference genome of Tifrunner, A and B genome-specific interspersed repeat (IR) oligonucleotide probe sets (Multiplex #A or Multiplex #B) were developed (Du et al., 2023). Staining radiation-induced M_1 and M_3 seedlings with oligonucleotide Multiplex #A or Multiplex #B identified a large number of chromosomal translocations between A and B genome chromosomes (Figure 3.3A). Further utilization of Oligo probes such a as Multiplex #3 (including FAM-modified TIF-439, TIF-185-1, TIF-134-3, and TIF-165-3) and Multiplex #4 (including TAMRA-modified Ipa-1162, Ipa-1137, DP-1, and DP-5 allowed the identification of chromosome B03 as those with an intergenomic translocation (Fu et al., 2021; Du et al., 2023) (Figure 3.3B).

Single-copy oligonucleotides include single-gene and single-copy genomic oligonucleotides. They are usually chromosome-specific, and are thus used to identify homoeologous chromosomes in closely related species (Han et al., 2015), haplotype analysis of homologous chromosomes (Beliveau et al., 2015) and meiotic crossovers (Do Vale Martins et al., 2019). The genome sequencing of cultivated peanut and its wild donor species, *A. duranensis* and *A. ipaënsis*, assembled 20 pseudomolecules, ten for the A and ten for the B genomes (Bertioli et al., 2016, 2019; Zhuang et al., 2019). However, the allocated chromosome numbers in the reference map do not correspond to the actual chromosomes in cytogenetic karyotypes. To correlate a sequenced chromosome with a cytologically identified chromosome, a single-copy oligonucleotide library was developed from a ~ 8.5 Mb distal region of A06 and used to paint metaphase chromosomes of *A. duranensis*. Sequential FISH revealed that the target signals were present in cytologically designated chromosome A3 (Du et al., 2018). More recently, single-copy oligonucleotide probe libraries were developed for individual chromosomes. With this approach, probe library paints developed from the whole chromosome A8 hybridizes onto the characteristic A chromosome (Figure 3.4A and B) that is most used to identify A genome species.

The development of Oligo-FISH has promoted the identification of peanut interspecific hybridization materials more efficiently and accurately. Using Oligo-FISH identification, a collection of alien chromosome variation lines was developed from a synthetic allohexaploid derived from the peanut Silihong cultivar and *A. duranensis* (Wang, Li, Wang et al., 2021).

3.6 GENOME SIZE

Genome size is a useful descriptor for characterizing plant genetic resources and for inferring patterns of genomic differentiation in eukaryotes (Ozias Akins and Jarret, 1994; Hendrix and Stewart, 2005; Loureiro et al., 2007; Bennett, 1982; Leitch and Bennett, 2004). Early studies using the Feulgen densitometry method overestimated the size of genomes of *Arachis* species by two-fold due to technical artifacts (Resslar et al., 1981; Singh et al., 1996; Lavia and Fernández, 2008; Temsch and Greilhuber, 2000, 2001). Flow cytometry analysis established a very uniform genome size of 5.8–6.0 pg for *A. hypogaea* and *A. monticola* (Temsch and Greilhuber, 2000). However, these authors observed that genomic size varies from 2.49 to 2.87 pg in different populations of *A. duranensis* and that this variation presented a high correlation with the latitude and altitude at which the analyzed materials are distributed (Temsch and Greilhuber, 2001).

A flow cytometry study conducted on 26 diploid species of section *Arachis* with $2n = 20$ showed that the 2C values of these species ranged from 2.55 to 3.22 pg (Samoluk, Robledo et al., 2015). Annual species belonging to different genomes tend to show significant differences in genome sizes, the lowest found in *A. duranensis* (A genome), the intermediate ones in the species of the D, K, and F genomes, and the highest in those of the B genome. However, the 2C values of perennial A genome species are almost continuously distributed throughout the variation range of 2.69 to 3.18 pg.

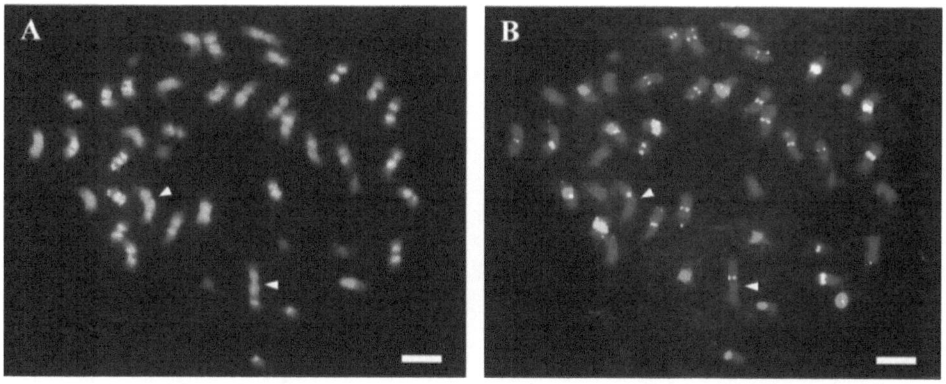

FIGURE 3.3 Chromosome variants (FS2020-4-1) detected in peanut via sequential Oligo GISH/FISH using repetitive oligonucleotides. A. Oligo GISH using Multiplex #A (green) and Multiplex #B (red); B. Oligo FISH using Multiplex #3 (green) and Multiplex #4 (red). Multiplex #3 included FAM-modified TIF-439, TIF-185-1, TIF-134-3, and TIF-165-3; and Multiplex #4 included TAMRA-modified Ipa-1162, Ipa-1137, DP-1, and DP-5. The two translocation in FS2020-4-1 were associated to chromosome B03. Arrowheads indicate translocated chromosomes. Scale bar = 5 μm.

Comparison of 2C values with karyotypic parameters suggests that changes in DNA content were proportionally distributed among chromosomal arms. It was also evident that the species with the highest proportion of heterochromatin in their karyotypes were those with the smallest genomic sizes, so it was proposed that variations in this fraction would not explain the variation in genomic size (Samoluk, Robledo et al., 2015). Within the A genome, annual species have lower DNA content than perennial species, which is consistent with the predictions of the nucleotype hypothesis (Bennett, 1982). However, the lack of significant relationships with geoclimatic variables suggests that there are many intrinsic factors that determine particular ecological roles of DNA content in different lineages of section *Arachis* (Samoluk et al., 2022). Critical analysis of DNA content of other *Arachis* species is still needed to address the direction of genome change during the evolution of the genus as a whole.

From an evolutionary perspective, the relationship between genome size variation and the amount of different genome fractions of repetitive sequences seems to be complex in *Arachis*. However, the largest genomes tend to present a higher accumulation of LTR elements and lower contents of satDNAs. In contrast, species with lowest genome sizes tend to accumulate satDNAs in detriment of LTR elements (Samoluk et al., 2022).

3.7 CYTOGENETIC CONTRIBUTIONS TO DISENTANGLE THE ORIGIN OF PEANUT

The presence of chromosomes with different morphology suggested the existence of two distinct chromosomal complements in cultivated peanut, and therefore of its allotetraploid constitution with $2n = 4x = 40$ (Husted, 1933, 1936; Smartt and Gregory, 1967; Stalker and Wynne, 1979; Singh, 1986). Interspecific hybridization studies between cultivated peanut and different wild diploid species confirmed this hypothesis and led to the proposal of a genomic formula AABB (Smartt et al., 1978; Fernández and Krapovickas, 1994; Robledo and Seijo et al., 2010).

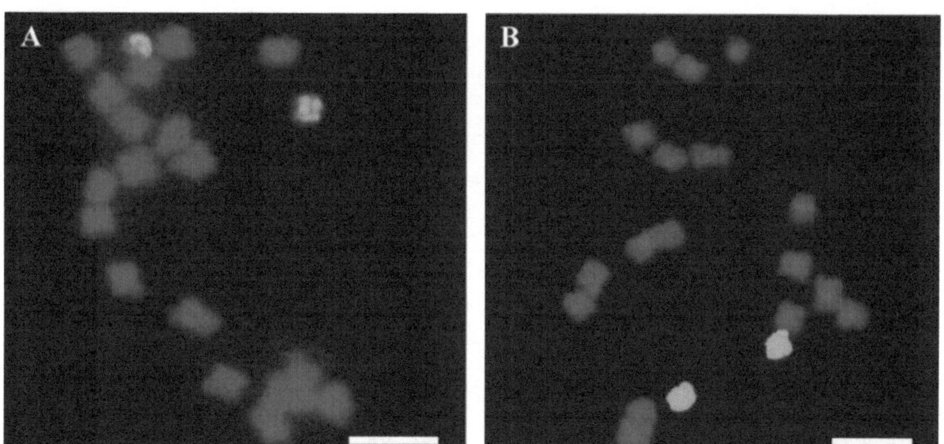

FIGURE 3.4 Chromosome painting (red) using a specific probe to identify the A08 chromosomes. A. *Arachis duranensis*. B. *A. cardenasii*. Scale bar = 5 μm.

Simultaneously, several research groups advocated for the identification of the most probable diploid species involved in the origin of cultivated peanut (reviewed in Singh, 1986; Seijo et al., 2004, 2007; Grabiele et al., 2012). Different A genome species like *A. duranensis*, *A. villosa*, *A. correntina*, and *A. cardenasii* were proposed as the most likely ancestors of cultivated peanut (Kochert et al., 1991, 1996; Raina et al., 2001; Moretzsohn et al., 2004; Milla et al., 2005). Similarly, *A. batizocoi* (K genome) was for a long time the best candidate until the proposal of *A. ipaënsis* (B genome) as the other B genome donor (Fernández and Krapovickas, 1994; Kochert et al., 1996). Even today, after different genome sequencing initiatives, the debate persists. One of the most consulted genome sequences available at PeanutBase, that of cv. Tiffrunner, clearly documents a recent origin of *A. hypogaea* from *A. ipaënsis* and *A. duranensis*, while those based on cv. Shitoqui suggest a spontaneous origin around 450,000 years ago and denies, from the sequenced accessions, that *A. duranensis* is the right donor of the A genome of peanut.

The implementation of molecular cytogenetic techniques in the study of *Arachis* species, determining heterochromatin patterns and the number and position of rRNA genes by FISH, provided strong data to identify the best candidates to be the wild diploid progenitors of cultivated peanut (Raina and Mukai, 1999; Seijo et al., 2004; Robledo et al., 2009; Robledo and Seijo, 2010; Custódio et al., 2013). Analysis of the distribution of rDNA loci and heterochromatin as well as double GISH using genomic DNA from diploid ($2n = 2x = 20$) candidate *Arachis* species identified *A. duranensis* as the most likely donor of the A genome and *A. ipaënsis* as the donor of the B genome of *A. hypogaea* (Figure 3.2C, D) (Seijo et al., 2007). The chloroplast and ribosomal sequences as well as SNPs analysis of a diverse set of A genome species, including many *A. duranensis* populations identified those from north Argentina as the most similar ones to those that participated in the origin of cultivated peanut (Grabiele et al., 2012; Bertioli et al., 2019, 2020).

The high degree of ribosomal gene homology detected between *A. monticola* and *A. hypogaea* provided support for the hypothesis that peanut is a domesticated variant of *A. monticola* (Seijo et al., 2004, 2007). It is the only wild allotetraploid within section *Arachis* and is currently only known from three closely related localities in northwest Argentina (Seijo, pers. obs.). This wild species is interfertile with cultivated peanut and, in all dendrograms developed from molecular markers, appears among the variability of the peanut landraces (Kirti et al., 1983; Gimenes et al., 2002; Grabiele et al., 2012; Moretzsohn et al., 2013). On this basis, *A. monticola* had been considered by different authors as the direct progenitor of peanut or as an introgressive derivative between peanut and wild species (Stalker and Moss, 1987; Grabiele et al., 2012). Spontaneous hybridization between *A. hypogaea* (AABB) and diploid species was never observed in nature (Krapovickas and Gregory, 1994, Seijo, pers. obs.). Moreover, the AABB genomic constitution of *A. monticola*, revealed by FISH and GISH, is identical to that of cultivated peanut (Seijo et al., 2004, 2007; Grabiele et al., 2012). These results ruled out the hypothesis that considers *A. monticola* as an introgressive derivative of peanut with wild species. The artificial resynthesis of amphidiploids from *A. ipaënsis* and *A. duranensis* produced plants morphologically very similar to *A. monticola* and supports the hypothesis that considers the latter as a non-domesticated version of *A. hypogaea* (Fávero et al., 2006). However, the fruit structure found in *A. monticola* is alike with the wild species (in which each seed has its own shell separated by an isthmus), and its persistence in natural population, unlike the cultivated peanut, supports its maintenance as a separate taxonomic species.

The similarity in the patterns of ribosomal loci, heterochromatin, genomic in situ hybridization (GISH), and other cytogenetic markers derived from repeated sequences among all botanical varieties of peanut provided strong evidence for a single origin from an ancestral allotetraploid population (Seijo et al., 2004, 2007, 2018). The common ancestry of all intraspecific taxa of *A. hypogaea* is supported by the low genetic variability detected so far with most molecular markers in cultivated peanut and by the SNPs analysis of the main peanut collections (Halward et al., 1991; Kochert et al., 1996; Herselman, 2003; Grabiele et al., 2012; Moretszohn et al., 2013; Pandey et al., 2017; Otyama et al., 2020).

The single origin is also supported by the genetic exchanges analyzed in the AABB allotetraploids. Genetic exchanges between the A and B genomes and insertions/deletions were inferred by studying more than 200 diverse genotypes comprising *A. monticola*, landraces and cultivars of *A. hypogaea*, and new amphidiploids made from the ancestral species. This analysis was done by mapping densities of short-read sequences onto the combined sequenced diploid ancestral species genomes, and through analysis of the short-read whole-genome sequences for single-nucleotide polymorphisms (SNPs) that consistently differentiate representatives of A and B genome diploid species. From these analyses, several genetic exchanges between ancestral genomes were inferred, mostly towards the ends of chromosomes. Analyzing co-linear pairs of homeologous chromosomes, the detection of regions AAAA or BBBB, that is, 'tetrasomic' conformations instead of the expected disomic AABB, was related to chromosome exchanges. The fine-scale patterns of these SNPs represent substantially fixed, or fossilized, genetic signals ("fingerprints") from past events. Their uniformity in all six botanical varieties and the wild counterpart of peanut *A. monticola* favors a single polyploid origin for the two species.

From a functional point of view, one of the most evident chromosomes markers is the number and location of active nucleolar organizers. The analysis of extended ribosomal DNA loci and their association with nucleolar organizing regions in the six botanical varieties in *A. monticola* and in a synthetic amphidiploid (BBAA) revealed a rapid establishment of nucleolar dominance in favor of the A genome in all the materials analyzed, regardless of the direction of crossing (Seijo et al., 2018).

3.8 FINAL CONSIDERATIONS

Despite the huge progress made since the availability of genomic resources in the comprehension of peanut origin and evolution, there is still an important set of questions to be addressed at the chromosomal organization level that cannot be fully afforded from the genome sequence. The availability of genomic resources has potentiated the development of genome, chromosomes, or even chromosomal discrete region-specific probes that may be implemented to shed light more deeply on the evolution of the tetraploid genome, the patterns of genome diversification among diploids, and the restriction that karyotype differences impose on hybridization and production of fertile offspring. Moreover, the implementation of routine and simple methods of chromosome identification may be very useful for the development of chromosomal lines, introgression of particular segments, and chromosome engineering.

REFERENCES

Alix, K., and J. S. P. Heslop-Harrison. "The Diversity of Retroelements in Diploid and Allotetraploid *Brassica* Species." *Plant Molecular Biology* 54, no. 6 (2004): 895–909. https://doi.org/10.1007/s11103-004-0391-z.

Bechara, M. D., M. C. Moretzsohn, D. A. Palmier, J. P. Monteiro, M. Bacci, J. Martins, J. F. M. Valls, C. R. Lopes, and M. A. Gimenes. "Phylogenetic Relationships in Genus *Arachis* Based on ITS and 5.8S rDNA Sequences." *BMC Plant Biology* 10 (2010): 255. https://doi.org/10.1186/1471-2229-10-255.

Beliveau, B. J., A. N. Boettiger, M. S. Avendaño, R. Jungmann, R. B. McCole, E. F. Joyce, and C. Kim-Kiselak. "Single-Molecule Super-Resolution Imaging of Chromosomes and *in Situ* Haplotype Visualization Using Oligopaint FISH Probes." *Nature Communications* 6 (2015): 7147. https://doi.org/10.1038/ncomms8147.

Bennett, M. D. "Nucleotypic Basis of the Spatial Ordering of Chromosomes in Eukaryotes and the Implications of the Order for Genome Evolution and Phenotypic Variation." In *Genome Evolution*, edited by G. A. Dover and R. B. Flavel, 239–261. Academic Press, 1982.

Bertioli, D. J., B. Abernathy, G. Seijo, J. Clevenger, and S. T. Cannon. "Evaluating Two Different Models of Peanut's Origin." *Nature Genetics* 52, no. 6 (2020): 557–559. https://doi.org/10.1038/s41588-020-0626-1.

Bertioli, D. J., S. B. Cannon, L. Froenicke, G. Huang, A. D. Farmer, E. K. S. Cannon, X. Liu, et al. "The Genome Sequences of *Arachis duranensis* and *Arachis ipaensis*, the Diploid Ancestors of Cultivated Peanut." *Nature Genetics* 48, no. 4 (2016): 438–446. https://doi.org/10.1038/ng.3517.

Bertioli, D. J., J. Jenkins, J. Clevenger, O. Dudchenko, D. Gao, G. Seijo, S. C. M. Leal-Bertioli, et al. "The Genome Sequence of Segmental Allotetraploid Peanut *Arachis hypogaea*." *Nature Genetics* 51, no. 5 (2019): 877–884. https://doi.org/10.1038/s41588-019-0405-z.

Bertioli, D. J., B. Vidigal, S. Nielen, M. B. Ratnaparkhe, T.-H. Lee, S. C. M. Leal-Bertioli, C. Kim, et al. "The Repetitive Component of the a Genome of Peanut (*Arachis hypogaea*) and Its Role in Remodeling Intergenic Sequence Space Since Its Evolutionary Divergence from the B Genome." *Annals of Botany* 112, no. 3 (2013): 545–559. https://doi.org/10.1093/aob/mct128.

Burow, M. D., C. E. Simpson, M. W. Faries, J. L. Starr, and A. H. Paterson. "Molecular Biogeographic Study of Recently Described B- and A-Genome *Arachis* Species, Also Providing New Insights into the Origins of Cultivated Peanut." *Genome* 52, no. 2 (2009): 107–119. https://doi.org/10.1139/G08-094.

Charlesworth, B., P. Sniegowski, and W. Stephan. "The Evolutionary Dynamics of Repetitive DNA in Eukaryotes." *Nature* 371, no. 6494 (1994): 215–220. https://doi.org/10.1038/371215a0.

Chen, X., H. Li, M. K. Pandey, Q. Yang, X. Wang, V. Garg, H. Li, et al. "Draft Genome of the Peanut A-Genome Progenitor (*Arachis duranensis*) Provides Insights into Geocarpy, Oil Biosynthesis, and Allergens." *Proceedings of the National Academy of Sciences* 113, no. 24 (2016): 6785–6790. https://doi.org/10.1073/pnas.1600899113.

Clevenger, J., Y. Chu, C. Chavarro, G. Agarwal, D. J. Bertioli, S. C. M. Leal-Bertioli, M. K. Pandey, et al. "Genome-Wide SNP Genotyping Resolves Signatures of Selection and Tetrasomic Recombination in Peanut." *Molecular Plant* 10, no. 2 (2017): 309–322. https://doi.org/10.1016/j.molp.2016.11.015.

Cuadrado, Á., and N. Jouve. "Chromosomal Detection of Simple Sequence Repeats (SSRs) Using Nondenaturing FISH (ND-FISH)." *Chromosoma* 119, no. 5 (2010): 495–503. https://doi.org/10.1007/s00412-010-0273-x.

Custódio, A. R., J. G. Seijo, and J. F. M. Valls. "Characterization of Brazilian Accessions of Wild *Arachis* Species of Section *Arachis* (Fabaceae) Using Heterochromatin Detection and Fluorescence *In Situ* Hybridization (FISH)." *Genetics and Molecular Biology* 36, no. 3 (2013): 364–370. https://doi.org/10.1590/S1415-47572013000300011.

Custódio, A. R., C. E. Simpson, and J. F. M. Valls. "Crossability Between *Arachis gregoryi* (Fabaceae) and Wild *Arachis* Species with Distinct Genome." *Bonplandia* 30 (2021). http://dx.doi.org/10.30972/bon.3014669.

De Oliveira Bustamante, F., T. H. do Nascimento, C. Montenegro, S. Dias, L. do Vale Martins, G. T. Braz, A. M. Benko-Iseppon, et al. "Oligo-FISH Barcode in Beans: A New Chromosome Identification System." *Theoretical and Applied Genetics* 134, no. 11 (2021): 3675–3686. https://doi.org/10.1007/s00122-021-03921-z.

Dhillon, S. S., A. V. Rake, and J. P. Miksche. "Reassociation Kinetics and Cytophotometric Characterization of Peanut (*Arachis hypogaea* L.) DNA." *Plant Physiology* 65, no. 6 (1980): 1121–1127. https://doi.org/10.1104/pp.65.6.1121.

Do Vale Martins, L., F. Yu, H. Zhao, T. Dennison, N. Lauter, H. Wang, Z. Deng, et al. "Meiotic Crossovers Characterized by Haplotype-Specific Chromosome Painting in Maize." *Nature Communications* 10 (2019): 4604. https://doi.org/10.1038/s41467-019-12646-z.

Du, P., C. Cui, H. Liu, L. Fu, L. Li, X. Dai, L. Qin, et al. "Development of an Oligonucleotide Dye Solution Facilitates High Throughput and Cost-Efficient Chromosome Identification in Peanut." *Plant Methods* 15 (2019): 69. https://doi.org/10.1186/s13007-019-0451-7.

Du, P., L. Fu, Q. Wang, T. Lang, H. Liu, S. Han, C. Li, et al. "Development of Oligo-GISH Kits for Efficient Detection of Chromosomal Variants in Peanut." *The Crop Journal* 11 (2023): 238–246. https://doi.org/10.1016/j.cj.2022.03.014.

Du, P., L. Li, H. Liu, L. Fu, Z. Sun, L. Qin, Z. Zhang, et al. "High-Resolution Chromosome Painting with Repetitive and Single Copy Oligonucleotides in *Arachis* Species Identifies Structural Rearrangements and Genome Differentiation." *BMC Plant Biology* 18 (2018): 240. https://doi.org/10.1186/s12870-018-1468-1.

Du, P., L. Zhuang, Y. Wang, L. Yuan, Q. Wang, D. Wang, Dawadondup, et al. "Development of Oligonucleotides and Multiplex Probes for Quick and Accurate Identification of Wheat and *Thinopyrum bessarabicum* Chromosomes." *Genome* 60, no. 2 (2017): 93–103. https://doi.org/10.1139/gen-2016-0095.

Fávero, A. P., A. R. Custodio, N. B. Dinato, I. J. Godoy, J. G. Seijo, and M. D. Michelotto. "Transference of Multiple Resistance to Peanut Through the Development of Cross-Compatible Complex Hybrids of Wild *Arachis*." *Genetics and Molecular Biology* 43 (2020). https://doi.org/10.1590/1678-4685-gmb-2019-0099.

Fávero, A. P., R. F. dos Santos, C. E. Simpson, J. F. M. Valls, and N. A. Vello. "Successful Crosses Between Fungal-Resistant Wild Species of *Arachis* (Section *Arachis*) and *Arachis hypogaea*." *Genetics and Molecular Biology* 38, no. 3 (2015): 353–365. https://doi.org/10.1590/S1415-475738320140376.

Fávero, A. P., C. E. Simpson, J. F. M. Valls, and N. A. Vello. "Study of the Evolution of Cultivated Peanut Through Crossability Studies Among *Arachis ipaensis*, *A. duranensis*, and *A. hypogaea*." *Crop Science* 46, no. 4 (2006): 1546–1552. https://doi.org/10.2135/cropsci2005.09-0331.

Fernández, A., and A. Krapovickas. "Cromosomas y Evolución en *Arachis* (Leguminosae)." *Bonplandia* 8 (1994): 187–220.

Friend, S. A., D. Quandt, S. P. Tallury, H. T. Stalker, and K. W. Hilu. "Species, Genomes, and Section Relationships in the Genus *Arachis* (Fabaceae): A Molecular Phylogeny." *Plant Systematics and Evolution* 290, no. 1 (2010): 185–199. https://doi.org/10.1007/s00606-010-0360-8.

Fu, L., Q. Wang, L. Li, T. Lang, J. Guo, S. Wang, Z. Sun, et al. "Physical Mapping of Repetitive Oligonucleotides Facilitates the Establishment of a Genome Map-Based Karyotype to Identify Chromosomal Variations in Peanut." *BMC Plant Biology* 21, no. 1 (2021): 107. https://doi.org/10.1186/s12870-021-02875-0.

Gao, D., A. C. G. Araujo, E. F. M. B. Nascimento, et al. "ValSten: A New Wild Species-Derived Allotetraploid for Increasing Genetic Diversity of the Peanut Crop (*Arachis hypogaea* L.)." *Genetic Resources and Crop Evolution* 68 (2021): 1471–1485. https://doi.org/10.1007/s10722-020-01076-2.

García, A. V. "Determination of the Mechanism of Origin of *Arachis hypogaea* L. Cultigen by Meiotic Analysis of the Interspecific Hybrid Between the Most Probable Diploid Progenitor Species." Final Graduation Paper for Bachelor's Degree in Biological Sciences, Universidad Nacional del Nordeste, 2015.

García, A. V., A. M. Ortiz, M. C. Silvestri, A. R. Custodio, M. C. Moretzsohn, and G. I. Lavia. "Occurrence of 2n Microspore Production in Diploid Interspecific Hybrids Between the Wild Parental Species of Peanut (*Arachis hypogaea* L., Leguminosae) and Its Relevance in the Genetic Origin of the Cultigen." *Crop Science* 60, no. 5 (2020): 2420. https://doi.org/10.1002/csc2.20233.

García, A. V., M. C. Silvestri, M. A. Vandecaveye, A. R. Custodio, M. C. Moretzsohn, and G. I. Lavia. "Genomic Affinity in Hybrids of B-Genome *Arachis* Species: New Genetic Resources Toward Peanut Improvement." *Crop Breeding Applied Biotechnology* 21 (2021): e38292139. https://doi.org/10.1590/1984-70332021v21n3a48.

Gao, D., E. F. M. B. Nascimento, S. C. M. Leal Bertioli, B. Abernathy, S. A. Jackson, A. C. G. Araujo, and D. J. Bertioli. "TAR30, a Homolog of the Canonical Plant TTTAGGG Telomeric Repeat, Is Enriched in the Proximal Chromosome Regions of Peanut (*Arachis hypogaea* L.)." *Chromosome Research* 30, no. 1 (2022): 77–90. https://doi.org/10.1007/s10577-022-09684-7.

Gimenes, M. A., C. Lopes Romero, M. L. Galgaro, and J. F. M. Valls. "RFLP Analysis of Genetic Variation in Species of Section *Arachis*, Genus *Arachis* (Leguminosae)." *Euphytica* 123, no. 3 (2002): 421–429. https://doi.org/10.1023/A:1015033700110.

Gindullis, F., C. Desel, I. Galasso, and T. Schmidt. "The Large-Scale Organization of the Centromeric Region in Beta Species." *Genome Research* 11, no. 2 (2001): 253–265. https://doi.org/10.1101/gr.162301.

Gowda, M. V. C., R. S. Bhat, V. Sujay, P. Kusuma, V. Varshakumari, S. Bhat, and R. K. Varshney. "Characterization of AhMITE1 Transposition and Its Association with the Mutational and Evolutionary Origin of Botanical Types in Peanut (*Arachis* spp.)." *Plant Systematics and Evolution* 291, no. 1 (2011): 153–158. https://doi.org/10.1007/s00606-010-0373-3.

Grabiele, M., L. Chalup, G. Robledo, and J. G. Seijo. "Genetic and Geographic Origin of Domesticated Peanut as Evidenced by 5S rDNA and Chloroplast DNA Sequences." *Plant Systematics and Evolution* 298, no. 6 (2012): 1151–1165. https://doi.org/10.1007/s00606-012-0627-3.

Gregory, W. C., and M. P. Gregory. "Exotic Germplasm of *Arachis* L. Interspecific Hybrids." *Journal of Heredity* 70, no. 2 (1979): 185–193. https://doi.org/10.1093/oxfordjournals.jhered.a109231.

Halward, T. M., H. T. Stalker, E. A. Larue, and G. Kochert. "Genetic Variation Detectable with Molecular Markers Among Unadapted Germplasm Resources of Cultivated Peanut and Wild Species." *Genome* 34, no. 6 (1991): 1013–1020. https://doi.org/10.1139/g91-156.

Han, Y. H., T. Zhang, P. Thammapichai, Y. Q. Weng, and J. M. Jiang. "Chromosome-Specific Painting in *Cucumis* Species Using Bulked Oligonucleotides." *Genetics* 200, no. 2 (2015): 771–779. https://doi.org/10.1534/genetics.115.177642.

Hendrix, B., and J. M. Stewart. "Estimation of the Nuclear DNA Content of *Gossypium* Species." *Annals of Botany* 95, no. 5 (2005): 789–797. https://doi.org/10.1093/aob/mci078.

Herselman, L. "Genetic Variation Among Southern African Cultivated Peanut (*Arachis hypogaea* L.) Genotypes as Revealed by AFLP Analysis." *Euphytica* 133, no. 3 (2003): 319–327. https://doi.org/10.1023/A:1025769212187.

Hudakova, S., W. Michalek, G. G. Presting, R. ten Hoopen, K. dos Santos, Z. Jasencakova, and I. Schubert. "Sequence Organization of Barley Centromeres." *Nucleic Acids Research* 29, no. 24 (2001): 5029–5035. https://doi.org/10.1093/nar/29.24.5029.

Husted, L. "Cytological Studies on the Peanut, *Arachis*. I. Chromosome Number and Morphology." *Cytologia* 5, no. 1 (1933): 109–117.

Husted, L. "Cytological Studies on the Peanut, *Arachis*. II. Chromosome Number and Morphology and Behavior, and Their Application to the Problem of the Origin of the Cultivated Forms." *Cytologia* 7, no. 4 (1936): 396–423.

Jiang, J. "Fluorescence *In Situ* Hybridization in Plants: Recent Developments and Future Applications." *Chromosome Research* 27, no. 2 (2019): 153–165. https://doi.org/10.1007/s10577-019-09607-z.

Kirti, P. B., M. Bharathi, U. R. Murty, and N. G. P. Rao. "Chromosome Morphology in Three Diploid Species of *Arachis* and Its Bearing on the Genomes of Groundnut (*Arachis hypogaea* L.)." *Cytologia* 48, no. 2 (1983): 139–151. https://doi.org/10.1508/cytologia.48.139.

Kochert, G., T. Halward, W. D. Branch, and C. E. Simpson. "RFLP Variability in Peanut (*Arachis hypogaea* L.) Cultivars and Wild Species." *Theoretical and Applied Genetics* 81, no. 5 (1991): 565–570. https://doi.org/10.1007/BF00226719.

Kochert, G., H. T. Stalker, M. Gimenes, L. Galgaro, C. R. Lopes, and K. Moore. "RFLP and Cytogenetic Evidence on the Origin and Evolution of Allotetraploid Domesticated Peanut, *Arachis hypogaea* (Leguminosae)." *American Journal of Botany* 83, no. 10 (1996): 1282–1291. https://doi.org/10.1002/j.1537-2197.1996.tb13912.x.

Krapovickas, A., and W. C. Gregory. "Taxonomía del Género *Arachis* (Leguminosae)." *Bonplandia* 8 (1994): 1–186.

Lanfredi, M., L. Congiu, M. A. Garrido-Ramos, R. de la Herrán, M. Leis, M. Chicca, R. Rossi, et al. "Chromosomal Location and Evolution of a Satellite DNA Family in Seven Sturgeon Species." *Chromosome Research* 9, no. 1 (2001): 47–52. https://doi.org/10.1023/A:1026739616749.

Lavia, G. I. "Chromosome Characterization of Peanut Germplasm." PhD diss., Universidad Nacional de Córdoba, 1999.

Lavia, G. I. "Estudios Cromosómicos en *Arachis* (Leguminosae)." *Bonplandia* 9, no. 1 (1996): 111–120.

Lavia, G. I. "Karyotypes of *Arachis palustris* and *A. praecox* (Section *Arachis*), Two Species with Basic Chromosome Number x = 9." *Cytologia* 63, no. 2 (1998): 177–181. https://doi.org/10.1508/cytologia.63.177.

Lavia, G. I., and A. Fernández. "Genome Size in Wild and Cultivated Peanut Germplasm." *Plant Systematics and Evolution* 272, no. 1 (2008): 1–10. https://doi.org/10.1007/s00606-007-0632-0

Lavia, G. I., and A. Fernández. "Karyotypic Studies in *Arachis hypogaea* L. Varieties." *Caryologia* 57, no. 4 (2004): 353–359. https://doi.org/10.1080/00087114.2004.10589416.

Lavia, G. I., A. Fernández, C. E. Simpson, and G. Seijo. "Meiotic Analysis in Wild Diploid *Arachis* Species." *Cytologia* 66, no. 3 (2001): 293–298. https://doi.org/10.1508/cytologia.66.293.

Lavia, G. I., A. M. Ortiz, and A. Fernández. "Karyotypic Studies in Wild Germplasm of *Arachis* (Leguminosae)." *Genetic Resources and Crop Evolution* 56, no. 6 (2009): 755–764. https://doi.org/10.1007/s10722-008-9399-6.

Leal-Bertioli, S. C. M., S. P. Santos, K. M. Dantas, P. W. Inglis, S. Nielen, A. C. Araujo, J. P. Silva, et al. "*Arachis batizocoi*: A Study of Its Relationship to Cultivated Peanut (*A. hypogaea*) and Its Potential for Introgression of Wild Genes into the Peanut Crop Using Induced Allotetraploids." *Annals of Botany* 115, no. 2 (2015b): 237–249.

Leal-Bertioli, S. C. M., K. Shirasawa, B. Abernathy, M. C. Moretzsohn, C. Chavarro, J. Clevenger, P. Ozias-Akins, S. Jackson, and D. J. Bertioli. "Tetrasomic Recombination Is Surprisingly Frequent in Allotetraploid *Arachis*." *Genetics* 199, no. 4 (2015a): 1093–1105. https://doi.org/10.1534/genetics.115.174607.

Leitch, I. J., and M. D. Bennett. "Genome Downsizing in Polyploidy Plants." *Biological Journal of the Linnean Society* 82, no. 4 (2004): 651–663. https://doi.org/10.1111/j.1095-8312.2004.00349.x.

Loureiro, J., E. Rodriguez, A. Costa, and C. Santos. "Nuclear DNA Content Estimations in Wild Olive (*Olea europaea* L. ssp. *europaea* var. *sylvestris* Brot.) and Portuguese Cultivars of *O. europaea* Using Flow Cytometry." *Genetic Resources and Crop Evolution* 54, no. 1 (2007): 21–25. https://doi.org/10.1007/s10722-006-9115-3.

Macas, J., D. Požárková, A. Navrátilová, M. Nouzová, and P. Neumann. "Two New Families of Tandem Repeats Isolated from Genus *Vicia* Using Genomic Self-Priming PCR." *Molecular and General Genetics* 263, no. 4 (2000): 741–751. https://doi.org/10.1007/s004380000245.

Milla, R., T. G. Isleib, and H. T. Stalker. "Taxonomic Relationships Among *Arachis* Sect. *Arachis* Species as Revealed by AFLP Markers." *Genome* 48, no. 1 (2005): 1–11. https://doi.org/10.1139/g04-089.

Moretzsohn, M. C., E. G. Gouvea, P. W. Inglis, S. C. M. Leal-Bertioli, J. F. M. Valls, and D. J. Bertioli. "A Study of the Relationships of Cultivated Peanut (*Arachis hypogaea*) and Its Most Closely Related Wild Species Using Intron Sequences and Microsatellite Markers." *Annals of Botany* 111, no. 1 (2013): 113–126. https://doi.org/10.1093/aob/mcs237.

Moretzsohn, M. C., M. S. Hopkins, S. E. Mitchell, S. Kresovich, J. F. M. Valls, and M. E. Ferreira. "Genetic Diversity of Peanut (*Arachis hypogaea* L.) and Its Wild Relatives Based on the Analysis of Hypervariable Regions of the Genome." *BMC Plant Biology* 4, no. 11 (2004): 11. https://doi.org/10.1186/1471-2229-4-11.

Mukai, Y., B. Friebe, and B. S. Gill. "Comparison of C-Banding Patterns and *In Situ* Hybridization Sites Using Highly Repetitive and Total Genomic Rye DNA Probes of 'Imperial' Rye Chromosomes Added to 'Chinese Spring' Wheat." *The Japanese Journal of Genetics* 67, no. 1 (1992): 71–83. https://doi.org/10.1266/jjg.67.71.

Nielen, S., F. Campos-Fonseca, S. C. M. Leal-Bertioli, P. Guimarães, J. G. Seijo, C. Town, R. Arrial, and D. Bertioli. "FIDEL—A Retrovirus-Like Retrotransposon and Its Distinct Evolutionary Histories in the A- and B-Genome Components of Cultivated Peanut." *Chromosome Research* 18, no. 3 (2010): 227–246. https://doi.org/10.1007/s10577-009-9109-z.

Nielen, S., B. S. Vidigal, S. C. M. Leal-Bertioli, M. Ratnaparkhe, A. H. Paterson, O. Garsmeur, A. D'Hont, P. M. Guimarães, and D. J. Bertioli. "Matita, a New Retroelement from Peanut: Characterization and Evolutionary Context in the Light of the *Arachis* A-B Genome Divergence." *Molecular Genetics and Genomics* 287, no. 1 (2012): 21–38. https://doi.org/10.1007/s00438-011-0656-6.

Otyama, P. I., R. Kulkarni, K. Chamberlin, P. Ozias-Akins, Y. Chu, L. M. Lincoln, G. E. MacDonald, et al. "Genotypic Characterization of the U.S. Peanut Core Collection." *G3: Genes|Genomes|Genetics* 10, no. 11 (2020): 4013–4026. https://doi.org/10.1534/g3.120.401306.

Ozias-Akins, P., and R. L. Jarret. "Nuclear DNA Content and Ploidy Levels in the Genus *Ipomoea*." *Journal of the American Society for Horticultural Science* 119, no. 1 (1994): 110–115. https://doi.org/10.21273/JASHS.119.1.110.

Pandey, M. K., G. Agarwal, S. M. Kale, J. Clevenger, S. N. Nayak, M. Sriswathi, A. Chitikineni, C. Chavarro, et al. "Development and Evaluation of a High-Density Genotyping 'Axiom_Arachis' Array with 58K SNPs for Accelerating Genetics and Breeding in Groundnut." *Scientific Reports* 7 (2017): 40577. https://doi.org/10.1038/srep40577.

Paula, A. F., N. B. Dinato, B. B. Z. Vigna, and A. P. Fávero. "Characterization of Diploid *Arachis* Interspecific Hybrids for Pest Resistance." *Bragantia* 78, no. 1 (2019): 96–108. https://doi.org/10.1590/1678-4499.2018130.

Peñaloza, A. P. S., and J. F. M. Valls. "Chromosome Number and Satellited Chromosome Morphology of Eleven Species of *Arachis* (Leguminosae)." *Bonplandia* 14, no. 1 (2005): 65–72.

Peñaloza, A. P. S., and J. F. M. Valls. "Contagem do Número Cromossômico em Acessos de *Arachis decora* (Leguminosae)." Poster presented at the Simpósio Latino-Americano de Recursos Genéticos Vegetais, August 18–22, 1997. (In Portuguese).

Plohl, M., A. Luchetti, N. Meštrović, and B. Mantovani. "Satellite DNAs Between Selfishness and Functionality: Structure, Genomics, and Evolution of Tandem Repeats in Centromeric Heterochromatin." *Gene* 409, no. 1–2 (2008): 72–82. https://doi.org/10.1016/j.gene.2007.11.013.

Raina, S. N., and Y. Mukai. "Detection of a Variable Number of 18S–5.8S–26S and 5S Ribosomal DNA Loci by Fluorescent *In Situ* Hybridization in Diploid and Tetraploid *Arachis* Species." *Genome* 42, no. 1 (1999): 52–59. https://doi.org/10.1139/g98-092.

Raina, S. N., V. Rani, T. Kojima, Y. Ogihara, K. P. Singh, and R. M. Devarumath. "RAPD and ISSR Fingerprints as Useful Genetic Markers for Analysis of Genetic Diversity, Varietal Identification, and Phylogenetic Relationships in Peanut (*Arachis hypogaea*) Cultivars and Wild Species." *Genome* 44, no. 5 (2001): 763–772. https://doi.org/10.1139/g01-064.

Raman, V. S. "Nature of Chromosome Pairing in Allopolyploids of *Arachis* and Their Stability." *Cytologia* 46, no. 2 (1981): 307–321. https://doi.org/10.1508/cytologia.46.307.

Resslar, P. M., and W. C. Gregory. "A Cytological Study of Three Diploid Species of the Genus *Arachis* L." *Journal of Heredity* 70, no. 1 (1979): 13–16. https://doi.org/10.1093/oxfordjournals.jhered.a109180.

Resslar, P. M., J. M. Stucky, and J. P. Miksche. "Cytophometric Determination of the Amount of DNA in *Arachis* (Leguminosae)." *American Journal of Botany* 68, no. 2 (1981): 149–153. https://doi.org/10.1002/j.1537-2197.1981.tb12373.x.

Robledo, G. "Genomic Characterization of *Arachis* Species (Section *Arachis*, Leguminosae)." PhD diss., Universidad Nacional de la Plata, 2010. (In Spanish).

Robledo, G., G. I. Lavia, and J. G. Seijo. "Species Relations Among Wild *Arachis* Species with the a Genome as Revealed by FISH Mapping of rDNA Loci and Heterochromatin Detection." *Theoretical and Applied Genetics* 118, no. 7 (2009): 1295–1307. https://doi.org/10.1007/s00122-009-0981-x.

Robledo, G., and J. G. Seijo. "Characterization of the *Arachis* (Leguminosae) D Genome Using Fluorescence *In Situ* Hybridization (FISH) Chromosome Markers and Total Genome DNA Hybridization." *Genetics and Molecular Biology* 31, no. 4 (2008): 717–724. https://doi.org/10.1590/S1415-47572008000400019.

Robledo, G., and G. J. Seijo. "Species Relationships Among the Wild B Genome of *Arachis* Species (Section *Arachis*) Based on FISH Mapping of rDNA Loci and Heterochromatin Detection: A New Proposal for Genome Arrangement." *Theoretical and Applied Genetics* 121, no. 6 (2010): 1033–1046. https://doi.org/10.1007/s00122-010-1369-7.

Rodríguez, V., J. G. Seijo, G. I. Lavia, A. Fernández, and C. E. Simpson. "Meiotic Behavior in Wild Diploid *Arachis* (Leguminosae) Species." *Cytologia* 69, no. 3 (2004): 209–214. https://doi.org/10.1508/cytologia.69.209.

Samoluk, S. S., L. M. I. Chalup, C. Chavarro, G. Robledo, D. J. Bertioli, S. A. Jackson, and J. G. Seijo. "Heterochromatin Evolution in *Arachis* Investigated through Genome-Wide Analysis of Repetitive DNA." *Planta* 249, no. 5 (2019): 1405–1415. https://doi.org/10.1007/s00425-019-03096-4.

Samoluk, S. S., L. Chalup, G. Robledo, and G. Seijo. "Genome Sizes in Diploid and Allopolyploid *Arachis* L. Species (Section *Arachis*)." *Genetic Resources and Crop Evolution* 62, no. 5 (2015): 747–763. https://doi.org/10.1007/s10722-014-0193-3.

Samoluk, S. S., G. Robledo, D. Bertioli, and J. G. Seijo. "Evolutionary Dynamics of an AT-Rich Satellite DNA and Its Contribution to Karyotype Differentiation in Wild Diploid *Arachis* Species." *Molecular Genetics and Genomics* 292, no. 2 (2017): 283–296. https://doi.org/10.1007/s00438-016-1271-3.

Samoluk, S. S., G. Robledo, M. Podio, L. Chalup, J. P. A. Ortiz, S. C. Pessino, and J. G. Seijo. "First Insight into Divergence, Representation, and Chromosome Distribution of Reverse Transcriptase Fragments from L1 Retrotransposons in Peanut and Wild Relative Species." *Genetica* 143, no. 1 (2015): 113–125. https://doi.org/10.1007/s10709-015-9820-y.

Samoluk, S. S., M. Vaio, A. M. Ortíz, L. M. I. Chalup, G. Robledo, D. J. Bertioli, and J. G. Seijo. "Comparative Repeatome Analysis Reveals New Evidence on Genome Evolution in Wild Diploid *Arachis* (Fabaceae) Species." *Planta* 256 (2022): 50. https://doi.org/10.1007/s00425-022-03961-9.

Santana, S. H., and J. F. M. Valls. "*Arachis veigae* (Fabaceae), the Most Dispersed Wild Species of the Genus, and Yet Taxonomically Overlooked." *Bonplandia* 24, no. 1 (2015): 139–150.

Schmidt, T., and J. S. Heslop-Harrison. "Genomes, Genes, and Junk: The Large-Scale Organization of Plant Chromosomes." *Trends in Plant Science* 3, no. 6 (1998): 195–199. https://doi.org/10.1016/S1360-1385(98)01223-0.

Seijo, J. G., M. Atahuachi, C. E. Simpson, and A. Krapovickas. "*Arachis inflata*: A New B Genome Species of *Arachis* (Fabaceae)." *Bonplandia* 30 (2021): 169–174.

Seijo, J. G., G. I. Lavia, A. Fernández, A. Krapovickas, D. Ducasse, D. J. Bertioli, and E. A. Moscone. "Genomic Relationships Between the Cultivated Peanut (*Arachis hypogaea*, Leguminosae) and Its Close Relatives Revealed by Double GISH." *American Journal of Botany* 94, no. 12 (2007): 1963–1971. https://doi.org/10.3732/ajb.94.12.1963.

Seijo, J. G., G. I. Lavia, A. Fernández, A. Krapovickas, D. Ducasse, and E. A. Moscone. "Physical Mapping of the 5S and 18S–25S rRNA Genes by FISH as Evidence That *Arachis duranensis* and *A. ipaensis* Are the Wild Diploid Progenitors of *A. hypogaea* (Leguminosae)." *American Journal of Botany* 91, no. 9 (2004): 1294–1303. https://doi.org/10.3732/ajb.91.9.1294.

Seijo, J. G., E. I. Kovalsky, L. M. I. Chalup, S. S. Samoluk, A. Fávero, and G. Robledo. "Karyotype Stability and Genome-Specific Nucleolar Dominance in Peanut, Its Wild 4× Ancestor, and a Synthetic AABB Polyploid." *Crop Science* 58, no. 4 (2018): 1671–1683. https://doi.org/10.2135/cropsci2018.02.0088.

Shirasawa, K., P. Koilkonda, K. Aoki, H. Hirakawa, S. Tabata, M. Watanabe, M. Hasegawa, et al. "In Silico Polymorphism Analysis for the Development of Simple Sequence Repeat and Transposon Markers and Construction of Linkage Map in Cultivated Peanut." *BMC Plant Biology* 12 (2012): 80. https://doi.org/10.1186/1471-2229-12-80.

Silvestri, M. C., A. M. Ortiz, and G. I. Lavia. "rDNA Loci and Heterochromatin Positions Support a Distinct Genome Type for 'x = 9 Species' of Section *Arachis* (*Arachis*, Leguminosae)." *Plant Systematics and Evolution* 301, no. 2 (2015): 555–562. https://doi.org/10.1007/s00606-014-1092-y.

Singh, A. K. "Utilization of Wild Relatives in the Genetic Improvement of *Arachis hypogaea* L. 8. Synthetic Amphidiploids and Their Importance in Interspecific Breeding." *Theoretical and Applied Genetics* 72, no. 3 (1986): 433–439. https://doi.org/10.1007/BF00289523.

Singh, A. K., and J. P. Moss. "Utilization of Wild Relatives in Genetic Improvement of *Arachis hypogaea* L." *Theoretical and Applied Genetics* 61, no. 4 (1982): 305–314. https://doi.org/10.1007/BF00272846.

Singh, A. K., and J. P. Moss. "Utilization of Wild Relatives in the Genetic Improvement of *Arachis hypogaea* L. 5. Genome Analysis in Section *Arachis* and Its Implications in Gene Transfer." *Theoretical and Applied Genetics* 68, no. 4 (1984): 355–364. https://doi.org/10.1007/BF00267889.

Singh, K. P., S. N. Raina, and A. K. Singh. "Variation in Chromosomal DNA Associated with the Evolution of *Arachis* Species." *Genome* 39, no. 5 (1996): 890–897. https://doi.org/10.1139/g96-112.

Slamovits, C. H., J. A. Cook, E. P. Lessa, and M. S. Rossi. "Recurrent Amplifications and Deletions of Satellite DNA Accompanied Chromosomal Diversification in South American Tuco-Tucos (Genus *Ctenomys*, Rodentia: Octodontidae): A Phylogenetic Approach." *Molecular Biology and Evolution* 18, no. 9 (2001): 1708–1719. https://doi.org/10.1093/oxfordjournals.molbev.a003959.

Smartt, J. "Cross-Compatibility Relationship Between the Cultivated Peanut *Arachis hypogaea* L. and Other Species of the Genus *Arachis*." PhD dissertation, North Carolina State University, 1964.

Smartt, J., and W. C. Gregory. "Interspecific Cross-Compatibility Between the Cultivated Peanut *Arachis hypogaea* L. and Other Members of the Genus *Arachis*." *Oléagineux* 22, no. 11 (1967): 455–459.

Smartt, J., W. C. Gregory, and M. P. Gregory. "The Genome of *Arachis hypogaea*. 1. Cytogenetic Studies of Putative Genome Donors." *Euphytica* 27, no. 3 (1978): 665–675. https://doi.org/10.1007/BF00023701.

Smartt, J., and H. T. Stalker. "Speciation and Cytogenetics of *Arachis*." In *Peanut Science and Technology*, edited by H. E. Pattee and C. T. Young, 21–49. Peanut Research and Education Society, 1982.

Stalker, H. T. "A New Species in Section *Arachis* of Peanuts with a D Genome." *American Journal of Botany* 78, no. 5 (1991): 630–637. https://doi.org/10.1002/j.1537-2197.1991.tb12587.x.

Stalker, H. T., J. S. Dhesi, D. C. Parry, and J. H. Hahn. "Cytological and Interfertility Relationships of *Arachis* Section *Arachis*." *American Journal of Botany* 78, no. 2 (1991): 238–246. https://doi.org/10.1002/j.1537-2197.1991.tb15751.x.

Stalker, H. T., and J. P. Moss. "Speciation, Cytogenetics, and Utilization of *Arachis* Species." *Advances in Agronomy* 41 (1987): 1–40. https://doi.org/10.1016/S0065-2113(08)60801-9.

Stalker, H. T., and J. C. Wynne. "Cytology of Interspecific Hybrids in Section *Arachis* of Peanuts." *Peanut Science* 6, no. 2 (1979): 110–114. https://doi.org/10.3146/i0095-3679-6-2-12.

Stebbins, G. L. "Types of Polyploids: Their Classification and Significance." *Advances in Genetics* 1 (1947): 403–429. https://doi.org/10.1016/S0065-2660(08)60490-3.

Tallury, S. P., K. W. Hilu, S. R. Milla, S. A. Friend, M. Alsaghir, H. T. Stalker, and D. Quandt. "Genomic Affinities in *Arachis* Section *Arachis* (Fabaceae): Molecular and Cytogenetic Evidence." *Theoretical and Applied Genetics* 111, no. 7 (2005): 1229–1237. https://doi.org/10.1007/s00122-005-0017-0.

Temsch, E. M., and J. Greilhuber. "Genome Size Variation in *Arachis hypogaea* and *A. monticola* Re-Evaluated." *Genome* 43, no. 3 (2000): 449–451. https://doi.org/10.1139/g99-130.

Temsch, E. M., and J. Greilhuber. "Genome Size in *Arachis duranensis*: A Critical Study." *Genome* 44, no. 5 (2001): 826–830. https://doi.org/10.1139/g01-081.

Urdampilleta, J. D., A. Pereira de Souza, D. R. S. Schneider, A. L. L. Vanzela, M. S. Ferrucci, and E. R. F. Martins. "Molecular and Cytogenetic Characterization of an AT-Rich Satellite DNA Family in *Urvillea chacoensis* Hunz. (*Paullinieae*, *Sapindaceae*)." *Genetica* 136, no. 2 (2009): 171–177. https://doi.org/10.1007/s10709-008-9332-0.

Valls, J. F. M., and C. E. Simpson. "New Species of *Arachis* (*Leguminosae*) from Brazil, Paraguay, and Bolivia." *Bonplandia* 14, no. 1 (2005): 35–63.

Valls, J. F. M., and C. E. Simpson. "A New Species of *Arachis* (*Fabaceae*) from Mato Grosso, Brazil, Related to *A. matiensis*." *Bonplandia* 26, no. 1 (2017): 143–149. https://doi.org/10.30972/bon.2622575.

Valls, J. F. M., L. C. Da Costa, and A. R. Custodio. "A Novel Trifoliolate Species of *Arachis* (*Fabaceae*) and Further Comments on the Taxonomic Section *Trierectoides*." *Bonplandia* 22, no. 1 (2013): 91–97. https://doi.org/10.30972/bon.2211257.

Wang, S. Y., L. N. Li, L. Y. Fu, H. Liu, L. I. Qin, C. H. Cui, L. J. Miao, et al. "Development and Characterization of New Allohexaploid Resistant to Web Blotch in Peanut." *Journal of Integrative Agriculture* 20, no. 1 (2021): 55–64. https://doi.org/10.1016/S2095-3119.

Wang, M., J. Li, P. Wang, F. Liu, Z. Liu, G. Zhao, Z. Xu, L. Pei, C. E. Grover, et al. "Comparative Genome Analyses Highlight Transposon-Mediated Genome Expansion and the Evolutionary Architecture of 3D Genomic Folding in Cotton." *Molecular Biology and Evolution* 38, no. 9 (2021): 3621–3636. https://doi.org/10.1093/molbev/msab128.

Wynne, J. C., and T. Halward. "Cytogenetics and Genetics of *Arachis*." *Critical Reviews in Plant Sciences* 8, no. 3 (1989): 189–220. https://doi.org/10.1080/07352688909382275.

Zhang, L., C. Xu, and W. Yu. "Cloning and Characterization of Chromosomal Markers from a Cot-1 Library of Peanut (*Arachis hypogaea* L.)." *Cytogenetic and Genome Research* 137, no. 1 (2012): 31–41. https://doi.org/10.1159/000339455.

Zhang, L., X. Yang, L. Tian, L. Chen, and W. Yu. "Identification of Peanut (*Arachis hypogaea*) Chromosomes Using a Fluorescence *In Situ* Hybridization System Reveals Multiple Hybridization Events During Tetraploid Peanut Formation." *New Phytologist* 211, no. 4 (2016): 1424–1439. https://doi.org/10.1111/nph.13999.

Zhuang, W., H. Chen, M. Yang, J. Wang, M. K. Pandey, C. Zhang, W. Chang, et al. "The Genome of Cultivated Peanut Provides Insight into Legume Karyotypes, Polyploid Evolution, and Crop Domestication." *Nature Genetics* 51, no. 5 (2019): 865–876. https://doi.org/10.1038/s41588-019-0402-2.

4 Fine Structure and Organization of Peanut Genome

Dongmei Yin, Xingguo Zhang, Sheidu Abdullaziz, Xiyin Wang, and Weijian Zhuang

4.1 INTRODUCTION

Since the first genome sequencing of the model plant *Arabidopsis thaliana* in 2000, the complete genomes of additional diploid plant species with small genome sizes, such as rice, corn, and soybean, have been sequenced due to their economic importance (Michael and Jackson, 2013). The advent of next-generation sequencing (NGS) technologies has revolutionized the field of plant genomics, leading to rapid, cost-effective, and high-throughput sequencing capabilities. This technological leap has particularly benefited the study of economically important and complex genomes, such as that of the peanut (*Arachis hypogaea* L.), an allotetraploid crop with a genome size of approximately 2.6 Gb and an AABB genome configuration (Zhuang et al., 2019; Bertioli et al., 2019; Wang et al., 2024). The genomic complexity of peanut, including its polyploid nature, initially delayed its sequencing compared to simpler genomes like that of rice. However, the collaborative efforts facilitated by both public and private sectors using advanced genomic techniques and international consortia have significantly accelerated peanut genome research.

In 2001, 26 legume scientists collaborated with the United Soybean Board, National Peanut Foundation, USA Dry Pea and Lentil Council, and USDA-ARS to develop a strategy for genomics research of five economically important legume species, including peanut. In 2004, the book *Legume Crop Genomics* was published, and a national strategic plan for the Peanut Genome Initiative (PGI) was developed, both aiming to advance genomics research in legume crops. This marks a significant milestone in legume genomics, highlighting the power of international collaboration in advancing agricultural biotechnology.

In 2006, the International Conference on Aflatoxin Management and Genomics established an open dialogue for cooperative research. In 2010, the International Peanut Genome Project (IPGI) made significant progress in peanut genome sequencing. A proposal was put forward to collaborate and pool resources for sequencing both the tetraploid and diploid peanut genomes. The project was discussed at the inaugural Peanut Genome Project meeting in Atlanta, where the executive committee agreed to proceed with an international collaborative effort. They decided to sequence four cultivars and 200 recombinant inbred line progenies. After meetings with Chinese collaborators, BGI proposed an integrated strategy combining whole-genome sequencing with bacterial artificial chromosomes, resequencing, and calling SNPs. Subsequently, the Peanut Genome Consortium (PGC) was formally established as an extension of the International Peanut Genome Initiative (IPGI), bringing together a coalition of international scientists and stakeholders involved in the Peanut Genome Project (PGP). Finally, the IPGI and PGC released the two diploid sequences for public use in April 2014 (http://www.peanutbase.org/node/618) and published the two genome sequences of the diploid ancestors of cultivated peanut in February 2016 (Bertioli et al., 2016). These foundational genomes have provided invaluable insights into disease resistance, oil biosynthesis, and allergens, crucial for the development of elite peanut cultivars (Qin et al., 2012; Bertioli et al., 2016). However, due to its allotetraploid nature, large genome size, and high rate of repetitive sequences, the genome of the cultivated peanut experienced significant delays in sequencing efforts for several years. However, through international collaborative efforts, three important peanut varieties were finally sequenced and published by three separate research teams (Zhuang et al., 2019; Bertioli et al., 2019; Chen et al., 2019). These advances have enormously promoted peanut functional genomics and genetics toward breeding application.

In this chapter, we primarily focus on the genomic complexities of cultivated peanuts (*Arachis hypogea*). We detail the collaborative efforts that led to sequencing the peanut genome, starting with the sequencing of diploid progenitors and culminating in the release of whole-genome sequences for cultivated varieties. The sequencing, as well as the genome structure and organization, are discussed in detail. Emphasizing the significance of three-dimensional chromatin organization in understanding gene regulation and its implications for peanut biology, we also discuss the historical progression of the peanut genome project and its impact on functional genomics and system biology for peanut breeding.

4.2 BREAKTHROUGH OF PEANUT GENOME SEQUENCING AND ASSEMBLY

The genus *Arachis*, which comprises 81 species, is predominantly diploid, with a chromosome count of $2n = 2x =$

DOI: 10.1201/9781003360988-4

20. Among these species, *Arachis hypogea*, the cultivated peanut, is an exception as an allotetraploid ($2n = 4x = 40$), formed from the hybridization of two diploid species, *A. duranensis* (AA genome) and *A. ipaensis* (BB genome). This hybridization event doubled the genome size compared to its wild diploid progenitors, contributing to the genomic complexity and challenges associated with sequencing the cultivated peanut. The genome assembly is particularly difficult due to the close relationship between the subgenomes and a high proportion of repeated sequences (Bertioli et al., 2014).

Evidence suggests that since becoming polyploid, the changes within the ancestral genomes have been minimal. Thus, studying the combined genetic content of these diploid ancestors provides a valuable approximation of the allotetraploid peanut's genome. The successful sequencing of *A. duranensis* and *A. ipaensis* in 2014 marked a significant milestone. These sequences are now publicly available (Table 4.1) via the National Center for Biotechnology Information (NCBI) (www.ncbi.nlm.nih.gov); this offers crucial insights into the genetic makeup and evolution of cultivated peanuts.

The divergence of the A and B genomes occurred about three million years ago. The allotetraploid nature of the cultivated peanut genome introduces complexities in sequencing and assembly efforts. This is due to the close similarity between the subgenomes, making it a formidable challenge to distinguish between sequence reads originating from each subgenome (Bertioli et al., 2014).

In 2016, the peanut research community had access to reference genome sequences for these diploid progenitors, alongside a high-throughput genotyping array featuring 56,000 single nucleotide polymorphisms (SNPs). This array facilitates deep genomic studies, enabling researchers to identify disease-resistance genes, develop new molecular markers, assist in the assembly of tetraploid transcripts, and explore genetic exchanges between the subgenomes of cultivated peanut (Bertioli et al., 2016).

In May 2016, a draft genome of *Arachis duranensis*, the A-genome progenitor of cultivated peanut, was released (Chen et al., 2016). This genomic study reveals that the peanut lineage has undergone at least three polyploidization events since the origin of eudicots, indicating significant evolutionary changes. Additionally, resequencing of synthetic *Arachis* tetraploids has shown extensive gene conversion events since their formation. The genome of *A. duranensis* is particularly valuable because it provides a rich source of candidate genes relevant to fruit development, oil biosynthesis, and allergens, thereby broadening our understanding of these critical areas in plant biology and the impacts on human health. Furthermore, this study has identified millions of structural variations that can serve as genetic markers, crucial for developing improved peanut varieties through genomics-assisted breeding.

In December 2017, the genome of *Arachis hypogaea* subsp. *fastigiata*, cv. Shitouqi (STQ), representing the Spanish ecotype and Valencia variant, was released on the Peanut Genome Database website (http://peanutgr.fafu.edu.cn), featuring detailed annotations of its 20 chromosomes. Shortly afterward, the genome sequence of *Arachis hypogaea* subsp. *hypogaea*, cv. Tifrunner, a runner-type peanut and hypogaea variant with the registration number CV-93, PI 644011, was also released on PeanutBase. These were the first two cultivated peanut genomes sequenced using whole-genome shotgun sequencing technology, with the detailed findings published in *Nature Genetics* at the same time (Zhuang et al., 2019; Bertioli et al., 2019). The Tifunner genome sequencing effort, led by David Bertioily in the United States, successfully produced sequences for 20 chromosomes, named Arahy.01 through Arahy.20. The chromosomes of the A subgenome are represented by Arahy.01–Arahy.10, and those of the B subgenome by Arahy.11–Arahy.20. Together, these chromosome sequences account for 99.3% of the assembled genome, which totals 2.54 Gb, approximately 93% of the genome size estimated by flow cytometry. This detailed sequencing revealed genetic deletions and exchanges between the subgenomes, providing insights into genetic variations that likely contributed to the domestication of *A. hypogea* over its diploid ancestors. These findings underscore the significant role of these genetic mechanisms in the prevalence of polyploidy among domesticated plant species.

The STQ genome sequencing, led by Weijian Zhuang from Fujian Agriculture and Forestry University in China, together with other institutes worldwide, provided the whole genome sequencing and assembly of 20 chromosomes, named Ahchr01–Ahchr20, totally 2.54 Gb. In this study, PacBio SMRT sequencing combined with Hi-C technology and high-density genetic mapping was utilized to achieve precise assembly of the A and B subgenomes, resulting in high-quality reference genomes (Table 4.1; Figure 4.1A and 4.1B). In addition, 52 peanut samples from 12 different species were selected for resequencing. The results provided new insights into the genome structure, biological characteristics, polyploid evolution, and crop domestication in peanut. The comparative genomics first revealed the legumes karyotype evolution, diploids as well as tetraploid peanut chromosome origins, and cultivated peanut origin and domestication. It also highlighted the important genes involved in the evolution of cultivated peanut yield, lipid metabolism, disease resistance, and nitrogen symbiosis. The quality genome-assisted gene discovery of agronomic important traits like seed size and color, rust, and LLS-resistant alleles and fine locations (Zhuang et al., 2019). These provide a firm foundation for peanut genetics and breeding and system biology.

In 2019, the whole-genome sequencing and gene-function analysis of the peanut cultivar Fuhuasheng, a renowned breeding parent in China, were also completed. This research provides valuable insights into peanut origin, evolution, and oil-content improvement (Chen et al., 2019). The project combined second and third-generation sequencing technologies, BioNano, high-density genetic mapping, and other technologies to successfully assemble

TABLE 4.1
Characteristics of Diploid and Tetraploid *Arachis* Genomes

	A. duranensis (accession PI475845 (Chen et al., 2016)	*A. duranensis* (accession V14167) (Bertioli et al., 2016)	*A. ipaensis* (accession K30076) (Bertioli et al., 2016)	*A.ipaensis* (accession ICG_8206) (Lu et al., 2018)	*A.monticola* Line PI 263393 (Xue et al., 2024)	*A.hypogaea* cv. Tifrunner (Bertioli et al., 2019)	*A.hypogaea* cv. Fuhuasheng (Chen et al., 2019)	*A.hypogaea* cv. Shitouqi (Zhuang et al., 2019)
No. contigs	90,568	765,406	869,435	1,008,989	80	7,698	–	7,232
Contig N50(kb)	29.584	22.293	23.492	8.067	68,296.015	1,500	211	1,509.43
Total contig length (Mb)	–	1,074.450	1,3886.399	–	2,620	2,552.5	2,580	2,538.411
No. scaffolds	8,173	635,392	759,499	79,408	46	384	486	1,297
Scaffold N50(kb)	649.840	947.955	5343.284	170.050	137,590	9,000	5,656.284	135,085.854
Total scaffold length (Mb)	1,051.524	1211.483	1,512.090	1,391700,926	2557.5	2,556.3	2,551.679	2,539.035
Total genes	50,324	36,734	41,840	39,704	75,226	66,469	108,604	83,709
Mean length of gene	–	1,100.72	1,048.69	–	–	876.51	–	1,164.51
Total transposable elements, Mb (TEs)	643.886	580.237	861.106	1,125.924	2123.0	1,890.048	1,386.549	1,971.142
Transposable element percent in genome	59.77	61.73	68.50	75	83.96	74.03	54.34	77.65

20 chromosomes. It provided abundant candidate gene resources for peanut-oil metabolism and other related research, which was conducive to peanut genetic improvement and genetic diversity research. The decoding of the cultivated peanut genome and the analysis of related biological challenges are expected to accelerate the shift from traditional genetic improvement to genome-assisted breeding in peanuts.

As a crucial wild resource donor for enhancing the agronomic traits of cultivated peanuts, the genomes of wild tetraploid peanuts have become a significant focus of research. The successful decoding of the genome of wild tetraploid peanut will help scientists and breeders understand the origin and domestication of *A. hypogea*. Researchers successfully sequenced and assembled the genome of *Arachis monticola* (Yin et al., 2018, 2020, 2024). To tackle the complexity of the highly homologous heterologous tetraploid genome in wild peanut, researchers employed Nanopore ultra-long reads, Hi-C reads, MGI Illumina sequencing platforms, and the most advanced genome assembly technology. Finally, researchers obtained a near-complete genome assembly for *Arachis monticola* (Amon2.0). The final genome assembly contained only 34 gaps, with a total size of 2.56 Gbps and an N50 value of 137.6 Mbps. In total, 11 chromosomes have both complete telomeric sequences, and the remaining 9 chromosomes have one complete telomeric sequences. All chromosomes have centromeric sequences, including 13 chromosomes with complete centromeric sequences. LTR assembly index (LAI) assessment demonstrated that Amon2.0 achieved gold-standard quality for repeat regions with an LAI score of 21.27. Researchers predicted a total of 75,226 protein-coding genes from 35,556 gene families, functionally annotating 74.6% of these genes, with 91.2% containing at least one known structural domain. The Benchmarking Universal Single-Copy Orthologs (BUSCO) scores were 99.3% for complete genes and 91.5% for duplicated genes, corresponding to 96.2% and 96.7% in the A and B subgenomes, respectively. The publication of the wild peanut genome had important scientific value for understanding the evolution of peanut and leguminous crops.

A three-dimensional (3D) chromatin organization map was depicted using Hi-C, ATAC-seq, multi-tissue expression spectrum, and 3C experiments, revealing the spatial structure characteristics of chromatin and its influence on plant type (Zhang et al., 2021). The 3D genome architecture map illustrated A/B compartments, topologically associated domains (TADs), and widespread chromatin interactions. Most peanut chromosomal arms (52.3%) had active regions (A compartment) with relatively high gene density and high transcriptional levels. About 2.0% of chromosomal regions switched from inactive to active (B-to-A) in a mutant line harboring 58 differentially expressed genes enriched in flavonoid biosynthesis and circadian rhythm functions. This study has expanded our understanding of the relationship between the 3D genome of peanuts and gene regulation, has promoted the research of peanut functional genomics, and finally has provided important theoretical innovation and technical support for the genetic improvement and germplasm creation of peanuts.

4.3 FINE STRUCTURE AND ORGANIZATION OF PEANUT GENOME

4.3.1 Peanut Gene Models

Multiple projects have completed the cultivated tetraploid genomes annotated with diverse gene models (Zhuang et al., 2019; Bertioli et al., 2019; Chen et al., 2019; Wang et al., 2024). The genome of *A. hypogea* cultivar Shitouqi, an important landrace, and breeding parent from southern China, was found to contain 83,709 predicted protein-coding genes (Table 4.1), including 37,059 genes in the At subgenome and 46,650 genes in the Bt subgenome. Interestingly, the Bt subgenome was found to be more similar to *A. ipaensis* than the At subgenome was to *A. duranensis* (Chen et al., 2019, Zhuang et al., 2019). In *A. hypogea* cv. Tifrunner, a runner-type peanut, there were 66,469 genes annotated and assigned to the pseudo-chromosomes, with 35,110 predicted genes in the Bt subgenome and 31,359 genes in the At subgenome (Bertioli et al., 2019). In another genome of *A. hypogea* cv. Fuhuasheng, PacBio technologies were combined with BioNano genome mapping and high-density genetic maps. The genome sequence was assembled into 20 pseudo-molecules, spanning a total of 2,552 Mb. The *Arachis hypogaea* genome sequence contained 108,604 predicted gene models, including 83,087 high-confidence genes. Gene numbers are 4.11% higher in the Bt subgenome, with 41,526 predicted genes compared to 39,888 genes in the At subgenome (Bertioli et al., 2019; Chen et al., 2019; Zhuang et al., 2019).

Peanut genome complexity combines allotetraploidy with other gene duplication mechanisms. The study identified 30,596 nonredundant peanut genes in the STQ genome, including 24,208 (79.12%) with homeologs and 6,338 (20.88%) without homeologs. Additionally, the study identified 27,913 duplicated genes within the peanut genome, with 10,590 located in subgenome A and 17,323 in subgenome B. Among these, 2,402 are tandem duplications (located consecutively), while 25,511 are dispersed duplications occurring on different chromosomes or spaced apart on the same chromosome (Zhuang et al., 2019). The average gene length is about 1,100 bp in peanut (Zhuang et al., 2019; Bertioli et al., 2016).

4.3.2 Peanut Repeats Sequences

The amplification of transposable elements (TEs) is widely regarded as a key mechanism for increasing plant genome size and driving plant genome evolution (Casacuberta et al., 2016; Liu and El-Kassaby, 2019). Peanut contains abundant TEs in its genome, and LTRs (long terminal repeats) retrotransposons are the dominant TEs in the *Arachis* genome. Different species seem to contain diverse TE contents.

In *A. duranensis* genome, ~59.77% sequence is TE. Further, LTR retrotransposons, the main TE type, account for ~40% of the genome. *Gypsy* and *Copia* account for 36.9% and 2.68% genome, respectively (Chen et al., 2016). TEs account for approximately 75.97% of the *A. ipaensis* genome, and LTRs cover 64.15% of the nuclear genome (Lu et al., 2018). TEs account for 61.7% and 68.5% and LTR for 44.7% and 45.72% of the *A. duranensis* and *A. ipaensis* genomes, respectively (Bertioli et al., 2016). The higher transposon content and local duplications in *Arachis ipaensis* are likely the primary factors contributing to the larger pseudomolecules in *A. ipaensis* compared to their counterparts in *Arachis duranensis*. LTR accounts for 63.94%, 59.15%, and 64.97% in the *A. hypogea*, *A. duranensis*, and *A. ipaensis* genomes, respectively. Gypsy comprised 40.59%, 31.32%, and 41.79% of the three genomes, respectively (Zhuang et al., 2019).

In the *A. monticola* genome, a total of 2,147,185,085 bps (83.96%) were annotated as repeat elements, which were mainly long terminal repeat (LTR) retroelements (65.3%) and DNA transposons (12.03%). The highest proportion of LTR retroelements were Gypsy sequences (67.1%). Genes, DNA transposons, and LTR/Copia elements were more concentrated near the telomeres, while LTR/Gypsy elements primarily accumulated near the centromeres (Yin et al., 2024).

The *A. hypogea* (cultivar Fuhuasheng) genome has about 54.34% repeat region and 52.3% LTR retrotransposons. Gypsy accounts for 49.29% of the genome and Copia for 2.83%. The percent of DNA transposon is 0.69%. Gypsy lineage is a higher proportion of TEs in the subgenome At (52.11%) than in the A-progenitor genome (42.07%) (Chen et al., 2019). LTR retrotransposons are 61.83% in the *A. hypogea* genome and highly abundant in pericentromeric regions like diploid genome (Bertioli et al., 2016, 2019), whereas DNA transposons are more frequent in euchromatic arms (Bertioli et al., 2019).

Moreover, the copy number of TEs is frequently higher in polyploids than in their related diploid species. For example, 77.65% of genome sequences were repetitive, including 64.74% of retrotransposons and 4.49% of DNA transposons in *A. hypogea* (Zhuang et al., 2019). Of genome sequences, 77.03% were TE in *A. hypogea* genome (Bertioli et al., 2019). Subgenome At had a higher proportion of TEs (52.11%) than the A-progenitor genome (42.07%) (Chen et al., 2019). In *the A. monticola* genome, TEs account for about 83.96%, including 43.93% LTR/Gypsy (Yin et al., 2020). In summary, the At subgenome contains higher TE than its diploid A genome, and Bt contain a paramount TE as to its diploid B genome.

Most transposable elements expanded after tetraploidization (Chen et al., 2016). Research focused on LTR expansion time with different views. LTR in the subgenome At in *A. hypogea* expanded for a high proportion in ~0.25 Mya but in Bt subgenome with less expansion (Zhuang et al., 2019). LTRs of the subgenome Bt and the two diploids, *A. duranensis* and *A. ipaensis*, expanded 0.89, 1.12, and 1.00 Mya, respectively, and At subgenome did not expand at this time (Zhuang et al. 2019). The At subgenome of *A. hypogea* with higher TEs and LTRs than the A diploid genome was shown to result from the expansion of LTR after tetroploidy at ~0.25 Mya (Zhuang et al., 2019, 2020). Different types of transposable elements show some specific distribution patterns, with Gypsy forming dense accumulations in pericentromeres, but Copia positioning more towards the ends of chromosomes (Figure 4.1; Davashuk et al., 2016; Zhuang et al., 2019). As repeats play a major part in peanut genome, with dynamics in amplification, structure, and components, further characterization of repeats will lead to the clarification of Arachis origin and domestication.

4.3.3 Peanut Noncoding RNAs

Noncoding RNAs (ncRNAs) are integral to the genomic architecture of the peanut (*Arachis hypogea*), playing pivotal roles in gene regulation without encoding proteins. These ncRNAs include microRNAs (miRNAs), small interfering RNAs (siRNAs), and long noncoding RNAs (lncRNAs), each serving distinct functional roles. MiRNAs, typically 21–24 nucleotides in length, are crucial for post-transcriptional gene regulation, targeting mRNAs for degradation or translational repression. Recent studies have identified over 200 conserved and novel miRNAs in the peanut genome, linked to important biological processes such as nodulation and oil production (Zhang et al., 2021).

SiRNAs, which are abundant in peanuts, arise from double-stranded RNA precursors and are primarily involved in silencing genes, transposons, and viral elements, particularly in heterochromatic regions associated with repetitive elements. Estimates suggest thousands of siRNA clusters distributed throughout the peanut genome (Prasad et al., 2023). LncRNAs, more than 200 nucleotides in length, regulate gene expression through mechanisms like chromatin modification and transcriptional control. Recent cataloging efforts have identified around 1,000 lncRNAs across all peanut chromosomes, indicating a wide range of functions from development to stress responses (Wang et al., 2023).

Evolutionary and functional constraints influence the genomic distribution of ncRNAs in peanuts. Many ncRNAs are located in gene-rich regions, enabling efficient gene regulation. SiRNAs are concentrated in repetitive DNA regions, including centromeres and telomeres, contributing to heterochromatin formation and genome stability. LncRNAs often originate from intergenic regions, potentially regulating adjacent genes through cis-regulation mechanisms. This diverse array of ncRNAs underscores their significance in peanut biology, providing insights that could enhance peanut breeding and cultivation strategies. Further research is essential to understand the complex interactions and functions of ncRNAs within the peanut genome.

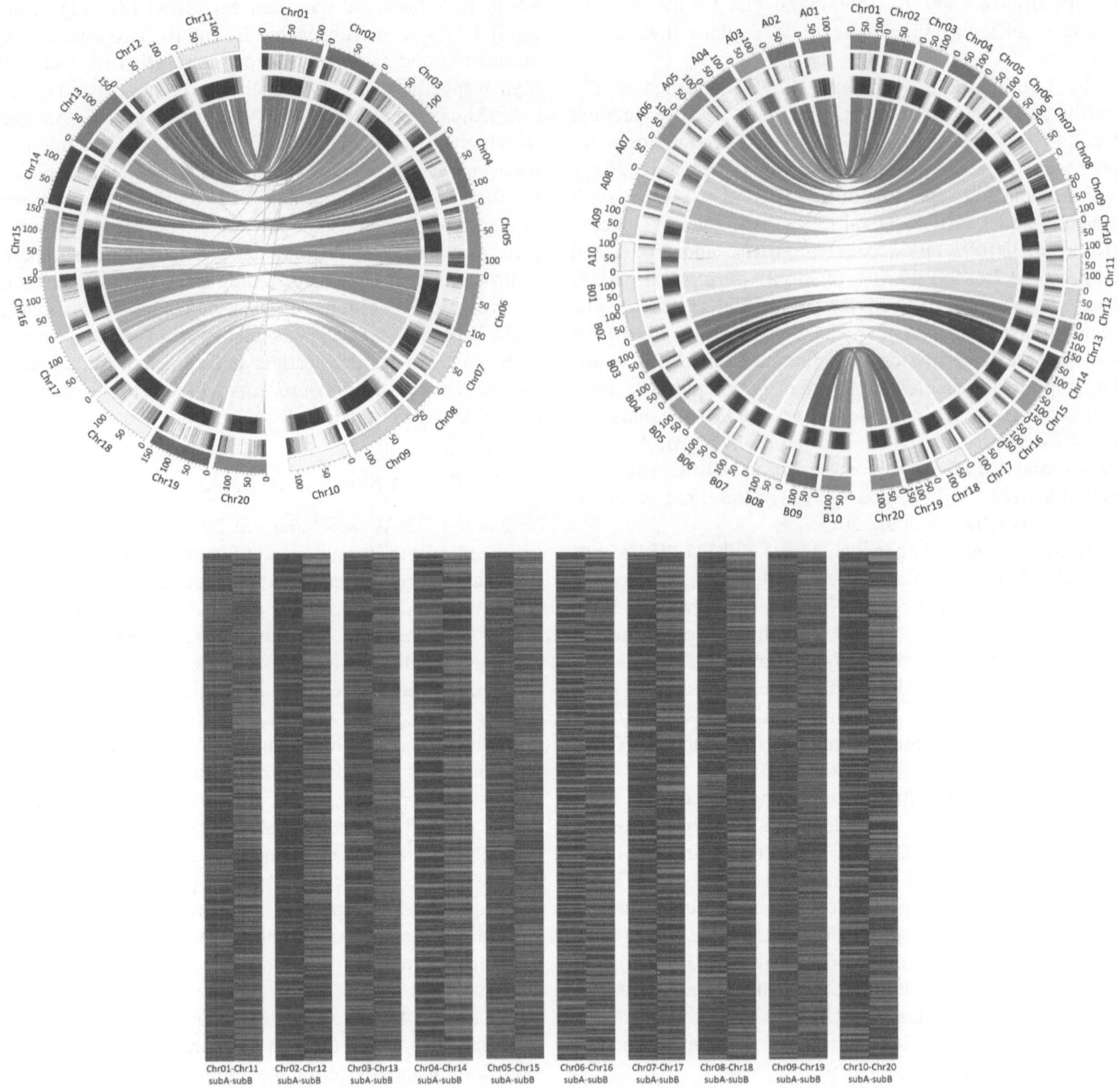

FIGURE 4.1 Peanut genome structure and gene expressions. A. Circos diagram depicting relationships of A and B subgenome chromosomal pseudomolecules. The scale for the chromosomes (outer bars) is megabases; colors represent the density of nonautonomous LTR retrotransposons and Ty3-gypsy elements (blue) and genes (green). Homeologous gene pairs between chromsomes are connected with lines. B. Syntenic comparisons between peanut subgenomes and diploid A and B genomes. The outer three circles are chromosomes, density of genes. and of Ty3-gypsy and nonautonomous LTR retrotransposons (as shown in a) between the A and B subgenomes and *A. duranensis* and *A. ipaensis* genomes, respectively, based on BLASTN. C. Differential expression of homeologous gene pairs between two subgenomes is shown.

4.4 EXPRESSION DIFFERENTIATION BETWEEN HOMOLOGOUS GENES

Unequal expression of homologous genes in allopolyploids is recognized as a significant consequence, potentially contributing to genetic novelty that facilitates domestication, as noted in several studies (Grover et al., 2012; Li et al., 2021; Powell et al., 2017; Wu et al., 2018; Yoo et al., 2013). Research focusing on three sequenced peanut cultivars has explored the expression patterns of homologous gene pairs from the At and Bt subgenomes (Bertioli et al., 2019; Chen et al., 2019; Zhuang et al., 2019). In the cultivar Tifrunner, analyses revealed that while many homologous pairs showed no asymmetry in expression, a certain proportion exhibited no expression at all. The count of gene pairs with expression bias was nearly balanced between the

At and Bt subgenomes, with 3,648 and 3,759 showing At and Bt biased expression, respectively. Notably, reproductive tissues and roots displayed a significant predisposition ($P < 0.05$) for more Bt subgenome-biased gene expression.

The genes with the most pronounced asymmetry in expression were primarily involved in critical biological processes such as oxidation reduction, pollen recognition, lipid and chitin metabolism, and responses to biotic stimuli (Bertioli et al., 2019). Additionally, pathways like mannose metabolism, nitrate assimilation, and phosphorylation signal transduction often involve genes with biased expression. Specifically, Bt-biased genes played significant roles in responding to biotic stimuli and inorganic ion transport (Bertioli et al., 2019). In the cultivar Fuhuasheng, both subgenomes show similar levels of gene function and a relatively even dominance in genome-wide expression, according to recent studies (Chen et al., 2019). Approximately 57% of high-confidence (HC) genes show expression across various plant parts, including roots, stems, leaves, and flowers, and throughout multiple pod developmental stages. Notably, the At and Bt subgenomes contribute about equally to the total number of expressed genes. However, a small fraction, about 6.2% (2029 homologs), exhibit biased expression in various tissues and developmental stages, with 1,027 showing a Bt bias and 1,002 showing an At bias (Chen et al., 2019). When homologous gene pairs are both active, the expression levels of Bt copies are generally comparable to those of At copies in roots, stems, leaves, flowers, and whole pods, though they tend to be slightly higher in seed developmental stages (Chen et al., 2019). Further investigations into the proteomics of *A. hypogea* cv. Tifrunner reveals that homologous proteins from different subgenomes display biased abundances in distinct biological processes. For instance, in leaves, At subgenome homologs are predominantly involved in biosynthetic and metabolic processes, while those from the Bt subgenome are more involved in iron ion homeostasis (Pan et al., 2022). These findings underscore the complexity of the peanut genome, which involves not only allotetraploidy but also other gene duplication mechanisms, contributing to a nuanced genetic architecture.

In the Shitouqi cultivar, researchers identified a total of 30,596 nonredundant peanut genes, of which 24,208 (79.12%) have corresponding homologs across the two subgenomes (Zhuang et al., 2019). Analysis of 29 RNA samples revealed that these 24,208 homologous pairs exhibit widespread differential expression; specifically, 13,180 pairs displayed significant dominant expression, while 7,156 pairs showed no differential expression (Figure 4.1c). Overall, these results indicate that although Bt homologs generally exhibit a higher level of expression across different cultivars, there is a noticeable variation in the degree of subgenome bias in homologous gene expression among them. The predominance of Bt-biased expression in homologous genes is largely attributed to long terminal repeat (LTR) expansions that occurred following the tetraploidization event, suggesting that these genomic changes have significantly influenced gene expression patterns (Wendel et al., 2018; Zhuang et al., 2019, 2020).

4.5 GENOME PROMOTES SCALE CHARACTERIZATION OF KEY GENES IN ARACHIS

Peanut genome provides insights into oil synthesis, nitrogen fixation, and genetic resources. Key findings include the identification of genes involved in fatty acid and triacylglycerol biosynthesis, root-nodule development, lipid metabolism, and so on. The genome also harbors numerous gene families crucial for plant growth, development, and stress responses.

4.5.1 Genome Provides Collective Analysis of Important Metabolic Genes

A. hypogea, an oilseed crop, has many excellent genetic resources. Approximately 57% of high-confidence (HC) genes show expression across various plant parts, including roots, stems, leaves, and flowers, and throughout multiple pod developmental stages. Below, we will describe disease-resistance genes, oil synthesis genes, and nodulation-related genes.

Peanut oil, primarily composed of triacylglycerol (TAG), is obtained by pressing the kernel cotyledons and provides essential nutrients for human health. Peanut breeding programs primarily target seed oil content and quality. A total of 1,613 genes related to the biosynthesis of fatty acids and triacylglycerols have been identified, including 116 nonredundant homologs potentially involved in lipid biosynthesis in *A. ipaensis*. These include acyl-ACP thioesterase A and B (FATA and FATB), FAD2, LACS, and KAS, the key enzymes leading to the synthesis of fatty acid or triacylglycerol synthesis (Lu et al., 2018). In the *Arachis hypogaea* genome, researchers identified 2,559 genes involved in fatty acid carbon flux and lipid storage, while 1,918 genes were classified under lipid metabolism. Among the 2,559 oil metabolism genes, 1,767 clustered into three major groups, representing relatively high expression at early, intermediate, and late stages. Gene families such as MCMT, FATB, CK, CCT, DAGTA, and FAD2, which encode crucial lipid biosynthesis enzymes, exhibited low expression during seed development. Oil-related gene loci clustered near the distal regions of chromosomes. The Ka/Ks ratios for oil metabolism genes indicate that only 12 are under positive selection (Chen et al., 2019). A total of 1,944 acyl-lipid orthologs were identified in peanut cv. Shitouqi, including fatty acid synthesis, lipid signaling, and triacylglycerol (TAG) biosynthesis genes (Zhuang et al., 2019).

Nitrogen is one of the essential elements for plant growth. Peanut can utilize nitrogen fixation with the help of rhizobia (Charpentier and Oldroyd, 2010). Researchers have identified 16 and 38 root-nodule developmental genes in the *A. ipaensis* and *A. duranensis* genomes, respectively. These genes include homologous LRR receptor kinase genes, LysM NF ion channel genes, and CaCa-dependent kinase genes (Lu et al., 2018). Peanut retained 95 families of 169 nodulation-related genes, with four families missing and 40 experiencing more gene loss (29 families) or gain (11 families) than the sequenced *A. duranensis* and *A. ipaensis* genomes (Zhuang et al., 2019) (Table 4.2).

4.5.2 Genome-Aid Important Gene Family Characterization

Genome-wide analyses have recently characterized important gene families in various plant species. Researchers finely characterized several gene families involved in plant growth, development, and resistance within the peanut genome. These studies have provided comprehensive insights into gene family number, length, chromosomal location, structure and functions, conserved domains and motifs, sequence diversity, phylogenesis and classification, evolution and duplication, and expression differentiation, as well as functional characterization and so on. All of these further refine the annotation of the peanut genome structure and promote peanut functional genomics. Recent advancements in peanut genomics have led to the extensive characterization of various gene families, with numbers surpassing the typical range of 60–70.

The WRKY gene family, a major transcription factor family involved in stress responses and development, includes 158 genes in cultivated peanut (*A. hypogea*) (Zhao et al., 2020), and 75 and 77 genes in its ancestors *A. ipaensis* and *A. duranensis*, respectively (Song et al., 2016). WRKY genes play crucial roles in drought adaptation and other stress responses. Similarly, the basic helix-loop-helix (bHLH) genes, involved in various biological processes and environmental stimuli, have 132 and 129 members in *A. duranensis* and *A. ipaensis*, respectively, and have been reported to play roles in pod development (Gao et al., 2017). Basic leucine zipper (bZIP) genes, another important group of transcription factors, which regulate seed germination, flower development, and stress responses, include 50 and 45 genes in *A. duranensis* and *A. ipaensis*, respectively (Wang et al., 2019).

TABLE 4.2
Summary of Genetic Resources, Oil Synthesis, and Nitrogen Fixation Genes in Peanut Crops

Category	Methods	Results	Identified Genes	Reference
Oil synthesis genes	Identification of genes related to fatty acid and triacylglycerol biosynthesis. Analysis of gene expression during seed development	1,613 genes related to fatty acid and triacylglycerol biosynthesis identified in *A. ipaensis* • 2,559 genes involved in fatty acid carbon flux and lipid storage identified in *A. hypogea* • 1,918 genes classified into lipid metabolism in *A. hypogea* • Expression profiles of oil metabolism genes during different stages of seed development • Clustering of oil-related gene loci near the distal chromosome regions	Acyl-ACP thioesterase A (FATA), Acyl-ACP thioesterase B (FATB), FAD2, LACS, KAS, MCMT, CK, CCT, DAGTA, FAD2	Chen et al., 2019
Nitrogen fixation genes	• Identification of root-nodule developmental genes in *A. ipaensis* and *A. duranensis* genomes. • Comparative analysis of nodulation-related genes in peanut genomes	• 16 root-nodule developmental genes identified in *A. ipaensis* • 38 root-nodule developmental genes identified in *A. duranensis* • Retention of 95 families of nodulation-related genes in peanut, with 4 families missing and 40 families experiencing gene loss or gain compared to sequenced genomes of *A. duranensis* and *A. ipaensis*	LRR receptor kinase genes, LysM NF ion channel genes, CaCa-dependent kinase genes	Lu et al., 2018; Zhuang et al., 2019
Genetic resources	• Assignment of homologous genes to biological process GO categories	• Homologous genes in A. hypogea assigned to 347 biological process GO categories. • Identification of 1,944 acyl-lipid orthologs in peanut, including genes related to fatty acid synthesis, lipid signaling, and triacylglycerol biosynthesis	650 FAR1 genes from four *Arachis* species, Peanut, 246 AhFAR1 genes (Ahy_A10g049543, Ahy_A06g026579, Ahy_A10g048401)	Lu et al., 2022

TABLE 4.3
Summary of Key Gene Families and Their Functions

Gene Family	Number Identified	Description	Functions/Association	References
MYB	129	DoMYB22	Potentially influences soluble tannin content during fruit development	Jie et al., 2023
MLPs (major latex-like proteins)	135	Divided into five evolutionary groups in peanuts.	Defense mechanisms	Jie et al., 2023
PLDs (phospholipase Ds)	22, 46	Identified in diploid progenitors and allotetraploid peanuts; expanded due to duplications under purifying selection.	Involved in signal transduction and membrane trafficking.	Zhang et al., 2023
OMTs (O-methyltransferases)	116	Play roles in development and stress tolerance.	Play roles in development, stress tolerance, and secondary metabolite pathways.	Tiecheng et al., 2023
AhAPX (ascorbate peroxidases)	11 main groups	Identified in the peanut genome with 38 transcription factor families targeting these genes.	Involved in the detoxification of reactive oxygen species.	Ali et al., 2022
LEA (late embryogenesis abundant)	126	Divided into eight groups involved in abiotic stress response.	Implicated in the protection of cellular structures during abiotic stress.	Ruo et al., 2022
SNARE	21 subfamilies	Characterized based on phylogenetic relationships; some play a critical role in stress responses.	Play a critical role in vesicle fusion and stress responses.	Lu et al., 2023
AhNACs	132	Classified into eight subgroups based on phylogenetic relationships.	Involved in various plant developmental processes and stress responses.	Pengxiang et al., 2021
FAD (fatty acid desaturases)	36	Includes six sub-families and four ω-3 AhFAD3 members.	Associated with fatty acid metabolism.	Peng et al., 2020
Aux/AhIAA (auxin/indole-3-acetic acid)	44	Located on 16 chromosomes, characterized during seed development.	Play a role in auxin signaling during seed development.	Zhang et al., 2022

Plant nucleotide-binding–leucine-rich repeat (NB-LRR) genes play important roles in their resistance ability against various pathogens (Jiang et al., 2018; Idehara et al., 2019; Noman et al., 2019). *A. duranensis* and *A. ipaensis* possess 345 and 397 of these genes, respectively (Bertioli et al., 2016). The *A. ipaensis* genomic assembly contains 1,437 putative disease NB-LRR genes (Lu et al., 2018). Researchers identified 2,294 disease-resistance gene analogs (RGAs) in the *Arachis monticola* genome. They also detected 661 genes with nucleotide-binding site (NBS) domains in the tetraploid peanut genome of *Arachis hypogea* cv. Shitouqi (Zhuang et al., 2019). Moreover, these NB-LRR genes tend to cluster on the distal regions of different scaffolds or chromosomes and are abundant in the distal regions of chromosomes (Bertioli et al., 2016; Lu et al., 2018; Yin et al., 2020). Many young NBS-LRRs were detected in *A. hypogea* cv. Tifrunner, indicating that *A. hypogea* cv. Tifrunner had greater disease susceptibility than *A. duranensis* and *A. ipaensis* (Song, H., Guo, Z., et al., 2019; Song, H., Sun, J., et al., 2019).

The MYB gene family in *A. hypogea* consists of 196 R2R3-MYB genes (Wang et al., 2023; Xue et al., 2024). Some key genes like DoMYB22 potentially influence soluble tannin content during fruit development in *Diospyros oleifera* (Kang et al., 2023). In cultivated peanuts and their ancestor species, researchers have identified 135 genes associated with major latex-like proteins (MLPs), encompassing five evolutionary groups, playing a role in plant defense and stress responses (Jie et al., 2023). The Phospholipase D (PLD) gene family includes 22, 22, and 46 genes in diploid progenitors and allotetraploid peanuts, respectively, with the expansion of the PLD gene family being a result of duplications under purifying selection. PLD plays a critical role in growth, development, and stress responses (Zhang et al., 2023). The peanut genome extensively represents O-Methyltransferase (OMT) genes, Ascorbate Peroxidase (AhAPX) genes, late embryogenesis abundant (LEA) proteins, SNARE family genes, and fatty acid desaturase (FAD) genes, all of which contribute to various stress responses and developmental processes. The Auxin/Indole-3-Acetic

Acid (Aux/AhIAA) gene families, with 44 genes located on 16 chromosomes, are crucial during seed development (Liu et al., 2022). Genome-wide analyses of these gene families in cultivated peanuts and their diploid progenitors have provided valuable insights into their evolutionary characteristics, expression patterns, and functional roles. This foundational work is instrumental for further research into their biological functions and potential applications in peanut breeding programs (Table 4.3).

Transcription factor (TF) participates in the regulation of a variety of physiological responses and molecular responses (Amorim et al., 2017; Baillo et al., 2019; Wang et al., 2018). Arachis genomes contain many TF genes. In *A. duranensis*, 5,251 putative TF genes were identified, spanning 57 families (Chen et al., 2016), while in *A. ipaensis*, 2,530 putative TF genes from 58 families have been identified. A total of 5,161 predictive TF genes from 58 families were identified in *A. hypogea* (Chen et al., 2016, 2019; Lu et al., 2018). Notably, researchers identified transcription factors associated with drought adaptation, such as MYB, MYB-related TFs, WRKY, ERF, and bHLH, in *Arachis ipaensis* (Lu et al., 2018).

4.6 FUTURE PROSPECTS OF PEANUT GENOME

The future of peanut genomics is set to transform peanut breeding and cultivation through a combination of advanced genomic technologies and innovative breeding practices. Enhanced understanding of genetic diversity will be crucial, particularly through the exploration of Arachis's wild relatives, to introduce beneficial traits into cultivated varieties. Combining sequencing with precise phenotyping of segregating populations and germplasm collections is expected to accurately measure the genetic diversity within the germplasm and identify the genes or nucleotides associated with specific traits. This approach will help develop better technologies and strategies for discovering functional sequence variations in the genome linked to desirable traits in peanuts.

The rapid evolution of NGS technologies and advances in next-generation breeding approaches will enable the dissection of desirable traits in peanuts. The availability of such information will facilitate faster alterations of multiple genes in the genome using modern molecular breeding technologies, including genome editing. More information about the functional variation for genes controlling traits of interest will eventually help in manipulating genes using appropriate technologies such as marker-assisted selection/back-crossing, genomic selection, and genome editing in peanut improvement. Looking forward, the data from the Peanut Genome Project offers vast potential for enhancing crop productivity and safety. The ongoing development of genomic tools and techniques continues to open new avenues for the precise genetic improvement of the peanut crop, aiming to address global challenges of food security and sustainability.

REFERENCES

Ali, Raza, Yasir Sharif, Kun Chen, Lihui Wang, Huiwen Fu, Yuhui Zhuang, Annapurna Chitikineni, Hua Chen, Cheng Zhang, Rajeev K. Varshney, and Weijian Zhuang. "Genome-Wide Characterization of Ascorbate Peroxidase Gene Family in Peanut (*Arachis hypogaea* L.) Revealed Their Crucial Role in Growth and Multiple Stress Tolerance." *Frontiers in Plant Science* (2022). https://doi.org/10.3389/fpls.2022.962182.

Amorim, L.L.B., R. da Fonseca Dos Santos, J.P.B. Neto, M. Guida-Santos, S. Crovella, and A.M. Benko-Iseppon. "Transcription Factors Involved in Plant Resistance to Pathogens." *Current Protein & Peptide Science* 18 (2017): 335–351.

Baillo, E.H., R.N. Kimotho, Z. Zhang, and P. Xu. "Transcription Factors Associated with Abiotic and Biotic Stress Tolerance and Their Potential for Crop Improvement." *Genes* 30 (2019): 771.

Bertioli, D.J., A.C.G. Araujo, S. Nielen, et al. "An Overview of Peanut Genome Structure." In *Genetics, Genomics and Breeding of Peanuts*, edited by N. Mallikarjuna and R.K. Varshney, 266p. Boca Raton, FL: CRC Press, 2014.

Bertioli, D.J., S.B. Cannon, L. Froenicke, et al. "The Genome Sequences of *Arachis duranensis* and *Arachis ipaensis*, the Diploid Ancestors of Cultivated Peanut." *Nature Genetics* 48 (2016): 438–446.

Bertioli, D.J., J. Jenkins, J. Clevenger, et al. "The Genome Sequence of Segmental Allotetraploid Peanut *Arachis hypogaea*." *Nature Genetics* 51 (2019): 877–884.

Casacuberta, J.M., S. Jackson, O. Panaud, M. Purugganan, and J. Wendel. "Evolution of Plant Phenotypes, from Genomes to Traits." *G3 (Bethesda)* 6 (2016): 775–778.

Charpentier, M., and G. Oldroyd. "How Close Are We to Nitrogen-Fixing Cereals?" *Current Opinion in Plant Biology* 13 (2010): 556–564.

Chen, X., H. Li, M.K. Pandey, Q. Yang, X. Wang, V. Garg, H. Li, X. Chi, D. Doddamani, Y. Hong, et al. "Draft Genome of the Peanut A-Genome Progenitor (*Arachis duranensis*) Provides Insights into Geocarpy, Oil Biosynthesis, and Allergens." *Proceedings of the National Academy of Sciences of the United States of America* 113 (2016): 6785–6790.

Chen, X., Q. Lu, H. Liu, J. Zhang, Y. Hong, H. Lan, H. Li, J. Wang, H. Liu, S. Li, et al. "Sequencing of Cultivated Peanut, *Arachis hypogaea*, Yields Insights into Genome Evolution and Oil Improvement." *Molecular Plant* 12 (2019): 920–934.

Divashuk, M. G., T. M. L. Khuat, P. Y. Kroupin, I. V. Kirov, D. V. Romanov, A. V. Kiseleva, L. I. Khrustaleva, D. G. Alexeev, A. S. Zelenin, M. V. Klimushina, and O. V. Razumova. "Variation in Copy Number of Ty3/Gypsy Centromeric Retrotransposons in the Genomes of Thinopyrum intermedium and Its Diploid Progenitors." *PLoS One* 11, no. 4 (2016): e0154241. https://doi.org/10.1371/journal.pone.0154241.

Gao, C., J. Sun, C. Wang, Y. Dong, S. Xiao, X. Wang, and Z. Jiao. "Genome-Wide Analysis of Basic/Helix-Loop-Helix Gene Family in Peanut and Assessment of Its Roles in Pod Development." *PLoS One* 12 (2017): e0181843.

Grover, C. E., J. P. Gallagher, E. P. Szadkowski, M. J. Yoo, L. E. Flagel, and J. F. Wendel. "Homoeolog Expression Bias and Expression Level Dominance in Allopolyploids." *New Phytologist* 196, no.4 (2012): 966–971. https://doi.org/10.1111/j.1469-8137.2012.04365.x.

Idehara, K., R. Tomita, K.T. Sekine, M. Nishiguchi, and K. Kobayashi. "Random Mutagenesis of Virus Gene for the Experimental Evaluation of the Durability of NB-LRR Class Plant Virus Resistance Gene." *Methods in Molecular Biology* 2028 (2019): 97–113.

Jiang, R., J. Li, Z. Tian, J. Du, M. Armstrong, K. Baker, J.T.-Y. Lim, J. Vossen, H. He, L. Portal, et al. "Potato Late Blight Field Resistance from QTL dPI09c Is Conferred by the NB-LRR Gene R8." *Journal of Experimental Botany* 69 (2018): 1545–1555.

Jie, L., Ruier Zeng, Ziwen Huang, He Gao, Shiyuan Liu, Yuzhu Gao, Suzhe Yao, Ying Wang, Hui Zhang, Lei Zhang, and Tingting Chen. "Genome-Wide Characterization of Major Latex Protein Gene Family in Peanut and Expression Analyses Under Drought and Waterlogging Stress." *Frontiers in Plant Science* (2023). https://doi.org/10.3389/fpls.2023.1152824.

Kang, J., Liu Cuiyu, Wu Kai, Yi Dong, Bang-chu Gong, and Xu Yang. "Genome-Wide Characterization of the R2R3-MYB Gene Family in *Diospyros oleifera*." *Agriculture* (2023). https://doi.org/10.3390/agriculture13050955.

Kyriacou, C. P., and E. Tauber. "Genes and Genomic Searches." In *Encyclopedia of Animal Behavior*, edited by M. D. Breed and J. Moore, 2 (2010): 12–20. Netherlands: Elsevier. https://doi.org/10.1016/B978-0-08-045337-8.00166-2.

Li, M., W. Sun, F. Wang, X. Wu, and J. Wang. "Asymmetric Epigenetic Modification and Homoeolog Expression Bias in the Establishment and Evolution of Allopolyploid Brassica napus." *New Phytologist* 232, no. 2 (2021): 898–913. https://doi.org/10.1111/nph.17695.

Liu, Nuoxuan, Li Li, Xiling Chen, Yanlong Zhang, and Tengxun Zhang. "The Auxin/Indole-3-Acetic Acid (Aux/IAA) Gene Family Analysis of Four Rosaceae Genomes and Expression Patterns of PmIAAs in Prunus mume." Horticulturae 8, no. 10 (2022): 899.

Liu, Y., and El-Kassaby, Y.A. "Novel Insights Into Plant Genome Evolution and Adaptation as Revealed Through Transposable Elements and Non-Coding RNAs in Conifers." *Genes* (Basel) 10 (2019): 228.

Lu, C., Z. Peng, Y. Liu, G. Li, and S. Wan. "Genome-Wide Analysis of the SNARE Family in Cultivated Peanut (*Arachis hypogaea* L.) Reveals That Some Members Are Involved in Stress Responses." *International Journal of Molecular Sciences* 24, no. 8 (2023): 7103. https://doi.org/10.3390/ijms24087103.

Lu, Q., H. Li, Y. Hong, G. Zhang, S. Wen, X. Li, G. Zhou, S. Li, H. Liu, H. Liu, et al. "Genome Sequencing and Analysis of the Peanut B-Genome Progenitor (*Arachis ipaensis*)." *Frontiers in Plant Science* 9 (2018): 604.

Lu, Q., H. Liu, Y. Hong, X. Liang, S. Li, H. Liu, H. Li, R.-Q. Wang, Q. Deng, H. Jiang, R.K. Varshney, M.K. Pandey, and X. Chen. "Genome-Wide Identification and Expression of FAR1 Gene Family Provide Insight Into Pod Development in Peanut (*Arachis hypogaea*)." *Frontiers in Plant Science* (2022). https://doi.org/10.3389/fpls.2022.893278.

Michael, T.P., and S. Jackson. "The First 50 Plant Genomes." *Plant Genome* 6 (2013): 1–7.

Noman, A., M. Aqeel, and Y. Lou. "PRRs and NB-LRRs: From Signal Perception to Activation of Plant Innate Immunity." *International Journal of Molecular Sciences* 20 (2019).

Pan, Y., Y. Zhuang, T. Liu, H. Chen, L. Wang, R. K. Varshney, W. Zhuang, and X. Wang. "Deciphering Peanut Complex Genomes Paves a Way to Understand Its Origin and Domestication." *Plant Biotechnology Journal* 21, no. 11 (2023): 2173–2181. https://doi.org/10.1111/pbi.14112.

Peng, Z., J. Ruan, H. Tian, L. Shan, J. Meng, F. Guo, Z. Zhang, H. Ding, S. Wan, and X. Li. "The Family of Peanut Fatty Acid Desaturase Genes and a Functional Analysis of Four ω-3 AhFAD3 Members." *Plant Molecular Biology Reporter* 38, no. 2 (2020): 209–221.

Pengxiang, L., Zhenying Peng, Pingli Xu, Guiying Tang, Changle Ma, Jieqiong Zhu, Lei Shan, and Shubo Wan. "Genome-Wide Identification of NAC Transcription Factors and Their Functional Prediction of Abiotic Stress Response in Peanut." *Frontiers in Genetics* (2021). https://doi.org/10.3389/FGENE.2021.630292.

Powell, A.F. *Two to Tango: Allopolyploidy and Root Nodule Symbiosis in Glycine Subgenus Glycine*. PhD diss., Cornell University, 2017.

Prasad, K., K. Yogendra, H. Sanivarapu, K. Rajasekaran, J. W. Cary, K. K. Sharma, and P. Bhatnagar-Mathur. "Multiplexed Host-Induced Gene Silencing of Aspergillus flavus Genes Confers Aflatoxin Resistance in Groundnut." *Toxins* 15, no. 5 (2023): 319. https://doi.org/10.3390/toxins15050319.

Ruo, H., Dong Xiao, Xin Wang, Jie Zhan, Ai-Qin Wang, and Long-Fei He. "Genome-Wide Identification, Evolutionary and Expression Analyses of LEA Gene Family in Peanut (*Arachis hypogaea* L.)." *BMC Plant Biology* (2022). https://doi.org/10.1186/s12870-022-03462-7.

Song, H., Z. Guo, X. Hu, L. Qian, F. Miao, X. Zhang, and J. Chen. "Evolutionary Balance Between LRR Domain Loss and Young NBS-LRR Genes Production Governs Disease Resistance in *Arachis hypogaea* cv. Tifrunner." *BMC Genomics* 20 (2019): 844.

Song, H., J. Sun, and G. Yang. "Old and Young Duplicate Genes Reveal Different Responses to Environmental Changes in *Arachis duranensis*." *Molecular Genetics and Genomics* 294 (2019): 1199–1209.

Song, H., P. Wang, J.-Y. Lin, C. Zhao, B.Y., and X. Wang. "Genome-Wide Identification and Characterization of WRKY Gene Family in Peanut." *Frontiers in Plant Science* 7 (2016): 534.

Tiecheng, C., Yasir Sharif, Yuhui Zhuang, Qiang Yang, Xiangyu Chen, Kun Chen, Yuting Chen, Meijiang Gao, H. Dang, Yi Yun Pan, Ali Raza, Cheng Zhang, Hua Chen, and Weijian Zhuang. "In-Silico Identification and Characterization of O-Methyltransferase Gene Family in Peanut (*Arachis hypogaea* L.) Reveals Their Putative Roles in Development and Stress Tolerance." *Frontiers in Plant Science* (2023). https://doi.org/10.3389/fpls.2023.1145624.

Valentine, H., E. Grabau, P. Ozias-Akins, et al. *Biotech Peanut White Paper: Benefits and Issues*. Alexandria, VA: American Peanut Council, 2006.

Wang, S., Z. Xu, Y. Yang, W. Ren, J. Fang, and L. Wan. "Genome-Wide Analysis of R2R3-MYB Genes in Cultivated Peanut (*Arachis hypogaea* L.): Gene Duplications, Functional Conservation, and Diversification." *Frontiers in Plant Science* 14 (2023): 1102174.

Wang, P., C. Yang, H. Chen, L. Luo, Q. Leng, S. Li, Z. Han, et al. "Exploring Transcription Factors Reveals Crucial Members and Regulatory Networks Involved in Different Abiotic Stresses in *Brassica napus* L." *BMC Plant Biology* 18 (2018): 202.

Wang, X., H. Fan, B. Wang, and F. Yuan. "Research Progress on the Roles of lncRNAs in Plant Development and Stress Responses." *Frontiers in Plant Science* 14 (2023): 1138901. https://doi.org/10.3389/fpls.2023.1138901.

Wang, Z., L. Yan, L. Wan, D. Huai, Y. Kang, L. Shi, H. Jiang, and B. Liao. "Genome-Wide Systematic Characterization of bZIP Transcription Factors and Their Expression Profiles During Seed Development and in Response to Salt Stress in Peanut." *BMC Genomics* 20 (2019): 1–14.

Wang, X., Z. Sun, F. Qi, Z. Zhou, P. Du, L. Shi, W. Dong et al. "A Telomere-to-telomere Genome Assembly of the Cultivated Peanut." *Molecular Plant* 18, no. 1 (2024): 5–8.

Wendel, J.F., D. Lisch, G. Hu, and A.S. Mason. "The Long and Short of Doubling Down: Polyploidy, Epigenetics, and the Temporal Dynamics of Genome Fractionation." *Current Opinion in Genetics & Development* 49 (2018): 1–7. https://doi.org/10.1016/j.gde.2018.01.004.

Wenlong, Y., Chunguang Xu, Xiaohui Zhang, Jianping Song, Huixia Jia, Qin H., Feng S., Chen C., et al. "An Integrated Genetic Linkage Map of Cultivated Peanut (*Arachis hypogaea* L.) Constructed from Two RIL Populations." *Theoretical and Applied Genetics* 124 (2012): 653–664.

Wu, J., L. Lin, M. Xu, P. Chen, D. Liu, Q. Sun, L. Ran, and Y. Wang. "Homoeolog Expression Bias and Expression Level Dominance in Resynthesized Allopolyploid Brassica napus." *BMC Genomics* 19, no. 1 (2018): 586. https://doi.org/10.1186/s12864-018-4964-0.

Xue, H., K. Zhao, K.K. Zhao, et al. "A Near Complete Genome of *Arachis monticola*, an Allotetraploid Wild Peanut." *Plant Biotechnology Journal* 22, no. 8 (2024): 2110.

Yin, D., Ji C., Ma X., et al. "Genome of an Allotetraploid Wild Peanut *Arachis monticola*: A De Novo Assembly." *GigaScience* 7 (2018): 1–9.

Yin, D., Ji C., Song Q., et al. "Comparison of *Arachis monticola* with Diploid and Cultivated Tetraploid Genomes Reveals Asymmetric Subgenome Evolution and Improvement of Peanut." *Advanced Science* 7 (2020): 1901672.

Yoo, M.J., E. Szadkowski, and J.F. Wendel. "Homoeolog Expression Bias and Expression Level Dominance in Allopolyploid Cotton." *Heredity* 110, no. 2 (2013): 171–180.

Zhang, X., M.K. Pandey, J. Wang, et al. "Chromatin Spatial Organization of Wild Type and Mutant Peanuts Reveals High-Resolution Genomic Architecture and Interaction Alterations." *Genome Biology* 22 (2021): 315.

Zhang, H., Yu Yu, S. Wang, J. Yang, X. Ai, N. Zhang, X. Zhao, X. Liu, C. Zhong, and H. Yu. "Phospholipase D Family Genes in Allotetraploid Peanut and Its." In *Peanut Genomics and Biotechnology in Breeding Applications* 14 (2023): 40.

Zhang, X., K. Zhang, L. Luo, Y. Lv, Y. Li, S. Zhu, B. Luo, Y. Wan, X. Zhang, and F. Liu. "Identification of Peanut Aux/IAA Genes and Functional Prediction during Seed Development and Maturation." *Plants* 11, no. 4 (2022): 472. https://doi.org/10.3390/plants11040472.

Zhao, Y., J. Ma, M. Li, et al. "Whole-Genome Resequencing-Based QTL-seq Identified AhTc1 Gene Encoding a R2R3-MYB Transcription Factor Controlling Peanut Purple Testa Colour." *Plant Biotechnology Journal* 18, no. 1 (2020): 96–105.

Zhuang, W., H. Chen, M. Yang, et al. "The Genome of Cultivated Peanut Provides Insight into Legume Karyotypes, Polyploid Evolution and Crop Domestication." *Nature Genetics* 51 (2019): 865–876.

5 Advances in Peanut Genomics, Evolution, and Domestication

Xiaoping Chen, Lu Huang, Qing Lu, Muhammad Jawad Umer, Sheidu Abdullaziz, Xiyin Wang, and Weijian Zhuang

5.1 INTRODUCTION

Peanut or groundnut (*Arachis hypogea* L.), a pivotal legume crop, originates from South America and has been a vital agricultural staple for approximately 6,000 years. This crop not only enriches the soil through nitrogen fixation (Nzepang et al., 2023) but also provides significant economic and nutritional value due to its high oil and protein contents (Mondal et al., 2020). Today, peanuts are cultivated extensively across the globe, with major production in China and India (Torres et al., 2014), reflecting its adaptation to diverse climates and its role in feeding a growing global population.

The history of plant genomics dates to the early 2000s when the first genome sequencing efforts began on *Arabidopsis thaliana* (*Arabidopsis* Genome Initiative, 2000). Since 2016, researchers have published the genomes of the cultivated peanut's probable wild progenitors (*A. duranensis* and *A. ipaënsis*) and its wild tetraploid counterpart (*A. monticola*) (Bertioli et al., 2016; Chen et al., 2016; Lu et al., 2018; Yin et al., 2018). In 2019, they successfully deciphered the complete genomes of three cultivated peanut genotypes, including *A. hypogea* cv. Fuhuasheng (Chen et al., 2019), *A. hypogea* cv. Shitouqi (Zhuang et al., 2019), and *A. hypogea* cv. Tifrunner (Bertioli et al., 2019). The exploration of peanut genomics has accelerated significantly due to the introduction of next-generation sequencing (NGS) technologies, which have dramatically influenced agricultural research and breeding strategies. Unlike traditional breeding methods, which can be lengthy and unpredictable, genomic tools provide a means for precise and efficient genetic enhancements. This is especially important for crops like peanuts, where the complexity of the allotetraploid genome poses unique challenges for sequence assembly and genetic improvement (Chen et al., 2019).

This introduction to peanut genomics charts the course from early domestication events, marked by significant genetic and morphological changes, to contemporary genomic studies that unravel the complex genetic architecture of this crop. By focusing on the genomic milestones and the technological strides in sequencing and genetic mapping, this section sets the stage for discussing the broader implications of genomic insights in peanut evolution, domestication, and breeding advancements.

In this chapter, we delve into the intricate relationship between peanut's genetic makeup and its agricultural traits, highlighting how genomic studies have not only illuminated its evolutionary process but also paved the way for future enhancements. The integration of genomic data has catalyzed a transformative shift in peanut breeding, moving towards more sustainable and productive agricultural practices.

5.2 PROGRESS IN PEANUT GENOMICS

5.2.1 Genome Features of Cultivated Peanut and Its Wild Relatives

The genus *Arachis*, belonging to Leguminosae, has evolved into species which were arranged in nine taxonomic sections based on morphological characteristics, cross incompatibility, and karyotype analyses (Krapovickas & Gregory, 1994). Cross-compatibility, meiotic, and karyotypic analyses revealed that there are six different genomes in the genus *Arachis*, designated as genome A, B, D, F, G, and K respectively (Robledo & Seijo, 2008; Samoluk & Chalup, 2015; Silvestri et al., 2015; Stalker, 1991), of which genomes A and B are pervasive (Jung et al., 2003; Robledo et al., 2009; Robledo & Seijo, 2010). The number of haploid chromosomes in most diploid wild species is 10 ($2n = 2x = 20$); however, there are four diploid wild species with a haploid chromosome number of 9 ($2n = 2x = 18$), that is, *A. palustris*, *A. praecox*, and *A. decora* in the section *Arachis* as well as *A. porphyrocalyx* in the section *Erectoidesn* (Silvestri et al., 2015). Over centuries, selective breeding by indigenous peoples enhanced desirable traits such as larger seed size, easier shell removal, and high yield. The process of domestication likely involved the hybridization of two wild species, *Arachis duranensis* and *Arachis ipaënsis*, resulting in the tetraploid *A. hypogaea*. According to cytological and PCR-based molecular marker analyses, the cultivated peanut (*A. hypogea*) is an allotetraploid ($2n = 2x = 40$), formed from two diploid wild species ($2n = 2x = 20$) after a natural cross-pollination followed by chromosome doubling (Jung et al., 2003; Smartt et al., 1978). The diploid wild species got a 2C value of between 2.55 and 3.22 pg, with an estimated genome size of 1.4 Gb (Samoluk & Chalup, 2015). The 2C value of the tetraploid cultivated peanut is 5.7 pg with an estimated genome size of 2.8 Gb (Samoluk & Chalup, 2015; Temsch & Greilhuber, 2000). The cultivated peanut has a large and complicated genome

DOI: 10.1201/9781003360988-5

with a high level of repeat sequence (>50%), posing substantial challenge in *de novo* whole-genome sequencing and assembly (Kyriakidou et al., 2018).

With the development of sequencing technology, *de novo* whole-genome assembly of peanut became possible in the past decade. Researchers first attempted to decipher the peanut genome by focusing on the putative wild diploid ancestors of cultivated peanut, *A. duranensis* (A genome) and *A. ipaënsis* (B genome) (Bertioli et al., 2016; Chen et al., 2016; Lu et al., 2018). The assembled genome sequences of *A. duranensis* V14167 and *A. ipaënsis* K30076 were generated by the International Peanut Genome Initiative (IPGI), while whole-genome sequencing and assembly of *A. duranensis* PI475845 and *A. ipaënsis* ICG8206 was performed by Chen et al. (2016) and Lu et al. (2018), respectively. So far, several peanut research groups have sequenced eight accessions from four diploid wild species containing A or B genomes (Table 5.1). The assembled genome size of A-genome diploid species was in the range of 1.07 gigabase (Gb) (*A. duranensis* V14167 v2) to 1.38 Gb (*A. duranensis* PI475845), resulting in a total of 33,810 predicted protein-coding genes in the former species, but 50,324 in the latter one. The GC content in the genome assemblies of A-genome species ranged from 31.79% (*A. duranensis* PI475845) to 37.55% (*A. stenosperma* V10309).

Compared to A-genome assemblies, B-genome diploid species demonstrate fewer variations in genome size (ranging from 1.35 Gb to 1.51 Gb), predicted gene count (between 38,783 and 41,840), and GC content (approximately 36%). B-genome assemblies typically exhibit larger average sizes than their A-genome counterparts. The genomes from four diploid peanut species are not only essential for decoding tetraploid peanut genomes but also represent a largely untapped source of genetic diversity, offering significant potential for future peanut molecular breeding.

In recent years, innovative sequencing technologies have emerged, such as single-molecule real-time (SMRT) sequencing, high-throughput chromosome conformation capture (Hi-C), and BioNano optical genome mapping. These technologies have significantly transformed the genomics field by offering powerful tools for investigating polyploid genomes (Kyriakidou et al., 2018). Benefiting from these sequencing technologies, researchers have sequenced the genomes of the allotetraploid cultivated peanut *A. hypogea* and its wild counterpart *A. monticola*. The wild tetraploid *A. monticola* PI263393 ($2n = 2x = 40$) was the first characterized polyploid peanut species. The final assembled genome size of *A. monticola* is 2.62 Gb, including 1.06 Gb and 1.45 Gb for the A subgenome and the B subgenome, respectively. Yin et al. (2018) predicted 74,907 protein-encoding gene models in the *A. monticola* genome. Transposable elements contributed 73.2% of the *A. monticola* genome.

Within the past 5 years, the genomes of cultivated peanut from five genotypes including cvs. Fuhuasheng (Chen et al., 2019), Shitouqi (Zhuang et al., 2019), Tifrunner (Bertioli et al., 2019), Haihua1, and Tennessee Red have been completed (Table 5.1). *A. hypogea* cvs. Fuhuasheng and Shitouqi have been widely used as breeding parents in Chinese peanut breeding programs during past half century, while cv. Tifrunner is a classic runner-type peanut adapted for the southeast of the United States. It is notable that the sizes of four assembled genomes are almost the same (approximately 2.55 Gb), with the exception of the genome of cv. Haihua1 (2.49 Gb). The number of protein-encoding gene models was approximately similar between the two varieties: 83,087 for Fuhuasheng and 83,709 for Shitouqi.

There was a significant reduction in the number of predicted gene models in the Tifrunner genome, totaling 66,469 genes. Similarly, repeat sequences constituted a substantial portion of the genomes, with 77.65% in Shitouqi and 74.03% in Tifrunner, while Fuhuasheng showed a lower repeat content at 55.34%. Despite these differences in repeat content, long-terminal repeats (LTR) were the predominant type of repetitive DNA in all three genomes. Additionally, the distribution of predicted genes was uneven across the A and B subgenomes, with the B subgenome containing more genes than the A subgenome.

Researchers observed a high degree of conformity not only between the subgenomes of cultivated peanut and its progenitors but also between the homologous chromosomes. Zhuang et al. (2019) observed an average ~97% and >99.5% identity between the subgenomes of *A. hypogea* cv. Shitouqi and its corresponding progenitors, respectively (Zhuang et al., 2019). Synteny analysis revealed many syntenic blocks between the subgenomes of *A. hypogea* cv. Fuhuasheng and its progenitors without large rearrangements. Furthermore, researchers intensively observed syntenic disruptions in chromosomes 07 and 08, suggesting the existence of complex rearrangements after tetraploidization (Chen et al., 2019). In *A. hypogea* cv. Tifrunner, most of the homologous chromosomes were in one-to-one correspondence, with varying amounts of inversions and reciprocal translocations (Bertioli et al., 2019). Despite a high degree of similarity, researchers have observed asymmetric evolution in the subgenomes of tetraploid peanuts following polyploidization.

Although a few peanut genome sequences are available, our understanding of the three-dimensional chromatin architecture in peanuts remains limited, particularly in terms of how it varies across cultivars and lines, in different tissues, and throughout various developmental stages. Zhang et al. (2021) made the initial effort to decipher the 3D chromatin architecture of peanut. By integrating high-throughput chromosome conformation capture (Hi-C) technology with assay for transposase-accessible chromatin sequencing (ATAC-seq), Zhang et al. compared the genome organization between two cultivated tetraploid peanut lines, wild-type H2014 and its dwarf mutant H1314 (Zhang et al., 2021). Researchers observed widespread A/B compartment switching and changes in topologically associated domains (TAD) in both peanut lines. The A compartment, characterized by positive eigenvectors, comprises 52.3% of the genomic regions and exhibits higher levels of gene density

TABLE 5.1
Basic Statistics on the Genome of Wild Relatives of Cultivated Peanut

Species	Genome	Genome Size (Gb)	Number of Coding Genes	Contig N50 (bp)	Scaffold N50 (bp)	Unplaced Scaffolds	GC Content (%)	Reference
A. duranensis PI475845	AA	1.38	50,324	29,584	649,840	NA	31.79	(Chen et al., 2016)
A. duranensis V14167	AA	1.21	36,734	24,685	1,068,268	1497	34.00	(Bertioli et al., 2016)
A. duranensis V14167 (v2)	AA	1.07	33,810	24,535	109,374,422	1,077	35.81	(Garg et al., 2022)
A. duranensis K30065	AA	1.11	NA	1,014,563	115,523,635	106	36.12	www.ncbi.nlm.nih.gov/assembly/GCA_014805325.1 (2020)
A. duranensis K30060	AA	1.10	NA	29,271,072	29,271,072	112	35.92	www.ncbi.nlm.nih.gov/assembly/GCA_018207795.1 (2021)
A. stenosperma V10309	AA	1.33	NA	1,268,048	141,817,970	20	37.55	www.ncbi.nlm.nih.gov/assembly/GCA_014773155.1 (2020)
A. cardenasii K10017	AA	1.24	NA	280,607	63,554,988	726	37.25	www.ncbi.nlm.nih.gov/assembly/GCA_018493915.1 (2021)
A. ipaënsis K30076	BB	1.51	41,840	24,488	6,169,993	538	35.49	(Bertioli et al., 2016)
A. ipaënsis ICG 8206	BB	1.39	39,704	8,067	170,050	NA	36.70	(Lu et al., 2018)
A. ipaënsis K30076 (v2)	BB	1.35	38,783	24,433	136,726,255	485	36.84	(Garg et al., 2022)
A. monticola PI 263393	AABB	2.62	74,907	106,661	124,915,013	6889	35.99	(Yin et al., 2018)
A. hypogaea cv. Fuhuasheng	AABB	2.55	83,087	213,557	137,243,429	9	36.33	(Chen et al., 2019)
A. hypogaea cv. Shitouqi	AABB	2.54	83,709	1,509,423	135,108,068	0	36.53	(Zhuang et al., 2019)
A. hypogaea cv. Tifrunner	AABB	2.55	66,469	1,498,096	135,150,084	364	36.21	(Bertioli et al., 2019)
A. hypogaea cv. Haihua1	AABB	2.49	NA	9,813,388	131,078,573	1,549	36.29	www.ncbi.nlm.nih.gov/assembly/GCA_016103905.1 (2020)
A. hypogaea cv. Tennessee Red	AABB	2.54	NA	2,585,762	132,678,025	1,549	36.41	www.ncbi.nlm.nih.gov/assembly/GCA_022829005.1 (2022)

and expression. They detected A/B compartment switching in approximately 2% of the genomic regions between the two peanut lines. Zhang et al. (2021) discovered that the peanut chromatin architecture and its effects on gene expression and underlying loci might be responsible for the dwarf phenotype in peanuts. Additionally, by comparing the chromatin architecture with the reference genome of *A. hypogaea* cv. Tifrunner, they identified several scaffolding errors, highlighting the need for further refinement of the shotgun sequencing-based genome sequences of cultivated allotetraploid peanuts. As high-throughput and low-cost sequencing technologies continue to advance, the sequencing of thousands of peanut germplasm resources and further exploration of their genome structure and chromatin architecture will likely become more accessible and feasible in future studies.

Peanut reference genomes have been fundamental for the advancement of many aspects in both basic and applied research fields in peanut. However, the access to a perfect reference genome of peanut is still some way in the future. Different lengths and numbers of gaps or unplaced scaffolds exist in current available assembled genomes (Bertioli et al., 2016, 2019; Chen et al., 2016; Garg et al., 2022; Chen et al., 2019; Lu et al., 2018; Yin et al., 2018, 2020; Zhuang et al., 2019). Moreover, extensive variations can be observed in the sequenced genomes in many aspects such as genome size and number of coding genes, as well as repeat content, for example, the repeat content of different assembled genomes of *A. hypogea* ranging from 54.34% to 74.03% (Bertioli et al., 2019; Chen et al., 2019; Zhuang et al., 2019). Such variations may be the result of either within-species variation in genome sequences, or the inherent defects in short-read NGS platforms (Su et al., 2011). When *de novo* assembling a genome using short reads, complex regions (e.g., tandem repeats, GC-rich regions, highly polymorphism regions, and large structure variations) can lead to erroneous rearrangements, deletions, collapsed repeats and other assembly errors (Treangen & Salzberg, 2012; Zavodna et al., 2014). In an effort to overcome the difficulties in short-read assembly, several sequencing innovations, such as Pacbio HiFi and Nanopore Ultra-long reads, have emerged (Jain et al., 2018; Wenger et al., 2019). They have been proved effective in the construction of telomere-to-telomere (T2T) chromosome assemblies in human, rice, barley, and watermelon (Cechova & Miga, 2023; Deng et al., 2022; Navrátilová et al., 2022; Nurk et al., 2022; Zhang et al., 2022). With the advancement of sequencing technology, peanut genomes with unprecedented accuracy and more comprehensive annotation may be expected in the near future, which will provide an impetus that drives studies on structural, functional, and applied genomics in peanut.

5.2.2 Genomic Resources

Over the past few years, numerous genome sequencing projects targeting various peanut species, combined with extensive sequencing efforts, have produced a significant amount of sequence data and genomic resources. These resources are instrumental in advancing peanut genomics and enhancing genetic improvement efforts. PeanutBase (https://peanutbase.org/), the primary repository for peanut genomic data for IPGI, is one of the best-known peanut web repository databases by far. It provides not only genome browsers for the genome sequence and annotation of *A. hypogea* cv. Tifrunner and its putative progenitors, but also convenient access to the gene expression data, germplasm information, genetic maps, QTLs, and a series of tools for browsing, searching, and interrelating these datasets (Bertioli et al., 2016, 2019). Another important database, Peanut Genome Resource (https://peanutgr.fafu.edu.cn), developed by Zhuang et al., holds the genome sequence of cultivated peanut *A. hypogea* cv. Shitouqi and related annotations (Zhuang et al., 2019). It has also collected peanut resources such as genome resequencing data from wild species and cultivated peanut, germplasm and variety information, gene expression data, genetic maps, and mapped traits collected from literature. Besides the above-mentioned peanut-specific genomic databases, all the genome sequences of peanut (including those from unpublished research) are available from NCBI (http://www.ncbi.nlm.nih.gov/). These publicly available genomic resources, together with the bioinformatics tools in web repository databases, would provide valuable information for analysis of the various features of peanut genomes.

5.3 CHROMOSOMAL EVOLUTION IN *ARACHIS*

5.3.1 Peanut Chromosome Origin and Karyotype Evolution

The cultivated peanut (*A. hypogea* L.) belongs to the genus *Arachis* that was established by Linnaeus in 1735 and placed in tribes *Hedysareae* and *Aeschynomenea* by Bentham (1865) and Schulze-Menz (1964), respectively. Both of them also placed this genus in the subtribe *Stylosanthinae*. Evolutionary analyses have revealed that the *Arachis* genus has undergone at least three polyploidization events. These include three significant polyploidization events. The first is the core eudicot paleo-hexaploidy event, which occurred approximately 130 million years ago and is shared with grape and *Theobroma cacao* (Tang et al., 2008). The second event is a more recent pan-legume duplication that took place around 55 million years ago, involving other legume species (Young et al., 2011). Finally, the third event is a duplication specific to closely related *Arachis* species (Bertioli et al., 2016; Chen et al., 2016, 2019; Zhuang et al., 2019). Polyploidization with genome doubling brings heterosis and genome redundancy, which promotes the adaptation to ecological niches with enhanced phenotypic plasticity (Leal-Bertioli et al., 2021; Pan et al., 2023; Schiavinato et al., 2021). Pan et al. (2023) discovered traces of former polyploidization events in peanut genomes, which have evolved through point mutation, gene conversion, gene loss, mobile-element activity, and chromosomal rearrangement. These findings provide valuable insights for evaluating the evolutionary history of peanuts.

The availability of genome sequences from diploid and tetraploid *Arachis* species serves as a valuable resource for understanding *Arachis* chromosome evolution. Comparative genomic analyses of peanuts with other eudicot and legume species have revealed traces of legume-common tetraploidy (~59 million years ago) (W7), and core-eudicot-common hexaploidy (~130 million years ago), which had remained in the peanut genome. Post-hexaploidy genome structures have been well preserved in *Vitis vinifera* (Jaillon et al., 2007; Schnable et al., 2011) and post-tetraploidy structures in *Phaseolus vulgaris*. A significant number of colinear gene pairs, that is, a total of 2,067 gene pairs with 1,289 from LCT (legume common tetraploidy) and 1,198 from ECH (some derived from both), within the A subgenome. In the B subgenome, a total of 2,283 collinear gene pairs, 1,508 from LCT and 1,372 from ECH, remained in tetraploid peanut. These existing collinear gene pairs made up 6.6% and 6.1% of the gene content in the two subgenomes, respectively (Zhuang et al., 2019). Therefore, researchers identified the origin and evolution from both tetraploidy events by analyzing collinearity with other legumes to infer the preservation of ancestral arrangements (Wang et al., 2017).

The peanut karyotype evolved independently of other legume lineages. To aid inference, researchers adapted a "top-down" grape–legume comparison strategy, identifying 5 independent chromosome fusions that produced 16 basic legume ancestral chromosomes before the LCT. Using a bottom-up approach, they revealed 5 common bean chromosomes and 11 chromosomal blocks that are largely preserved across different legumes, identifying 16 post-LCT ancestral chromosomes. Comparison of diploid peanut genomes with the 16 post-LCT chromosomes, reconstructed using common bean genes, revealed that peanut ancestral chromosomes A1, A3, A4, A5, A6, and A7 consist of chromosomal segments derived from Lu chromosomes through six fusions, resulting in a reduction in chromosome number (Figure 5.1). Two crossovers of Lu chromosomes formed the ancestral peanut chromosomes A2, A8, A9, and A10 (Figure 5.1). Post separating from the ancestral A genome, there was crossing-over in the peanut B genome, which produced its specific chromosomes 7 and 8. These results further clarified the origin and evolution of peanut genomes (Zhuang et al., 2019).

5.3.2 Peanut Chromosome Recombination Post Polyploidy

Polyploidy resulted in rapid, dynamic alterations to the genomic structure, which played a significant role in plant evolution and speciation (Cai et al., 2021; Wang et al., 2021; Zhao et al., 2017). Notably, in peanut chromosomes, pairs 05, 06, and 09 had substantial inversions on one arm, whereas chromosome pair 01 features extensive inversions on both arms. Furthermore, intricate rearrangements between chromosomes 07 and 08 had resulted in the transfer of repeat-rich DNA to A07 and gene-rich DNA to A08 (Bertioli et al., 2016). In *Arachis hypogea* (Tifrunner), four pairs of homologous chromosomes (Arahy.02/12, 03/13, 04/14, and 10/20) displayed collinearity. There were multiple large inversions across these chromosomes: one each on 06/16 and 09/19, two on 05/15, and three on 01/11. Additionally, chromosomes 17/18 showed reciprocal translocations compared to 07/08. Comparison with two sequenced ancestral diploid species revealed three additional inversions in the *A. hypogea* Tifrunner genome: two in the At subgenome (on Arahy.05 and Arahy.07) and one in the Bt subgenome (on Arahy.11) (Bertioli et al., 2019). Genetic interactions between the At and Bt subgenomes included six exchanges or substitutions, particularly evident between chromosomes 03 and 10. Eleven genomic regions showed inversions between the At and Bt subgenomes and between the *A. hypogea* and the diploid A genome, while only four smaller inversions were seen between the Bt subgenome and the diploid B genome (Zhuang et al., 2019).

Asymmetric evolution was notably distinct among the diploid genomes and the tetraploid subgenomes. The Bt subgenome more closely resembled *A. ipaënsis* compared to the similarity of At subgenome to *A. duranensis*. The diploid genome B and the subgenome Bt shared more colinear genes than the diploid A genome and At subgenome (Bertioli et al., 2019; Chen et al., 2019; Zhuang et al., 2019). The At subgenome, measuring 1,159 Mb, was larger than the A-progenitor genome (1,068 Mb), whereas the Bt subgenome, at 1,349 Mb, matched the size of the B-progenitor genome (Chen et al., 2019).

The gene sets of At might have evolved faster than those of Bt, indicated by their distinct K_a and K_s rates. The Bt subgenome contained 11% more genes, numbering 35,110, compared to 31,359 in the At subgenome (Bertioli et al., 2019). In *A. hypogea*, the number of orthologous gene groups identified were 19,961 in the At subgenome and 20,206 in the Bt subgenome (Chen et al., 2019). The At subgenome showed significantly more positively selected genes (395) compared to Bt (299), with gene conversion occurring much more frequently in At. In At, 3,747 genes possess at least two conversion sites, compared to 640 in Bt. Additionally, At genes convert to Bt alleles at a rate more than five times higher than the reverse (Chen et al., 2019). Results also showed conversions from the Bt to At were far more common through comparison of a synthetic tetraploid peanut line and its parents (Chen et al., 2016).

Additionally, six well-defined exchanges or substitutions occurred between the subgenomes At and Bt. Eleven big regions of chromosomal incongruence were detected between subgenomes, alongside significant crossover specific to the B-genome occurring in the Bt genome lineage, impacting chromosomes 7 and 8 (Bertioli et al., 2019; Zhuang et al., 2019).

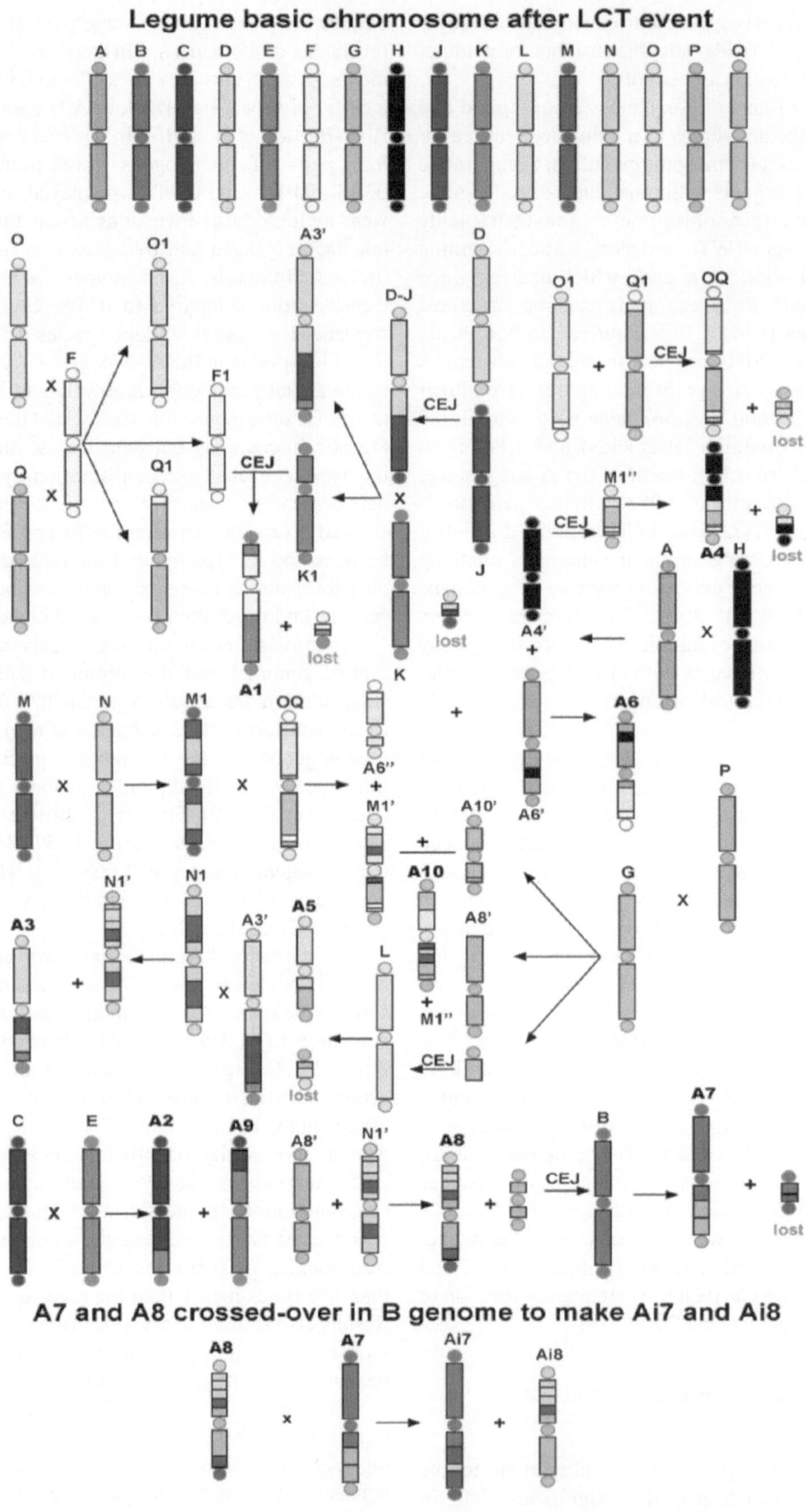

FIGURE 5.1 Schematic overview of chromosomal evolution in peanut (*Arachis hypogaea* L.) and highlights for major chromosomal rearrangements, including inversions, reciprocal translocations, end-to-end joining, and exchanges between the At and Bt subgenomes. These structural variations illustrate the dynamic changes associated with polyploidization and subsequent evolutionary diversification. CEJ: chromosome end-end joining; X: crossing-over.

Structural variations (SV), including gene deletions, insertions, single nucleotide polymorphisms, and copy number variations, contributed significantly to genetic diversity and trait regulation within and across species (Long et al., 2021; Yuan et al., 2021). A comparison between the wild and cultivated tetraploid genomes revealed that *A. hypogea* had a higher enrichment of deletions and insertions in upstream regions compared to *A. monticola*. In the wild tetraploid, the At subgenome showed higher frequencies of SNPs, deletions, and insertions, as well as intra-subgenomic insertions, than the Bt subgenome. The median K_a/K_s values for SV genes in the At subgenome were significantly lower than those in the Bt subgenome in *A. monticola*, contrasting with *A. hypogea*, where the Bt subgenome exhibited higher SV (Yin et al., 2020).

5.4 SPECIATION AND EVOLUTION OF *A. HYPOGEA*: A FRESH PERSPECTIVE FROM GENOME

5.4.1 Genomic Insights into the Speciation of Cultivated Peanut

The details regarding the genetic origin of cultivated peanut have yet to be definitively determined. It is commonly acknowledged that cultivated peanut is the result of hybridization between two diploid wild species (Bertioli et al., 2016, 2019; Chen et al., 2019; Moretzsohn et al., 2013; Zhuang et al., 2019). Archeological evidence for probable progenitor species of *A. hypogea* has been reported (Hammons et al., 2016; Simpson et al., 2001). Currently, there are several major hypotheses regarding the speciation and evolutionary history of peanut. Prior to the availability of whole genome sequencing, several studies have utilized molecular markers such as AFLP, SSR, RAPD, ISSR, and the highly variable plastid region (for example, *trn*T-F) to examine the evolutionary dynamics of tetraploid peanut. These researches have proposed the involvement of B genome species, *A. ipaënsis*, as well as several A genome species such as *A. helodes*, *A. simpsonii*, *A. villosa*, *A. cardenasii*, and *A. duranensis*, in the formation and evolution of tetraploid peanut (Jung et al., 2003; Koppolu et al., 2010; Milla et al., 2005; Raina et al., 2001; Tallury et al., 2005). Investigations of these markers have revealed genetic relationships, genetic diversity, and polyploidization events within the *Arachis* genus, providing valuable insights into the evolutionary history and genomic contributions of different species to tetraploid peanuts. However, another assumption regarding the origin of *A. hypogaea* is explained by Stalke (2017), who suggests that *A. hypogaea* evolved about four thousand years ago in the region from southern Bolivia to northern Argentina, resulting from a cross between *A. duranensis* Krapov. and W. C. Gregory and *A. ipaënsis* Krapov. and W. C. Gregory, considered the diploid progenitors of cultivated peanut (Kochert et al., 1996). Cultivated peanut exhibits relatively less genetic variation compared to diploid species.

Recent advancements in whole genome sequencing have provided an opportunity to explore the speciation and evolution of tetraploid peanut in detail and with higher resolution. Through analysis of whole-chromosome-scale DNA divergence rates and the uniformity of patterns of homologous recombination at the ends of chromosomes, strong evidence suggests that tetraploid peanut originated from a single hybridization event that occurred relatively recently (<10,000 years ago) in South America. This hybridization involved the wild diploid species *A. duranensis* as the A genome donor and *A. ipaënsis* as the B genome donor (Bertioli et al., 2019). This conclusion is further supported by experimental hybridization studies, which reinforce the notion of recent genomic evolution (Gregory & Gregory, 1979). Furthermore, the findings suggest that human-mediated transportation likely contributed to the formation of tetraploid peanut (Bertioli et al., 2020; Bertioli et al., 2016, 2019).

However, formation of allotetraploid peanut appears to have been more complex than a single hybridization. In addition to the single origin hypothesis, some studies propose alternative scenarios that involve recurrent hybridization processes and subsequent genomic rearrangements between *A. duranensis* and *A. ipaënsis*, which lead to the formation of tetraploid peanut. For instance, comparative karyotyping analysis revealed that multiple hybridization events of *A. ipaënsis* with a population or different varieties of *A. duranensis* might have occurred during the origination of peanut (Zhang et al., 2016). Evolution analysis suggested that the B subgenome of *A. hypogea* more consistently resembled the ancestral condition, but the A subgenome underwent more genomic changes (Chen et al., 2019). Moreover, Zhuang et al., through analyses of divergence at synonymous codon sites in paired genes from corresponding chromosomes (Ks analyses) in *A. hypogea* cv. Shitouqi and two diploid peanuts, proposed that the tetraploid peanut originated approximately 450,000 years ago. This finding contrasts with the previously held belief that the tetraploid peanut originated less than 10,000 years ago (<10,000 years ago) (Bertioli et al., 2020; Zhuang et al., 2019, 2020). They suggest that the result of a recent origin might have arisen due to inadequate utilization of variation data and potential biases in alignment algorithms (Zhuang et al., 2020). These findings highlight the complexity of peanut speciation, necessitating further research and empirical evidence to establish the accurate evolutionary model. According to Yin et al., it is proposed that hybridization between diploid wild species occurred approximately ~11,690 years ago, resulting in the polyploidization event that gave rise to the tetraploid wild species *A. monticola* and domesticated into the cultivated peanut in around ~4500 years ago (Yin et al., 2020). Another K_s estimation by Chen et al. revealed incongruent divergence times between each subgenome of *A. hypogea* cv. Fuhuasheng and corresponding genome donors (0.25 Mya for the At-*A. duranensis* divergence and

0.18 Mya for the Bt-*A. ipaënsis* divergence), suggesting the existence of occasional outcrossing events in the speciation of extant cultivated peanut (Chen et al., 2019).

5.4.2 Asymmetric Evolution in the Subgenomes of *A. Hypogea*

Asymmetric evolution occurred in the two subgenomes of *A. hypogea*. Compared to the B subgenome, the A subgenome of cultivated peanut exhibits less similarity to its ancestral counterparts in terms of sequence divergence, number of homologous genes, and structural variations. Both *A. hypogea* and *A. monticola*, the other tetraploid in the genus *Arachis*, display a higher occurrence of homologous sequence exchanges from the A to B subgenomes, as opposed to the B to A subgenomes (Chen et al., 2019; Yin et al., 2020). The genome of *A. hypogea* cv. Shitouqi shares 34,266 and 38,417 genes with the corresponding diploid genomes in the A and B subgenomes, respectively, suggesting a better-preserved conformity in the B subgenome. Among the 23 independent inversions with a size of more than 3 Mb, 91.3% (21/23) were identified in the A lineage (Zhuang et al., 2019). In *A. hypogea* cv. Fuhuasheng, a higher number of syntenic blocks was observed between the B subgenome and *A. ipaënsis* than between the A subgenome and *A. duranensis*. However, the A subgenome displayed more disrupted genes (frameshift or premature stop codons) compared to the B subgenomes (Chen et al., 2019). In addition to the imbalanced structural variation and gene content, asymmetry is also evident in the pattern of evolution and unequal expression of homologous genes in the subgenomes (Yin et al., 2020). The B subgenome of allotetraploid peanut is exceptionally well conserved by structure and sequence compared with the A subgenome. This asymmetric evolution has played a significant role in the functional divergence of subgenomes during domestication and has raised questions regarding the true donor of the A subgenome of cultivated peanut (Chen et al., 2019; Tian et al., 2021; Wang et al., 2019; Zhuang et al., 2019). Several studies showed the genetic erosion of the ancestral A genome of cultivated peanut following polyploidy by invasion of B genome alleles into the A subgenome (Bertioli et al., 2019, 2021; Lamon et al., 2020).

In the genomic evolutionary process, gene loss and conversion are important genomic changes that can facilitate genome evolution by deletions which alter a species' genetic information (Lawrence & Roth, 2014). In allotetraploid peanut, only few genes have been lost, with no significant difference between A and B subgenomes (Chen et al., 2019). However, frameshifts or premature stop codons disrupted more than one-third of *A. hypogea* genes compared to their orthologous progenitor genes (Chen et al., 2019). The higher number of disrupted genes than lost genes implied the recent origin of allotetraploid peanut (Schnable et al., 2011). It has been discovered that large-scale transfers of chromosomal segments in *A. hypogea*, cause significant gene translocation or loss as well as, multiple consecutive genes were lost on a few chromosomal segments. Extensive gene conversion between A and B subgenomes, a possible contributor to the transgressive properties of polyploids relative to their progenitors, has occurred not only in *A. hypogea* but also in synthetic tetraploids. Some large chromosomal segment exchanges between subgenomes followed gene conversion, which generated a significant number of SNPs that are characteristics of the corresponding subgenome (Bertioli et al., 2019). Researchers mainly found the conversion from the A genome to the B genome in *A. hypogea*, while the conversion from B to A occurred primarily in synthetic tetraploids (Chen et al., 2016, 2019). The contrary results suggest that DNA sequence changes in the allotetraploid progeny of artificial crosses are completely different from those in natural allotetraploid, suggesting the difficulty of mimicking the speciation processes of natural evolution.

At the chromosome level, most of the homologous chromosomes of *A. hypogea* show a high collinearity with few homologous chromosomes differentiated by large inversions and reciprocal translocations (Bertioli et al., 2019). Each subgenome of *A. hypogea* transmits to progeny via conventional disomic inheritance, rendering the allotetraploid-cultivated peanut genetically diploid. However, chromosome associations during meiosis and genetic changes have indeed occurred in homologous chromosomes between A and B genomes even at a very low frequency. A few segments of the A genome, most of which are at the very distal ends of chromosomes, has been transferred into B chromosomes, and vice versa (Bertioli et al., 2019).

5.5 TRACING THE PATH OF PEANUT DOMESTICATION

The genus *Arachis* is endemic to South America, where it is still naturally confined in the countries of Brazil, Bolivia, Paraguay, Uruguay, and Argentina (Gregory & Gregory, 1979). Species similarity within the genus make them practically indistinguishable without all of the plant parts, although a few keys, such as leaflet characters, branching patterns, corolla color, peg position, and fruit morphology, were used to separate species in the genus (Krapovickas & Gregory, 2007). The definition of the sections was based on the results of diallelic crosses (Gregory & Gregory, 1979). The first interspecific crosses were made between *A. hypogea* and *A. correntina* by V. A. Rigoni (Krapovickas & Rigoni, 1949). The cross-compatibility data suggest that the fragmentation caught significant genomic portions among sections, including portions of lost ancestor in some sections (Gregory & Gregory, 1979). With a combination of their morphology, karyotype, and sexual compatibility, all species in the genus *Arachis* were finally arranged into nine taxonomic sections: *Arachis*, *Trierectoides*, *Erectoides*, *Extranervosae*, *Triseminatae*, *Heteranthae*, *Caulorrhizae*, *Procumbentes*, and *Rhizomatosae* (Krapovickas & Gregory,

1994; Krapovickas & Gregory, 2007). However, it is anticipated that additional sections within the genus *Arachis* may be established in the future, given that some taxa currently have only provisional classifications. The placement of certain species remains uncertain due to insufficient chromosomal, genetic, and geographic data. Most *Arachis* species are diploid, whereas the cultivated peanut (*A. hypogaea*) and its wild relative (*A. monticola*) are allotetraploids.

All species in the genus *Arachis* produce subterranean fruit. A great deal of intraspecific variability unrelated to the geographic distribution of the accessions were found in the genus *Arachis* (Krapovickas & Gregory, 2007). Several *Arachis* species have been cultivated for their edible seeds, historically and to the present day. The diploid species that have a history of cultivation show relatively small signatures of domestication. Genomic analysis revealed that genetic changes following tetraploidization help to favor the domestication of *A. hypogea* over other diploid *Arachis* species cultivated by humans (Bertioli et al., 2019). The tetraploid cultivated peanut (*A. hypogea*) has evolved into highly domesticated forms and has become a worldwide crop due to improvement in major agronomic traits, and thus enhanced attractiveness for cultivation (Leal-Bertioli et al., 2017).

Cultivated peanut has a domestication history of at least 3,500 years (Simpson et al., 2001). The original allotetraploid that gave rise to cultivated peanut may have been attractive because of an increase in plant size, different transpiration characteristics, higher photosynthetic capacity, or other characteristics (Leal-Bertioli et al., 2017). The cultivated peanut has larger leaves, stomata, and epidermal cells than its diploid parents. In addition, the cultivated peanut produces more photosynthetic pigments. These traits have the same trend across the different hybrid combinations, suggesting that the changes are more likely due to ploidy rather than hybridity (Leal-Bertioli et al., 2017). Even though *A. hypogea* has considerable morphological variation, its suspected recent origin is thought to account for the paucity of genetic diversity among naturally occurring peanut genotypes, and the genetic vulnerability of natural tetraploids is in contrast to rich diversity among diploid *Arachis* (Chen et al., 2016).

All wild species in the genus are found only in South America, the unquestioned place of original cultivation. The diversity center of *Arachis* lies in Mato Grosso, Brazil. Evidence from early ceramics to the west suggests that peanuts were domesticated east of the Andes (Stalker et al., 2016). The earliest publication records the wide distribution of this important crop plant that had occurred before the discovery of America (Hammons et al., 2016). All species in the genus *Arachis* exhibit a unique fruiting type, characterized by a high rate of self-pollination and the production of subterranean fruits. As a result, peanut seed dispersal is typically limited to within one meter of the parent plant. Broader dispersal relies on external factors such as transportation by floodwaters, animals, or human activity. The earliest recorded introduction indicated that the peanut was probably carried to Europe by a Seville physician in the sixteenth century (Stalker et al., 2016). Several pieces of evidence indicated that peanuts were found in China 2,000 years ago, or even earlier, approximately 4,900 years ago (Simpson et al., 2001). The domesticated peanut was cultivated widely throughout the tropical and subtropical regions of South America and then was disseminated to Europe, to both coasts of Africa, to Asia, and to the Pacific Islands by New World explorers (Holbrook, 2001). To date, this species is adapted to a wide ecological range of tropical and subtropical regions and is cultivated under diverse agricultural production systems in Asia, Africa, and America (Holbrook, 2001). Farmers grow cultivated peanut in tropical and subtropical regions across 100 countries on six continents between 40°N and 40°S.

5.6 CONCLUSION AND PERSPECTIVES

Genomics yields insights into the relationships of genotype and phenotype and the applications of molecular and quantitative genetics for major agronomic traits. In the past decade, several key milestones in peanut genomics have been achieved, such as *de novo* genome sequencing of the wild tetraploid, the cultivated tetraploid, and its probable wild diploid progenitors. New methods for high-throughput genotyping and phenotyping will make population genomics possible, such as studies of quantitative variations in populations on a genomic scale (Lu et al., 2024). The *de novo* sequencing of genomes of the genus *Arachis* and whole genome resequencing of germplasms will be beneficial for a detailed understanding of peanut evolution and domestication from a new perspective. Availability of increasing genomic resources has brought low-cost genotyping to the peanut community that will facilitate the use of modern genetics and breeding approaches such as genome-wide association studies for trait mapping and genomic selection for peanut improvement. Furthermore, bioinformatics and new theoretical frameworks will drive future studies to deepen our understanding of peanut origin, evolution, domestication, and association with various agronomic traits.

REFERENCE

Arabidopsis Genome Initiative genomeanalysis@ tgr. org genomeanalysis@ gsf. de. "Analysis of the genome sequence of the flowering plant *Arabidopsis thaliana*." *Nature* 408, no. 6814 (2000): 796–815.

Bentham, George, and Joseph D. Hooker. Genera Plantarum ad exemplaria imprimis in herbariis kewensibus servata definita. Vol. 1. London: Reeve & Co., 1865. "Stylosantheae." In Genera Plantarum, vol. 1: 449. Reeve & Co., London.

Bertioli, David J., Brian Abernathy, Guillermo Seijo, Josh Clevenger, and Steven B. Cannon. "Evaluating two different models of peanut's origin." *Nature Genetics* 52, no. 6 (2020): 557–559.

Bertioli, David J., Steven B. Cannon, Lutz Froenicke, Guodong Huang, Andrew D. Farmer, Ethalinda K. S. Cannon, Xin Liu et al. "The genome sequences of *Arachis duranensis* and *Arachis ipaensis*, the diploid ancestors of cultivated peanut." *Nature Genetics* 48, no. 4 (2016): 438–446.

Bertioli, David J., Josh Clevenger, Ignacio J. Godoy, H. T. Stalker, Shona Wood, Joáo F. Santos, Carolina Ballén-Taborda et al. "Legacy genetics of Arachis cardenasii in the peanut crop shows the profound benefits of international seed exchange." *Proceedings of the National Academy of Sciences* 118, no. 38 (2021): e2104899118.

Bertioli, David J., Jerry Jenkins, Josh Clevenger, Olga Dudchenko, Dongying Gao, Guillermo Seijo, Soraya C. M. Leal-Bertioli et al. "The genome sequence of segmental allotetraploid peanut Arachis hypogaea." *Nature Genetics* 51, no. 5 (2019): 877–884.

Cai, Xu, Lichun Chang, Tingting Zhang, Haixu Chen, Lei Zhang, Runmao Lin, Jianli Liang, Jian Wu, Michael Freeling, and Xiaowu Wang. "Impacts of allopolyploidization and structural variation on intraspecific diversification in *Brassica rapa*." *Genome Biology* 22, no. 1 (2021): 166.

Cechova, Monika, and Karen H. Miga. "Comprehensive variant discovery in the era of complete human reference genomes." *Nature Methods* 20, no. 1 (2023): 17–19.

Chen, Xiaoping, Hongjie Li, Manish K. Pandey, Qingli Yang, Xiyin Wang, Vanika Garg, Haifen Li et al. "Draft genome of the peanut A-genome progenitor (*Arachis duranensis*) provides insights into geocarpy, oil biosynthesis, and allergens." *Proceedings of the National Academy of Sciences* 113, no. 24 (2016): 6785–6790.

Chen, Xiaoping, Qing Lu, Hao Liu, Jianan Zhang, Yanbin Hong, Haofa Lan, Haifen Li et al. "Sequencing of cultivated peanut, *Arachis hypogaea*, yields insights into genome evolution and oil improvement." *Molecular Plant* 12, no. 7 (2019): 920–934.

Deng, Yun, Shoucheng Liu, Yilin Zhang, Jingsheng Tan, Xiaopeng Li, Xiao Chu, Binghua Xu et al. "A telomere-to-telomere gap-free reference genome of watermelon and its mutation library provide important resources for gene discovery and breeding." *Molecular Plant* 15, no. 8 (2022): 1268–1284.

Garg, Vanika, Olga Dudchenko, Jinpeng Wang, Aamir W. Khan, Saurabh Gupta, Parwinder Kaur, Kai Han et al. "Chromosome-length genome assemblies of six legume species provide insights into genome organization, evolution, and agronomic traits for crop improvement." *Journal of Advanced Research* 42 (2022): 315–329.

Gregory, M. Pfluge, and W. C. Gregory. "Exotic germ plasm of Arachis L. interspecific hybrids." *Journal of Heredity* 70, no. 3 (1979): 185–193.

Hammons, Ray O., Danielle Herman, and H. Thomas Stalker. "Origin and early history of the peanut." In *Peanuts: Genetics, Processing, and Utilization*, edited by H. Thomas Stalker and Richard F. Wilson, 1–26. Urbana, IL: Academic PressAocs Press, 2016.

Holbrook, C. C. "Status of the Arachis germplasm collection in the United States." *Peanut Science* 28, no. 2 (2001): 84–89.

Jaillon, Olivier, Jean-Marc Aury, Benoît Noël, Attilio Policriti, Claire Clepet, Angela Casagrande, Nathalie Choisne, et al. "The grapevine genome sequence suggests ancestral hexaploidization in major angiosperm phyla." Nature 449, no. 7161 (2007): 463–467.

Jain, Miten, Sergey Koren, Karen H. Miga, Josh Quick, Arthur C. Rand, Thomas A. Sasani, John R. Tyson et al. "Nanopore sequencing and assembly of a human genome with ultra-long reads." *Nature Biotechnology* 36, no. 4 (2018): 338–345.

Jung, S., P. L. Tate, R. Horn, G. Kochert, K. Moore, and A. G. Abbott. "The phylogenetic relationship of possible progenitors of the cultivated peanut." *Journal of Heredity* 94, no. 4 (2003): 334–340.

Kochert, Gary, H. Thomas Stalker, Marcos Gimenes, Leticia Galgaro, Catalina Romero Lopes, and Kim Moore. "RFLP and cytogenetic evidence on the origin and evolution of allotetraploid domesticated peanut, *Arachis hypogaea* (Leguminosae)." *American Journal of Botany* 83, no. 10 (1996): 1282–1291.

Koppolu, Ravi, Hari D. Upadhyaya, Sangam L. Dwivedi, David A. Hoisington, and Rajeev K. Varshney. "Genetic relationships among seven sections of genus Arachis studied by using SSR markers." *BMC Plant Biology* 10 (2010): 1–12.

Krapovickas, Antonio, and Walton C. Gregory. "'Taxonomia del genero' arachis (leguminosae)." *Bonplandia* 8 1- 4 (1994): 1–186.

Krapovickas, Antonio, Walton C. Gregory, David E. Williams, and Charles E. Simpson. "Taxonomy of the genus Arachis (Leguminosae)." *Bonplandia* 16 (2007): 7–205.

Krapovickas, A., and V. A. Rigoni. "Cromosomas de una especie silvestre de Arachis." *Idia* 2, no. 24 (1949): 23–24.

Kyriakidou, Maria, Helen H. Tai, Noelle L. Anglin, David Ellis, and Martina V. Strömvik. "Current strategies of polyploid plant genome sequence assembly." *Frontiers in Plant Science* 9 (2018): 1660.

Lamon, Samuele, Ye Chu, Larissa A. Guimaraes, David J. Bertioli, Soraya C. M. Leal-Bertioli, João F. Santos, Ignácio J. Godoy, Albert K. Culbreath, C. Corley Holbrook, and Peggy Ozias-Akins. "Characterization of peanut lines with interspecific introgressions conferring late leaf spot resistance." *Crop Science* 61, no. 3 (2021): 1724–1738.

Lawrence, Jeffrey G., and John R. Roth. 2014. "Genomic flux: Genome evolution by gene loss and acquisition." In Organization of the Prokaryotic Genome, edited by R. Charlebois, 263–289. Hoboken, NJ: Wiley, 2014.

Leal-Bertioli, Soraya C. M., Márcio C. Moretzsohn, Silvio P. Santos, Ana CM Brasileiro, Patrícia M. Guimarães, David J. Bertioli, and Ana Claudia G. Araujo. "Phenotypic effects of allotetraploidization of wild Arachis and their implications for peanut domestication." *American Journal of Botany* 104, no. 3 (2017): 379–388.

Leal-Bertioli, Soraya C. M., Eliza F. M. B. Nascimento, M. Carolina F. Chavarro, Adriana R. Custódio, Mark S. Hopkins, Márcio C. Moretzsohn, David J. Bertioli, and Ana Claudia G. Araújo. "Spontaneous generation of diversity in Arachis neopolyploids (*Arachis ipaënsis*× *Arachis duranensis*) 4x replays the early stages of peanut evolution." *G3* 11, no. 11 (2021): jkab289.

Long, Yuexuan, Zhenping Liu, Pengcheng Wang, Hang Yang, Yuejin Wang, Sainan Zhang, Xianlong Zhang, and Maojun Wang. "Disruption of topologically associating domains by structural variations in tetraploid cottons." *Genomics* 113, no. 5 (2021): 3405–3414.

Lu, Qing, Lu Huang, Hao Liu, Vanika Garg, Sunil S. Gangurde, Haifen Li, Annapurna Chitikineni et al. "A genomic variation map provides insights into peanut diversity in China and associations with 28 agronomic traits." *Nature Genetics* 56, no. 3 (2024): 530–540.

Lu, Qing, Haifen Li, Yanbin Hong, Guoqiang Zhang, Shijie Wen, Xingyu Li, Guiyuan Zhou et al. "Genome sequencing and analysis of the peanut B-genome progenitor (*Arachis ipaensis*)." *Frontiers in Plant Science* 9 (2018): 604.

Milla, S. R., T. G. Isleib, and H. T. Stalker. "Taxonomic relationships among Arachis sect. Arachis species as revealed by AFLP markers." *Genome* 48, no. 1 (2005): 1–11.

Mondal, Mousumi, Milan Skalicky, Sourav Garai, Akbar Hossain, Sukamal Sarkar, Hirak Banerjee, Rajib Kundu et al. "Supplementing nitrogen in combination with rhizobium inoculation and soil mulch in peanut (*Arachis hypogaea* L.) production system: Part II. Effect on phenology, growth, yield attributes, pod quality, profitability and nitrogen use efficiency." *Agronomy* 10, no. 10 (2020): 1513.

Moretzsohn, Márcio C., Ediene G. Gouvea, Peter W. Inglis, Soraya C. M. Leal-Bertioli, José F. M. Valls, and David J. Bertioli. "A study of the relationships of cultivated peanut (*Arachis hypogaea*) and its most closely related wild species using intron sequences and microsatellite markers." *Annals of Botany* 111, no. 1 (2013): 113–126.

Navrátilová, Pavla, Helena Toegelová, Zuzana Tulpová, Yi-Tzu Kuo, Nils Stein, Jaroslav Doležel, Andreas Houben, Hana Šimková, and Martin Mascher. "Prospects of telomere-to-telomere assembly in barley: Analysis of sequence gaps in the MorexV3 reference genome." *Plant Biotechnology Journal* 20, no. 7 (2022): 1373–1386.

Nurk, Sergey, Sergey Koren, Arang Rhie, Mikko Rautiainen, Andrey V. Bzikadze, Alla Mikheenko, Mitchell R. Vollger et al. "The complete sequence of a human genome." *Science* 376, no. 6588 (2022): 44–53.

Nzepang, Darius T., Djamel Gully, Joël R. Nguepjop, Arlette Zaiya Zazou, Hodo-Abalo Tossim, Aissatou Sambou, Jean-François Rami et al. "Mapping of QTLs associated with biological nitrogen fixation traits in peanuts (*Arachis hypogaea* l.) using an interspecific population derived from the cross between the cultivated species and its wild ancestors." *Genes* 14, no. 4 (2023): 797.

Pan, Yuxin, Yuhui Zhuang, Tao Liu, Hua Chen, Lihui Wang, Rajeev K. Varshney, Weijian Zhuang, and Xiyin Wang. "Deciphering peanut complex genomes paves a way to understand its origin and domestication." *Plant Biotechnology Journal* 21, no. 11 (2023): 2173–2181.

Raina, S. N., V. Rani, T. Kojima, Y. Ogihara, K. P. Singh, and R. M. Devarumath. "RAPD and ISSR fingerprints as useful genetic markers for analysis of genetic diversity, varietal identification, and phylogenetic relationships in peanut (*Arachis hypogaea*) cultivars and wild species." *Genome* 44, no. 5 (2001): 763–772.

Robledo, Dobladez, Germán Ariel, and José Guillermo Seijo. "Characterization of the Arachis (Leguminosae) D genome using fluorescence in situ hybridization (FISH) chromosome markers and total genome DNA hybridization." *Genetics and Molecular Biology* 31 (2008): 717–724.

Robledo, Dobladez, Germán Ariel, and José Guillermo Seijo. "Species relationships among the wild B genome of Arachis species (section Arachis) based on FISH mapping of rDNA loci and heterochromatin detection: a new proposal for genome arrangement." *Theoretical and Applied Genetics* 121, no. 6 (2010): 1033–1046.

Robledo, G., G. I. Lavia, and G. Seijo. "Species relations among wild Arachis species with the A genome as revealed by FISH mapping of rDNA loci and heterochromatin detection." *Theoretical and Applied Genetics* 118, no. 7 (2009): 1295–1307.

Samoluk, Sergio Sebastián, Laura Chalup, Germán Robledo, and José Guillermo Seijo. "Genome sizes in diploid and allopolyploid Arachis L. species (section Arachis)." *Genetic Resources and Crop Evolution* 62 (2015): 747–763.

Schiavinato, Matteo, Alexandrina Bodrug-Schepers, Juliane C. Dohm, and Heinz Himmelbauer. "Subgenome evolution in allotetraploid plants." *The Plant Journal* 106, no. 3 (2021): 672–688.

Schnable, James C., Nathan M. Springer, and Michael Freeling. "Differentiation of the maize subgenomes by genome dominance and both ancient and ongoing gene loss." *Proceedings of the National Academy of Sciences* 108, no. 10 (2011): 4069–4074.

Schulze-Menz, G. K. *'Rosaceae,' in Engler's Syllabus der Pflanzenfamilien*, 12th ed., vol. II, edited by H. Melchior, 209–218. Berlin: Gebrüder Borntraeger, 1964.

Silvestri, María Celeste, Alejandra Marcela Ortiz, and Graciela Inés Lavia. "rDNA loci and heterochromatin positions support a distinct genome type for 'x= 9 species' of section Arachis (Arachis, Leguminosae)." *Plant Systematics and Evolution* 301, no. 2 (2015): 555–562.

Simpson, C. E., A. Krapovickas, and J. F. M. Valls. "History of Arachis including evidence of A. hypogaea L. progenitors." *Peanut Science* 28, no. 2 (2001): 78–80.

Smartt, J., W. C. Gregory, and M. Pfluge Gregory. "The genomes of Arachis hypogaea. 1. Cytogenetic studies of putative genome donors." *Euphytica* 27 (1978): 665–675.

Stalker, H. T. "A new species in section Arachis of peanuts with ad genome." *American Journal of Botany* 78, no. 5 (1991): 630–637.

Stalker, H. Thomas. "Utilizing wild species for peanut improvement." *Crop Science* 57, no. 3 (2017): 1102–1120.

Stalker, H. Thomas, Shyamalrau P. Tallury, Guillermo R. Seijo, and Soraya C. Leal-Bertioli. "Biology, speciation, and utilization of peanut species." In *Peanuts: Genetics, Processing and Utilization*, edited by H. Thomas Stalker and Richard F. Wilson, 27–66. Urbana, IL: Academic Press and AOCS Press, 2016.

Su, Zhenqiang, Baitang Ning, Hong Fang, Huixiao Hong, Roger Perkins, Weida Tong, and Leming Shi. "Next-generation sequencing and its applications in molecular diagnostics." *Expert Review of Molecular Diagnostics* 11, no. 3 (2011): 333–343.

Tallury, S. P., K. W. Hilu, S. R. Milla, S. A. Friend, M. Alsaghir, H. T. Stalker, and D. Quandt. "Genomic affinities in Arachis section Arachis (Fabaceae): molecular and cytogenetic evidence." *Theoretical and Applied Genetics* 111 (2005): 1229–1237.

Tang, Haibao, Xiyin Wang, John E. Bowers, Ray Ming, Maqsudul Alam, and Andrew H. Paterson. "Unraveling ancient hexaploidy through multiply-aligned angiosperm gene maps." *Genome Research* 18, no. 12 (2008): 1944–1954.

Temsch, Eva M., and Johann Greilhuber. "Genome size variation in Arachis hypogaea and A. monticola re-evaluated." *Genome* 43, no. 3 (2000): 449–451.

Tian, Xiangyu, Luye Shi, Jia Guo, Liuyang Fu, Pei Du, Bingyan Huang, Yue Wu, Xinyou Zhang, and Zhenlong Wang. "Chloroplast phylogenomic analyses reveal a maternal hybridization event leading to the formation of cultivated peanuts." *Frontiers in Plant Science* 12 (2021): 804568.

Torres, Adriana Mabel, Germán Gustavo Barros, Sofia Alejendra Palacios, Sofia Noemi Chulze, and Paola Battilani. "Review on pre-and post-harvest management of peanuts to minimize aflatoxin contamination." *Food Research International* 62 (2014): 11–19.

Treangen, Todd J., and Steven L. Salzberg. "Repetitive DNA and next-generation sequencing: computational challenges and solutions." *Nature Reviews Genetics* 13, no. 1 (2012): 36–46.

Wang, Juan, Yuan Li, Chunjuan Li, Caixia Yan, Xiaobo Zhao, Cuiling Yuan, Quanxi Sun, Chengren Shi, and Shihua Shan. "Twelve complete chloroplast genomes of wild peanuts: great genetic resources and a better understanding of Arachis phylogeny." *BMC Plant Biology* 19 (2019): 1–18.

Wang, Jinpeng, Pengchuan Sun, Yuxian Li, Yinzhe Liu, Jigao Yu, Xuelian Ma, Sangrong Sun et al. "Hierarchically aligning 10 legume genomes establishes a family-level genomics platform." *Plant Physiology* 174, no. 1 (2017): 284–300.

Wang, Longfei, Guanghong Jia, Xinyu Jiang, Shuai Cao, Z. Jeffrey Chen, and Qingxin Song. "Altered chromatin architecture and gene expression during polyploidization and domestication of soybean." *The Plant Cell* 33, no. 5 (2021): 1430–1446.

Wenger, Aaron M., Paul Peluso, William J. Rowell, Pi-Chuan Chang, Richard J. Hall, Gregory T. Concepcion, Jana Ebler et al. "Accurate circular consensus long-read sequencing improves variant detection and assembly of a human genome." *Nature Biotechnology* 37, no. 10 (2019): 1155–1162.

Yin, Dongmei, Changmian Ji, Xingli Ma, Hang Li, Wanke Zhang, Song Li, Fuyan Liu et al. "Genome of an allotetraploid wild peanut Arachis monticola: a de novo assembly." *Gigascience* 7, no. 6 (2018): giy066.

Yin, Dongmei, Changmian Ji, Qingxin Song, Wanke Zhang, Xingguo Zhang, Kunkun Zhao, Charles Y. Chen et al. "Comparison of Arachis monticola with diploid and cultivated tetraploid genomes reveals asymmetric subgenome evolution and improvement of peanut." *Advanced Science* 7, no. 4 (2020): 1901672.

Young, Nevin D., Frédéric Debellé, Giles ED Oldroyd, Rene Geurts, Steven B. Cannon, Michael K. Udvardi, Vagner A. Benedito et al. "The Medicago genome provides insight into the evolution of rhizobial symbioses." *Nature* 480, no. 7378 (2011): 520–524.

Yuan, Yuxuan, Philipp E. Bayer, Jacqueline Batley, and David Edwards. "Current status of structural variation studies in plants." *Plant Biotechnology Journal* 19, no. 11 (2021): 2153–2163.

Zavodna, Monika, Andrew Bagshaw, Rudiger Brauning, and Neil J. Gemmell. "The accuracy, feasibility and challenges of sequencing short tandem repeats using next-generation sequencing platforms." *PLoS One* 9, no. 12 (2014): e113862.

Zhang, Laining, Xiaoyu Yang, Li Tian, Lei Chen, and Weichang Yu. "Identification of peanut (*Arachis hypogaea*) chromosomes using a fluorescence in situ hybridization system reveals multiple hybridization events during tetraploid peanut formation." *New Phytologist* 211, no. 4 (2016): 1424–1439.

Zhang, Xingguo, Manish K. Pandey, Jianping Wang, Kunkun Zhao, Xingli Ma, Zhongfeng Li, Kai Zhao et al. "Chromatin spatial organization of wild type and mutant peanuts reveals high-resolution genomic architecture and interaction alterations." *Genome Biology* 22 (2021): 1–21.

Zhang, Yilin, Jun Fu, Kai Wang, Xue Han, Tianze Yan, Yanning Su, Yanfeng Li et al. "The telomere-to-telomere gap-free genome of four rice parents reveals SV and PAV patterns in hybrid rice breeding." *Plant Biotechnology Journal* 20, no. 9 (2022): 1642–1644.

Zhao, Meixia, Biao Zhang, Damon Lisch, and Jianxin Ma. "Patterns and consequences of subgenome differentiation provide insights into the nature of paleopolyploidy in plants." *The Plant Cell* 29, no. 12 (2017): 2974–2994.

Zhuang, Weijian, Hua Chen, Meng Yang, Jianping Wang, Manish K. Pandey, Chong Zhang, Wen-Chi Chang et al. "The genome of cultivated peanut provides insight into legume karyotypes, polyploid evolution and crop domestication." *Nature Genetics* 51, no. 5 (2019): 865–876.

Zhuang, Weijian, Xiyin Wang, Andrew H. Paterson, Hua Chen, Meng Yang, Chong Zhang, Pengchuan Sun et al. "Reply to: Evaluating two different models of peanut's origin." *Nature Genetics* 52, no. 6 (2020): 560–563.

6 Genebank Genomics and Genetic Diversity in Peanut

Priya Shah, Sunil S Gangurde, Baozhu Guo, Xingjun Wang, Weijian Zhuang, Rajeev K. Varshney, and Manish K. Pandey

6.1 INTRODUCTION

The peanut (*Arachis hypogaea* L.) is classified as an allotetraploid ($2n = 4x = 40$) leguminous crop. Its AABB genomic constitution originated from a single hybridization and spontaneous tetraploidization of two ancestral diploid species, *A. duranensis* (A genome) and *A. ipaensis* (B genome) (Bertioli et al., 2016; Halward et al., 1992; Moretzsohn et al., 2013). One of the main causes of the limited genetic diversity in domesticated peanuts is their brief evolutionary history and the existence of hybridization barriers between diploid and tetraploid species. The mining of large-scale polymorphic markers, such as simple-sequence repeats (SSRs), for the creation of high-density genetic maps and Quantitative traits loci (QTL) investigations has been significantly affected as the result of this low level of diversity (Pandey et al., 2016; Qin et al., 2012; Varshney et al., 2013). The construction of high-density genetic linkage maps, which are necessary for fine mapping and candidate gene discovery for traits of interest in crop species, has been facilitated by whole genome re-sequencing (WGRS) of biparental and multi-parental mapping populations. A SNP-based high-density genetic map was created using WGRS data on a population of recombinant inbred lines, and fine mapping, candidate gene discovery, and marker validation for disease resistance were all performed.

Germplasm refers to the collection of genetic resources, such as seeds, pollen, or other reproductive materials, used in crop breeding and improvement. Its diversity is fundamental to peanut improvement for several reasons. Genetic variation within germplasm represents the pool of diversity available within a species, which provides the foundation for developing new varieties with desirable traits. By utilizing diverse germplasm, breeders can tap into a wider spectrum of genetic attributes, including resistance to diseases, pests, drought, and other environmental stresses (Pandey et al., 2012). Thus, germplasm serves as a critical resource for identifying and incorporating beneficial traits into breeding programs, enabling the development of improved and resilient peanut varieties. By screening diverse germplasm collections, breeders can identify accessions or varieties that possess desirable traits, such as higher yield potential, improved nutritional quality, or enhanced oil content. These traits can then be introgressed into elite cultivars through traditional breeding techniques or biotechnological approaches. The germplasm also helps in improving disease resistance; germplasm diversity is particularly crucial for managing diseases that affect peanut crops. Cultivated peanuts are susceptible to various diseases, including fungal, viral, and bacterial infections. By incorporating genes for disease resistance from diverse germplasm sources, breeders can develop peanut varieties that are more resilient to these diseases. This approach helps in reducing yield losses, minimizing the need for chemical inputs, and enhancing the overall sustainability of peanut production (Guo et al., 2013; Pandey et al., 2012; Varshney et al., 2013).

Additionally, in terms of adaptation to environmental conditions, the germplasm collections encompass peanut varieties from different geographic regions, each adapted to specific environmental conditions. By studying and utilizing diverse germplasm, breeders can identify genotypes that thrive under specific climatic conditions, such as high temperatures, drought, or salinity. This information is vital for developing peanut varieties that can perform well in different regions, thereby ensuring stable and sustainable production across diverse agro-ecological zones.

Germplasm diversity serves as a valuable reservoir of genetic traits that may become increasingly important in the face of evolving challenges, such as climate change or emerging pests and diseases. Having a broad range of genetic resources available ensures that breeders have access to the necessary tools to tackle these future challenges and develop peanut varieties that are resilient, productive, and adapted to changing conditions. Hence, germplasm, its diversity, and the evolution sweep, or selected regions are of great importance in peanut improvement. By harnessing the genetic variation present in germplasm collections, breeders can develop peanut varieties with improved traits, enhanced disease resistance, adaptability to diverse environments, and the capacity to meet future agricultural challenges.

DOI: 10.1201/9781003360988-6

6.2 GLOBAL GENEBANK DIVERSITY AND DIFFERENT DIVERSITY PANELS

The collection of genetic resources is one of the most crucial prebreeding activities for generating data on the variety of germplasm for a species with economically important traits. The cornerstone for continued crop improvement is the wise utilization of the genetic variety that is already there. Utilizing the genetic diversity that is currently available is crucial for crop improvement because cultivated peanuts originated very recently in evolutionary history and have a limited genetic diversity when compared to other crop species. Genetic resources hold the key to success in boosting the genetic gain and quality of production in farmers' fields since genetic variety in a given crop is the key asset for subsequent crop improvement. Four countries—India, Brazil, the United States, and China—maintain large peanut germplasm collections (Pandey et al., 2012).

The genebank at ICRISAT, India, holds the largest collection (15,445 accessions from 93 countries) followed by the National Bureau of Plant Genetic Resources (NBPGR) (14,585 accessions) and the Directorate of Groundnut Research (DGR) of the Indian Council of Agricultural Research (ICAR) (9024 accessions) in India. Apart from India, several institutions hold significant peanut germplasm collections. In China, Agricultural Sciences (CAAS) holds 8,083 accessions and the Crops Research Institute of the Guangdong Academy of Agricultural Sciences holds 4,210 accessions. In the United States, the US Department of Agriculture (USDA) holds 9,917 accessions, along with smaller to medium collections at Texas A&M University (TAMU) and North Carolina State University (NCSU). EMBRAPACENARGEN and the Instituto Agronomico de Campinas in Brazil, as well as the Instituto Nacional de Tecnología Agropecuaria (INTA) and the Instituto de Botánica del Nordeste (IBONE) in Argentina, also maintain peanut accessions. Major centers holding wild *Arachis* species include Texas A&M University (1,200 accessions), the USDA (607 accessions), and the NCSU (406 accessions) in the United States; EMBRAPA-CENARGEN, Brazil (1,220 accessions); ICRISAT in India (477 accessions); and IBONE (472 accessions) in Argentina (Figure 6.1) (Pandey et al., 2012; Gangurde et al., 2019).

The genus *Arachis* can be divided into nine sections, which consist of 81 species including different ploidy and an enormous diversity of genomes (A, B, AB, D, F, K, EX, T, PR, H, C, T, E, R1, and R2) (Figure 6.2) (Stalker, 2017). The *A. hypogaea* species holds greater relevance among these 81 species because it is the only one currently farmed for commercial purposes. These germplasm collections include various diversity panels used in breeding and characterized for significant agronomic traits. For instance, the core collections with 1,704 (ICRISAT), 831 (United States), and 576 (China) genotypes led to the development of mini-core collections with 184, 112, and 298 genotypes, respectively, for easier maintenance. To increase variety, ICRISAT also established other panels, such as the composite collection and reference set, which serve as excellent resources for trait discovery and breeding. Researchers should assess these various panels in numerous hotspot locations to identify present-day and future features significant for marker discovery and crop breeding (Pandey et al., 2020). Resequencing whole germplasm collections along with precise phenotyping data for current and future need-based features may also help redefine diverse sets for various uses in peanut research and genetic improvement.

With so many cultivated accessions readily available, choosing the best lines for crop breeding raises several concerns. A reasonable and affordable entry point into germplasm collections for finding candidate genotypes with new sources of disease and pest resistance or abiotic stress tolerance is the "core collection," which comprises 10% of the whole germplasm collection (Brown, 1989). ICRISAT created a core collection of 1,704 *A. hypogaea* genotypes as a result, which is comparable to the US core collection (831 accessions) (Holbrook et al., 1993). Jiang et al. (2008) used the same process to generate the 576 accessions of the core collection in China. Breeders cannot easily make use of these core collections because of their size. As a result, "mini-core collections" of 184, 112, and 298 accessions have been produced at ICRISAT (Upadhyaya et al., 2002), USDA/ARS (Holbrook & Dong, 2005), and China (JIANG et al., 2010), respectively. These collections make up 10% of the core collections and 1% of the whole germplasm collection. In addition to the collections mentioned above, researchers created a composite collection of 1,000 accessions using phenotypic characterization, geographic origin, and taxonomic information. Twenty SSR loci were used to genotype these accessions.

A group of 300 accessions has been designated as a "reference set" based on genetic diversity analyses and after taking population structure into consideration (Upadhyaya et al., 2002). For tetraploid peanut, researchers are also establishing induced mutant collections in addition to germplasm collections that represent naturally occurring variety. In the past 10 years, the peanut research community has made great strides towards obtaining the best genetic resources. Only three (A, B, AB) of the 15 genomes in the *Arachis* genus have reference genomes accessible. For the implementation of translational genomics activities and various applications in agriculture, genomic resources such as reference genomes, genome-wide genetic markers, genome-wide gene expression atlases, genotyping assays (low, mid, and high density), quality control (QC) panels, and diagnostic markers are crucial.

First-generation sequencing techniques, which generate longer reads, faced challenges due to the enormous genome size and allopolyploidy of the peanut, making assembly complex and whole genome sequencing prohibitively expensive. In contrast, second-generation sequencing generates short reads, offering high-throughput capabilities and reduced costs. The scientific advancements of the past two decades have led to the development of various high-throughput sequencing methods for short reads,

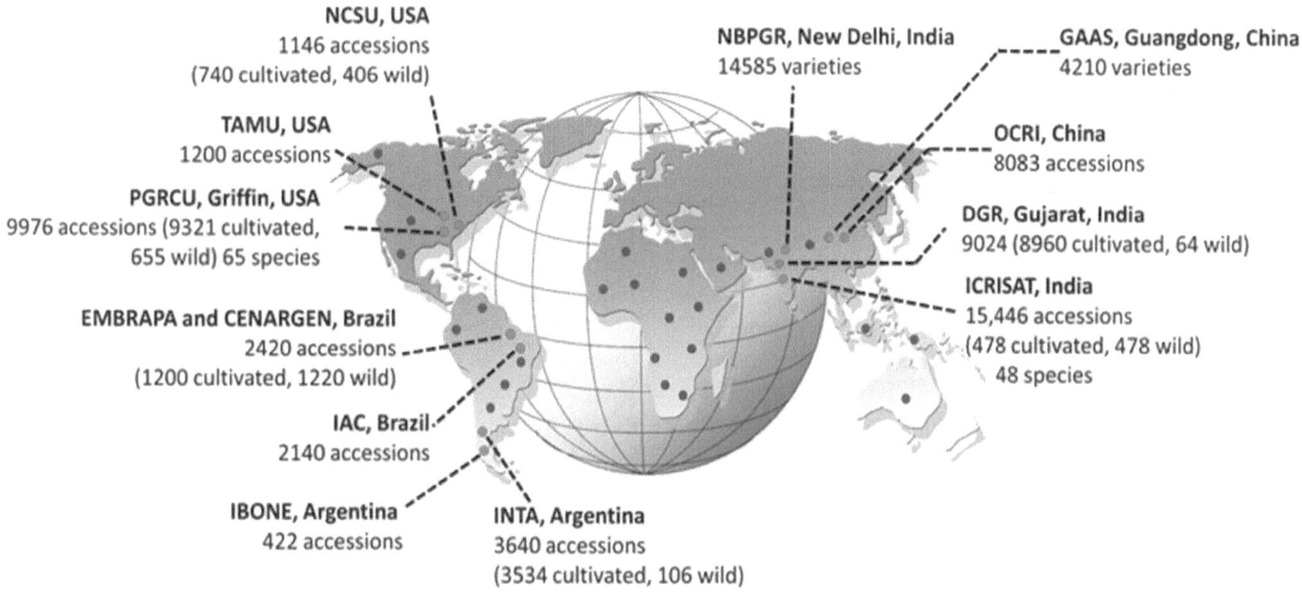

FIGURE 6.1 Distribution of peanut diversity across more than 100 countries on four continents, comprising approx. 72,000 accessions representing 81 species in the genus *Arachis*.

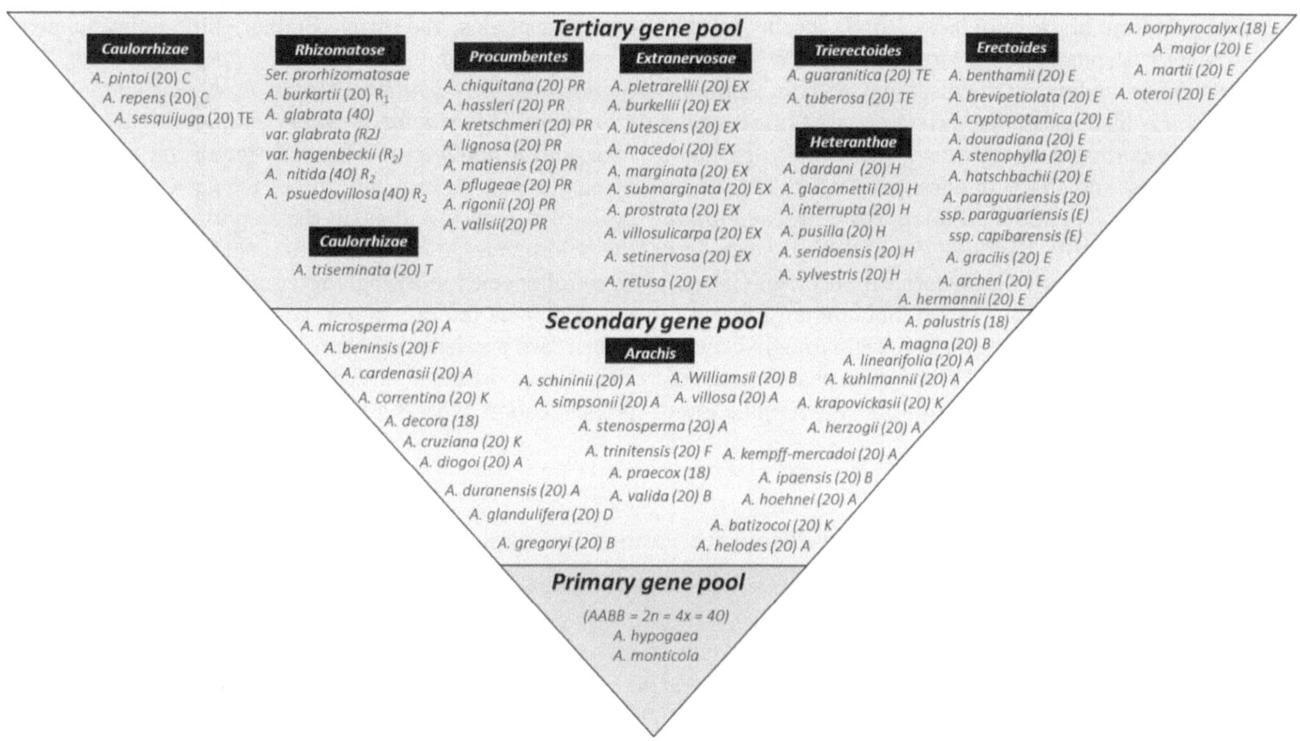

FIGURE 6.2 Gene pool of *Arachis* species. The *primary gene pool* includes cultivated *A. hypogaea* and its tetraploid relative *A. monticola* (AABB). The *secondary gene pool* comprises diploid wild relatives with A, B, D, F, and K genomes, closely related to the cultivated species. The *tertiary gene pool* consists of more distantly related wild species from sections such as *Caulorrhizae*, *Rhizomatosae*, *Procumbentes*, and others, carrying diverse genomes (C, R, PR, EX, TE, H, E). These species represent reservoirs of genetic diversity for crop improvement, although introgression is limited by crossing barriers.

significantly simplifying and lowering the cost of sequencing. As a result, researchers have generated and utilized a wealth of improved genetic materials in breeding programs over the past 10 years, marking a golden era for peanut research.

Despite this progress, the demand for resources remains high, making it a top priority for the community as it continues to grow in the coming years. By effectively leveraging resources such as genome-wide genetic markers, gene expression atlases, high-density genotyping assays, diagnostic markers, reference genomes (for progenitors, wild tetraploids, and cultivated tetraploids), and tetraploid progenitors, researchers can plan and execute breeding programs more efficiently.

6.2.1 Beneficial Alleles from Wild *Arachis* Species

The genus *Arachis* comprises 81 wild and cultivated species, all of which harbor valuable alleles for resistance to diseases and insect pests. For example, the widely used donor GPBD 4, employed in breeding for late leaf spot (LLS) and rust resistance, derives its resistance primarily from *A. cardenasii* (Gowda et al., 2002). Additionally, *A. cardenasii* and *A. duranensis* exhibit resistance to in vitro seed colonization (IVSC) and aflatoxin production (AP) (Xue et al., 2004). *A. benensis* is resistant to peanut bud necrosis disease (PBND), LLS diseases, corn earworm, and leaf miner insect pest of peanut (Reddy et al., 2000). *A. batizocoi* showed resistance to early leaf spot (ELS), LLS, leaf rust, tomato spotted wilt virus (TSWV) diseases, and pests such as corn earworm, aphids, leafhoppers, nematodes, and thrips (Pande & Rao, 2001; Reddy et al., 2000; Stalker, 2017; Subrahmanyam et al., 1983). *A. cardenasii* is resistant to cylindrocladium black rot (CBR), PBND, peanut mottle virus (PMV), rust, peanut stripe virus (PSV), TSWV, IVSC, AP, ELS, and groundnut rosette disease (GRD). In addition, *A. cardinasii* is also resistant to broad range of insect pests such as armyworms, corn earworm, leaf miner, leafhoppers, nematodes, and southern corn room worm (Pande & Rao, 2001; Reddy et al., 2000; Stalker, 2017). *A. correntina* is resistant to rust, CBR, LLS, PMV, and PSV diseases and showed resistance to armyworm, corn earworm, leaf miner, leafhoppers, spider mites, and thrips (Subrahmanyam et al., 1983, 1985). A wild relative *A. diogoi* originating in central Brazil is resistant to ELS, LLS, PBND, rust, peanut ringspot virus (PRV), PSV, and GRD. In addition, *A. diogoi* shows resistance to insect pests such as armyworm, corn earworm, aphids, leafhoppers, nematodes, and thrips (Abdou et al., 1974; Stalker, 2017; Subrahmanyam et al., 1983, 1985, 2001). The two progenitors are also rich in disease resistant alleles for several fungal and bacterial diseases as well as resistance to insect pests. For instance, *A. duranensis* represents A genome of peanut resistant to LLS, rust, PSV, and aflatoxin and showed insect resistance against armyworms, corn earworms, aphids, leaf miner, leafhoppers, and nematodes thrips. *A ipaensis* represents the B subgenome of peanut and is resistant to LLS, rust, and GRD diseases and insects such as corn earworm and leaf hoppers (Pande & Rao, 2001; Stalker, 2017). *A. monticola* is an intermediate between diploid species, and cultivated tetraploid *A hypogaea* is resistant to ELS, rust, and CBR and insects such as corn earworm, leaf miner, leafhopper, and thrips (Pande & Rao, 2001; Sharma et al., 2003; Stalker et al., 2013).

Researchers have recently reported a new source of root knot nematode resistance from *A. stenosperma* incorporated into allotetraploid *A. hypogaea*. Earlier reports suggested that *A. stenosperma* is resistant to ELS, LLS, rust, TSWV, and GRD. *A. stenosperma* also shows resistance to insect pests such as corn earworm, aphids, leaf miner, leafhoppers, nematodes, and thrips (Stalker, 2017; Subrahmanyam et al., 1983, 2001). In addition, *A. glandulifera*, *A. helodes*, *A. hoehnei*, *A. kempff-mercades*, *A. kempff-mercadoi*, *A. kuhnmannii*, *A. magna*, *A. microsperma*, *A. valida*, *A. vellosa*, and *A. decora* are also very important wild relatives of peanut rich in resistant alleles for various diseases and pests of peanut. To identify the important alleles from all the wild species, it is essential to include the variable region from all species in a single reference genome. Future peanut genomics requires a pan genome or super pangenome to increase precision (Table 6.1). Pangenome investigations clearly support the significance of structural differences in domestication and crop breeding (Zhao et al., 2018). As a

TABLE 6.1
Resistance Sources from Wild relatives of Peanut for Various Diseases

Species	Disease Tesistance	Insect Pest Resistance	Citation
A. benensis	PBND, LLS	Corn earworm (*Helicoverpa armigera*), leaf miner (*Aproaerena modicella*)	(Reddy et al., 2000; Stalker, 2017)
A. batizocoi	ELS, LLS, TSWV, leaf rust	Corn earworm (*Heloithis zea*), groundnut aphids (*Aphis crassivora*), leaf hopper (*Empoasca fabae*), nematodes (*Meloidogyne spp.*), thrips (*Frankliniella fusca*)	(Lyerly et al., 2002; Pande & Rao, 2001; Stalker, 2017; Subrahmanyam et al., 1983)
A. cardenasii	CBR, PBND, PMV, leaf rust, peanut stripe virus, TSWV, IVSC, AP, ELS, GRD	Army worm (*Spodoptera litura*), corn earworm (*Helicoverpa armigera*), leaf miner (*Aproaerena modicella*), leaf hopper (*Empoasca fabae*), nematodes (*Meloidogyne spp.*), southern corn root worm (*Diabrotica undecimpunctata howardi*)	(Nigam et al., 1991; Pande & Rao, 2001; Reddy et al., 2000; Stalker, Beute, Shew, & Barker, 2002; Stalker, Beute, Shew, & Isleib, 2002; Stalker, 2017; Subrahmanyam et al., 1983, 2001)
A. correntina	Leaf rust, CBR, LLS, peanut mottle virus, peanut stunt virus	Army worm (*Spodoptera litura*), corn earworm (*Heloithis zea*), groundnut aphids (*Aphis crassivora*), leaf hoppers (*Empoasca fabae*), lesser cornstalk borer (*Elasmopalpus lignosellus*), spider mites (*Tetranychus urticae*), thrips (*Frankliniella fusca*)	(Stalker, 2017; Subrahmanyam et al., 1983, 1985)

(*Continued*)

TABLE 6.1 (CONTINUED)
Resistance Sources from Wild relatives of Peanut for Various Diseases

Species	Disease Tesistance	Insect Pest Resistance	Citation
A. decora	GRD, LLS		(H Thomas Stalker, 2017; Pande & Rao, 2001; Subrahmanyam et al., 2001)
A. diogoi	ELS, LLS, PBND, leaf rust, peanut ringspot virus, peanut stripe virus, peanut mottlle virus, GRD	Army worm (*Spodoptera litura*), corn earworm (*Heloithis zea*), groundnut aphids (*Aphis crassivora*), leaf hoppers (*Empoasca fabae*), nematodes (*Meloidogyne spp.*), thrips (*Frankliniella fusca*), chilli thrips (*Scirtothrips dorsalis*)	(Abdou et al., 1974; Nigam et al., 1991; Stalker, 2017; Subrahmanyam et al., 1983, 1985, 2001)
A. duranensis	Peanut rust, Peanut stunt virus, Aflatoxin resistance, LLS	Army worm (*Spodoptera litura*), corn earworm (*Heloithis zea*), corn earworm (*Helicoverpa armigera*), groundnut aphids (*Aphis crassivora*), leaf miner (*Aproaerena modicella*), leaf hopper (*Empoasca fabae*), nematodes (*Meloidogyne spp.*), thrips (*Frankliniella fusca*, *Frankliniella schultzei*), chilli thrips (*Scirtothrips dorsalis*)	(Pande & Rao, 2001; Stalker, 2017; Subrahmanyam et al., 1983, 1985)
A. glandulifera	Sclerotinia blight		(Stalker, 2017; Tallury et al., 2011)
A. helodes	Peanut stripe virus, sclerotia blight, CBR, peanut clump virus, peanut rust	Nematodes (*Meloidogyne spp.*)	(Culver et al., 1987; Stalker, 2017; Subrahmanyam et al., 1983; Tallury et al., 2011)
A. hoehnei	LLS, peanut rust, GRD	Corn earworm (*Helicoverpa armigera*), leaf hopper (*Empoasca fabae*), nematodes (*Meloidogyne spp.*)	(Pande & Rao, 2001; Stalker, 2017; Subrahmanyam et al., 2001)
A. ipaensis	LLS, ELS, peanut rust	Army worm (*Spodoptera litura*), corn earworm (*Helicoverpa armigera*), leaf hoppers (*Empoasca fabae*)	(Pande & Rao, 2001; Stalker, 2017)
A. kempff-mercades	CRD		(Fitzner et al., 1985; Stalker, 2017)
A. kempff-mercadoi	ELS, LLS, peanut rust	Army worm (*Spodoptera litura*), corn earworm (*Helicoverpa armigera*), groundnut aphids (*Aphis crassivora*), leaf miner (*Aproaerena modicella*), leaf hopper (*Empoasca fabae*)	(Mallikarjuna et al., 2004; Stalker, 2017; Subrahmanyam et al., 1983)
A. kuhlmannii	Peanut rust, CBR, GRD, ELS, LLS		(Pande & Rao, 2001; Stalker, 2017; Subrahmanyam et al., 2001)
A. magna	ELS		(Stalker, 2017)
A. microsperma	CBR		
A. monticola	ELS, peanut rust, CBR	Corn earworm (*Helicoverpa armigera*), leaf miner (*Aproaerena modicella*), leaf hopper (*Empoasca fabae*), thrips (*Frankliniella fusca*)	(Pande & Rao, 2001; H. C. Sharma et al., 2003; Stalker, 2017)
A. stenosperma	ELS, LLS, peanut rust, TSWV, GRD	Corn earworm (*Heloithis zea*), corn earworm (*Helicoverpa armigera*), groundnut aphids (*Aphis crassivora*), leaf miner (*Aproaerena modicella*), leaf hopper (*Empoasca fabae*), nematodes (*Meloidogyne spp.*), thrips (*Frankliniella fusca*)	(Nigam et al., 1991; Stalker, 2017; Subrahmanyam et al., 1983, 2001)
A. valida	ELS, LLS, peanut rust, CBR		(Pande & Rao, 2001; Stalker, 2017)
A. villosa	GRD, LLS, PBND, peanut rust, peanut stunt virus, CBR	Army worm (*Spodoptera litura*), corn earworm (*Heloithis zea*), groundnut aphids (*Aphis crassivora*), leaf hopper (*Empoasca fabae*), lesser cornstalk borer (*Elasmopalpus lignosellus*), spider mites (*Tetranychus urticae*), nematodes (*Meloidogyne spp.*), thrips (*Frankliniella fusca*)	(Reddy et al., 2000; Stalker, 2017; Subrahmanyam et al., 2001)

ELS, early leaf spot; LLS, late leaf spot; PBND, peanut bud necrosis disease; CBR, cylindrocladium black rot.

result, creating pangenomes preserves genetic varieties that might otherwise be missed or lost during domestication. Marker-assisted breeding in *Arachis* has benefited greatly from the Axiom_*Arachis* 58K SNP array (Clevenger et al., 2017; Pandey, Agarwal, et al., 2017), which has made it easier to identify the genes and genomic regions linked to desired traits. In a similar vein, building "super-pangenomes," which expand the pangenome strategy now in use to a genus-level approach, offers a more thorough representation of the diverse array of features seen in crop wild relative (CWR) genomes (Khan et al., 2020).

6.3 SEQUENCING OF DIVERSE GERMPLASM, GENETIC DIVERSITY, AND PANGENOMICS

Over the past decade, significant progress in genome sequencing has rapidly advanced high-resolution trait mapping, the identification of candidate genes, and their application in breeding across various legume species (Pandey et al., 2016). Single nucleotide polymorphisms (SNPs) have become the preferred genetic markers for numerous genetic and breeding activities due to their widespread presence across the genome. The recent publication of reference genomes for both progenitor species, namely *A. duranensis* (Bertioli et al., 2016; Chen et al., 2016) and *A. ipaensis* (Bertioli et al., 2016), has facilitated the practical integration of next-generation sequencing (NGS) techniques in peanut research (Table 6.2).

These seminal references illuminate genome architecture and evolutionary trends while paving the way for the development of essential genomic tools and technologies tailored for a wide array of genetic and breeding applications. For example, these references enabled the construction of a high-density genotyping array, incorporating more than 58,000 notably informative SNPs (Clevenger et al., 2017; Pandey, Khan et al., 2017). They have also enabled the identification of SNPs through RNA-seq data (Nayak et al., 2017; Gangurde et al., 2021), substantially broadening the scope for in-depth exploration of the genome and the discovery of new genes. Moreover, the rapid evolution of sequencing technologies has rendered the sequencing process more economically feasible by increasing data generation capacity and intensifying market competition.

Molecular markers, among all genomic resources, hold direct utility for germplasm characterization, trait mapping, and molecular breeding. Over the past three decades, various marker systems have been developed, including restriction fragment length polymorphisms (RFLPs), random amplified polymorphic DNAs (RAPDs), amplified fragment length polymorphisms (AFLPs), and diversity arrays technology (DArT) markers (Gupta et al., 2010; Varshney et al., 2006). However, for plant genetics and breeding applications, simple sequence repeats (SSRs) or microsatellites and single nucleotide polymorphism (SNP) markers are generally preferred. While SSR markers are multiallelic, codominant, and user-friendly, SNP markers are highly compatible with high-throughput genotyping techniques. Yet, the development and application of SNP markers, especially in less technologically advanced laboratories, remain less routine.

In recent years, genome sequences have become accessible for several crop species, including rice (2005), maize (2009), sorghum (2009), and soybean (2010). The peanut genome is approximately 20 times larger than that of *Arabidopsis thaliana* and 2–6 times larger than the genomes of rice, sorghum, or soybean. Due to its considerable genome size and the amphidiploid nature of its genome, peanut genome sequencing has only recently commenced. The Peanut Genome Consortium (PGC) set specific goals, including creating a high-quality draft of the reference genome sequence for the cultivated tetraploid species, organized on a chromosomal scale. This initiative also involves the development of detailed maps for both the progenitor and synthetic amphidiploid genomes.

Additionally, the PGC aims to conduct high-throughput transcriptome analysis of the reference tetraploid cultivar and explore the gene space within the germplasm of amphidiploid and diploid progenitors. This effort seeks to establish correlations between phenotypic traits and genetically mapped markers, ultimately creating interactive bioinformatics tools for data curation and practical application in breeding programs. This comprehensive approach will facilitate molecular breeding strategies aimed at enhancing peanut yield, improving resistance to diseases and pests, increasing tolerance to environmental stresses, and refining overall quality characteristics.

In this context, the Peanut Genome Project (PGP) is collaborating with BGI-Shenzhen in China to embark on the sequencing of the peanut genome. The approach involves a comprehensive strategy that combines whole-genome shotgun sequencing for both tetraploid and diploid genotypes. Additionally, there will be highly multiplexed BAC-by-BAC sequencing undertaken on a specific tetraploid cultivar, Tifrunner, utilizing HiSeq2000 technology. Researchers will achieve genetic ordering by employing low-coverage sequencing of recombinant inbred lines. Concurrently, a separate initiative has been initiated at the University of California-Davis, the United States, led by Richard Michelmore (pers. comm.), aimed at generating ultra-high-density genetic maps. This endeavor will involve low-coverage, shotgun sequencing of diploid and tetraploid mapping populations, as well as the reference germplasm set (Froenicke et al., 2011).

These populations include combinations such as *A. duranensis* crossed with *A. stenosperma* (AA genome), *A. ipaënsis* crossed with *A. magna* (BB genome), and *A. hypogaea* cv. IAC Runner crossed with the synthetic amphidiploid of the two progenitor species. Furthermore, the reference set encompasses 300 accessions chosen based on SSR genotyping data to represent the diversity of the international peanut germplasm collection (Upadhyaya et al., 2002). The US mini-core collection (Holbrook & Dong, 2005) and historical and contemporary US cultivars are also incorporated.

TABLE 6.2
Information about Reference Genomes of Different Species of Peanut

Species	*Arachis duranensis*	*Arachis duranensis*	*Arachis ipaensis*	*Arachis ipaensis*	*Arachis monticola*	*Arachis hypogaea* ssp. *hypogaea*	*Arachis hypogaea* ssp. *fastigiata*	*Arachis hypogaea* ssp. *fastigiata*
Ploidy level	Diploid	Diploid	Diploid	Diploid	Tetraploid	Tetraploid	Tetraploid	Tetraploid
Genome	AA	AA	BB	BB	AABB	AABB	AABB	AABB
Wild/cultivated	Wild	Wild	Wild	Wild	Wild	Cultivated	Cultivated	Cultivated
Accessions	PI475845	V14167	K30076	ICG 8206	PI 263393	Tifrunner	Shitouqi	Fuhuasheng
Sequencing platform	Illumina HiSeq2500	Illumina HiSeq2000	Illumina HiSeq2000	Illumina HiSeq2500	Pacbio, BioNano optics and Hi-C	PacBio and Illumina HiSeq 2500	PacBio and Illumina HiSeq 2000	Illumina HiSeq and PacBio
Genome size	1.38 Gb	1.081 Gb	1.371 Gb	1.39 Gb	2.62 Gb	2.556 Gb	2.54 Gb	2.58 Gb
Number of scaffolds	8,173	635,392	759,499	79,408	4,135	384	1297	491
GC content (%)	31.79	34.00	35.49	36.70	35.99	36.21	36.53	36.33
Protein coding genes	50,324	36,734	41,840	39,704	43,961	66,469	83,709	81,414
Reference	(Chen et al., 2016)	(Bertioli et al., 2016)	(Bertioli et al., 2016)	(Lu et al., 2018)	(Yin et al., 2018)	(Bertioli et al., 2019)	(Zhuang et al., 2019)	(Chen et al., 2019)

Researchers will meticulously analyze the SNPs discovered within this diversity panel for linkage disequilibrium (LD) patterns and then use these findings to refine the genetic bins generated from segregation data across various mapping populations. These collective efforts provides a vital assistance in complementing the process of assembling the reference genome sequence for peanuts. The approach will validate the accuracy of genome assemblies and supply essential chromosomal genetic coordinates for the contigs and scaffolds. As a result, it is anticipated that in the coming years, a preliminary draft of the genome sequence will be made available, accompanied by an extensive repository of genome and transcriptome data, effectively serving the peanut research community's needs.

6.4 INTEGRATION OF GENEBANK DIVERSITY, PREBREEDING, AND GENOMICS FOR DIVERSIFYING PRIMARY GENEPOOL

The integration of peanut genebank diversity, prebreeding, and genomics holds significant promise for diversifying the primary genepool of peanuts, thereby enhancing their adaptability, resilience, and overall sustainability. The genebank houses a rich repository of genetic resources encompassing diverse peanut varieties, landraces, and wild relatives. By strategically leveraging these resources, breeders can initiate prebreeding efforts to introgress valuable traits from wild relatives and unexplored landraces into cultivated varieties (Pattee & Stalker, 1995). This process involves targeted hybridization, followed by rigorous selection, with the aim of broadening the genetic base of cultivated peanuts.

Peanuts are an important self-pollinating tetraploid legume produced for its high-quality edible oil and easily digestible protein. About two-thirds of the world's peanut production, predominantly grown as an oilseed crop, undergoes crushing to produce vegetable oil. The sustainability of mixed crop-livestock farming depends on these crops, which also offer genetic resources for adaptability to the harsh and marginal conditions that are typical in the semi-arid tropics. Due to a variety of distinct qualities and advantages, these crops are crucial in sustainable agriculture systems in rain-fed areas. In the genus *Arachis*, which includes 81 wild species and domesticated peanuts, there is a tremendous amount of genetic diversity. Breeders can exploit wild species to create cultivars with higher levels of resistance to biotic and abiotic stresses (Dwivedi, 2018). To employ these species, they must use ploidy manipulations, bridge crossings, and embryo or ovule rescue. Researchers have created several synthetics (amphidiploids and autotetraploids) using A- and B-genome accessions with high levels of resilience to numerous stresses to effectively utilize diploid wild species from section *Arachis*. Breeders utilize these synthetics in crossing programs with cultigens to create prebreeding populations and introgression lines (ILs) with high frequencies of beneficial genes and alleles in good agronomic backgrounds.

The evaluation of two populations—ICGV 91114 × ISATGR 121250 (a synthetic derived from *A. duranensis* Krapov. & Greg. × *A. ipaensis* Krapov. & Greg.) and ICGV 87846 × ISATGR 265-5 (*A. kempf-mercadoi* Greg. & Simpson × *A. hoehnei* Krapov. & Greg.)—revealed promising results in genotyping with markers for rust and LLS resistance. The findings suggest potential novel allele introgression and resistance sources from wild species other than the commonly used *A. cardenasii* Krapov. & Greg., which could be highly beneficial for peanut improvement (Sharma et al., 2017).

The integration of genomics provides a cutting-edge approach to identify specific genes and markers associated with desirable traits for prebreeding. Through techniques like marker assisted selection (MAS), breeders can efficiently track and transfer these traits across different genetic backgrounds, expediting the breeding process. Genomic tools also facilitate the identification of novel genetic variations that may confer resistance to diseases, pests, and environmental stresses, thereby bolstering the adaptability of peanut varieties (Varshney et al., 2014).

The utilization of genebank resources in prebreeding endeavors requires comprehensive phenotypic and genotypic characterization of the collected materials (Pandey, Khan et al., 2017). Advanced phenotyping techniques allow for the accurate assessment of various traits, including yield, stress tolerance, and nutritional content. This information, coupled with genotypic data from high-throughput sequencing, enables breeders to make informed decisions and prioritize materials for hybridization and subsequent selection. Moreover, the integration of genebank diversity, prebreeding, and genomics supports the conservation of locally adapted landraces and wild species, which may possess unique genes for stress resistance and other beneficial traits (Varshney et al., 2013). This approach ensures that valuable genetic resources are preserved and utilized to tackle emerging challenges, such as climate change and evolving pathogens.

Furthermore, the integration of these strategies aligns well with the goals of sustainable agriculture, as it fosters genetic diversity and reduces the risk of vulnerability to single stress factors (Hoisington et al., 1999). By incorporating diverse genetic traits into improved varieties, the agricultural sector becomes better prepared to withstand a range of challenges, leading to stable and resilient peanut production. As global demand for peanuts increases and climate variability becomes more pronounced, the integration of genebank diversity, prebreeding, and genomics becomes crucial for ensuring food security. These strategies offer a holistic and forward-looking approach to peanut breeding, providing the means to develop adaptable, high-yielding, and resilient varieties that can meet the needs of a changing world (Bertioli et al., 2019). In conclusion, the integration of peanut genebank diversity, prebreeding, and genomics represents a holistic strategy to diversify the primary genepool. This approach capitalizes on the wealth of genetic resources available, accelerates the transfer of desirable traits, and strengthens peanut breeding programs against emerging challenges.

6.5 CONCLUSION AND FUTURE PROSPECTS

The peanut crop holds diverse applications for the global population, society, and the environment. It serves nutritional needs and fights malnutrition due to its affordability and accessibility. In the context of global markets and trade, crops that align with farmers' interests and meet industry and consumer preferences will thrive. Over centuries, the peanut crop has become integral to global food production. The research community has created pivotal genomic resources and successfully applied them to molecular breeding. Integrating advanced technologies into agriculture is pivotal for augmenting production amidst evolving climate conditions and diminishing natural resources.

However, contemporary peanut cultivation faces challenges from both biotic and abiotic stresses, suboptimal crop management, and an unstable seed system, further exacerbated by erratic weather events. The expanding use of peanuts for specialized food products adds complexity to breeding programs. Accelerating genetic improvements remains a central challenge, alongside refining crop management and the seed system. Consequently, investing in genetic breeding tools and technologies like speed breeding, genomic selection, and gene editing is crucial for expediting variety development.

Addressing the issue of limited genetic diversity in peanuts has involved the creation of numerous molecular markers. However, the development of cost-effective platforms for SNP genotyping of tetraploid peanut germplasm collections and mapping populations is still ongoing. As a result, SSR markers continue to be the favored choice for genetic and breeding purposes in cultivated peanuts. Despite the construction of genetic maps for both diploid and tetraploid genomes, only a few hundred SSR markers have been precisely located on the genetic map. A major challenge for the peanut research community is to incorporate a maximal number of markers onto genetic maps. While researchers have established a physical map for AA-genome species, the absence of a similar map for BB-genome species remains a critical gap. This gap hampers the merging of genetic and physical maps, which would greatly facilitate gene cloning and molecular breeding.

Advancements in genomics for peanut breeding have predominantly targeted trait mapping and molecular breeding, especially for resistance against specific biotic stresses like leaf rust and root nematodes. Moreover, it is important to address other challenges, such as production and quality constraints including drought stress and aflatoxin contamination, which pose health risks. Additionally, considering stagnant yields, utilizing molecular breeding to enhance oil content offers a viable avenue to boost farmers' profitability. Improving nutritional quality is also imperative, particularly by increasing the oleic: linoleic acid ratio to extend shelf life and provide health benefits, such as reducing LDL cholesterol levels and countering inflammatory diseases.

As sequencing technologies and bioinformatics advance, it is foreseeable that peanut genome sequences or gene spaces will become more accessible. However, this will necessitate a concerted effort to implement high-throughput and precise phenotyping for various traits crucial to peanut breeding. Overall, the future of peanut genomics and molecular tools in breeding holds promise, fortifying peanuts against a range of production and quality constraints. With optimal genomic resources, peanuts have the potential to withstand climate change and satisfy consumer and trade preferences. Strategic research investments should prioritize sequencing the entire germplasm, developing marker assays, enhancing the mapping of complex traits, broadening the genepool, and modernizing the peanut value chain. Collaborative endeavors will sustain the notable achievements of the peanut research community, ultimately benefiting end-users and stakeholders.

REFERENCES

Abdou, Yousef A. M., W. C. Gregory, and W. E. Cooper. "Sources and nature of resistance to Cercospora arachidicola Hori and Cercosporidium personatum (Beck & Curtis) Deighton in Arachis species." *Peanut Science* 1, no. 1 (1974): 6–11.

Bertioli, David John, Steven B. Cannon, Lutz Froenicke, Guodong Huang, Andrew D. Farmer, Ethalinda K. S. Cannon, Xin Liu et al. "The genome sequences of Arachis duranensis and Arachis ipaensis, the diploid ancestors of cultivated peanut." *Nature Genetics* 48, no. 4 (2016): 438–446.

Bertioli, David J., Jerry Jenkins, Josh Clevenger, Olga Dudchenko, Dongying Gao, Guillermo Seijo, Soraya C. M. Leal-Bertioli et al. "The genome sequence of segmental allotetraploid peanut Arachis hypogaea." *Nature Genetics* 51, no. 5 (2019): 877–884.

Brown, A. H. D. "Core collections: a practical approach to genetic resources management." *Genome* 31, no. 2 (1989): 818–824.

Chen, Xiaoping, Hongjie Li, Manish K. Pandey, Qingli Yang, Xiyin Wang, Vanika Garg, Haifen Li et al. "Draft genome of the peanut A-genome progenitor (Arachis duranensis) provides insights into geocarpy, oil biosynthesis, and allergens." *Proceedings of the National Academy of Sciences* 113, no. 24 (2016): 6785–6790.

Chen, Xiaoping, Qing Lu, Hao Liu, Jianan Zhang, Yanbin Hong, Haofa Lan, Haifen Li et al. "Sequencing of cultivated peanut, Arachis hypogaea, yields insights into genome evolution and oil improvement." *Molecular Plant* 12, no. 7 (2019): 920–934.

Clevenger, Josh, Ye Chu, Carolina Chavarro, Gaurav Agarwal, David J. Bertioli, Soraya CM Leal-Bertioli, Manish K. Pandey et al. "Genome-wide SNP genotyping resolves signatures of selection and tetrasomic recombination in peanut." *Molecular Plant* 10, no. 2 (2017): 309–322.

Culver, J. N., J. L. Sherwood, and H. A. Melouk. "Resistance to peanut stripe virus in Arachis germ plasm." (1987): 640.

Dwivedi, Sangam L. "Hari Deo Upadhyaya: Plant breeder, geneticist and genetic resources specialist." *Plant Breeding Reviews* 41 (2018): 1–53.

Fitzner, M. S., S. C. Alderman, and H. T. Stalker. "Greenhouse evaluation of cultivated and wild peanut species for resistance to Cylindrocladium black rot." *Proceedings of American Peanut Research and Education Society* 17 (1985): 28.

Froenicke, L., M. Pandey, H. Upadhyaya, M. C. Moretzsohn, P. Guimarães, S. Leal-Bertioli, R. K. Varshney, D. Bertioli, and R. W. Michelmore. "Towards ultra-dense genetic maps of peanut generated by sequencing diploid and tetraploid RIL populations and a peanut diversity panel." *Proc Advances in Arachis through Genomics and Biotechnology (AAGB)* 13 (2011): 15.

Gangurde, Sunil S., Rakesh Kumar, Arun K. Pandey, Mark Burow, Haydee E. Laza, Spurthi N. Nayak, Baozhu Guo et al. "Climate-smart groundnuts for achieving high productivity and improved quality: current status, challenges, and opportunities." *Genomic Designing of Climate-smart Oilseed Crops* (2019): 133–172.

Gangurde, Sunil S., Spurthi N. Nayak, Pushpesh Joshi, Shilp Purohit, Hari K. Sudini, Annapurna Chitikineni, Yanbin Hong et al. "Comparative transcriptome analysis identified candidate genes for late leaf spot resistance and cause of defoliation in groundnut." *International Journal of Molecular Sciences* 22, no. 9 (2021): 4491.

Gowda, M. V. C., B. N. Motagi, G. K. Naidu, S. B. Diddimani, and R. Sheshagiri. "GPBD 4: a spanish bunch groundnut genotype resistant to rust and late leaf spot." *International Arachis Newsletter* 22 (2002): 29–32.

Guo, Baozhu, Manish K. Pandey, Guohao He, Xinyou Zhang, Boshou Liao, Albert Culbreath, Rajeev K. Varshney, Victor Nwosu, Richard F. Wilson, and H. Thomas Stalker. "Recent advances in molecular genetic linkage maps of cultivated peanut." *Peanut Science* 40, no. 2 (2013): 95–106.

Gupta, P. K., Jitendra Kumar, R. R. Mir, and Ajay Kumar. "4 Marker-assisted selection as a component of conventional plant breeding." *Plant Breeding Reviews* 33, no. 4 (2010): 145–217.

Halward, Tracy, Tom Stalker, Elizabeth LaRue, and Gary Kochert. "Use of single-primer DNA amplifications in genetic studies of peanut (Arachis hypogaea L.)." *Plant Molecular Biology* 18 (1992): 315–325.

Hoisington, David, Mireille Khairallah, Timothy Reeves, Jean-Marcel Ribaut, Bent Skovmand, Suketoshi Taba, and Marilyn Warburton. "Plant genetic resources: what can they contribute toward increased crop productivity?." *Proceedings of the National Academy of Sciences* 96, no. 11 (1999): 5937–5943.

Holbrook, C. Corley, William F. Anderson, and Roy N. Pittman. "Selection of a core collection from the US germplasm collection of peanut." *Crop Science* 33, no. 4 (1993): 859–861.

Holbrook, C. Corley, and Weibo Dong. "Development and evaluation of a mini core collection for the US peanut germplasm collection." *Crop Science* 45, no. 4 (2005): 1540–1544.

Jiang, HuiFang., Ren XiaoPing, Liao BoShou, Huang JiaQuan, Lei Yong, Chen BenYin, B. Z. Guo, C. C. Holbrook, and H. D. Upadhyaya. "Peanut core collection established in China and compared with ICRISAT mini core collection." *Acta Agronomica Sinica* 34, no. 1 (2008): 25–30.

Jiang, Hui-Fang, Ren Xiao-Ping, Xiao-Jie Zhang, Jia-Quan Huang, Yong Lei, Li-Ying Yan, Bo-Shou Liao, Hari D. Upadhyaya, and Corley C. Holbrook. "Comparison of genetic diversity based on SSR markers between peanut mini core collections from China and ICRISAT." Acta Agronomica Sinica 36, no. 7 (2010): 1084–1091.

Khan, Aamir W., Vanika Garg, Manish Roorkiwal, Agnieszka A. Golicz, David Edwards, and Rajeev K. Varshney. "Super-pangenome by integrating the wild side of a species for accelerated crop improvement." *Trends in Plant Science* 25, no. 2 (2020): 148–158.

Lu, Qing, Haifen Li, Yanbin Hong, Guoqiang Zhang, Shijie Wen, Xingyu Li, Guiyuan Zhou et al. "Genome sequencing and analysis of the peanut B-genome progenitor (Arachis ipaensis)." *Frontiers in Plant Science* 9 (2018): 604.

Lyerly, Jeanette H., H. T. Stalker, J. W. Moyer, and K. Hoffman. "Evaluation of Arachis species for resistance to tomato spotted wilt virus." *Peanut Science* 29, no. 2 (2002): 79–84.

Mallikarjuna, N., Suresh Pande, Deepak R. Jadhav, D. C. Sastri, and J. Narayan Rao. "Introgression of disease resistance genes from Arachis kempff-mercadoi into cultivated groundnut." *Plant breeding* 123, no. 6 (2004): 573–576.

Moretzsohn, M. C., E. G. Gouvea, P. W. Inglis, S. C. M. Leal-Bertioli, J. F. M. Valls, and D. J. Bertioli. "A study of the relationships of cultivated peanut (Arachis hypogaea) and its most closely related wild species using intron sequences and microsatellite markers." *Annals of Botany* 111, no. 1 (2013): 113–126.

Nayak, S. N., G. Agarwal, M. K. Pandey, H. K. Sudini, A. S. Jayale, S. Purohit, A. Desai, L. Wan, B. Guo, and B. Liao. "Aspergillus flavus infection triggered immune responses and host-pathogen cross-talks in groundnut during in-vitro seed colonization." Scientific Reports 7, no. 1 (2017): 9659.

Nigam, S. N., S. L. Dwivedi, and R. W. Gibbons. "Groundnut breeding: Constraints, achievements and future possibilities." Plant Breeding Abstracts 61, no. 10 (1991): 112–1136.

Pande, S., and J. Narayana Rao. "Resistance of wild Arachis species to late leaf spot and rust in greenhouse trials." *Plant Disease* 85, no. 8 (2001): 851–855.

Pandey, Manish K., Gaurav Agarwal, Sandip M. Kale, Josh Clevenger, Spurthi N. Nayak, Manda Sriswathi, Annapurna Chitikineni et al. "Development and evaluation of a high density genotyping 'Axiom_ Arachis' array with 58 K SNPs for accelerating genetics and breeding in groundnut." *Scientific Reports* 7, no. 1 (2017): 40577.

Pandey, Manish K., Aamir W. Khan, Vikas K. Singh, Manish K. Vishwakarma, Yaduru Shasidhar, Vinay Kumar, Vanika Garg et al. "QTL-seq approach identified genomic regions and diagnostic markers for rust and late leaf spot resistance in groundnut (A rachis hypogaea L.)." *Plant Biotechnology Journal* 15, no. 8 (2017): 927–941.

Pandey, Manish K., Emmanuel Monyo, Peggy Ozias-Akins, Xuanquiang Liang, Patricia Guimarães, Shyam N. Nigam, Hari D. Upadhyaya et al. "Advances in Arachis genomics for peanut improvement." *Biotechnology Advances* 30, no. 3 (2012): 639–651.

Pandey, Manish K., Arun K. Pandey, Rakesh Kumar, Chogozie Victor Nwosu, Baozhu Guo, Graeme C. Wright, Ramesh S. Bhat et al. "Translational genomics for achieving higher genetic gains in groundnut." *Theoretical and Applied Genetics* 133 (2020): 1679–1702.

Pandey, Manish K., Manish Roorkiwal, Vikas K. Singh, Abirami Ramalingam, Himabindu Kudapa, Mahendar Thudi, Anu Chitikineni, Abhishek Rathore, and Rajeev K. Varshney. "Emerging genomic tools for legume breeding: current status and future prospects." *Frontiers in Plant Science* 7 (2016): 455.

Pattee, H. E., and H. T. Stalker. *Advances in Peanut Science*. American Peanut Research and Education Society, Stillwater, OK, USA, 1995.

Qin, Hongde, Suping Feng, Charles Chen, Yufang Guo, Steven Knapp, Albert Culbreath, Guohao He et al. "An integrated genetic linkage map of cultivated peanut (Arachis hypogaea L.) constructed from two RIL populations." *Theoretical and Applied Genetics* 124 (2012): 653–664.

Reddy, A. S., L. J. Reddy, N. Mallikarjuna, M. D. Abdurahman, Y. V. Reddy, P. J. Bramel, and D. V. R. Reddy. "Identification of resistance to peanut bud necrosis virus (PBNV) in wild Arachis germplasm." Annals of Applied Biology 137, no. 2 (2000): 135–139.

Sharma, H. C., G. Pampapathy, S. L. Dwivedi, and L. J. Reddy. "Mechanisms and diversity of resistance to insect pests in wild relatives of groundnut." Journal of Economic Entomology 96, no. 6 (2003): 1886–1897.

Sharma, S., M. K. Pandey, H. K. Sudini, H. D. Upadhyaya, and R. K. Varshney. "Harnessing genetic diversity of wild Arachis species for genetic enhancement of cultivated peanut." Crop Science 57, no. 3 (2017): 1121–1131.

Stalker, H. Thomas. "Utilizing wild species for peanut improvement." Crop Science 57, no. 3 (2017): 1102–1120.

Stalker, H. T., M. K. Beute, B. B. Shew, and K. R. Barker. "Registration of two root-knot nematode-resistant peanut germplasm lines." Crop Science 42, no. 1 (2002): 312–314.

Stalker, H. T., M. K. Beute, B. B. Shew, and T. G. Isleib. "Registration of five leaf spot-resistant peanut germplasm lines." Crop Science 42, no. 1 (2002): 314–317.

Stalker, H. T., S. P. Tallury, B. Ozias-Akins, D. Bertioli, and S. C. L. Bertioli. "The value of diploid peanut relatives for breeding and genomics." Peanut Science 40, no. 2 (2013): 70–88.

Subrahmanyam, P., R. Anaidu, L. J. Reddy, P. L. Kumar, and M. E. Ferguson. "Resistance to groundnut rosette disease in wild Arachis species." Annals of Applied Biology 139, no. 1 (2001): 45–50.

Subrahmanyam, P., J. P. Moss, D. McDonald, P. V. Subba Rao, and V. R. Rao. "Resistance to leaf spot caused by Cercosporidium personatum in wild Arachis species." Plant Disease 69, no. 11 (1985): 951–954.

Subrahmanyam, P., J. P. Moss, and V. R. Rao. "Resistance to peanut rust in wild Arachis species." *Plant Disease* 67, no. 2 (1983): 209–212.

Tallury, S. P., J. Hollowell, and T. G. Isleib. "Greenhouse evaluation of section Arachis wild species for Sclerotinia blight and CBR resistance." Proceedings of the American Peanut Research and Education Society 43 (2011): 69.

Upadhyaya, H. D., P. J. Bramel, R. Ortiz, and S. Singh. "Developing a mini core of peanut for utilization of genetic resources." Crop Science 42, no. 6 (2002): 2150–2156.

Varshney, R. K., D. A. Hoisington, and A. K. Tyagi. "Advances in cereal genomics and applications in crop breeding." Trends in Biotechnology 24, no. 11 (2006): 490–499.

Varshney, R. K., S. M. Mohan, P. M. Gaur, N. Gangarao, M. K. Pandey, A. Bohra, S. L. Sawargaonkar, A. Chitikineni, P. K. Kimurto, and P. Janila. "Achievements and prospects of genomics-assisted breeding in three legume crops of the semi-arid tropics." Biotechnology Advances 31, no. 8 (2013): 1120–1134.

Varshney, R. K., M. K. Pandey, P. Janila, S. N. Nigam, H. Sudini, M. V. C. Gowda, M. Sriswathi, T. Radhakrishnan, S. S. Manohar, and P. Nagesh. "Marker-assisted introgression of a QTL region to improve rust resistance in three elite and popular varieties of peanut (Arachis hypogaea L.)." Theoretical and Applied Genetics 127 (2014): 1771–1781.

Xue, H. Q., T. G. Isleib, H. T. Stalker, G. A. Payne, and G. OBrian. "Evaluation of Arachis species and interspecific tetraploid lines for resistance to aflatoxin production by Aspergillus flavus." Peanut Science 31, no. 2 (2004): 134–141.

Yin, D., C. Ji, X. Ma, H. Li, W. Zhang, S. Li, F. Liu, K. Zhao, F. Li, and K. Li. "Genome of an allotetraploid wild peanut Arachis monticola: A de novo assembly." Gigascience 7, no. 6 (2018): giy066.

Zhao, Q., Q. Feng, H. Lu, Y. Li, A. Wang, Q. Tian, Q. Zhan, Y. Lu, L. Zhang, and T. Huang. "Pan-genome analysis highlights the extent of genomic variation in cultivated and wild rice." Nature Genetics 50, no. 2 (2018): 278–284.

Zhuang, W., H. Chen, M. Yang, J. Wang, M. K. Pandey, C. Zhang, W.-C. Chang, L. Zhang, X. Zhang, and R. Tang. "The genome of cultivated peanut provides insight into legume karyotypes, polyploid evolution and crop domestication." Nature Genetics 51, no. 5 (2019): 865–876.

7 Development of Molecular Markers and Its Application in Peanut Breeding

Charles Chen, Jinesh Patel, Xinlei Yang, Hui Zhang, Li Li, Mingli Wang, and Xiaoping Chen

7.1 INTRODUCTION

Grown throughout the world, extensively in Asia, Africa, and the Americas, cultivated peanut (*Arachis hypogaea* L.) ranks as the world's 12th most important food crop and 4th most important oilseed crop. While its major producers are China, India, and the United States, over 48.8 million metric tons (in shell) are currently being produced on approximately 29.6 million hectares on six different continents (FAO, 2022). In addition to being a rich source of oil (44%–50%), protein (20%–50%), and carbohydrates (10%–20%), peanut seeds contain vitamin E, niacin, folacin, calcium, phosphorus, magnesium, zinc, iron, riboflavin, thiamine, and potassium (Varshney et al., 2016). *Arachis hypogaea* is a self-pollinated, annual allotetraploid ($2n = 4x = 40$, AABB), although most of its wild relatives are diploid species (Bertioli et al., 2016). A member of the Fabaceae family, the species itself is comprised of two subspecies (*hypogaea* and *fastigiata*), which are further classified into two (*hypogaea* and *hirsuta*) and four (*fastigiata*, *vulgaris*, *aequatoriana*, and *peruviana*) botanical varieties. While it has a large nuclear genome ($n = 2.8$ Gbp), since cultivated peanut is believed to be monophyletic in origin, the germplasm study shows significantly less genetic diversity than most other cultivated crops, which makes molecular marker development and QTL identification much more difficult and challenging (Zhao et al., 2012).

With recent efforts in peanut genomic studies, the availability of a tremendous number of molecular markers and genetic and physical maps have greatly assisted the mapping of QTL and/or genes (Vishwakarma et al., 2017). Many QTLs have been identified for disease resistance (Han et al., 2018), seed and pod quality traits (Chavarro et al., 2020; D.P. Chen et al., 2016), seed dormancy (Wang et al., 2022; Patel et al., 2022), nutritional quality traits (Wilson et al., 2017; Zhang et al., 2021) and drought resistance (Ravi et al., 2011; Pandey et al. 2020), providing possibilities for maker assisted selection and genomic selection for peanut improvement. In this chapter, we will focus on the development of molecular markers and their application in peanut breeding.

7.2 DEVELOPMENT OF MOLECULAR MARKERS

Molecular markers are essential for constructing genetic linkage maps, dissecting QTLs, and identifying genes for traits of interest. Such markers can also be used to understand genome evolution with complex genetic inheritance. Several types of markers have been used in different studies on cultivated peanuts, including isozymes, restriction fragment length polymorphisms (RFLP), random amplified polymorphic DNAs (RAPD), amplified fragment length polymorphisms (AFLP), simple sequence repeats (SSR), and single nucleotide polymorphisms (SNP) (Stalker and Mozingo, 2001).

7.2.1 Isozyme

Isozymes are protein-based markers that are used in plant breeding studies. These are forms of enzyme markers that have the same function but vary in molecular weight between different accessions due to differences in amino acid sequence (Weeden, 1989). Some of the isozymes used in these studies are lipoxygenases, phenolics, peroxidases, triose phosphate isomerase (TPI), phosphohexose isomerase (PHI), isocitrate dehydrogenase (IDH), and glutamate oxaloacetate transaminase (GOT) (Sanders et al., 1975; Grieshammer and Wynne, 1990b). Isozyme markers have been used to study the genetic diversity, hybridization, and inheritance of traits in peanuts. A total of 17 enzymes were used to evaluate the genetic diversity among 131 accessions of *Arachis* species consisting of diploid and tetraploid biotypes, including cultivated peanut (*A. hypogaea*). From these studies, it was concluded that there was a low amount of variation between cultivated species, whereas diploids showed a high level of variation. Additionally, this research ruled out the possibility of *A. batizocoi* being a possible progenitor (B genome donor) of *A. hypogaea* (Stalker et al., 1994). Two separate studies testing several isozymes on different accessions of *Arachis* determined only three isozymes, PHI, IDH, and GOT, as promising

DOI: 10.1201/9781003360988-7

isozymes for studying genetic variation in peanut accessions (Grieshammer and Wynne, 1990a; Lacks and Stalker, 1993). This indicates that peanut, like many plant species, also shows a lack of variability in isozymes.

7.2.2 RFLP, AFLP, and RAPD

RFLP was the first DNA-based marker system that generated a high amount of polymorphisms between two accessions in plant species that had been used in the construction of genetic linkage maps (Helentjaris et al., 1986; Nam et al., 1989). In general, the digestion of DNA using restriction enzymes results in different DNA fragments, which can be separated by size using gel electrophoresis. Genetic variation between individuals will confer the presence or absence of the restriction site, thus producing different band patterns between the individuals. This information can be used to evaluate genetic diversity, develop genetic maps, or identify genes associated with traits of interest. Similarly to isozyme markers, RFLP markers produce a good amount of variation between different species of *Arachis*, especially the diploid species (Kochert et al., 1991; Paik-Ro et al., 1992). RFLPs were used to develop a genetic linkage map in an interspecific cross between diploid peanut species, resulting in 11 linkage groups spanning about 1,063 cM (Halward Stalker and Kochert, 1993). However, RFLPs were still not effective in revealing variation among accessions within *A. hypogaea*. Additionally, this research also failed to reveal polymorphisms between *A. hypogaea*, exotic germplasm lines, and its wild relative (*A. monticola*) (Halward et al., 1991; Kochert et al., 1991; Paik-Ro et al., 1992).

The polymerase chain reaction (PCR) technique allows for the development of easy-to-use, inexpensive, and more reliable markers like RAPD and AFLP (Singh et al., 2015). RAPD uses a different size of primers to amplify DNA fragments that are randomly distributed throughout the genome. At the same time, AFLP markers are based on a combination of different sets of restriction enzymes and PCR to generate unique DNA fingerprinting for each accession. RAPDs were considered easier to use and inexpensive, while AFLPs are more reliable than RAPDs and produce a greater number of bands to evaluate for polymorphisms. It can be challenging to determine the heterozygous or homozygous nature of progenies using RAPD and AFLP markers, as they tend to be dominant in nature. These markers are better suited for screening backcross populations (Amiteye, 2021). In peanut, these markers were utilized not only for creating genetic linkage maps and analyzing genetic variations but also for recognizing regions of the genome that relate to desirable traits.

A genetic linkage map of 11 linkage groups consisting of 167 RAPD and 39 RFLP markers was developed in a backcross population generated by the cross *A. stenosperma* × (*A. stenosperma* × *A. cardenasii*) (Garcia et al., 2005). RAPD markers were used to identify three linked regions to root-knot nematode resistance using a bulk segregation analysis in BC_4F_2 population (Burow et al., 1996). In a study of 22 USDA *A. hypogaea* accessions, 27 RAPD primer sets revealed 80 polymorphic bands. These bands were useful in distinguishing genetically diverse and closely related accessions. This study highlighted the value of using RAPD markers in diversity studies for crops with a very limited number of markers (Al-Saghir and Abdel-Salam, 2015).

The first attempt to develop a reliable technique for detecting DNA variation in peanut cultivars using AFLP markers involved evaluating six diverse peanut germplasm accessions distributed across three botanical varieties (He and Prakash, 1997). Later, two separate studies used AFLP markers to determine genetic variation in 21 commercially grown South African peanut cultivars and 44 peanut cultivars representing six botanical varieties (He and Prakash, 2001; Herselman, 2003). Studies have also identified genomic regions through the use of AFLP markers for resistance to biotic stress (early leaf spot, late leaf spot, rust, and groundnut rosette disease) (Dwivedi et al., 2002; Herselman et al., 2004). Although AFLP, RFLP, and RAPD were used extensively for peanut research, the use of these markers significantly decreased with the availability of more reliable and easy-to-use markers such as SSR and SNP.

7.2.3 SSR Marker

Microsatellites, a ubiquitous class of simple repetitive DNA sequences, thus also known as simple sequence repeats (SSRs), represent tandem repeats of 1–6 bp in length that are widely distributed across genomes (Mason, 2015). SSR is one of the most indicative markers to study polymorphism between DNA sequences due to its high polymorphism, high abundance, co-dominant inheritance, multi-allelic nature, and ease of scoring (McCouch, Chen, Panaud et al., 1997). In one early study,^{32}P-labeled dinucleotide repeats were used to discover SSRs in peanut. This involved hybridizing the repeats with clones containing peanut genome sequences. Out of the total of 26 SSR primer pairs identified through this technique, 5 were found to be polymorphic in nature based on screening of wild and cultivated peanut (Hopkins et al., 1999). Another study used DNA fragments from AFLP marker protocol and hybridized them with biotinylated SSR probes, which identified 19 polymorphic SSRs (He et al., 2003). Later, several studies discovered new sets of SSR markers from various genetic sequences, such as ESTs, transcriptome sequences, and genomic sequences. The use of a combination of EST and RNAseq dataset revealed over 6,000 SSRs with a subset of 380 validated through PCR (Peng et al., 2016). Recent research on the genome-wide search for SSR has identified approximately 136,000 SSRs from the A genome (*A. duranensis*) and 200,000 SSRs from the B genome (*A. ipaensis*) (Zhao, Qiu et al., 2017). In addition, high-quality genome assembly of the tetraploid peanut genomes has helped to identify around one million potential SSR markers (Lu et al., 2019).

SSRs have been used for constructing detailed genetic maps, fine-mapping desired traits, and conducting genome-wide association studies. Out of all marker systems, SSR markers have been widely used in peanut research to construct genetic maps using biparental populations. The first SSR-based *Arachis* diploid genetic map was published using an F_2 population obtained from a cross between *A. duranensis* and *A. stenosperma* (AA genome) employing 204 polymorphic SSRs (Moretzsohn et al., 2005). Meanwhile, the first SSR-based map in tetraploid *A. hypogaea* L. was developed using 318 recombinant inbred lines (RILs) from the cross TAG 24 × ICGV 86031 using 191 polymorphic SSRs (Varshney et al., 2016). With the advancement of sequencing technology and the availability of more sequence information, the development and use of SSRs have significantly increased in peanut. This influx of SSR markers has enabled the creation of denser genetic maps, incorporating over 700 SSR markers to identify QTLs for various agronomically important traits (Shirasawa, Koilkonda et al., 2012; Huang et al., 2016; Luo et al., 2017; Liu et al., 2019). Additionally, consensus genetic maps were also developed with different populations and using SSR markers (Hong et al., 2010; Gautami et al., 2012a, Gautami et al., 2012b; Qin et al., 2012; Shirasawa et al., 2013). A total of ten RIL mapping populations and one backcross population were evaluated for drought tolerance, late leaf spot, rust, oil and protein content, tomato spotted wilt virus, and aflatoxin contamination. These populations were genotyped with SSR markers, and a consensus genetic map of these 11 populations was constructed through information provided from common markers used across different genetic maps. These consensus maps consist of 895 SSR marker loci distributed across 20 linkage groups (Gautami et al., 2012a). Another study integrated 16 linkage maps and developed a consensus map consisting of 3,693 SSR markers and covering 2,651 cM of tetraploid peanut genome (Shirasawa et al., 2013). These dense reference consensus maps using SSR-based markers helped peanut researchers align new marker information on the genetic map and precisely identify genomic regions for the traits of interest. SSRs are widely dispersed throughout the genome and are commonly used to analyze diversity panels, such as the peanut mini-core collection, and to carry out genome-wide association studies (GWAS) for agronomical traits (Wang et al., 2011; Jiang et al., 2014; Zhao, Li et al., 2017; Daudi et al., 2021). However, the accessibility of SNP arrays and cost-effective sequencing methods have resulted in SNP markers replacing or complimenting SSRs, especially in GWAS studies.

7.2.4 SNP Markers- and Sequencing-Based Genotyping and SNP Arrays

SNP markers are highly effective and can reveal changes in a single nucleotide (A, C, G, or T) in the genome. These markers are unequivocally superior to other marker systems that rely on discerning band size. SNP markers can offer greater affordability, require significantly less effort and time, and minimize the likelihood of errors. They are also ubiquitous and located in genetic regions linked to the desired traits. As a result, they can significantly improve the effectiveness of breeding programs and are a valuable option for marker-assisted selection (MAS) in breeding programs.

There are three alternatives to utilizing SNP marker information for high throughput genotyping: 1) whole genome resequencing (WGRS), 2) reduced representation sequencing (RRS), and 3) SNP arrays (Pavan et al., 2020). Whole genome resequencing is an effective but costly method that can detect all variations between two accessions. It can identify different types of variations like SNP, structural variation, and copy number variation. However, for studies that do not require a lot of polymorphism marker information, genotype-by-sequencing (a form of reduced representation sequencing) is often used to reduce costs. Although it may not be as effective at detecting indels and copying number variation, it can still accurately identify SNPs between accessions. The Affymetrix and Illumina platforms provide high throughput SNP array genotyping through sets of probes designed to bind to specific SNPs (Rasheed et al., 2017). Although SNP arrays are inexpensive and easy to use, they do not detect all forms of variation and novel mutation (Nowak et al., 2009). The high-throughput sequencing-based SNP markers on peanuts have been reported in several studies and were applied in genome-wide association studies (GWAS) and genome-wide selection (GWS).

The first high-density map in *Arachis duranensis* was developed using a combination of EST-based SSR markers and SNP markers (Alves et al., 2008; Nagy et al., 2012). The SNP data from this study was used to develop a 1536 Illumina GoldenGate SNP assay, which was further used to develop genetic maps using two RIL mapping populations, one diploid (*A. duranensis* × *A. stenosperma*) and one tetraploid [*A. hypogaea* cv. Runner IAC 886 × synthetic tetraploid (*A. ipaënsis* × *A. duranensis*) (Bertioli et al., 2014). SNP markers were validated using PCR-based methods. Khera et al. (2013) applied Kompetitive allele-specific polymerase chain reaction (KASP) assays to develop peanut-specific SNP markers and validated 73 polymorphic SNP markers in a panel of 94 genotypes that included cultivated peanut, synthetic autotetraploid/amphidiploid lines, and wild species. High-resolution melting (HRM) was also applied to validate SNPs in peanut. Analyzing 200 amplicons derived from expressed sequence tags (ESTs) and genome survey sequences (GSS), Hong et al. (2015) developed 62 SNP markers (18 EST-SNPs and 44 genomic SNPs) among 12 *A. hypogaea* varieties and two diploid wild species (*A. duranensis* and *A. ipaensis*).

Information from transcription sequences has proven effective in developing a large set of genetic markers in a complex polyploid genome. Specifically, studies have identified 189,705 SNPs (Chopra et al., 2015), 11,902 SNPs

(Peng et al., 2016), and 6,965 SNPs (Peng et al., 2017) from transcriptome sequences. Through double-digest restriction-site-associated DNA sequencing (ddRADseq), over 53,000 SNPs were identified between two peanut varieties involved in developing the RIL population. This information aids in creating the high-density SNP-based genetic map for *A. hypogaea* containing 1,621 SNPs, distributed in 20 linkage groups and covering a distance of 1,446.7 cM (Zhou et al., 2014). Another study involving the RAD sequencing technique found only 3,248 SNPs among tetraploid genotypes, which is substantially lower than in diploid peanuts and other crop species. The reason for this relatively low number of SNPs can be attributed to the limited genetic diversity in tetraploid genotypes and the computational difficulties in identifying SNPs in the tetraploid genome (Gupta et al., 2015). Several studies have been done involving RRS and SNP-based linkage mapping to identify genomic regions of important agronomic traits (Han et al., 2018; Li et al., 2019; Zhang et al., 2019; Khan et al., 2020; Wang et al., 2020). Compared to reduced representation sequencing (RRS), whole genome resequencing (WGRS) provides a higher resolution and comprehensive view of the genome, generating a much larger number of SNPs that can be used for genetic mapping and association studies. In a study conducted by Agarwal et al. (2018), WGRS was employed on a RIL population, resulting in the generation of a high-density map consisting of 8,869 SNPs. This dense SNP map was then utilized to identify SNP markers associated with early leaf spot (ELS), late leaf spot (LLS), and tomato spotted wilt virus (TSWV). Furthermore, WGRS was performed on 203 accessions from four different botanical classifications, leading to the discovery of 5,439,320 SNPs in cultivated accessions. These SNPs were subsequently used in GWAS to identify genomic regions and candidate genes associated with seed traits (Liu et al., 2022). A comprehensive analysis of WGRS reads from a dataset of 179 peanut accessions, available on the Sequence Read Archive (SRA) database (https://www.ncbi.nlm.nih.gov/sra), revealed the presence of 4,309,724 SNPs. Among these SNPs, a subset of 46,087 exhibited high polymorphic information content (PIC) values of 0.38, indicating their potential utility in genetic studies and breeding programs. The utilization of WGRS in these studies demonstrates its effectiveness in capturing a comprehensive set of genetic variations across the entire genome. The larger number of SNPs obtained through WGRS enhances the resolution and power of genetic mapping and association analyses, enabling the identification of specific genomic regions and candidate genes that may be associated with important traits. Furthermore, in peanut studies, the combination of WGRS with bulk segregant analysis (BSA), known as QTLseq, has been frequently employed. Luo et al. (2019) utilized QTLseq to identify SNPs for marker-assisted selection for shelling percentage, while Zhao et al. (2020) used it to identify SNPs associated with testa color. Similarly, QTLseq has been used to identify SNPs related to fresh seed dormancy (Kumar et al., 2020), seed weight (Wang, Yan et al., 2022), and various biotic stresses (Pandey, Khan et al., 2017; Clevenger et al., 2018; Luo, Pandey, Khan et al., 2019). This integration of QTLseq with WGRS enables the identification of SNPs with potential applications in marker-assisted selection and the study of complex traits in peanuts. By analyzing genotypic data obtained from this array, researchers have been able to study the genetic basis of important traits in peanuts, such as disease resistance, yield, quality, and abiotic stress tolerance. This has accelerated the development of improved peanut cultivars.

Zhang et al. (2017) reported 17,338 SNP markers in the cultivated peanut genome using the SLAFseq method based on 158 peanut accessions and analyzed SNPs associated with 11 agronomic traits using genome-wide association studies (GWAS). Through resequencing of 41 peanut samples with distinct genotypes, Pandey et al. identified 98,375 and 65,407 SNP markers from the A and B subgenomes, respectively. A total of 58,233 unique and informative SNPs were selected for the development of a high-density SNP array "Axiom_Arachis" for further high throughput detection and precise genotyping of SNP markers in peanut (Pandey, Agarwal et al., 2017). The Axiom_Arachis array has proven to be a valuable tool in genetic studies, including QTL mapping, GWAS, population structure analysis, and marker-assisted breeding (Chu et al., 2020; Gangurde et al., 2020; Pandey et al., 2020; Zou et al., 2020; Patel et al., 2022; Wang, Wang et al., 2022).

7.2.5 Miscellaneous Markers Sets

In addition to the marker systems mentioned above, peanut researchers have also sporadically used a few other systems. InDels are a common type of structural variation in plant genomes caused by replication slippage, transposable elements, and crossing-over. A total of 48 InDel markers were developed from the conserved sequence of functional genes. These markers were tested in a diverse panel of 118 accessions to identify markers that could be used to indicate resistance to disease (Liu et al., 2015). In a GWAS study, 49 peanut cultivars were analyzed using RNA-Seq data to identify differences in gene expression and InDels. The study discovered a total of 61,942 InDels, and 48 of them were significantly linked to seed oil content in various environments. This study also helped pinpoint crucial genomic regions on the A03 chromosome that can be used for MAS in peanut breeding programs (Wang, Xu et al., 2018).

Diversity arrays technology (DArT) is a high-throughput assay based on microarray hybridizations that analyze genetic variation across a genome without needing prior DNA sequence information. It involves two steps: complexity reduction and hybridization-based genotyping (Wenzl et al., 2004). The DArT platform has been developed for peanut, which contains more than 15,000 features (DArT Pty Ltd, Australia). This platform has been used in a few studies to identify genomic regions associated with disease resistance, yield traits, oil content, and fatty acids (Pandey et al., 2014; Shasidhar et al., 2017). Despite its high throughput capability, the DArT assay has proven to be ineffective in achieving a high level of polymorphism in tetraploid peanuts (Pandey et al., 2012). As a result, it has lost its appeal in both peanut breeding and genetic research.

Transposable elements (TEs), often called jumping genes, are segments of DNA that can move within a genome, ranging from 100 bp to several kilobases. Due to their widespread presence throughout genomes and their variable insertion sites, transposons have been utilized to create TE markers, which aid in studying genetic variation and evolutionary processes (Kumar and Hirochika, 2001; Bhat, Shirasawa et al., 2020). Transposable elements can disrupt the function of genes. One such example in peanut is the disruption of the fatty-acid desaturase-encoding gene ahFAD2B (Patel et al., 2004). Identifying variations in the transposon insertion site makes it possible to create a closely linked marker for this trait. In a study, 504 primer pairs were designed to target the AhMITE1 sequence. Out of these, 169 primer pairs showed polymorphic bands among the four accessions evaluated, suggesting that insertional polymorphism can be utilized to understand peanut diversity and develop trait-associated markers (Shirasawa, Hirakawa et al., 2012a). Furthermore, TE markers have been identified for rust resistance, high-oleic peanut, nutritional traits, productivity traits, LLS resistance, plant height, and pod width (Mondal et al., 2014; Hake et al., 2017; Wang, Song et al., 2022).

7.2.6 Epigenetic Signatures

Epigenetic signatures are changes to DNA methylation or histone modification that can be impacted by external factors. These changes can affect gene expression and modulation of chromatin structure, which can also be inherited by offspring. Such modifications can influence plant growth, morphology, plasticity, and ability to respond to biotic and abiotic stress (Kooke et al., 2015). Limited focus has been placed on identifying epigenetic signatures and their association with traits of interest. Research has shown that histone modification is linked to regulating the peanut allergen gene Ara h 3 expression in various peanut tissues, such as developing embryos (Fu et al., 2010). Another study also revealed that histone deacetylation plays a crucial role in controlling the seed storage protein (SSP) gene Ara h 2.02 expression (Yang et al., 2015). Similarly, a study on global methylation has discovered the connection between siRNAs and miRNAs and DNA methylation and the effect of methylation on gene expression during the development of peanut pods (Wang, Shi et al., 2018). These findings indicate that epigenetic modifications play an essential role in pod development and suppressing genes related to seed development while the peanut plant is in the vegetative growth stage. Through conducting a bisulfite sequencing and RNA-Seq experiment on 11 peanut accessions, researchers were able to identify various differentially methylated sites between the foliar disease-resistant and susceptible genotypes. These sites had an impact on gene expression (Bhat, Rockey et al., 2020). Similarly, Liu, Wu et al. (2022) theorized that changes in methylation can affect the biosynthesis of oil in peanut seeds by impacting the expression of lipid metabolic genes, which in turn can affect the oil content in the seeds. To identify differentially methylated regions (DMRs), whole-genome bisulfite sequencing (WGBS) is typically used, but it can be expensive. Cheaper alternatives like reduced representation bisulfite sequencing (RRBS), enrichment-based methods (such as MeDIP-seq), and methylation-sensitive restriction enzyme sequencing are also available (Chen et al., 2016). Additionally, array-based whole-genome DNA methylation profiling has also been developed in human genetics, which can be high-throughput and cost-efficient (Bibikova et al., 2009; Bibikova et al., 2011).

While peanut researchers have developed a well-established pipeline for DNA markers, it is time to also focus on developing epigenetic markers or assays and incorporating them into peanut breeding programs to avoid missing differentially expressed genes due to epigenetic modification. Peanut researchers should expand their DNA marker pipeline and develop epigenetic markers or assays. This will help incorporate epigenetic modification into peanut breeding programs that impact the phenotypic outcome by controlling gene expression and preventing missed opportunities.

7.3 GENETIC MAP DEVELOPMENT AND APPLICATION

7.3.1 Segregating Population Development

Genetic maps are the basis of QTL mapping, map-based cloning, and MAS breeding. The first step in constructing a genetic map is to establish suitable populations, such as F_2 population, backcross (BC) progenies, RILs, double haploids (DHs), near isogenic lines (NILs), and nested association mapping (NAM). According to the accuracy of QTL localization, mapping populations can be divided into primary localization populations and fine localization populations. The F_2 population, RIL, BC population, and DH population belong to the primary localization population. Peanuts are widely used as self-pollination crops, and F2 populations and RILs are also widely used. In particular, the RIL populations, which are homozygous and stable in population structure, are also known as permanently separated populations. Based on the primary population, the secondary population can also be derived for fine mapping, such as NILs, residual heterozygous lines (RHLs) and QTL isogenic recombinants (QIRs). There is also a class of substitution populations that are not associated with the primary localization population, like introgressive lines (ILs), single segment substitution lines (SSSLs), and chromosome segment substitute lines (CSSLs).

RILs have been extensively used in peanut mapping for various quantitative traits and have been proven to be effective. To overcome the limitations of a biparental population as a relatively narrow genetic base represented by the two parents, multiparental mapping populations have been created. Nested association mapping (NAM) and multi-parent advanced generation inter-crossing (MAGIC) are the two most important multi-parent mapping populations. A NAM population was created in peanut (Holbrook et al., 2013) in which two common parents, Tifrunner (Holbrook and Culbreath, 2007) and Florida-07 (Gorbet and Tillman, 2009), were selected as common females and crossed with eight unique male parents: N08082olJCT, C76-16, NC 3033 (Beute et al., 1976), GP-NC WS 16 (Tallury et al., 2014), SSD 6 (PI 576638), OLin (Simpson et al., 2003), New Mexico Valencia A (Hsi and Finkner, 1972), and Florunner (Norden et al., 1969), resulting in 16 structured populations. These populations provided significant genetic sources with maximum genetic diversity and target traits important to US peanut breeding programs to work with on QTL mapping and genomic research. A MAGIC population comprising 600 RILs together with eight founder parents was developed at ICRISAT (International Crops Research Institute for the Semi-Arid Tropics) by making all possible two-way inter-crosses between eight diverse parents (ICGV 91114, ICGV 06040, 55-437, ICGV 00440, ICGV 00308, GPBD 4, ICGV 05155, and ICGV 88145), followed by intercrossing the F_1s from the two-way crosses, and then intercrossing the F_1s' four-way crosses. Then, the F_1s obtained from four-way intercrosses were advanced to F_2 generation, and the generation advancement from F_2 to F_8 was carried out following single seed descent (SSD). With this MAGIC population, a total of 62 markers–trait associations (MTAs) were identified across the A and B subgenomes for LLS resistance in peanuts (Wankhade et al., 2023).

7.3.2 Genotyping of Population

To determine the genotype of each line in the population, some molecular markers were used, such as RFLP, RAPD, SSR, sequence-characterized amplified regions (SCAR), AFLP, cleaved amplified polymorphic sequences (CAPS), Kompetitive allele specific PCR (KASP), DArT, and SNP. Based on reliability, quality of DNA, polymorphism, mode of inheritance, costs, reproducibility, time, and so on, the polymorphic markers will be used to construct the linkage map. With the next-generation sequencing technology, sequencing-based markers, such as SNPs, are becoming more common for genetic mapping and QTL analysis as they are saturated on the genetic maps. The appropriate markers were selected first to survey the level of parental polymorphism. It is worth noting that finding polymorphic markers covering the genome is essential. These polymorphic markers are to be detected in the progenies of the given mapping population. Generally, markers segregate in a Mendelian fashion, but sometimes, there is partial separation. Calculating the recombination fractions can estimate the relative order and genetic distances between markers.

7.3.3 Linkage Analysis and Map Construction with High Throughput Software

Genotyping-by-sequencing (GBS) was employed to construct a highly saturated genetic linkage map or physical map in given populations. The genetic or physical map of mapping populations was constructed with SNP markers distributed on each chromosome. The linkage map can be constructed with software such as MAPMAKER/EXP, Join Map or R/qtl, MapMaker, and MapManager.

7.3.3 Bin Map Construction

Through the GBS approach, we can construct a high-resolution genetic map of the mapping population, but GBS includes a certain proportion of missing data, rarely corrected, and a percentage of SNP genotyping sequencing errors. Recently, researchers developed a pipeline for imputing missing SNP data in the sliding-window approach, where adjacent SNPs with the same genotype in an interval are combined into bins. For example, the maps of the RILs were aligned and compared for their genotypes for a 100kb interval. Adjacent 100kb intervals with the same genotype across all RILs were combined into a recombination bin. The linkage map was constructed from the recombination bins serving as genetic makers using MAPMAKER/ EXP V3.0. The bin-map method is demonstrated to be more powerful for detecting QTLs than traditional methods and has also been employed for fine-mapping QTLs in some crops.

7.4 MAPPING OF AGRONOMIC TRAIT LOCI

7.4.1 Linkage Mapping of QTLs

A QTL is a chromosomal region of DNA that is associated with variation of a quantitative trait in the phenotype of a population. QTLs are mapped by identifying which molecular markers correlate with an observed trait, and this is the first step in identifying the actual genes that cause the trait variation. QTL mapping is a genome-wide inference of the relationship between phenotypes for a set of quantitative traits and genotypes at various genomic locations in terms of the quantity, genomic positions, effects, and interaction of QTLs. Linkage mapping of QTLs can help us to understand the genetic architecture of quantitative traits in the population, such as how many QTLs significantly affect the trait variation in a population. With the development of statistical approaches and molecular linkage maps for QTL mapping, it is possible to detect the genetic loci controlling genetically complex traits.

7.4.2 Association Mapping Covering GWAS

Association mapping is a high-resolution method for mapping QTLs based on the linkage disequilibrium (LD) principle. It aims to identify specific functional genetic variants (loci, alleles) linked to phenotypic differences in a trait to facilitate the detection of trait-causing DNA sequence polymorphisms and the selection of genotypes that closely resemble the phenotype. The size of the experimental population, the density of markers used, the magnitude of the target allele effect, the rate of LD decay, errors in phenotyping and genotyping data, and the desired resultant statistical significance level can affect the power of association study. SNP and SSR markers have been widely used in plant breeding programs due to their high abundance, high degree of polymorphisms, codominant inheritance, ease of use, and relatively low cost.

Based on the association mapping, high-throughput genotype data coupled with phenotypic data can be used to identify marker–trait associations via GWAS. GWAS can potentially offer higher mapping resolution at a lower cost in terms of both time and money; therefore, it has emerged as a powerful tool for identifying markers closely linked to QTLs. It has been successfully conducted in many major agronomic crops, such as wheat, soybean, and cotton, to detect genes or markers responsible for various QTLs. In peanut, the first GWAS report was produced in 2014 by Pandey et al. (2014). In that study, 300 genotypes were tested for 36 traits, including biotic and abiotic resistances, seed quality, and yield. Subsequently, GWAS has been widely used in peanut for different traits, such as leaf spots, ionomic variation, seed composition traits, root-knot nematode resistance, growth habit–related traits, seed aspect ratio, and major agronomic traits related to domestication.

The software used for GWAS includes the SVS software package, GAPIT, Tassel, mrMLM, and so on. Compared with other software, the SVS model usually fits the data better. The phenotype data, genotype data, and genetic marker map must be imported into SVS. To efficiently correct population structure in the association test, efficient mixed-model association eXpedited (EMMAX) analyses in the SVS software package using the first ten principal components as covariates were conducted for GWAS. The model is listed as follows:

$$Y = Xb + Zu + e,$$

where Y is the vector of phenotype; X is the matrix of fixed effects including the first ten principal components; b is the vector representing the coefficients; Z is the matrix of random effect; u is the coefficient vector, $\mathrm{Var}(u) = \sigma^2 gG$, where $\sigma^2 g$ represents the additive genetic variance, and G stands for the genomic kinship matrix; and e is the vector of random residuals. The threshold P value for genome-wide significance was calculated based on Bonferroni correction with the estimated number of independent SNPs and LD blocks.

7.4.3 Mapping by Sequencing Covering BSAseq/QTLseq, BSRseq, and Mutseq

Besides GWAS, other methods such as BSAseq (bulked segregant analysis), RNAseq, QTLseq, BSRseq, and Mutseq could also detect agronomic trait loci. BSAseq, which only requires the genotyping of individuals with extreme phenotypes, has been widely used in QTL identification and gene mining related to specific traits. This approach has been successfully deployed in locating genes in many crops, including rice, soybean, maize, and peanut. QTLs of testa color, bacterial wilt resistance, chilling tolerance, and nodulation factors have been mapped in peanuts using BSAseq. Recently, a series of NGS-based BSA approaches were developed, including QTLseq, BSRseq, and Mutseq. QTLseq is a whole-genome resequencing-based approach, which firstly resequences two bulked DNA of progenies (each with 20–50 individuals) showing extreme phenotypic values and then identifies the candidate region or genes by counting and comparing the index SNPs between two bulks. Faster identification of genomic regions and candidate genes for the target traits is made possible by sequencing-based mapping approaches. The QTLseq approach is one of the successful methods for sequencing pooled samples for RILs with extreme phenotypes and parental genotypes. This strategy is cost-effective since it uses fewer samples for sequencing and facilitates the development of cost-effective markers in less time. QTLseq has been used in cultivated peanuts to identify QTLs or genes related to shelling percentage, leaf spot resistance, bacterial wilt, and seed weight (Clevenger et al., 2018; Luo, Pandey, Khan, Guo et al., 2019, 2020; Wang, Yan et al., 2022).

Based on BSA theory, Mutseq is a unique, highly parallel approach to defining functional residues of proteins based on their mutability alone. This approach, which maps coding sequences for residues that demonstrate statistically high or low rates of mutability, makes use of the depth provided by next-generation sequencing to evaluate complex pools of altered genomes and genes for functioning. Mutseq involves operationally the following steps: i) mutagenesis of a genome or gene; ii) recovery of a bank of mutagenized targets under a positive selection condition; iii) deep sequencing of the entire bank; and iv) alignment of sequence reads to identify and quantify base substitutions within the genomes or genes that represent mutations. MutMap is suitable for the analysis of recessive mutated genes mutated by EMS. After the homozygote is obtained by EMS mutagenesis and self-cross, the mutant and its parent are backcrossed to obtain F_1. The F_2 offspring obtained by F_1 self-cross will appear to have phenotypic separation, and wild-type phenotypic population and mutant super-type population will be obtained. The DNA of these two populations was mixed in equal quantities, the wild-type

DNA mixed was pooled, and the mutant DNA mixed pool was obtained. DNA sequencing was performed on the two mixed pools, data analysis was conducted using the MutMap pipeline, and the frequency of SNP occurrence in the mutant mixed pool and wild-type mixed pool was calculated to obtain the chromosome segment linked to the mutant phenotype and the possible mutation site.

7.4.4 Gene Expression Profiling: RNAseq

After obtaining the QTLs and candidate genes from the above methods, we need to check gene expression profiling by RNAseq, an easy and cost-effective method to identify differentially expressed genes (DEGs). RNAseq is a sequencing technique that uses next-generation sequencing (NGS) to reveal the presence and quantity of RNA in a biological sample, and it relies on the principle that read counts for each transcript from the NGS data reflect relative transcript concentrations. RNA extraction is the first step for RNAseq, and many factors need to be considered, such as treatments, sampling time, and tissue. RNA samples were pooled for three biological replications. Total RNA was extracted using the RNeasy Extraction Kit. Gel electrophoresis and Nanodrop 2000 were used to determine the quality and quantity of total RNA. High-quality RNA was used for transcriptome sequencing.

FastQC software was used to check raw read quality, and then Trimmomatic v.0.33 was performed to remove adapters and low-quality reads and sequences (Andrews, 2010). Tophat2 was used to align with all the parameters set to the default. The cultivated peanut genome Tifrunner and the annotation GFF file were used as a reference for alignment. The alignment files of the samples resulting from Tophat2 were input into Cufflinks for genome-guided assembly and differentially expressed gene analysis. The DEGs were identified using Cufflinks and Cuffdiff. The DEGs were obtained through the comparison of gene expression among different samples.

In terms of the status of QTL identified by all means, a group of scientists from France, Ethiopia, Senegal, and Cameroon compiled all efforts of QTL mapping and conducted a comprehensive review to summarize QTL mapping in peanuts (Kassie et al., 2023). The review indicated that 1,243 QTLs were identified at the present time, as shown in Table 7.1. The total number of QTL identified for agronomic and morphological traits was 556, followed by seed quality-related traits (413), disease and pest-resistance traits (134), and abiotic stress traits (140), respectively.

7.5 APPLICATION OF MAS IN PEANUT BREEDING

MAS, with modern technology applied to conventional breeding, can be used throughout the whole growth period of crops. Sexual hybridization is a significant approach to peanut conventional breeding. Because of the self-pollination style, the progress of hybridization requires artificial emasculation and pollination, which means extirpating the flowers with non-pollination to avoid the appearance of non-hybridized pegs. Incomplete removal of the stamen androecium in the female and other reasons may cause the appearance of false hybrids. Breeders generally select hybrids through the morphologically identified method, observing the phenotype of F_1 and the situation of trait separation in F_2 to eliminate false hybrids. However, demanding conditions and high costs were involved in phenotyping selection. MAS has become the more favored breeder choice.

Because of the specialty of the whole growth period in peanuts, the pods are under the soil, and the breeder needs to dig out pods in order to realize the agronomic traits, nutrition traits, and traits of stress tolerance, for example, weight and size of seed, content of oleic acid, and pods rot. Deploying MAS during the early stages of the peanut growth period will provide great support to make decisions on other important traits in future generations. MAS was used in combination with conventional breeding to obtain high-yield, high-quality, and highly resistant peanut varieties.

7.5.1 MAS in Agronomic Traits

Production of peanut has been unable to satisfy global demand with the rapid increase of human populations worldwide. There is a great potential to further enhance peanut production through the genetic improvement of high-yield varieties with MAS. Yield is one of the most important and complex traits in peanut, which is directly and indirectly influenced by agronomic traits, such as the height of the main stem, length of lateral branches, morphology of pods, and total branching number. The pod-related traits include pod length, pod width, seed length, and seed width, which directly influence peanut yield.

Agronomic and quality-related traits are almost always quantitative traits in plant species. Linkage mapping or QTL mapping based on segregating populations derived from bi-parental crossing is the most popular and successful method of identifying QTLs that control complex traits in plants. QTLseq was applied for a RIL population (ZH16×sd-H1), and three major QTLs on chromosomes A05, B02, and B06 were determined. Among these QTL regions, *qSWB06.3* exhibited stable expression with a large contribution to phenotypic variance across all environments. Three KASP markers (*Ah011475*, *Ah011476*, and *Ah011478*) in *qSWB06.3* were successfully validated in 60 natural peanut varieties, including 30 varieties with high seed weight and 30 varieties with low seed weight; they amplified the AA, TT, and TT alleles in the parent ZH16 of high seed weight, while CC, GG, and GG alleles in parent sd-H1of low seed weight (Wang et al., 2022). Researchers identified seven QTLs related to seed weight and shape on chromosome A05 by deploying QTL-seq in a segregating

TABLE 7.1
Summary of QTL Identified for the Important Traits in Peanut

Class	Traits Studied	Number of QTLs Identified	% of Phenotypic Variance Explained
Agronomic and morphological related traits (556)	Plant height	77	0.01–26.7
	Hundred pod weight	48	3.33–38.15
	Fresh seed/seed dormancy	54	69.3–74.7
	Days to flowering	31	1.15–21.82
	Pod weight	20	7.7–29.7
	Pod length	52	1.25–26.46
	Pod width	54	5.1–43.63
	Seed length	32	3.03–20.8
	Seed width	33	2.21–23.7
	Harvest index	15	11.0–18.1
	Hundred seed weight	42	5.68–35.9
	Haulm weight	11	2.9–33.36
	Pod number	24	3.91–14.2
	Total biomass	15	4.34–22.39
	Growth habit	48	4.55–27.14
Disease and pest-resistance traits (134)	Nematode	7	1.3–22.18
	Leaf spot	80	1.7–50.9
	Bacterial wilt	3	0.12–0.22
	Rust	34	7.24–48.7
	Smut	10	7.24–11.
Abiostress traits (140)	Shoot dry weight	16	4.2–22.09
	Transpiration efficiency	27	4.47–18.12
	Leaf area	26	5.0–16.2
	Transpiration rate	13	4.3–17.3
	Transpiration	16	4.36–18.17
	SPAD chlorophyll meter readings (SCMR)	29	4.00–19.53
	Stress tolerance index	13	10.4–20.1
Seed quality–related traits (413)	Oil content	80	0.76–27.19
	Protein content	22	0.76–26.99
	Oleic acid	58	0.13–57.56
	Linoleic acid	54	0.17–57.56
	Oleic/linoleic acid ratio	32	1.04–43.41
	Palmitic acid	32	0.3–34.35
	Arachidic acid	32	0.13–36.93
	Stearic acid	31	0;13–78.6
	Behenic acid	32	0.76–26.99
	Eicosanoid	1	0.2
	Lignoceric acid	15	2.89–12.61
	Gadoleic acid	16	2.55–15.11
	Arachidonic acid	8	0.76–26.99

Source: (Adapted from Kassie et al., 2023)

population obtained from the progeny of Tifrunner × IpaDur. They used molecular markers on chromosome 5 of peanuts reported in the literature and newly developed markers for verification. Four markers (A05A1344, AX-147223558, AX-147223576, and AX-147223670) showed a strong linkage with the weight and shape of peanut seeds (Huang et al., 2022).

In addition, pod size is a vital biological and economic element of peanuts, being the determinant of pod weight. Pod size is also correlated with grain quality traits and grain appearance. Desirable grain appearance varies in different regions of the world. Therefore, it is important to breed peanut cultivars with various grain sizes to meet the needs of different market classes. Forty-six putative candidate genes were found by using GBS to sequence 250 accessions of the Chinese peanut mini-core collection. The diagnostic marker for yield-related traits from non-synonymous SNP (Aradu-A06-107901527) was successfully validated, which showed that the GG genotype contained a significantly higher seed (pod)-related trait value than those having the TT genotype (Zhou et al., 2021). A RIL population (Chico×ICGV02251) of 385 lines was developed to identify candidate genes and develop markers by QTLseq and BSA. Four KASP markers (*SNPAh0031*, *SNPAh0033*, *SNPAh0037*, and *SNPAh0038*) were developed and successfully validated for deployment in breeding for large-seeded peanut varieties (Gangurde et al., 2022). The large pod variety Luhua11 and small pod variety BC54 were used as crossing parents, the F_1 generation was obtained, and those true hybrids were screened by KASP analysis. The screened true hybrid progeny was self-pollinated continuously to produce an RIL population containing 520 lines (Hu et al., 2022).

Furthermore, the growth habit of peanuts is another element that determines the yield. The pod number per plant is influenced by peanut growth habit, which can be categorized into four types: erect, bunch, spreading, and prostrate. Restricting pod development at the plant base would decrease pod yield, as would be the case for peanut plants with upright lateral branches. On the other hand, growth habits characterized by the spreading of lateral branches on the ground facilitate pod formation on the nodes, thereby increasing yield potential. QTLseq and analysis of resequencing data were applied to an RIL population containing 521 lines derived from a cross between YZ9102 and wt09-0023 to identify QTLs on chromosomes A05 and A06. Four KASP markers (*Arahy15.156854742*, *Arahy15.156931574*, *Arahy15.156976352*, and *Arahy06.111973258*) were developed and validated in a panel of 77 accessions with distinct growth habit. Among them, 55 spreading or prostrate RILs had consistent genotypes featuring chromosome 15 markers "C:C"/"A:A"/"-:-" and chromosome 6 INDEL marker $C(ATT)_3$; 31 erect RILs carried the "C:C"/"A:A"/"-:-" genotype and $C(ATT)_5$ INDEL genotype; and 79 erect RILs carried genotype "A:A"/"C:C"/"TA:TA" on chromosome 15 and either $C(ATT)_5$ or $C(ATT)_3$ as the INDEL genotype on chromosome 6 (Fang et al., 2023).

Fresh-seed germination (FSG) prior to harvest has become a growing problem, especially in regions where the harvest season coincides with wet weather. FSG-affected peanut seeds are susceptible to pathogen infection, which renders them vulnerable for seed contamination by aflatoxin or other mycotoxin detrimental to human health. Developing FSG-resistant cultivars by molecular breeding is expected to mitigate yield loss and quality impairment caused by FSG. An RIL population comprised of 521 lines was developed by single-seed descent from a cross between YZ9102, which is FSG-susceptible, and wt09-0023, which is FSG-resistant. QTLs associated with FSG were mapped. By designing the KASP marker to genotype the RIL population with 200 lines and a verification group comprised of 96 cultivars, one SNP and one InDel associated with FSG were identified. T:T (YZ9102 type) and G:G (wt09-0023 type) represented the genotype at Arahy.04:101089930, while T:T (YZ9102 type) and /:/ (wt09-0023) represent the InDel at Arahy.04:100184540 (Zhang et al., 2022).

The testa color is an important trait of peanut with many variations, such as white, pink, red, and black. Anthocyanin content and composition are crucial factors in determining the color of the testa. The testa was developed from the integument and with the same genotype as its maternal plant, which means the testa color phenotype appears one generation later than other traits. For conventional breeding methods, the phenotype of the testa color can only be identified after harvest. MAS, selecting the target progenies according to the genotype, can significantly accelerate the breeding process and improve breeding efficiency, especially for testa color selection. By screening 96 SNPs from genotyping data and polymorphism in 80 tetraploid lines, peanut KASP array markers (GKAMs) were constructed. A total of 73 GKAMs were validated and showed polymorphism across 94 genotypes (validation set) (Khera et al., 2013). Two F_4 populations were constructed through the cross of YZ9102 (pink testa) with ZH12 (red testa) and Zhanhong2 (red testa). The use of BSA-seq and linkage mapping identified *AhRt2* on chromosome 12. A KASP marker *KASP_AhRt2* was developed and used to detect sites in 220 lines of YZZH12 populations and 57 lines of YZZH2 populations, and found that the locus of almost all red testa lines was "T:T", and the pink testa lines were "C:C" or "T:C". Results suggested that *KASP_AhRt2* could be used as the diagnostic genotyping marker to predict red testa through MAS (Zhang et al., 2021).

7.5.2 MAS in Stress Tolerance

In recent years, the incidence of diseases and damage caused by disease in peanut-producing areas around the world are increasing due to the harsh environment caused by the intensification of global climate change and poor planting methods such as single planting varieties and continuous cropping.

One example of MAS in peanut breeding is the use of molecular markers to identify resistance to fungal diseases such as late leaf spot and rust. By identifying and selecting plants with resistance genes, breeders can develop peanut varieties that are less susceptible to these diseases. Peanut bacterial wilt, a soil-borne disease, seriously harms plants throughout the whole growth season, especially in the blooming period. The economic loss caused by bacterial disease is the most hazardous in peanut. Using marker-assisted selection to breed plant-resistant cultivars are the best approach to manage the disease. QTL-seq was applied for a RIL population (YZ9102 with resistance to bacterial wilt×Xuzhou68-4 with susceptibility to bacterial wilt), and a candidate genomic region associated with bacterial wilt resistance (BWR) was identified on chromosome B02. Three SNP markers were developed and validated on 21 resistant and 9 susceptible breeding lines; two markers (*Araip_B02_3740746* and *Araip_B02_5804063*) showed clear differentiation between resistant and susceptible breeding lines. *Araip_B02_3740746* amplified the CC alleles for the resistant lines while the AA alleles for the susceptible lines. *Araip_B02_5804063* amplified the GG alleles for the resistant lines, while the TT alleles for the susceptible lines (Luo et al., 2019). A population comprising 521 $F_{8:10}$ RILs was derived from a cross between YZ9102 and wt09-0023 to construct a genetic linkage map and identify QTLs for bacterial wilt resistance. Four major QTLs were determined on linkage groups LG03, LG12, LG14, and LG20. A major QTL on LG12 *qBWA12* showed a phenotypic variation explained (PVE) of 31.2%–44.8%. Forty-two polymorphic SNPs were converted to KASP markers, and 17 of them successfully genotyped the 521 RILs. These results implicated that *A12.4097252* was tightly linked with bacterial wilt resistance, and the resistant allele was G instead of C at the SNP site (Qi et al., 2022).

Foliar fungal diseases (rust and LLS) incur large yield losses and deterioration of fodder quality in peanut worldwide. The co-occurrence of two foliar fungal diseases, namely rust (caused by *Puccinia arachidis*) and LLS (caused by *Cercosporidium personatum*), causes chlorotic lesions leading to defoliation, which lowers crop yield and fodder quality. The application of fungicides can control these diseases. However, this increases the financial burden on farmers. The development and adoption of resistant cultivars are the most cost-efficient and effective ways to control the spread of the disease and reduce yield losses. MAS can speed the introgression of disease resistance. An RIL population, comprised of 94 lines, was developed by single-seed descent from a cross between GKBSPSc 30076 and GKSSc 30097 with high rust-resistance. A strong QTL linked to rust resistance was identified, and KASP markers suitable for foreground selection were developed; 19 of them successfully distinguished rust-resistant cultivars from non-resistant ones (Leal-Bertioli et al., 2015). By using polyploid-specific SNP identification pipelines identified robust parental SNPs, the problem of false positive SNP with high rates obscured significant QTLs can be solved. In addition, an LLS-resistant QTL was identified by deploying QTL-seq in a RIL population that developed from a cross between Florida-07 and GP-NC WS16. Three KASP markers were developed and used to phenotype the entire population; the results revealed a significant effect of each marker on leaf spot field resistance across all four years of data. Markers identified can then be deployed in breeding programs, increasing the efficiency of selection using molecular tools (Clevenger et al., 2018). Foliar disease resistance of three popular Indian groundnut varieties (GJG9, GG20, and GJGHPS1) was improved by marker-assisted backcrossing (MABC). Allele-specific markers for rust (*GMRQ517, GMRQ786*, and *GMRQ843*) and LLS (*GMLQ975*) were developed to perform foreground selection in progenies crossed by the genotype GPBD4 with foliar disease-resistant and Indian groundnut varieties. A total of 22 BC_3F_4 and 30 BC_2F_4 introgression lines (ILs) were developed. Up to 94% of genome recovery among second and third-backcross progenies was identified (Shasidhar et al., 2020).

Peanut root-knot nematode is an important pest of diseases because its narrow genetic base and limited gene flow lead to a lack of strong resistance alleles for pests and diseases in the primary gene pool. Using high-yielding and nematode-resistant cultivars in combination with rotation is the most efficient and effective way to control nematode populations and maintain yield while reducing the use of nematicides. RAPD markers (RKN410 and RKN440) and RFLP markers (R2430E, R2545E, and S1137E) were developed to breed cultivars with resistance to root-knot nematode initially; however, the validated results in other populations were unsuccessful (Garcia et al., 1996). SSR markers were used to test the relationship between DNA markers and resistance genes based on bulked segregant pools derived from F_2 progenies of the cross between Huayu22 and D099. SSR markers (*S32-380* and *S89-140*) linked to resistance to root-knot nematode were identified with genetic distance of 4.421cM and 7.404cM (Wang et al., 2008). Peanut root-knot nematode resistant variety *A. stenosperma* is used as a donor parent to incorporate resistance QTL into tetraploid lines. Sixteen reported SNPs were used to perform foreground selection and developed SNP markers to genotype breeding programs (Ballén-Taborda et al., 2022). Breeding aflatoxin resistance in peanuts is highly limited by many factors, such as complex interactions among the peanut varieties with the pathogen and lack of desirable parental genotypes with reliable and high-level resistance.

Moreover, the lack of rapid and accurate methods for phenotypic identification of aflatoxin contamination also represents a common technical drawback worldwide. Therefore, more precise identification of peanut materials resistant to aflatoxin contamination and the development of molecular markers applicable for MAS in breeding are of great significance. The 99 peanut accessions were also genotyped by restriction site-associated DNA sequencing (RAD-Seq) for GWAS. A total of 60 SNPs associated with aflatoxin production were detected, and they explained 16.87%–31.70 % of PVE. Among these markers, SNP02686 had the largest effect on AFB_2 content; accessions with genotype GG or GA of SNP02686 accumulated significantly lower AFB_1 and AFB_2 than AA genotypes (Yu et al., 2020).

Using Pearson correlation analysis and multiple linear stepwise regression analysis, 12 pairs of SSR markers and 48 pairs of InDel markers were used to detect 77 American peanut germplasm resources; the result revealed that four SSR markers (3A8, GM1760, PM163, and 14H6) and two InDel markers (InDel-004 and InDel-020) were significantly associated with peanut pod rot resistance. An F_2 segregating population was constructed to validate SSR markers, and the results revealed that markers can be applied for marker-assisted breeding to develop peanut varieties resistant to pod rot (Liu et al., 2019).

7.5.3 MAS in Nutrition Traits

Oil content is a polygenetic trait and is significantly influenced by the environment. In the current breeding programs, high oil content lines are selected based on the phenotyping data generated from multiple environments, which is an un-efficient and time-consuming process. Deployment of well-validated markers could accelerate the process and precision in genetic improvement. Allele-specific and CAPS markers are used for genotyping the F_2 progenies of the cross between SunOleic and ICGV05141. SSR markers were further used to screen recombinant lines and female parents. The majority of recombinant lines were in combinations of high oil content similar to ICGV05141 and high oleic acid content similar to SunOleic 95R, which suggested the success of marker-assisted selection (Bera et al., 2018). An F_2 population with 1,160 individual plants was constructed based on NILs derived from heterozygous residuals in the RIL population (XuHua13 with high oil content×ZhongHua6 with low oil content). Nine InDel markers with polymorphism between the parents were used to genotype the population and construct the genetic linkage map, demonstrating the feasibility of InDel markers for peanut QTL mapping (Tao et al., 2023). Using KASP markers, progenies of the genetic population related to oil content were detected by crossing with Yuhua 14 as a high-oil female parent and with LOP215 as a low-oil male parent. F_1 true hybrids were selected to obtain RILs (Qiu et al., 2022).

The nutritional quality, shelf life, and flavor of peanut oil and its products are reliant on different proportions of saturated fatty acids (SFAs), monounsaturated FA (MUFA), and polyunsaturated FA (PUFA). Peanut seeds, containing 22%–30% proteins and about 50% fatty acid, are a significant source of protein and healthy fatty acids. Unsaturated fatty acids, accounting for 80% of the total fatty acids in peanut, can lower blood lipids and cholesterol, preventing hardening of the arteries. High oleic peanut has a relatively longer shelf life. A diet with high oleic acid can reduce the risk of heart diseases by, for example, reducing systolic blood pressure and slowing down atherosclerosis. As a result, the holistic image and consumer perception of peanuts has shifted from high-energy foods to foods associated with improved health benefits.

High oleic oils are in great demand by various industries, including food and oleochemical industries, due to the low oxidative and frying stability of high oleic peanuts. The first natural high oleate mutant line, F435, was identified with about 80% oleic acid, much higher than the traditional peanut genotypes (Norden et al., 1987). High-oleic-acid mutants M2-225 and C458 were generated by chemical mutagenesis (Patel et al., 2004). The discovery of high-oleic-acid mutants provides the foundation for isolating and functional research on genes controlling oleic acid content in peanuts.

The fatty acid desaturase gene *AhFAD2* is the critical gene that controls the conversion of oleic acid to linoleic acid, determining peanut kernels' oleic and linoleic acid contents. Near-infrared reflectance spectroscopy (NIR) provided a low-cost and efficient method of measuring unbroken seeds' oil content and fatty acid composition. The KASP method was used for genotyping the seeds from the male parent, female parent, and F_1 crossed between high oleate donors, Kainong176, and normal oleate cultivars, Huayu31. The results demonstrated that the developed KASP could detect the genotypes of peanut *AhFAD2A* and *AhFAD2B* genes (Zhao et al., 2017). Four peanut cultivars of different market types were hybridized with high-oleic-acid donors and backcrossed for four generations as recurrent parent. Near-infrared reflectance spectroscopy was used to detect oleic acid and KASP markers, and fad2 mutant markers were screened to develop high-oleic-acid lines. Twenty-four lines were developed and genetic backgrounds estimated by KASP assay. The results indicated that KASP assay can detect fad2 mutations and evaluate genetic background in a peanut breeding program (Huang et al., 2019). In order to achieve introgression of the mutant ahfad2 alleles conferring high oleic content into an Indian cultivar GPBD4, which is also resistant to foliar fungal diseases like rust and late leaf spot (LLS), the AS-PCR and CAPS assays were used to identify the F_1 generation and segregation populations crossed between GPBD4 and SunOleic95R. Finally, 64 introgression lines were generated (Nawade et al., 2019). The high oleic acid content of three popular Indian groundnut varieties (GJG9, GG20, and GJGHPS1) was improved by MABC. Allele-specific markers (F435 and F435SUB) and CAPS markers (aF19F and 1956R) were developed to perform foreground selection in progenies crossed by the SunOleic 95R and Indian groundnut varieties. A total of 46 BC_3F_4 and 41 BC_2F_4 introgression lines were developed (Shasidhar et al., 2020).

7.6 CONCLUSIONS

The first type of DNA marker was RFLP, developed in 1974. As DNA sequencing and PCR technology advances, many different types of DNA markers have been developed (RAPD, AFLP, SSR, CAPS, STS, DArF, SNP, and KASP). DNA markers have been successfully used in many aspects (genetic mapping, physical mapping, QTL mapping, heterosis, GWAS, phylogeny, and evolution) of plant studies. But MAS in plant breeding (including parent selection based on genetic diversity, early generation selection, introgression selection, backcross selection, recurrent selection, and pyramiding selection) is one of the most important application aspects that can increase both accuracy and efficiency of plant breeding in a time- and cost-effective manner. This has been demonstrated by many successful examples in peanut, as discussed above. For any crops employing MAS, we need to mainly consider the following factors: species-specific (annual/perennial, self/cross-pollination, and diploid/tetraploid), breeding program–specific (single institution/international collaboration), improved trait–specific (single trait/multiple traits, qualitative/quantitative trait), types of markers (random/functional) available, lab platforms (low/high throughput) available, and resources available. The cultivated peanut is an annual, mainly self-pollinated, tetraploid species. The genomes from diploid progenitors and several tetraploid references of peanut have been sequenced. Thousands of polymorphic DNA markers were mined from DNA sequencing. Using these polymorphic DNA markers, many QTLs have been mapped on certain chromosomal regions. SNP arrays have been developed, commercialized, and used in genotyping and breeding programs. Although many random DNA markers are available, functional DNA markers are still very limited. Functional DNA markers were developed from the genes that directly control the traits of interest. Therefore, using functional DNA markers in MAS breeding will be more efficient and effective. However, only a few functional DNA markers (such as *FAD2* markers for the level of oleic acid and RKN440 markers for root-knot nematode resistance) have been developed and used successfully in peanut breeding programs. Development and validation of functional DNA markers is still the high priority and main task for promoting MAS in the peanut community. Functional DNA markers can be developed by using major QTLs, anchored genome regions of physical mapping, and QTL-seq. Once candidate genes are identified, functional DNA markers can be developed and validated for MAS. Some traits may be controlled by many minor QTLs. GS may hold promise for improving these kinds of traits. Genome editing (such as CRISPR) can specifically target the designated locus on the genome for improving a specific trait, but delivering CRISPR constructs into the genome by transformation is still a bottleneck in peanut. As new technology advances, the integration of GS for quantitative trait improvement and CRISPR for qualitative trait improvement may revolutionize peanut breeding programs in the future.

REFERENCES

Agarwal, G., J. Clevenger, M. K. Pandey, H. Wang, Y. Shasidhar, Y. Chu, J. C. Fountain et al. "High-density genetic map using whole-genome resequencing for fine mapping and candidate gene discovery for disease resistance in peanut." *Plant Biotechnology Journal* 16, no. 11 (2018): 1954–1967.

Al-Saghir, M. G., and A. G. Abdel-Salam. "Genetic diversity of peanut (Arachis hypogea L.) cultivars as revealed by RAPD markers." *American Journal of Plant Sciences* 6, no. 14 (2015): 2303–2308.

Alves, D. M. T., R. W. Pereira, S. C. de Macedo LEAL BERTIOLI, Márcio de Carvalho Moretzsohn, Patrícia M. Guimarães, and David John Bertioli. "Development and use of single nucleotide polymorphism markers for candidate resistance gene in wild peanuts (Arachis spp)." *Genetics and Molecular Research* 7, no. 3 (2008): 631–642.

Amiteye, S. "Basic concepts and methodologies of DNA marker systems in plant molecular breeding." *Heliyon* 7, no. 10 (2021).

Andrews, S. *FastQC: A Quality Control Tool for High Throughput Sequence Data.* 2010. http://www.bioinformatics.babraham.ac.uk/projects/fastq_screen/.

Ballén-Taborda, C., Y. Chu, P. Ozias-Akins, C. C. Holbrook, P. Timper, S. A. Jackson, D. J. Bertioli, and S. C. M. Leal-Bertioli. "Development and genetic characterization of peanut advanced backcross lines that incorporate root-knot nematode resistance from Arachis stenosperma." Frontiers in Plant Science 12 (2022): 785358.

Bera, S. K., J. H. Kamdar, S. V. Kasundra, P. Dash, A. K. Maurya, M. D. Jasani, and A. B. Chandrashekar. "Improving oil quality by altering levels of fatty acids through marker-assisted selection of Ahfad2 alleles in peanut (Arachis hypogaea L.)." Euphytica 214, no. 9 (2018): 162.

Bertioli, D. J., S. B. Cannon, L. Froenicke, G. Huang, A. D. Farmer, E. K. Cannon, X. Liu, D. Gao, J. Clevenger, S. Dash, et al. "The genome sequences of Arachis duranensis and Arachis ipaensis, the diploid ancestors of cultivated peanut." Nature Genetics 48, no. 4 (2016): 438–446.

Bertioli, D. J., P. O. Akins, Y. Chu, K. M. Dantas, S. P. Santos, E. Gouvea, P. M. Guimarães, S. C. M. Leal-Bertioli, S. J. Knapp, and M. C. Moretzsohn. "The use of SNP markers for linkage mapping in diploid and tetraploid peanuts." *G3: Genes, Genomes, Genetics* 4, no. 1 (2014): 89–96.

Beute, M. K., J. C. Wynne, and D. A. Emery. "Registration of NC3033 peanut germplasm (Reg. No. GP 9)." *Crop Science* 16 (1976): 887.

Bhat, R., K. Shirasawa, Y. Monden, H. Yamashita, and M. Tahara. "Developing transposable element marker system for molecular breeding." *Legume Genomics: Methods and Protocols* (2020): 233–251.

Bhat, R. S., J. Rockey, K. Shirasawa, I. S. Tilak, M. P. B. Patil, and V. B. R. Lachagari. "DNA methylation and expression analyses reveal epialleles for the foliar disease resistance genes in peanut (Arachis hypogaea L.)." *BMC Research Notes* 13 (2020): 1–7.

Bibikova, M., B. Barnes, C. Tsan, V. Ho, B. Klotzle, J. M. Le, D. Delano et al. "High density DNA methylation array with single CpG site resolution." *Genomics* 98, no. 4 (2011): 288–295.

Bibikova, M., J. Le, B. Barnes, S. S. Melnyk, L. Zhou, R. Shen, and K. L. Gunderson. "Genome-wide DNA methylation profiling using Infinium® assay." *Epigenomics* 1, no. 1 (2009): 177–200.

Burow, M. D., C. E. Simpson, A. H. Paterson, and J. L. Starr. "Identification of peanut (*Arachis hypogaea* L.) RAPD markers diagnostic of root-knot nematode (*Meloidogyne arenaria* (Neal) Chitwood) resistance." *Molecular Breeding* 2, no. 4 (1996): 369–379.

Chavarro, C., Y. Chu, C. Holbrook, T. Isleib, D. Bertioli, R. Hovav, C. Butts, M. Lamb, R. Sorensen, S. A. Jackson, and P. O. Akins. "Pod and seed trait QTL identification to assist breeding for peanut market preferences." *G3: Genes, Genomes, Genetics* 10, no. 7 (2020): 2297–2315.

Chen, D.-P., Y.-C. Lin, and C. S. Fann. "Methods for identifying differentially methylated regions for sequence- and array-based data." *Briefings in Functional Genomics* 15 (2016): 485–490.

Chen, W., Y. Jiao, L. Cheng, L. Huang, B. Liao, M. Tang, X. Ren, X. Zhou, Y. Chen, and H. Jiang. "Quantitative trait locus analysis for pod-and kernel-related traits in the cultivated peanut (Arachis hypogaea L.)." *BMC Genetics* 17 (2016): 1–9.

Chopra, R., G. Burow, A. Farmer, J. Mudge, C. E. Simpson, T. A. Wilkins, M. R. Baring, N. Puppala, K. D. Chamberlin, and M. D. Burow. "Next-generation transcriptome sequencing, SNP discovery and validation in four market classes of peanut, Arachis hypogaea L." *Molecular Genetics and Genomics* 290 (2015): 1169–1180.

Chu, Y., P. Chee, T. G. Isleib, C. C. Holbrook, and P. O. Akins. "Major seed size QTL on chromosome A05 of peanut (Arachis hypogaea) is conserved in the US mini core germplasm collection." *Molecular Breeding* 40 (2020): 1–16.

Clevenger, J., Ye Chu, C. Chavarro, S. Botton, A. Culbreath, Thomas G. Isleib, C. C. Holbrook, and Peggy Ozias-Akins. "Mapping late leaf spot resistance in peanut (Arachis hypogaea) using QTL-seq reveals markers for marker-assisted selection." *Frontiers in Plant Science* 9 (2018): 83.

Daudi, H., H. Shimelis, I. Mathew, R. O. Frimpong, C. Ojiewo, and R. K. Varshney. "Genetic diversity and population structure of groundnut (Arachis hypogaea L.) accessions using phenotypic traits and SSR markers: implications for rust resistance breeding." *Genetic Resources and Crop Evolution* 68 (2021): 581–604.

Dwivedi, S. L., S. Gurtu, and S. N. Nigam. "AFLP diversity among selected foliar diseases resistant groundnut (Arachis hypogaea L.) germplasm." *Indian Journal of Plant Genetic Resources* 15, no. 1 (2002): 46–50.

FAO. World Food and Agriculture – Statistical Yearbook 2022. Rome, 2022. https://doi.org/10.4060/cc2211en.

Fang, Y., X. Zhang, H. Liu, J. Wu, F. Qi, Z. Sun, Z. Zheng, W. Dong, and B. Huang. "Identification of quantitative trait loci and development of diagnostic markers for growth habit traits in peanut (Arachis hypogaea L.)." *Theoretical and Applied Genetics* 136, no. 5 (2023): 105.

Fu, G., Y. Zhong, C. Li, Y. Li, X. Lin, B. Liao, E. W. Tsang, K. Wu, and S. Huang. "Epigenetic regulation of peanut allergen gene Ara h 3 in developing embryos." *Planta* 231 (2010): 1049–1060.

Gangurde, S. S., A. W. Khan, P. Janila, M. T. Variath, S. S. Manohar, P. Singam, A. Chitikineni, R. K. Varshney, and M. K. Pandey. "Whole-genome sequencing based discovery of candidate genes and diagnostic markers for seed weight in groundnut." *The Plant Genome* (2022).

Garcia, G., H. Stalker, E. Schroeder, J. Lyerly, and G. Kochert. "A RAPD-based linkage map of peanut based on a backcross population between the two diploid species Arachis stenosperma and A. cardenasii." *Peanut Science* 32 (2005): 1–8.

Garcia, G. M., H. T. Stalker, E. Shroeder, and G. Kochert. "Identification of RAPD, SCAR, and RFLP markers tightly linked to nematode resistance genes introgressed from Arachis Cardenasii into Arachis Hypogaea." *Genome* 39, no. 5 (1996): 836–845.

Gautami, B., D. Foncéka, M. K. Pandey, M. C. Moretzsohn, V. Sujay et al. "An international reference consensus genetic map with 897 marker loci based on 11 mapping populations for tetraploid groundnut (Arachis hypogaea L.)." *PLoS One* 7 (2012a): e41:213.

Gautami, B., M. Pandey, V. Vadez, S. Nigam, P. Ratnakumar et al. "QTL analysis and consensus genetic map for drought tolerance traits based on three RIL populations of cultivated groundnut (Arachis hypogaea L.)." *Mol Breed* 32 (2012b): 757–772.

Gorbet, D. W., and B. L. Tillman. "Registration of 'Florida-07' peanut." *Journal of Plant Registrations* 3 (2009): 14–18.

Grieshammer, U., and J. Wynne. "Isoozyme variability in mature seeds of US peanut cultivars and collections." *Peanut Science* 17 (1990a): 72–75.

Grieshammer, U., and J. C. Wynne. "Mendelian and non-Mendelian inheritance of three isozymes in peanut (*Arachis hypogaea L.*)." *Peanut Science* 17 (1990b): 101–105.

Gupta, S. K., J. Baek, N. Carrasquilla-Garcia, and R. V. Penmetsa. "Genome-wide polymorphism detection in peanut using next-generation restriction-site-associated DNA (RAD) sequencing." *Molecular Breeding* 35 (2015): 1–10.

Hake, A., K. Shirasawa, A. Yadawad, S. Nayak, S. Mondal et al. "Identification of transposable element markers associated with yield and quality traits from an association panel of independent mutants in peanut (Arachis hypogaea L.)." *Euphytica* 213 (2017): 1–13.

Halward, T. M., H. T. Stalker, E. A. Larue, and G. Kochert. "Genetic variation detectable with molecular markers among unadapted germ-plasm resources of cultivated peanut and related wild species." *Genome* 34 (1991): 1013–1020.

Halward, T., H. Stalker, and G. Kochert. "Development of an RFLP linkage map in diploid peanut species." *Theoretical and Applied Genetics* 87 (1993): 379–384.

Han, S., M. Yuan, J. P. Clevenger, C. Li, A. Hagan et al. "A SNP-based linkage map revealed QTLs for resistance to early and late leaf spot diseases in peanut (Arachis hypogaea L.)." *Frontiers in Plant Science* 9 (2018): 1012.

He, G., and C. S. Prakash. "Identification of polymorphic DNA markers in cultivated peanut (Arachis hypogaea L.)." *Euphytica* 97 (1997): 143–149.

He, G., and C. Prakash. "Evaluation of genetic relationships among botanical varieties of cultivated peanut (Arachis hypogaea L.) using AFLP markers." *Genetic Resources and Crop Evolution* 48 (2001): 347–352.

He, G. H., R. H. Meng, M. Newman, G. Q. Gao, R. N. Pittman, and C. S. Prakash. "Microsatellites as DNA markers in cultivated peanut (*Arachis hypogaea* L.)." *BMC Plant Biology* 3 (2003): 3.

Helentjaris, T., M. Slocum, S. Wright, A. Schaefer, and J. Nienhuis. "Construction of genetic linkage maps in maize and tomato using restriction fragment length polymorphisms." *Theoretical and Applied Genetics* 72 (1986): 761–769.

Herselman, L. "Genetic variation among Southern African cultivated peanut (Arachis hypogaea L.) genotypes as revealed by AFLP analysis." *Euphytica* 133 (2003): 319–327.

Herselman, L., R. Thwaites, F. Kimmins, B. Courtois, P. Van Der Merwe et al. "Identification and mapping of AFLP markers linked to peanut (Arachis hypogaea L.) resistance to the aphid vector of groundnut rosette disease." *Theoretical and Applied Genetics* 109 (2004): 1426–1433.

Holbrook, C. C., and A. K. Culbreath. "Registration of 'Tifrunner' peanut." *Journal of Plant Registrations* 1 (2007): 124.

Holbrook, C. C., T. G. Isleib, P. Ozias-Akins, Y. Chu, S. Knapp, B. Tillman, B. Guo., R. Gill, and M. D. Burow. "Development and phenotyping of recombinant inbred line (RIL) populations for peanut (Arachis hypogaea)." *Peanut Sci.* 40 (2013): 89–94.

Hong, Y., X. Chen, X. Liang, H. Liu, G. Zhou, S. Li, S. Wen, C. C. Holbrook, and B. Guo. "A SSR-based composite genetic linkage map for the cultivated peanut (Arachis hypogaea L.) genome." *BMC Plant Biology* 10 (2010): 1–13.

Hong, Y., M. K. Pandey, Y. Liu, X. Chen, H. Liu, R. K. Varshney, X. Liang, and S. Huang. "Identification and evaluation of single-nucleotide polymorphisms in allotetraploid peanut (Arachis hypogaea L.) based on amplicon sequencing combined with high resolution melting (HRM) analysis." *Frontiers in Plant Science* 6 (2015): 1068.

Hopkins, M. S., A. M. Casa, T. Wang, S. E. Mitchell, R. E. Dean, G. D. Kochert, and S. Kresovich. "Discovery and characterization of polymorphic simple sequence repeats (SSRs) in peanut." *Crop Science* 39, no. 4 (1999): 1243–1247.

Hsi, D. C. H., and R. E. Finkner. 1972. "Registration of New Mexico Valencia a peanut." *Crop Science* 12 (1972): 256.

Hu, C., X. Li, J. Li, H. Ji, Y. Tang, X. Zou, D. Jiang, B. Guo, J. Wang, and L. Qiao. "Construction of RIL population on pod size by FAD2 marker in peanut (in Chinese)." *Molecular Plant Breeding* 20 (2022): 1–8.

Huang, B., F. Qi, Z. Sun, L. Miao, Z. Zhang, H. Liu, and Y. Fang. "Marker-assisted backcrossing to improve seed oleic acid content in four elite and popular peanut (*Arachis hypogaea L.*) cultivars with high oil content." *Breeding Science* 69, no. 2 (2019): 234–243.

Huang, H., M. Wang, C. Zhao, L. Hou, S. Fan, X. Wang, and H. Xia. "Variation analysis and marker screening of seed weight and shape of peanut germplasm (in Chinese)." *Molecular Plant Breeding* 20 (2022): 1–14.

Huang, L., X. Ren, B. Wu, X. Li, W. Chen, X. Zhou, Y. Chen et al. "Development and deployment of a high-density linkage map identified quantitative trait loci for plant height in peanut (Arachis hypogaea L.)." *Scientific Reports* 6, no. 1 (2016): 39478.

Jiang, H., L. Huang, X. Ren, Y. Chen, X. Zhou, Y. Xia, J. Huang et al. "Diversity characterization and association analysis of agronomic traits in a Chinese peanut (Arachis hypogaea L.) mini-core collection." *Journal of Integrative Plant Biology* 56, no. 2 (2014): 159–169.

Kassie, F. C., J. R. Nguepjop, H. B. Ngalle, D. V. M. Assaha, M. K. Gesese, W. G. Abtew, H.-A. Tossim, A. Sambou, M. Seye, J.-F. Rami, et al. "An overview of mapping quantitative trait loci in peanut (*Arachis hypogaea* L.)." *Genes* 14, no. 6 (2023): 1176.

Khan, S. A., H. Chen, Y. Deng, Y. Chen, C. Zhang, T. Cai, N. Ali et al. "High-density SNP map facilitates fine mapping of QTLs and candidate genes discovery for Aspergillus flavus resistance in peanut (Arachis hypogaea)." *Theoretical and Applied Genetics* 133 (2020): 2239–2257.

Khera, P., H. D. Upadhyaya, M. K. Pandey, M. Roorkiwal, M. Sriswathi, P. Janila, and Y. Guo. "Single nucleotide polymorphism-based genetic diversity in the reference set of peanut (*Arachis spp.*) by developing and applying cost-effective kompetitive allele-specific polymerase chain reaction genotyping assay." *The Plant Genome* 6, no. 3 (2013). https://doi.org/10.3835/plantgenome2013.06.0019.

Kochert, G., T. Halward, W. Branch, and C. Simpson. "RFLP variability in peanut (*Arachis hypogaea L.*) cultivars and wild species." *Theoretical and Applied Genetics* 81 (1991): 565–570.

Kooke, R., F. Johannes, R. Wardenaar, F. Becker, M. Etcheverry, V. Colot, D. Vreugdenhil, and J. J. Keurentjes. "Epigenetic basis of morphological variation and phenotypic plasticity in Arabidopsis thaliana." *The Plant Cell* 27, no. 2 (2015): 337–348.

Kumar, A., and H. Hirochika. "Applications of retrotransposons as genetic tools in plant biology." *Trends in Plant Science* 6 (2001): 127–134.

Kumar, R., P. Janila, M. K. Vishwakarma, A. W. Khan, S. S. Manohar, S. S. Gangurde, M. T. Variath, Y. Shasidhar, M. K. Pandey, and R. K. Varshney. "Whole-genome resequencing-based QTL-seq identified candidate genes and molecular markers for fresh seed dormancy in groundnut." *Plant Biotechnology Journal* 18, no. 4 (2020): 992–1003.

Lacks, G. D., and H. T. Stalker. "Isozyme analyses of Arachis species and interspecific hybrids." *Peanut Science* 20, no. 2 (1993): 76–81.

Leal-Bertioli, S. C. M., U. Cavalcante, E. G. Gouvea, C. Ballén-Taborda, K. Shirasawa, P. M. Guimarães, S. A. Jackson, D. J. Bertioli, and M. C. Moretzsohn. "Identification of QTLs for rust resistance in the peanut wild species *Arachis magna* and the development of KASP markers for marker-assisted selection." *G3 Genes|Genomes|Genetics* 5, no. 7 (2015): 1403–1413.

Li, L., X. Yang, S. Cui, X. Meng, G. Mu, M. Hou, M. He, H. Zhang, L. Liu, and C. Y. Chen. "Construction of high-density genetic map and mapping quantitative trait loci for growth habit-related traits of peanut (Arachis hypogaea L.)." *Frontiers in Plant Science* 10 (2019): 745.

Liu, L., P. M. Dang, and C. Y. Chen. "Development and utilization of InDel markers to identify peanut (Arachis hypogaea) disease resistance." *Frontiers in Plant Science* 6 (2015): 988.

Liu, N., B. Wu, M. K. Pandey, L. Huang, H. Luo, Y. Chen, X. Zhou et al. "Gene expression and DNA methylation altering lead to the high oil content in wild allotetraploid peanut (A. monticola)." *Frontiers in Plant Science* 13 (2022): 1065267.

Liu, Y., J. He, S. Cui, X. Yang, G. Mu, et al. "Screening and analysis of molecular markers for pod rot resistance in American peanut (*Arachis hypogaea*) germplasm resources (in Chinese)." *Journal Agricultural Biotechnology* 27, no. 4 (2019): 9.

Lu, Q., Y. Hong, S. Li, H. Liu, H. Li, J. Zhang, H. Lan et al. "Genome-wide identification of microsatellite markers from cultivated peanut (Arachis hypogaea L.)." *BMC genomics* 20 (2019): 1–9.

Luo, H., M. K. Pandey, A. W. Khan, J. Guo, B. Wu, Y. Cai, L. Huang et al. "Discovery of genomic regions and candidate genes controlling shelling percentage using QTL-seq approach in cultivated peanut (Arachis hypogaea L.)." *Plant Biotechnology Journal* 17, no. 7 (2019): 1248–1260.

Luo, H., M. K. Pandey, Y. Zhi, H. Zhang, S. Xu, J. Guo, B. Wu et al. "Discovery of two novel and adjacent QTLs on chromosome B02 controlling resistance against bacterial wilt in peanut variety Zhonghua 6." *Theoretical and Applied Genetics* 133 (2020): 1133–1148.

Luo, H., X. Ren, Z. Li, Z. Xu, X. Li, L. Huang, X. Zhou et al. "Co-localization of major quantitative trait loci for pod size and weight to a 3.7 cM interval on chromosome A05 in cultivated peanut (Arachis hypogaea L.)." *BMC Genomics* 18 (2017): 1–12.

Luo, H., M. K. Pandey, A. W. Khan, B. Wu, J. Guo, X. Ren, X. Zhou et al. "Next-generation sequencing identified genomic region and diagnostic markers for resistance to bacterial wilt on chromosome B02 in peanut (Arachis hypogaea L.)." *Plant Biotechnology Journal* 17, no. 12 (2019): 2356–2369.

Mason, A. S. "SSR genotyping." In *Plant* Genotyping: Methods and Protocols, edited by J. Batley. New York: Springer, 2015, 77–89.

McCouch, S. R., X. Chen, O. Panaud, S. Temnykh, Y. Xu, Y. G. Cho, N. Huang, T. Ishii, and M. Blair. "Microsatellite marker development, mapping and applications in rice genetics and breeding." *Plant Molecular Biology* 35, no. 1 (1997): 89–99.

Mondal, S., P. Hande, and A. M. Badigannavar. "Identification of transposable element markers for a rust (*Puccinia arachidis* Speg.) resistance gene in cultivated peanut." *Journal of Phytopathology* 162 (2014): 548–552.

Moretzsohn, M. C., L. Leoi, K. Proite, P. M. Guimaraes, S. C. M. Leal-Bertioli, M. A. Gimenes, W. S. Martins, J. F. M. Valls, D. Grattapaglia, and D. J. Bertioli. "A microsatellite-based, gene-rich linkage map for the AA genome of Arachis (Fabaceae)." *Theoretical and Applied Genetics* 111 (2005): 1060–1071.

Nagy, E. D., Y. Guo, S. Tang, J. E. Bowers, R. A. Okashah, C. A. Taylor, D. Zhang et al. "A high-density genetic map of Arachis duranensis, a diploid ancestor of cultivated peanut." *BMC Genomics* 13 (2012): 1–11.

Nam, H.-G., J. Giraudat, B. D. Boer, F. Moonan, W. D. B. Loos, B. M. Hauge, and H. M. Goodman. "Restriction fragment length polymorphism linkage map of Arabidopsis thaliana." *The Plant Cell* 1, no. 7 (1989): 699–705.

Nawade, B., G. P. Mishra, and R. Thankappan. "Development of high oleic peanut lines through marker-assisted introgression of mutant ahFAD2 alleles and its fatty acid profiles under open-field and controlled conditions." *3 Biotech* 9, no. 6 (2019): 243. https://doi.org/10.1007/s13205-019-1774-9.

Norden, A. J., D. W. Gorbet, D. A. Knauft, and C. T. Young. "Variability in oil quality among peanut genotypes in the Florida breeding program." *Peanut Science* 14, no. 1 (1987): 7–11.

Norden, A. J., R. W. Lipscomb, and W. A. Carver. "Registration of Florunner peanut." *Crop Science* 9, no. 850 (1969): 850.

Nowak, D., W.-K. Hofmann, and H. P. Koeffler. "Genome-wide mapping of copy number variations using SNP arrays." *Transfusion Medicine and Hemotherapy* 36 (2009): 246–251.

Paik-Ro, O., R. Smith, and D. Knauft. "Restriction fragment length polymorphism evaluation of six peanut species within the *Arachis* section." *Theoretical and Applied Genetics* 84 (1992): 201–208.

Pandey, Manish K., Gaurav Agarwal, Sandip M. Kale, Josh Clevenger, Spurthi N. Nayak, Manda Sriswathi, Annapurna Chitikineni et al. "Development and evaluation of a high density genotyping 'Axiom_ Arachis' array with 58 K SNPs for accelerating genetics and breeding in groundnut." *Scientific Reports* 7, no. 1 (2017): 40577.

Pandey, M. K., S. S. Gangurde, V. Sharma, S. K. Pattanashetti, G. K. Naidu, I. Faye, F. Hamidou et al. "Improved genetic map identified major QTLs for drought tolerance-and iron deficiency tolerance-related traits in groundnut." *Genes* 12, no. 1 (2020): 37.

Pandey, M. K., A. W. Khan, V. K. Singh, M. K. Vishwakarma, Y. Shasidhar, V. Kumar, V. Garg et al. "QTL-seq approach identified genomic regions and diagnostic markers for rust and late leaf spot resistance in groundnut (A rachis hypogaea L.)." *Plant Biotechnology Journal* 15, no. 8 (2017): 927–941.

Pandey, M. K., E. Monyo, P. O. Akins, X. Liang, P. Guimarães, S. N. Nigam, H. D. Upadhyaya et al. "Advances in Arachis genomics for peanut improvement." *Biotechnology Advances* 30, no. 3 (2012): 639–651.

Pandey, M. K., H. D. Upadhyaya, A. Rathore, V. Vadez, M. S. Sheshshayee, M. Sriswathi, M. Govil, A. Kumar, M. V. C. Gowda, S. Sharma, et al. "Genomewide association studies for 50 agronomic traits in peanut using the 'reference set'comprising 300 genotypes from 48 countries of the semi-arid tropics of the world." *PLoS One* 9, no. 8 (2014): 105228.

Patel, J. D., M. L. Wang, P. Dang, C. Butts, M. Lamb, and C. Y. Chen. "Insights into the genomic architecture of seed and pod quality traits in the US Peanut mini-core diversity panel." *Plants* 11, no. 7 (2022): 837.

Patel, M., S. Jung, K. Moore, G. Powell, C. Ainsworth, and A. Abbott. "High-oleate peanut mutants result from a MITE insertion into the FAD2 gene." *Theoretical and Applied Genetics* 108 (2004): 1492–1502.

Pavan, S., C. Delvento, L. Ricciardi, C. Lotti, E. Ciani, and N. D'Agostino. "Recommendations for choosing the genotyping method and best practices for quality control in crop genome-wide association studies." *Frontiers in Genetics* 11 (2020): 447.

Peng, Z., W. Fan, L. Wang, D. Paudel, D. Leventini, B. L. Tillman, and J. Wang. "Target enrichment sequencing in cultivated peanut (Arachis hypogaea L.) using probes designed from transcript sequences." *Molecular Genetics and Genomics* 292 (2017): 955–965.

Peng, Z., M. Gallo, B. L. Tillman, D. Rowland, and J. Wang. "Molecular marker development from transcript sequences and germplasm evaluation for cultivated peanut (Arachis hypogaea L.)." *Molecular Genetics and Genomics* 291 (2016): 363–381.

Qi, F., Z. Sun, H. Liu, Z. Zheng, L. Qin, L. Shi, Q. Chen et al. "QTL identification, fine mapping, and marker development for breeding peanut (Arachis hypogaea L.) resistant to bacterial wilt." *Theoretical and Applied Genetics* 135, no. 4 (2022): 1319–1330.

Qin, H., S. Feng, C. Chen, Y. Guo, S. Knapp, A. Culbreath, G. He et al. "An integrated genetic linkage map of cultivated peanut (Arachis hypogaea L.) constructed from two RIL populations." *Theoretical and Applied Genetics* 124 (2012): 653–664.

Qiu, X., J. Li, X. Li, L. Wu, Y. Tang, H. Zhu, J. Sui, X. Zhang, J. Wang, and L. Qiao. Construction of genetic population for oil content using AhFAD2 as a selectable marker in peanut (in Chinese). *Chinese Journal of Oil Crop Sciences* 45, no. 3 (2022): 1–7.

Rasheed, A., Y. Hao, X. Xia, A. Khan, Y. Xu, R. K. Varshney, and Z. He. Crop breeding chips and genotyping platforms: progress, challenges, and perspectives. *Molecular Plant* 10, no. 8 (2017): 1047–1064.

Ravi, K., V. Vadez, S. Isobe, R. R. Mir, Y. Guo, S. N. Nigam, M. V. C. Gowda et al. "Identification of several small main-effect QTLs and a large number of epistatic QTLs for drought tolerance related traits in groundnut (Arachis hypogaea L.)." *Theoretical and Applied Genetics* 122 (2011): 1119–1132.

Sanders, T. H., H. E. Pattee, and J. A. Singleton. "Lipoxygenase isozymes of peanut." *Lipids* 10, no. 11 (1975): 681–685.

Shasidhar, Y., M. T. Variath, M. K. Vishwakarma, S. S. Manohar, S. S. Gangurde, M. Sriswathi, H. K. Sudini et al. "Improvement of three popular Indian groundnut varieties for foliar disease resistance and high oleic acid using SSR markers and SNP array in marker-assisted backcrossing." *The Crop Journal* 8, no. 1 (2020): 1–15.

Shasidhar, Y., M. K. Vishwakarma, M. K. Pandey, P. Janila, M. T. Variath, S. S. Manohar, S. N. Nigam, B. Guo, and R. K. Varshney. "Molecular mapping of oil content and fatty acids using dense genetic maps in groundnut (Arachis hypogaea L.)." *Frontiers in Plant Science* 8 (2017): 794.

Shirasawa, K., H. Hirakawa, S. Tabata, M. Hasegawa, H. Kiyoshima, S. Suzuki, S. Sasamoto, A. Watanabe, T. Fujishiro, and S. Isobe. "Characterization of active miniature inverted-repeat transposable elements in the peanut genome." *Theoretical and Applied Genetics* 124 (2012): 1429–1438.

Shirasawa, K., P. Koilkonda, K. Aoki, H. Hirakawa, S. Tabata, M. Watanabe, M. Hasegawa et al. "In silico polymorphism analysis for the development of simple sequence repeat and transposon markers and construction of linkage map in cultivated peanut." *BMC Plant Biology* 12 (2012): 1–13.

Shirasawa, K. E. N. T. A., D. J. Bertioli, R. K. Varshney, M. C. Moretzsohn, S. C. M. Leal-Bertioli, M. A. H. E. N. D. A. R. Thudi, M. K. Pandey et al. "Integrated consensus map of cultivated peanut and wild relatives reveals structures of the A and B genomes of Arachis and divergence of the legume genomes." *DNA Research* 20, no. 2 (2013): 173–184.

Simpson, C. E., M. R. Baring, A. M. Schubert, H. A. Melouk, A., Lopez, J. S. Kirby. 2003. "Registration of 'Olin' peanut." *Crop Science* 43, no. 6 (2003): 1880–1881.

Singh, B. D., A. K. Singh, B. D. Singh, and A. K. Singh. "Polymerase chain reaction-based markers." *Marker-Assisted Plant Breeding: Principles and Practices* (2015): 47–75.

Stalker, H. T., and L. G. Mozingo. "Molecular markers of Arachis and marker-assisted selection." *Peanut Science* 28, no. 2 (2001): 117–123.

Stalker, H. T., T. D. Phillips, J. P. Murphy, and T. M. Jones. "Variation of isozyme patterns among Arachis species." *Theoretical and Applied Genetics* 87 (1994): 746–755.

Sun, H., L. Ren, F. Qi, H. Wang, S. Yu, Z. Sun, B. Huang et al. "BSA-seq approach identified candidate region and diagnostic marker for chilling tolerance of high oleic acid peanut at germination stage." *Agronomy* 13, no. 1 (2022): 18.

Tallury, S. P., T. G. Isleib, S. C. Copeland, P. Rosas-Anderson, M. Balota, D. Singh, and H. T. Stalker. "Registration of two multiple disease-resistant peanut germplasm lines derived from Arachis cardenasii Krapov. & WC Gregory, GKP 10017." *Journal of Plant Registrations* 8, no. 1 (2014): 86–89.

Tao, S. Y., B. Wu, N. Liu, H. Y. Luo, L. Huang, X. J. Zhou, W.-G. Chen et al. "Development and employment of InDel marker in peanut QTL mapping of oil content." *Acta Agronomica Sinica* 49, no. 5 (2023): 1222–1230.

Varshney, R. K. "Exciting journey of 10 years from genomes to fields and markets: some success stories of genomics-assisted breeding in chickpea, pigeonpea and groundnut." *Plant Science* 242 (2016): 98–107.

Vishwakarma, M. K., S. N. Nayak, B. Guo, L. Wan, B. Liao, R. K. Varshney, and M. K. Pandey. "Classical and molecular approaches for mapping of genes and quantitative trait loci in peanut." *The Peanut Genome,* 93–116. Springer; Berlin/Heidelberg, Germany, 2017.

Wang, C. T., G. S. Song, Z. W. Wang, H. J. Li, H. W. Han, X. Y. Chi, X. Z. Wang, and X. S. Sun. "Assessment of genetic diversity among Chinese high-oleic peanut genotypes using miniature inverted-repeat transposable element markers." *Genetic Resources and Crop Evolution* 69, no. 3 (2022): 949–958.

Wang, H., Y. M. Shi, Y. Ren, S. L. Li, K. Jiao, M. Yuan, and H. J. Li. "Development of SSR markers for root-knot nematode resistance in peanut." *Journal of Peanut Science* 37, no. 2 (2008): 14–17.

Wang, L., X. Yang, S. Cui, N. Zhao, L. Li, M. Hou, G. Mu, L. Liu, and Z. Li. "High-density genetic map development and QTL mapping for concentration degree of floret flowering date in cultivated peanut (Arachis hypogaea L.)." *Molecular Breeding* 40 (2020): 1–14.

Wang, M. L., S. Sukumaran, N. A. Barkley, Z. Chen, C. Y. Chen, B. Guo, R. N. Pittman et al. "Population structure and marker–trait association analysis of the US peanut (Arachis hypogaea L.) mini-core collection." *Theoretical and Applied Genetics* 123 (2011): 1307–1317.

Wang, M. L., H. Wang, C. Zhao, B. Tonnis, S. Tallury, X. Wang, J. Clevenger, and B. Guo. "Identification of QTLs for seed dormancy in cultivated peanut using a recombinant inbred line mapping population." *Plant Molecular Biology Reporter* 40, no. 1 (2022): 208–217.

Wang, P., S. Shi, J. Ma, H. Song, Y. Zhang, C. Gao, C. Zhao et al. "Global Methylome and gene expression analysis during early Peanut pod development." *BMC Plant Biology* 18 (2018): 1–13.

Wang, X., P. Xu, L. Yin, Y. Ren, S. Li, Y. Shi, T. D. Alcock et al. "Genomic and transcriptomic analysis identified gene clusters and candidate genes for oil content in peanut (Arachis hypogaea L.)." *Plant Molecular Biology Reporter* 36 (2018): 518–529.

Wang, Z., L. Yan, Y. Chen, X. Wang, D. Huai, Y. Kang, H. Jiang, K. Liu, Y. Lei, and B. Liao. "Detection of a major QTL and development of KASP markers for seed weight by combining QTL-seq, QTL-mapping and RNA-seq in peanut." *Theoretical and Applied Genetics* 135, no. 5 (2022): 1779–1795.

Wankhade, A. P., V. P. Chimote, K.P. Viswanatha, S. Yadaru, D. B. Deshmukh, S. Gattu, H. K. Sudini, M. P. Deshmukh, V. S. Shinde, A. K. Vemula et al. "Genome-wide association mapping for LLS resistance in a MAGIC population of groundnut (*Arachis hypogaea* L.)". *Theoretical and Applied Genetics*, 136, no. 3 (2023): 43.

Weeden, N. F. "Applications of isozymes in plant breeding." *Plant Breeding Reviews* 6 (1989): 11–54.

Wenzl, P., J. Carling, D. Kudrna, D. Jaccoud, E. Huttner, A. Kleinhofs, and A. Kilian. "Diversity Arrays Technology (DArT) for whole-genome profiling of barley." *Proceedings of the National Academy of Sciences* 101, no. 26 (2004): 9915–9920.

Wilson, J. N., R. Chopra, M. R. Baring, M. G. Selvaraj, C. E. Simpson, J. Chagoya, and M. D. Burow. "Advanced backcross quantitative trait loci (QTL) analysis of oil concentration and oil quality traits in peanut (Arachis hypogaea L.)." *Tropical Plant Biology* 10 (2017): 1–17.

Yang, P., F. Zhang, X. Luo, Y. Zhou, and J. Xie. "Histone deacetylation modification participates in the repression of peanut (Arachis hypogaea L.) seed storage protein gene Ara h 2.02 during germination." *Plant Biology* 17, no. 2 (2015): 522–527.

Yu, B., H. Jiang, M. K. Pandey, L. Huang, D. Huai, X. Zhou, Y. Kang et al. "Identification of two novel peanut genotypes resistant to aflatoxin production and their SNP markers associated with resistance." *Toxins* 12, no. 3 (2020): 156.

Zhang, H., M. L. Wang, P. Dang, T. Jiang, S. Zhao, M. Lamb, and C. Chen. "Identification of potential QTLs and genes associated with seed composition traits in peanut (Arachis hypogaea L.) using GWAS and RNA-Seq analysis." *Gene* 769 (2021): 145215.

Zhang, H., M. L. Wang, R. Schaefer, P. Dang, T. Jiang, and C. Chen. "GWAS and coexpression network reveal ionomic variation in cultivated peanut." *Journal of Agricultural and Food Chemistry* 67, no. 43 (2019): 12026–12036.

Zhang, K., M. Yuan, H. Xia, L. He, J. Ma, M. Wang, H. Zhao et al. "BSA-seq and genetic mapping reveals AhRt2 as a candidate gene responsible for red testa of peanut." *Theoretical and Applied Genetics* 135, no. 5 (2022): 1529–1540.

Zhang, X., J. Zhang, X. He, Y. Wang, X. Ma, and D. Yin. "Genome-wide association study of major agronomic traits related to domestication in peanut." *Frontiers in Plant Science* 8 (2017): 1611.

Zhao, C., J. Qiu, G. Agarwal, J. Wang, X. Ren, H. Xia, B. Guo et al. "Genome-wide discovery of microsatellite markers from diploid progenitor species, Arachis duranensis and *A. ipaensis*, and their application in cultivated peanut (*A. hypogaea*)." *Frontiers in Plant Science* 8 (2017): 1209.

Zhao, J., L. Huang, X. P. Ren, M. K. Pandey, B. Wu, Y. N. Chen, X. J. Zhou et al. "Genetic variation and association mapping of seed-related traits in cultivated peanut (Arachis hypogaea L.) using single-locus simple sequence repeat markers." *Frontiers in Plant Science* 8 (2017): 2105.

Zhao, Y., C. S. Prakash, and G. He. "Characterization and compilation of polymorphic simple sequence repeat (SSR) markers of peanut from public database." *BMC Research Notes* 5 (2012): 1–7.

Zhao, Y., J. Ma, M. Li, L. Deng, G. Li, H. Xia, S. Zhao et al. "Whole-genome resequencing-based QTL-seq identified AhTc1 gene encoding a R2R3-MYB transcription factor controlling peanut purple testa colour." *Plant Biotechnology Journal* 18, no. 1 (2020): 96–105.

Zhou, X., J. Guo, M. K. Pandey, R. K. Varshney, L. Huang, H. Luo, and N. Liu. Dissection of the genetic basis of yield-related traits in the Chinese peanut mini-core collection through genome-wide association studies. *Frontiers in Plant Science* 12 (2021): 637284.

Zhou, X., Y. Xia, X. Ren, Y. Chen, L. Huang, S. Huang, B. Liao, Y. Lei, L. Yan, and H. Jiang. "Construction of a SNP-based genetic linkage map in cultivated peanut based on large scale marker development using next-generation double-digest restriction-site-associated DNA sequencing (ddRADseq)." *BMC Genomics* 15 (2014): 1–14.

Zou, K., K. S. Kim, K. Kim, D. Kang, Y. H. Park, H. Sun, B. K. Ha, J. Ha, and T. H. Jun. "Genetic diversity and genome-wide association study of seed aspect ratio using a high-density SNP array in peanut (Arachis hypogaea L.)." *Genes* 12, no. 1 (2020): 2.

8 Functional Genomics Research Advancement in Peanut

Radhakrishnan Thankappan Mahesh K. Mahatma, Papa Rao Vaikuntappu, Manish K. Pandey, and Kamal Krishna Pal

8.1 INTRODUCTION

Peanut (*Arachis hypogaea*) is a major global source of both edible oil and protein, cultivated in 121 countries across the tropics and subtropics. However, its production is constrained by both abiotic stresses (primarily drought) and biotic stresses (such as diseases and pests), which are key breeding challenges. The main breeding objectives for peanut include improving yield, kernel quality, flavor, ensuring aflatoxin-free seeds, and enhancing oil quality, particularly increasing oleic acid content. Historically, the lack of a reference genome for the cultivated peanut, an autoallotetraploid (AABB), has made gene discovery, annotation, and functional validation difficult. However, the recent sequencing of the wild progenitors' genomes and the availability of the reference genome for *Arachis hypogaea* has significantly advanced peanut genomics. Researchers have now successfully annotated numerous functional genes and developed gene atlases, which can help guide breeding programs.

Peanut seeds are rich in oil (35.8%–54.2%) and protein (16.2%–36.0%). These kernels are primarily used either for oil extraction (approximately 60% of global production) or for direct consumption in forms such as raw, boiled, or roasted, or as part of confectionary products. The remaining 40% of global production is used for direct consumption and seed purposes (Birthal et al., 2010; Radhakrishnan et al., 2022). Given its role as a food crop, peanut kernel quality is of paramount importance. Various quality traits are governed by multiple genes, which, once functionally characterized and validated, can be incorporated into breeding practices for improved quality and yield. The basic understanding of gene function for key plant features is very important for their further genetic improvement and achieving desirable features in plant species. In peanut, huge genomic and transcriptomic data have been generated for different tissues and developmental stages as well as stress conditions. However, the focus needs to be shifted from generating omics data to analyzing and understanding the molecular mechanism and genes for better utilization in peanut improvement programs. In this context, the need for understating the relationship between the phenotype and the genome has also become apparent, which has led to the approach of functional genomics. Functional genomics involves the interaction of the genome, transcriptome, proteome, and metabolome. The genetic information is translated to the proteome through transcription and finally achieves the gene function, which may involve signalling as well as metabolic networks. Activation of the signalling pathways through various interactions may lead to the switching on/off of gene(s) while the proteome plays the enzymatic activities involved in the metabolic pathways. Both the substrates and products of enzymatic reactions, cofactors, and organic (non-peptide) molecules are involved in the signalling pathways. Thus, functional genomics essentially reflects the dynamics of genome expression (phenotype). In peanut, a crop characterized by a narrow genetic base, the development of genomic resources over the past decade has significantly advanced research, particularly in gene expression profiling and genetic regulation. While substantial information from expressed sequence tags (ESTs), unigenes, transcriptomes, full-length cDNA clones, and derived proteins has been available for nearly two decades, enabling potential exploitation for peanut improvement through targeted pathway interventions, the advent of whole-genome sequencing has revolutionized the field. With the availability of complete genome sequences, numerous functional genes have been annotated, and comprehensive gene atlases have been constructed. Integrating structural and functional genomics data offers powerful opportunities for gene discovery, functional characterization, and the identification of key genes modulating agronomically important traits in peanut. Moreover, the development of genetic transformation protocols provides a crucial tool for validating candidate gene function before their application in breeding programs.

8.2 ANALYSIS OF GENE EXPRESSION

Analysis of gene expression studies in peanut has evolved significantly alongside advancements in sequencing technologies. Prior to the availability of reference genomes for cultivated peanut and its progenitors, transcriptome analysis provided valuable genomic data (Chen, 2008; Zhang et al., 2012; Chopra, 2016), serving as a cost-effective alternative to whole-genome sequencing in various crops (Guimarães et al., 2017). These early transcriptome studies in peanut primarily employed direct assembly strategies

 DOI: 10.1201/9781003360988-8

for marker and gene discovery, identifying numerous genes expressed in response to biotic and abiotic stresses, as well as during key developmental processes such as geocarpic, embryo, and pod development. The advent of next-generation sequencing (NGS) and the recognition of peanut's large genome size (2.7 Gb) and high repetitive content (~64%; Dhillon, 1980) established high-throughput RNA sequencing (RNA-Seq) as the preferred technique for comprehensive gene expression profiling. RNA-Seq enabled the analysis of coding and non-coding RNAs across diverse tissues and developmental stages (Shen, 2015). Consequently, techniques like DNA microarrays, serial analysis of gene expression (SAGE) (Luo, 2005; Payton, 2009; Bi, 2010; Guo, 2011), TILLING arrays (Lin et al., 2012), and, increasingly, RNA-Seq itself (Zhang, 2012; Chen, 2013, 2016; Clevenger et al., 2016a, 2016b; Gupta, 2016; Nayak, 2017; Wan, 2017), have become widely adopted for gene expression studies in peanut. Leveraging the increasing availability of genomic resources and powerful tools like RNA-Seq, researchers have focused on identifying and cataloging functional genes in gene expression atlases spanning a wide range of tissues. One such atlas, developed for *A. hypogaea* subspecies, utilized 22 different tissues representing key organs and 260 Gb of data (encompassing full developmental stages of the tetraploid cultivar Tifrunner), identifying 8,816 putative homeologous genes, over 9,000 alternative splicing events, and more than 6,000 non-coding RNAs (Clevenger et al., 2016a). Another comprehensive transcriptome map, based on 39 samples from diverse tissues and conditions (including biotic and abiotic stresses and hormone treatments), covered 91.73% of total annotated genes (Zhuang, 2019). Furthermore, a gene atlas using different tissues from the drought-tolerant cultivar ICGS 91114 has also been reported (Sinha, 2020). Integrating proteome and metabolome data with these genomic resources will further enhance genome annotation and provide a more holistic understanding of trait development and gene expression regulation.

8.3 GROWTH AND DEVELOPMENT

Analysis of wild peanut progenitors (*A. duranensis* and *A. ipaensis*) identified 16 genes encoding phosphatidyl ethanolamine-binding proteins (PEBs), whose roles in photoperiod-dependent pathways were analyzed across various tissues and growth stages under short- and long-day conditions. Candidate genes regulating plant architecture were identified in *A. duranensis* (*AraduF950M/AraduWY2NX*) and *A. ipaensis* (*Aradu80YRY/AraduYY72S*). Key roles for *AraduEHZ9Y* (*A. duranensis*) and *AraipVEP8T* (*A. ipaensis*) in regulating flower initiation time were also reported (Jin, 2019). Transcriptome analysis of the semi-dwarf mutant *sdm2* revealed differential expression of genes involved in hormone biosynthesis and signalling, cell wall synthesis, and metabolic pathways (Guo et al., 2020). Specifically, genes in the brassinosteroid (BR) pathway and those related to cell elongation within cell wall synthesis and metabolic pathways were significantly downregulated in *sdm2* stems and leaves compared to the cultivar Fenghua1 (FH1).

Given the economic importance of pods and seeds, research has heavily focused on peg development, geocarpy (subterranean fruit development), pod development, seed morphology, and seed quality traits like lipid metabolism (Nayak, 2021). Peanut fruit development is unique due to geocarpy, where the fertilized ovule is pushed underground by the peg (gynophore) through gravitropic extension. *In vitro* gravitropic simulation studies of gynophore development identified 20 pathways involved in peg development and growth, including carbon fixation, aminoacyl-tRNA biosynthesis, the pentose phosphate pathway, starch and sucrose metabolism, and flavonoid biosynthesis (Li, Chen 2013). Geotropic movement is regulated by auxin concentration in the gynophore. Hormones like auxins, gibberellins (GA), and abscisic acid (ABA) are crucial for embryo development during peg growth. It is postulated that indole-3-acetic acid (IAA) may be degraded by peroxidase and laccase, while IAA may, in turn, repress these enzymes. Modulation of laccase expression during different growth stages regulates IAA concentration. Downregulation of ethylene-responsive transcription factors increases auxin response and contributes to peg gravitropism. The stress-responsive RD22-like protein (BURP-domain family) is also implicated in peg gravitropic growth. The sugar transporter gene *AhERD6*, involved in starch and sucrose metabolism, is upregulated at low aluminum concentrations (Bao, 2022). At high aluminum concentrations, hormone signaling transduction is enriched, and endogenous ABA content increases due to enhanced expression of 9-cis-epoxycarotenoid dioxygenase. RNA-Seq expression profiling generated over 70,000 unigenes, revealing modulation of genes involved in light signaling, hormone metabolism, cell division and growth, photosynthesis, carbon and nitrogen metabolism, and stress responses before and after peg penetration into the soil (Xia, 2013). Calcium plays a critical role in peg and pod development, influencing pod filling and ultimately yield (Poovaiah, 1993; Beringer and Taha, 1976; Cox, 1982). Calcium-dependent protein kinase (CDPK) activity is linked to peg and seed development (Jain, 2011), with studies showing differential expression of genes related to Ca^{2+}signaling, hormone signaling (auxin, ethylene, GA, cytokinin, and ABA), and development during peg and pod development (Li, 2017; Yang et al., 2017; Yang, 2020). Genes encoding phytochrome A (*phyA*), auxin response factor 9 (*IAA9*), and mechanosensitive ion channel protein have significant roles in geocarpy (Cui, 2022a). Studies on chloroplast-related genes proposed two signaling pathways: one originating from chloroplasts and signaling via phosphorylation, and another initiated during abiotic stress and signaling via phosphorylation, calcium signaling, and ubiquitination. Studies on gene action during pod and seed development, from pre-embryonic stages to mature pod formation, have been extensively reported in peanut. Analysis of 165,689 transcripts in shell and kernel tissues during pod development demonstrated the significant influence of photomorphogenesis-related genes (Liu, 2020). Transcriptome analysis of four peanut accessions revealed enrichment of mitogen-activated protein kinase (MAPK) signaling pathways, phytohormones, and transcription factors during pod development (Wu, 2022). It is hypothesized that MPK3 promotes integument cell division and regulates cell division through phosphorylation. This suggests the involvement of candidate genes regulating auxin synthesis and metabolism (*PNC*, *YUC*) and those regulating cell differentiation and proliferation (*IAA*, *CYCD*, *CYCU*, and *GASA*) in regulating seed size and pod elongation through the gibberellic acid pathway. A network of molecular pathways, formed by the interaction of these genes, regulates pod filling and size.

Seed coat (testa) color is a crucial trait influencing market preference. Pigments like anthocyanins (Chukwumah, 2009), flavonols, and proanthocyanidins (Iwashina, 2015a, 2015b) contribute to testa color. MYB proteins, basic helix-loop-helix (bHLH) proteins, and WD40 transcription factors have been identified as regulators of flavonoid biosynthesis (Falcone-Ferreyra, 2012). Studies using low-anthocyanin mutants demonstrated the differential expression of caffeoyl-CoA O-methyltransferase (*CCoAOMT*) and *R2R3-Myb* in seed color development (Wan, 2016), consistent with their roles in flavonoid accumulation in *Arabidopsis thaliana* (Zhang, 2014; Baudry, 2006). Analysis of flavonoid metabolism in peanut seed coats identified 5,452 differentially expressed unigenes involved in flavonoid biosynthesis, phenylalanine metabolism, phenylpropanoid biosynthesis, and hormone signaling pathways (Wan, 2018), suggesting roles for *PAL*, *C4H*, *CHS*, *CHI*, *MYB111*, *bHLH*, and *WD40* in seed coat color development, coordinated with auxin, ABA, ethylene, jasmonic acid, and C-metabolism pathways. Studies in black peanuts reported upregulation of transcriptional activators *AhMYB1*, *AhMYB2*, and *AhTT8*, leading to skin-specific anthocyanin accumulation (Huang et al., 2019). The black color in the peanut cultivar ZH9 was linked to reduced lignin and isoflavonoid biosynthesis, directing more substrate towards anthocyanin biosynthesis, along with the upregulation of *DFR* genes and three *R2R3MYB* transcription factors (Xia et al., 2020). qRT-PCR validated 15 key enzyme and transcription factor genes. Combined genetic analysis, tissue expression patterns, and differentially expressed gene annotations revealed the involvement of seed coat-specific genes *Araip.M7RY3*, *Aradu.R8PMF*, and *Araip.MHR6K* in white testa color formation, regulated by hormone synthesis, signaling, and the MBW complex (Wan, 2020). Transcriptome analysis of red and pink testa color varieties identified 198 flavonoid metabolites and confirmed the role of anthocyanins in red coloration (Xue, 2021). Differentially expressed genes in these genotypes were enriched in flavonoid biosynthesis, metabolic pathways, and secondary metabolite biosynthesis. Key functional genes controlling red and pink testa color development were identified as *F3'H* (*arahy.8F7PE4* and *arahy.K8H9R8*), *CHS* (*arahy.CM90T6*), and *DFR* (*arahy.LDV9QN* and *arahy.X8EVF3*). Three transcription factor families were also implicated: *MYB* (*arahy.A2IWKV*, *arahy.US2SKM*, *arahy.SJGE27*, *arahy.H8DJRL*, and *arahy.PR7AYB*), *bHLH* (*arahy.26781N*, *arahy.HM1IVV*, and *arahy.MP3D3D*), and *WD40* (*arahy.L6JJW9*).

Seed size is a crucial determinant of peanut yield and market value, particularly for direct consumption and confectionery uses. Several candidate genes have been implicated in controlling pod width and seed size, including cinnamyl alcohol dehydrogenase (*CAD*, *c26901_g1*) and 1-aminocyclopropane-1-carboxylate synthase (*ACS*, *c37339_g1*) (Wan, 2017). A comprehensive transcriptome analysis encompassing approximately 36,000 gene loci identified additional seed size–related genes, such as topless-related protein 2, IAA-amino acid hydrolase ILR1-like 5, transcription factor TGA7, and a putative pentatricopeptide repeat (PPR) protein, with the PPR protein having a major influence on large seed size (Li, 2021).

Peanut kernel nutritional quality is paramount due to its diverse uses in direct consumption, confectionery products, and oil extraction. Kernels are rich in fatty acids, proteins, carbohydrates, and minerals like copper (Cu), manganese (Mn), potassium (K), calcium (Ca), and phosphorus (P). They also contain essential vitamins (tocopherols/vitamin E, niacin, folic acid), dietary fibers, flavonoids, and phytosterols (resveratrol, beta-sitosterol, phenolic acids) (Francisco and Resurreccion, 2008; Settaluri, 2012; Griel, 2004; Mahatma, 2016). Among the 12 fatty acids present, palmitic (16:0), oleic (18:1), and linoleic (18:2) acids are particularly important (Orsavova, 2015). Oleic acid typically comprises around 80% of peanut oil, while palmitic acid makes up about 10% (Ahmed and Young, 1982; Treadwell, 1983). High-oleic peanut varieties, with oleic acid content similar to olive oil (approximately 80%), low linoleic acid (3%–4%), and palmitic acid (around 6%–9%), are considered healthier and have a longer shelf life (Rizzo, 1986; Vasiliou et al., 2009).

Seed development involves the coordinated expression of genes related to protein and oil biosynthesis, as well as seed coat color regulation. RNA-Seq studies on three peanut cultivars with varying oil contents annotated 125 metabolic pathways associated with 18,028 unigenes (Zhang, 2012). Transcriptome analysis of two genotypes at four developmental stages showed higher expression of energy and primary metabolite production genes in genotypes with greater pod filling (Gupta, 2016). Studies on lipid metabolism in developing seeds identified 18,028 differentially expressed unigenes associated with 125 KEGG pathways (Zhang et al., 2012), with 146 of 3,899 metabolic pathway genes linked to lipid biosynthesis (GO: 0008610) and 69 to fatty acid biosynthesis. RNA-Seq identified 1,500 unigenes related to lipid metabolism and fatty acid synthesis in developing seeds, with 654 involved in lipid transport (Yin et al., 2013). Key metabolic pathways identified in oil accumulation included the glyoxylate cycle, gluconeogenesis, glycolysis, pentose phosphate pathway, citric acid cycle, and oxidation.

Research isolating 12 new *AhFAD* genes from *A. hypogaea* and evaluating their expression under abiotic stress showed distinct expression patterns during seed development (Chi et al., 2017). *AhFAB2-2*, *AhFAD3-1*, *AhFAD7-1*, and *AhSLD-3* expression gradually increased, peaking before slowly declining. *AhFAB2-3*, *AhFAD3-2*, *AhFAD4*, *AhADS*, and *AhDES* showed high expression early in development, decreasing over time. *AhSLD-4* transcript levels peaked at the final stages. A transcriptome sequencing study revealed 2,243 upregulated and 3,215 downregulated genes involved in oil biosynthesis, with comparative genomics identifying 147 common gene clusters across 17 chromosomes (Wang et al., 2018). Studies on peanut β-ketoacyl-CoA synthase (*AhKCS*) genes, involved in fatty acid elongation, identified enzymes with different substrate specificities, confirming that endoplasmic reticulum-localized *AhKCS1* and *AhKCS28* regulate very long-chain fatty acid content in peanut seeds, further validated by overexpression in *Arabidopsis thaliana* (Huai et al., 2020).

Fresh seed dormancy, crucial for preventing viviparity, is linked to ethylene, abscisic acid (ABA), auxin, and gibberellic acid (GA) metabolism, as shown by transcriptome analysis (Chen, Jiang et al., 2015). Two key candidate genes, *AhNCED2* (maintaining dormancy) and *AhCYP707A1* (facilitating dormancy release), were identified and confirmed in Spanish bunch varieties (Silva et al., 2017). The seed coat, a protective barrier against pathogens and a maintainer of seed viability, is influenced by ethylene signalling and ABA metabolism, as shown in studies on a *pscb* mutant (Wan et al., 2016). Seed coat cracking may result from asynchronous growth between the epidermis and hypodermis due to downregulation of genes affecting ABA and ethylene signaling (e.g., ethylene receptor 1, ethylene-responsive transcription factor 1, *PYR/PYL* family ABA receptors). Interactions between stress hormones (ABA and ethylene) and kinesin-4-like proteins, which regulate cell division and growth, are thought to contribute to this cracking (Gazzarrini and McCourt, 2001; Wan et al., 2016).

8.4 RESPONSE TO STRESS

8.4.1 Abiotic Stress

Water deficit (drought) is a primary abiotic stress limiting peanut cultivation. Numerous studies have investigated gene action and expression profiles under drought conditions in both wild and cultivated peanut species. Studies on gene expression modulation under water-limited conditions in *A. duranensis* revealed modulation of 31 candidate genes, including expansins, nitrilase, P450 proteins, and transcription factors from the NAC, bZIP, late embryogenesis abundant (LEA), ERF, and MYB families (Brasileiro et al., 2015). In *A. magna*, genes involved in drought response signal transduction pathways included drought-induced proteins (DiP), carbonic anhydrase, Ca^{2+} /H^{+} exchanger, chloroplast drought-induced stress protein (CDSP), metallothionein, and aminomethyltransferase (Brasileiro et al., 2015).

Expansins, cell wall proteins responding to environmental stimuli at various developmental stages, have been the subject of genome-wide and RNA-Seq analyses in *Arachis*, revealing their modulation in response to both biotic and abiotic stresses (Guimaraes et al., 2017). The expansin-like gene *AraEXLB8* was found to be highly conserved across *Arachis* species, with eight identified haplotypes. Studies on *A. duranensis* root tissues under water stress identified differentially expressed genes related to drought perception and response, including a DOT1-like glycine-rich protein, MLP-like protein, chitinase-2, maturase K, and HSP70 and HSP17.3 (Carmo et al., 2019). Expression and abundance analysis of 14 of 15 identified proteins revealed negative regulation in roots under stress.

Transcriptome analysis of two peanut genotypes under induced drought identified 4,508 differentially expressed genes, including 1,554 transcription factors involved in ABA and sucrose metabolism pathways (Bhogireddy et al., 2020). Genome-wide analysis of NAC transcription factors in the wild species *A. duranensis* and *A. ipaensis* under drought and salinity stress revealed 43 differentially expressed NAC genes, with 17 responsive to both stresses (Yuan et al., 2020). Studies comparing tolerant and sensitive peanut cultivars under progressive drought stress highlighted the importance of osmotic potential maintenance, reactive oxygen species (ROS) elimination, and signal transduction in drought response (Jiang et al., 2021). Root transcriptome analysis of the drought-tolerant *A. duranensis* and drought-sensitive *A. ipaensis* under drought stress identified 1,464 differentially expressed genes, with *A. duranensis* showing more transcripts related to signal transduction, flavonoid biosynthesis, DNA methylation, and ethylene response (Thoppurathu et al., 2022). Genome-wide analysis identified 162 small auxin-upregulated RNA (SAUR) genes in peanut (Liu et al., 2022), with transcriptomic data revealing prominent expression in leaves, perianth, pistils, and peg tips. The gene *AhSAUR3* was shown to have a functionally negative role in drought tolerance. Zhang et al. (2020) conducted transcriptome sequencing of young peanut leaves subjected to 24-hour salt treatments and control conditions, generating six libraries. Comprehensive analysis of differentially expressed genes (DEGs) identified 3,425 DEGs, including 141 transcription factors responding to salinity stress. Upregulated DEGs were enriched in pathways regulating cell growth and antioxidant activity, while downregulated DEGs were enriched in metabolic processes. Fourteen DEGs were validated using real-time PCR. Jiang (2020) performed a comparative transcriptome analysis to investigate global gene expression of transcription factors (TFs) in two peanut varieties exhibiting contrasting cold tolerance. The study identified 87 TF families, comprising 2,328 TF genes, with 445 TF genes showing significant differential expression under cold stress. The TF families with the largest number of differentially expressed members included bHLH (basic helix-loop-helix protein),

C2H2 (Cys2/His2 zinc finger protein), ERF (ethylene-responsive factor), MYB (v-myb avian myeloblastosis viral oncogene homolog), NAC (NAM, ATAF1/2, CUC2), and WRKY TFs. Phylogenetic evolutionary analysis, temporal expression profiling, protein–protein interaction (PPI) network analysis, and functional enrichment of these differentially expressed TFs provided insights into the importance of plant hormone signal transduction and plant–pathogen interaction pathways and their potential roles in peanut cold tolerance.

8.4.2 Biotic Stress

Peanut is susceptible to various pathogens and insect pests, impacting yield and quality. Major fungal diseases affecting aerial parts include leaf spots and rust, while subterranean pods are affected by stem rot, pod rot, and aflatoxin contamination. Sucking pests and pod-boring insects also contribute to yield losses. Research on gene expression in response to biotic stress in peanut has primarily focused on aflatoxin contamination and nematode infestations, with comparatively less attention given to foliar fungal diseases at the functional genomics level. Using cDNA-AFLP, Kumar and Kirti (2011) demonstrated upregulation of several defense-responsive genes in *A. diogoi* challenged with *Phaeoisariopsis personata*, the causal agent of late leaf spot. Two isolated clones coded for cyclophilin-like proteins. In the resistant wild species *A. stenosperma*, Guimaraes et al. (2012) identified 7,723 unigenes from infected leaves, including transcription factors (20), defense-related genes, resistance gene analogs, and retrotransposon-related sequences. Upregulation of genes encoding oxygen-evolving enhancer protein, glyceraldehyde-3-phosphate dehydrogenase, F-box protein, dihydroflavonol reductase, sedoheptulose-1,7-biphosphatase, phytochrome A, and photosystem-II chlorophyll a/b binding protein upon pathogen infection were also reported (Kumar and Kirti, 2015a). Jasmonic acid and ethylene signaling pathways were implicated in the response to *P. personata* infection (Kumar et al., 2015). Han et al. (2017) found that the resistant mutant M14 became susceptible to fungal infection due to downregulation of chloroplast genes, upregulation of WRKY transcription factors, and suppression of growth-related phytohormones. Despite high levels of pathogenesis-related (PR) proteins, these negative effectors, triggered by fungal infection in the susceptible plant, reduced photosynthesis and phytohormone levels, leading to symptom development. Gong et al. (2020), using RNA-Seq, reported 133 differentially expressed genes upon infection by *Cercospora arachidicola*, the causal agent of early leaf spot (ELS). Two copies of the coiled-coil nucleotide-binding leucine-rich repeat (NLR) R gene were identified on chromosome B2, mediating the peanut phytoalexin deficient 4 (PAD4) regulators of effector-triggered immunity. Polyphenol oxidase (PPO) genes also played a significant role in the resistance response. Catto et al. (2021) studied differential gene expression in peanut cultivars susceptible and field-resistant to tomato spotted wilt virus (TSWV). The resistant cultivar Tifguard showed 2,579 differentially expressed genes, while the susceptible cultivar SunOleice showed 4,605. Despite having fewer differentially expressed genes, the resistant cultivar exhibited higher upregulation of defense-related genes, including disease resistance (R) proteins, leucine-rich repeats, stilbene synthase, dicer, and calmodulin.

Sequencing of seedlings infected with *Sclerotium rolfsii*, a major peanut fungal pathogen causing stem rot in India, identified 170 genes, including resistance genes with LRR, ERF, and zinc finger motifs, within a 5.2 Mb QTL region on chromosome B04 (Dodia et al., 2019). Further RNA-Seq studies, validated by qRT-PCR, demonstrated the involvement of jasmonic acid pathway enzymes, receptor-like kinases, and transcription factors (TFs) like WRKY, zinc finger protein, and C2-H2 zinc finger in the defense response of a resistant genotype upon pathogen infection (Bosamia et al., 2020). These studies proposed PAMP-triggered immunity (PTI) as a likely resistance mechanism and the jasmonic acid signaling pathway as a key defense mechanism in peanut. The subterranean nature of peanut pod development exposes pods and seeds to *Aspergilus flavus (A. flavus)* infection, leading to aflatoxin accumulation, a major constraint in peanut trade. The ubiquitous nature of the fungus allows infection pre- and post-harvest under conducive moisture conditions. The prominent roles of LOX homologs, LOX2-3 and LOX4, in seed colonization by *A. flavus* were established (Zeringue et al., 1996), while enhanced expression of *LOX1* and alpha-dioxygenase 1 (*DOX1*) upon fungal infection was reported (Burow et al., 2000). Research on the fungal infection process has revealed that resistance to *A. flavus* is manifested through three distinct mechanisms depending on the infection type: pre-harvest aflatoxin contamination (PAC), *in vitro* seed coat colonization (IVSC), and post-harvest seed infection (aflatoxin production, AP) (Nigam et al., 2009). Methyl-jasmonate's role in inhibiting fungal sporulation and reducing toxin production in *Aspergillus* was reported (Goodrich-Tanrikulu et al., 1995; Meimaroglou et al., 2009). The WRKY transcription factor's role in regulating jasmonate and salicylate production in response to stress, as part of the host immune system, was also documented (Gimenez-Ibanez and Solano, 2013). The inhibitory effect of resveratrol, a major antioxidant, on aflatoxin production was demonstrated (Wang et al., 2015), showing reduced production of hexanoate (aflatoxin biosynthesis precursor) through downregulation of aflatoxin biosynthesis pathway genes *aflA* and *aflB*. Resveratrol also suppressed regulatory genes *stuA*, *fluG*, and *flbC*, inhibiting mycelial and conidial formation. Infection studies revealed that defense signaling genes from mitogen-activated protein kinase (MAPK) pathways react with mycelial chitin, inducing pathogen-associated molecular pattern (PAMP)-triggered immunity (PTI) (Fountain et al., 2015). The equilibrium between antioxidant biosynthesis and ROS-scavenging enzymes is crucial for suppressing fungal growth and aflatoxin production. Transcriptomic shifts enriched pathways synthesizing antioxidants like phenylpropanoids, stilbenes, peroxidase, coumarins, cinnamic acid, ascorbate, and flavonoids (Wang et al., 2016). Studies on *A. flavus*-contaminated peanut kernels revealed expression of genes involved in cell wall restructuring and morphology, ABA signaling, fatty acid metabolism, nitrogen metabolism, sugar metabolism and transport, and sulfate metabolism, with fatty acid biosynthesis and ABA signaling gene clusters functionally validated as key regulators of PAC (Clevenger et al., 2016b). Transcriptome analysis of resistant and susceptible cultivars challenged with *A. flavus* reported 30,143 differentially expressed genes, including pathogenesis-related proteins, defense-related genes, and transcription factors (Wang et al., 2016). Highly responsive genes controlled phenylpropanoid-derived compounds and secondary metabolite synthesis. Nayak et al. (2017) reported 4,445 differentially expressed genes related to defense reactions, including senescence-related proteins, 9s-lipoxygenase, resveratrol synthase, PR proteins, and so on. Transcriptional level diversity assessment of pathways from *A. flavus*-infected resistant and susceptible genotypes overexpressing resveratrol synthase and chalcone synthase confirmed reduced fungal colonization and aflatoxin load. The defense mechanism involves oxidative homeostasis counteracting ROS produced by fungal infection, with enhanced expression of genes like chalcone synthase, phenylalanine ammonia lyase, and resveratrol synthase (antioxidant biosynthesis), catalases, glutathione-S-transferase, peroxidases, superoxide dismutase, senescence-associated proteins (stress response), chitinases, defensins, Mlp-like protein (defense), and linoleate 9s-lipoxygenases (lipid signaling). Transcription factors (TFs) like NAC, ERF, WRKY, and MYB were identified as key regulators of pathogenesis-related and antioxidant-producing genes, also controlling pathways involved in fungal colonization and toxin production, such as ethylene production and signaling and polyamine biosynthesis (Huang et al., 2009; Wang et al., 2017; Nayak et al., 2017; Majumdar et al., 2018). Transcriptomic comparison of resistant and susceptible cultivars revealed 4,272 differentially expressed genes, implicating protein processing in the endoplasmic reticulum, alpha-linolenic acid metabolism, carbon fixation, carbon metabolism spliceosome, ethylene, heat shock proteins, and NBS-LRR receptors in the defense response to *Aspergillus* infection, and demonstrating the significant role of the defense mechanism in aflatoxin production (Korani et al., 2018). Transcriptomic and proteomic studies on infected and uninfected peanut samples revealed a weak correlation between transcript and protein levels, suggesting the presence of post-transcriptional regulation in toxin production (Zhao et al., 2019). Differentially expressed genes in resistant genotypes included chitinases, PR proteins, and hevamine-A. *In vitro* fungistasis tests demonstrated the functional role of hevamine-A in inhibiting fungal growth. Soni et al. (2021) identified 45,336 transcripts, including genes encoding bHLH (719), MYB-related (479), NAC (437), FAR1 family protein (320), and other TFs (57) in infected and uninfected samples. Genes involved in senescence, resveratrol synthesis, PR proteins, peroxidase, seed linoleate content, ABA response, glutathione-S-transferases, chitinases, and chalcone synthase were differentially expressed in the host plant. Using transcriptomic comparison and weighted gene co-expression network analysis (WGCNA), Cui et al. (2022b) identified hub genes associated with *Aspergillus* resistance, including 18 genes encoding pathogenesis-related protein PR10, MAPK kinase, 1-aminocyclopropane-1-carboxylate oxidase (ACO1), serine/threonine kinase (STK), cytochrome P450, pattern recognition receptors (PRRs), phosphatidylinositol transfer protein, SNARE protein SYP121, pectinesterase, and pentatricopeptide repeat (PPR) proteins.

Meloidogyne arenaria, the root-knot nematode (RKN), is a major soilborne pest causing substantial economic losses in peanut by damaging pods. Studies have investigated the molecular basis of resistance to RKN. Proite et al. (2007) reported the involvement of candidate genes related to hypersensitive response (HR), pathogen defense-related secondary metabolites, resistance protein MG13, hormonal balance, auxin-repressed protein (ARP), patatin-like protein, metallothionein, DUF protein, resveratrol synthase, and catalase in the defense response to RKN infection. These findings were further validated through RT-qPCR and microarray analysis (Morgante et al., 2013). Guimaraes et al. (2010) identified auxin-repressed protein (*AsARP*), cytokinin oxidase (*AsCKX*), and metallothionein type 2 (*AsMET2*) as the most differentially expressed genes during expression profiling of the nematode-resistant *A. stenosperma* and the susceptible *A. hypogaea*. These genes contribute to maintaining plant hormonal balance and responding to the hyper sensitive reaction (HR) triggered by reactive oxygen species (ROS), with some expressed even before nematode challenge. An RNA-Seq study of *A. stenosperma* roots identified jasmonic acid pathways (patatin, allene oxidase cyclase), salicylic acid pathway (NBS-LRR, lipocalins, resveratrol synthase), hormonal balance (auxin-responsive protein, GH3), and cellular plasticity and signaling (tetraspanin, integrin, expansin) as differentially expressed and involved in the host plant's defense response during RKN infection (Guimaraes et al., 2015). Comparative transcriptome analysis of *A. stenosperma* (RKN resistant) and *A. duranensis* (RKN susceptible) revealed the involvement of genes responsible for secondary metabolite production, hormone signaling, and the NBS-LRR class of R genes (Mota et al., 2018). RT-qPCR validation showed contrasting expression patterns of most identified NBS-LRR genes between the resistant and susceptible *Arachis* species.

8.5 FUNCTIONAL GENE MUTATIONS

Mutation is a fundamental driver of biological evolution, resulting in changes in genomic sequence composition through single-base substitutions, small insertions or deletions (InDels), and large-scale duplications or deletions (Jankowicz-Cieslak et al., 2017). In peanut, understanding gene function is crucial for accelerating crop improvement through targeted breeding techniques. Historically, functional gene characterization in peanut has been limited by a lack of genomic information and resources. However, the recent publication of high-quality reference genomes (Bertioli et al., 2019; Zhuang, 2019) has facilitated the development of genetic and genomic resources that support both fundamental research and breeding efforts. Sequenced mutant populations, post-transcriptional gene silencing (PTGS) techniques (including RNA interference (RNAi), small interfering RNA (siRNA), virus-induced gene silencing (VIGS)), and structured natural populations (Chaudhary et al., 2019) now provide rapid approaches for directly examining gene function in peanut. While research in model plant species offers valuable insights, direct functional characterization in crop species like peanut is essential for effective crop improvement, as findings from model species do not always translate directly (Sikora et al., 2011).

8.5.1 Gene-Specific/Site-Directed Mutagenesis

Mutation breeding is a well-established approach for increasing genetic variability and improving desired traits in peanut. This is particularly valuable in peanut due to its narrow genetic base, which limits the variability available for traditional cross-breeding, especially for biotic and abiotic stress resistance. Two main classes of mutagens are commonly used: physical mutagens (X-rays, gamma rays, neutrons, etc.) and chemical mutagens (ethyl methane sulfonate (EMS), diethyl sulfate (DES), ethylene-imine (EI), N-nitroso-N-methyl urethane (NMUT), N-nitro-N-methyl urea (NMU), ethyl bromide (EB), sodium azide (SA), etc.). Physical mutagenesis typically involves drying seeds to low moisture content and then irradiating them in sealed packets for a specific duration depending on the radiation dose (C. T. Wang et al., 2012). Chemical mutagenesis involves soaking seeds in water for 2–4 hours, followed by immersion in a specific concentration of the chemical mutagen for 5–6 hours. Excess chemical is then removed by rinsing the seeds in running tap water for approximately 10 hours, followed by shade drying before sowing (C. T. Wang et al., 2012).

Both physical and chemical mutagens have been used to generate a broad spectrum of genetic diversity in peanut for use in crop improvement and inheritance studies. Mutations affecting a wide range of traits have been induced, including agronomic traits (large seed size, earliness, etc.), biotic and abiotic stress tolerance (late leaf spot, early leaf spot, rust, *sclerotium rolfsii*, collar rot, peanut bud necrosis disease, peanut stem necrosis disease, *Spodoptera*, thrips, bruchid beetle, phosphorus efficiency, acid soils, drought, heat, etc.), physical traits (photosynthesis, nitrogen fixation, harvest index, partitioning efficiency, water use efficiency, etc.) (Addai et al., 2022), biochemical traits (high oil, protein, amino acids, modified linoleic and oleic acid content, etc.) (M. C. Wang et al., 2020; Han et al., 2022), and other characteristics such as the size, shape, and colour of leaves, plant height, and blossom colour (Branch and Brown, 2019; Panda et al., 2022); these are only a few examples of the qualitative and quantitative attributes that have been shown to be impacted by various mutations in peanut in previous research (Murthy et al., 2004). Sodium azide (0.39%) mutagenesis produced a peanut plant with increased oleate content (Wang et al., 2011; Mondal et al., 2011). EMS and gamma radiation mutagenesis of the peanut cultivar GPBD-4 yielded 20 high-oleate mutants with stable fatty acid profiles across generations. Mutant GE-113 showed the greatest increases in oleic acid (74.48%) and the oleic/linoleic (O/L) ratio (8.12) compared to the parent (Kavera et al., 2013). The Bhabha Atomic Research Center (BARC) in India has developed numerous peanut varieties for commercial cultivation through mutant breeding, with mutant G2-214 exhibiting maximum pod production and a higher O/L ratio than its parent GPBD-4 (Kavera and Nadaf 2017). Combining 200 Gy gamma rays with 3 mM sodium azide successfully induced mutants with dominant rose testa and fused terminal leaflets (Mondal et al., 2007). The high-yielding peanut cultivar Huayu 40 was developed by injecting 0.3% EMS into Huayu 16 flowers (C. T. Wang et al., 2010). Treatment of Huayu 22 peanut seeds with 0.39% sodium azide (NaN_3) created line 11-L36, which produced higher yields than the control. Treating Huayu 22 with 0.39% diethyl sulfate (DES) produced lines 11-L39 and 11-L40, which produced more kernels per acre than the control (C. T. Wang et al., 2012). Pre-treatment of seeds with EMS and SA improved plant germination, height, branching, foliage, and reduced mortality (Jadhav et al., 2018). Variance analysis in M1 and M2 generations indicated significant differences between genotypes treated with sodium azide and gamma rays, with sodium azide concentrations significantly affecting the number of primary branches (Abdelrahem et al., 2021). Mutagenesis has also been employed to develop low-aflatoxin varieties. Inducing genetic variation in four peanut cultivars (Giza 6, Gregory, Giza 4, and Giza 5) using EMS and gamma rays resulted in ten *A. flavus*-resistant mutants with high β-1,3-glucanase activity (Sharaf et al., 2011). Several peanut varieties developed through mutation breeding (e.g., TMV 10, TG 1, MH 2, BG 1, BG 2, CO 2, Kaushal, TG 3, TG 17, ICGS 76, VRI 2, MH 4, Somnath, TAG 24, TG 22, TKG 19A, Dh 40, TLG45) have been successfully released for commercial cultivation in India (Murthy et al., 2004; Mondal et al., 2007).

8.5.2 RNA Interference (RNAi)

RNA interference (RNAi), a process whereby double-stranded RNA (dsRNA) triggers the degradation of messenger RNA (mRNA), has become a prominent biotechnological tool for crop improvement due to its high accuracy and fidelity (Rajam, 2020). Since its discovery, RNAi has evolved into a powerful technique for functional genomics and genetic engineering, providing a precise method for suppressing specific genes without affecting the expression of others. This technology has been effectively used in various plant species to modify desirable traits (Tiwari and Rajam, 2022). In peanut, RNAi has been successfully applied to address several key challenges. For example, transgenic peanuts resistant to peanut stem necrosis disease (PSND) were developed by transforming the peanut cultivar TMV-7 with an RNAi vector containing inverted repeats of the tomato spotted wilt virus (TSWV) coat protein (CP) gene (Reddy et al., 2016; Senthilraja et al., 2018). RNA interference has also been proposed as a strategy to reduce allergenic proteins in peanut (Singh et al., 2021). Transgenic lines with reduced levels of the major peanut allergens Ara h1 and Ara h6 have been created using RNAi, leading to the development of hypoallergenic peanut varieties (Dodo et al., 2005; Chu et al., 2008). Furthermore, RNAi has been employed to control aflatoxin contamination caused by *A. flavus* and *A. parasiticus.* Silencing five aflatoxin biosynthesis-related genes (*aflR*, *aflS*, *aflC*, *aflP*, and *pes1*) involved in aflatoxin biosynthesis, transport, or non-ribosomal peptide synthetase (NRPS) activity, RNAi significantly reduced aflatoxin B1 and B2 accumulation in transgenic peanut plants following *A. flavus* inoculation. Notably, RNAi-expressing peanut lines 288-72 and 288-74 showed up to 100% reduction in aflatoxin B1 and B2 compared to control plants when inoculated with aflatoxigenic *A. flavus* (Arias et al., 2015). Finally, RNAi-mediated silencing of the *AhFAD2* gene resulted in a 15%–36% increase in seed oleic acid content (Xu et al., 2018). This demonstrates the versatility of RNAi in peanut improvement, effectively addressing issues ranging from disease resistance and allergenicity to seed quality traits.

8.5.3 Small Interference RNAs (siRNA)

Small RNAs (sRNAs), including microRNAs (miRNAs) and small interfering RNAs (siRNAs), are key agents of RNA interference in plants. Increasing evidence supports the intercellular, systemic, and even interspecies movement of sRNAs (Kamthan et al., 2015). These non-coding sRNAs can move symplasmically through plasmodesmata (PD) and via the phloem from source to sink tissues to silence target genes. siRNAs play a crucial role in mediating plant defense against biotic stresses (viruses, bacteria, fungi, oomycetes) and abiotic stresses (drought, salinity, cold, etc.) (Kong et al., 2022). Altering siRNA-producing loci could enhance stress tolerance. Next-generation high-throughput sequencing, combined with bioinformatics, has revealed a complex small RNA population in peanut, identifying 14 novel species-specific miRNA families and 75 conserved miRNAs (Zhao et al., 2010). miRNAs regulate diverse genes in response to the soilborne pathogen *Sclerotium rolfsii*, establishing a systemic defense network against this devastating peanut disease. Studies have identified sets of known and novel miRNAs differentially expressed upon *S. rolfsii* infection, with predicted target genes including receptor kinases, transcription factors, and genes involved in metabolite production and transport (Xu et al., 2022). Integrated analysis of the transcriptome, miRNAome, and degradome of susceptible and resistant peanut varieties in response to *A. flavus* revealed that the miR482/2118 family may regulate the *NBS-LRR* gene, which shows higher expression in resistant genotypes (Zhao et al., 2020). High-throughput small RNA sequencing of control and transformed peanut lines expressing an inverted repeat targeting five aflatoxin biosynthesis pathway genes showed up to 100% reduction in aflatoxin B1 compared to controls. Analysis of small RNAs from aflatoxin-resistant peanut led to the discovery of 10 novel miRNAs, 89 known miRNAs, and 2 siRNAs associated with aflatoxin suppression (Power et al., 2017). Transient expression analysis in *Nicotiana benthamiana* suggested that miR160, miR482, and miR2118 may positively regulate peanut cold tolerance, while miR396, miR162, and miR1511 may act as negative regulators (Zhang et al., 2022). Tissue-specific miRNA expression has also been observed in peanut, with miR3 and miR7 more highly expressed in leaves, flowers, and roots than in seeds, and miR156 more highly expressed in leaves, roots, and stems than in flowers and seeds (Chi et al., 2011). Upon infection with *Ralstonia solanacearum*, miR2118 was upregulated in both cultivated and wild peanut, while miR482 was upregulated in cultivated peanut but downregulated in wild peanut, suggesting roles for these miRNAs in bacterial wilt (BW) resistance (Zhao et al., 2015). Small RNA transcriptome sequencing of peanut infected with pod rot revealed 334 miRNAs, with 97 downregulated and 27 upregulated after infection. Functional validation showed downregulation of ahy-miR396e-5 and upregulation of its target gene *c39419-g1-i1* after infection. Similarly, ahy-miR3509-5p, ahy-miR166f, and ahy-miR159b were downregulated, while their corresponding target genes *c40055-g1-i3*, *c31393-g1-i1*, and *c41016-g4-i1* were upregulated (Wang et al., 2020). Genome-wide analysis of peanut transcriptomes identified 1,442 long non-coding RNAs (lncRNAs), with *TCONS-00292946* showing downregulation in roots within 12 hours of salinity stress but upregulation at 24 hours. *TCONS 00176941* was upregulated in roots and downregulated in leaves under salinity stress within 12 hours, while *TCONS 00011551* was upregulated (Tian et al., 2020). Small RNA and degradome deep sequencing have revealed roles for miRNAs in cold and aluminum stress responses, with identified miRNA–stress response relationships potentially serving as markers for breeding future-ready peanut varieties (Zhang et al., 2022; Tong et al., 2022). Increasing evidence highlights the role of non-coding RNAs (miRNAs, siRNAs) in determining biotic and abiotic stress responses, alongside protein-coding genes. MicroRNAs (miRNAs) are among the most studied non-coding RNAs for functional characterization of their regulatory effects on crop traits (Tang and Chu, 2017). High-throughput RNA sequencing in peanut identified approximately 33 miRNA families and 25 peanut-specific novel miRNAs (Chi et al., 2011). Studies on embryo and pod development reported 70 known and 24 novel miRNA families expressed during these processes (Gao et al., 2017).

Pod initiation from the peanut peg involves expression of 10 known miRNAs (miR1520, miR2199, miR2628, miR4414, miR5221, miR5227, miR5234, miR5244, miR6300, and miR7696), which are either legume- or peanut-specific. Peg swelling into a pod, essential for seed enlargement, requires the action of miR3512 (Gao et al., 2017). Similar to *Arabidopsis* and rice, several genes in peanut, such as *squamosa promoter binding-like (SPL), teosinte branched1/cycloidea/proliferating cell factor1 (TCP), MYB, ARF, NAC, GRAS,* and *apetala2 (AP2),* key regulators of growth and development, are targeted by distinct miRNA classes (Gao et al., 2017). For instance, *MYB* TFs are regulated by miR159 (controlling pigment accumulation during seed development), *AP2* by miR172, *SPL* by miR156, and *NAC* by miR164. MiR167 and miR1088 regulate *PPRP* cleavage, and miR390 reduces *ARF* transcripts, with these TFs known to participate in embryogenesis (Sauer et al., 2013). During peanut pod initiation, *TAS3*, regulated by miR390, induces phased ta-siRNA formation, which regulates auxin signaling and may control pod swelling, embryo growth, and thus seed size (Gao et al., 2017). These studies provide transcriptome-wide insights into miRNA-mediated regulation of embryo and pod development, key determinants of peanut seed quality.

8.5.4 Virus-Induced Gene Silencing (VIGs)

Virus-induced gene silencing (VIGS) is a powerful tool for high-throughput gene function analysis in plants. Advancements in genomics and transcriptomics have enabled the discovery of novel genes linked to complex metabolic processes, and VIGS has emerged as a potent method for rapid *in planta* functional evaluation. A key advantage of VIGS is that it does not require the creation of stable transgenic lines, which can be technically challenging and time-consuming in some crops. VIGS also offers advantages over other reverse genetic techniques like RNAi-mediated suppression. The VIGS methodology involves creating virus-derived vectors, selecting the optimal size and orientation of the target gene sequence, and employing an appropriate inoculation technique (Ramegowda et al., 2014). Several studies have demonstrated the effectiveness of VIGS in peanut. Silencing both *PeaeIF4E* and *PeaeIF(iso)4E* genes using VIGS significantly reduced peanut stripe virus (PStV) accumulation, whereas silencing either gene individually did not have a significant impact (Xu et al., 2017). The crucial role of the peanut drought-induced gene *HSP70* in drought tolerance was confirmed by silencing it in *Nicotiana benthamiana* using a VIGS approach (Govind et al., 2009). VIGS has also been used to investigate other drought-responsive genes in peanut (Huang et al., 2012). Whole-plant VIGS was employed to silence two *ABI4*-encoding genes cloned from the peanut cultivar Fenghua2 (Fh2, Spanish type). Under salt stress, the *AhABI4s*-silenced plants exhibited significantly higher survival rates, biomass accumulation, and root/shoot ratios compared to mock-inoculated control plants. This silencing also induced significant changes in ion transporter/channel expression under salt stress (Luo et al., 2021). Another approach, known as host-induced gene silencing (HIGS), targeting specific genes in the aflatoxin production pathway, such as aflM and aflP, was found to reduce the amount of aflatoxin production in peanut seeds (Sharma et al., 2018).

8.5.5 Targeting Induced Local Lesions in Genomes (TILLING)

Targeting induced local lesions in genomes (TILLING) is a valuable reverse genetic approach for functional genomics and agronomic applications, applicable to diverse genomes. This technique involves high-throughput screening for point mutations in specific genomic regions following chemical random mutagenesis. TILLING is a PCR-based method that detects single-nucleotide polymorphisms (SNPs) or small insertions/deletions (InDels) using CEL1 nuclease mismatch cleavage (Till et al., 2007). Next-generation sequencing (NGS) is often the preferred method for mutation identification due to its capacity for rapid analysis of multiple amplicons. In peanut, two homeologous *AhFAD2* genes, *AhFAD2A* and *AhFAD2B*, have been identified (Moore and Knauft, 1989). While reduced *AhFAD2* activity increases the oleic to linoleic acid ratio, only one functional allele is sufficient for a typical oleate phenotype. TILLING has been used to identify mutations in both *AhFAD2* genes (Fang et al., 2012; Nadaf et al., 2017; Karaman et al., 2021). This technique has also been explored for developing peanut cultivars with reduced allergenicity (Knoll et al., 2011). In one study, researchers attempted to induce point mutations in the *AhFAD2* and *Ara h 1* genes using EMS followed by TILLING. Two heterozygous mutations were identified in the *AhFAD2B* and *AhFAD2A* gene regions in populations treated with 1.2% and 0.4% EMS, respectively. However, no mutations were found in *Ara h 1.01* and *Ara h 1.02* for either EMS treatment (Karaman et al., 2021). The TILLING by sequencing approach has also been successfully applied to identify stress resistance genes in peanut. These findings indicate that sequencing-based TILLING is effective for detecting mutations in both single-copy and multi-copy genes (Guo et al., 2015).

8.6 PROTEOME ANALYSIS

Proteome analysis is a direct method for defining gene functions. Combined with genomic sequence information, it plays a crucial role in functional genomics. Cataloging and identifying proteins across different proteome states and environments can provide valuable insights. Two widely used proteomics techniques for this purpose are two-dimensional electrophoresis (2-DE) and mass spectrometry (MS). These methods allow researchers to better understand how proteins, produced by genomes, respond to various environmental stresses and biological activities (Graves and Haystead, 2002; Jamshidi Goharrizi, Baghizadeh et al., 2020; Jamshidi Goharrizi, Fatehi et al. 2020). Proteins involved in chemical reactions and regulation are often organized in pathway databases. Similar to gene ontology (GO) analysis, protein or gene lists can also be analyzed for pathway abundance. There are many available resources and databases for extracting pathway constraints from biological data. Proteomics has been applied in various studies, including those investigating the development of peanut seeds (Kang et al., 2007), gynophores (Zhao et al., 2015), leaves (Katam et al., 2010), allergens (Porterfield et al., 2009), and drought tolerance (Kottapalli et al., 2013; Carmo et al., 2019). Additionally, proteomics has been used to study seed proteins related to aflatoxin contamination in resistant peanut genotypes (T. Wang, et al., 2010; C. T. Wang, et al., 2012; Zhao et al., 2019), revealing differentially expressed proteins (DEPs) associated with pre-harvest aflatoxin contamination. Non-targeted proteomics, combining 2D gel electrophoresis with LC-MS/MS, has also been employed to explore the resistance mechanisms against *A. flavus* (Ouakhssase et al., 2019). Proteomic techniques have also identified DEPs linked to specific traits such as storage proteins, anti-nutritional factors, allergens, and cultivar identification (Kottapalli et al., 2008). Other studies have focused on peanut germin-like proteins (AhGLPs) involved in plant development and defense (Wang et al., 2013), proteins related to pod swelling (Zhu et al., 2013), and oleic acid accumulation during seed development (Liu et al., 2018). Proteomic research has even examined gene expression under stress conditions to a limited extent (T. Wang et al., 2010; C. T. Wang, et al., 2012; Kottapalli et al., 2013; Katam et al., 2016). Global proteome profiling can be used to quantify protein dynamics, post-translational modifications, and biological pathways that contribute to immune function. This approach has revealed various resistance mechanisms, such as oxidative stress responses, the induction of PR proteins, and activation of secondary metabolic pathways (Balotf et al., 2021; Elmore et al., 2021). To investigate the diverse functions of proteins in peanut gynophores, Zhao, Zhao et al. (2015) performed gene ontology (GO) analysis on proteins from three stages (S1, S2, and S3) of gynophore samples. A total of 2,584, 2,335, and 2,119 proteins were identified in the S1, S2, and S3 stages, respectively, and categorized into 45 functional groups. Some proteins were assigned to multiple categories, indicating their involvement in multiple biological processes. Proteins related to cell functions, metabolic processes, cellular activities, organelles, binding, catalytic activity, and response to stimuli were highly represented in all three samples. To further assess the proteins, cluster of orthologous groups (COG) analysis was conducted, categorizing the proteins into 23 groups. The largest group across all stages was the general function prediction category, followed by categories related to post-translational modification, energy production and conversion, carbohydrate transport, metabolism, and translation. The smallest groups were related to nuclear structure, chromatin structure and dynamics, and RNA processing. Additionally, the identified proteins were mapped to reference pathways in the Kyoto Encyclopedia of Genes and Genomes (KEGG) database. This revealed the biological pathways associated with peanut gynophore development at different stages, including metabolic pathways, plant–pathogen interaction pathways, plant hormone signal transduction pathways, and fatty acid biosynthesis pathways. These annotations provide valuable insights into the processes and pathways involved in peanut gynophore and pod development.

A comprehensive protein expression profile of developing peanut seeds was generated at seven stages of development (seed size in mm): R1 and R2 (1–2 mm), R3 (2–4 mm), R4 (4–6 mm), R5 (6–8 mm), R6 (8–10 mm), and R7 (10–12 mm) to understand underground seed development (Li et al., 2020). High-resolution two-dimensional electrophoresis (2-DE) revealed that peanut seed development can be divided into early stages (R1-R4) and late stages (R5-R7), with the proteome being highly dynamic during seed development. More than 800 reproducible protein spots were observed, suggesting their involvement in seed development. A total of 264 proteins were identified through MALDI-TOF/TOF-MS analysis, yielding at least one unique peptide with 1% false discovery rate (FDR). Approximately 80% of these proteins consisted of more than one peptide. The proteins were classified into five primary functional categories: metabolism, cellular processes and signaling, poorly characterized proteins, energy production and conversion, and information storage and processing. The metabolism category, which contained the most identified proteins (IPs), was further subdivided into carbohydrate transport and metabolism (42 IPs), amino acid transport and metabolism (32 IPs), inorganic ion transport and metabolism (14 IPs), secondary metabolite biosynthesis (14 IPs), lipid transport and metabolism (11 IPs), nucleotide transport and metabolism (6 IPs), and coenzyme transport and metabolism (5 IPs). These results highlighted the importance of carbohydrate, amino acid, and lipid metabolism in underground peanut seed development. The cellular processes and signalling category, containing 84 IPs, included proteins related to post-translational modification, protein turnover, and chaperones (43 IPs). These findings provide a clearer understanding of the molecular mechanisms driving peanut seed development during both early and late stages.

8.6.1 Peanut Allergens, Storage, Anti-Nutritional Proteins

During seed development, Li et al. (2020) identified allergen proteins using translated allergen sequences from the A genome (Arachis duranensis) and B genome (Arachis ipaensis) of peanuts. A total of 14 allergen-like proteins were identified through the BLAST method. The relative abundance of these allergens was analyzed, revealing that most allergens expressed in the R1 to R4 stages had low accumulation or were undetectable by 2-DE (two-dimensional electrophoresis). As the seeds developed, six allergen proteins showed significant accumulation starting from the R5 stage. Three of these proteins were identified from Arachis ipaensis: XP_015936536.1 (arachin Ahy-3), XP_016174800.1 (allergen Ara h 1, clone P41B), and XP_016205863.1 (arachin Ahy-3). The remaining three, identified from Arachis duranensis, were XP_015948729.1 (legumin type B isoform X1), XP_015968272.1 (11S globulin subunit beta), and XP_015968165.1 (arachin Ahy-3). These proteins began to accumulate at the R5 stage and continued to increase in abundance as the seeds matured. By the R6 stage, the accumulation of allergen proteins such as XP_015968162.1 (arachin Ahy-3), XP_015968164.1 (legumin type B), and XP_020964607.1 (conglutin-7 isoform X2: Arachis ipaensis) reached their peak, with the highest levels observed at the R7 stage. At this stage, other allergen proteins like XP_015968160.1 (arachin Ahy-3), XP_015968312.1 (arachin Ahy-3), XP_016205798.1 (arachin Ahy-3), and XP_020959918.1 (arachin Ahy-3-like) also accumulated significantly. Generally, most of these allergen proteins served as seed storage proteins, accumulating during the late stages of seed development. A total seed protein profile was also generated by analyzing mature seeds from four peanut cultivars: New Mexico Valencia C (NM Valencia C), Tanspan 90, Georgia Green, and NC-7, using two-dimensional gel electrophoresis coupled with nano-electrospray ionization liquid chromatography-tandem mass spectrometry (LC-MS/MS). The majority of the seed storage proteins were classified within the globulin fraction, which included arachin (glycinin and Arah3/4) and conarachin, along with other allergenic proteins. Cultivar-specific differences were observed in the expression of these proteins. For instance, allergens Arah3/Arah4 and conarachin were detected only in Tanspan 90 and NC-7, while Gly1 protein was specific to NM Valencia C and NC-7. Additionally, a galactose-binding lectin protein with anti-nutritional properties was found only in Tanspan 90. There were also differences in the relative abundance of proteins such as 13-lipoxygenase, fructose diphosphate aldolase, and glyceraldehyde-3-phosphate dehydrogenase among the four cultivars. These identified proteins can serve as valuable markers in peanut breeding programs to distinguish cultivars based on sensory and nutritional characteristics (Kottapalli et al., 2008).

8.6.2 Growth, Development, and Stress Responsive Proteins

Proteomic analysis of peanut leaves was conducted using two-dimensional gel electrophoresis combined with MALDI/TOF mass spectrometry to identify proteins associated with growth, development, and responses to environmental stresses (Katam et al., 2010). This analysis revealed 300 distinct polypeptides with isoelectric points (pI) ranging from 3.5 to 8.0 and molecular weights between 12 and 100 kDa. A comprehensive leaf protein profile was established, highlighting proteins that were consistently expressed across the samples. Using GPS software and the Viridiplantae database (NCBI), a total of 133 unique proteins were identified. These proteins were categorized into ten groups based on their cellular compartmentalization and eight groups based on their biological functions. The majority of the identified leaf proteins were involved in crucial processes such as metabolism, transport (e.g., ATPase synthase), energy production, protein synthesis, signal transduction (e.g., calcium-binding proteins, calreticulin: CRT), and photosynthesis (e.g., chaperonin, heat shock proteins, chlorophyll a/b binding proteins, and RuBisCO). Additionally, proteins involved in cellular biogenesis (e.g., actin and tubulin) were also present. Enzymes critical for carbohydrate metabolism and photosynthesis dominated the protein set. The proteomic maps generated from drought-tolerant varieties like Vemana can serve as a foundation for future studies on peanut physiology, especially in response to biotic and abiotic stresses. Furthermore, these leaf proteome maps could assist in the development of protein markers for cultivar identification during the seedling stage of plant growth. In addition, key candidate proteins involved in oleic acid (OA) biosynthesis were identified using iTRAQ-based proteomics. A total of 389 differentially expressed proteins (DEPs) were found between the high-oleic cultivar Kainong176 and the low-oleic cultivar Kainong70. These DEPs were primarily involved in unsaturated fatty acid biosynthesis during the early stages of seed development, with some proteins linked to lipid oxidation pathways found during seed maturation. In Kainong176, the expression of FAB2 (stearoyl-acyl carrier protein desaturase), a key enzyme in OA biosynthesis, was significantly elevated in early seed development but declined during the later stages. Conversely, the expression pattern in Kainong70 (low oleic acid cultivar) was the opposite. This comparative proteomic analysis provides valuable insights into the molecular dynamics of OA biosynthesis throughout peanut seed development (Liu et al., 2018). Another significant study by Bhatnagar-Mathur et al. (2021) explored the molecular mechanisms underlying the interaction between Aspergillus and peanuts, with a focus on proteomic analysis of A. flavus-responsive proteins. The study found that differentially expressed proteins (DEPs) were primarily involved in stress responses, such as cell wall biosynthesis, hypersensitive response, secondary metabolic pathways (including phenylpropanoids and flavonoids), lipid metabolism, and plant–pathogen interactions. In transgenic samples, cell wall biosynthesis enzymes, including chitinase (108.30-fold change) and pectin esterase (10.51-fold change), were upregulated, while β-1,3-glucanase (14.37-fold change) was more abundant in wild-type (WT) peanut plants. Disease resistance genes (R genes) such as NBS-LRR proteins (2.0–89.00 FC), TIR-NBS-LRR (2.5–20.00 FC), PR protein 4A (7.26 FC), and NPR1 protein (75.00 FC) were significantly more abundant in the transgenic lines compared to WT controls. Secondary metabolite biosynthesis proteins, including phenylpropanoids (e.g., phenylalanine ammonia-lyase 2-like, caffeic acid methyltransferase) and flavonoids (e.g., chalcone synthase, resveratrol synthase), were also upregulated in transgenic lines. In the fatty acid biosynthesis pathway, proteins such as acyl carrier proteins (397.56 FC), lipoxygenase (131.89 FC), and diacylglycerol-glycerol-3-phosphate acyltransferase (153.72 FC) were upregulated in transgenic lines as well. Putative susceptibility-associated proteins (SAPs) in WT controls compared to OE-Def transgenic lines were identified, revealing infection progression over time. Higher expression of several plant–pathogen interaction proteins, such as annexin (4.20 FC), syntaxin (9.72 FC), and mildew resistance locus O-like proteins, were found at early stages of infection (24 hours post-infection, hpi) in compatible *A. flavus*-peanut interactions. Several fatty acid pathway proteins and MYB transcription factors (e.g., MYB4 and MYB9) were also differentially expressed. The identified proteins, including PR proteins, phytohormones, and transcription factors regulating secondary metabolic pathways, play crucial roles in suppressing pathogen development in peanuts. These proteins can serve as biomarkers for plant breeding programs aimed at selecting aflatoxin-free crops. Furthermore, these findings contribute to understanding the genetic mechanisms behind peanut resistance to A. flavus and the resulting aflatoxin contamination, offering novel targets for precision breeding in peanuts (Bhatnagar-Mathur et al., 2021).

8.7 GENETIC TRANSFORMATION

Genetic transformation is a crucial technique for introducing desired genes across species, enabling the enhancement of plant traits. In peanuts, successful transformation has been achieved using various methods, including Agrobacterium tumefaciens, A. rhizogenes, and particle bombardment. These techniques have led to the development of several effective genetic transformation protocols (Chen, Yang et al., 2015). However, peanut remains a challenging species for regeneration and genetic modification, which presents several obstacles to successful transformation. Several factors influence the efficiency of genetic transformation in peanuts, including the choice of genotypes, explants, co-cultivation time, virulence-inducing agents, hormonal combinations, and selectable markers. Among the different explants, cotyledonary nodes (CNs) regenerate more rapidly than other types of nodes, which improves transformation success (Mallikarjuna et al., 2016). The last decade has seen significant advancements in gene-editing research, leading to the development of novel lines with improved traits and enhanced plant functionality (Zhang et al., 2022). The first transgenic peanut plants stably expressing Cry genes derived from Bacillus thuringiensis, which impart resistance to lepidopteran pests, were reported by Ozias-Akins et al. (1993). This was achieved using the biolistic approach. Subsequently, several protocols employing Agrobacterium and high-frequency regeneration from cotyledons have been developed, leading to increased transformation success (Mehta et al., 2013; Sarkar et al., 2014, 2016, 2019; Patel et al., 2016, 2017; Patil et al., 2017). Over the years, genetic transformation has also enabled the development of transgenic peanut plants expressing genes that provide resistance to fungal and viral pathogens as well as insect pests. These advancements have been achieved through the incorporation of gene constructs that include PR proteins, genes encoding cell wall components, bacterial chloroperoxidase, oxalate oxidase, coat proteins, RNA interference (RNAi), crystal proteins, and various transcription factors like DREB, mtlD, PDH45, NHX, and NAC (Krishna et al., 2015; Mallikarjuna et al., 2016). A comprehensive list of genes expressed in peanuts through genetic transformation of explants is summarized in Table 8.1.

8.8 CONCLUSIONS

The advent of high-throughput sequencing technologies and advanced bioinformatic tools has significantly improved the accessibility and affordability of annotating and validating genes that govern key traits in peanuts. While several functional genes were reported before the high-throughput era, the availability of reference genomes for both wild and cultivated peanut species has revolutionized the understanding of peanut genetics. This has enabled the development of gene expression atlases and the discovery of numerous candidate genes. The use of RNA sequencing (RNA-Seq) in combination with reference genomes has made transcriptome analysis more precise. This approach has facilitated the identification and validation of additional functional genes, pathways, and regulatory elements with greater accuracy. Moreover, proteomic techniques have been applied to identify differentially expressed proteins and protein interaction networks associated with various traits, such as storage and allergenic proteins, defense mechanisms, pod development, drought tolerance, and resistance to *A. flavus*. In addition, efficient genetic transformation protocols, combined with high-frequency regeneration systems, have enabled the in planta validation of several functional genes in peanuts. However, more work is needed to further dissect the functional roles of genes identified through transcriptomics and gene expression analysis. This can be achieved through the use of different mutant lines and genes modulating identified pathways. Many of the candidate genes discovered are involved in key traits, such as resistance to biotic and abiotic stresses and addressing critical issues like aflatoxin contamination. These candidate genes can be utilized in the development of functional markers, which could be applied in marker-assisted selection (MAS) to achieve complex phenotypes that are challenging to obtain through conventional breeding methods. Although some of these candidate genes, such as those responsible for high oleic acid content in groundnuts, are already being used in MAS, more efforts are required to fully exploit the wealth of information generated from functional genomics research in peanuts.

TABLE 8.1
Comprehensive List of Genes Expressed in Planta through Genetic Transformation of Peanut Tissues

Trait Targeted	Gene (s) Used	Method Used	Peanut Variety Used	Reference
Abiotic stress				
Drought tolerance	*DREB1A* of *Arabidopsis*	*Agrobacterium*	JL 24	Bhatnagar- Mathur et al. (2007, 2009, 2014)
Drought and salinity tolerance	*AtNHXI* of *Arabidopsis*	*Agrobacterium*	Golden BARI- 2000	Asif et al. (2011)
Drought tolerance	Isopentenyl transferase of *Arabidopsis*	*Agrobacterium*	New Mexico Valencia A	Qin et al. (2011)
Drought tolerance	*AhAREB1* of *Arachis hypogaea*	*Agrobacterium*	Shanyou 523	X. Li et al. (2013)
Drought tolerance	*MuNAC4* of Horse gram	In planta	Narayani	Pandurangaiah et al. (2014)
Drought tolerance	*AtNAC2* of *Arabidopsis*	In planta	K134	Patil et al. (2014)
Drought tolerance	*PDH45* of pea	In planta	K 134	Manjulatha et al. (2014)
Drought tolerance	*DREB1A* of *Arabidopsis*	*Agrobacterium*	GG 20	Sarkar et al. (2014)
Salinity tolerance	*SbpAPX* of Salicornia brachiata	*Agrobacterium*	GG20	Singh et al. (2014)
Drought and salinity tolerance	*SbASR-1* of Salicornia brachiata	*Agrobacterium*	GG20	Tiwari et al. (2015)
Drought tolerance	*mtlD* of *E. coli*	*Agrobacterium*	GG20	Patel et al. (2016)
Biotic stress				
Resistance to lesser crosstalk borer	*CryIA(c)* of *Bacillus thuringiensis*	Biolistics	Toalson Florunner MARC-1	Singsit et al. (1997)
Resistance to TSWV	*TSWV N* of tomato spotted wilt virus	*Agrobacterium*	New Mexico Valencia A	Li et al. (1997)
Resistance to TSWV	*TSWV N* of tomato spotted wilt virus	Biolistics	Florunner, Georgia Runner MARC-1	Yang et al. (1998)
Resistance TSWV	*TSWV N* of tomato spotted wilt virus	Biolistics	VC1 AT120	Magbanua et al. (2000)
Leaf spots (*C. arachidicola*)	Chitinase of tobacco	*Agrobacterium*	TNV2	Rohini and Rao (2001)
Resistance to PStV	Coat protein CP2 of PStV and CP4 of potato spindle tuber viroid	Biolistics	Gajah NC7	Higgins et al. (2004)
Resistance to *Sclerotinia minor*	Oxalate oxidase	Biolistics	Wilson Perry NC-7	Livingstone et al. (2005), Partridge-Telenko et al. (2011)
Resistance to Sclerotinia blight	Chitinase of rice and glucanase of alfalfa	Biolistics	Okrun	Chenault et al. (2006)
Resistance to Leaf spots	*RsAFP1* and *RsAFP2* of mustard	*Agrobacterium*	JL 24	Swathi Anuradha et al. (2008)
Resistance to GBNV	*GBNV-N* of groundnut bud necrosis virus	*Agrobacterium*	Kadiri 6 (K6), Kadiri 134 (K134)	Yadav (2009)
Resistance to Cercospora arachidicola and *A. flavus*	β 1–3 glucanase of *Arabidopsis*	In planta	TMV 2	Sundaresha et al. (2010)
Resistance to *Spodoptera litura*	Cry1EC of *Bacillus thuringiensis*	*Agrobacterium*	JL 24	Tiwari et al. (2011)
Resistance to *Cercospora arachidicola*	Oleosin, chitinase-3 of Rice	*Agrobacterium*	Golden BARI- 2000	Iqbal et al. (2012)
Resistance to *Holotrichia parallela*	*cry8Ea1* of *Bacillus thuringiensis*	*A. rhizogenes*	Luhua11 Huayu16 Huayu28 Baisha1016 Yuanhua8 Xinhua1 Xinhua5	Geng et al. (2012)
Resistance to late leaf spots	*PR5, Rs-AFP2* of *Raphanus sativum*	*Agrobacterium*	JL 24	Vasavirama and Kirti (2012)
Resistance to TSV	CP of tobacco streak virus	*Agrobacterium*	Kadiri 6 (K6), Kadiri 134 (K134)	Mehta et al. (2013)

(*Continued*)

TABLE 8.1 (CONTINUED)
Comprehensive List of Genes Expressed in Planta through Genetic Transformation of Peanut Tissues

Trait Targeted	Gene (s) Used	Method Used	Peanut Variety Used	Reference
Resistance to late leaf spots, rust, and *A. flavus*	Rchit (class 1 chitinase) of Rice	*Agrobacterium*	JL 24, ICGV 89104, ICGV 86031	Prasad et al. (2013)
Resistance to PBN	*PBNV- N* of peanut bud necrosis virus, *uidA* of *E. coli*.	Biolistics	JL 24	Rao et al. (2013)
Resistance to *Spodoptera litura*	*Cry1AcF* of *Bacillus thuringiensis*	In planta	K 143	Keshavareddy et al. (2013)
Resistance to *Cercospora personata*	β-1–3 glucanase of *Arabidopsis*	*Agrobacterium*	Huaya 20, Huaya 22	Qiao et al. (2014)
Resistance to Late leaf spot	*AdSGT1* of *Arachis diogoi*	*Agrobacterium*	JL 24	Kumar and Kirti (2015b)
Early and late leaf spots	Fusion defensin gene from radish and fenugreek	*Agrobacterium*	GG 20	Bala et al. (2016)
Resistance to TSV and GBNV	CP of tobacco streak virus + *GBNV-N* of groundnut bud necrosis virus	*Agrobacterium*	Kadiri 6 (K6)	Patil et al. (2017)
Others				
Edible vaccine against Bluetongue virus	*BT-VP2* of Bluetongue virus	Biolistics	JL 24	Athmaram et al. (2006)
Edible vaccine against *Helicobacter pylori*	*Ure B* of *Helicobacter pylori*	*Agrobacterium*		Yang et al. (2011)
Silencing *Arah 2* and *Arah 6* genes which produce allergens	*Arah 2 RNAi* of peanut	Biolistics	Georgia Green	Chu et al. (2008)
Silencing *Arah 2* and *Arah 6* genes which produce allergens	*Arah 2 RNAi* of peanut	*Agrobacterium*	Georgia Green	Dodo et al. (2008)

REFERENCES

Abdelrahem, A. A., and H. S. Seham. "Genetic Improvement for Yield and Its Components through Induced Chemical and Physical Mutagenesis in Peanut (*A. hypogaea* L.)." *Journal of Plant Production Sciences* 10, no. 1 (2021): 49–66. https://doi.org/10.21608/jpps.2021.178524.

Addai, I. K., A. Bawa, and B. N-Yanyi. "Mutagenesis for Improvement of Growth and Yield of Groundnut (*A. hypogaea* L.) in the Guinea Savannah Agro-Ecology of Ghana." *International Journal of Plant Breeding and Genetics* 16, no. 1 (2022): 10–17. https://doi.org/10.3923/ijpbg.2022.10.17.

Ahmed, E. M., and C. T. Young. "Composition, Quality and Flavour of Peanuts." In *Peanut Science and Technology*, edited by H. E. Pattee and C. T. Young, 655–688. Yoakum, TX: Peanut Res, Education and Society, 1982.

Arias, R. S., P. M. Dang, and V. S. Sobolev. "RNAi-Mediated Control of Aflatoxins in Peanut: Method to Analyze Mycotoxin Production and Transgene Expression in the Peanut/Aspergillus Pathosystem." *Journal of Visualized Experiments* (2015): e53398. https://doi.org/10.3791/53398.

Asif, M. A., Y. Zafar, J. Iqbal, M. M. Iqbal, U. Rashid, G. M. Ali, A. Arif, and F. Nazir. "Enhanced Expression of AtNHX1 in Transgenic Groundnut (*A. hypogaea* L.) Improves Salt and Drought Tolerance." *Molecular Biotechnology* 49, no. 3 (2011): 250–256. https://doi.org/10.1007/s12033-011-9399-1.

Athmaram, T. N., G. Bali, and K. M. Devaiah. "Integration and Expression of Bluetongue VP2 Gene in Somatic Embryos of Peanut Through Particle Bombardment Method." *Vaccine* 24, no. 15 (2006): 2994–3000. https://doi.org/10.1016/j.vaccine.2005.11.019.

Bala, M., T. Radhakrishnan, A. Kumar, G. P. Mishra, J. R. Dobraia, and P. B. Kirti. "Overexpression of a Fusion Defensin Gene from Radish and Fenugreek Improves Resistance Against Leaf Spot Diseases Caused by *Cercospora arachidicola* and *Phaeoisariopsis personata* in Peanut." *Turkish Journal of Biology* 40, no. 1 (2016): 139–149. https://doi.org/10.3906/biy-1902-25.

Balotf, S., R. Wilson, R. S. Tegg, D. S. Nichols, and C. R. Wilson. "Quantitative Proteomics Provides an Insight into Germination-Related Proteins in the Obligate Biotrophic Plant Pathogen *Spongospora subterranea*." *Environmental Microbiology Reports* 13, no. 4 (2021): 521–532. https://doi.org/10.1111/1758-2229.12955.

Bao, G., S. Li, Q. Zhou, U. Ashraf, J. Qiao, X. Li, X. Wan, and Y. Zheng. "Transcriptomic Analysis Provides Insights into the Differential Effects of Aluminum on Peanut (*A. hypogaea* L.)." *Genes* 13, no. 10 (2022): 1830. https://doi.org/10.3390/genes13101830.

Baudry, A., M. Caboche, and L. Lepiniec. "TT8 Controls Its Own Expression in a Feedback Regulation Involving TTG1 and Homologous MYB and bHLH Factors, Allowing a Strong and Cell-Specific Accumulation of Flavonoids in

Arabidopsis thaliana." *Plant Journal* 46, no. 5 (2006): 768–779. https://doi.org/10.1111/j.1365-313X.2006.02733.x.

Bera, S. K., J. H. Kamdar, S. V. Kasundra, S. V. Patel, M. D. Jasani, A. K. Maurya, P. Dash, et al., "Steady Expression of High Oleic Acid in Peanut Bred by Marker-Assisted Backcrossing for Fatty Acid Desaturase Mutant Alleles and Its Effect on Seed Germination Along with Other Seedling Traits." *PLoS One* 14, no. 12 (2019): e0226252. https://doi.org/10.1371/journal.pone.0226252.

Beringer, H., and M. A. Taha. "45 Calcium Absorption by Two Cultivars of Groundnut (*Arachis hypogaea*)." *Experimental Agriculture* 12, no. 1 (1976): 1–7. https://doi.org/10.1017/S0014479700006992.

Bertioli, D. J., J. Jenkins, J. Clevenger, O. Dudchenko, D. Gao, G. Seijo, S. C. M. Leal-Bertioli, et al., "The Genome Sequence of Segmental Allotetraploid Peanut *Arachis hypogaea*." *Nature Genetics* 51, no. 5 (2019): 877–884. https://doi.org/10.1038/s41588-019-0405-z.

Bhatnagar-Mathur, P., M. J. Devi, D. S. Reddy, M. Lavanya, V. Vadez, R. Serraj, K. Yamaguchi-Shinozaki, and K. K. Sharma. "Stress-Inducible Expression of AtDREB1A in Transgenic Peanut (*A. hypogaea* L.) Increases Transpiration Efficiency Under Water-Limiting Conditions." *Plant Cell Reports* 26, no. 12 (2007): 2071–2082. https://doi.org/10.1007/s00299-007-0406-8.

Bhatnagar-Mathur, P., M. J. Devi, V. Vadez, and K. K. Sharma. "Differential Antioxidative Responses in Transgenic Peanut Bear No Relationship to Their Superior Transpiration Efficiency Under Drought Stress." *Journal of Plant Physiology* 166, no. 11 (2009): 1207–1217. https://doi.org/10.1016/j.jplph.2009.01.001.

Bhatnagar-Mathur, P., J. S. Rao, V. Vadez, S. R. Dumbala, A. Rathore, K. Yamaguchi-Shinozaki, and K. K. Sharma. "Transgenic Peanut Overexpressing the DREB1A Transcription Factor Has Higher Yields Under Drought Stress." *Molecular Breeding* 33, no. 2 (2014): 327–340. https://doi.org/10.1007/s11032-013-9952-7.

Bhatnagar-Mathur, P., K. Yogendra, S. Parankusam, H. Sanivarapu, K. Prasad, S. B. Lingampali, and K. K. Sharma. "Comparative Proteomics Provide Insights on the Basis of Resistance to *A. flavus* Infection and Aflatoxin Production in Peanut (*A. hypogaea* L.)." *Journal of Plant Interactions* 16, no. 1 (2021): 494–509. https://doi.org/10.1080/17429145.2021.1995058.

Bhogireddy, S., A. Xavier, V. Garg, N. Layland, R. Arias, P. Payton, S. N. Nayak, M. K. Pandey, N. Puppala, and R. K. Varshney. "Genome-Wide Transcriptome and Physiological Analyses Provide New Insights into Peanut Drought Response Mechanisms." *Scientific Reports* 10, no. 1 (2020): 4071. https://doi.org/10.1038/s41598-020-60187-z.

Bi, Y. P., W. Liu, H. Xia, L. Su, C. Z. Zhao, S. B. Wan, and X. J. Wang. "EST Sequencing and Gene Expression Profiling of Cultivated Peanut (*A. hypogaea* L.)." *Genome* 53, no. 10 (2010): 832–839. https://doi.org/10.1139/g10-074.

Birthal, P. S., P. P. Rao, S. N. Nigam, C. S. Bantilan, and S. Bhagavatulu. *Groundnut and Soybean Economies in Asia: Facts, Trends and Outlook.* International Crops Research Institute for the Semi-Arid Tropics, 2010. http://oar.icrisat.org/id/eprint/190.

Bosamia, T. C., S. M. Dodia, G. P. Mishra, S. Ahmad, B. Joshi, P. P. Thirumalaisamy, N. Kumar, et al., "Unravelling the Mechanisms of Resistance to *Sclerotium rolfsii* in Peanut (*A. hypogaea* L.) Using Comparative RNA-Seq Analysis of Resistant and Susceptible Genotypes." *PLoS One* 15, no. 8 (2020): e0236823. https://doi.org/10.1371/journal.pone.0236823.

Branch, W. D., and N. Brown. "Inheritance of an Albino-Virescent Leaf Mutant in the Cultivated Peanut (*A. hypogaea* L.)." *Peanut Science* 46, no. 2 (2019): 203–205. https://doi.org/10.3146/PS19-3.1.

Brasileiro, A. C., C. V. Morgante, A. C. Araujo, S. C. Leal-Bertioli, A. K. Silva, A. C. Martins, C. C. Vinson, et al., "Transcriptome Profiling of Wild *Arachis* from Water-Limited Environments Uncovers Drought Tolerance Candidate Genes." *Plant Molecular Biology Reporter* 33 (2015): 1876–1892. https://doi.org/10.1007/s11105-015-0882-x.

Burow, G. B., H. W. Gardner, and N. P. Keller. "A Peanut Seed Lipoxygenase Responsive to *Aspergillus* Colonization." *Plant Molecular Biology* 42, no. 5 (2000): 689–701. https://doi.org/10.1023/a:1006361305703.

Carmo, L. S. T., A. C. Q. Martins, C. C. C. Martins, M. A. S. Passos, L. P. Silva, A. C. G. Araujo, A. C. M. Brasileiro, et al., "Comparative Proteomics and Gene Expression Analysis in *Arachis duranensis* Reveal Stress Response Proteins Associated to Drought Tolerance." *Journal of Proteomics* 192 (2019): 299–310. https://doi.org/10.1016/j.jprot.2018.09.011.

Catto, M. A., A. Shrestha, M. R. Abney, D. E. Champagne, A. K. Culbreath, S. C. M. Leal-Bertioli, B. G. Hunt, and R. Srinivasan. "Defense-Related Gene Expression Following an Orthotospovirus Infection Is Influenced by Host Resistance in *Arachis hypogaea*." *Viruses* 13, no. 7 (2021): 1303. https://doi.org/10.3390/v13071303.

Chaudhary, J., R. Deshmukh, and H. Sonah. "Mutagenesis Approaches and Their Role in Crop Improvement." *Plants* 8, no. 11 (2019): 467. https://doi.org/10.3390/plants8110467.

Chen, J., L. Jiang, C. M. Wang, X. H. Hu, H. Q. Zhai, and J. M. Wan. "Expression Analysis of Genes Involved in Peanut Seed Dormancy Release (*A. hypogaea* L.)." *Acta Agronomica Sinica* 41, no. 6 (2015): 845–860. https://doi.org/10.3724/SP.J.1006.2015.00845.

Chen, M., Q. Yang, T. Wang, N. Chen, L. Pan, X. Chi, Z. Yang, M. Wang, and S. Yu. "Agrobacterium-Mediated Genetic Transformation of Peanut and the Efficient Recovery of Transgenic Plants." *Canadian Journal of Plant Science* 95, no. 4 (2015): 735–744. https://doi.org/10.4141/cjps-2014-012.

Chen, X., Q. Yang, H. Li, et al., "Transcriptome-Wide Sequencing Provides Insights into Geocarpy in Peanut (*A. hypogaea* L.)." *Plant Biotechnology Journal* 14, no. 5 (2016): 1215–1224. https://doi.org/10.1111/pbi.12487.

Chen, X., B. Scully, X. Liang, C. Holbrook, J. Yu, and A. Culbreath. "Peanut Gene Expression Profiling in Developing Seeds at Different Reproduction Stages during *Aspergillus parasiticus* Infection." *BMC Developmental Biology* 8 (2008): 12. https://doi.org/10.1186/1471-213X-8-12.

Chen, X., W. Zhu, S. Azam, H. Li, F. Zhu, H. Li, Y. Hong, H. Liu, E. Zhang, H. Wu, S. Yu, G. Zhou, S. Li, N. Zhong, S. Wen, X. Li, S. J. Knapp, P. Ozias-Akins, R. K. Varshney, and X. Liang. "Deep Sequencing Analysis of the Transcriptomes of Peanut Aerial and Subterranean Young Pods Identifies Candidate Genes Related to Early Embryo Abortion." *Plant Biotechnology Journal* 11, no. 1 (2013): 115–127. https://doi.org/10.1111/pbi.12018.

Chenault, K., H. Melouk, and M. Payton. "Effect of Anti-Fungal Transgene(s) on Agronomic Traits of Transgenic Peanut Lines Grown under Field Conditions." *Peanut*

Science 33 (2006): 12–19. https://doi.org/10.3146/0095-3679(2006)33[12:EOATOA]2.0.CO;2.

Chi, X., Q. Yang, X. Chen, J. Wang, L. Pan, M. Chen, Z. Yang, Y. He, X. Liang, and S. Yu. "Identification and Characterization of microRNAs from Peanut (*A. hypogaea* L.) by High-Throughput Sequencing." *PLoS One* 6, no. 11 (2011): e27530. https://doi.org/10.1371/journal.pone.0027530.

Chi, X., Z. Zhang, N. Chen, X. Zhang, M. Wang, M. Chen, T. Wang, L. Pan, J. Chen, Z. Yang, X. Guan, and S. Yu. "Isolation and Functional Analysis of Fatty Acid Desaturase Genes from Peanut (*A. hypogaea* L.)." *PLoS One* 12, no. 12 (2017): e0189759. https://doi.org/10.1371/journal.pone.0189759.

Chopra, R., G. Burow, C. E. Simpson, J. Chagoya, J. Mudge, and M. D. Burow. "Transcriptome Sequencing of Diverse Peanut (*Arachis*) Wild Species and the Cultivated Species Reveals a Wealth of Untapped Genetic Variability." *G3* 6, no. 12 (2016): 3825–3836. https://doi.org/10.1534/g3.115.026898.

Chu, Y., P. Faustinelli, M. L. Ramos, M. Hajduch, S. Stevenson, J. J. Thelen, S. J. Maleki, H. Cheng, and P. Ozias-Akins. "Reduction of IgE Binding and Nonpromotion of *A. flavus* Fungal Growth by Simultaneously Silencing Ara h 2 and Ara h 6 in Peanut." *Journal of Agricultural and Food Chemistry* 56, no. 23 (2008): 11225–11233. https://doi.org/10.1021/jf802600r.

Chukwumah, Y., L. T. Walker, and M. Verghese. "Peanut Skin Color: A Biomarker for Total Polyphenolic Content and Antioxidative Capacities of Peanut Cultivars." *International Journal of Molecular Sciences* 10, no. 11 (2009): 4941–4952. https://doi.org/10.3390/ijms10114941.

Clevenger, J., Y. Chu, B. Scheffler, and P. Ozias-Akins. "A Developmental Transcriptome Map for Allotetraploid *Arachis hypogaea*." *Frontiers in Plant Science* 7 (2016a): 1446. https://doi.org/10.3389/fpls.2016.01446.

Clevenger, J., K. Marasigan, V. Liakos, V. Sobolev, G. Vellidis, C. Holbrook, and P. Ozias-Akins. "RNA Sequencing of Contaminated Seeds Reveals the State of the Seed Permissive for Pre-Harvest Aflatoxin Contamination and Points to a Potential Susceptibility Factor." *Toxins* 8, no. 11 (2016b): 317. https://doi.org/10.3390/toxins8110317.

Cox, F. R., F. Adams, and B. B. Tucker. "Liming, Fertilization, and Mineral Nutrition." In *Peanut Science and Technology*, edited by H. E. Pattee and C. T. Young, 139–163. Yoakum, TX: APRES, Inc., 1982.

Cui, Y., J. Bian, Y. Lv, J. Li, X. W. Deng, and X. Liu. "Analysis of the Transcriptional Dynamics of Regulatory Genes during Peanut Pod Development Caused by Darkness and Mechanical Stress." *Frontiers in Plant Science* 13 (2022a): 904162. https://doi.org/10.3389/fpls.2022.904162.

Cui, M., S. Han, D. Wang, M. S. Haider, J. Guo, Q. Zhao, P. Du, Z. Sun, F. Qi, Z. Zheng, B. Huang, W. Dong, P. Li, and X. Zhang. "Gene Co-Expression Network Analysis of the Comparative Transcriptome Identifies Hub Genes Associated with Resistance to *A. flavus* L. in Cultivated Peanut (*A. hypogaea* L.)." *Frontiers in Plant Science* 13 (2022b): 899177. https://doi.org/10.3389/fpls.2022.899177.

Dhillon, S. S., A. V. Rake, and J. P. Miksche. "Reassociation Kinetics and Cytophotometric Characterization of Peanut (*A. hypogaea* L.) DNA." *Plant Physiology* 65, no. 6 (1980): 1121–1127. https://doi.org/10.1104/pp.65.6.1121.

Dodia, S. M., B. Joshi, S. S. Gangurde, P. P. Thirumalaisamy, G. P. Mishra, D. Narandrakumar, P. Soni, A. L. Rathnakumar, J. R. Dobaria, C. Sangh, A. Chitikineni, S. V. Chanda, M. K. Pandey, R. K. Varshney, and R. Thankappan. "Genotyping-by-Sequencing Based Genetic Mapping Reveals Large Number of Epistatic Interactions for Stem Rot Resistance in Groundnut." *TAG. Theoretical and Applied Genetics* 132, no. 4 (2019): 1001–1016. https://doi.org/10.1007/s00122-018-3255-7.

Dodo, H. W., K. N. Konan, F. C. Chen, M. Egnin, and O. M. Viquez. "Alleviating Peanut Allergy Using Genetic Engineering: The Silencing of the Immunodominant Allergen Ara h 2 Leads to Its Significant Reduction and a Decrease in Peanut Allergenicity." *Plant Biotechnology Journal* 6, no. 2 (2008): 135–145. https://doi.org/10.1111/j.1467-7652.2007.00292.x.

Dodo, H., K. Konan, and O. Viquez. "A Genetic Engineering Strategy to Eliminate Peanut Allergy." *Current Allergy and Asthma Reports* 5, no. 1 (2005): 67–73. https://doi.org/10.1007/s11882-005-0058-0.

Elmore, J. M., B. D. Griffin, and J. W. Walley. "Advances in Functional Proteomics to Study Plant–Pathogen Interactions." *Current Opinion in Plant Biology* 63 (2021): 102061. https://doi.org/10.1016/j.pbi.2021.102061.

Falcone Ferreyra, M. L., S. P. Rius, and P. Casati. "Flavonoids: Biosynthesis, Biological Functions, and Biotechnological Applications." *Frontiers in Plant Science* 3 (2012): 222. https://doi.org/10.3389/fpls.2012.00222.

Fang, C. Q., C. T. Wang, P. W. Wang, Y. Y. Tang, X. Z. Wang, F. G. Cui, and S. L. Yu. "Identification of a Novel Mutation in *FAD2B* from a Peanut EMS Mutant with Elevated Oleate Content." *Journal of Oleo Science* 61, no. 3 (2012): 143–148. https://doi.org/10.5650/jos.61.143.

Fountain, J. C., P. Khera, L. Yang, S. N. Nayak, B. T. Scully, R. D. Lee, Z.-Y. Chen, R. C. Kemerait, R. K. Varshney, and B. Guo. "Resistance to *A. flavus* in Maize and Peanut: Molecular Biology, Breeding, Environmental Stress, and Future Perspectives." *Crop Journal* 3, no. 3 (2015): 229–237. https://doi.org/10.1016/j.cj.2015.02.003.

Francisco, M. L., and A. V. Resurreccion. "Functional Components in Peanuts." *Critical Reviews in Food Science and Nutrition* 48, no. 8 (2008): 715–746. https://doi.org/10.1080/10408390701640718.

Gao, C., P. Wang, S. Zhao, C. Zhao, H. Xia, L. Hou, Z. Ju, Y. Zhang, C. Li, and X. Wang. "Small RNA Profiling and Degradome Analysis Reveal Regulation of microRNA in Peanut Embryogenesis and Early Pod Development." *BMC Genomics* 18, no. 1 (2017): 220. https://doi.org/10.1186/s12864-017-3587-8.

Gazzarrini, S., and P. McCourt. "Genetic Interactions between ABA, Ethylene, and Sugar Signaling Pathways." *Current Opinion in Plant Biology* 4, no. 5 (2001): 387–391. https://doi.org/10.1016/s1369-5266(00)00190-4.

Geng, L., L. Niu, P. M. Gresshoff, C. Shu, F. Song, D. Huang, and J. Zhang. "Efficient Production of *Agrobacterium rhizogenes*-Transformed Roots and Composite Plants in Peanut (*A. hypogaea* L.)." *Plant Cell, Tissue and Organ Culture (PCTOC)* 109, no. 3 (2012): 491–500. https://doi.org/10.1007/s11240-012-0113-1.

Gimenez-Ibanez, S., and R. Solano. "Nuclear Jasmonate and Salicylate Signaling and Crosstalk in Defense against Pathogens." *Frontiers in Plant Science* 4 (2013): 72. https://doi.org/10.3389/fpls.2013.00072.

Gong, L., S. Han, M. Yuan, X. Ma, A. Hagan, and G. He. "Transcriptomic Analyses Reveal the Expression and Regulation of Genes Associated with Resistance to Early

Leaf Spot in Peanut." *BMC Research Notes* 13, no. 1 (2020): 381. https://doi.org/10.1186/s13104-020-05225-9.

Goodrich-Tanrikulu, M., N. E. Mahoney, and S. B. Rodriguez. "The Plant Growth Regulator Methyl Jasmonate Inhibits Aflatoxin Production by *A. flavus*." *Microbiology* 141, no. 11 (1995): 2831–2837. https://doi.org/10.1099/13500872-141-11-2831.

Govind, G., V. T. Harshavardhan, J. K. Patricia, R. Dhanalakshmi, M. Senthil Kumar, N. Sreenivasulu, and M. Udayakumar. "Identification and Functional Validation of a Unique Set of Drought Induced Genes Preferentially Expressed in Response to Gradual Water Stress in Peanut." *Molecular Genetics and Genomics* 281, no. 6 (2009): 591–605. https://doi.org/10.1007/s00438-009-0432-z.

Graves, P. R., and T. A. Haystead. "Molecular Biologist's Guide to Proteomics." *Microbiology and Molecular Biology Reviews* 66, no. 1 (2002): 39–63. https://doi.org/10.1128/MMBR.66.1.39-63.2002.

Griel, A. E., B. Eissenstat, V. Juturu, G. Hsieh, and P. M. Kris-Etherton. "Improved Diet Quality with Peanut Consumption." *Journal of the American College of Nutrition* 23, no. 6 (2004): 660–668. https://doi.org/10.1080/07315724.2004.10719408.

Guimaraes, L. A., A. P. Z. Mota, A. C. G. Araujo, L. F. de Alencar Figueiredo, B. M. Pereira, M. A. de Passos Saraiva, R. B. Silva, E. G. J. Danchin, P. M. Guimaraes, and A. C. M. Brasileiro. "Genome-Wide Analysis of Expansin Superfamily in Wild *Arachis* Discloses a Stress-Responsive Expansin-Like B Gene." *Plant Molecular Biology* 94, no. 1–2 (2017): 79–96. https://doi.org/10.1007/s11103-017-0594-8.

Guimaraes, P. M., A. C. Brasileiro, C. V. Morgante, A. C. Martins, G. Pappas, O. B. Silva Jr., R. Togawa, S. C. Leal-Bertioli, A. C. Araujo, M. C. Moretzsohn, and D. J. Bertioli. "Global Transcriptome Analysis of Two Wild Relatives of Peanut under Drought and Fungi Infection." *BMC Genomics* 13 (2012): 387. https://doi.org/10.1186/1471-2164-13-387.

Guimaraes, P. M., A. C. M. Brasileiro, K. Proite, A. C. G. de Araújo, S. C. M. Leal-Bertioli, A. Pic-Taylor, F. R. da Silva, C. V. Morgante, S. d. G. Ribeiro, and D. J. Bertioli. "A Study of Gene Expression in the Nematode Resistant Wild Peanut Relative, *Arachis stenosperma*, in Response to Challenge with *Meloidogyne arenaria*." *Tropical Plant Biology* 3, no. 4 (2010): 183–192. https://doi.org/10.1007/s12042-010-9056-z.

Guimaraes, P. M., L. A. Guimaraes, C. V. Morgante, O. B. Silva Jr, A. C. Araujo, A. C. Martins, M. A. Saraiva, T. N. Oliveira, R. C. Togawa, S. C. Leal-Bertioli, D. J. Bertioli, and A. C. Brasileiro. "Root Transcriptome Analysis of Wild Peanut Reveals Candidate Genes for Nematode Resistance." *PLoS One* 10, no. 10 (2015). https://doi.org/10.1371/journal.pone.0140937.

Guo, B., N. D. Fedorova, X. Chen, C. H. Wan, W. Wang, W. C. Nierman, D. Bhatnagar, and J. Yu. "Gene Expression Profiling and Identification of Resistance Genes to *A. flavus* Infection in Peanut through EST and Microarray Strategies." *Toxins* 3, no. 7 (2011): 737–753. https://doi.org/10.3390/toxins3070737.

Guo, F., J. Ma, L. Hou, S. Shi, J. Sun, G. Li, C. Zhao, H. Xia, S. Zhao, X. Wang, and Y. Zhao. "Transcriptome Profiling Provides Insights into Molecular Mechanism in Peanut Semi-Dwarf Mutant." *BMC Genomics* 21, no. 1 (2020): 211. https://doi.org/10.1186/s12864-020-6614-0.

Guo, Y., B. Abernathy, Y. Zeng, and P. Ozias-Akins. "TILLING by Sequencing to Identify Induced Mutations in Stress Resistance Genes of Peanut (*Arachis hypogaea*)." *BMC Genomics* 16, no. 1 (2015): 157. https://doi.org/10.1186/s12864-015-1348-0.

Gupta, K., G. Kayam, A. Faigenboim-Doron, J. Clevenger, P. Ozias-Akins, and R. Hovav. "Gene Expression Profiling During Seed-Filling Process in Peanut with Emphasis on Oil Biosynthesis Networks." *Plant Science* 248 (2016): 116–127. https://doi.org/10.1016/j.plantsci.2016.04.014.

Han, H. W., Z. W. Wang, X. Z. Wang, X. S. Sun, C. Q. Fang, and C. T. Wang. "Identification of High-Oleic Peanut Chemical Mutants and Functional Analysis of Mutated *FAD2B* Gene." *Plant Genetic Resources: Characterization and Utilization* 20, no. 1 (2022): 15–21. https://doi.org/10.1017/S1479262122000053.

Han, S., H. Liu, M. Yan, F. Qi, Y. Wang, Z. Sun, B. Huang, W. Dong, F. Tang, X. Zhang, and G. He. "Differential Gene Expression in Leaf Tissues between Mutant and Wild-Type Genotypes Response to Late Leaf Spot in Peanut (*A. hypogaea* L.)." *PLoS One* 12, no. 8 (2017): e0183428. https://doi.org/10.1371/journal.pone.0183428.

Higgins, C. M., R. M. Hall, N. Mitter, A. Cruickshank, and R. G. Dietzgen. "Peanut Stripe Potyvirus Resistance in Peanut (*A. hypogaea* L.) Plants Carrying Viral Coat Protein Gene Sequences." *Transgenic Research* 13, no. 1 (2004): 59–67. https://doi.org/10.1023/b:trag.0000017166.29458.74.

Huai, D., X. Xue, Y. Li, P. Wang, J. Li, L. Yan, Y. Chen, X. Wang, N. Liu, Y. Kang, Z. Wang, Y. Huang, H. Jiang, Y. Lei, and B. Liao. "Genome-Wide Identification of Peanut KCS Genes Reveals That AhKCS1 and AhKCS28 Are Involved in Regulating VLCFA Contents in Seeds." *Frontiers in Plant Science* 11 (2020): 406. https://doi.org/10.3389/fpls.2020.00406.

Huang, C. J., Y. J. Qian, Z. H. Li, and X. P. Zhou. "Virus-Induced Gene Silencing and Its Application in Plant Functional Genomics." *Science China. Life Sciences* 55, no. 2 (2012): 99–108. https://doi.org/10.1007/s11427-012-4280-4.

Huang, J. Q., H. F. Jiang, Y. Q. Zhou, Y. Lei, S. Y. Wang, and B. S. Liao. "Ethylene Inhibited Aflatoxin Biosynthesis Is Due to Oxidative Stress Alleviation and Related to Glutathione Redox State Changes in *A. flavus*." *International Journal of Food Microbiology* 130, no. 1 (2009): 17–21. https://doi.org/10.1016/j.ijfoodmicro.2008.12.027.

Huang, J., M. Xing, Y. Li, F. Cheng, H. Gu, C. Yue, and Y. Zhang. "Comparative Transcriptome Analysis of the Skin-Specific Accumulation of Anthocyanins in Black Peanut (*A. hypogaea* L.)." *Journal of Agricultural and Food Chemistry* 67, no. 4 (2019): 1312–1324. https://doi.org/10.1021/acs.jafc.8b05915.

Iqbal, M. M., F. Nazir, S. Ali, M. A. Asif, Y. Zafar, J. Iqbal, and G. M. Ali. "Over Expression of Rice Chitinase Gene in Transgenic Peanut (*A. hypogaea* L.) Improves Resistance Against Leaf Spot." *Molecular Biotechnology* 50, no. 2 (2012): 129–136. https://doi.org/10.1007/s12033-011-9426-2.

Iwashina, T. "Contribution to Flower Colors of Flavonoids Including Anthocyanins: A Review." *Natural Product Communications* 10, no. 3 (2015a): 529–544. https://doi.org/10.1177/1934578X1501000335.

Iwashina, T. "Flavonoid Properties in Plant Families Synthesizing Betalain Pigments (Review)." *Natural Product Communications* 10, no. 6 (2015b): 1103–1114. https://doi.org/10.1177/1934578X1501000675.

Jadhav, M. V., R. K. Zote, and A. D. Sable. "Studies on Induced Chemical Mutagenesis in Groundnut (*A. hypogaea* L.)." *International Journal of Current Microbiology and Applied Sciences* 7, no. 10 (2018): 1662–1669. https://doi.org/10.20546/ijcmas.2018.710.190.

Jain, M., B. P. Pathak, A. C. Harmon, B. L. Tillman, and M. Gallo. "Calcium Dependent Protein Kinase (CDPK) Expression During Fruit Development in Cultivated Peanut (*Arachis hypogaea*) Under Ca2+ -Sufficient and -Deficient Growth Regimens." *Journal of Plant Physiology* 168, no. 18 (2011): 2272–2277. https://doi.org/10.1016/j.jplph.2011.07.005.

Jamshidi Goharrizi, K., A. Baghizadeh, M. Kalantar, and F. Fatehi. "Assessment of Changes in Some Biochemical Traits and Proteomic Profile of UCB-1 Pistachio Rootstock Leaf Under Salinity Stress." *Journal of Plant Growth Regulation* 39, no. 2 (2020): 608–630. https://doi.org/10.1007/s00344-019-10004-3.

Jamshidi Goharrizi, K., F. Fatehi, M. Nazari, F. Salehi, and M. Maleki. "Assessment of Changes in the Content of Sulforaphane and Expression Levels of CYP79F1 and Myrosinase Genes and Proteomic Profile of *Lepidium draba* Plant Under Water-Deficit Stress Induced by Polyethylene Glycol." *Acta Physiologiae Plantarum* 42, no. 6 (2020): 1–18. https://doi.org/10.1007/s11738-020-03085-1.

Jankowicz-Cieslak, J., C. Mba, and B. J. Till. "Mutagenesis for Crop Breeding and Functional Genomics." In *Biotechnologies for Plant Mutation Breeding: Protocols*, edited by J. Jankowicz-Cieslak, T. H. Tai, J. Kumlehn, and B. J. Till, 3–18. Springer International Publishing, 2017. https://doi.org/10.1007/978-3-319-45021-6_1.

Jiang, C., X. Li, J. Zou, J. Ren, C. Jin, H. Zhang, H. Yu, and H. Jin. "Comparative Transcriptome Analysis of Genes Involved in the Drought Stress Response of Two Peanut (*A. hypogaea* L.) Varieties." *BMC Plant Biology* 21, no. 1 (2021): 64. https://doi.org/10.1186/s12870-020-02761-1.

Jiang, C., H. Zhang, J. Ren, J. Dong, X. Zhao, X. Wang, J. Wang, C. Zhong, S. Zhao, X. Liu, S. Gao, and H. Yu. "Comparative Transcriptome-Based Mining and Expression Profiling of Transcription Factors Related to Cold Tolerance in Peanut." *International Journal of Molecular Sciences* 21, no. 6 (2020): 1921. https://doi.org/10.3390/ijms21061921.

Jin, H., X. Tang, M. Xing, H. Zhu, J. Sui, C. Cai, and S. Li. "Molecular and Transcriptional Characterization of Phosphatidyl Ethanolamine-Binding Proteins in Wild Peanuts *Arachis duranensis* and *Arachis ipaensis*." *BMC Plant Biology* 19, no. 1 (2019): 484. https://doi.org/10.1186/s12870-019-2113-3.

Kamthan, A., A. Chaudhuri, M. Kamthan, and A. Datta. "Small RNAs in Plants: Recent Development and Application for Crop Improvement." *Frontiers in Plant Science* 6 (2015): 208. https://doi.org/10.3389/fpls.2015.00208.

Kang, I. -H., and M. Gallo. "Cloning and Characterization of a Novel Peanut Allergen Ara h 3 Isoform Displaying Potentially Decreased Allergenicity." *Plant Science* 172, no. 2 (2007): 345–353. https://doi.org/10.1016/j.plantsci.2006.09.014.

Kang, I. H., P. Srivastava, P. Ozias-Akins, and M. Gallo. "Temporal and Spatial Expression of the Major Allergens in Developing and Germinating Peanut Seed." *Plant Physiology* 144, no. 2 (2007): 836–845. https://doi.org/10.1104/pp.107.096933.

Karaman, K., S. Kizil, M. Başak, B. Uzun, and E. Yol. "Development of EMS-Induced Mutagenized Groundnut Population and Discovery of Point Mutations in the *ahFAD2* and *Ara h 1* Genes by Tilling." *Journal of Oleo Science* 70, no. 11 (2021): 1631–1640. https://doi.org/10.5650/jos.ess21075.

Katam, R., S. M. Basha, P. Suravajhala, and T. Pechan. "Analysis of Peanut Leaf Proteome." *Journal of Proteome Research* 9, no. 5 (2010): 2236–2254. https://doi.org/10.1021/pr901009n.

Katam, R., K. Sakata, P. Suravajhala, T. Pechan, D. M. Kambiranda, K. S. Naik, B. Guo, and S. M. Basha. "Comparative Leaf Proteomics of Drought-Tolerant and -Susceptible Peanut in Response to Water Stress." *Journal of Proteomics* 143 (2016): 1–13. https://doi.org/10.1016/j.jprot.2016.05.031.

Kavera, D., and H. L. Nadaf. "Genetic Improvement for Yield Through Induced Mutagenesis in Groundnut (*A. hypogaea* L.)." *Legume Research* 40 (2017): 32–35. https://doi.org/10.18805/lr.v0i0.7019.

Kavera, N. H. L., H. L. Nadaf, and R. R. Hanchinal. "Genetic Improvement for Oil Quality Through Induced Mutagenesis in Groundnut (*A. hypogaea* L.)." *Indian Journal of Genetics and Plant Breeding* 73, no. 4 (2013): 378–385. https://doi.org/10.5958/j.0975-6906.73.4.057.

Keshavareddy, G., S. Rohini, S. V. Ramu, S. Sundaresha, A. R. Kumar, P. A. Kumar, and M. Udayakumar. "Transgenics in Groundnut (*A. hypogaea* L.) Expressing Cry1AcF Gene for Resistance to *Spodoptera litura* (F.)." *Physiology and Molecular Biology of Plants* 19, no. 3 (2013): 343–352. https://doi.org/10.1007/s12298-013-0182-6.

Knoll, J. E., M. L. Ramos, Y. Zeng, C. C. Holbrook, M. Chow, S. Chen, S. Maleki, A. Bhattacharya, and P. Ozias-Akins. "TILLING for Allergen Reduction and Improvement of Quality Traits in Peanut (*A. hypogaea* L.)." *BMC Plant Biology* 11 (2011): 81. https://doi.org/10.1186/1471-2229-11-81.

Kong, X., M. Yang, B. H. Le, W. He, and Y. Hou. "The Master Role of siRNAs in Plant Immunity." *Molecular Plant Pathology* 23, no. 10 (2022): 1565–1574. https://doi.org/10.1111/mpp.13250.

Korani, W., Y. Chu, C. C. Holbrook, and P. Ozias-Akins. "Insight into Genes Regulating Postharvest Aflatoxin Contamination of Tetraploid Peanut from Transcriptional Profiling." *Genetics* 209, no. 1 (2018): 143–156. https://doi.org/10.1534/genetics.118.300478.

Kottapalli, K. R., P. Payton, R. Rakwal, G. K. Agrawal, J. Shibato, M. Burow, and N. Puppala. "Proteomics Analysis of Mature Seed of Four Peanut Cultivars Using Two-Dimensional Gel Electrophoresis Reveals Distinct Differential Expression of Storage, Anti-Nutritional, and Allergenic Proteins." *Plant Science* 175, no. 3 (2008): 321–329. https://doi.org/10.1016/j.plantsci.2008.05.005.

Kottapalli, K. R., M. Zabet-Moghaddam, D. Rowland, W. Faircloth, M. Mirzaei, P. A. Haynes, and P. Payton. "Shotgun Label-Free Quantitative Proteomics of Water-Deficit-Stressed Midmature Peanut (*A. hypogaea* L.) Seed." *Journal of Proteome Research* 12, no. 11 (2013): 5048–5057. https://doi.org/10.1021/pr400936d.

Krishna, G., B. K. Singh, E. K. Kim, V. K. Morya, and P. W. Ramteke. "Progress in Genetic Engineering of Peanut (*A. hypogaea* L.)—A Review." *Plant Biotechnology Journal* 13, no. 2 (2015): 147–162. https://doi.org/10.1111/pbi.12339.

Kumar, D., and P. B. Kirti. "Pathogen-Induced SGT1 of *Arachis diogoi* Induces Cell Death and Enhanced Disease Resistance in Tobacco and Peanut." *Plant Biotechnology Journal* 13, no. 1 (2015a): 73–84. https://doi.org/10.1111/pbi.12237.

Kumar, D., and P. B. Kirti. "Transcriptomic and Proteomic Analyses of Resistant Host Responses in *Arachis diogoi* Challenged with *Phaeoisariopsis personata*, the Late Leaf Spot Pathogen." *PLoS One* 10, no. 2 (2015b): e0117559. https://doi.org/10.1371/journal.pone.0117559.

Kumar, D., S. Rampuria, N. K. Singh, P. Shukla, and P. B. Kirti. "Characterization of a Vacuolar Processing Enzyme Expressed in *Arachis diogoi* in Resistance Responses Against the Late Leaf Spot Pathogen, *Phaeoisariopsis personata*." *Plant Molecular Biology* 88, no. 1–2 (2015): 177–191. https://doi.org/10.1007/s11103-015-0318-x.

Kumar, K. R., and P. B. Kirti. "Differential Gene Expression in *Arachis diogoi* Upon Interaction with Peanut Late Leaf Spot Pathogen, *Phaeoisariopsis personata* and Characterization of a Pathogen-Induced Cyclophilin." *Plant Molecular Biology* 75, no. 4–5 (2011): 497–513. https://doi.org/10.1007/s11103-011-9747-3.

Li, H. F., X. P. Chen, F. H. Zhu, H. Y. Liu, Y. B. Hong, and X. Q. Liang. "Transcriptome Profiling of Peanut (*Arachis hypogaea*) Gynophores in Gravitropic Response." *Functional Plant Biology* 40, no. 12 (2013): 1249–1260. https://doi.org/10.1071/FP13075.

Li, H., X. Liang, B. Zhou, X. Chen, Y. Hong, R. Zhou, S. Li, H. Liu, Q. Lu, H. Liu, and H. Wu. "A Proteomic Analysis of Peanut Seed at Different Stages of Underground Development to Understand the Changes of Seed Proteins." *PLoS One* 15, no. 12 (2020): e0243132. https://doi.org/10.1371/journal.pone.0243132.

Li, X. Y., X. Liu, Y. Yao, Y. H. Li, S. Liu, C. Y. He, J. M. Li, Y. Y. Lin, and L. Li. "Overexpression of *A. hypogaea* AREB1 Gene Enhances Drought Tolerance by Modulating ROS Scavenging and Maintaining Endogenous ABA Content." *International Journal of Molecular Sciences* 14, no. 6 (2013): 12827–12842. https://doi.org/10.3390/ijms140612827.

Li, Y., J. Meng, S. Yang, F. Guo, J. Zhang, Y. Geng, L. Cui, S. Wan, and X. Li. "Transcriptome Analysis of Calcium- and Hormone-Related Gene Expressions During Different Stages of Peanut Pod Development." *Frontiers in Plant Science* 8 (2017): 1241. https://doi.org/10.3389/fpls.2017.01241.

Li, Z., R. L. Jarret, and J. W. Demski. "Engineered Resistance to Tomato Spotted Wilt Virus in Transgenic Peanut Expressing the Viral Nucleocapsid Gene." *Transgenic Research* 6, no. 4 (1997): 297–305. https://doi.org/10.1023/A:1018462729127.

Li, Z., X. Zhang, K. Zhao, K. Zhao, C. Qu, G. Gao, F. Gong, X. Ma, and D. Yin. "Comprehensive Transcriptome Analyses Reveal Candidate Genes for Variation in Seed Size/Weight During Peanut (*A. hypogaea* L.) Domestication." *Frontiers in Plant Science* 12 (2021): 666483. https://doi.org/10.3389/fpls.2021.666483.

Lin, J., F. M. Bruni, Z. Fu, J. Maloney, L. Bardina, A. L. Boner, G. Gimenez, and H. A. Sampson. "A Bioinformatics Approach to Identify Patients with Symptomatic Peanut Allergy Using Peptide Microarray Immunoassay." *Journal of Allergy and Clinical Immunology* 129, no. 5 (2012): 1321–1328.e5. https://doi.org/10.1016/j.jaci.2012.02.012.

Liu, H., H. Li, J. Gu, L. Deng, L. Ren, Y. Hong, Q. Lu, X. Chen, and X. Liang. "Identification of the Candidate Proteins Related to Oleic Acid Accumulation During Peanut (*A. hypogaea* L.) Seed Development Through Comparative Proteome Analysis." *International Journal of Molecular Sciences* 19, no. 4 (2018): 1235. https://doi.org/10.3390/ijms19041235.

Liu, H., X. Liang, Q. Lu, H. Li, H. Liu, S. Li, R. Varshney, Y. Hong, and X. Chen. "Global Transcriptome Analysis of Subterranean Pod and Seed in Peanut (*A. hypogaea* L.) Unravels the Complexity of Fruit Development Under Dark Condition." *Scientific Reports* 10, no. 1 (2020): 13050. https://doi.org/10.1038/s41598-020-69943-7.

Liu, Y., L. Xiao, J. Chi, R. Li, Y. Han, F. Cui, Z. Peng, S. Wan, and G. Li. "Genome-Wide Identification and Expression of SAUR Gene Family in Peanut (*A. hypogaea* L.) and Functional Identification of AhSAUR3 in Drought Tolerance." *BMC Plant Biology* 22, no. 1 (2022): 178. https://doi.org/10.1186/s12870-022-03564-2.

Livingstone, D. M., J. L. Hampton, P. M. Phipps, and E. A. Grabau. "Enhancing Resistance to *Sclerotinia minor* in Peanut by Expressing a Barley Oxalate Oxidase Gene." *Plant Physiology* 137, no. 4 (2005): 1354–1362. https://doi.org/10.1104/pp.104.057232.

Luo, L., Q. Wan, K. Zhang, X. Zhang, R. Guo, C. Wang, C. Zheng, F. Liu, Z. Ding, and Y. Wan. "AhABI4s Negatively Regulate Salt-Stress Response in Peanut." *Frontiers in Plant Science* 12 (2021): 741641. https://doi.org/10.3389/fpls.2021.741641.

Luo, M., X. Q. Liang, P. Dang, C. C. Holbrook, M. G. Bausher, R. D. Lee, and B. Z. Guo. "Microarray-Based Screening of Differentially Expressed Genes in Peanut in Response to *Aspergillus parasiticus* Infection and Drought Stress." *Plant Science* 169, no. 4 (2005): 695–703. https://doi.org/10.1016/j.plantsci.2005.05.020.

Magbanua, Z. V., H. D. Wilde, J. K. Roberts, K. Chowdhury, J. Abad, J. W. Moyer, H. Y. Wetzstein, and W. A. Parrott. "Field Resistance to Tomato Spotted Wilt Virus in Transgenic Peanut (*A. hypogaea* L.) Expressing an Antisense Nucleocapsid Gene Sequence." *Molecular Breeding* 6, no. 2 (2000): 227–236. https://doi.org/10.1023/A:1009649408157.

Mahatma, M. K., L. K. Thawait, S. K. Bishi, N. Khatediya, A. L. Rathnakumar, H. B. Lalwani, and J. B. Misra. "Nutritional Composition and Antioxidant Activity of Spanish and Virginia Groundnuts (*A. hypogaea* L.): A Comparative Study." *Journal of Food Science and Technology* 53 (2016): 2279–2286. https://doi.org/10.1007/s13197-016-2187-y.

Majumdar, R., M. Lebar, B. Mack, R. Minocha, S. Minocha, C. Carter-Wientjes, C. Sickler, K. Rajasekaran, and J. W. Cary. "The *A. flavus* Spermidine Synthase (spds) Gene, Is Required for Normal Development, Aflatoxin Production, and Pathogenesis During Infection of Maize Kernels." *Frontiers in Plant Science* 9 (2018): 317. https://doi.org/10.3389/fpls.2018.00317.

Mallikarjuna, G., T. S. R. B. Rao, and P. B. Kirti. "Genetic Engineering for Peanut Improvement: Current Status and Prospects." *Plant Cell, Tissue and Organ Culture* 125, no. 3 (2016): 399–416. https://doi.org/10.1007/s11240-016-0966-9.

Manjulatha, M., R. Sreevathsa, A. M. Kumar, C. Sudhakar, T. G. Prasad, N. Tuteja, and M. Udayakumar. "Overexpression of a Pea DNA Helicase (PDH45) in Peanut (*A. hypogaea* L.) Confers Improvement of Cellular Level Tolerance and Productivity Under Drought Stress." *Molecular Biotechnology* 56, no. 2 (2014): 111–125. https://doi.org/10.1007/s12033-013-9687-z.

Mehta, R., T. Radhakrishnan, A. Kumar, R. Yadav, J. R. Dobaria, P. P. Thirumalaisamy, R. K. Jain, and P. Chigurupati. "Coat Protein-Mediated Transgenic Resistance of Peanut (*A. hypogaea* L.) to Peanut Stem Necrosis Disease Through Agrobacterium-Mediated Genetic Transformation." *Indian*

Journal of Virology 24, no. 2 (2013): 205–213. https://doi.org/10.1007/s13337-013-0157-9.

Meimaroglou, D. M., D. Galanopoulou, and P. Markaki. "Study of the Effect of Methyl Jasmonate Concentration on Aflatoxin B(1) Biosynthesis by *Aspergillus parasiticus* in Yeast Extract Sucrose Medium." *International Journal of Microbiology* (2009): 842626. https://doi.org/10.1155/2009/842626.

Mondal, S., A. M. Badigannavar, and S. F. D'Souza. "Induced Variability for Fatty Acid Profile and Molecular Characterization of High Oleate Mutant in Cultivated Groundnut (*A. hypogaea* L.)." *Plant Breeding* 130, no. 2 (2011): 242–247. https://doi.org/10.1111/j.1439-0523.2010.01787.x.

Mondal, S., A. M. Badigannavar, D. M. Kale, and G. S. S. Murthy. "Induction of Genetic Variability in a Disease-Resistant Groundnut Breeding Line." *BARC Newsletter* 285 (2007): 237–247.

Moore, K. M., and D. A. Knauft. "The Inheritance of High Oleic Acid in Peanut." *Journal of Heredity* 80, no. 3 (1989): 252–253. https://doi.org/10.1093/oxfordjournals.jhered.a110845.

Morgante, C. V., A. C. M. Brasileiro, P. A. Roberts, L. A. Guimaraes, A. C. G. Araujo, L. N. Fonseca, S. C. M. Leal-Bertioli, D. J. Bertioli, and P. M. Guimaraes. "A Survey of Genes Involved in *Arachis stenosperma* Resistance to *Meloidogyne Arenaria* Race 1." *Functional Plant Biology* 40, no. 12 (2013): 1298–1309. https://doi.org/10.1071/FP13096.

Mota, A. P. Z., B. Vidigal, E. G. J. Danchin, R. C. Togawa, S. C. M. Leal-Bertioli, D. J. Bertioli, A. C. G. Araujo, A. C. M. Brasileiro, and P. M. Guimaraes. "Comparative Root Transcriptome of Wild *Arachis* Reveals NBS-LRR Genes Related to Nematode Resistance." *BMC Plant Biology* 18, no. 1 (2018): 159. https://doi.org/10.1186/s12870-018-1373-7.

Murthy, G. S. S., A. M. Badigannavar, S. Mondal, and D. M. Kale. "Research and Impact of Groundnut Mutation Breeding in India." In *Groundnut Research in India*, edited by M. S. Basu and N. B. Singh, 57–69. Junagadh, India: NRCG, 2004.

Nadaf, H. L., K. Biradar, G. S. S. Murthy, P. U. Krishnaraj, R. S. Bhat, M. A. Pasha, and A. S. Yerimani. "Novel Mutations in Oleoyl-PC Desaturase (ahFAD2B) Identified from New High Oleic Mutants Induced by Gamma Rays in Peanut." *Crop Science* 57, no. 5 (2017): 2538–2546. https://doi.org/10.2135/cropsci2017.01.0039.

Nayak, S. N., G. Agarwal, M. K. Pandey, H. K. Sudini, A. S. Jayale, S. Purohit, A. Desai, L. Wan, B. Guo, B. Liao, and R. K. Varshney. "*A. flavus* Infection Triggered Immune Responses and Host–Pathogen Cross-Talks in Groundnut During In-Vitro Seed Colonization." *Scientific Reports* 7, no. 1 (2017): 9659. https://doi.org/10.1038/s41598-017-09260-8.

Nayak, S. N., V. Hebbal, P. Soni, R. Kumar, A. K. Pandey, L. Wan, M. K. Pandey, L. Boshou, and R. K. Varshney. "Groundnut Kernel Transcriptome." In *Comprehensive Foodomics*, edited by A. Cifuentes, 1:528–543. Elsevier, 2021. ISBN: 9780128163955.

Nigam, S. N., F. Waliyar, R. Aruna, S. V. Reddy, P. L. Kumar, P. Q. Craufurd, A. T. Diallo, B. R. Ntare, and H. D. Upadhyaya. "Breeding Peanut for Resistance to Aflatoxin Contamination at ICRISAT." *Peanut Science* 36, no. 1 (2009): 42–49. https://doi.org/10.3146/AT07-008.1.

Orsavova, J., L. Misurcova, J. V. Ambrozova, R. Vicha, and J. Mlcek. "Fatty Acids Composition of Vegetable Oils and Its Contribution to Dietary Energy Intake and Dependence of Cardiovascular Mortality on Dietary Intake of Fatty Acids." *International Journal of Molecular Sciences* 16, no. 6 (2015): 12871–12890. https://doi.org/10.3390/ijms160612871.

Ouakhssase, A., A. Chahid, H. Choubbane, A. Aitmazirt, and E. A. Addi. "Optimization and Validation of a Liquid Chromatography/Tandem Mass Spectrometry (LC-MS/MS) Method for the Determination of Aflatoxins in Maize." *Heliyon* 5, no. 5 (2019): e01565. https://doi.org/10.1016/j.heliyon.2019.e01565.

Ozias-Akins, P., J. A. Schnall, W. F. Anderson, C. Singsit, T. E. Clemente, M. J. Adang, and A. K. Weissinger. "Regeneration of Transgenic Peanut Plants from Stably Transformed Embryogenic Callus." *Plant Science* 93, no. 1–2 (1993): 185–194. https://doi.org/10.1016/0168-9452(93)90048-5.

Panda, P., D. B. Pradhan, and D. G. R. Rout. "Comparative Mutagenic Frequency, Efficiency, Effectiveness and Rate of Ethyl Methane Sulphonate and Sodium Azide in Groundnut (*A. hypogaea* L.)." *Pharma Innovation* 11, no. 4 (2022): 24–28. https://doi.org/10.22271/tpi.2022.v11.i4a.12264.

Pandurangaiah, M., G. Lokanadha Rao, O. Sudhakarbabu, A. Nareshkumar, K. Kiranmai, U. Lokesh, G. Thapa, and C. Sudhakar. "Overexpression of Horsegram (*Macrotyloma uniflorum* Lam.Verdc.) NAC Transcriptional Factor (MuNAC4) in Groundnut Confers Enhanced Drought Tolerance." *Molecular Biotechnology* 56, no. 8 (2014): 758–769. https://doi.org/10.1007/s12033-014-9754-0.

Partridge-Telenko, D. E., J. Hu, D. M. Livingstone, B. B. Shew, P. M. Phipps, and E. A. Grabau. "Sclerotinia Blight Resistance in Virginia-Type Peanut Transformed with a Barley Oxalate Oxidase Gene." *Phytopathology®* 101, no. 7 (2011): 786–793. https://doi.org/10.1094/PHYTO-10-10-0266.

Patel, K. G., V. B. Mandaliya, G. P. Mishra, J. R. Dobaria, and R. Thankappan. "Transgenic Peanut Overexpressing mtlD Gene Confers Enhanced Salinity Stress Tolerance via Mannitol Accumulation and Differential Antioxidative Responses." *Acta Physiologiae Plantarum* 38, no. 7 (2016): 181. https://doi.org/10.1007/s11738-016-2200-0.

Patel, K. G., R. Thankappan, G. P. Mishra, V. B. Mandaliya, A. Kumar, and J. R. Dobaria. "Transgenic Peanut (*A. hypogaea* L.) Overexpressing mtlD Gene Showed Improved Photosynthetic, Physio-Biochemical, and Yield-Parameters under Soil-Moisture Deficit Stress in Lysimeter System." *Frontiers in Plant Science* 8 (2017): 1881. https://doi.org/10.3389/fpls.2017.01881.

Patil, A. S., R. Thankappan, R. Mehta, R. Yadav, A. Kumar, G. P. Mishra, J. R. Dobaria, P. P. Thirumalaisamy, and R. K. Jain. "Evaluation of Transgenic Peanut Plants Encoding Coat Protein and Nucleocapsid Protein Genes for Resistance to Tobacco Streak Virus and Peanut Bud Necrosis Virus." *Journal of Environmental Biology* 38, no. 2 (2017): 187–196. https://doi.org/10.22438/jeb/38/2/MS-191.

Patil, M., S. V. Ramu, P. Jathish, R. Sreevathsa, P. C. Reddy, T. G. Prasad, and M. Udayakumar. "Overexpression of AtNAC2 (ANAC092) in Groundnut (*A. hypogaea* L.) Improves Abiotic Stress Tolerance." *Plant Biotechnology Reports* 8, no. 2 (2014): 161–169. https://doi.org/10.1007/s11816-013-0305-0.

Payton, P., K. R. Kottapalli, D. Rowland, W. Faircloth, B. Guo, M. Burow, N. Puppala, and M. Gallo. "Gene Expression Profiling in Peanut Using High Density Oligonucleotide

Microarrays." *BMC Genomics* 10, no. 1 (2009): 265. https://doi.org/10.1186/1471-2164-10-265.

Poovaiah, B. W., A. S. N. Reddy, and L. Feldman. "Calcium and Signal Transduction in Plants." *Critical Reviews in Plant Sciences* 12, no. 3 (1993): 185–211. https://doi.org/10.1080/07352689309701901.

Porterfield, H. S., K. S. Murray, D. G. Schlichting, X. Chen, K. C. Hansen, M. W. Duncan, and S. C. Dreskin. "Effector Activity of Peanut Allergens: A Critical Role for Ara h 2, Ara h 6, and Their Variants." *Clinical and Experimental Allergy* 39, no. 7 (2009): 1099–1108. https://doi.org/10.1111/j.1365-2222.2009.03273.x.

Power, I. L., P. M. Dang, V. S. Sobolev, V. A. Orner, J. L. Powell, M. C. Lamb, and R. S. Arias. "Characterization of Small RNA Populations in Non-Transgenic and Aflatoxin-Reducing-Transformed Peanut." *Plant Science* 257 (2017): 106–125. https://doi.org/10.1016/j.plantsci.2016.12.013.

Prasad, K., P. Bhatnagar-Mathur, F. Waliyar, and K. K. Sharma. "Overexpression of a Chitinase Gene in Transgenic Peanut Confers Enhanced Resistance to Major Soil-Borne and Foliar Fungal Pathogens." *Journal of Plant Biochemistry and Biotechnology* 22 (2013): 222–233. https://doi.org/10.1007/s13562-012-0155-9.

Proite, K., S. C. Leal-Bertioli, D. J. Bertioli, M. C. Moretzsohn, F. R. da Silva, N. F. Martins, and P. M. Guimaraes. M. "ESTs from a Wild *Arachis* Species for Gene Discovery and Marker Development." *BMC Plant Biology* 7 (2007): 7. https://doi.org/10.1186/1471-2229-7-7.

Qiao, L. X., X. Ding, H. C. Wang, J. M. Sui, and J. S. Wang. "Characterization of the Beta-1,3-Glucanase Gene in Peanut (*A. hypogaea* L.) by Cloning and Genetic Transformation." *Genetics and Molecular Research* 13, no. 1 (2014): 1893–1904. https://doi.org/10.4238/2014.March.17.17.

Qin, H., Q. Gu, J. Zhang, L. Sun, S. Kuppu, Y. Zhang, M. Burow, P. Payton, E. Blumwald, and H. Zhang. "Regulated Expression of an Isopentenyltransferase Gene (IPT) in Peanut Significantly Improves Drought Tolerance and Increases Yield Under Field Conditions." *Plant and Cell Physiology* 52, no. 11 (2011): 1904–1914. https://doi.org/10.1093/pcp/pcr125.

Radhakrishnan, T., A. L. Rathnakumar, M. K. Mahatma, S. Chandramohan, and S. Patel. "Genetic Resources of Groundnut." In *Cash Crops*, edited by P. Priyadarshan and S. M. Jain, 344–406. Springer, 2022. https://doi.org/10.1007/978-3-030-74926-2_10.

Rajam, M. V. "RNA Silencing Technology: A Boon for Crop Improvement." *Journal of Biosciences* 45 (2020): 118. https://doi.org/10.1007/s12038-020-00082-x.

Ramegowda, V., K. S. Mysore, and M. Senthil-Kumar. "Virus-Induced Gene Silencing Is a Versatile Tool for Unraveling the Functional Relevance of Multiple Abiotic-Stress-Responsive Genes in Crop Plants." *Frontiers in Plant Science* 5 (2014): 323. https://doi.org/10.3389/fpls.2014.00323.

Rao, S. C., P. Bhatnagar-Mathur, P. L. Kumar, A. S. Reddy, and K. K. Sharma. "Pathogen-Derived Resistance Using a Viral Nucleocapsid Gene Confers Only Partial Non-Durable Protection in Peanut Against Peanut Bud Necrosis Virus." *Archives of Virology* 158, no. 1 (2013): 133–143. https://doi.org/10.1007/s00705-012-1483-8.

Reddy, M. G., C. Senthilraja, R. Adhithya, V. K. Satya, E. Kokiladevi, D. Sudhakar, R. Rabindran, and R. Velazhahan. "Integration of a Hairpin RNA-Encoding Gene Derived from Tobacco Streak Virus Confers Resistance in Groundnut (*A. hypogaea* L.) Against Peanut Stem Necrosis Disease." *Journal of Plant Diseases and Protection* 123, no. 5 (2016): 205–214. https://doi.org/10.1007/s41348-016-0039-7.

Rizzo, W. B., P. A. Watkins, M. W. Phillips, D. Cranin, B. Campbell, and J. Avigan. "Adrenoleukodystrophy: Oleic Acid Lowers Fibroblast Saturated C22-26 Fatty Acids." *Neurology* 36, no. 3 (1986): 357–361. https://doi.org/10.1212/wnl.36.3.357.

Rohini, V. K., and K. S. Sankara Rao. "Transformation of Peanut (*A. hypogaea* L.) with Tobacco Chitinase Gene: Variable Response of Transformants to Leaf Spot Disease." *Plant Science* 160, no. 5 (2001): 889–898. https://doi.org/10.1016/s0168-9452(00)00462-3.

Sarkar, T., R. Thankappan, A. Kumar, G. P. Mishra, and J. R. Dobaria. "Heterologous Expression of the AtDREB1A Gene in Transgenic Peanut Conferred Tolerance to Drought and Salinity Stresses." *PLoS One* 9, no. 12 (2014): e110507. https://doi.org/10.1371/journal.pone.0110507.

Sarkar, T., R. Thankappan, A. Kumar, G. P. Mishra, and J. R. Dobaria. "Stress Inducible Expression of AtDREB1A Transcription Factor in Transgenic Peanut (*A. hypogaea* L.) Conferred Tolerance to Soil-Moisture Deficit Stress." *Frontiers in Plant Science* 7 (2016): 935. https://doi.org/10.3389/fpls.2016.00935.

Sarkar, T., R. Thankappan, G. P. Mishra, and B. D. Nawade. "Advances in the Development and Use of DREB for Improved Abiotic Stress Tolerance in Transgenic Crop Plants." *Physiology and Molecular Biology of Plants* 25, no. 6 (2019): 1323–1334. https://doi.org/10.1007/s12298-019-00711-2.

Sauer, M., S. Robert, and J. Kleine-Vehn. "Auxin: Simply Complicated." *Journal of Experimental Botany* 64, no. 9 (2013): 2565–2577. https://doi.org/10.1093/jxb/ert139.

Senthilraja, C., M. Gurivi Reddy, J. Rajeswaran, E. Kokiladevi, and R. Velazhahan. "RNA Interference-Mediated Resistance to Tobacco Streak Virus in Transgenic Peanut." *Australasian Plant Pathology* 47, no. 2 (2018): 227–230. https://doi.org/10.1007/s13313-018-0549-9.

Settaluri, V. S., C. V. K. Kandala, N. Puppala, and J. Sundaram "Peanuts and Their Nutritional Aspects—A Review." *Food and Nutrition Sciences* 3, no. 12 (2012): 1644–1650. https://doi.org/10.4236/fns.2012.312215.

Sharaf, A. N., A. Gaber, and A. A. Abdelhadi. "Induction, Characterization and Genetic Analysis of *A. flavus* Resistant Mutants in *Arachis hypogaea*." *African Journal of Biotechnology* 10, no. 75 (2011): 17095–17105. https://doi.org/10.5897/AJB11.2574.

Sharma, K. K., A. Pothana, K. Prasad, D. Shah, J. Kaur, D. Bhatnagar, Z. Y. Chen, Y. Raruang, J. W. Cary, K. Rajasekaran, H. K. Sudini, and P. Bhatnagar-Mathur. "Peanuts That Keep Aflatoxin at Bay: A Threshold That Matters." *Plant Biotechnology Journal* 16, no. 5 (2018): 1024–1033. https://doi.org/10.1111/pbi.12846.

Shen, Y., E. Zhiguo, X. Zhang, Y. Liu, and Z. Chen. "Screening and Transcriptome Analysis of Water Deficiency Tolerant Germplasms in Peanut (*Arachis hypogaea*)." *Acta Physiologiae Plantarum* 37 (2015). https://doi.org/10.1007/s11738-015-1840-9.

Sikora, P., A. Chawade, M. Larsson, J. Olsson, and O. Olsson. "Mutagenesis as a Tool in Plant Genetics, Functional Genomics, and Breeding." *International Journal of Plant Genomics* 2011 (2011): 314829. https://doi.org/10.1155/2011/314829.

Silva, M. F. C., C. R. C. Silva, L. M. Lima, J. P. C. Ramos, and R. C. Santos. "Differential Expression of Dormancy-Associated Genes in Fastigiata and Hypogaea Peanut." *Genetics and Molecular Research* 16, no. 4 (2017).

Singh, A., S. N. Raina, M. Sharma, M. Chaudhary, S. Sharma, and V. R. Rajpal. "Functional Uses of Peanut (*A. hypogaea* L.) Seed Storage Proteins." In *Peanut Science*, edited by J. C. Jimenez-Lopez, 8. IntechOpen, 2021. https://doi.org/10.5772/intechopen.96871.

Singh, N., A. Mishra, and B. Jha. "Ectopic Over-Expression of Peroxisomal Ascorbate Peroxidase (SbpAPX) Gene Confers Salt Stress Tolerance in Transgenic Peanut (*Arachis hypogaea*)." *Gene* 547, no. 1 (2014): 119–125. https://doi.org/10.1016/j.gene.2014.06.037.

Singsit, C., M. J. Adang, R. E. Lynch, W. F. Anderson, A. Wang, G. Cardineau, and P. Ozias-Akins. "Expression of a *Bacillus thuringiensis* cryIA(c) Gene in Transgenic Peanut Plants and Its Efficacy Against Lesser Cornstalk Borer." *Transgenic Research* 6, no. 2 (1997): 169–176. https://doi.org/10.1023/a:1018481805928.

Sinha, P., P. Bajaj, L. T. Pazhamala, S. N. Nayak, M. K. Pandey, A. Chitikineni, D. Huai, A. W. Khan, A. Desai, H. Jiang, W. Zhuang, B. Guo, B. Liao, and R. K. Varshney. "*A. hypogaea* Gene Expression Atlas for Fastigiata Subspecies of Cultivated Groundnut to Accelerate Functional and Translational Genomics Applications." *Plant Biotechnology Journal* 18, no. 11 (2020): 2187–2200. https://doi.org/10.1111/pbi.13374.

Soni, P., A. K. Pandey, S. N. Nayak, M. K. Pandey, P. Tolani, S. Pandey, H. K. Sudini, P. Bajaj, J. C. Fountain, P. Singam, B. Guo, and R. K. Varshney. "Global Transcriptome Profiling Identified Transcription Factors, Biological Processes, and Associated Pathways for Pre-Harvest Aflatoxin Contamination in Groundnut." *Journal of Fungi* 7, no. 6 (2021): 413. https://doi.org/10.3390/jof7060413.

Sundaresha, S., A. M. Manoj Kumar, S. Rohini, S. A. Math, E. Keshamma, S. C. Chandrashekar, and M. Udayakumar. "Enhanced Protection against Two Major Fungal Pathogens of Groundnut, *Cercospora arachidicola* and *A. flavus* in Transgenic Groundnut Over-Expressing a Tobacco β 1–3 Glucanase." *European Journal of Plant Pathology* 126, no. 4 (2010): 497–508. https://doi.org/10.1007/s10658-009-9556-6.

Anuradha, T. S. Swathi, K. Divya, S. K. Jami, and P. B. Kirti. "Transgenic Tobacco and Peanut Plants Expressing a Mustard Defensin Show Resistance to Fungal Pathogens." *Plant Cell Reports* 27, no. 11 (2008): 1777–1786. https://doi.org/10.1007/s00299-008-0596-8.

Tang, J., and C. Chu. "MicroRNAs in Crop Improvement: Fine-Tuners for Complex Traits." *Nature Plants* 3, no. 7 (2017): 17077. https://doi.org/10.1038/nplants.2017.77.

Thoppurathu, F. J., Z. Ghorbanzadeh, A. K. Vala, R. Hamid, and M. Joshi. "Unravelling the Treasure Trove of Drought-Responsive Genes in Wild-Type Peanut through Transcriptomics and Physiological Analyses of Root." *Functional and Integrative Genomics* 22, no. 2 (2022): 215–233. https://doi.org/10.1007/s10142-022-00833-z.

Tian, H., F. Guo, Z. Zhang, H. Ding, J. Meng, X. Li, Z. Peng, and S. Wan. "Discovery, Identification, and Functional Characterization of Long Noncoding RNAs in *A. hypogaea* L." *BMC Plant Biology* 20, no. 1 (2020): 308. https://doi.org/10.1186/s12870-020-02510-4.

Till, B. J., L. Comai, and S. Henikoff. "Tilling and Ecotilling for Crop Improvement." In *Genomics-Assisted Crop Improvement*, edited by Name of Editor(s), 333–49. Springer Netherlands, 2007.

Tiwari, R., and M. V. Rajam. "RNA- and MiRNA-Interference to Enhance Abiotic Stress Tolerance in Plants." *Journal of Plant Biochemistry and Biotechnology* 31, no. 4 (2022): 689–704. https://doi.org/10.1007/s13562-022-00770-9.

Tiwari, S., D. K. Mishra, K. Chandrasekhar, P. K. Singh, and R. Tuli. "Expression of Delta-Endotoxin Cry1EC from an Inducible Promoter Confers Insect Protection in Peanut (*A. hypogaea* L.) Plants." *Pest Management Science* 67, no. 2 (2011): 137–145. https://doi.org/10.1002/ps.2041.

Tiwari, V., A. K. Chaturvedi, A. Mishra, and B. Jha. "Introgression of the SbASR-1 Gene Cloned from a Halophyte *Salicornia brachiata* Enhances Salinity and Drought Endurance in Transgenic Groundnut (*Arachis hypogaea*) and Acts as a Transcription Factor."*PLoS One* 10, no. 7 (2015): e0131567. https://doi.org/10.1371/journal.pone.0131567.

Tong, B., Y. Shi, A. Ntambiyukuri, X. Li, J. Zhan, A. Wang, D. Xiao, and L. He. "Integration of Small RNA and Degradome Sequencing Reveals the Regulatory Network of Al-Induced Programmed Cell Death in Peanut." *International Journal of Molecular Sciences* 23, no. 1 (2022): 246. https://doi.org/10.3390/ijms23010246.

Treadwell, K., C. T. Young, and J. C. Wynne. "Evaluation of Fatty Acid Content of Forty Peanut Cultivars." *Oleagineux* 38 (1983): 381–385.

Vasavirama, K., and P. B. Kirti. "Increased Resistance to Late Leaf Spot Disease in Transgenic Peanut Using a Combination of PR Genes." *Functional & Integrative Genomics* 12 (2012): 625–634. https://doi.org/10.1007/s10142-012-0298-8.

Vassiliou, E. K., A. Gonzalez, C. Garcia, J. H. Tadros, G. Chakraborty, and J. H. Toney. "Oleic Acid and Peanut Oil High in Oleic Acid Reverse the Inhibitory Effect of Insulin Production of the Inflammatory Cytokine TNF-α Both in Vitro and in Vivo Systems." *Lipids in Health and Disease* 8, no. 1 (2009): 25. https://doi.org/10.1186/1476-511X-8-25.

Wan, L., Y. Lei, L. Yan, Y. Liu, M. K. Pandey, X. Wan, R. K. Varshney, J. Fang, and B. Liao. "Transcriptome and Metabolome Reveal Redirection of Flavonoids in a White Testa Peanut Mutant." *BMC Plant Biology* 20, no. 1 (2020): 161. https://doi.org/10.1186/s12870-020-02383-7.

Wan, L., B. Li, Y. Lei, L. Yan, D. Huai, Y. Kang, H. Jiang, J. Tan, and B. Liao. "Transcriptomic Profiling Reveals Pigment Regulation during Peanut Testa Development." *Plant Physiology and Biochemistry* 125 (2018): 116–125. https://doi.org/10.1016/j.plaphy.2018.01.029.

Wan, L., B. Li, Y. Lei, L. Yan, X. Ren, Y. Chen, X. Dai, H. Jiang, J. Zhang, W. Guo, A. Chen, and B. Liao. "Mutant Transcriptome Sequencing Provides Insights into Pod Development in Peanut (*A. hypogaea* L.)." *Frontiers in Plant Science* 8 (2017): 1900. https://doi.org/10.3389/fpls.2017.01900.

Wan, L., B. Li, M. K. Pandey, Y. Wu, Y. Lei, L. Yan, X. Dai, H. Jiang, J. Zhang, G. Wei, R. K. Varshney, and B. Liao. "Transcriptome Analysis of a New Peanut Seed Coat Mutant for the Physiological Regulatory Mechanism Involved in Seed Coat Cracking and Pigmentation." *Frontiers in Plant Science* 7 (2016): 1491. https://doi.org/10.3389/fpls.2016.01491.

Wang, C. T., Y. Tang, Z. Wang, W. Zhang, J. Li, C. Zhang, J. Yu, Y. F. Luo, and S. Lin. "Sodium Azide Mutagenesis Resulted in a Peanut Plant with Elevated Oleate Content." *Electronic Journal of Biotechnology* 14 (2011): 1–7. https://doi.org/10.2225/vol14-issue2-fulltext-4.

Wang, C. T., X. Z. Wang, Y. Y. Tang, D. X. Chen, J. C. Zhang, F. G. Cui, and S. L. Yu. "High Yielding Mutants Achieved by Injecting EMS into Peanut Flower Organs." *Journal of Nuclear Agricultural Sciences* 24 (2010): 239–242. https://doi.org/10.11869/hnxb.2010.02.0239.

Wang, T., E. Zhang, X. Chen, L. Li, and X. Liang. "Identification of Seed Proteins Associated with Resistance to Pre-Harvested Aflatoxin Contamination in Peanut (*A. hypogaea* L.)." *BMC Plant Biology* 10 (2010): 267. https://doi.org/10.1186/1471-2229-10-267.

Wang, C. T., Y. Y. Tang, X. Z. Wang, Q. Wu, H. Y. Gao, T. Feng, J. W. Su, S. T. Yu, X. L. Fang, W. L. Ni, Y. S. Jiang, L. Qian, and D. Q. Hu. "Mutagenesis: A Useful Tool for the Genetic Improvement of the Cultivated Peanut (*A. hypogaea* L.)." In *Mutagenesis*, edited by R. Mishra, 1. IntechOpen, 2012. https://doi.org/10.5772/50514.

Wang, H., Y. Lei, L. Wan, L. Yan, J. Lv, X. Dai, X. Ren, W. Guo, H. Jiang, and B. Liao. "Comparative Transcript Profiling of Resistant and Susceptible Peanut Post-Harvest Seeds in Response to Aflatoxin Production by *A. flavus*." *BMC Plant Biology* 16 (2016): 54. https://doi.org/10.1186/s12870-016-0738-z.

Wang, H., Y. Lei, L. Yan, K. Cheng, X. Dai, L. Wan, W. Guo, L. Cheng, and B. Liao. "Deep Sequencing Analysis of Transcriptomes in *A. flavus* in Response to Resveratrol." *BMC Microbiology* 15 (2015): 182. https://doi.org/10.1186/s12866-015-0513-6.

Wang, J., L. Shi, Y. Liu, M. Zhao, X. Wang, L. Qiao, J. Sui, G. Li, H. Zhu, and S. Yu. "Development of Peanut Varieties with High Oil Content by In Vitro Mutagenesis and Screening." *Journal of Integrative Agriculture* 19 (2020): 2974–2982. https://doi.org/10.1016/S2095-3119(20)63182-3.

Wang, M., C. Zhang, L. Pan, T. Wang, N. Chen, M. Chen, Z. Yang, X. Guo, S. Yu, and X. Chi. "Small RNA Profiling Reveal Regulation of MicroRNAs in Field Peanut Pod Rot Pathogen Infection." *Biologia* 75 (2020): 1779–1788. https://doi.org/10.2478/s11756-020-00485-z.

Wang, S., Y. S. Park, Y. Yang, E. J. Borrego, T. Isakeit, X. Gao, and M. V. Kolomiets. "Seed-Derived Ethylene Facilitates Colonization but Not Aflatoxin Production by *A. flavus* in Maize." *Frontiers in Plant Science* 8 (2017): 415. https://doi.org/10.3389/fpls.2017.00415.

Wang, T., X. Chen, F. Zhu, H. Li, L. Li, Q. Yang, X. Chi, S. Yu, and X. Liang. "Characterization of Peanut Germin-Like Proteins, AhGLPs in Plant Development and Defense." *PLoS One* 8 (2013): e61722. https://doi.org/10.1371/journal.pone.0061722.

Wang, X., P. Xu, L. Yin, Y. Ren, S. Li, Y. Shi, T. D. Alcock, Q. Xiong, W. Qian, X. Chi, M. K. Pandey, R. K. Varshney, and M. Yuan. "Genomic and Transcriptomic Analysis Identified Gene Clusters and Candidate Genes for Oil Content in Peanut (*A. hypogaea* L.)." *Plant Molecular Biology Reporter* 36 (2018): 518–529. https://doi.org/10.1007/s11105-018-1088-9.

Wang, Z., S. Yan, C. Liu, F. Chen, and T. Wang. "Proteomic Analysis Reveals an Aflatoxin-Triggered Immune Response in Cotyledons of *A. hypogaea* Infected with *A. flavus*." *Journal of Proteome Research* 11 (2012): 2739–2753. https://doi.org/10.1021/pr201105d.

Wu, Y., Z. Sun, F. Qi, M. Tian, J. Wang, R. Zhao, X. Wang, X. Wu, X. Shi, H. Liu, W. Dong, B. Huang, Z. Zheng, and X. Zhang. "Comparative Transcriptomics Analysis of Developing Peanut (*A. hypogaea* L.) Pods Reveals Candidate Genes Affecting Peanut Seed Size." *Frontiers in Plant Science* 13 (2022): 958808. https://doi.org/10.3389/fpls.2022.958808.

Xia, H., C. Zhao, L. Hou, A. Li, S. Zhao, Y. Bi, J. An, Y. Zhao, S. Wan, and X. Wang. "Transcriptome Profiling of Peanut Gynophores Revealed Global Reprogramming of Gene Expression During Early Pod Development in Darkness." *BMC Genomics* 14 (2013): 517. https://doi.org/10.1186/1471-2164-14-517.

Xia, H., L. Zhu, C. Zhao, K. Li, C. Shang, L. Hou, M. Wang, J. Shi, S. Fan, and X. Wang. "Comparative Transcriptome Analysis of Anthocyanin Synthesis in Black and Pink Peanut." *Plant Signaling and Behavior* 15 (2020): 1721044. https://doi.org/10.1080/15592324.2020.1721044.

Xu, M., H. Xie, J. Wu, L. Xie, J. Yang, and Y. Chi. "Translation Initiation Factor eIF4E and eIFiso4E Are Both Required for Peanut Stripe Virus Infection in Peanut (*A. hypogaea* L.)." *Frontiers in Microbiology* 8 (2017): 338. https://doi.org/10.3389/fmicb.2017.00338.

Xu, P., G. Tang, Y. Bi, Z. Liu, and L. Shan. "Analysis of the Peanut Transgenic Offspring with Depressing AhFAD2 Gene." *Sheng Wu Gong Cheng Xue Bao = Chinese Journal of Biotechnology* 34 (2018): 1469–1477. https://doi.org/10.13345/j.cjb.170508.

Xu, Y., X. Zhang, R. Hou, X. Zhang, S. Li, F. Yue, X. Zhang, and X. Zhu. "Multiple MicroRNAs Are Involved in Regulating Peanut (*A. hypogaea* L.) Resistance to Sclerotium rolfsii at the Early Stage." *Tropical Plant Biology* 15 (2022): 276–287. https://doi.org/10.1007/s12042-022-09321-x.

Xue, Q., X. Zhang, H. Yang, H. Li, Y. Lv, K. Zhang, Y. Liu, F. Liu, and Y. Wan. "Transcriptome and Metabolome Analysis Unveil Anthocyanin Metabolism in Pink and Red Testa of Peanut (*A. hypogaea* L.)." *International Journal of Genomics* 2021 (2021): 5883901. https://doi.org/10.1155/2021/5883901.

Yadav, R. "Agrobacterium Mediated Genetic Transformation of Groundnut for Nucleocapsid Mediated Resistance to Bud Necrosis Virus Disease." PhD thesis, Saurashtra University, 2009. http://hdl.handle.net/10603/15973.

Yang, C. Y., S. Y. Chen, and G. C. Duan. "Transgenic Peanut (*A. hypogaea* L.) Expressing the Urease Subunit B Gene of *Helicobacter pylori*." *Current Microbiology* 63 (2011): 387–391. https://doi.org/10.1007/s00284-011-9991-4.

Yang, H., C. Singsit, A. Wang, D. Gonsalves, and P. Ozias-Akins. "Transgenic Peanut Plants Containing a Nucleocapsid Protein Gene of Tomato Spotted Wilt Virus Show Divergent Levels of Gene Expression." *Plant Cell Reports* 17 (1998): 693–699. https://doi.org/10.1007/s002990050467.

Yang, S., L. Li, J. Zhang, Y. Geng, F. Guo, J. Wang, J. Meng, N. Sui, S. Wan, and X. Li. "Transcriptome and Differential Expression Profiling Analysis of the Mechanism of Ca2+ Regulation in Peanut (Arachis hypogaea) Pod Development." *Frontiers in Plant Science* 8 (2017): 1609. https://doi.org/10.3389/fpls.2017.01609.

Yang, S., J. Wang, Z. Tang, F. Guo, Y. Zhang, J. Zhang, J. Meng, L. Zheng, S. Wan, and X. Li. "Transcriptome of Peanut Kernel and Shell Reveals the Mechanism of Calcium on Peanut Pod Development." *Scientific Reports* 10 (2020): 15723. https://doi.org/10.1038/s41598-020-72893-9.

Yin, D., Y. Wang, X. Zhang, H. Li, X. Lu, J. Zhang, W. Zhang, and S. Chen. "De Novo Assembly of the Peanut (*A. hypogaea* L.) Seed Transcriptome Revealed Candidate Unigenes for Oil Accumulation Pathways." *PLoS One* 8 (2013): e73767. https://doi.org/10.1371/journal.pone.0073767.

Yuan, C., C. Li, X. Lu, X. Zhao, C. Yan, J. Wang, Q. Sun, and S. Shan. "Comprehensive Genomic Characterization of NAC Transcription Factor Family and Their Response to Salt and Drought Stress in Peanut." *BMC Plant Biology* 20 (2020): 454. https://doi.org/10.1186/s12870-020-02678-9.

Zeringue, H. J., R. L. Brown, J. N. Neucere, and T. E. Cleveland. "Relationships Between C6 –C12 Alkanal and Alkenal Volatile Contents and Resistance of Maize Genotypes to *A. flavus* and Aflatoxin Production." *Journal of Agricultural and Food Chemistry* 44 (1996): 403–407. https://doi.org/10.1021/jf950313r.

Zhang, J., S. Liang, J. Duan, J. Wang, S. Chen, Z. Cheng, Q. Zhang, X. Liang, and Y. Li. "De Novo Assembly and Characterisation of the Transcriptome During Seed Development, and Generation of Genic-SSR Markers in Peanut (*A. hypogaea* L.)." *BMC Genomics* 13 (2012): 90. https://doi.org/10.1186/1471-2164-13-90.

Zhang, G., Y. Zhang, J. Xu, X. Niu, J. Qi, A. Tao, L. Zhang, P. Fang, L. Lin, and J. Su. "The CCoAOMT1 Gene from Jute (*Corchorus Capsularis* L.) Is Involved in Lignin Biosynthesis in *Arabidopsis thaliana*." *Gene* 546 (2014): 398–402. https://doi.org/10.1016/j.gene.2014.05.011.

Zhang, H., X. Zhao, Q. Sun, C. Yan, J. Wang, C. Yuan, C. Li, S. Shan, and F. Liu. "Comparative Transcriptome Analysis Reveals Molecular Defensive Mechanism of *A. hypogaea* in Response to Salt Stress." *International Journal of Genomics* 2020 (2020): 6524093. https://doi.org/10.1155/2020/6524093.

Zhang, R. X., B. B. Li, Z. G. Yang, J. Q. Huang, W. H. Sun, N. Bhanbhro, W. T. Liu, and K. M. Chen. "Dissecting Plant Gene Functions Using CRISPR Toolsets for Crop Improvement." *Journal of Agricultural and Food Chemistry* 70 (2022): 7343–7359. https://doi.org/10.1021/acs.jafc.2c01754.

Zhang, X., C. Ren, Y. Xue, Y. Tian, H. Zhang, N. Li, C. Sheng, H. Jiang, and D. Bai. "Small RNA and Degradome Deep Sequencing Reveals the Roles of MicroRNAs in Peanut (*A. hypogaea* L.) Cold Response." *Frontiers in Plant Science* 13 (2022): 920195. https://doi.org/10.3389/fpls.2022.920195.

Zhao, C., T. Li, Y. Zhao, B. Zhang, A. Li, S. Zhao, L. Hou, H. Xia, S. Fan, J. Qiu, P. Li, Y. Zhang, B. Guo, and X. Wang. "Integrated Small RNA and mRNA Expression Profiles Reveal miRNAs and Their Target Genes in Response to *A. flavus* Growth in Peanut Seeds." *BMC Plant Biology* 20 (2020): 215. https://doi.org/10.1186/s12870-020-02426-z.

Zhao, C., H. Xia, T. Cao, Y. Yang, S. Zhao, L. Hou, Y. Zhang, C. Li, X. Zhang, and X. Wang. "Small RNA and Degradome Deep Sequencing Reveals Peanut MicroRNA Roles in Response to Pathogen Infection." *Plant Molecular Biology Reporter* 33 (2015): 1013–1029. https://doi.org/10.1007/s11105-014-0806-1.

Zhao, C., S. Zhao, L. Hou, H. Xia, J. Wang, C. Li, A. Li, T. Li, X. Zhang, and X. Wang. "Proteomics Analysis Reveals Differentially Activated Pathways That Operate in Peanut Gynophores at Different Developmental Stages." *BMC Plant Biology* 15 (2015): 188. https://doi.org/10.1186/s12870-015-0582-6.

Zhao, C. Z., H. Xia, T. P. Frazier, Y. Y. Yao, Y. P. Bi, A Q. Li, M. J. Li, C. S. Li, B. H. Zhang, and X. J. Wang. "Deep Sequencing Identifies Novel and Conserved MicroRNAs in Peanuts (Arachis hypogaea L.)." *BMC Plant Biology* 10, no. 1 (2010): 3. https://doi.org/10.1186/1471-2229-10-3.

Zhao, X., C. Li, C. Yan, J. Wang, C. Yuan, H. Zhang, and S. Shan. "Transcriptome and Proteome Analyses of Resistant Preharvest Peanut Seed Coat in Response to *A. flavus* Infection." *Electronic Journal of Biotechnology* 39 (2019): 82–90. https://doi.org/10.1016/j.ejbt.2019.03.003.

Zhu, W., E. Zhang, H. Li, X. Chen, F. Zhu, Y. Hong, B. Liao, S. Liu, and X. Liang. "Comparative Proteomics Analysis of Developing Peanut Aerial and Subterranean Pods Identifies Pod Swelling Related Proteins." *Journal of Proteomics* 91 (2013): 172–187. https://doi.org/10.1016/j.jprot.2013.07.002.

Zhuang, W., H. Chen, M. Yang, J. Wang, M. K. Pandey, C. Zhang, W. C. Chang, L. Zhang, X. Zhang, R. Tang, V. Garg, X. Wang, J. Tang, C. N. Chow, J. Wang, Y. Deng, D. Wang, A. W. Khan, Q. Yang, et al., "The Genome of Cultivated Peanut Provides Insight into Legume Karyotypes, Polyploid Evolution and Crop Domestication." *Nature Genetics* 51 (2019): 865–876. https://doi.org/10.1038/s41588-019-0402-2.

9 Molecular Mechanisms Underlying Peanut Growth, Development, and Stress Tolerance

Sheidu Abdullaziz, Chong Zhang, Yuhui Zhuang, Yasir Sharif, Hua Chen, Xingjun Wang, Rajeev R. Varshney, and Weijian Zhuang

9.1 INTRODUCTION

Peanut (Arachis hypogaea) is an important legume crop worldwide, offering a rich source of protein, oil, and essential nutrients. Beyond being a dietary staple in many regions, peanuts significantly impact the agricultural economies of countries in Africa, Asia, and South America. Despite their importance, peanut production is challenged by diseases, drought, salinity, and extreme temperatures (Gangurde et al., 2019; Ana et al., 2014; 2017).

At the molecular level, peanuts are fascinating for genetic research due to their allotetraploid nature, which presents both challenges and opportunities for breeding programs aimed at enhancing yield, quality, and stress resistance. Advances in genomics, transcriptomics, and proteomics have greatly expanded our understanding of the molecular mechanisms governing peanut growth and stress responses. Recent breakthroughs in next-generation sequencing and functional genomics have identified key genes involved in processes such as flowering, root architecture, nodule formation, and defense against biotic and abiotic stress. For instance, studies on transcription factors like AhMYB30 and AhGLK1b have revealed their roles in improving tolerance to salinity and drought, respectively (Chen et al., 2023; Ali et al., 2020).

Signal transduction pathways, including MAPK and calcium-mediated signaling, are crucial for coordinating peanut responses to environmental challenges, while epigenetic mechanisms provide an additional layer of regulatory control (Pirayesh et al., 2021). Moreover, tools like CRISPR/Cas9 are increasingly used to introduce desirable traits, paving the way for peanut varieties with enhanced stress tolerance and higher yields (Jin et al., 2022).

This chapter offers a comprehensive review of the molecular mechanisms underlying peanut growth and development, focusing on genetic regulation, hormonal pathways, and defense responses to environmental stresses. It also explores recent advances in molecular breeding technologies that are contributing to the development of peanut cultivars capable of thriving in challenging environments.

9.2 MOLECULAR PATHWAYS IN PEANUT GROWTH, DEVELOPMENT, AND STRESS RESILIENCE

9.2.1 Genetic Insights into the MAPK Pathway in Peanut Development and Disease Stress

The MAPK (mitogen-activated protein kinase) pathway is a conserved signaling cascade that regulates various cellular processes in response to external stimuli, such as pathogen attacks, drought, salinity, and temperature changes (Sinha et al., 2011). This pathway transduces signals from the environment into the nucleus, where it activates transcription factors that control genes involved in stress tolerance and growth regulation (Tena et al., 2001; Jonak et al., 2002). The cascade consists of three tiers: MAPKKK, MAPKK, and MAPK. When activated, MAPKKKs initiate phosphorylation events, ultimately activating MAPKs like MAPK3, which translocate to the nucleus and phosphorylate transcription factors, including WRKY31 and MYB30 (Mishra et al., 2006). These factors regulate stress-responsive genes involved in drought tolerance and defense against pathogens, such as chitinases and pathogenesis-related proteins (Dan et al., 2001; Wang and Gou, 2020).

A study by Wu et al. (2022) analyzed differentially expressed genes (DEGs) in the WRKY family during key stages of peanut pod development. The study found significant enrichment of MAPK signaling pathways and highlighted several key genes involved in auxin synthesis and metabolism, such as *PNC*, *YUC*, and *IAA*. Notably, *WRKY40* showed lower expression in large-pod genotypes, indicating its negative regulatory role in pod development, while *MYB20* and *MAPKKK17* also displayed reduced expression during seed and pod growth in large-pod varieties. Key MAPK pathway-related genes, including *MPK3*, *WRKY33*, *WRKY40*, and *NPR1*, were identified as central regulators of pod and seed development. *MPK3*, in particular, played an important role in pod formation, with its expression increasing in shells as development progressed. Genes like *CYCD* and *CYCU* contributed to cell differentiation and proliferation, while *GASA* was linked to

DOI: 10.1201/9781003360988-9

seed size and pod elongation via the gibberellin pathway. Additionally, two variant sites, *GCP4* and *RPPL1*, associated with seed size characteristics, were identified within stable QTL intervals, further suggesting their role in regulating microtubule nucleation and seed size. These findings shed light on the complex regulatory mechanisms of pod and seed development in peanuts, with these DEGs holding potential for improving pod size and yield in cultivated varieties. Interestingly, the MAPK signaling pathway is also targeted during biotic stress. For example, *R. solanacearum*, the causal agent of bacterial wilt, delivers the toxic effector RipAU, which interferes with MAPK signaling to promote disease progression.

9.2.2 Role of Calcium Signaling in Peanut Development and Disease Resistance

Calcium (Ca^{2+}) plays a crucial role as a secondary messenger in a wide range of cellular processes, including gene expression, cell division, and stress responses. In peanuts, calcium signaling is essential for regulating developmental processes such as seed formation, root development, and adaptation to abiotic stresses like drought and salinity (Zhang et al., 2007). Calcium signaling is integral to regulating nodule formation in peanuts, a process critical for nitrogen fixation and peanut seed development. Calcium deficiency has been linked to embryo abortion in peanuts, primarily through apoptosis and cell wall degradation, producing empty pods. This is common phenomenon in peanut production especially in sandy and acidic areas. A study by Chen et al. (2016) identified 62 differentially expressed genes under low calcium conditions, including AhCYP707A4, a key enzyme of abscisic acid (ABA) catabolism, which reduces ABA levels and triggers apoptosis (Petal et al., 2024). The role of *AhCYP707A4* in regulating seed abortion was further validated through its overexpression in *Nicotiana benthamiana*, which resulted in similar ABA-mediated effects, causing a low ABA level and high degree of sterile or abortive seeds. The *AhHsfA4a* gene plays a crucial role in regulating cellular functions and biological pathways in peanuts, with significant downregulation over 10-fold under calcium (Ca^{2+}) deficiency, particularly linked to early embryo abortion. RT-PCR analysis revealed that while *AhHsfA4a* expression increases in embryos under sufficient Ca^{2+} conditions, its expression significantly decreases during Ca^{2+} deficiency, correlating with embryo contraction and subsequent abortion. Zhang et al. (2007) reveal that peanut gynophores and pods exhibit differential gene expression in response to sufficient and deficient free calcium (Ca^{2+}), with upregulation of transcription factors like WRKY, NAC, and bHLH in gynophores and distinct regulation of calcium signaling and hormone-related genes in pods. *AhCCaMK* plays a critical role in proper nodule organogenesis, endosymbiont dissemination, and symbiosome formation during peanut nodule development (Sinharoy and DasGupta, 2009). *AhCYCLOPS* is essential for nodulation. It successfully complemented *Medicago truncatula* mutants (*ipd3-1* and *ipd3-2*) when expressed under a native promoter, demonstrating functional conservation across species (Das et al., 2019). Silencing *AhCYCLOPS* using RNA interference (RNAi) delayed nodulation and prevented rhizobia from differentiating into functional bacteroids, which remained in their rod-shaped form. This silencing also affected the expression of other key symbiotic genes: *AhCCaMK* levels remained unchanged, *AhHK1* increased by 1.5-fold, and the transcription of *AhENOD40* and *AhNIN* decreased moderately (Das et al., 2019). These findings suggest that *AhCYCLOPS* plays a crucial role in rhizobia differentiation, likely through the regulation of symbiotic signaling components. *AhENOD40*, which is expressed early in nodule development, is also involved in forming the nodule structure (Ganguly et al., 2021). Moreover, acetylene reduction assays revealed a substantial decrease in nitrogenase activity in RNAi nodules compared to control plants (Sinharoy et al., 2009). It is pertinent to acknowledge that *AhCYCLOPS* and related transcription factors coordinate nodule development, ensuring proper cellular differentiation and the establishment of symbiotic relationships.

Additionally, resistance to bacterial wilt has been linked to calcium-mediated signaling pathways in peanuts. The *AhCML69* gene, which encodes a calmodulin-like protein, is involved in regulating defense responses against *Ralstonia solanacearum*. Calcium signaling activates *AhCML69*, which interacts with downstream kinases to induce the expression of resistance-related genes (Yang et al., 2024). Overexpression of *AhCML69* in peanut plants has been shown to enhance resistance to bacterial wilt, reducing the incidence and severity of infection. Another key mechanism in peanut disease resistance is the activation of transcription factors involved in regulating defense-related gene expression. The WRKY transcription factor family, particularly AhWRKY31, has been shown to play a crucial role in peanut defense against fungal pathogens. AhWRKY31 is activated in response to pathogen attack and regulates the expression of genes involved in the production of antimicrobial compounds and cell wall fortification (Yan et al., 2022). By modulating the activity of AhWRKY31, peanut plants can mount a robust defense against fungal infections, minimizing crop losses due to diseases. Continued advances in functional genomics are revealing numerous candidate genes and transcription factors involved in root nodule symbiosis (RNS) in legumes, offering deeper insights into the molecular mechanisms underlying this complex process (Karmakar et al., 2019; Peng et al., 2017).

9.2.3 ROS Signaling and Stress Responses in Peanut

Reactive oxygen species (ROS), such as hydrogen peroxide (H_2O_2) and superoxide anions (O_{2}-), are generated as byproducts of metabolic processes, particularly under stress conditions like drought, salinity, and pathogen attack (Mittler, 2002). While excessive ROS can cause oxidative damage to cellular structures, they also serve as essential signaling molecules that trigger defense responses and activate stress-tolerance pathways in plants (Gill and Tuteja, 2010). The balance between ROS production and scavenging is crucial for stress adaptation, and this balance is tightly regulated by antioxidant enzymes such as superoxide dismutase (SOD), catalase (CAT), and peroxidases (POD). These enzymes prevent ROS accumulation, thereby reducing oxidative damage and maintaining redox homeostasis (Nakagami et al., 2006). Recent studies have identified the gene *AhUGT75A*, encoding a UDP-glycosyltransferase that enhances ROS scavenging by modulating the levels of antioxidant compounds like flavonoids. Overexpression of *AhUGT75A* has been shown to improve peanut tolerance to drought and salinity by reducing ROS levels and protecting cellular structures from oxidative damage (Ouyang et al., 2023). The *AhPRR1* gene encodes a receptor-like kinase in peanuts that plays a crucial role in PAMP recognition. Activation of *AhPRR1* induces downstream defense responses, including the production of pathogenesis-related proteins like *AhPR5*, which have antifungal activity (Hao et al., 2017).

The salt overly sensitive (SOS) pathway is essential in regulating salinity tolerance. The *SOS1* gene encodes a plasma membrane sodium/hydrogen antiporter that prevents the accumulation of toxic sodium ions (Na^{2+}) in root cells, thereby maintaining ion balance under saline conditions (Zhu, 2001). The *SOS1* pathway is regulated by the calcium-binding protein CBL4, which interacts with the protein kinase CIPK24 to initiate the SOS signaling cascade. This cascade ensures efficient sodium removal from the cytoplasm, reducing ionic toxicity and allowing plants to survive in saline soils (Pirayesh et al., 2021). Additionally, the transcription factor *AhMYB30* has been shown to enhance salinity tolerance by regulating genes involved in ion transport and osmoprotection. Overexpression of *AhMYB30* promotes the accumulation of osmoprotectants such as glycine betaine, which helps maintain cellular turgor under saline stress (Chen et al., 2023). During heat stress, heat shock proteins (HSPs) such as *AhHSP70* act as molecular chaperones, stabilizing proteins and preventing denaturation. The expression of *AhHSP70* is regulated by heat shock transcription factors like *AhHSFA2*, which modulate the plant's heat stress response and enhance thermotolerance (Wang et al., 2019). Cold stress, on the other hand, induces oxidative stress by increasing ROS production, which can damage cellular membranes and proteins. In response, peanut plants upregulate antioxidant enzymes such as SOD and CAT to scavenge excess ROS and protect cells from oxidative damage (Su et al., 2021). Lipid metabolism also plays a critical role in cold tolerance, as maintaining membrane fluidity is essential for cellular integrity at low temperatures. The *AhFAD2* gene, which encodes fatty acid desaturase, regulates the synthesis of unsaturated fatty acids, ensuring membrane fluidity under cold conditions and allowing peanut cells to function effectively (Zhao et al., 2022; Zhang et al., 2024).

In summary, peanut plants employ a wide range of molecular mechanisms to defend against both biotic and abiotic stresses. ROS signaling, mediated by transcription factors and antioxidant enzymes, plays a central role in regulating stress responses. Additionally, hormone signaling pathways, ion transport mechanisms, and transcription factors like WRKY and MYB contribute to peanut's ability to tolerate environmental stress. Understanding these molecular responses is critical for breeding peanut varieties that are more resilient and capable of thriving in diverse environmental conditions.

9.3 EPIGENETIC REGULATION IN PEANUT GROWTH AND DEVELOPMENT

9.3.1 DNA Methylation in Peanut Development

DNA methylation is a crucial epigenetic mechanism in plants, characterized by the addition of a methyl group to cytosine residues in DNA. This modification generally leads to the repression of gene expression by hindering transcription factors from accessing their target DNA sequences. In peanuts, DNA methylation plays a vital role in regulating essential growth processes, including seed development and stress adaptation. Wang et al. (2016) demonstrated that methylation patterns change dynamically throughout seed development. Notably, specific genes involved in seed maturation, particularly those that govern embryo growth and nutrient accumulation, exhibit differential methylation patterns corresponding to distinct stages of seed formation. Investigations into cytosine-5 DNA methyltransferases and demethylases in wild peanut species underscore the complexity of epigenetic regulation during peanut development. Wang et al. (2018) reported that during the early stages of peanut pod development, substantial changes in DNA methylation were observed, with specific genes related to growth and stress responses showing differential expression. Methyltransferase genes such as DRM2 and MET1 were found to be upregulated, indicating their essential role in establishing DNA methylation patterns crucial for pod development. Furthermore, Ramesh et al. (2020) revealed significant variations in DNA methylation patterns among diverse peanut genotypes, influencing gene expression and potentially contributing to phenotypic diversity. Moreover, studies on O-methyltransferase genes in peanuts suggest their involvement in development and stress tolerance, highlighting a potential role for these genes in DNA methylation during peanut development (Cai et al., 2023). DNA methylation also plays a significant role in regulating oil content and pod formation in peanuts. A recent study by Nian et al. (2022) indicates that variations in DNA methylation between wild and cultivated peanuts markedly affect gene expression related to lipid metabolism, thereby influencing oil accumulation. Specific gene clusters associated with lipid metabolism demonstrate divergent expression patterns, suggesting that DNA methylation is a critical factor in both oil biogenesis and degradation processes. In a recent report, m6A methylation arround splicing junctions was found to affect gene alternative splice and expression in peanut; a higher level of m6A causes a lower level of gene expression and lower alternative splicing (Wang et al., 2024). Overall, the variation in DNA methylation patterns across peanut genotypes significantly influences gene expression, emphasizing the importance of epigenetic regulation in peanut growth and development.

9.3.2 Histone Modifications and Chromatin Remodeling

Histone modifications are another key epigenetic mechanism regulating peanut growth. Histones, the proteins around which DNA is wrapped, can undergo various post-translational modifications, such as acetylation, methylation, and phosphorylation. These modifications influence chromatin structure, either relaxing it to promote gene expression or condensing it to repress gene activity. In peanut, histone modifications are involved in regulating developmental processes like flowering and seed maturation. During seed formation, histone methylation at specific loci has been linked to the activation of genes responsible for nutrient storage and embryo development. These modifications ensure that genes involved in seed maturation are turned on at the appropriate developmental stages, ensuring the proper formation and quality of peanut seeds (Kundu and DasGupta, 2017). Studies by Yang et al. (2015) show that histone acetylation and deacetylation play significant roles in regulating stress responses, essential for repressing seed storage protein genes, such as *Ara h 2.02*, during germination. This process prevents the expression of embryonic traits in vegetative tissues. The mutation of histone deacetylase (HDA19) leads to hyperacetylation and ectopic expression of *Ara h 2.02*, indicating that histone modifications are critical for maintaining seed dormancy and regulating developmental transitions. Furthermore, Su et al. (2021) report that *AhHDA1*, a gene encoding a histone deacetylase, has been shown to influence flavonoid and phenylpropanoid metabolism in peanut roots. Overexpression of *AhHDA1* enhances the biosynthesis of these secondary metabolites, which are critical for plant defense against pathogens and abiotic stresses. By modulating histone acetylation, *AhHDA1* controls the accessibility of stress-responsive genes, allowing peanut plants to fine-tune their defense mechanisms. Histone modifications, including acetylation and nucleosome loss, regulate oleosin gene expression in developing peanut embryos, showcasing dynamic chromatin remodeling during maturation stages (Li et al., 2009). The interplay between histone variants and DNA methylation is also significant, as it influences epigenetic inheritance and gene regulation in plants (Adams et al., 2023). Chromatin remodeling complexes, which include ATP-dependent factors, facilitate the accessibility of DNA for transcription by altering nucleosome positioning. This is vital for the expression of genes necessary for seed development and response to environmental cues (Lucchesi, 2019). During periods of water deficit, plants utilize histone modifications to control the expression of drought-responsive genes, allowing the plant to conserve resources and survive under adverse conditions (Berr and Shen, 2010; Amar et al., 2014). These epigenetic adjustments are essential for maintaining peanut productivity, even in challenging environments.

9.3.3 Non-Coding RNAs in Peanut Growth and Stress Responses

In recent years, non-coding RNAs (ncRNAs) have emerged as important regulators of gene expression in plants. ncRNAs do not code for proteins, but they play crucial roles in silencing or activating specific genes, particularly those involved in developmental processes and stress responses. The regulatory landscape of peanut flowering is influenced by various genetic and epigenetic factors, including long non-coding RNAs (lncRNAs) that interact with flowering-related genes. The study identified lncRNAs linked to MADS-box transcription factor, influencing flowering and fruit development in peanut, indicating their role in regulating flowering (Zhang et al., 2016). LncRNAs have been shown to modulate the expression of genes involved in seed maturation and response to nutrient availability. By influencing the expression of these genes, lncRNAs contribute to the regulation of peanut yield and seed quality (Zhao et al., 2019). Small interfering RNAs (siRNAs) and microRNAs (miRNAs) have been shown to regulate key genes involved in root development, flowering, and response to abiotic stresses like drought and salinity. For example, miR156, a conserved microRNA across plant species, targets *AhSPL* (SQUAMOSA promoter binding protein-like) genes in peanuts. These *AhSPL* genes are involved in regulating root development and flowering time, ensuring that peanuts adapt their growth to environmental conditions (Sun et al., 2024). Furthermore, during stress conditions such as drought, specific miRNAs are upregulated to modulate the expression of genes involved in water retention and root architecture. By silencing genes that inhibit root elongation, these miRNAs enhance root growth, allowing peanut plants to access deeper soil layers and improve water uptake. This epigenetic regulation via ncRNAs is critical for the plant's ability to adapt to fluctuating environmental conditions (Su et al., 2021).

In summary, epigenetic modifications, including DNA methylation, histone alterations, and non-coding RNA regulation, play a crucial role in controlling peanut growth and development. These mechanisms allow peanuts to rapidly adjust gene expression in response to environmental challenges, optimizing growth and improving stress resilience. Harnessing the power of epigenetics in peanut breeding offers a promising avenue for developing cultivars that can better withstand biotic and abiotic stresses, ensuring stable yields in varying conditions.

9.4 FLOWERING AND SEED DEVELOPMENT IN PEANUT

9.4.1 Molecular Regulation of Flowering in Peanut

Flowering in plants is tightly controlled by both internal hormonal signals and external environmental cues such as photoperiod, temperature, and water availability. Several genes and pathways have been identified that regulate the timing and initiation of flowering, allowing peanut plants to synchronize their reproductive cycle with favorable environmental conditions. The molecular regulation of flowering in peanut involves complex genetic pathways and specific genes like Phy A, Phy B, Phy E, FT2a, FT5a, and COL2, influenced by photoperiod conditions, with higher expression in short days (Jannat et al., 2022). Phosphatidyl ethanolamine-binding proteins (PEBPs) are primarily regulated by hormonal signals and environmental cues, which are essential for the adaptation of peanuts to diverse growing conditions (Jin et al., 2019). Recent studies by Zhong et al. (2022) have elucidated the roles of these genes (PEBPs) in flowering time regulation, revealing their response to environmental factors such as photoperiod and temperature. A total of 33 PEBP genes have been identified in cultivated peanuts, categorized into four subgroups: FT, TFL, MFT, and FT-like. These genes exhibit distinct expression patterns between early and late-flowering genotypes, indicating their critical role in flowering regulation. Sun et al. (2023) reported that overexpression of gibberellin biosynthesis genes like *AhGA20ox* and *AhGA3ox* has been shown to accelerate flowering, making these genes potential targets for breeding programs aimed at manipulating flowering times for different growing environments. These genes (*AhGA20ox* and *AhGA3ox*) show high expression in flowers, indicating their involvement in molecular regulation of flowering in peanut. Exogenous application of GA3 also positively influences morphological traits, including seed size and pod number, thereby increasing overall yield (Oldroyd et al., 2001). Additionally, drought and heat can delay or inhibit flowering by altering hormonal balances, particularly through the accumulation of abscisic acid (ABA). The ABA signaling pathway interacts with flowering genes, repressing their activity under stress conditions to conserve resources and protect the plant from reproductive failure (Rouhi and Sepehri, 2020).

In summary, the flowering process in peanuts is orchestrated by a complex interplay of genetic pathways, hormones, and environmental signals. Key gene families like PEBPs regulate flowering timing in response to factors such as photoperiod and temperature. Gibberellins (GAs) further accelerate flowering by activating growth-related genes, while environmental stresses like drought alter hormonal balances, delaying the process. Understanding these molecular mechanisms offers exciting opportunities for breeding programs to optimize flowering times and improve peanut yields under diverse conditions.

9.4.2 Hormonal Control of Seed Development

Seed development in peanuts is a complex process influenced by a combination of hormones, transcription factors, and environmental signals, all of which play crucial roles in determining the final yield. The peanut shell, which forms first, provides essential protection and support for seed growth during this process (Liu et al., 2020). After fertilization, peanuts develop pegs, and pod growth begins only after the pegs successfully penetrate the soil (Chen et al., 2016). To better understand the growth patterns of peanut pods, Lv et al. (2023) studied the length and width of pods from the Tif and Lps varieties at five different stages after pegging. Their analysis, using a logistic growth model, identified three distinct phases of pod expansion: an initial slow growth phase, followed by a rapid-growth phase between 10 and 15 days after pegging (DAP), and finally, a slower phase as the pods reached their full size. This research indicated that the number of cells during the rapid-growth phase was a major factor in determining the differences in pod size between the Tif and Lps varieties. Earlier studies, however, had suggested that cell area was the primary factor influencing pod size (Wang et al., 2022; Wan et al., 2017).

Phytohormones, including auxin, cytokinin (CK), and gibberellin (GA), play a critical role in regulating the early stages of peanut pod development (Yang et al., 2020; Held et al., 2014; Reinecke et al., 2013). These hormones primarily affect pod size by promoting cell division and expansion. Research suggests that during the rapid-growth phase, when the cell number is most crucial to pod size, the concentration of phytohormones in the pods is at its peak. Consequently, it is hypothesized that variations in phytohormone levels during this stage are the main cause of differences in cell number and therefore pod size between the Tif and Lps varieties (Lv et al., 2023). Auxins, particularly indole-3-acetic acid (IAA), play a pivotal role during the early stages of peanut seed development. They regulate the expression of genes involved in embryo patterning and seed coat formation. Auxin-responsive factors (ARFs) control key processes, including the formation of the seed coat and the establishment of nutrient pathways from the mother plant to the developing embryo, ensuring proper seed compartment differentiation (Lv et al., 2023; Li et al., 2021). Studies have shown that auxin-related genes are differentially expressed during various stages of pod and seed development, underscoring auxin's critical role in promoting cell division and expansion, which are essential for seed growth and pod formation (Li et al., 2017). Auxin biosynthesis, which significantly influences cell division and expansion, begins with tryptophan as a key precursor (Zhao, 2010). Tryptophan is converted by the TAA family of aminotransferases into indole-3-pyruvic acid (IPA), and then the YUC family of flavin monooxygenases transforms IPA into indole-3-acetic acid (IAA) (Zhao, 2012; Won et al., 2011; Mashiguchi et al., 2011). In peanuts, the expression pattern of gene-LOC112801162 (TAA1) aligned with changes in IAA levels, suggesting that differential expression of TAA1 may be responsible for the variation in IAA content between the Tif and Lps varieties, ultimately contributing to the differences in pod size (Lv et al., 2023). This study also reported that two IPT genes (gene-LOC112741152 and gene-LOC112795845) were significantly upregulated in the Lps variety at 10 DAP, resulting in higher cytokinin accumulation in Lps. This elevated CK concentration likely promotes increased cell division and contributes to enhanced pod and seed growth (Frugier et al., 2008). The balance between auxins and cytokinins is crucial for regulating seed enlargement and the allocation of resources between the seed and other plant tissues. Cytokinin biosynthesis genes such as IPT (ISOPENTENYLTRANSFERASE) are highly expressed during seed development, promoting cell proliferation and increasing seed size (Takei et al., 2004). Gibberellins (GAs) are involved in multiple aspects of plant growth and development (Reinecke et al., 2013). Gibberellic acid (GA3) has been shown to enhance seed vigor in peanut (Rouhi and Sepehri, 2020). However, Wang et al. (2022) reported that GA20ox genes were significantly downregulated in a peanut mutant with smaller pods. Similarly, Lv et al. (2023) observed that several genes involved in GA biosynthesis, including gene-LOC112756333 (KAO), gene-LOC112706011 (GA20ox), gene-LOC112711704 (GA20ox), gene-LOC112709954 (GA3ox), and gene-LOC112727565 (GA3ox), were significantly upregulated in the Lps variety at 10DAP. These findings suggest that further research into GA20ox genes in peanuts is necessary, as these key differentially expressed genes (DEGs) may play a role in regulating pod size through modulation of GA biosynthesis.

By understanding the molecular mechanisms governing hormonal signaling, breeders can exploit these pathways to optimize pod and seed development in peanuts, contributing to more efficient peanut production.

9.4.3 The Role of SWEET and AhARF6 in Seed Development

Seed development in peanuts is governed by a complex network of genes, with the *AhARF6* and SWEET gene families playing pivotal roles in regulating growth and nutrient accumulation. These genetic mechanisms ensure that developing seeds receive the necessary energy and nutrients for proper development. Recent research has highlighted the critical role of sugar transporters, particularly the SWEET (sugars will eventually be exported transporter) gene family, in seed development. A total of 94 SWEET genes have been identified across three peanut species, with 47 present in *Arachis hypogaea*. Phylogenetic analysis categorized these genes into seven distinct groups, indicating both evolutionary conservation and functional diversity (Li et al., 2024). Transcriptomic analyses revealed that 12 *AhSWEET* genes are particularly crucial for seed development, displaying diverse expression patterns across different peanut varieties. For instance, *AhSWEET24* was found to be highly expressed in the endosperm and embryo during seed maturation. This gene plays a functional role in loading sucrose into the developing seed, ensuring proper nutrient accumulation and seed maturation. Notably, the overexpression of *AhSWEET24* has been associated with increased seed size and enhanced seed filling, making it a promising target for breeding programs aimed at improving peanut yield and seed quality (Li et al., 2024). While *AhSWEET* genes are integral to seed development, they do not work in isolation. Their interactions with other gene families, such as *AhUGT* (UDP-glycosyltransferases) and *AhLEA* (late embryogenesis abundant proteins), suggest a more complex regulatory network that also responds to abiotic stresses like drought. These interactions highlight the multifaceted role of the SWEET genes in peanut growth, resilience, and productivity (Huang et al., 2022). Understanding how these genes coordinate with hormonal pathways opens up new avenues for improving seed development and productivity in peanuts.

In addition to SWEET genes, *AhARF6* (auxin response factor 6) plays a significant role in pod development. According to Li et al. (2021), *AhARF6* shows differential expression across various peanut tissues and is particularly linked to pod development. The gene is highly expressed in the stem and branches of both wild and cultivated peanut species but shows lower expression levels in roots and leaves. Transcriptome analyses further revealed that *AhARF6* expression was more pronounced in larger pods, such as those in cultivars H8107 and ZP06, compared to medium-sized pods in H103 and smaller pods in *Arachis monticola*. Additionally, *AhARF6-5* was found to be localized in the nucleus, which aligns with the role of AtARF6 in Arabidopsis, further supporting its potential involvement in regulating pod development in peanuts (Lv et al., 2023). The regulation of peanut pegs and pods by auxin underscores the significance of auxin signaling in peanut development. Studies have shown that applying indole-3-acetic acid (IAA) or an auxin polar transport inhibitor (TIBA) can significantly increase pod weight and yield per pod, further supporting the role of auxin in enhancing peanut productivity (Li et al., 2021). Conversely, high temperatures can accelerate seed maturation but may reduce seed quality by affecting the accumulation of storage proteins and oils. In peanuts, genes involved in lipid biosynthesis, such as *AhFAD2* (fatty acid desaturase), are particularly sensitive to temperature fluctuations. These genes control the composition of fatty acids in the seed, which in turn affects oil quality.

Understanding the molecular and genetic control mechanisms, such as *AhARF6* and *AhSWEET* genes, allows breeders to explore new strategies for improving pod size, seed quality, and overall yield in peanuts, ultimately contributing to more efficient and resilient peanut production.

9.5 ROOT AND NODULE DEVELOPMENT

9.5.1 Genetic and Hormonal Regulation of Root Architecture

The architecture of the peanut root system plays a fundamental role in determining the plant's ability to explore the soil for water and nutrients. The regulation of root development involves a complex interplay of genetic and hormonal factors, with auxins, cytokinins, and brassinosteroids being the primary drivers of processes like root elongation, lateral root branching, and root hair formation. Auxin plays a central role in root architecture by promoting cell division and elongation, especially through auxin accumulation in the root tip. This accumulation stimulates root growth, allowing the plant to better absorb water and nutrients from the soil. In response to environmental stimuli such as drought or nutrient deficiency, auxin signaling is adjusted to modulate root architecture and increases the plant's adaptive capacity to stress (Lou et al., 2022).

Auxin signaling is tightly regulated by a group of transcription factors known as auxin response factors (ARFs). In peanuts, the ARF family, particularly genes such as *AhARF2*, regulates gene expression by binding to auxin response elements (AuxREs) in the promoters of auxin-responsive genes, controlling root cell division and elongation (Li et al., 2021). Additionally, auxin-responsive gene expression is influenced by the degradation of Aux/IAA proteins, which inhibit ARF activity in the absence of auxin. Once auxin is perceived, Aux/IAA proteins are degraded, allowing ARFs to activate genes involved in root development (Luo et al., 2018). Research has identified 63 *AhARF* genes in peanuts, including specific ARFs such as *AhARF14*, *AhARF26*, and *AhARF45*, which correlate positively with root length and surface area, highlighting the crucial role of these genes in peanut root architecture (Luo et al., 2022). ABA signaling inhibits root growth and promotes the formation of deeper, more drought-resistant root systems (Patel et al., 2024). Studies have shown that ABA treatment in peanut leads to increased root length and lateral root density, enhancing water uptake from deeper soil layers (Chen et al., 2020).

The interaction between auxin and other phytohormones, such as cytokinins and brassinosteroids, further modulates root development. These hormones work either synergistically or antagonistically to regulate root organogenesis and lateral root formation, both of which are essential for the uptake of water and nutrients (Singla and Kaur, 2018; Sharma et al., 2021; Held et al., 2014). While auxins generally promote root formation, cytokinins inhibit lateral root growth by directing resources toward the primary root system. The balance between auxin and cytokinin signaling is therefore critical for determining the overall structure of the root system. In Arabidopsis, for example, the cytokinin receptor *AhCRE1* modulates cytokinin levels in response to environmental conditions, ensuring optimal root development (Inoue et al., 2001;Gonzalez-Rizzo et al., 2006). In peanuts, this balance becomes particularly important during periods of drought, where deeper root growth is favored to access water from lower soil layers. By fine-tuning the auxin–cytokinin balance, peanuts are better able to adapt their root systems to environmental stresses, such as drought or nutrient limitations (Li et al., 2023).

Furthermore, nitrogen deficiency significantly impacts root morphology in peanuts, as shown by transcriptomic analyses. This highlights the critical role of genetic expression in response to nutrient availability, particularly in regulating root growth and development (Li et al., 2023). While significant progress has been made in understanding the genetic and hormonal networks governing peanut root architecture, gaps remain in fully elucidating these complex interactions. More research is necessary to optimize root traits for improved agricultural outcomes, particularly through breeding strategies that target key genetic pathways linked to root traits. Advanced phenotyping technologies provide valuable insights into these traits, enhancing the potential for developing peanut varieties with optimized root architectures that can thrive in various environmental conditions. By understanding the hormonal regulation of root development through auxins, cytokinins, and their interactions, breeders can explore new avenues for improving peanut root systems. This will contribute to enhanced water and nutrient uptake, better stress resilience, and, ultimately, increased crop yields.

9.5.2 Nodule Development and Symbiotic Nitrogen Fixation

9.5.2.1 Root Nodule Symbiosis (RNS)

Root nodule symbiosis (RNS) is characterized by complex molecular interactions between the host plant and symbiotic bacteria, which ultimately result in the formation of functional nodules. In these nodules, differentiated bacteria known as bacteroids convert nitrogen gas into ammonium ions (Broughton et al., 2000). The initial signaling event in this process is the release of flavonoids by leguminous roots, which accumulate in the rhizosphere and activate the bacterial transcriptional regulator NodD. This activation induces the transcription of nodulation-related genes (nod, nol, and noe), leading to the production and secretion of bacterial nodulation signals called nod factors (NF) or lipochitooligosaccharides (LCOs) (Spaink, 2000). Peanut has been predominantly nodulated by Bradyrhizobium strains carrying nodABC genes, but there is evidence that it can also be nodulated by nodABC-lacking Bradyrhizobium strains such as Btai1 (Noisangiam et al., 2012). This suggests that nod factors may not always be indispensable for RNS in peanuts, leading to the hypothesis that two modes of RNS might exist: i) an NF-dependent mode and ii) an NF-independent mode (Ibáñez and Fabra, 2011; Guha et al.,

2016). In the NF-dependent mode, nod factors are required to initiate the cell divisions necessary for nodule primordium development (Ibáñez and Fabra, 2011). Peanut root cracks are intercellularly infected by rhizobia, leading to the formation of aeschynomenoid nodules through repeated divisions of infected cells. Two alternative models have been proposed to explain NF perception and the infection process. The first model suggests that nod factors are necessary for the intracellular infection of the basal cortical cell that forms the nodule primordium. The second model posits that rhizobia can infect cells intracellularly without nod factors, but the nod factors are essential for reactivating meristematic activity in the infected cells. However, the exact mechanism of colonization by Bradyrhizobium nodC mutants, whether intercellular or intracellular, remains unclear, and both possibilities require further investigation (Ibáñez and Fabra, 2011).

In peanuts, the *AhHK1* gene encodes a receptor with an extracellular CHASE domain, two transmembrane domains, a cytosolic kinase domain, and a C-terminal receiver domain. RNA interference (RNAi) targeting *AhHK1* significantly downregulated type-A response regulators such as *AhRR5* and *AhRR3*, as well as symbiotic genes like *AhNIN* and *AhENOD40*. Plants treated with *AhHK1*-RNAi showed a marked reduction in nodulation, and the few nodules that formed were inactive. Microscopic analysis revealed an elevated mitotic index, indicating undifferentiated nodules with rod-shaped rhizobia in spherical symbiosomes. Increased expression of the meristem maintenance factor Wuschel-Related Homeobox 5 (WOX5) in these nodules suggests a link between *AhHK1* downregulation and disrupted nodule development (Kundu and DasGupta, 2017). These findings underscore the importance of *AhHK1*-mediated cytokinin signaling in initiating nodulation and ensuring proper nodule differentiation and development. It remains unclear whether factors like retinoblastoma-related protein (RBR) (Perilli et al., 2013) and Wuschel-related homeobox 5 (WOX5)(Franssen et al., 2015), which are conserved in root nodule meristems, are involved.

The molecular perception of NFs triggers symbiotic signaling that is critical for establishing nitrogen-fixing root nodule symbiosis. In the model legume *Lotus japonicus*, NFs are recognized by two LysM domain-containing receptor-like kinases (RLKs), LjNFR1 and LjNFR5, which together form a heteromeric complex (Madsen et al., 2011). In peanut, *LjNFR1* and *LjNFR5* have been reported as orthologues of *AhNFR1* and *AhNFP*, respectively, based on phylogenetic analyses (Ibañez et al., 2015). Expressed sequence tag (EST) contigs that showed significant similarity to *LjNFR1* contained extracellular LysM domains in the predicted protein sequences, along with characteristic CXC motifs in the interspacer domains between LysM1-LysM2 and LysM2-LysM3. Additionally, PCR-based strategies retrieved an amplicon with high identity to *LjNFR5*-like RLKs, encoding the extracellular region of *AhNFP*. A three-dimensional homology model of the LysM2 domain of NFP demonstrated conserved proximity of glutamine and isoleucine residues associated with NF binding. However, transcriptional analysis of post-infection root samples showed no significant induction of *AhNFP* (Ibañez et al., 2015). Earlier studies have demonstrated the inability of *Bradyrhizobium SEMIA* nodC mutants to nodulate, indicating the NF dependence of peanut nodulation (Ibáñez and Fabra, 2011). Nevertheless, research by Noisangiam et al. (2012) provided evidence of NF-independent nodulation by *Bradyrhizobium* BTAi1, suggesting that peanut can engage in both NF-dependent and NF-independent nodulation, depending on the bacterial strain involved. In summary, while peanut possesses NF receptors, the dependence on NFs for nodulation appears to vary and may be strain-specific for different *Bradyrhizobium* partners.

9.5.2.2 Symbiotic Receptor Kinase

Symbiotic receptor kinase (SYMRK) plays a critical role in amplifying the signal from NFs during the establishment of legume-rhizobia symbiosis (Stracke et al., 2002). In peanuts, *AhSYMRK* is a conserved receptor kinase comprising an N-terminal signal peptide, a NORK-like (Malectin-like) domain, three leucine-rich repeats (LRRs), a transmembrane domain, and a cytoplasmic kinase domain with a C-terminal tail. Notably, *AhSYMRK* functions as a dual-specific serine/threonine/tyrosine (S/T/Y) receptor kinase, with autophosphorylation occurring at a crucial "gatekeeper" residue (Y670) (Samaddar et al., 2013). *AhSYMRK* has demonstrated cross-species functionality by complementing *Medicago truncatula* SYMRK-null mutants (TR25), restoring nodulation and facilitating intracellular accommodation of *Sinorhizobium*. However, when the kinase domain (KD) of *AhSYMRK* was overexpressed in *M. truncatula*, it activated nodule organogenesis without rhizobial infection, a phenomenon not observed in peanuts, likely due to the requirement of bacterial invasion for nodulation (Saha et al., 2014). Mutations at the gatekeeper tyrosine (Y670) residue, substituting it with phenylalanine (Phe) or alanine (Ala), significantly impaired symbiotic function in *M. truncatula*, halting rhizobial infection and preventing it from penetrating the epidermal–cortical barrier. These findings suggest that precise phosphorylation of SYMRK is crucial for facilitating rhizobial infection and promoting intracellular invasion in the nodule cortex (Saha et al., 2016). However, how this epidermal–cortical regulation operates specifically in peanuts requires further investigation.

9.6 FUTURE PERSPECTIVES IN MOLECULAR MECHANISM IN PEANUT GROWTH AND DEVELOPMENT

The genetic improvement of peanut has seen significant advancements; however, there remains limited progress in identifying and functionally characterizing key genes associated with essential agronomic traits such as growth, development, yield, and resistance to diseases and environmental stresses. Currently, only a few genes have been cloned and studied in depth, and even fewer mechanisms have been disclosed that directly relate to the molecular basis of these critical traits. Omics studies including genomics, transcriptomics, proteomics, and metabolomics have provided insights through gene enrichment analyses and preliminary functional evaluations, yet these findings alone do not sufficiently reveal the precise molecular mechanisms needed for effective applications in peanut breeding.

To address these gaps, future research should prioritize the following key areas:

1. Genetic mechanism discovery: Rapidly identifying and characterizing genes associated with vital agronomic traits, such as seed development, drought tolerance, and disease resistance, is crucial. This involves using positional cloning and other high-throughput approaches, such as RNA sequencing (RNA-seq) and genome-wide association studies (GWAS), to map and verify candidate genes. These discoveries will establish a genetic foundation essential for further molecular studies in peanut breeding.
2. Molecular mechanism elucidation: Functional characterization of these identified genes should be conducted through transgenic methods to disclose the molecular pathways governing essential traits. A deep understanding of these pathways will help clarify how specific genes influence the expression of key traits under varying environmental conditions, shedding light on the broader molecular processes affecting peanut resilience and productivity.
3. Genetic enhancement: Utilizing CRISPR/Cas9 genome-editing technology holds significant potential to improve peanut yield, quality, and disease resistance by enabling targeted genetic modifications. Further exploration is needed to optimize genotype-dependent transformation efficiencies, especially within the complex allotetraploid peanut genome. Advances in bioinformatics and genomic libraries will aid in identifying alleles from wild relatives for introgression, enhancing genetic diversity. Additionally, efforts to establish non-GMO regulatory frameworks for gene-edited crops will be essential to support commercialization and facilitate market entry.

By concentrating on these areas—genetic mechanism discovery, molecular mechanism elucidation, and genetic enhancement—research can drive the development of high-performance peanut varieties that are well-equipped to meet global agricultural demands and adapt to changing environmental conditions. Integrating traditional breeding with molecular approaches offers a pathway toward more resilient and productive peanut crops, aligning with sustainable agriculture goals.

REFERENCES

Adams, D. W., J. J. Ipsaro, J. Cahn, J. Lynn, H. S. Kim, B. Berube, V. Major, J. P. Calarco, C. LeBlanc, S. Bhattacharjee, U. Ramu, D. Grimanelli, Y. Jacob, P. Voigt, L. Joshua-Tor, and R. A. Martienssen. "Chromatin Remodeling of Histone H3 Variants Underlies Epigenetic Inheritance of DNA Methylation." *bioRxiv* (2023). https://doi.org/10.1101/2023.07.11.548598.

Ali, N., H. Chen, C. Zhang, S. A. Khan, M. Gandeka, D. Xie, and W. Zhuang. "Ectopic Expression of AhGLK1b (GOLDEN2-like Transcription Factor) in Arabidopsis Confers Dual Resistance to Fungal and Bacterial Pathogens." *Genes* 11, no. 3 (2020): 343. https://doi.org/10.3390/genes11030343.

Amar, R., A. Kumari, and D. M. Pandey. "Annotation of Stress Responsive Candidate Genes in Peanut ESTs." *Interdisciplinary Sciences: Computational Life Sciences* (2014). https://doi.org/10.1007/S12539-015-0010-5.

Ana, F., E. Bianucci, S. Castro, and K.-J. Dietz. "Metabolic Features Involved in Drought Stress Tolerance Mechanisms in Peanut Nodules and Their Contribution to Biological Nitrogen Fixation." *Plant Science* (2017). https://doi.org/10.1016/J.PLANTSCI.2017.06.009.

Ana, F., E. Bianucci, M. D. C. Tordable, S. Castro, and K.-J. Dietz. "Antioxidant Enzyme Activities and Gene Expression Patterns in Peanut Nodules During a Drought and Rehydration Cycle." *Functional Plant Biology* (2014). https://doi.org/10.1071/FP13311.

Berr, A., and W. H. Shen. "Molecular Mechanisms in Epigenetic Regulation of Plant Growth and Development." In *Plant Developmental Biology - Biotechnological Perspectives*, edited by E. Pua and M. Davey, 301–320. Springer, 2010. https://doi.org/10.1007/978-3-642-04670-4_17.

Broughton, W. J., S. Jabbouri, and X. Perret. "Keys to Symbiotic Harmony." *Journal of Bacteriology* 182, no. 20 (2000): 5641–5652. https://doi.org/10.1128/JB.182.20.5641-5652.2000.

Cai, T., Y. Sharif, Y. Zhuang, Q. Yang, X. Chen, K. Chen, Y. Chen, M. Gao, H. Dang, Y. Pan, A. Raza, C. Zhang, H. Chen, and W. Zhuang. "In-Silico Identification and Characterization of O-Methyltransferase Gene Family in Peanut (Arachis hypogaea L.) Reveals Their Putative Roles in Development and Stress Tolerance." *Frontiers in Plant Science* 14 (2023): 1145624. https://doi.org/10.3389/fpls.2023.1145624.

Chen, H., C. Zhang, T. C. Cai, Y. Deng, S. Zhou, Y. Zheng, and W. Zhuang. "Identification of Low Ca2+ Stress-Induced Embryo Apoptosis Response Genes in Arachis hypogaea by SSH-Associated Library Lift (SSHaLL)." *Plant Biotechnology Journal* 14, no. 2 (2016): 682–698. https://doi.org/10.1111/pbi.12415.

Chen, N., L. Pan, Z. Yang, M. Su, J. Xu, X. Jiang, X. Yin, T. Wang, F. Wan, and X. Chi. "A MYB-Related Transcription Factor from Peanut, AhMYB30, Improves Freezing and Salt Stress Tolerance in Transgenic Arabidopsis Through Both DREB/CBF and ABA-Signaling Pathways." *Frontiers in Plant Science* 14 (2023): 1136626. https://doi.org/10.3389/fpls.2023.1136626.

Chen, X., Y. Zhang, X. Wang, and H. Zhang. "Abscisic Acid Levels and Root Architecture in Groundnut (Arachis hypogaea L.) Cultivars Under Drought Stress." *Plant Physiology and Biochemistry* 150 (2020): 123–132. https://doi.org/10.1016/j.plaphy.2020.02.037.

Dan, I., N. M. Watanabe, and A. Kusumi. "The Ste20 Group Kinases as Regulators of MAP Kinase Cascades." *Trends in Cell Biology* 11, no. 5 (2001): 220–230.

Das, D. R., B. Horváth, A. Kundu, P. Kaló, and M. DasGupta. "Functional Conservation of CYCLOPS in Crack Entry Legume *Arachis hypogaea*." *Plant Science* 281 (2019): 232–241. https://doi.org/10.1016/j.plantsci.2019.01.002.

Franssen, H. J., T. T. Xiao, O. Kulikova, X. Wan, T. Bisseling, B. Scheres, and R. Heidstra. "Root Developmental Programs Shape the *Medicago truncatula* Nodule Meristem." *Development* 142, no. 17 (2015): 2941–2950. https://doi.org/10.1242/dev.123273.

Frugier, F., S. Kosuta, J. D. Murray, M. Crespi, and K. Szczyglowski. "Cytokinin: Secret Agent of Symbiosis." *Trends in Plant Science* 13, no. 3 (2008): 115–120. https://doi.org/10.1016/j.tplants.2008.01.007.

Ganguly, P., D. Roy, T. Das, A. Kundu, F. Cartieaux, Z. Ghosh, and M. DasGupta. "The Natural Antisense Transcript DONE40 Derived from the lncRNA ENOD40 Locus Interacts with SET Domain Protein ASHR3 During Inception of Symbiosis in *Arachis hypogaea*." *Molecular Plant-Microbe Interactions* 34, no. 9 (2021): 1057–1070. https://doi.org/10.1094/MPMI-12-20-0357-R.

Gangurde, S. S., R. Kumar, A. K. Pandey, M. Burow, H. E. Laza, S. N. Nayak, et al. "Climate-Smart Groundnuts for Achieving High Productivity and Improved Quality: Current Status, Challenges, and Opportunities." In *Genomic Designing of Climate-Smart Oilseed Crops*, edited by Chittaranjan Kole, 133–172. Cham: Springer International Publishing, 2019. https://doi.org/10.1007/978-3-319-93536-2_5.

Gill, S. S., and N. Tuteja. "Reactive Oxygen Species and Antioxidant Machinery in Abiotic Stress Tolerance in Crop Plants." *Plant Physiology and Biochemistry* 48 (2010): 909–930. http://dx.doi.org/10.1016/j.plaphy.2010.08.016.

Gonzalez-Rizzo, S., M. Crespi, and F. Frugier. "The *Medicago truncatula* CRE1 Cytokinin Receptor Regulates Lateral Root Development and Early Symbiotic Interaction with *Sinorhizobium meliloti*." *The Plant Cell* 18, no. 10 (2006): 2680–2693. https://doi.org/10.1105/tpc.106.042184.

Guha, S., M. Sarkar, P. Ganguly, R. Uddin, S. Mandal, and M. DasGupta. "Segregation of Nod-Containing and Nod-Deficient *Bradyrhizobia* as Endosymbionts of *Arachis hypogaea* and as Endophytes of *Oryza sativa* in Intercropped Fields of Bengal Basin, India." *Environmental Microbiology* 18, no. 12 (2016): 2575–2590. https://doi.org/10.1111/1462-2920.13426.

Hao, K., F. Wang, X. Nong, M. R. McNeill, S. Liu, G. Wang, G. Cao, and Z. Zhang. "Response of Peanut *Arachis hypogaea* Roots to the Presence of Beneficial and Pathogenic Fungi by Transcriptome Analysis." *Scientific Reports* 7, no. 1 (2017): 964. https://doi.org/10.1038/s41598-017-01029-3.

Held, M., H. Hou, M. Miri, C. Huynh, L. Ross, S. Hossain, S. Sato, S. Tabata, J. Perry, and T. L. Wang. "*Lotus japonicus* Cytokinin Receptors Work Partially Redundantly to Mediate Nodule Formation." *Plant Cell* 26, no. 2 (2014): 678–694. https://doi.org/10.1105/tpc.113.122675.

Huang, R., D. Xiao, X. Wang, J. Zhan, A. Q. Wang, and L. F. He. "Genome-Wide Identification, Evolutionary and Expression Analyses of LEA Gene Family in Peanut (*Arachis hypogaea* L.)." *BMC Plant Biology* 22, no. 1 (2022): 1–15. https://doi.org/10.1186/s12870-022-03462-7.

Ibañez, F., J. Angelini, M. S. Figueredo, V. Muñoz, M. L. Tonelli, and A. Fabra. "Sequence and Expression Analysis of Putative *Arachis hypogaea* (Peanut) Nod Factor Perception Proteins." *Journal of Plant Research* 128, no. 5 (2015): 709–718. https://doi.org/10.1007/s10265-015-0776-7.

Ibáñez, F., and A. Fabra. "Rhizobial Nod Factors Are Required for Cortical Cell Division in the Nodule Morphogenetic Programme of the Aeschynomeneae Legume *Arachis*." *Plant Biology* 13, no. 6 (2011): 794–800. https://doi.org/10.1111/j.1438-8677.2011.00468.x.

Inoue, T., K. Takei, H. Sakakibara, and N. Sugiyama. "Identification of Genes Expressed in Response to Cytokinins in *Arabidopsis thaliana*." *Gene* 279, no. 1 (2001): 43–50. https://doi.org/10.1016/S0378-1119(01)00679-2

Jannat, S., M. U. Hassan, G. Ortiz, M. K. Nawaz Shah, M. Ahmed, A. Hussain Shah, A. Qayyum, and A. Qayyum. "Genetic Characterization of Flowering and Phytochrome Genes in Peanut (*Arachis hypogaea* L.) for Early Maturity." *Molecular Biology Reports* 49, no. 10 (2022): 1–10. https://doi.org/10.1007/s11033-022-07362-9.

Jin, F. J., B. T. Wang, Z. D. Wang, L. Jin, and P. Han. "CRISPR/Cas9-Based Genome Editing and Its Application in Aspergillus Species." *Journal of Fungi* 8, no. 5 (2022): 467. https://doi.org/10.3390/jof8050467.

Jin, H., X. Tang, M. Xing, H. Zhu, J. Sui, C. Cai, and S. Li. "Molecular and Transcriptional Characterization of Phosphatidyl Ethanolamine-Binding Proteins in Wild Peanuts *Arachis duranensis* and *Arachis ipaensis*." *BMC Plant Biology* 19, no. 1 (2019): 1–14. https://doi.org/10.1186/s12870-019-2113-3.

Jonak, C., L. Okrész, L. Bögre, and H. Hirt. "Complexity, Cross Talk and Integration of Plant MAP Kinase Signalling." *Current Opinion in Plant Biology* 5, no. 5 (2002): 415–424.

Karmakar, K., A. Kundu, A. Z. Rizvi, E. Dubois, D. Severac, P. Czernic, F. Cartieaux, and M. DasGupta. "Transcriptomic Analysis with the Progress of Symbiosis in 'Crack-Entry' Legume *Arachis hypogaea* Highlights Its Contrast with 'Infection Thread' Adapted Legumes." *Molecular Plant-Microbe Interactions* 32, no. 3 (2019): 271–285. https://doi.org/10.1094/MPMI-11-18-0347-R.

Kundu, A., and M. DasGupta. "Silencing of Putative Cytokinin Receptor Histidine Kinase1 Inhibits Both Inception and Differentiation of Root Nodules in *Arachis hypogaea*." *Molecular Plant-Microbe Interactions* 31, no. 2 (2017): 187–199. https://doi.org/10.1094/MPMI-05-16-0105-R.

Li, C., K. Wu, G. Fu, Y. Li, Y. Zhong, X. Lin, Y. Zhou, L. Tian, and S. Huang. "Regulation of Oleosin Expression in Developing Peanut (*Arachis hypogaea* L.) Embryos through Nucleosome Loss and Histone Modifications." *Journal of Experimental Botany* 60, no. 11 (2009): 3173–3184. https://doi.org/10.1093/jxb/erp275.

Li, L., X. Cheng, X. Kong, P. Jia, X. Wang, L. Zhang, X. Zhang, Y. Zhang, Z. Zhang, and B. Zhang. "Comparative Transcriptomic Analysis Reveals the Negative Response Mechanism of Peanut Root Morphology and Nitrate Assimilation to Nitrogen Deficiency." *Plants* 12, no. 4 (2023): 732. https://doi.org/10.3390/plants12040732.

Li, P., Q. Ma, C. Qu, S. Zhu, K. Zhao, X. Ma, ... and D. Yin. "Genome-Wide Identification and Expression Analysis of Auxin Response Factors in Peanut (*Arachis hypogaea* L.)." *PeerJ* 9 (2021): e12319. https://doi.org/10.7717/peerj.12319.

Li, Y., M. Fu, J. Li, J. Wu, Z. Shua, T. Chen, W. Yao, and D. Huai. "Genome-Wide Identification of SWEET Genes Reveals Their Roles during Seed Development in Peanuts." *BMC Genomics* 25, no. 1 (2024): 1–14. https://doi.org/10.1186/s12864-024-10173-w.

Li, Y., J. J. Meng, Y. Sha, F. Guo, J. Zhang, Y. Geng, L. Cui, S. Wan, and X. Li. "Transcriptome Analysis of Calcium- and Hormone-Related Gene Expressions during Different

Stages of Peanut Pod Development." *Frontiers in Plant Science* 8 (2017): 1241. https://doi.org/10.3389/fpls.2017.01241.

Liu, H., X. Liang, Q. Lu, H. Li, H. Liu, S. Li, et al. "Global Transcriptome Analysis of Subterranean Pod and Seed in Peanut (*Arachis hypogaea* L.) Unravels the Complexity of Fruit Development under Dark Conditions." *Scientific Reports* 10, no. 1 (2020): 13050. https://doi.org/10.1038/s41598-020-69713-3.

Lucchesi, J. C. "Chromatin Modification and Remodeling." In *Oxford Scholarship Online*, 2019. https://doi.org/10.1093/oso/9780198831204.003.0004.

Luo, J., J. J. Zhou, and J. Z. Zhang. "Aux/IAA Gene Family in Plants: Molecular Structure, Regulation, and Function." *International Journal of Molecular Sciences* 19, no. 1 (2018): 259. https://doi.org/10.3390/ijms19010259.

Luo, L., Q. Wan, Z. Yu, K. Zhang, X. Zhang, S. Zhu, Y. Wan, Z. Ding, and F. Liu. "Genome-Wide Identification of Auxin Response Factors in Peanut (*Arachis hypogaea* L.) and Functional Analysis in Root Morphology." *International Journal of Molecular Sciences* 23, no. 10 (2022): 5309. https://doi.org/10.3390/ijms23105309.

Lv, Z., D. Zhou, X. Shi, J. Ren, H. Zhang, C. Zhong, ... and C. Wang. "The Determination of Peanut (*Arachis hypogaea* L.) Pod-Sizes during the Rapid-Growth Stage by Phytohormones." *BMC Plant Biology* 23, no. 1 (2023): 371. https://doi.org/10.1186/s12870-023-04298-5.

Madsen, E. B., M. Antolín-Llovera, C. Grossmann, J. Ye, S. Vieweg, A. Broghammer, L. Krusell, S. Radutoiu, O. N. Jensen, J. Stougaard, and M. Parniske. "Autophosphorylation Is Essential for the In Vivo Function of the *Lotus japonicus* Nod Factor Receptor 1 and Receptor-Mediated Signalling in Cooperation with Nod Factor Receptor 5." *The Plant Journal* 65, no. 3 (2011): 404–417. https://doi.org/10.1111/j.1365-313X.2010.04432.x.

Mashiguchi, K., K. Tanaka, T. Sakai, S. Sugawara, H. Kawaide, M. Natsume, et al. "The Main Auxin Biosynthesis Pathway in *Arabidopsis*." *Proceedings of the National Academy of Sciences USA* 108, no. 45 (2011): 18512–18517. https://doi.org/10.1073/pnas.1108355108.

Mishra, N. S., R. Tuteja, and N. Tuteja. "Signaling through MAP Kinase Networks in Plants." *Archives of Biochemistry and Biophysics* 452, no. 1 (2006): 55–68.

Mittler, R. "Oxidative Stress, Antioxidants and Stress Tolerance." *Trends in Plant Science* 7, no. 9 (2002): 405–410.

Nakagami, H., H. Soukupova, A. Schikora, V. Zarsky, and H. Hirt. "A Mitogen-Activated Protein Kinase Kinase Kinase Mediates Reactive Oxygen Species Homeostasis in *Arabidopsis*." *Journal of Biological Chemistry* 281, no. 50 (2006): 38697–38704.

Nian, L., B. Wei, K. P. Manish, L. Huang, H. H. Luo, Y. Chen, X. Zhou, W. G. Chen, D. X. Huai, B. L. Yu, H. Chen, J. B. Guo, Y. Lei, B. S. Liao, R. K. Varshney, and H. F. Jiang. "Gene Expression and DNA Methylation Altering Lead to the High Oil Content in Wild Allotetraploid Peanut (A. monticola)." *Frontiers in Plant Science* 13 (2022): 1065267. https://doi.org/10.3389/fpls.2022.1065267.

Noisangiam, R., K. Teamtisong, P. Tittabutr, N. Boonkerd, T. Ueda, K. Minamisawa, and N. Teaumroong. "Genetic Diversity, Symbiotic Evolution, and Proposed Infection Process of Bradyrhizobium Strains Isolated from Root Nodules of *Aeschynomene americana* L. in Thailand." *Applied and Environmental Microbiology* 78, no. 17 (2012): 6236–6250. https://doi.org/10.1128/AEM.01187-12.

Oldroyd, G. E., E. M. Engstrom, and S. R. Long. "Ethylene Inhibits the Nod Factor Signal Transduction Pathway of *Medicago truncatula*." *The Plant Cell* 13, no. 8 (2001): 1835–1849. https://doi.org/10.1105/tpc.13.8.1835.

Ouyang, L., Y. Liu, R. Yao, D. He, L. Yan, Y. Chen, D. Huai, Z. Wang, B. Yu, Y. Kang, H. Jiang, Y. Lei, B. Liao, and X. Wang. "Genome-Wide Analysis of UDP-Glycosyltransferase Gene Family and Identification of a Flavonoid 7-O-UGT (AhUGT75A) Enhancing Abiotic Stress in Peanut (*Arachis hypogaea* L.)." *BMC Plant Biology* 23, no. 1 (2023): 1–16. https://doi.org/10.1186/s12870-023-04656-3.

Patel, K., K. R. Sahoo, S. R. Mondal, and M. L. Sharma. "Molecular Characterization and Functional Analysis of the ABA Receptor Gene Family in Peanut (*Arachis hypogaea* L.)." *BMC Plant Biology* 24, no. 1 (2024): 1–13. https://doi.org/10.1186/s12870-024-04635-2.

Peng, Z., F. Liu, L. Wang, H. Zhou, D. Paudel, L. Tan, J. Maku, M. Gallo, and J. Wang. "Transcriptome Profiles Reveal Gene Regulation of Peanut (*Arachis hypogaea* L.) Nodulation." *Scientific Reports* 7, no. 1 (2017): 40066. https://doi.org/10.1038/srep40066.

Perilli, S., J. M. Pérez-Pérez, R. Di Mambro, C. L. Peris, S. Díaz-Trivino, M. Del Bianco, et al. "Retinoblastoma-Related Protein Stimulates Cell Differentiation in the Arabidopsis Root Meristem by Interacting with Cytokinin Signaling." *Plant Cell* 25, no. 11 (2013): 4469–4478. https://doi.org/10.1105/tpc.113.118474.

Pirayesh, N., M. Giridhar, A. Ben Khedher, U. C. Vothknecht, and F. Chigri. "Organellar Calcium Signaling in Plants: An Update." *Biochimica et Biophysica Acta (BBA) - Molecular Cell Research* 1868, no. 4 (2021): 118948. https://doi.org/10.1016/j.bbamcr.2021.118948.

Ramesh, S. B., J. Rockey, K. Shirasawa, M. P. Tilak, B. V. Patil, and V. B. Lachagari. "DNA Methylation and Expression Analyses Reveal Epialleles for the Foliar Disease Resistance Genes in Peanut (*Arachis hypogaea* L.)." *BMC Research Notes* 13, no. 1 (2020): 1–8. https://doi.org/10.1186/s13104-020-4883-y.

Reinecke, D. M., A. D. Wickramarathna, J. A. Ozga, L. V. Kurepin, A. L. Jin, A. G. Good, et al. "Gibberellin 3-Oxidase Gene Expression Patterns Influence Gibberellin Biosynthesis, Growth, and Development in Pea." *Plant Physiology* 163, no. 2 (2013): 929–945. https://doi.org/10.1104/pp.113.221113.

Rouhi, H. R., and A. Sepehri. "Hormonal Priming to Overcome Drought Stress and Aging Damage in Groundnut Seed (*Arachis hypogaea* L.)." *International Journal of Plant Production* 14, no. 2 (2020): 1–10. https://doi.org/10.22034/ijpp.2020.672575.

Saha, S., A. Dutta, A. Bhattacharya, and M. Dasgupta. "Gatekeeper Tyrosine Phosphorylation of SYMRK Is Essential for Synchronizing the Epidermal and Cortical Responses in Root Nodule Symbiosis." *Plant Physiology* 171, no. 1 (2016): 71–81. https://doi.org/10.1104/pp.16.00437.

Saha, S., A. Dutta, A. Bhattacharya, S. Samaddar, M. B. Goshe, and M. Das Gupta. "Autophosphorylation of Gatekeeper Tyrosine by Symbiosis Receptor Kinase." *FEBS Letters* 587, no. 18 (2014): 2972–2979. https://doi.org/10.1016/j.febslet.2013.08.023.

Samaddar, S., A. Dutta, S. Sinharoy, A. Paul, A. Bhattacharya, S. Saha, K. Y. Chien, M. B. Goshe, and M. Dasgupta. "Autophosphorylation of Gatekeeper Tyrosine by Symbiosis

Receptor Kinase." *FEBS Letters* 587, no. 19 (2013): 2972–2979. https://doi.org/10.1016/j.febslet.2013.07.030.

Sharma, M., D. Singh, H. Bhat, P. Saksena, M. Sharma, A. Tiwari, P. Awasthi, K. B. Botta, B. N. Shukla, and A. Laxmi. "Understanding the Intricate Web of Phytohormone Signalling in Modulating Root System Architecture." *International Journal of Molecular Sciences* 22, no. 11 (2021): 5508. https://doi.org/10.3390/ijms22115508.

Singla, P., and S. Kaur. "Insights into Pivotal Role of Phytohormonal Cross Talk in Tailoring Underground Plant Root System Architecture." In *Phytohormones and Stress Responsive Secondary Metabolites*, edited by S. Kaur and P. Singla, 1–22. Springer, 2018. https://doi.org/10.1007/978-3-319-75910-4_1.

Sinha, A. K., M. Jaggi, B. Raghuram, and N. Tuteja. "Mitogen-Activated Protein Kinase Signaling in Plants under Abiotic Stress." *Plant Signaling and Behavior* 6, no. 2 (2011): 196–203. https://doi.org/10.4161/psb.6.2.14701.

Sinharoy, S., and M. DasGupta. "RNA Interference Highlights the Role of CCaMK in Dissemination of Endosymbionts in the Aeschynomeneae Legume *Arachis*." *Molecular Plant-Microbe Interactions* 22, no. 11 (2009): 1466–1475. https://doi.org/10.1094/MPMI-22-11-1466.

Spaink, H. P. "Root Nodulation and Infection Factors Produced by Rhizobial Bacteria." *Annual Review of Microbiology* 54 (2000): 257–288. https://doi.org/10.1146/annurev.micro.54.1.257.

Stracke, S., C. Kistner, S. Yoshida, L. Mulder, S. Sato, T. Kaneko, et al. "A Plant Receptor-Like Kinase Required for Both Bacterial and Fungal Symbiosis." *Nature* 417, no. 6887 (2002): 959–962. https://doi.org/10.1038/nature00882.

Su, L., S. Liu, X. Liu, B. Zhang, M. Li, L. Zeng, and L. Li. "Transcriptome Profiling Reveals Histone Deacetylase 1 Gene Overexpression Improves Flavonoid, Isoflavonoid, and Phenylpropanoid Metabolism in *Arachis hypogaea* Hairy Roots." *PeerJ* 9 (2021): e10976. https://doi.org/10.7717/peerj.10976.

Sun, J., X. Zhang, C. Fu, N. Ahmad, C. Zhao, L. Hou, M. Naeem, J. Pan, X. Wang, and S. Zhao. "Genome-Wide Identification and Expression Analysis of GA20ox and GA3ox Genes during Pod Development in Peanut." *PeerJ* 11 (2023): e16279. https://doi.org/10.7717/peerj.16279.

Sun, X., L. Zhang, W. Xu, J. Zheng, M. Yan, M. Zhao, X. Wang, and Y. Yin. "A Comprehensive Analysis of the Peanut SQUAMOSA Promoter Binding Protein-Like Gene Family and How AhSPL5 Enhances Salt Tolerance in Transgenic Arabidopsis." *Plants* 13, no. 8 (2024): 1057. https://doi.org/10.3390/plants13081057.

Takei, K., M. Taniguchi, and N. Sugiyama. "Regulatory Role of Cytokinin Oxidase/Dehydrogenase in Cytokinin Homeostasis." *Journal of Biological Chemistry* 279, no. 39 (2004): 40935–40943. https://doi.org/10.1074/jbc.M406750200.

Tena, G., T. Asai, W. L. Chiu, and J. Sheen. "Plant Mitogen-Activated Protein Kinase Signaling Cascades." *Current Opinion in Plant Biology* 4, no. 5 (2001): 392–400.

Wan, L., B. Li, Y. Lei, L. Yan, X. Ren, Y. Chen, et al. "Mutant Transcriptome Sequencing Provides Insights into Pod Development in Peanut (*Arachis hypogaea* L.)." *Frontiers in Plant Science* 8 (2017): 1900. https://doi.org/10.3389/fpls.2017.01900.

Wang, L., H. Chen, Y. Zhuang, K. Chen, C. Zhang, T. Cai, et al. "Multiple Strategies, Including 6mA Methylation, Affecting Plant Alternative Splicing in Allopolyploid Peanut." *Plant Biotechnology Journal* 22, no. 6 (2024): 1681–1702.

Wang, P., C. Gao, X. Bian, S. Zhao, C. Zhao, H. Xia, H. Song, L. Hou, S. Wan, and X. Wang. "Genome-Wide Identification and Comparative Analysis of Cytosine-5 DNA Methyltransferase and Demethylase Families in Wild and Cultivated Peanut." *Frontiers in Plant Science* 7 (2016): 7. https://doi.org/10.3389/fpls.2016.00007.

Wang, P., S. Shi, J. Ma, H. Song, Y. Zhang, C. Gao, C. Zhao, S. Zhao, L. Hou, J. Lopez-Baltazar, S. Fan, H. Xia, X. Wang, and X. Wang. "Global Methylome and Gene Expression Analysis During Early Peanut Pod Development." *BMC Plant Biology* 18, no. 1 (2018): 1–14. https://doi.org/10.1186/s12870-018-1546-4.

Wang, Q., X. Zhao, Q. Sun, Y. Mou, J. Wang, C. Yan, C. Yuan, C. Li, and S. Shan. "Genome-Wide Identification of the LRR-RLK Gene Family in Peanut and Functional Characterization of AhLRR-RLK265 in Salt and Drought Stresses." *International Journal of Biological Macromolecules* 254 (2024): 127829. https://doi.org/10.1016/j.ijbiomac.2023.127829.

Wang, X., Y. Guo, and W. Zhu. "Molecular Mechanisms of Peanut Resistance to Biotic and Abiotic Stresses." *Frontiers in Plant Science* 10 (2019): 1–12. https://doi.org/10.3389/fpls.2019.00001.

Wang, Y., M. Zhang, P. Du, H. Liu, Z. Zhang, J. Xu, et al. "Transcriptome Analysis of Pod Mutant Reveals Plant Hormones Are Important Regulators in Controlling Pod Size in Peanut (*Arachis hypogaea* L.)." *PeerJ* 10 (2022): e12965. https://doi.org/10.7717/peerj.12965.

Wang, Z., and X. Gou. "Receptor-Like Protein Kinases Function Upstream of MAPKs in Regulating Plant Development." *International Journal of Molecular Sciences* 21, no. 20 (2020): 7638. https://doi.org/10.3390/ijms21207638.

Won, C., X. Shen, K. Mashiguchi, Z. Zheng, X. Dai, Y. Cheng, et al. "Conversion of Tryptophan to Indole-3-Acetic Acid by Tryptophan Aminotransferases of Arabidopsis and YUCCAs in Arabidopsis." *Proceedings of the National Academy of Sciences USA* 108, no. 45 (2011): 18518–18523. https://doi.org/10.1073/pnas.1108436108.

Wu, Y., Z. Sun, F. Qi, M. Tian, J. Wang, R. Zhao, X. Wang, X. Wu, X. Shi, H. Liu, W. Dong, B. Huang, Z. Zheng, and X. Zhang. "Comparative Transcriptomics Analysis of Developing Peanut (*Arachis hypogaea* L.) Pods Reveals Candidate Genes Affecting Peanut Seed Size." *Frontiers in Plant Science* 13 (2022): 958808. https://doi.org/10.3389/fpls.2022.958808.

Yan, L., H. Jin, A. Raza, Y. Huang, D. Gu, and X. Zou. "WRKY Genes Provide Novel Insights into Their Role Against *Ralstonia solanacearum* Infection in Cultivated Peanut (*Arachis hypogaea* L.)." *Frontiers in Plant Science* 13 (2022): 986673. https://doi.org/10.3389/fpls.2022.986673.

Yang, D., T. Chen, Y. Wu, H. Tang, J. Yu, X. Dai, Y. Zheng, X. Wan, Y. Yang, and X. Tan. "Genome-Wide Analysis of the Peanut CaM/CML Gene Family Reveals That the AhCML69 Gene Is Associated with Resistance to *Ralstonia solanacearum*." *BMC Genomics* 25, no. 1 (2024): 1–14. https://doi.org/10.1186/s12864-024-10108-5.

Yang, P., F. Zhang, X. Luo, Y. Zhou, and J. K. Xie. "Histone Deacetylation Modification Participates in the Repression of Peanut (*Arachis hypogaea* L.) Seed Storage Protein Gene Ara h 2.02 During Germination." *Plant Biology* 17, no. 4 (2015): 782–789. https://doi.org/10.1111/plb.12268.

Yang, S., J. Wang, Z. Tang, F. Guo, Y. Zhang, J. Zhang, et al. "Transcriptome of Peanut Kernel and Shell Reveals the Mechanism of Calcium on Peanut Pod Development." *Scientific Reports* 10, no. 1 (2020): 15723. https://doi.org/10.1038/s41598-020-72379-6.

Zhang, J. C., N. B. Cai, X. W. Zhang, and W. J. Zhuang. "Isolation and Identification of Specific Expressed Proteins from Peanut (*Arachis hypogaea*) Development/Abortion Embryo Mediated by Ca^{2+}." *Acta Agronomica Sinica* 33 (2007): 814–819. https://doi.org/10.3724/SP.J.1006.2007.00814.

Zhang, X., Y. Zhou, Y. Liu, B. Li, S. Tian, and Z. Zhang. "Research Progress on the Mechanism and Function of Histone Acetylation Regulating the Interaction Between Pathogenic Fungi and Plant Hosts." *Journal of Fungi* 10, no. 8 (2024): 522. https://doi.org/10.3390/jof10080522.

Zhang, Y., P. Wang, H. Xia, C. Zhao, L. Hou, C. Li, et al. "Comparative Transcriptome Analysis of Basal and Zygote-Located Tip Regions of Peanut Ovaries Provides Insight into the Mechanism of Light Regulation in Peanut Embryo and Pod Development." *BMC Genomics* 17, no. 1 (2016): 606. https://doi.org/10.1186/s12864-016-2928-7.

Zhao, S., J. Sun, J. Sun, X. Zhang, C. Zhao, J. Pan, L. Hou, R. Tian, and X. Wang. "Insights into the Novel FAD2 Gene Regulating Oleic Acid Accumulation in Peanut Seeds with Different Maturity." *Genes* 13, no. 11 (2022): 2076. https://doi.org/10.3390/genes13112076.

Zhao, X., L. Gan, C. Yan, C. Li, Q. Sun, J. Wang, C. Yuan, H. Zhang, S. Shan, and J. N. Liu. "Genome-Wide Identification and Characterization of Long Non-Coding RNAs in Peanut." *Genes* 10, no. 7 (2019): 536. https://doi.org/10.3390/genes10070536.

Zhao, Y. "Auxin Biosynthesis and Its Role in Plant Development." *Annual Review of Plant Biology* 61 (2010): 49–64. https://doi.org/10.1146/annurev-arplant-042809-112226.

Zhao, Y. "Auxin Biosynthesis: A Simple Two-Step Pathway Converts Tryptophan to Indole-3-Acetic Acid in Plants." *Molecular Plant* 5, no. 2 (2012): 334–338. https://doi.org/10.1093/mp/ssr098.

Zhong, C., Z. Li, Y. Cheng, H. Zhang, Y. Liu, X. Wang, C. Jiang, X. Zhao, S. Zhao, J. Wang, H. Zhang, X. Liu, and H. Yu. "Comparative Genomic and Expression Analysis Insight into Evolutionary Characteristics of PEBP Genes in Cultivated Peanuts and Their Roles in Floral Induction." *International Journal of Molecular Sciences* 23, no. 20 (2022): 12429. https://doi.org/10.3390/ijms232012429.

Zhu, J. K. "Plant Salt Tolerance." *Trends in Plant Science* 6, no. 2 (2001): 66–71. https://doi.org/10.1016/S1360-1385(00)01837-6.

10 Genomics-Assisted Breeding in Peanut

Recent Advances and Future Perceptions

Chuanzhi Zhao and Xingjun Wang

10.1 INTRODUCTION

Improved varieties play an indispensable part in supporting the sustainable development of agriculture and food demands globally. In general, breeding technology development has experienced three main stages: original domestication and breeding, traditional breeding, and molecular breeding (or biotechnology breeding). In a broad sense, molecular breeding may be defined as improving target traits of interest in plants and animals by using of genetic manipulation at DNA molecular levels, including genetic engineering or gene manipulation, molecular marker-assisted selection, genomic selection, and so on. More often, molecular breeding only refers to molecular marker-assisted breeding, which is defined as the application of molecular markers combined with linkage maps and genomics to alter and improve plant or animal traits on the basis of genotypic assays (Jiang, 2013).

Molecular breeding is a new breeding method that combines traditional cross breeding, molecular biology, high-throughput sequencing, gene chip, and bioinformatics. It is the best choice for crop breeding in the future. Its precision and efficiency will lead crop breeding into a new era. For molecular breeding, there are several other definitions and upgraded versions, including marker-assisted selection (MAS), genetically modified organism breeding (GMOB), gene editing breeding (GEB), molecular module breeding (MMB), molecular design breeding (MDB), intelligent breeding (IB), and genome breeding or genome-assisted breeding or genomics-assisted breeding (GAB).

MAS, GMOB, and GEB focus on single gene or few quality traits or primary quantitative traits (QTLs); however, many important agronomic traits of crops are controlled by multiple quantitative loci, and thus it is usually hard to combine multiple complex traits because they have only a certain degree of correlation. MMB, MDB, IB, and GAG are updated MAS approaches that integrate genome resources, phenotyping data, and QTL/gene information. These approaches aim to breed superior varieties or lines with enhanced biotic or abiotic stress tolerance and improved yield and quality. They can aggregate a certain quantitative trait or multiple agronomic traits, providing alternatives to current challenges (Varshney et al., 2013). For example, using the MMB strategy, researchers successfully pyramided 15 favorable genes from five parent lines to develop a new elite rice variety, Molecular Design 1 (MD1). MD1 exhibited higher yield potential, superior quality, the desired grain shape, and high resistance to the brown plant hopper, demonstrating that MMB is a powerful strategy to improve multiple complex traits and provides a reference for future commercial breeding (Liu et al., 2022).

10.2 MARKER-ASSISTED SELECTION (MAS)

10.2.1 Definition and Development of MAS

Traditional breeding strategy relies on the selection of phenotypes, which requires rich experience and a long selection period. In addition, many conditions limit the direct identification of some special traits, so improving the efficiency of selection and reducing the uncertainty in the breeding process are key to advancing breeding efforts. With the development of genetics, people have long been aware of the use of easily distinguishable genetic markers to assist selection. Molecular marker–assisted breeding uses the closely linked characteristics of molecular markers and genes to track and select target traits. Detecting molecular markers allows researchers to identify the existence of target genes, enabling the selection of desired traits. This approach is fast, accurate, and free from interference by environmental conditions. It can be used as an auxiliary means to identify parental relationship, transfer of quantitative and recessive traits in backcross breeding, selection of hybrid offspring, prediction of heterosis, and identification of variety purity.

In the early stages, most of the markers used in plant genetics and breeding were based on morphological, cytological, and protein markers. These markers are not suitable for direct selection, and their numbers are limited. For example, researchers have developed and used isoenzymes as near-neutral genetic markers; however, different isoenzymes often require analysis through different detection procedures. Additionally, the expression of some isoenzymes is tissue- and development-specific. Moreover, the biggest disadvantage of isoenzyme markers is their limited number and low polymorphism in most breeding populations (Tanksley and Orton, 1983).

DNA sequence is the carrier of genetic information and its variation is the basis of genetic diversity of species. Although DNA can copy itself precisely in the process of transmission of genetic information, many natural and human factors can lead to DNA sequence variation in the

DOI: 10.1201/9781003360988-10

form of single nucleotide changes and insertion or deletion of DNA fragments, accumulating rich DNA sequence polymorphisms. DNA molecular markers are the direct reflection of genetic polymorphisms at the DNA level, reflecting the nucleotide sequence differences among individuals. Compared to morphological, cytological, and protein markers, DNA molecular markers have the following advantages: most of them have no effect on phenotypes, they are co-dominant, and it is convenient to select recessive traits. Researchers can use DNA from different tissues at various stages of development for marker analysis, enabling early selection of plant genotypes (Fowler et al., 1994; Davis and DeNise, 1998). Therefore, the emergence of DNA molecular markers has aroused the interest of breeders.

There are two important conditions for using DNA molecular marker-assisted selection, one is the molecular marker closely linked to the target trait, and the second is that the molecular marker has polymorphism in breeding population. In 1980, Botstein first developed the restrictive fragment length polymorphism (RFLP) marker (Botstein et al., 1980). Over the past 40 years, with the continuous development of PCR technology and genomics, especially the rapid development of functional genomics in the past 10 years, different types of molecular markers have been developed, including RAPD, SSR, ISSR, AFLP, SCAR, SNP, RGAP, DArT, SRAP, CAPS, and KASP. Chapter 7 of this book provides detailed information on the principles, history, development, and application of these DNA markers.

10.2.2 Single or Few Genes/Traits Selection

Marker-assisted selection (MAS) is one of the simplest and most widely used molecular breeding methods, especially for single trait/gene screening, which can efficiently track and screen the target genes that confer the desired traits such as disease resistance, high yield, and high quality from populations (Dekkers and Hospital, 2002). Involving screening and tracking of a single gene or few genes and their transfer into a specific line, MAS is now widely used. For example, using polymorphic markers, eight lines containing a gene resistant to bacterial leaf blight were quickly obtained from a large number of rice hybrids (Davierwala et al., 2001).

Researchers can combine MAS with backcross selection to transfer target genes or traits into recurrent parents. Backcrossing, a common crop breeding method, involves using parents with excellent overall traits, though they may have disadvantage in a few traits. To address this, breeders integrate one or more genes into the parent, known as the recurrent parent. Plant breeders have widely used backcrossing method for nearly a century (Collard and Mackill, 2008). The use of molecular markers significantly increases the efficiency of backcross selection. In general, breeders can apply three levels of marker-assisted backcrossing (MAB): foreground selection, recombinant selection, and background selection.

Foreground selection refers to screening the target gene or QTLs using molecular markers from the offspring. Breeders can use marker-assisted selection for reproductive-stage traits in the seedling stage, allowing the removal of plants without the desired gene/QTL early on. This approach is useful for screening traits that are laborious or time-consuming to assess using phenotypic screening procedures and for traits that are susceptible to environmental factors. Additionally, it effectively selects recessive alleles, which are difficult to distinguish using conventional methods (Zhou et al., 2003).

Foreground selection has been widely used in peanut molecular breeding programs. For example, *ahFAD2A* and *ahFAD2B*, located on chromosomes A09 and B09, respectively, primarily control the conversion of oleic acid to linoleic acid. The simultaneous mutation of these two genes resulted in the accumulation of high levels of oleic acid in peanut (Jung et al., 2000, Barkley et al., 2013). At present, the high oleic acid mutant *F435*, widely used in peanut breeding programs, is partially or completely inactivated by the ORF-448 bp G/A replacement of *ahFAD2A* and the "A" insertion of ORF-441_442 of *ahFAD2B* (Jung et al., 2000). To accurately screen the target genes (*ahFAD2A* and *ahFAD2B*), CAPS, allele-specific (AS) -PCR and Kompetitive allele specific PCR (KASP) markers–based methods were developed (Barkley Noelle et al., 2010). Since 2013, with funding from MARS and other projects, several teams in China have started to develop high-oleic acid peanut varieties through the marker-assisted backcrossing strategy. Up to now, more than 300 new commercial peanut cultivars with high oleic acid have been registered (Pan et al., 2019; Tang, Qiu et al., 2022).

In addition, breeders also used foreground selection in breeding peanuts for disease resistance. Through a marker-assisted backcrossing approach, a new resistant QTL derived from wild peanut *Arachis stenosperma* has been successfully incorporated into cultivated peanut, which represented an important step toward the development of high-yielding and nematode-resistant peanut cultivars (Ballen-Taborda et al., 2022). Moreover, marker-assisted introgression of wild chromosome segments has also conferred resistance to fungal foliar diseases in peanut (Moretzsohn et al., 2023; Rajarathinam et al., 2023).

Recombinant selection involves selecting BC progeny with the target gene and recombination events between the target loci and linked flanking markers. The transfer of the target gene from the donor parent may bring along many undesirable genes into the recurrent parent. These undesirable genes (or linkage drag) might negatively affect crop performance. It is necessary to reduce the size of the donor chromosome segment containing the target locus to break the linkage drag and generate novel allele combinations (Young and Tanksley, 1989). Using traditional breeding methods, the donor segment can remain very large, and it is difficult to remove the linkage drag even in many BC generations. By using molecular markers that are closely linked to target gene (e.g. less than 5cM on either side), the linkage drag can be minimized (Parkin and Lydiate, 1997; Frisch and Melchinger, 2001; Hospital, 2001). Kunitz trypsin inhibitor (KTI)-free soybean is crucial for the soy-food industry. Researchers performed molecular marker-assisted recombinant selection to eliminate KTI from two popular soybean genotypes, DS9712 and DS9814, which led to the identification of plants with minimal linkage drag and the rapid development of six KTI-free breeding lines. This result suggested that marker-assisted recombinant selection is an effect approach for introgression of the target gene/QTL with minimal linkage drag (Kumar et al., 2015; Maranna et al., 2016).

Background selection refers to selecting backcross generations with the greatest proportion of recurrent parent genome using markers that are unlinked to the target locus. In general, researchers often combine background selection and foreground selection. They use tightly linked flanker markers for recombination selection, while employing non-linked markers for recurrent parental selection (Hospital, 2001; Hiebert et al., 2007; Neeraja et al., 2007; Ribaut and Ragot, 2007). The markers for background selection do not relate to the target gene or QTL. In other words, researchers use these markers to select loci that characterize the recurrent parent. In wheat, researchers employed marker-assisted background selection and foreground selection to transfer four drought-tolerant QTLs into a popular high-yielding, drought-sensitive variety, HD2733. This process resulted in 23 improved lines carrying combinations of the four drought-tolerant QTLs, with background recovery ranging from 85.35% to 95.79%. Among these lines, they finally developed and released a promising genotype, HD3411, after multi-location coordinating trials (Kumar et al., 2023). In rice, through marker-based foreground and background selection, four Pi genes (*Piz*, *Pib*, *Pita*, and *Pik*) were introgressed into a renowned japonica rice variety, with 95.65% of the recurrent genome being recovered (Zampieri et al., 2023).

In peanut, researchers applied foreground selection combined with background selection to increase oleic acid content and enhance resistance to foliage diseases. For example, researchers modified Huayu22, an elite variety in China, to a high oleic acid variety YH61 using backcrossing and KASP markers for FAD2 genes screening. *Axiom_Arachis* 58 K SNP array analysis showed that the average background recovery rate in BC_4F_4 lines was 92.34% (Tang, Qiu et al., 2022). Using a similar strategy, researchers obtained 24 high-oleic-acid lines from the BC_4F_4 and BC_4F_5 populations. All these lines exhibited morphological features and agronomic traits similar to those of the recurrent parents. KASP assay showed that the recovery rate of the genetic background of BC_4F_5 lines was 79.49%–92.31% (Huang et al., 2019). Researchers employed eight SSR markers linked with disease-resistant QTLs for foreground screening to facilitate the transfer of QTLs for late leaf spot and rust resistance from the donor variety GPBD4 using the marker-assisted backcross method. Rajarathinam et al. (2023) utilized a comprehensive set of 217 SSR markers spanning the entire peanut genome for background screening of BC_3F_2 lines.

10.2.3 Multi-Genes/Traits Pyramiding

To obtain new germplasms with superior agronomic traits, breeders need to select multiple genes or loci simultaneously. Multiple genes or loci control many important agronomic traits, necessitating the pyramiding of these genes or traits into a single genotype. In traditional breeding approaches, distinguishing plants containing more than one gene or QTL is challenging. However, using DNA molecular markers, breeders can test multiple specific genes from a single DNA sample, greatly facilitating the selection process.

Many rice genotypes cannot survive under deep standing water and cannot germinate under oxygen-limiting conditions. Researchers identified two QTLs linked to resistance to submergence, Submergence 1 (Sub1), and associated Anaerobic Germination-9-2 (AG1) with anaerobic germination. The QTLs of Sub1 and AG1 were pyramided together in several genotypes, such as Ciherang, IR64, Swarna, NSIC Rc222, and Bg358, using marker-assisted backcrossing, leading to multi-stress-tolerant rice varieties (Sartaj et al., 2016). To increase resistance to rice blast caused by the pathogenic fungus *Magnaporthe oryzae*, researchers used marker-assisted selection to transfer three resistance genes, Pigm, Pi48, and Pi49, into the male sterile line Chuang5S. The gene pyramiding lines (*Pigm* + *Pi48* + *Pi49*) showed a higher level of rice-blast resistance. The newly improved lines, with broad spectrum blast resistance, can be used for variety development (Peng et al., 2023). Using a similar approach, researchers employed a marker-assisted backcrossing breeding strategy to transfer three major bacterial blight resistance genes (xa21, xa13, and xa5) into a rice variety. The three-gene pyramided lines exhibited a significant yield advantage over the control under pathogen infection (Pradhan et al., 2015).

In peanut breeding, researchers have used multi-gene/trait pyramiding for the genetic improvement of high oleic acid content and leaf spot resistance. The selection for mutant alleles of *ahFAD2A* and *ahFAD2B* using molecular markers lead to the creation of many new high oleic acid peanuts (Huang et al., 2019; Tang, Qiu et al., 2022). In addition, molecular screening for the QTLs of late leaf spot and rust combined with background screening resulted in five advanced BC_3F_2 lines with more than 90% recurrent parent genome (Rajarathinam et al., 2023).

10.3 TRANSGENIC BREEDING

Plant genetic transformation methods mainly include agrobacterium-mediated genetic transformation, particle gun bombardment, pollen-mediated gene transfer (flower dipping), and other systems using various vectors (Kamo et al., 1995; Wu et al., 2003). Among them, agrobacterium-mediated genetic transformation is simple to operate, cost-effective, and exhibits a high conversion rate, with successful tests conducted in many plant species. Since 1990, researchers around the world have carried out genetic transformation of peanuts, primarily using the agrobacterium infection method. They have used peanut cotyledons, cotyledon nodules, apical meristems, young leaves, immature embryos, and embryonic calli as explants for transformation (Krishna et al., 2015; Gantait and Mondal, 2018).

In comparison with rice, corn, and soybean, peanut genetic transformation is still very difficult and not very efficient. Even so, research groups from several countries have obtained transgenic peanut plants using agrobacterium-mediated transformation. For example, researchers performed agrobacterium-mediated transformation of peanut cotyledons and immature leaves to overexpress the tobacco streak virus (TSV) coat protein gene in peanut. As a result, researchers obtained six homozygous transgenic plants. Phenotypic identification showed that the transgenic lines remained symptomless, and there was only traces of or no systemic accumulation of virus after inoculation with TSV (Mehta et al., 2013). In peanut, expressing an *Arabidopsis* homeodomain–leucine zipper transcription factor (*AtHDG11*) increased drought and salt tolerance, and increased the yield of peanuts in drought and salt stress conditions (Banavath et al., 2018). In addition, the transgenic peanut line expressing *AtDREB1A* displayed better physio-biochemical characteristics and growth parameters under drought stress over the wild-type plants (Bhalani et al., 2019). The transgenic peanuts expressing synthetic *cry1EC* gene showed 100% killing of larvae at the 2(nd) instar stage of *Spodoptera litura*, which is the first report of transgenic peanut plants with resistance to *S. litura* and provided materials for breeding insect-resistant peanuts (Tiwari et al., 2008). Moreover, overexpression of a pea DNA helicase (*PDH45*) gene in peanut increased cellular-level tolerance and productivity under drought stress (Manjulatha et al., 2014). Expression of *AtLEC1* increased oil content and altered fatty acid composition in seeds of transgenic peanuts (Tang et al., 2018).

In addition, researchers have made significant efforts to alter the expression of target genes derived from peanuts. For instance, researchers cloned a fatty acid elongase gene, KCS1, from the drought-tolerant cultivar K-9 and overexpressed it in the drought-sensitive groundnut cultivar K-6 under the control of CaMV35S promoter. The transgenic peanuts exhibited increased cuticular wax content, reduction of water loss, lower membrane damage, decreased MDA content, and high proline content. These results suggested that *ahKCS1* play an important role in plant response to drought stress (Lokesh et al., 2019). Overexpression of histone deacetylases in peanut showed that hairy root growth was retarded after 20 d in vitro cultivation with a greater accumulation of superoxide anions and hydrogen peroxide in comparison with that of wild type (Su et al., 2021).

Despite significant progress in genetic transformation, peanut transformation remains difficult, inefficient, and unstable. There are many factors may affect the gene engineering of peanut, including the complex tetraploid genome and low transformation efficiency. As researchers focus more on peanut molecular breeding, especially with the development of gene editing technology, they expect an increase in the demand for peanut genetic transformation, which will drive greater progress in the field.

10.4 GENE EDITING BREEDING

Gene editing, also known as genome editing, is an emerging genetic engineering technology that can accurately modify specific target genes in the genome. Gene editing will have broad application prospects in the field of crop breeding. More importantly, gene editing does not introduce foreign genes, and only results in the replacement or deletion of few nucleotides of endogenous gene. Therefore, gene-editing products could be easily accepted by customers and government supervision departments.

Recently, researchers have made significant progress in creating new germplasms using gene-editing methods. There are three widely used gene editing strategies: zinc finger nucleases (ZFNs) (Townsend et al., 2009), transcription activator-like effector nucleases (TALENs) (Miller et al., 2011), and clustered regularly interspaced short palindromic repeat (CRISPR) (Jinek et al., 2012; Larson et al., 2013). All these gene-editing techniques can accurately carry out target knockout, nucleotide insertion, and replacement of plant genes. Among them, the CRISPR/Cas system, known as the third-generation genome editing technology, can carry out gene knockout, single base editing, and insertion, and prime editing (PE) has become an important tool and research focus for crop gene function verification and genetic improvement in rice, wheat, maize, and soybean.

Gene editing shows great potential for increasing crop yields. In rice, gene-editing mutations in three members of *OsPYLs* genes result in increased biomass and yield (Miao et al., 2018). Gene editing knockout of MIR396ef (MIR396e and MIR396f) enhanced both grain size and panicle branching, resulting in increased grain yield (Zhang et al., 2020). Moreover, gene editing knockout lines of amino acid transporter gene *OsAAP3* significantly increased grain yield (Lu et al., 2018). In wheat, gene editing mutant of cytokinin response 1 repressor 1 (*ARE1*) gene enhanced tolerance to N starvation, and showed delayed senescence and increased grain yield in field conditions (Zhang et al., 2021).

Gene editing technology can effectively improve the resistance of crops to biotic and abiotic stresses. In plants, pathogens leverage the well-characterized S gene, mildew-resistance locus (MLO), to achieve successful infection. In wheat, gene editing generated loss-of-function mutations in *TaMLO* genes and resulted in durable and broad-spectrum resistance to powdery mildew (Wang et al., 2014). Inactivating the wheat protein kinase (TaPsIPK1) gene provides broad-spectrum resistance to stripe rust (Wang et al., 2022). In rice, disease susceptibility requires the expression of the host sucrose transporter gene, mutating the promoter region of the rice sucrose transporter family gene *OsSWEET* destroy the binding element of its susceptible effector protein and improve the resistance of rice to bacterial blight (Oliva et al., 2019). The valine-glutamine (VQ) motif-containing protein of rice, *OsVQ25*, plays an important role in balancing broad-spectrum disease resistance and plant growth. The knockout of *OsVQ25* significantly improved the broad-spectrum resistance of rice-to-rice blast and bacterial blight (Hao et al., 2022).

In addition, gene editing has shown great potential in crop quality improvement. In rice, aroma is a key grain quality trait that directly influences the market price of rice globally. The main component of rice aroma, 2-acetyl-1-pyrroline (2AP), is synthesized from the precursor 4-aminobutyraldehyde, which can be oxidized by the aldehyde dehydrogenase 2 (OsBADH2) protein. Loss of function of OsBADH2 affects the biosynthesis of 2-AP, which is responsible for aroma in fragrant rice (Hui et al., 2022). CRISPR-Cas9 technology was used to edit *OsBadh2* in Zhonghua 11 and other conventional varieties respectively, and successfully obtained mutant lines with increased 2-AP content and significantly improved fragrance traits (Imran et al., 2023). Amylose content is a key component of rice food quality, and the Waxy (*Wx*) gene directly controls amylose synthesis in rice. CRISPR/Cas9-mediated editing of *Wx* and *BADH2* genes lead to create glutinous and aromatic rice (Zhang et al., 2023). CRISPR/Cas9 targeted mutagenesis of *OsAAP10* and *OsAAP10* can rapidly reduce grain protein content and improve cooking quality of rice (Wang et al., 2020). The above research provided new approaches for breeding cultivars with desired qualities through a gene editing strategy.

In recent years, gene editing has also made important progress in peanuts (Neelakandan, Subedi et al., 2022). Gene editing mutant of two nod factor receptors (NFRs) genes showed nod-phenotype, suggesting that CRISPR-Cas9 could be used in peanut hairy root transformation system for peanut functional genomic studies (Shu et al., 2020). As an important oil crop, peanut oil stability and quality are key traits, determined by the relative proportions of saturated fatty acids (SFAs) and unsaturated fatty acids (UFAs). Acyl-acyl carrier protein thioesterases (FATs) play an essential role in chain termination during de novo fatty acid synthesis and determine the types and levels of fatty acids. Gene editing mutant of *AhFatB* gene Arahy.4E7QKU resulted in low palmitic acid and high oleic acid, suggesting great potential for improving peanut oil quality (Tang, Huang et al., 2022). Using CRISPR/Cas9 system, the hot spot of natural mutation in *FAD2* genes was selected as the target region, and three mutations were identified—G448A in *ahFAD2A*, and 441_442insA and G451T in *ahFAD2B* (Yuan et al., 2019). A CRISPR/Cas9-based mutant of FAD2 cis-regulatory motifs can also increase the oleic acid content, providing a promising approach for peanut breeding (Neelakandan, Wright et al., 2022).

10.5 GENOMICS SELECTION BREEDING

10.5.1 Theory of Genomic Selection Breeding

Genomic selection (GS), also referred to as genome-wide selection (GWS), was first proposed in 2007 as a novel approach to marker-based breeding (Meuwissen 2007). Unlike conventional marker-assisted selection, genomic selection (GS) simultaneously utilizes tens to hundreds of thousands of markers distributed across the entire genome. Genomic selection (GS) is a method that simultaneously estimates all marker effects (or genetic loci) across the whole genome to calculate genomic estimated breeding values (GEBVs). Genomic selection breeding (GSB) is based on predicted breeding value calculated using an innovative method based on genome-wide dense DNA markers (Meuwissen et al., 2001). Although GS is not perfect, it is a potent, attractive, and valuable approach for plant breeding (Nakaya and Isobe, 2012). In traditional marker-assisted selection, QTLs, target genes, or linked markers typically originate from a specific parent or germplasm, and they may not be effective in other populations or hybrid combinations. GS does not need significant testing and identification of a subset of markers associated with the trait, and QTL mapping with populations derived from specific crosses can be avoided (Meuwissen et al., 2001). In addition, traditional marker-assisted breeding techniques are only effective for traits controlled by large effect genes or major QTLs, and it is difficult to improve quantitative traits controlled by small effect polygenes. Selection on genetic values predicted from markers could substantially increase the rate of genetic gain (Meuwissen et al., 2001; Jiang, 2013).

Genomic estimated breeding value (GEBV) is the basic criteria for the selection of desirable individuals in the breeding phase. Therefore, the effect value of each molecular marker covering the whole genome according to the phenotypes and genome-wide genotypes of the training population needs to be estimated, and then these marker effect values are accumulated in the test population genotype to obtain the GEBV. Thus, the biggest difference between GSB and MAB is that there is no need to locate the gene loci controlling the traits because the genetic information of the whole genome is used to select the individuals, which has high accuracy (Nakaya and Isobe, 2012). When GSB technology was first developed, high-throughput sequencing technology was not popular and the cost of genotyping of molecular markers was very high, and the application of GSB was limited. The year 2006 saw the first employment of GSB on dairy cows with breeding value estimate accuracy of up to 0.80. The cost of GSB in breeding was reduced by 92%, and genetic improvement speed increased two-fold (Schaeffer, 2006). Due to its remarkable results, GSB has become the standard method of dairy breeding (Schaeffer, 2006; Hayes et al., 2009). Partly because the economic value of individual crops is much lower than that of animals, the application of GSB in crops lags behind. However, crops have unique advantages for the use of GSB, such as low planting cost, short generation cycle, and large seed amount. At present, the use of GSB has achieved remarkable results in commercial crop breeding companies, and the in-depth application of GS in crop breeding will become an inevitable trend.

10.5.2 Tools for High Throughput Genomic Selection

Genomic selection breeding relies on high throughput genotyping techniques, including whole genome resequencing selection techniques, genotyping-by-sequencing (GBS) selection techniques, SNP microchips with different scales of selective markers, and high throughput selection by KASP genotyping.

Compared with other methods, whole genome resequencing could identify more variation sites from different individuals including SNPs, InDels, copy number variations (CNVs) and structure variation (SV). In addition, whole genome resequencing covers the whole genome of the species without bias, providing a robust and reliable tool for accelerating the development of varieties with desirable target traits in many crops, including peanut. With the rapid improvement of sequencing technology and the continuous reduction of sequencing cost, whole genome resequencing will provide more help for genomic selection breeding.

GBS refers to methods that obtain genotype data from samples by using restriction enzyme digestion followed by high-throughput sequencing. GBS has simple experimental steps, is low cost, and can obtain genome-wide genetic variation sites without relying on the reference genome. Classic GBS, RAD-seq (restriction-site associated DNA sequence), 2bRAD, and ddGBS are the most commonly used GBS technologies. The above four methods have similar principles.

Gene chipping is a nucleotide sequence hybridization method based on the principle of complementary base pairing. First, the designed and synthesized nucleotide probe cover the treated solid phase carrier with high density by a special method, and then the sample to be tested is labeled with fluorescent dye (generally labeled with Cy3 and Cy5) and hybridized with the probe on the chip under appropriate conditions. If the sequence of the sample tested is completely complementary to the probe sequence, the fluorescence signal will be strong; otherwise, the signal will be very weak. After hybridization, the chip scanner detects the fluorescence signal and converts it into a digital signal for analysis. Currently, commercial SNP chips are available for most crops, including peanut.

Kompetitive allele-specific PCR (KASP) relies on bi-allelic scoring of SNPs and InDels at specific loci coupled with endpoint fluorescence. KASP is an advanced PCR-based homogenous fluorescent genotyping method that allows rapid and precise identification of co-dominant alleles associated with SNPs and InDels at a specific locus in parent and offspring populations. Compared to GBS and SNP array, KASP is cost-effective. Moreover, KASP is also a high-throughput genotyping platform customized for a wide range of experimental designs involving different target loci and different numbers of samples.

10.5.3 Application of Genomic Selection in Crop Breeding

GSB is a powerful, attractive, and valuable approach for plant breeding. Studies on staple crops such as wheat, rice, and maize have primarily focused on creating models for genomic estimated breeding values (GEBV). To validate the practical application of GS for wheat breeding, a total of 376 elite wheat varieties were genotyped using 3,046 DArT, SNP, and gene markers. Seven traits were phenotyped over two years in different environments. Testing results showed that GSB represented an exciting opportunity to select key traits, and that optimization of the training population is crucial to its successful implementation (Bentley et al., 2014). In addition, GSB showed great potential to increase grain Zn and Fe concentrations in bread wheat (Velu et al., 2016). Using BayesR and Bayesian ridge regression (BRR) models, grain yield and the concentration of 13 nutritional elements in grains (B, Ca, Co, Cu, Fe, K, Mg, Mn, Mo, Na, Ni, P, and Zn) were predicted using a population of 1,470 spring wheat lines. The testing results demonstrated that genomic selection has great potential for enriching wheat grain with nutritional elements (Joukhadar et al., 2021). In rice, using genome-wide association mapping and selection, several superior alleles were identified and led to the development of elite lines with excellent taste, high yield, and a broad spectrum of blast resistance, suggesting the value of GSB in rice breeding (Xiao et al., 2021). Compared with traditional breeding, GSB showed the characteristics of accuracy and intelligence in breeding, which could greatly reduce workload and cost. However, there are still many challenges for GSB, especially in terms of its prediction accuracy. It is necessary to use a large number of multi-year multi-point data, multi-trait data, crossbreeding unit data, and early-accumulated genealogical data. Phenotypic data mostly rely on manual investigation, which is time-consuming and labor-intensive with a lot of room for error, and lacks intelligent and efficient equipment suitable for various trait identification. Most GSB studies have focused on staple crops (wheat, maize, and rice), and it has rarely been used in other crops. Due to the significant early-stage investment required, particularly for creating GEBV models, this technology has primarily been applied in developed countries, such as the United States, Germany, France, the United Kingdom, and Australia. To facilitate its application

TABLE 10.1
Partial Available Genome Sequences of *Arachis* Species

Species	Accession Number	Subgenome	NCBI Taxonomy ID	Link Website	Description of the Species	Reference
A. duranensi	V14167	AA	130453	https://legacy.peanutbase.org/gb2/gbrowse/Aradu1.0	*A. duranensis*,a wild herb native to Argentina, Bolivia, and Paraguay, is thought to be one of the diploid ancestors of cultivated peanut. *A. duranensis* strain V14167 originates in a region east of the Andes, in the vicinity of northern Argentina and southern Bolivia. This accession is thought to be one of the diploid ancestors of cultivated peanut.	(Bertioli et al., 2016)
A. ipaensis	K30076	BB	130454	https://legacy.peanutbase.org/gb2/gbrowse/Araip1.0	*A. ipaensis* accession K30076 is the only accession collected from what is thought to be the origin of tetraploid peanut, Villa Montes, Bolivia.	(Bertioli et al., 2016)
A. hypogaea	Tifrunner	AABB	3818	https://legacy.peanutbase.org/gb2/gbrowse/arahy.Tifrunner.gnm2/	Cultivar Tifrunner is a runner-type peanut bred in the United States, widely planted as a variety and widely used in breeding programs.	(Bertioli et al., 2019)
	Fuhuasheng	AABB		https://www.peanutbase.org/tools/jbrowse2/?config=config.json&session=local-j1iTd_zw5	The mid-twentieth-century landrace Fuhuasheng, from North China, is a widely used parent for the derivation of Chinese peanut cultivars released during the past half century.	(Chen et al., 2019)
	Shitouqi	AABB		http://peanutgr.fafu.edu.cn/; https://www.peanutbase.org/tools/jbrowse2/?config=config.json&session=local-eaHPYHcdA	Shitouqi (zh.h0235), the most widely cultivated peanut ecotype in the world, is a Chinese cultivar and breeding parent belonging to subspecies *fastigiata*, botanical type vulgaris and agronomic type Spanish with heterozygosity only 1/6,537 nucleotides on average.	(Di Giammartino et al., 2011)
	BaileyII	AABB		https://www.peanutbase.org/tools/jbrowse2/?config=config.json&session=local-qpcHfiWVD	A Virginia-type peanut cultivar developed by the NCSU breeding program, Bailey II is a near-isogenic line to Bailey, possessing high-oleic trait and resistance to several important pathogens, derived from *A. cardenasii* introgressions.	
A. cardenasii	K10017	AA	51121	https://www.peanutbase.org/tools/jbrowse2/?config=config.json&session=local-r1ETV3fHj	A wild peanut relative native to Bolivia. It has been used by peanut breeders and geneticists in interspecific hybrids with cultivated peanut as a source of pest and disease resistance. *A. cardenasii* is a diploid species in the A-genome group of *Arachis*, similar to *A. duranensis*. K10017 is perennial plant. Taproot deep, mainstem erect, with long procumbent lateral branches growing from base. Leaves tetrafoliolate, fruit geocarpic.	
A. stenosperma	V10309		217475	https://www.peanutbase.org/tools/jbrowse2/?config=config.json&session=local-WXLjSHOrD	A wild peanut relative native to central Brazil; in the past it was cultivated by native peoples of South America, and was carried to the Atlantic coast, where populations persist to the present day. It is a source of strong pest and disease resistance and has been used by peanut breeders and geneticists in interspecific hybrids. *A. stenosperma* is a diploid species in the A-genome group of *Arachis*, which has similarity to the A genome of tetraploid cultivated peanut.	

in developing countries, it is necessary to reduce the technical threshold and cost.

10.6 SUMMARY AND FUTURE PERSPECTIVE

In general, the traditional breeding approach relies on hybridization, chromosome recombination, and exchange to create good varieties by optimizing the combination of related genes. However, it has some disadvantages, such as low recombination exchange frequency, uneven distribution of recombination sites, and harmful allelic linkage. Genomics-assisted breeding shows more outstanding advantages than traditional breeding approach, greatly improving breeding efficiency, and has become a new development direction in crop breeding. For GSB, genomic sequences, large-scale molecular markers, and phenotypic data are necessary. In recent years, peanut genome sequencing has made great progress. In 2016, the genomes of diploid ancestor species *Arachis duranensis* ($2n = 2x = 20$, A genome) and *Arachis ipanesis* ($2n = 2x = 20$, A genome) were first reported (Bertioli et al., 2016; Chen, Li et al., 2016). In 2018, the genome of *Arachis monticola* ($2n = 4x = 40$, AABB), the tetraploid wild species of section *Arachis* was reported (Yin et al., 2018). In 2019, the genomes of three cultivated peanut cultivars ($2n = 4x = 40$, AABB), Shitouqi, Tifrunner and Fuhuasheng, were reported (Bertioli et al., 2019; Chen et al., 2019; Zhuang et al., 2019). In addition, the genomes of two other diploid wild peanuts, *Arachis cardenasii* and *Arachis stenosperma*, are present on PeanutBase (www.peanutbase.org). The following table lists the genome information of the *Arachis* species (Table 10.1), with most of the genome sequence information combined in PeanutBase. In addition to the genome, transcriptome, proteome, metabolome, and other omics, researchers integrate the information from Table 10.2 and Table 10.3 into whole-genome selection. These genome resources have laid a foundation for applying genome selection breeding (GSB) in peanuts. In GSB, GEBV models play a central role in genome-wide selection and directly influence its accuracy and efficiency. Additionally, the successful application of GSB relies on the following advances in research.

10.6.1 The Regulation Network Underlying Complex Agronomic traits

Many of the important traits of peanut, such as yield, quality, disease resistance, and nutrient absorption and utilization, are controlled by multiple loci/genes and are susceptible to environmental factors. In addition, many excellent traits might have mutual restriction or inhibition. Therefore, it is interesting to underpin the regulation network for complex agronomic traits of peanut through analyzing the regulatory factors and molecular networks of complex traits by comprehensive application of omics, genetics, systems biology, and computational biology. Using genetic resource materials and Big Data to study the genomic variation and epigenetic variation of complex traits and their mechanisms to cope with environmental variation, establish the accurate control theory of complex traits, and obtain the accurate design of high-quality and high-yield, resource-efficient, environment-intelligent responsive, and quality-healthy seeds.

10.6.2 Systematic Study and Identify Important Haplotypes

Haplotypes information is also important for constructing the GEBV model. However, available haplotypes are lacking in the case of peanut. Obtaining haplotype information involves applying poly-recombination populations (NAM MAJIC), pan-omics, artificial intelligence, and systems biology. Applying comprehensive multidimensional omics will reveal the formation and evolution laws of important traits in the development process of peanut from wild species to local varieties to modern varieties, dig out key regulatory genes in the formation of excellent traits, and analyze the regulatory networks that they participated in. In addition, systematic study of the evolution paths of important haplotypes, structural variation, and epigenetic variation during domestication and breeding of major varieties will reveal the transfer rules of important genes during domestication and breeding of major varieties, and clarify the genetic regulation mechanism during domestication and improvement of germplasm resources. Research progress in the above directions will further promote the progress of GSB of peanut.

Moreover, the potential of artificial intelligence and Big Data is huge, and there are important application scenarios for these in peanut breeding. For example, based on Big Data, AI can be used to model and accurately predict relevant genetic loci to provide target points for accurate GEBV. Currently, researchers rarely use GSB in peanut breeding. However, with the progress of peanut genomics, genetics, Big Data and AI, GSB will be one of the leading technologies of peanut breeding in the future.

TABLE 10.2
Reported Transcriptome Research of *Arachis* Species

Species	Tissue or Organ	Treatment/Different Stages	Description	Methods	Reference
Cultivar FlavRunner 458	Pod, peg, leaf, stem, and root	-	Developed a high-density oligonucleotide microarray using 49,205 publicly ESTs and compared transcript levels in pod, peg, leaf, stem, and root tissues	Oligonucleotide microarrays	(Payton et al., 2009)
Cultivar Luhua-14	Seeds, leaves, stems, roots, flowers, and gynophores	15, 25, 35, 45, and 70 days after pegging	Developed 17,000 ESTs and analyzed the expression profile of 5066 ESTs	cDNA microarray	(Bi et al., 2010)
-	-	-	PeanutDB: an integrated bioinformatics web portal for *Arachis hypogaea* transcriptomics	-	(Duan et al., 2012)
Cultivar Jihua 4, Kaixuan01-6, Te21	Immature seeds	-	Transcriptomes of the immature seeds of three different peanut varieties with different oil contents. Denove assembly of 59,077 unigenes.	Illumina HiSeq™ 2000	(Zhang et al., 2012)
Cultivar Luhua-14	Gynorphore	Stage 1, 2, 3	Assembled into 72,527 unigenes and revealed the gene express profile of gynorphore before and after soil penetration.	Illumina HiSeq™ 2000	(Xia et al., 2013)
Cultivar YJ-1 and Yueyou 7	Mature seeds	Well-watered, drought stress and both *A. flavus* and drought stress	Transcriptome identification of the resistance-associated genes (RAGs) to Aspergillus flavus infection in pre-harvested peanut	NimbleGen microarray	(Wang et al., 2013)
Spanish group variety, Spanco	Leaf, stem, root, flower, peg, and pod	10-day-old to seed-filled stage	A total of 26,048 unigenes were generated, providing valuable tools and resources for genomic research in legumes and other plant species.	Illumina HiSeq™ 2000	(Wu et al., 2013)
Cultivar Yueyou 7	Pegs	Aerial pods 8, 10, 12, 16, and 20 DAF and subterranean pods 2, 4, 6, 8, 10, and 12 days after soil penetration (DASP)	Generated 74,974 transcript assembly contigs, providing evidence for some functional genes as robust candidates underlying aerial and subterranean pod development and contributes to an elucidation of the evolutionary implications resulting from fruit development under light and dark conditions.	Roche 454 sequencing	(Chen et al., 2013)
A. stenosperma (V10309) and *A. duranensis* (K7988)	Leaves	Fungi inoculation, water-limited stress	Generated 7,723 contigs for *A. stenosperma* and 12,792 for *A. duranensis*, and helped to elucidate genes involved in responses to biological processes such as fungal diseases and water-limited stress.	Roche 454 sequencing	(Guimarães et al., 2012)

(*Continued*)

TABLE 10.2 (CONTINUED)
Reported Transcriptome Research of *Arachis* Species

Species	Tissue or Organ	Treatment/Different Stages	Description	Methods	Reference
Cultivar Yueyou 7	Pods	Aerial pods and subterranean pods at 1, 2, 4, 8 days after marked (DAM)	Identification of 6,203 differentially expressed genes, 4,732 stage-specific expressed genes, and 2,401 specific expressed genes only in aerial or subterranean pods	NimbleGen microarray	(Zhu et al., 2014)
U606 and U12	Seeds	30 DAF, 40 DAF, 50 DAF, 60 DAF, and 70 DAF (days after flowering)	Generated 59,236 unique sequences and revealed candidate unigenes for oil accumulation pathways. U606 and U12 originated from the same male (Yuhua15) and female (Huayu17) parent. These two subline cultivars differed greatly in oil content (U606 > 58% and U12 < 45%) and fatty acid composition in mature seeds.	Illumina Genome Analyzer II	(Yin et al., 2013)
Cultivar Olin and *A. ipaënsis* KGBSPSc30076 (BB), and *A. duranensis* K38901 (AA)	Leaf, root, and pod (yellow, brown, and black stages of maturity) tissues		Denerated de novo transcriptome assemblies, with about 38 million reads in the tetraploid cultivar OLin and 16 million reads in each of the diploids.	Illumina HiSeq™ 2000	(Chopra et al., 2014)
Two peanut genotypes, R genotype J04 and S genotype J62	Root tips	0, 6, 12, 24, 48, and 72 hours inoculated with R. solanacearum strain race 1	The transcriptome profiling provided the largest data set that explores the dynamic in crosstalk between peanut and *R. solanacearum*, contributing to a better understandin g of the molecular interaction between plant and *R. solanacearum*.	Illumina HiSeq™ 2000	(Chen et al., 2014)
Cultivar Yueyou 7	Pegs and pods	Two stages of whole pods (aerial pod, P0 and not swelling subterranean pod, P1) and nine stages of isolated pod walls (P2SH–P10SH)	Identified 110 217 transcripts across developmental stages and quantified their abundance along a pod developmental gradient in pod wall	Illumina HiSeq™ 2000	(Chen, Yang et al., 2016)
The seed coat crack and brown testa mutant pscb and Zhonghua 16	Seed coat	20, 40, and 60 DAF	Examined the seed coat transcriptome in three stages of seed development in the wild type and the *pscb* mutant; provided greater insights and molecular genetic control of peanut seed-coat cracking and color variation.	Illumina HiSeqTM2500	(Wan et al., 2016)
Cultivar Tifrunner	22 different tissue types	-	Producing a gene atlas for cultivated peanut, 22 different tissue types and ontogenies that represent the full development of peanut were sequenced, including a complete reproductive series from flower to peg elongation and peg tip immersion in the soil to fully mature seed. The data has been integrated into PeanutBase.	Illumina HiSeq 2500	(Clevenger et al., 2016)

(Continued)

TABLE 10.2 (CONTINUED)
Reported Transcriptome Research of *Arachis* Species

Species	Tissue or Organ	Treatment/Different Stages	Description	Methods	Reference
Arachis glabrata	Root, stem, leaf, flower, peg, and pod	Drought, salt, cold, and leaf spot disease infection	Denove generated 109,747 unique transcripts and revealed that 1229, 1039, 1671, 3923, 1521, and 1799 transcripts expressed specifically in the root, stem, leaf, flower, peg, and pod, respectively. Identified thousands of differentially expressed transcripts in response to drought, salt, cold, and leaf spot disease.	PacBio combined with BGISEQ-500 RNA-seq	(Zhao et al., 2021)
Tifrunner (susceptible), C76-16 (tolerant), 587 (tolerant), and 506 (susceptible)	Leaf	Drought treatment	Identification of 2,457 differentially expressed genes (DEGs).	Illumina HiSeq 4000	(Wang, Yang et al., 2021)

TABLE 10.3
Reported Proteomic and Metabolomic Research of *Arachis* Species

Species	Tissue or Organ	Treatment/Different Stages	Description	Methods	Reference
Cultivar YJ-1 and Yueyou 7	Mature seeds	Well-watered, drought, *A. flavus* inoculation	Identify resistance-associated proteins in response to A. flavus infection under drought stress using	Two-dimensional electrophoresis and MALDI-TOF–TOF MS	(Wang et al., 2010)
Cultivar Yueyou 7	Elongating aerial pegs	-	Identified 31 distinct potential candidate proteins and played critical roles in regulating pod swelling and development	Two-dimensional electrophoresis and MALDI-TOF–TOF MS	(Zhu et al., 2013)
The wsc mutant (white testa) and Zhonghua 16 (WT)	Seed coat	20 days after flowering (20DAF), 40DAF, and 60DAF	Metabolomes revealed flavonoids were redirected in *wsc*, while multi-omics analyses of *wsc* mutant seeds and testa uncovered that WSC influenced the flavonoids biosynthesis in testa as well as suberin formation, glycolysis, the TCA cycle, and amino acid metabolism.	LC-ESI-MS/MS system	(Wan et al., 2020)
Cultivars Shanhua 15 with pink testa and Zhonghua 12 with red testa	Seed coat	R4 and R7 stage	A total of 198 flavonoid metabolites were detected. Transcriptome analysis showed that there were 6,059 and 3,153 differentially expressed genes between Shanhua 15 and Zhonghua 12 in different growth periods.	LC-ESI-MS/MS system for metabolite identification and HiSeq 4000 platform for RNA-seq	(Xue et al., 2021)
The cold-resistant peanut cultivar "Silihong" (SLH) and cold-sensitive "Zhonghua 12" (ZH12)	Leaf	0, 3, 24, and 48 h cold treatment	Obtained a set of cold-responsive metabolites, including several carbohydrates and polyamines and 3,620 cold tolerance genes; is a valuable resource for study of cold resistance in peanut.	LC-MS/MS system for metabolite identification and Illumina NovaSeq for RNA-seq	(Wang, Liu et al., 2021)

REFERENCES

Ballen-Taborda C., Y. Chu, P. Ozias-Akins, C. C. Holbrook, P. Timper, S. A. Jackson, D. J. Bertioli, S. C. M. Leal-Bertioli. 2021. "Development and genetic characterization of peanut advanced backcross lines that incorporate root-knot nematode resistance from *Arachis stenosperma*." *Front Plant Sci* 12:785358.

Banavath J. N., T. Chakradhar, V. Pandit, S. Konduru, K. K. Guduru, C. S. Akila, S. Podha, C. O. R. Puli. 2018. "Stress inducible overexpression of AtHDG11 leads to improved drought and salt stress tolerance in peanut (*Arachis hypogaea* L.)." *Front Chem* 6:34.

Barkley N. A., M. Li Wang, R. N. Pittman. 2001. "A real-time PCR genotyping assay to detect FAD2A SNPs in peanuts (*Arachis hypogaea* L.)." *Electron J Biotechnol* 14.

Barkley N. A., T. G. Isleib, M. L. Wang, R. N. Pittman. 2013. "Genotypic effect of ahFAD2 on fatty acid profiles in six segregating peanut (*Arachis hypogaea* L) populations." *BMC Genet* 14:62.

Bentley A. R., M. Scutari, N. Gosman, S. Faure, F. Bedford, P. Howell, J. Cockram, G. A. Rose, et al. 2014. "Applying association mapping and genomic selection to the dissection of key traits in elite European wheat." *Theor Appl Genet* 127:2619–2633.

Bertioli D. J., J. Jenkins, J. Clevenger, O. Dudchenko, D. Gao, G. Seijo, S. C. M. Leal-Bertioli, L. Ren, et al. 2019. "The genome sequence of segmental allotetraploid peanut *Arachis hypogaea*." *Nat Genet* 51:877–884.

Bertioli D. J., S. B. Cannon, L. Froenicke, G. Huang, A. D. Farmer, E. K. Cannon, X. Liu, D. Gao, et al. 2016. "The genome sequences of *Arachis duranensis* and *Arachis ipaensis*, the diploid ancestors of cultivated peanut." *Nat Genet* 48:438–446.

Bhalani H., R. Thankappan, G. P. Mishra, T. Sarkar, T. C. Bosamia, J. R. Dobaria. 2019. "Regulation of antioxidant mechanisms by AtDREB1A improves soil-moisture deficit stress tolerance in transgenic peanut (*Arachis hypogaea* L.)." *PLoS One* 14:e0216706.

Bi Y. P., W. Liu, H. Xia, L. Su, C. Z. Zhao, S. B. Wan, X. J. Wang. 2010. "EST sequencing and gene expression profiling of cultivated peanut (*Arachis hypogaea* L.)." *Genome* 53:832–839.

Botstein D., R. L. White, M. Skolnick, R. W. Davis. 1980. "Construction of a genetic linkage map in man using restriction fragment length polymorphisms." *Am J Hum Genet* 32:314–331.

Chen X., H. Li, M. K. Pandey, Q. Yang, X. Wang, V. Garg, H. Li, X. Chi, et al. 2016a. "Draft genome of the peanut A-genome progenitor (*Arachis duranensis*) provides insights into geocarpy, oil biosynthesis, and allergens." *Proc Natl Acad Sci U S A* 113:6785–6790.

Chen X., Q. Lu, H. Liu, J. Zhang, Y. Hong, H. Lan, H. Li, J. Wang, et al. 2019. "Sequencing of cultivated peanut, arachis hypogaea, yields insights into genome evolution and oil improvement." *Mol Plant* 12: 920–934.

Chen X., Q. Yang, H. Li, H. Li, Y. Hong, L. Pan, N. Chen, F. Zhu, et al. 2016b. "Transcriptome-wide sequencing provides insights into geocarpy in peanut (*Arachis hypogaea* L.)." *Plant Biotechnol J* 14:1215–1224.

Chen X., W. Zhu, S. Azam, H. Li, F. Zhu, H. Li, Y. Hong, H. Liu, et al. 2013. "Deep sequencing analysis of the transcriptomes of peanut aerial and subterranean young pods identifies candidate genes related to early embryo abortion." *Plant Biotechnol J* 11:115–127.

Chen Y., X. Ren, X. Zhou, L. Huang, L. Yan, Y. Lei, B. Liao, J. Huang, et al. 2014. "Dynamics in the resistant and susceptible peanut (*Arachis hypogaea* L.) root transcriptome on infection with the Ralstonia solanacearum." *BMC Genomics* 15:1078.

Chopra R., G. Burow, A. Farmer, J. Mudge, C. E. Simpson, M. D. Burow. 2014. "Comparisons of de novo transcriptome assemblers in diploid and polyploid species using peanut (*Arachis* spp.) RNA-Seq data." *PLoS One* 9:e115055.

Clevenger J., Y. Chu, B. Scheffler, P. Ozias-Akins. 2016. "A developmental transcriptome map for Allotetraploid *Arachis hypogaea*." *Front Plant Sci* 7:1446.

Collard B. C., D. J. Mackill. 2008. "Marker-assisted selection: an approach for precision plant breeding in the twenty-first century." *Philos Trans R Soc Lond B Biol Sci* 363:557–572.

Davierwala A. P., A. P. Reddy, M. D. Lagu, P. K. Ranjekar, V. S. Gupta. 2001. "Marker assisted selection of bacterial blight resistance genes in rice." *Biochem Genet* 39:261–278.

Davis G. P., S. K. DeNise. 1998. "The impact of genetic markers on selection." J Anim Sci 76:2331–2339.

Dekkers J. C., F. Hospital. 2002. "The use of molecular genetics in the improvement of agricultural populations." *Nat Rev Genet* 3:22–32.

Di Giammartino D. C., K. Nishida, J. L. Manley. 2011. "Mechanisms and consequences of alternative polyadenylation." *Mol Cell* 43:853–866.

Ditt R. F., E. W. Nester, L. Comai. 2001. "Plant gene expression response to *Agrobacterium tumefaciens*." *Proc Natl Acad Sci U S A* 98:10954–10959.

Duan X., E. Schmidt, P. Li, D. Lenox, L. Liu, C. Shu, J. Zhang, C. Liang. 2012. "PeanutDB: an integrated bioinformatics web portal for *Arachis hypogaea* transcriptomics." *BMC Plant Biol* 12:94.

Fowler J. C., J. M. Kijas, M. R. Thomas. 1994. "Current molecular methods for plant genome identification." *Australas Biotechnol* 4:153–157.

Frisch M., A. E. Melchinger. 2001. "The length of the intact donor chromosome segment around a target gene in marker-assisted backcrossing." *Genetics* 157:1343–1356.

Gantait S., S. Mondal. 2018. "Transgenic approaches for genetic improvement in groundnut (*Arachis hypogaea* L.) against major biotic and abiotic stress factors." *J Genet Eng Biotechnol* 16:537–544.

Guimarães P. M., A. C. Brasileiro, C. V. Morgante, A. C. Martins, G. Pappas, O. B. Silva, Jr., R. Togawa, S. C. Leal-Bertioli, et al. 2012. "Global transcriptome analysis of two wild relatives of peanut under drought and fungi infection." *BMC Genomics* 13:387.

Hao Z., J. Tian, H. Fang, L. Fang, X. Xu, F. He, S. Li, W. Xie, et al. 2022. "A VQ-motif-containing protein fine-tunes rice immunity and growth by a hierarchical regulatory mechanism." *Cell Rep* 40:111235.

Hayes B. J., P. J. Bowman, A. J. Chamberlain, M. E. Goddard. 2009. "Invited review: Genomic selection in dairy cattle: progress and challenges." *J Dairy Sci* 92:433–443.

Heuberger S. M. (2009) Understanding pollen-mediated transgene flow in agricultural seed production. Esa Convention.

Hiebert C. W., J. B. Thomas, D. J. Somers, B. D. McCallum, S. L. Fox. 2007. "Microsatellite mapping of adult-plant leaf rust resistance gene Lr22a in wheat." *Theor Appl Genet* 115:877–884.

Hospital F. 2001. "Size of donor chromosome segments around introgressed loci and reduction of linkage drag in marker-assisted backcross programs." *Genetics* 158:1363–1379.

Huang B., F. Qi, Z. Sun, L. Miao, Z. Zhang, H. Liu, Y. Fang, W. Dong, et al. 2019. "Marker-assisted backcrossing to improve seed oleic acid content in four elite and popular peanut (*Arachis hypogaea* L.) cultivars with high oil content." *Breed Sci* 69:234–243.

Hui S., H. Li, A. M. Mawia, L. Zhou, J. Cai, S. Ahmad, C. Lai, J. Wang, et al. 2022. "Production of aromatic three-line hybrid rice using novel alleles of BADH2." *Plant Biotechnol J* 20:59–74.

Imran M., S. Shafiq, X. Tang. 2023. "CRISPR-Cas9-mediated editing of BADH2 gene triggered fragrance revolution in rice." *Physiol Plant* 175:e13871.

Jiang G. L. (2013) "Molecular markers and marker-assisted breeding in plants." In Plant Breeding from Laboratories to Fields.

Jinek M., K. Chylinski, I. Fonfara, M. Hauer, J. A. Doudna, E. Charpentier. 2012. "A programmable dual-RNA-guided DNA endonuclease in adaptive bacterial immunity." *Science* 337:816–821.

Joukhadar R., R. Thistlethwaite, R. M. Trethowan, M. J. Hayden, J. Stangoulis, S. Cu, H. D. Daetwyler. 2021. "Genomic selection can accelerate the biofortification of spring wheat." *Theor Appl Genet* 134:3339–3350.

Jung S., G. Powell, K. Moore, A. Abbott. 2000. "The high oleate trait in the cultivated peanut [*Arachis hypogaea* L]. II. Molecular basis and genetics of the trait." *Mol Gen Genet* 263:806–811.

Kamo K., A. Blowers, F. Smith, J. V. J. P. e. Eck. 1995. "Stable transformation of Gladiolus by particle gun bombardment of cormels." 110(1):105–111.

Krishna G., B. K. Singh, E. K. Kim, V. K. Morya, P. W. Ramteke. 2015. "Progress in genetic engineering of peanut (*Arachis hypogaea* L.)—a review." *Plant Biotechnol J* 13:147–162.

Kumar P. K. C., A. Bellundagi, H. Krishna, M. G. Mallikarjuna, R. K. Thimmappa, N. Rai, P. Shashikumara, N. Sinha, et al. 2023. "Development of bread wheat (*Triticum aestivum* L) variety HD3411 following marker-assisted backcross breeding for drought tolerance." *Front Genet* 14:1046624.

Kumar V., A. Rani, R. Rawal, V. Mourya. 2015. "Marker assisted accelerated introgression of null allele of kunitz trypsin inhibitor in soybean." Breed Sci 65:447–452.

Larson M. H., L. A. Gilbert, X. Wang, W. A. Lim, J. S. Weissman, L. S. Qi. 2013. "CRISPR interference (CRISPRi) for sequence-specific control of gene expression." *Nat Protoc* 8:2180–2196.

Liu M., F. Fan, S. He, Y. Guo, G. Chen, N. Li, N. Li, H. Yuan, et al. 2022. "Creation of Elite Rice with High-Yield, Superior-Quality and High Resistance to Brown Planthopper Based on Molecular Design." *Rice* 15:17.

Lokesh U., B. Venkatesh, K. Kiranmai, A. Nareshkumar, V. Amarnathareddy, G. L. Rao, A. M. Anthony Johnson, M. Pandurangaiah, et al. 2018. "Overexpression of ß-Ketoacyl Co-A Synthasel gene improves tolerance of drought susceptible groundnut (*Arachis hypogaea* L.) cultivar K-6 by increased leaf epicuticular wax accumulation." *Front Plant Sci* 9:1869.

Lu K., B. Wu, J. Wang, W. Zhu, H. Nie, J. Qian, W. Huang, Z. Fang. 2018. "Blocking amino acid transporter OsAAP3 improves grain yield by promoting outgrowth buds and increasing tiller number in rice." *Plant Biotechnol J* 16:1710–1722.

Manjulatha M., R. Sreevathsa, A. M. Kumar, C. Sudhakar, T. G. Prasad, N. Tuteja, M. Udayakumar. 2014. "Overexpression of a pea DNA helicase (PDH45) in peanut (*Arachis hypogaea* L.) confers improvement of cellular level tolerance and productivity under drought stress." *Mol Biotechnol* 56:111–125.

Maranna S., K. Verma, A. Talukdar, S. K. Lal, A. Kumar, K. Mukherjee. 2016. "Introgression of null allele of Kunitz trypsin inhibitor through marker-assisted backcross breeding in soybean (Glycine max L. Merr.)." *BMC Genet* 17:106.

Mariel A., U. Toledo, J. Carlos, I. Ignacio, E. M. J. P. B. Septiningsih, Biotechnology. 2015. "Development of Improved Ciherang-Subl Having Tolerance to Anaerobic Germination Conditions." *Breeding and Biotechnology* 3:77–87.

Mehta R., T. Radhakrishnan, A. Kumar, R. Yadav, J. R. Dobaria, P. P. Thirumalaisamy, R. K. Jain, P. Chigurupati. 2013. "Coat protein-mediated transgenic resistance of peanut (*Arachis hypogaea* L.) to peanut stem necrosis disease through Agrobacterium-mediated genetic transformation." *Indian J Virol* 24:205–213.

Meuwissen T. 2007. "Genomic selection: marker assisted selection on a genome wide scale." *J Anim Breed Genet* 124:321–322.

Meuwissen T. H., B. J. Hayes, M. E. Goddard. 2001. "Prediction of total genetic value using genome-wide dense marker maps." *Genetics* 157:1819–1829.

Miao C., L. Xiao, K. Hua, C. Zou, Y. Zhao, R. A. Bressan, J. K. Zhu. 2018. "Mutations in a subfamily of abscisic acid receptor genes promote rice growth and productivity." *Proc Natl Acad Sci U S A* 115:6058–6063.

Miller J. C., S. Tan, G. Qiao, K. A. Barlow, J. Wang, D. F. Xia, X. Meng, D. E. Paschon, et al. 2011. "A TALE nuclease architecture for efficient genome editing." *Nat Biotechnol* 29:143–148.

Moretzsohn M. C., J. F. Dos Santos, A. R. A. Moraes, A. R. Custodio, M. D. Michelotto, N. Mahrajan, S. C. M. Leal-Bertioli, I. J. Godoy, et al. 2023. "Marker-assisted introgression of wild chromosome segments conferring resistance to fungal foliar diseases into peanut (*Arachis hypogaea* L.)." *Front Plant Sci* 14:1139361.

Nakaya A., S. N. Isobe. 2012. "Will genomic selection be a practical method for plant breeding?" *Ann Bot* 110:1303–1316.

Neelakandan A. K., B. Subedi, S. M. Traore, P. Binagwa, D. A. Wright, G. He. 2022a. "Base editing in peanut using CRISPR/nCas9." *Front Genome Ed* 4:901444.

Neelakandan A. K., D. A. Wright, S. M. Traore, X. Chen, M. H. Spalding, G. He. 2022b. "CRISPR/Cas9 based site-specific modification of FAD2 cis-regulatory motifs in peanut (*Arachis hypogaea* L)." *Front Genet* 13:849961.

Neeraja C. N., R. Maghirang-Rodriguez, A. Pamplona, S. Heuer, B. C. Collard, E. M. Septiningsih, G. Vergara, D. Sanchez, et al. 2007. "A marker-assisted backcross approach for developing submergence-tolerant rice cultivars." *Theor Appl Genet* 115:767–776.

Noelle, A., Barkley, Kelly, D., Chenault, Chamberlin, Ming, et al. 2010. "Development of a real-time PCR genotyping assay to identify high oleic acid peanuts (*Arachis hypogaea* L.)." Molecular Breeding.

Okuley J., J. Lightner, K. Feldmann, N. Yadav, E. Lark, J. Browse. 1994. "Arabidopsis FAD2 gene encodes the enzyme that is essential for polyunsaturated lipid synthesis." *Plant Cell* 6:147–158.

Oliva R., C. Ji, G. Atienza-Grande, J. C. Huguet-Tapia, A. Perez-Quintero, T. Li, J. S. Eom, C. Li, et al. 2019. "Broad-spectrum resistance to bacterial blight in rice using genome editing." *Nat Biotechnol* 37:1344–1350.

Pan L., Y. Jiang, W. Zhou, P. Jiang, L. Wu, A. Chen, H. Zhu, J. Sui, et al. 2019. "[Breeding on a new peanut variety Yuhua91 with high oleic acid content]." *Sheng Wu Gong Cheng Xue Bao* 35:1698–1706.

Parkin I. A., D. J. Lydiate. 1997. "Conserved patterns of chromosome pairing and recombination in Brassica napus crosses." *Genome* 40:496–504.

Pattee H. E., T. G. Isleib, D. W. Gorbet, K. M. Moore, Y. Lopez, M. R. Baring, C. E. Simpson. 2002. "Effect of the high-oleic trait on roasted peanut flavor in backcross-derived breeding lines." *J Agric Food Chem* 50:7362–7365.

Payton P., K. R. Kottapalli, D. Rowland, W. Faircloth, B. Guo, M. Burow, N. Puppala, M. Gallo. 2009. "Gene expression profiling in peanut using high density oligonucleotide microarrays." *BMC Genomics* 10:265.

Peng P., H. Jiang, L. Luo, C. Ye, Y. Xiao. 2023. "Pyramiding of multiple genes to improve rice blast resistance of photo-thermo sensitive male Sterile Line, without yield penalty in hybrid rice production." *Plants (Basel)* 12:1389.

Pradhan S. K., D. K. Nayak, S. Mohanty, L. Behera, S. R. Barik, E. Pandit, S. Lenka, A. Anandan. 2015. "Pyramiding of three bacterial blight resistance genes for broad-spectrum resistance in deepwater rice variety, Jalmagna." *Rice* 8:51.

Rafael A. B., W. V. Barroga, P. N. M. Marcelo, E. M. Septiningsih, O. E. Manangkil. 2015. "Introgression of Sub1 and AG1 tolerance genes to NSIC Rc222 through marker-assisted backcrossing."

Rajarathinam P., G. Palanisamy, R. P, M. Narayana, M. Alagirisamy. 2023. "Marker assisted backcross to introgress late leaf spot and rust resistance in groundnut (*Arachis hypogaea* L.)." *Mol Biol Rep* 50:2411–2419.

Ribaut J. M., M. Ragot. 2007. "Marker-assisted selection to improve drought adaptation in maize: the backcross approach, perspectives, limitations, and alternatives." *J Exp Bot* 58:351–360.

Sartaj A. B., K. S. Udawela, H. M. V. G. Herath. 2016. "Marker assisted backcross breeding of Bg 358 (*Oryza sativa* L.) for the anaerobic germination tolerant QTL AG1." *Trop Agric Res* 27:287.

Schaeffer L. R. 2006. "Strategy for applying genome-wide selection in dairy cattle." *J Anim Breed Genet* 123:218–223.

Shu H., Z. Luo, Z. Peng, J. Wang. 2020. "The application of CRISPR/Cas9 in hairy roots to explore the functions of AhNFR1 and AhNFR5 genes during peanut nodulation." *BMC Plant Biol* 20:417.

Su L., S. Liu, X. Liu, B. Zhang, M. Li, L. Zeng, L. Li. 2021. "Transcriptome profiling reveals histone deacetylase 1 gene overexpression improves flavonoid, isoflavonoid, and phenylpropanoid metabolism in *Arachis hypogaea* hairy roots." *PeerJ* 9:e10976.

Tang G., P. Xu, W. Ma, F. Wang, Z. Liu, S. Wan, L. Shan. 2018. "Seed-specific expression of AtLEC1 increased oil content and altered fatty acid composition in seeds of peanut (*Arachis hypogaea* L.)." *Front Plant Sci* 9:260.

Tang Y., J. Huang, H. Ji, L. Pan, C. Hu, X. Qiu, H. Zhu, J. Sui, et al. 2022a. "Identification of AhFatB genes through genome-wide analysis and knockout of AhFatB reduces the content of saturated fatty acids in peanut (*Arichis hypogaea* L.)." *Plant Sci* 319:111247.

Tang Y., X. Qiu, C. Hu, J. Li, L. Wu, W. Wang, X. Li, X. Li, et al. 2022b. "Breeding of a new variety of peanut with high-oleic-acid content and high-yield by marker-assisted backcrossing." *Mol Breed* 42:42.

Tanksley S. D., T. J. Orton. 1983. "Isozymes in plant genetics and breeding."

Tiwari S., D. K. Mishra, A. Singh, P. K. Singh, R. Tuli. 2008. "Expression of a synthetic cry1EC gene for resistance against Spodoptera litura in transgenic peanut (*Arachis hypogaea* L.)." *Plant Cell Rep* 27:1017–1025.

Townsend J. A., D. A. Wright, R. J. Winfrey, F. Fu, M. L. Maeder, J. K. Joung, D. F. Voytas. 2009. "High-frequency modification of plant genes using engineered zinc-finger nucleases." *Nature* 459:442–445.

Varshney R. K., S. M. Mohan, P. M. Gaur, N. V. Gangarao, M. K. Pandey, A. Bohra, S. L. Sawargaonkar, A. Chitikineni, et al. 2013. "Achievements and prospects of genomics-assisted breeding in three legume crops of the semi-arid tropics." *Biotechnol Adv* 31:1120–1134.

Velu G., J. Crossa, R. P. Singh, Y. Hao, S. Dreisigacker, P. Perez-Rodriguez, A. K. Joshi, R. Chatrath, et al. 2016. "Genomic prediction for grain zinc and iron concentrations in spring wheat." *Theor Appl Genet* 129:1595–1605.

Wan L., B. Li, M. K. Pandey, Y. Wu, Y. Lei, L. Yan, X. Dai, H. Jiang, et al. 2016. "Transcriptome Analysis of a New Peanut Seed Coat Mutant for the Physiological Regulatory Mechanism Involved in Seed Coat Cracking and Pigmentation." Front *Plant Sci* 7:1491.

Wan L., Y. Lei, L. Yan, Y. Liu, M. K. Pandey, X. Wan, R. K. Varshney, J. Fang, et al. 2020. "Transcriptome and metabolome reveal redirection of flavonoids in a white testa peanut mutant." *BMC Plant Biol* 20:161.

Wang N., C. Tang, X. Fan, M. He, P. Gan, S. Zhang, Z. Hu, X. Wang, et al. 2022. "Inactivation of a wheat protein kinase gene confers broad-spectrum resistance to rust fungi." *Cell* 185:2961–2974.e2919.

Wang S., Y. Yang, M. Guo, C. Zhong, S. J. T. C. J. Sun. 2020. "Targeted mutagenesis of amino acid transporter genes for rice quality improvement using the CRISPR/Cas9 system." 8.

Wang T., E. Zhang, X. Chen, L. Li, X. Liang. 2010. "Identification of seed proteins associated with resistance to pre-harvested aflatoxin contamination in peanut (*Arachis hypogaea* L)." *BMC Plant Biol* 10:267.

Wang T., X. P. Chen, H. F. Li, H. Y. Liu, Y. B. Hong, Q. L. Yang, X. Y. Chi, Z. Yang, et al. 2013. "Transcriptome identification of the resistance-associated genes (RAGs) to Aspergillus flavus infection in pre-harvested peanut (*Arachis hypogaea*)." *Funct Plant Biol* 40:292–303.

Wang X., X. Yang, Y. Feng, P. Dang, W. Wang, R. Graze, J. P. Clevenger, Y. Chu, et al. 2021b. "Transcriptome profile reveals drought-induced genes preferentially expressed in response to water deficit in cultivated peanut (*Arachis hypogaea* L.)." *Front Plant Sci* 12:645291.

Wang X., Y. Liu, Z. Han, Y. Chen, D. Huai, Y. Kang, Z. Wang, L. Yan, et al. 2021a. "Integrated transcriptomics and metabolomics analysis reveal key metabolism pathways contributing to cold tolerance in peanut." *Front Plant Sci* 12:752474.

Wang Y., X. Cheng, Q. Shan, Y. Zhang, J. Liu, C. Gao, J. L. Qiu. 2014. "Simultaneous editing of three homoeoalleles in hexaploid bread wheat confers heritable resistance to powdery mildew." *Nat Biotechnol* 32:947–951.

Wu H., C. Sparks, B. Amoah, H. D. Jones. 2003. "Factors influencing successful Agrobacterium-mediated genetic transformation of wheat." *Plant Cell Rep* 21:659–668.

Wu N., K. Matand, H. Wu, B. Li, Y. Li, X. Zhang, Z. He, J. Qian, et al. 2013. "De novo next-generation sequencing, assembling and annotation of *Arachis hypogaea* L. Spanish botanical type whole plant transcriptome." *Theor Appl Genet* 126:1145–1149.

Xia H., C. Zhao, L. Hou, A. Li, S. Zhao, Y. Bi, J. An, Y. Zhao, et al. 2013. "Transcriptome profiling of peanut gynophores revealed global reprogramming of gene expression during early pod development in darkness." *BMC Genomics* 14:517.

Xiao N., C. Pan, Y. Li, Y. Wu, Y. Cai, Y. Lu, R. Wang, L. Yu, et al. 2021. "Genomic insight into balancing high yield, good quality, and blast resistance of japonica rice." *Genome Biol* 22:283.

Xue Q., X. Zhang, H. Yang, H. Li, Y. Lv, K. Zhang, Y. Liu, F. Liu, et al. 2021. "Transcriptome and Metabolome Analysis Unveil Anthocyanin Metabolism in Pink and Red Testa of Peanut (*Arachis hypogaea* L.)." *Int J Genomics* 2021:5883901.

Yin D., C. Ji, X. Ma, H. Li, W. Zhang, S. Li, F. Liu, K. Zhao, et al. 2018. "Genome of an allotetraploid wild peanut *Arachis monticola*: a de novo assembly." *Gigascience* 7.

Yin D., Y. Wang, X. Zhang, H. Li, X. Lu, J. Zhang, W. Zhang, S. Chen. 2013. "De novo assembly of the peanut (*Arachis hypogaea* L.) seed transcriptome revealed candidate unigenes for oil accumulation pathways." *PLoS One* 8:e73767.

Young N. D., S. D. Tanksley. 1989. "RFLP analysis of the size of chromosomal segments retained around the Tm-2 locus of tomato during backcross breeding." *Theor Appl Genet* 77:353–359.

Yuan M., J. Zhu, L. Gong, L. He, C. Lee, S. Han, C. Chen, G. He. 2019. "Mutagenesis of FAD2 genes in peanut with CRISPR/Cas9 based gene editing." *BMC Biotechnol* 19:24.

Zampieri E., A. Volante, C. Marè, G. Orasen, F. Desiderio, C. Biselli, M. Canella, L. Carmagnola, et al. 2023. "Marker-assisted pyramiding of blast-resistance genes in a japonica elite rice cultivar through forward and background selection." *Plants* 12:757.

Zhang C., P. Yun, J. Xia, K. Zhou, L. Wang, J. Zhang, B. Zhao, D. Yin, et al. 2023. "CRISPR/Cas9-mediated editing of Wx and BADH2 genes created glutinous and aromatic two-line hybrid rice." *Mol Breed* 43:24.

Zhang J., H. Zhang, S. Li, J. Li, L. Yan, L. Xia. 2021. "Increasing yield potential through manipulating of an ARE1 ortholog related to nitrogen use efficiency in wheat by CRISPR/Cas9." *J Integr Plant Biol* 63:1649–1663.

Zhang J., S. Liang, J. Duan, J. Wang, S. Chen, Z. Cheng, Q. Zhang, X. Liang, et al. 2012. "De novo assembly and characterisation of the transcriptome during seed development, and generation of genic-SSR markers in peanut (*Arachis hypogaea* L.)." *BMC Genomics* 13:90.

Zhang J., Z. Zhou, J. Bai, X. Tao, L. Wang, H. Zhang, J. K. Zhu. 2020. "Disruption of MIR396e and MIR396f improves rice yield under nitrogen-deficient conditions." *Natl Sci Rev* 7:102–112.

Zhao C., L. He, H. Xia, X. Zhou, Y. Geng, L. Hou, P. Li, G. Li, et al. 2021. "De novo full length transcriptome analysis of *Arachis glabrata* provides insights into gene expression dynamics in response to biotic and abiotic stresses." *Genomics* 113:1579–1588.

Zhou Y. C., W. R. Wu, J. M. Qi. 2003. "Approximate estimation of minimal sample size required for marker-assisted backcross breeding." *Yi Chuan Xue Bao* 30:625–630.

Zhu W., E. Zhang, H. Li, X. Chen, F. Zhu, Y. Hong, B. Liao, S. Liu, et al. 2013. "Comparative proteomics analysis of developing peanut aerial and subterranean pods identifies pod swelling related proteins." *J Proteomics* 91:172–187.

Zhu W., X. Chen, H. Li, F. Zhu, Y. Hong, R. K. Varshney, X. Liang. 2014. "Comparative transcriptome analysis of aerial and subterranean pods development provides insights into seed abortion in peanut." *Plant Mol Biol* 85:395–409.

Zhuang W., H. Chen, M. Yang, J. Wang, M. K. Pandey, C. Zhang, W. C. Chang, L. Zhang, et al. 2019. "The genome of cultivated peanut provides insight into legume karyotypes, polyploid evolution and crop domestication." *Nat Genet* 51:865–876.

11 Developments in Omics Research on the Tolerance and Response to Biotic Stresses in Peanut

Ruihua Huang, Hongqing Li, Sheidu Abdullaziz, Shengchun Zhang, and Weichang Yu

11.1 INTRODUCTION

The study of plant genomes has greatly advanced our understanding of gene function and has aided in the identification of new genes. On the basis of omics' technical advancements, scientists can use omics to regulate complex global biological systems (Mochida and Shinozaki, 2010). The rapid advancement of research and the possibility of exploring new applications beyond genome sequencing have been enabled by recent developments in DNA sequencing technology (Lister et al., 2009). The development of next-generation sequencing in particular has made the following applications feasible: RNA sequencing analysis (RNA-seq) for transcriptome and noncoding RNAome, whole-genome resequencing for variant analysis focusing on transcriptional regulatory networks, quantitative detection analysis (Chip-seq) for epigenome dynamics, and DNA–protein interactions (Lister et al., 2009). Other methods have also been developed, such as hormonomic analysis for plant hormone signaling, interactomic analysis for protein–protein interactions, and methabolomic analysis for metabolic products (Kojima et al., 2009; Saito and Matsuda, 2010). Researchers will be able to mine and test certain genes involved in crop development with the use of omics technology. Furthermore, integrated network analysis enhances our comprehension of the links between genotype and phenotype by revealing the molecular pathways connecting genes and metabolites (Shinozaki and Sakakibara, 2009; Vadivel, 2015; Kumar et al., 2017). The past several decades have seen a significant growth in the research of the molecular aspects of peanut–biotic factor interactions owing to technological advancements in the fields of genomes, transcriptomics, metabolomics, and proteomic studies. As a result, several omics-based studies have made an effort to identify the molecular pathways that regulate peanut resistance to bacteria, viruses, and pests as well as those that lead to yield loss.

This chapter primarily elucidates the fundamentals of various omics analyses, with a focus on studies that provide pertinent information on peanut defense mechanisms and resilience to biotic stressors such as bacteria, pathogens, and insect pests.

11.2 INFLUENCE OF SEVERAL BIOTIC STRESSORS ON PEANUT

11.2.1 Microbial Pathogens' Effect on Peanut

11.2.1.1 Fungi

Numerous biotic stressors impede the growth and development of peanuts, which include fungal diseases (leaf spot, rust, stem rot, and *Aspergillus flavus*). Leaf spot comprises the late leaf spot (LLS) caused by *Phaeoisariopsispersonata* and the early leaf spot (ELS) produced by *Cercosporaarachidicola*. According to McDonald et al. (1985), both leaf spot diseases can cause lesions up to one centimeter (1 cm) in diameter on peanut leaves, petioles, stems, and pegs. According to Tshilenge-Lukanda et al. (2012), leaves infected with ELS often have brown lesions surrounding them with a yellow ring on the top side, but leaves infected with LLS fungal disease typically have dark brown to black patches. When circumstances are right, LLS fungal disease can cause up to 70% of planted peanut yield losses (Grichar et al., 1998).

Another significant fungal disease limiting peanut output is rust, which is brought on by *Puccinia arachidis* Speg. (Subrahmanyam et al., 1993; Varshney et al., 2014). The illness can spread swiftly and widely because of the rust-infected leaves' propensity to remain connected to the plant and the pathogen's brief life cycle. More significantly, rust-infected peanuts lower agricultural production and have an impact on haulm, oxidizer yield, and seed oil quality (Leal-Bertioli et al., 2015).

The worst disease that affects peanuts is stem rot, which causes a noticeable production reduction (Punja, 1985). According to Thiessen and Woodward (2012), *Sclerotium rolfsii*, which grows thick, white hyphae that resemble silk, is the source of this illness, which is common in regions where peanuts are grown. *S. rolfsii*–infected peanuts typically show dark-brown lesions on the stem at or below the soil's surface, which are followed by the leaves progressively becoming yellow and drooping (Termorshuizen et al., 2007). *S. rolfsii* infections in peanuts can result in up to 80% yield loss as well as rot on the stem, peg, and

DOI: 10.1201/9781003360988-11

pod (Mehan et al., 1994). However, it is difficult to manage because it originates from *S. rolfsii* overwinter in the soil and attacks peanuts in the following season (Le et al., 2012; Mayee et al., 1988). Peanut plants may exhibit branch withering and maybe entire plant wilting following *S. rolfsii* infection (Le et al., 2012; Mayee et al., 1988).

The peanut sector is at risk from aflatoxin, which is produced by the fungus *Aspergillus flavus* (Krishna et al., 2015). When eaten, aflatoxin-contaminated peanuts have an impact on both human and animal health (Pittet, 1998; Kew, 2013). According to Torres et al. (2014), aflatoxin is formed prior to harvest as well as throughout the drying, storage, and transporting processes following harvest. It is also possible for peanut pods and seeds to become contaminated.

11.2.1.2 Viruses

Tomato spotted wilt virus (TSWV) is a significant pathogenic virus causing damage to peanuts in the southeastern United States (Garcia et al., 2000). It is transmitted by at least ten thrips species (Pappu et al., 2009; Riley et al., 2011), with *Frankliniella fusca* and *F. occidentalis* being predominant (Riley et al., 2011). *TSWV* causes stunting in peanuts and chlorosis, necrosis, or ring spots in leaves (Culbreath et al., 2003). The disease costs the United States $12.3 million annually (Riley et al., 2011).

Peanut stripe virus (PStV) is a pervasive plant-infecting virus, limiting peanut yield (Singh et al., 2009). It is a member of the genus *Potyvirus* and is one of the largest groups of viruses that infect plants. PStV can be spread by aphids in a non-sustained manner and is 10%–100% prevalent in fields and 1%–50% in peanut seeds (Chen et al., 1990; Xu et al., 1991; Bi et al., 1999; Xu, 2002). The principal infection source in the field is infected peanut seeds. Symptoms of PStV include stripes, light mottle, and blotches encircled by necrotic or chlorotic ringspots (Middleton and Saleh, 1988).

11.2.1.3 Bacteria

There are several bacteria that interact with peanuts. Different bacterial populations can live on the surfaces of seeds, roots, stems, and leaves. Because of the warm, humid weather, peanuts produced in tropical and subtropical regions are susceptible to several bacterial illnesses, the most common of which is bacterial wilt (Jiang et al., 2017).

One of the most destructive diseases affecting peanuts is peanut bacterial wilt, which is brought on by the soil-borne bacterium *Ralstonia solanacearum* (Salanoubat et al., 2002). Due to its global expansion and long soil survival time, *R. solanacearum* ranks second among the top ten harmful bacteria in plant pathology (Mansfield et al., 2012). *R. solanacearum* often infects peanuts via the root system and uses the vascular system to propagate to the aboveground portions (see Chapter 17). The plant will begin to wilt and eventually die if the germs multiply to an excessive degree (Genin, 2010). Furthermore, according to Jiang et al. (2017), bacterial wilt disease can cause yield losses of 10%–30% and, in extreme cases, 50%–100%.

11.2.2 Insects Pests' Effect on Peanut

Peanut, a significant crop in semi-arid tropical regions, is a key part of both developed and emerging nations' diets. However, due to insect pests and diseases, farmed fields yield little, leading to disastrous consequences such as incidental feeding, plant destruction, and yield loss (Wightman and Rao, 1994). Over 350 types of insects, including root-knot nematodes, *Aphids craccivora*, *Helicoverpa armigera*, and *Sodoptera litura*, are harmful to peanut (Campbell, 1983).

Root-knot nematodes (RKNs) are destructive plant-parasitic nematodes, with around 100 species worldwide (Jones et al., 2013). They can infest almost all cultivated plant species (Trudgill and Blok, 2001), with *Meloidogyne incognita*, *M. arenaria*, *M. javanica*, and *M. hapla* being the most common (Agrios, 2005). RKNs hatch second-stage juveniles that mechanically breach the root tip epidermis and develop giant cells (GCs), which impair root activity. The significant modification of the cell wall in multinucleated GCs is similar to transfer cells in cereal endosperm development. Infected plants show reduced growth, withering, and increased sensitivity to other infections (Jones, 1981).

Aphis craccivora is a significant insect group worldwide, causing approximately 20% peanut yield loss (Jagtap et al., 1984). It damages peanut plants from seedlings to mature green plants, removing sap; causing irritation, toxicity, honeydew accumulation, and sooty mold growth; and spreading the rosette virus (Murant, 1990). Aphids also spread the catjang disease (Raha and Raychaudhuri, 1981). At least seven viruses, including the peanut stripe virus and groundnut rosette virus, are transmitted by Aphids to damage peanut (Metcalf and Flint, 1975; Metcalf and Flint, 1975). Severe infestations significantly decrease peanut yields (Tarimo and Karel, 1987).

Helicoverpa armigera, a major agricultural pest, costs the US economy $2 billion annually (Tay et al., 2013) and affects over 200 plant species, including peanut (Sharma et al., 2005; Pratissoli et al., 2015). Its direct feeding behavior on flowering and fruiting bodies significantly reduces agricultural productivity (McGahan et al., 1991).

One of the most damaging species is *Spodoptera litura*, whose larvae devour leaves inordinately, severely defoliating the plant until only the midrib veins remain. This can lead to yield losses of 35%–55% (Srinivasa et al., 2012). Maximum damage is caused by *S. litura* during the blooming and fruiting stages (Prasad and Gowda, 2006).

11.3 OMICS DDVANCES IN UNDERSTANDING PEANUT RESPONSES TO BIOTIC STRESS

11.3.1 Peanut Responses to Pests

11.3.1.1 Root-Knot Nematode

The root-knot nematode (*Meloidogyne arenaria*) poses a significant threat to peanut yield in India, China, and the United States (Dong et al., 2008). Wild *Arachis* species show high resistance (Holbrook and Stalker 2003), and gene segments from these species were introduced into peanut cultivars to improve resistance (Garcia et al., 1995; Stalker et al., 2002). NemaTAM, a root-knot nematode-resistant variety, was developed in the United States using marker-assisted backcrossing (Tirumalaraju et al., 2011). Pathogenesis-related proteins, patatin-like proteins, and universal stress-related proteins were all detected by suppression subtractive hybridization to be expressed in RKN-infected NemaTAM roots. Because of its remarkable resistance to RKN, *Arachis stenosperma* has great potential in peanut breeding (Leal-Bertioli et al., 2015). In the peanut cultivar, nematode reproduction is reduced by up to 98% by two loci affecting resistance on chromosomes A02 and A09 (Ballén-Taborda et al., 2022). Genes producing TIR-NBS-LRR proteins are enriched in the large-effect QTL on A02 (Ballén-Taborda et al., 2019).

NBS-LRR class plant disease resistance genes, hormone signaling, and the synthesis of secondary metabolites were identified as resistance genes against root-knot nematode infection in transcriptome study of two wild relatives of peanuts (Mota et al., 2018). Genes involved in the regulation of auxin balance and signaling pathways for both jasmonic and salycilic acids were consistently found in the transcriptome analysis of RKN-resistant *Arachis* genotypes (Guimaraes et al., 2015). These results suggest a role for phytohormones in nematode resistance against root knots.

To investigate the molecular elements of RKN resistance in wild *Arachis* species (*A. stenosperma)*, researchers have employed transcriptomics and genomics. Resistance genes (R), which are primarily NBS-LRR–type genes, are discovered to be the source of high levels of resistance (Proite et al., 2008; Guimaraes et al., 2015; Dangl et al., 2013; Meyers et al., 1999; Bertioli et al., 2003). Numerous peanut cultivars were discovered to have hundreds of RGAs, whereas over 300 representative genes were identified in wild peanut species (Bertioli et al., 2016; Song et al., 2017; Ratnaparkhe et al., 2011; Wang et al., 2012; Yuksel et al., 2005). Among the seven genes that were differently expressed in RKN-infected wild *Arachis*, one of them encoded the resistance protein MG13. The generation of secondary metabolites and plant defense are among the conserved immune response genes that are induced by RKN infection, according to comparative genomics and differential gene expression studies conducted on 22 plant species, including peanuts (Morgante et al., 2013; Mota et al., 2020).

Numerous techniques, such as transcriptomics, genomics, metabolomics, and miRNAomics, are being used to further peanut RKN research. During RKN infection, 430 mRNAs, 111 miRNAs, 4453 lncRNAs, and 123 circRNAs were found to have differential expression by miRNAomics; these mRNAs, 10 lncRNAs, 4 circRNAs, 5 miRNAs, and 13 mRNAs were shown to be involved in oxidation reduction and biological metabolic processes (Xu et al., 2022). During root-knot nematode infection, differentially expressed proteins and genes were found using protease combined with transcriptome analysis. The majority of these proteins were linked to plant responses to pathogens. According to Martins et al. (2020), AsMLP34 was a potential candidate for peanut RKN resistance.

11.3.1.2 *Helicoverpa Armigera*

Over 200 plant species, including peanuts, are attacked by the significant agricultural pest *H. armigera*, which results in an annual crop loss of $2 billion (Tay et al., 2013; Pratissoli et al., 2015; McGahan et al., 1991). Twelve resistance lines were chosen under field and greenhouse conditions following the evaluation of 30 *Arachis* species. Morphological features are associated with resistance to the pest (Harsulkar et al., 1999; Harsulkar et al., 1999; Sharma et al., 2003). The protective mechanisms of salicylic acid (SA) and jasmonic acid (JA) in peanuts against *H. armigera* infestation have been found. Pretreating peanuts with JA greatly enhanced their POD and PPO activity as well as their high concentrations of MDA, H_2O_2, and total phenols. According to War et al. (2011), the exogenous administration of JA and SA increased resistance to *H. armigera*, with sensitive genotypes exhibiting greater levels of enzyme activity and secondary metabolites in peanuts. The phenol fingerprinting of peanut plants with varying degrees of resistance to *H. armigera* infestation was analyzed using metabolomics and high performance liquid chromatography (HPLC) (War et al., 2011). Chlorogen, clove, quercetin, and ferulic acid are among the common substances seen during *H. armigera* infestations (War et al., 2011).

11.3.1.3 *Aphids Craccivora*

Major peanut pest *A. craccivora* reduces output by propagating viruses and absorbing phloem sap (Feakin, 1973; Wightman & Amin, 1988). Phenomics was used to identify resistance lines in order to stop infestation (Padgham et al., 1990). *A. craccivora* resistance is controlled by a single recessive gene, according to genomic study (Merwe et al., 2001). Using amplified fragment length polymorphism and bulk segregation analysis, DNA markers associated with aphid resistance were created (Herselman et al., 2004).

High performance liquid chromatography (HPLC) and metabolomics were utilized to examine the phenol fingerprinting of peanut plants with varying degrees of resistance. During an aphid infestation, common substances such as ferulic acid, syringic acid, quercetin, and chlorogenic acid were found; the amounts of these compounds varied according to the genotypes and feeding habits of the aphids (War et al., 2016).

11.3.1.4 *Spodoptera Litura*

In nations that produce peanuts, *S. litura* is a significant insect pest that causes severe defoliation and a 35%–55% loss in production (Ahmad et al., 2007; Tojo et al., 2008). According to Sharma et al. (2003), researchers have examined peanut resistance to *S. litura* and discovered relationships between morphological traits and damage in both field and greenhouse settings. Increased CO_2 in peanuts slows pest growth, decreases *S. litura* pupation, and reduces insect digesting efficiency (Srinivasa et al., 2012). Application of JA can also increase resistance to *S. litura*. These results emphasize how critical it is to comprehend and control peanut pests (War et al., 2011). Insect pests can be effectively controlled by genetic engineering. Peanuts that expressed *AhMPK3A* ectopically exhibited resistance to *S. litura* larvae and produced more defense response genes (Kumar et al., 2009). Cry1AcF, a chimeric Bt toxin protein, was used to create peanut resistance against *S. litura*. According to Keshavareddy et al. (2013), the cry1AcF transgenic peanut was efficient against *S. litura* but also markedly raised the death rate of larvae.

11.3.2 Peanut Responses to Microbial Pathogens

11.3.2.1 Fungi

11.3.2.1.1 Leaf Spot

Leaf spot diseases are severe fungal foliar diseases that create dark or black lesions on the underside of infected leaves. Examples of these diseases include *Phaeoisariopsis personata*, causing late leaf spot (LLS); and *Cercospora arachidicola*, causing early leaf spot (ELS) (Tshilenge et al., 2012). Many studies have been conducted in the last 10 years with the goal of identifying the key genes responsible for peanut leaf spot resistance, and a large number of QTLs have been found. Through genetic mapping of GPBD4 (resistant varieties), 11 QTLs were identified on a genetic map encompassing 56 SSR marker loci, resulting in a recombinant inbred line (RIL) population for LLS resistance (Khedikar et al., 2010). An enhanced genetic map with 260 SSR loci in the same RIL revealed 28 QTLs (Sujay et al., 2012). Based on mapping with 139 additional SSR and transposable element markers, two significant QTLs for LLS resistance were characterized (Pandey et al., 2016). In marker-assisted breeding, 9 QTLs linked to ELS resistance and 22 QTLs linked to LLS resistance were identified (Pandey, Khan et al., 2017). Double-digested RAD-seq markers were used to identify six QTLs on various chromosomes of the ELS-resistant parent Tx964117 (Zhou et al., 2016; Khera et al., 2016; Liang et al., 2017; Han et al., 2018; Chu et al., 2019; Zhang et al., 2020; Ahmad et al., 2020). Consensus QTLs across various genetic backgrounds are essential and important in marker-assisted selection (Shirasawa et al., 2018). In 1.4-Mb genomic areas on A02, Shirasawa et al. (2018) discovered QTL candidates for LLS, while Lu et al. (2018) found one large meta-QTL including 26 candidate genes for LLS in an area of around 0.38 cM.

A highly susceptible M14 mutant was created, and RNA-Seq analysis revealed 2,219 differentially expressed genes, including upregulated pathogenesis-related proteins, WRKY transcription factor genes, downregulated chloroplast genes, and depressed plant hormones. Genes involved in recognition events and early signaling responses to the pathogen were identified (Han et al., 2017), and Dang et al. (2019) verified a group of 214 R-genes expressed in peanut leaves induced with leaf spot pathogen infection.

Gong et al. (2020) found 133 differentially expressed genes (DEGs) in peanut lines that were ELS sensitive and resistant. The LLS resistance genes *Aradu.P20JR* and *Aradu.Z87JB* were shown to be resistant. Dang et al. (2021) discovered that 36 R-genes, mostly receptor-like kinases and receptor-like proteins, which are involved in sensing the presence of pathogens and triggering defense responses, were associated with and differently expressed in resistant lines.

Additionally, the mechanism of the peanut response to LLS has been investigated by the application of metabolomics. Higher concentrations of secondary metabolites, such as phenolic acid, flavanols, stilbenes, and terpenoids, are seen in genotypes of peanuts that are resistant to LLS (Mahatma et al., 2021).

11.3.2.2 Rust

Puccinia arachidis Speg is the source of rust, a common foliar disease that greatly reduces peanut output (Subrahmanyam et al., 1993; Leal-Bertioli et al., 2015). In order to find candidate genes, mapping efforts have mostly concentrated on SNP enrichment mapping and genomic sequence-based analysis. Using QTL analysis of 268 recombinant inbred lines from the rust segregation mapping population (TAG 24 x GPBD 4), 12 QTLs for rust were found. Using single-marker analysis and composite interval mapping, a large QTL linked to rust resistance was found and confirmed (Khedikar et al., 2010; Pasupuleti et al., 2016). Using a QTL-Seq method, nine candidate genes for rust resistance on chromosome A03 were found (Pandey, Wang et al., 2017).

Additionally, two SSR markers (SSR_GO340445 and FRS 72) flanking a 1.2 cM region on chromosome A03 containing five potential genes were discovered (Mondal and Badigannavar, 2018). Using whole-genome resequencing and ddRAD-Seq, it was shown that the rust-resistance genes are located in the same genomic area of chromosome A03, at a distance of 2.7 MB (Shirasawa et al., 2018). Under rust disease pressure, nine genotypes resistant to rust showed a 77%–120% increase in pod yield above control, indicating substantial interactions between genotype and environment as well as genotype × environment (Chaudhari et al., 2019).

To find genomic areas and diagnostic markers for rust resistance in peanuts, a QTL-Seq method was employed. Three allele-specific SNP diagnostic markers were created, and 30 nonsynonymous SNPs impacting 25 potential genes were found (Pandey, Wang et al., 2017). Using competitive allele-specific polymerase chain reaction markers with single-nucleotide polymorphism, QTLs for rust resistance in the wild species *Arachis magna* were created (Leal-Bertioli et al., 2015). Gene expression in susceptible and resistant peanut genotypes was shown to vary by RNA-Seq analysis; resistant genotypes had upregulated genes, whereas susceptible genotypes had downregulated genes (Rathod et al., 2020; Mondal et al., 2022).

11.3.2.2.1 Stem Rot

Stem rot disease, caused by *Sclerotium rolfsii*, is a severe disease in peanuts, causing significant yield loss (Punja, 1985). One hundred and thirty-two isolates from Ibaraki, Japan, were found to have four mycelial compatibility groups (MCG) (Okabe and Matsumoto, 2000). Researchers found that *S. rolfsii*-derived elicitors can induce various systemic acquired resistance, including salicylic acid, defense-related signal molecule, peroxidase, marker enzymes, and polymer lignin (Nandini et al., 2010). Using Omics methods, the resistance of peanuts to *S. rolfsii* has mostly been studied. Jogi et al. (2016) discovered 12 differently expressed genes, 3 of which were linked to fungal virulence and 7 of which were associated with defense in plants. According to Dodia et al. (2019), there are 44 significant epistatic QTLs that account for 14.32%–67.95% of the variance in phenotype. Genes encoding enzymes involved in the jasmonic acid pathway, receptor-like kinases, and transcription factors (TFs) were shown to have high expression in genotypes of resistant peanuts by Bosamia et al. (2020). While the jasmonic acid signaling pathway was thought to be a potential peanut defense mechanism, PAMP-triggered immunity was thought to be a potential mechanism of stem rot resistance. The genetic foundation of *S. rolfsii* varying levels of aggression was uncovered by de novo genome sequencing of two separate pathogen strains (ZY and GP3) (Yan et al., 2021).

11.3.2.2.2 Aspergillus Flavus

The global peanut sector is seriously threatened by the carcinogenic fungus *A. flavus*, which produces aflatoxin (Torres et al., 2014). A number of QTLs in peanuts that exhibit resistance to *A. flavus* infection have been found in recent investigations. Two QTLs with 7.96% and 12.16% PVE on chromosomes A03 and A10, respectively, were found by Yu et al. (2019). A significant QTL with 18.11% PVE on A03 and a minor QTL with 4.4% PVE on B04 were found by Khan et al. (2020). Six unique QTLs with 5.03%–10.87% PVE were found on chromosomes A05, A08, B01, B03, and B10 by Jiang et al. in 2021.

In order to create efficient defenses against aflatoxin contamination after harvest, the study focuses on the molecular processes of peanut–*Aspergillus* interactions and peanut resistance to aflatoxin formation. Researchers have found 22 potential *Aspergillus*-resistance genes with high expression in the resistant cultivar and 62 genes with increased expression in resistant cultivars (Guo et al., 2011; Wang et al., 2010). Using a differential proteomics approach, a number of aflatoxins-responsive proteins were identified in peanut cotyledons infected with aflatoxin-producing (toxigenic) but not non-aflatoxinogenic (toxigenic) *A. flavus* strains. These proteins included those involved in immune signaling, innate immunity, induction of defense, PAMP perception, penetration resistance, hypersensitive response, DNA and RNA stabilization, biosynthesis of phytoalexins, condensed tannin synthesis, cell wall responses, peptidoglycan assembly, detoxification, and metabolic regulation.

Transcriptome and proteome studies conducted during *A. flavus* infection of J11 peanut (Zhao et al., 2019) found 663 DEGs and 314 differentially expresed proteins. A number of potential genes implicated in the resistance response to *A. flavus* were identified by transcriptomic network analysis using publicly accessible RNA-seq datasets of resistant and susceptible peanut cultivars infected with the pathogen (Jayaprakash et al., 2021). Weighted gene coexpression network analysis and the comparative transcriptome technique were used to identify the core genes that are positively linked with peanut resistance to *A. flavus* (Cui et al., 2022).

A. flavus resistance in peanuts is influenced by a number of genes, including cytochrome P450, pectinesterase, phosphatidylinositol transfer protein, SNARE protein SYP121, pattern recognition receptors (PRRs), MAPK kinase, STK, ACO1, PR10, phosphatidylinositol transfer protein, and cytochrome P450. When resistant and susceptible cultivars were treated with *A. flavus*, correlation analysis of small RNAs, transcriptomes, and degradationomes showed 447 genes, 30 miRNAs, and 21 possible miRNA/mRNA pairings exhibiting differential expression. MiR156/SPL pairings may govern the accumulation of flavonoids in susceptible and resistant genotypes, and the miR482/2118 family may regulate the expression level of the NBS-LRR gene in resistant genotypes (Zhao et al., 2020).

11.3.2.3 Viruses

Tomato spotted wilt virus (TSWV), cucumber mosaic virus (CMV), peanut stripe virus (PSTV), peanut stunting virus (PSV), peanut bud necrosis virus (PBNV), peanut mottle virus (PEMOV), and peanut ringspot virus (PRSV) are among the viruses that attack peanuts. One of the most harmful single-stranded RNA viruses in peanuts is TSWV, which can limit peanut output and induce spotted wilt disease (Garcia et al., 2000). The identification of TSWV-resistant QTLs has been accomplished by quantitative trait locus (QTL) mapping. A significant QTL with a phenotypic variation explained (PVE) of 35.8% connected to resistance to TSWV on linkage group A01 was found in the first genetic linkage map based on NC94022-derived population (Qin et al., 2012; Khera et al., 2016).

11.3.2.4 Bacterial

Rastonia solanacearum, the bacterium that causes bacterial wilt, is a severe disease that has a big effect on the peanut business (Salanoubat et al., 2002). It is challenging to eradicate because of its broad host range and high long-term survival capabilities. Genetic behavior, trait mapping, gene identification, and diagnostic indicators of peanut bacterial wilt resistance have advanced throughout the last 10 years. Four SSR primers and one AFLP primer were shown to be useful in assessing the genetic connections of 31 genotypes of peanuts with varying levels of resistance to *R. solanacearum* (Jiang et al., 2007). According to Zhao et al. (2016), linkage map analysis identified two significant QTLs with PVEs of 12%–21% connected to *R. solanacearum* resistance on linkage groups LG01 and LG10.

A high-density SNP map using a RIL population from the hybrid Yuanza YZ9012 with Xuzhou68-4 verified a significant and consistent QTL for bacterial wilt resistance (qBWB02.1) on chromosome B02 (Wang et al., 2018). For a recombinant inbred line (RIL) resulting from a hybrid between cultivars Yuanza 9102 and Xuzhou 68-4, a QTL was verified using sequencing-based trait mapping and QTL-seq (Luo et al., 2019). Diagnostic markers confirmed and validated two significant QTLs on chromosome B02 (Luo et al., 2020). Using polymorphic KASP markers, the major QTL for peanut resistance to bacterial wilt was fine-mapped to a 216.7 kb area (Qi et al., 2022) (see Chapter 17).

Numerous peanut genes, including those involved in cell structure, defense, energy, signal transduction, metabolism, cell development, and transcription, have been linked by researchers to interactions with *R. solanacearum*. Additionally, studies have discovered 40 transcript-derived fragments (TDFs) which are connected to BW resistance and exhibit variable expression (Peng et al., 2011). Chen et al. (2014) discovered that resistance variations between susceptible and resistant genotypes are influenced by down-regulation of primary metabolism and genotype-specific expression patterns of defense-related DEGs. Direct cloning of peanuts has produced a number of candidate genes with potential resistance to bacterial wilt that has been functionally investigated.

11.4 CONCLUSION AND PERSPECTIVE

This chapter highlights peanut resilience to several biotic pests and diseases by summarizing several omics research conducted in the past 20 years on peanut biotic stress. The area of peanut response to biotic stress is making significant strides, and a significant quantity of omics data has been generated to interpret the molecular cues between the invading pathogens and the peanut. Finding the genomic areas in wild animals that offer disease resistance is the first step in using marker-assisted selection, which can facilitate the introgression of wild disease resistance genes with reduced linkage drag. In order to create disease-resistant cultivars and lower yield loss in the production of peanuts, the discovered disease resistance genes can also be used in genomics- and biotechnology-assisted breeding. But the information that has been gathered over the years is jumbled and unstructured, and it has not been logically sorted out. Our capacity to evaluate and make use of the vast amounts of omics data generated by technology has been superseded. These days, Big Data is a typical occurrence. However, it is vital to build technology to deal with omics data, and, more specifically, to use omics data in crop breeding to improve tolerance to biotic and abiotic stressors.

Research on peanut omics is comparatively behind that of other crops and model plant species because of the complexity of the peanut genome. Peanut biotic stressors yield little data when analyzed with omics methods, including transcriptomics, proteomics, and other technologies. The most significant pests and diseases that affect peanuts are primarily focused on, such as fungus, insects, and root knot nematodes; bacterial and viral pathogens receive less attention. Therefore, additional work is required in omics research in the future about biotic stresses that influence peanuts.

Furthermore, before Omics data products—such as potential QTLs, molecular markers, candidate genes, metabolites, proteins, and so on—can be trusted fully applied in academic studies and commercial breeding, they must be cross-checked with results from wet lab research. In order to precisely map the coarsely mapped QTLs to a smaller area on the relevant chromosome and ultimately clone them, more technologies could be required. Functional genetics is required to overexpress candidate genes that may be involved in peanut disease and insect resistance in transgenic research, or CRISPR genome editing technology may be used to induce their mutations in order to confirm their associated roles. However, omics data is useful for genome editing to create effective CRISPR targets that minimize off-target mutations in crop breeding.

Omics data may be used in peanut molecular breeding for pest and disease resistance, thanks to the success of peanut genome editing using CRISPR technology (Zhang et al., 2021).

REFERENCES

Agrios, George N. "Plant diseases caused by nematodes." *Plant Pathology* 5 (2005): 825–874.

Ahmad, Mushtaq, M. Iqbal Arif, and Munir Ahmad. "Occurrence of insecticide resistance in field populations of Spodoptera litura (Lepidoptera: Noctuidae) in Pakistan." *Crop Protection* 26, no. 6 (2007): 809–817.

Ahmad, Suhail, Bhagwat Nawade, Chandramohan Sangh, Gyan P. Mishra, Tejas C. Bosamia, Narendra Kumar, Jentilal R. Dobaria, and Harsukh P. Gajera. "Identification of novel QTLs for late leaf spot resistance and validation of a major rust QTL in peanut (Arachis hypogaea L.)." *3 Biotech* 10 (2020): 1–13.

Ballén-Taborda, Carolina, Ye Chu, Peggy Ozias-Akins, C. Corley Holbrook, Patricia Timper, Scott A. Jackson, David J. Bertioli, and Soraya CM Leal-Bertioli. "Development and genetic characterization of peanut advanced backcross lines that incorporate root-knot nematode resistance from Arachis stenosperma." *Frontiers in Plant Science* 12 (2022): 785358.

Ballén-Taborda, Carolina, Ye Chu, Peggy Ozias-Akins, Patricia Timper, C. Corley Holbrook, Scott A. Jackson, David J. Bertioli, and Soraya CM Leal-Bertioli. "A new source of root-knot nematode resistance from Arachis stenosperma incorporated into allotetraploid peanut (Arachis hypogaea)." *Scientific Reports* 9, no. 1 (2019): 17702.

Bertioli, David John, Steven B. Cannon, Lutz Froenicke, Guodong Huang, Andrew D. Farmer, Ethalinda KS Cannon, Xin Liu et al. "The genome sequences of Arachis duranensis and Arachis ipaensis, the diploid ancestors of cultivated peanut." *Nature Genetics* 48, no. 4 (2016): 438–446.

Bi, Y., X. Li, H. Wang, and J. Liu. "Survey of peanut stripe virus in Chinese peanut seed production regions." Plant Protection 25, no. 2 (1999): 45–49.

Bosamia, Tejas C., Sneha M. Dodia, Gyan P. Mishra, Suhail Ahmad, Binal Joshi, Polavakkalipalayam P. Thirumalaisamy, Narendra Kumar et al. "Unraveling the mechanisms of resistance to Sclerotium rolfsii in peanut (Arachis hypogaea L.) using comparative RNA-Seq analysis of resistant and susceptible genotypes." *PLoS One* 15, no. 8 (2020): e0236823.

Bertioli, David J., Jerry Jenkins, Josh Clevenger, Olga Dudchenko, Dongying Gao, Guillermo Seijo, Soraya CM Leal-Bertioli et al. "The genome sequence of segmental allotetraploid peanut Arachis hypogaea." *Nature Genetics* 51, no. 5 (2019): 877–884.

Bertioli, David J., Soraya C. M. Leal-Bertioli, M. B. Lion, V. L. Santos, G. Pappas, S. B. Cannon, and P. M. Guimaraes. "A large scale analysis of resistance gene homologues in Arachis." *Molecular Genetics and Genomics* 270 (2003): 34–45.

Chaudhari, Sunil, Dhirendra Khare, Sudam C. Patil, Subramaniam Sundravadana, Murali T. Variath, Hari K. Sudini, Surendra S. Manohar, Ramesh S. Bhat, and Janila Pasupuleti. "Genotype× environment studies on resistance to late leaf spot and rust in genomic selection training population of peanut (Arachis hypogaea L.)." *Frontiers in Plant Science* 10 (2019): 1338.

Chen, Xiaoping, He Li, Manish K. Pandey, Qingli Yang, Xingjun Wang, Vanika Garg, Hongjie Li, et al. "Draft genome of the peanut a-genome progenitor (Arachis duranensis) provides insights into geocarpy, oil biosynthesis, and allergens. Proceedings of the National Academy of Sciences 113, no. 24 (2014): 6785–6790.

Chen, Z., Z. Xu, Z. Zhang, and Y. Wu. 1990. "Incidence and seed transmission of peanut stripe virus in china." Plant Disease 74, no. 10 (1990): 834–837.

Chu, Ye, Peng Chee, Albert Culbreath, Thomas G. Isleib, C. Corley Holbrook, and Peggy Ozias-Akins. "Major QTLs for resistance to early and late leaf spot diseases are identified on chromosomes 3 and 5 in peanut (Arachis hypogaea)." *Frontiers in Plant Science* 10 (2019): 883.

Cui, Mengjie, Suoyi Han, Du Wang, Muhammad Salman Haider, Junjia Guo, Qi Zhao, Pei Du et al. "Gene Co-expression Network Analysis of the Comparative Transcriptome Identifies Hub Genes Associated With Resistance to Aspergillus flavus L. in Cultivated Peanut (Arachis hypogae a L.)." *Frontiers in Plant Science* 13 (2022): 899177.

Culbreath, A. K., J. W. Todd, and S. L. Brown. "Epidemiology and management of tomato spotted wilt in peanut." Annual Review of Phytopathology 41 (2003): 53–75.

Dangl, Jeffery L., Diana M. Horvath, and Brian J. Staskawicz. "Pivoting the plant immune system from dissection to deployment." *Science* 341, no. 6147 (2013): 746–751.

Dang, Phat M., Marshall C. Lamb, Kira L. Bowen, and Charles Y. Chen. "Identification of expressed R-genes associated with leaf spot diseases in cultivated peanut." *Molecular Biology Reports* 46 (2019): 225–239.

Dang, Phat M., Marshall C. Lamb, and Charles Y. Chen. "Association of differentially expressed R-gene candidates with leaf spot resistance in peanut (Arachis hypogaea L.)." *Molecular Biology Reports* 48, no. 1 (2021): 323–334.

Dong, W. B., C. C. Holbrook, P. Timper, T. B. Brenneman, Y. Chu, and P. Ozias-Akins. "Resistance in peanut cultivars and breeding lines to three root-knot nematode species." *Plant Disease* 92, no. 4 (2008): 631–638.

Dodia, Sneha M., Binal Joshi, Sunil S. Gangurde, Polavakkalipalayam P. Thirumalaisamy, Gyan P. Mishra, Dayama Narandrakumar, Pooja Soni et al. "Genotyping-by-sequencing based genetic mapping reveals large number of epistatic interactions for stem rot resistance in groundnut." *Theoretical and Applied Genetics* 132 (2019): 1001–1016.

Feakin, Susan D., ed. "Pest control in groundnuts" (3rd ed.). *PANS Manual No. 2* (pp. iii–197). London: Centre for Overseas Pest Research, 1973.

Garcia, G. M., H. T. Stalker, and G. Kochert. "Introgression analysis of an interspecific hybrid population in peanuts (Arachis hypogaea L.) using RFLP and RAPD markers." *Genome* 38, no. 1 (1995): 166–176.

Garcia, L. E., R. L. Brandenburg, and J. E. Bailey. "Incidence of Tomato spotted wilt virus (Bunyaviridae) and tobacco thrips in Virginia-type peanuts in North Carolina." *Plant Disease* 84, no. 4 (2000): 459–464.

Genin, Stéphane. "Molecular traits controlling host range and adaptation to plants in Ralstonia solanacearum." *New Phytologist* 187, no. 4 (2010): 920–928.

Gong, Limin, Suoyi Han, Mei Yuan, Xingli Ma, Austin Hagan, and Guohao He. "Transcriptomic analyses reveal the expression and regulation of genes associated with resistance to early leaf spot in peanut." *BMC Research Notes* 13 (2020): 1–6.

Grichar, W. J., B. A. Besler, and A. J. Jaks. "Peanut (Arachis hypogaea L.) cultivar response to leaf spot disease development under four disease management programs." *Peanut Science* 25, no. 1 (1998): 35–39.

Guimaraes, Patricia M., Larissa A. Guimaraes, Carolina V. Morgante, Orzenil B. Silva Jr, Ana Claudia G. Araujo, Andressa CQ Martins, Mario AP Saraiva et al. "Root transcriptome analysis of wild peanut reveals candidate genes for nematode resistance." *PLoS One* 10, no. 10 (2015): e0140937.

Guo, Baozhu, Natalie D. Fedorova, Xiaoping Chen, Chun-Hua Wan, Wei Wang, William C. Nierman, Deepak Bhatnagar, and Jiujiang Yu. "Gene expression profiling and identification of resistance genes to Aspergillus flavus infection in peanut through EST and microarray strategies." *Toxins* 3, no. 7 (2011): 737–753.

Han, Suoyi, Hua Liu, Mei Yan, Feiyan Qi, Yaqi Wang, Ziqi Sun, Bingyan Huang et al. "Differential gene expression in leaf tissues between mutant and wild-type genotypes response to late leaf spot in peanut (Arachis hypogaea L.)." *PLoS One* 12, no. 8 (2017): e0183428.

Han, Suoyi, Mei Yuan, Josh P. Clevenger, Chun Li, Austin Hagan, Xinyou Zhang, Charles Chen, and Guohao He. "A SNP-based linkage map revealed QTLs for resistance to early and late leaf spot diseases in peanut (Arachis hypogaea L.)." *Frontiers in Plant Science* 9 (2018): 1012.

Harsulkar, Abhay M., Ashok P. Giri, Aparna G. Patankar, Vidya S. Gupta, Mohini N. Sainani, Prabhakar K. Ranjekar, and Vasanti V. Deshpande. "Successive use of non-host plant proteinase inhibitors required for effective inhibition of Helicoverpa armigera gut proteinases and larval growth." *Plant Physiology* 121, no. 2 (1999): 497–506.

Herselman, L., R. Thwaites, F. M. Kimmins, Brigitte Courtois, P. J. A. Van Der Merwe, and Sue E. Seal. "Identification and mapping of AFLP markers linked to peanut (Arachis hypogaea L.) resistance to the aphid vector of groundnut rosette disease." *Theoretical and Applied Genetics* 109, no. 7 (2004): 1426–1433.

Holbrook, C. Corley, and H. Thomas Stalker. "Peanut breeding and genetic resources." *Plant Breeding Reviews* 22 (2003): 297–356.

Jiang, Gaofei, Zhong Wei, Jin Xu, Huilan Chen, Yong Zhang, Xiaoman She, Alberto P. Macho, Wei Ding, and Boshou Liao. "Bacterial wilt in China: history, current status, and future perspectives." *Frontiers in Plant Science* 8 (2017): 1549.

Jayaprakash, A., A. Roy, R. R. Thanmalagan, A. Arunachalam, and L. Ptv. Immune response gene co-expression network analysis of Arachis hypogaea infected with Aspergillus flavus. Genomics 113, no. 5 (2021): 2977–2988.

Jiang, Yifei, Huaiyong Luo, Bolun Yu, Yingbin Ding, Yanping Kang, Li Huang, Xiaojing Zhou et al. "High-density genetic linkage map construction using whole-genome resequencing for mapping QTLs of resistance to aspergillus flavus infection in peanut." *Frontiers in Plant Science* 12 (2021): 745408.

Jiang, Huifang, Boshou Liao, Xiaoping Ren, Yong Lei, Emma Mace, Tingdong Fu, and J. H. Crouch. "Comparative assessment of genetic diversity of peanut (Arachis hypogaea L.) genotypes with various levels of resistance to bacterial wilt through SSR and AFLP analyses." *Journal of Genetics and Genomics* 34, no. 6 (2007): 544–554.

Jones, John T., Annelies Haegeman, Etienne GJ Danchin, Hari S. Gaur, Johannes Helder, Michael GK Jones, Taisei Kikuchi et al. "Top 10 plant-parasitic nematodes in molecular plant pathology." *Molecular Plant Pathology* 14, no. 9 (2013): 946–961.

Jones, M. G. K. "Host cell responses to endoparasitic nematode attack: structure and function of giant cells and syncytia." *Annals of Applied Biology* 97, no. 3 (1981): 353–372.

Jogi, Ansuya, John W. Kerry, Timothy B. Brenneman, James H. Leebens-Mack, and Scott E. Gold. "Identification of genes differentially expressed during early interactions between the stem rot fungus (Sclerotium rolfsii) and peanut (Arachis hypogaea) cultivars with increasing disease resistance levels." *Microbiological Research* 184 (2016): 1–12.

Jagtap, A. B., B. D. Ghule, and A. B. Deokar. "Assessment of losses in yield of'Phule Pragati'groundnut caused by insect pests." Journal of Maharashtra Agricultural Universities 9 (1984): 697–698.

Keshavareddy, G., S. Rohini, S. V. Ramu, S. Sundaresha, A. R. V. Kumar, P. Ananda Kumar, and M. Udayakumar. "Transgenics in groundnut (Arachis hypogaea L.) expressing cry1AcF gene for resistance to Spodoptera litura (F.)." *Physiology and Molecular Biology of Plants* 19 (2013): 343–352.

Kew, Michael C. "Aflatoxins as a cause of hepatocellular carcinoma." *Journal of Gastrointestinal & Liver Diseases* 22, no. 3 (2013): 305–310.

Khedikar, Y. P., M. V. C. Gowda, C. Sarvamangala, K. V. Patgar, H. D. Upadhyaya, and R. K. Varshney. "A QTL study on late leaf spot and rust revealed one major QTL for molecular breeding for rust resistance in groundnut (Arachis hypogaea L.)." *Theoretical and Applied Genetics* 121 (2010): 971–984.

Khera, Pawan, Manish K. Pandey, Hui Wang, Suping Feng, Lixian Qiao, Albert K. Culbreath, Sandip Kale et al. "Mapping quantitative trait loci of resistance to tomato spotted wilt virus and leaf spots in a recombinant inbred line population of peanut (Arachis hypogaea L.) from SunOleic 97R and NC94022." *PLoS One* 11, no. 7 (2016): e0158452.

Khan, Shahid Ali, Hua Chen, Ye Deng, Yuhua Chen, Chong Zhang, Tiecheng Cai, Niaz Ali et al. "High-density SNP map facilitates fine mapping of QTLs and candidate genes discovery for Aspergillus flavus resistance in peanut (Arachis hypogaea)." *Theoretical and Applied Genetics* 133 (2020): 2239–2257.

Kojima, Mikiko, Tomoe Kamada-Nobusada, Hirokazu Komatsu, Kentaro Takei, Takeshi Kuroha, Masaharu Mizutani, Motoyuki Ashikari et al. "Highly sensitive and high-throughput analysis of plant hormones using MS-probe modification and liquid chromatography–tandem mass spectrometry: an application for hormone profiling in Oryza sativa." *Plant and Cell Physiology* 50, no. 7 (2009): 1201–1214.

Krishna, Gaurav, Birendra K. Singh, Eun-Ki Kim, Vivek K. Morya, and Pramod W. Ramteke. "Progress in genetic engineering of peanut (Arachis hypogaea L.)—A review." *Plant Biotechnology Journal* 13, no. 2 (2015): 147–162.

Kumar, Koppolu Raja Rajesh, Tantravahi Srinivasan, and Pulugurtha Bharadwaja Kirti. "A mitogen-activated protein kinase gene, AhMPK3 of peanut: molecular cloning, genomic organization, and heterologous expression conferring resistance against Spodoptera litura in tobacco." *Molecular Genetics and Genomics* 282 (2009): 65–81.

Kumar, Rakesh, Abhishek Bohra, Arun K. Pandey, Manish K. Pandey, and Anirudh Kumar. "Metabolomics for plant improvement: status and prospects." *Frontiers in Plant Science* 8 (2017): 1302.

Pasupuleti, Janila, Manish K. Pandey, Surendra S. Manohar, Murali T. Variath, Premalatha Nallathambi, Hajisaheb L. Nadaf, Harikishan Sudini, and Rajeev K. Varshney. "Foliar fungal disease-resistant introgression lines of groundnut (Arachis hypogaea L.) record higher pod and haulm yield in multilocation testing." *Plant Breeding* 135, no. 3 (2016): 355–366.

Le, C. N., R. Mendes, M. Kruijt, and J. M. Raaijmakers. "Genetic and phenotypic diversity of Sclerotium rolfsii in groundnut fields in central Vietnam." *Plant Disease* 96, no. 3 (2012): 389–397.

Leal-Bertioli, Soraya C. M., Ana Carolina V. F. José, Dione M. T. Alves-Freitas, Márcio C. Moretzsohn, Patrícia M. Guimarães, Stephan Nielen, Bruna S. Vidigal et al. "Identification of candidate genome regions controlling disease resistance in Arachis." *BMC Plant Biology* 9 (2009): 1–12.

Leal-Bertioli, Soraya C. M., Uiara Cavalcante, Ediene G. Gouvea, Carolina Ballén-Taborda, Kenta Shirasawa, Patrícia M. Guimarães, Scott A. Jackson, David J. Bertioli, and Márcio C. Moretzsohn. "Identification of QTLs for rust resistance in the peanut wild species Arachis magna and the development of KASP markers for marker-assisted selection." *G3: Genes, Genomes, Genetics* 5, no. 7 (2015): 1403–1413.

Luo, Huaiyong, Manish K. Pandey, Aamir W. Khan, Bei Wu, Jianbin Guo, Xiaoping Ren, Xiaojing Zhou et al. "Next-generation sequencing identified genomic region and diagnostic markers for resistance to bacterial wilt on chromosome B02 in peanut (Arachis hypogaea L.)." *Plant Biotechnology Journal* 17, no. 12 (2019): 2356–2369.

Luo, Huaiyong, Manish K. Pandey, Ye Zhi, Huan Zhang, Siliang Xu, Jianbin Guo, Bei Wu et al. "Discovery of two novel and adjacent QTLs on chromosome B02 controlling resistance against bacterial wilt in peanut variety Zhonghua 6." *Theoretical and Applied Genetics* 133 (2020): 1133–1148.

Lu, Qing, Hao Liu, Yanbin Hong, Haifen Li, Haiyan Liu, Xingyu Li, Shijie Wen et al. "Consensus map integration and QTL meta-analysis narrowed a locus for yield traits to 0.7 cM and refined a region for late leaf spot resistance traits to 0.38 cM on linkage group A05 in peanut (Arachis hypogaea L.)." *BMC Genomics* 19 (2018): 1–10.

Liang, Yuya, Michael Baring, Shichen Wang, and Endang M. Septiningsih. "Mapping QTLs for leafspot resistance in peanut using SNP-based next-generation sequencing markers." *Plant Breeding and Biotechnology* 5, no. 2 (2017): 115–122.

Lister, Ryan, Brian D. Gregory, and Joseph R. Ecker. "Next is now: new technologies for sequencing of genomes, transcriptomes, and beyond." *Current Opinion in Plant Biology* 12, no. 2 (2009): 107–118.

Martins, Andressa CQ, Angela Mehta, André M. Murad, Ana PZ Mota, Mário AP Saraiva, Ana CG Araújo, Robert NG Miller, Ana CM Brasileiro, and Patrícia M. Guimarães. "Proteomics unravels new candidate genes for Meloidogyne resistance in wild Arachis." *Journal of Proteomics* 217 (2020): 103690.

Mayee, C. D., and V. V. Datar. "Diseases of groundnut in the Tropics." In Diseases of Field Crops in the Tropics, edited by S. P. Raychaudhuri and J. P. Verma, 85–118. Oxford & IBH Publishing Co.

Mansfield, John, Stephane Genin, Shimpei Magori, Vitaly Citovsky, Malinee Sriariyanum, Pamela Ronald, M. A. X. Dow et al. "Top 10 plant pathogenic bacteria in molecular plant pathology." *Molecular Plant Pathology* 13, no. 6 (2012): 614–629.

Mahatma, M. K., Lokesh Kumar Thawait, K. S. Jadon, P. P. Thirumalaisamy, S. K. Bishi, Khyati J. Rathod, Aman Verma, Narendra Kumar, and B. A. Golakiya. "Metabolic profiling for dissection of late leaf spot disease resistance mechanism in groundnut." *Physiology and Molecular Biology of Plants* 27, no. 5 (2021): 1027–1041.

Meyers, Blake C., Allan W. Dickerman, Richard W. Michelmore, Subramoniam Sivaramakrishnan, Bruno W. Sobral, and Nevin D. Young. "Plant disease resistance genes encode members of an ancient and diverse protein family within the nucleotide-binding superfamily." *The Plant Journal* 20, no. 3 (1999): 317–332.

McDonald, D., P. W. G. R. S. Subrahmanyam, R. W. Gibbons, and D. H. Smith. "Early and late leaf spots of groundnut." Information Bulletin No. 21. Patancheru, Andhra Pradesh, India: International Crops Research Institute for the Semi-Arid Tropics (ICRISAT) (1985): 19861315652.

McGahan, P., R. J. Lloyd, and K. P. Rynne. "The cost of Helicoverpa in Queensland crops." In *A Review of Heliothis Research in Australia*. Conference and Workshop Series QC91006. Queensland Department of Primary Industries, Brisbane, Australia, pp. 11–28. 1991.

Mehan, V. K., C. D. Mayee, and D. McDonald. "Management of Sclerotium rolfsii-caused stem and pod rots of groundnut—a critical review." *International Journal of Pest Management* (1994): 313–320.

Metcalf, R. L., and W. P. Flint. *Destructive and Useful Insects, their Habit & Control*. 4th ed. McGraw Hill, 1975: 646.

Middleton, K. J., and N. Saleh. "Surveys for peanut stripe virus in East and Southeast Asia." In Coordination of Research on Peanut Stripe Virus: Summary Proceedings of the First Meeting to Coordinate Research on Peanut Stripe Virus Disease of Groundnut, June 9–12, 1987, Malang, Indonesia, edited by D. V. R. Reddy, 95–102. Patancheru, India: ICRISAT, 1988.

Mochida, Keiichi, and Kazuo Shinozaki. "Genomics and bioinformatics resources for crop improvement." *Plant and Cell Physiology* 51, no. 4 (2010): 497–523.

Mondal, Suvendu, and Anand M. Badigannavar. "Mapping of a dominant rust resistance gene revealed two R genes around the major Rust_QTL in cultivated peanut (Arachis hypogaea L.)." *Theoretical and Applied Genetics* 131 (2018): 1671–1681.

Mondal, Suvendu, K. Mohamed Shafi, Avi Rạizada, Hui Song, Anand M. Badigannavar, and Ramanathan Sowdhamini. "Development of candidate gene-based markers and map-based cloning of a dominant rust resistance gene in cultivated groundnut (Arachis hypogaea l.)." *Gene* 827 (2022): 146474.

Moretzsohn, Márcio C., Ediene G. Gouvea, Peter W. Inglis, Soraya CM Leal-Bertioli, José FM Valls, and David J. Bertioli. "A study of the relationships of cultivated peanut (Arachis hypogaea) and its most closely related wild species using intron sequences and microsatellite markers." *Annals of Botany* 111, no. 1 (2013): 113–126.

Morgante, Carolina V., Ana CM Brasileiro, Philip A. Roberts, Larissa A. Guimaraes, Ana CG Araujo, Leonardo N. Fonseca, Soraya CM Leal-Bertioli, David J. Bertioli, and Patricia M. Guimaraes. "A survey of genes involved in Arachis stenosperma resistance to Meloidogyne arenaria race 1." *Functional Plant Biology* 40, no. 12 (2013): 1298–1309.

Mota, Ana Paula Zotta, Diana Fernandez, Fabricio BM Arraes, Anne-Sophie Petitot, Bruno Paes de Melo, Maria E. Lisei de Sa, Priscila Grynberg et al. "Evolutionarily conserved plant genes responsive to root-knot nematodes identified by comparative genomics." *Molecular Genetics and Genomics* 295 (2020): 1063–1078.

Mota, Ana Paula Zotta, Bruna Vidigal, Etienne GJ Danchin, Roberto Coiti Togawa, Soraya CM Leal-Bertioli, David John Bertioli, Ana Claudia Guerra Araujo, Ana Cristina Miranda Brasileiro, and Patricia Messenberg Guimaraes. "Comparative root transcriptome of wild Arachis reveals NBS-LRR genes related to nematode resistance." *BMC Plant Biology* 18 (2018): 1–16.

Murant, A. F. "Dependence of groundnut Rosette virus on its satellite RNA as well as on groundnut Rosette assist or luteovirus for tr̥ansmission by Aphis craccivora." Journal of General Virology 71, no. 9 (September 1990): 2163–2166.

Nandini, Durgesh, Jampala Siva Satya Mohan, and Gunjan Singh. "Induction of systemic acquired resistance in Arachis hypogaea L. by Sclerotium rolfsii derived elicitors." *Journal of Phytopathology* 158, no. 9 (2010): 594–600.

Okabe, Ikuko, and Naoyuki Matsumoto. "Population structure of Sclerotium rolfsii in peanut fields." *Mycoscience* 41, no. 2 (2000): 145–148.

Padgham, D. E., F. M. Kimmins, and GV Ranga Rao. "Resistance in groundnut (Arachis hypogaea L.) to Aphis craccivora (Koch)." *Annals of Applied Biology* 117, no. 2 (1990): 285–294.

Pandey, Manish K., Aamir W. Khan, Vikas K. Singh, Manish K. Vishwakarma, Yaduru Shasidhar, Vinay Kumar, Vanika Garg et al. "QTL-seq approach identified genomic regions and diagnostic markers for rust and late leaf spot resistance in groundnut (A rachis hypogaea L.)." *Plant Biotechnology Journal* 15, no. 8 (2017): 927–941.

Pandey, Manish K., Hui Wang, Pawan Khera, Manish K. Vishwakarma, Sandip M. Kale, Albert K. Culbreath, C. Corley Holbrook, Xingjun Wang, Rajeev K. Varshney, and Baozhu Guo. "Genetic dissection of novel QTLs for resistance to leaf spots and tomato spotted wilt virus in peanut (Arachis hypogaea L.)." *Frontiers in Plant Science* 8 (2017): 25.

Pappu, H. R., P. C. Coy, and R. K. K. Fock. "Thrips vectors of tospoviruses." In Thrips and Tospovirus Management, edited by H. R. Pappu and R. A. C. Jones, 123–139. St. Paul, MN: APS Press, 2009.

Peng, Wen-Fang, Jian-Wei Lv, Xiao-Ping Ren, Li Huang, Xin-Yan Zhao, Qi-Gen Wen, and Hui-Fang Jiang. "Differential expression of genes related to bacterial wilt resistance in peanut (Arachis hypogaea L.)." *Yi Chuan= Hereditas* 33, no. 4 (2011): 389–396.

Pittet, A. "Natural occurrence of mycotoxins in foods and feeds-an updated review." Revue de Médecine Vétérinaire 149, no. 6 (1998): 479–492.

Prasad, M. N Rajendra, and M. V. C. Gowda. "Mechanisms of resistance to tobacco cutworm (Spodoptera litura F.) and their implications to screening for resistance in groundnut." *Euphytica* 149 (2006): 387–399.

Pratissoli, Dirceu, Victor L. S Lima, Victor D. Pirovani, and Wallace L. Lima. "Occurrence of Helicoverpa armigera (Lepidoptera: Noctuidae) on tomato in the Espírito Santo state." *Horticultura Brasileira* 33 (2015): 101–105.

Proite, K., R. Carneiro, R. Falcão, A. Gomes, S. Leal-Bertioli, P. Guimarães, and D. Bertioli. "Post-infection development and histopathology of Meloidogyne arenaria race 1 on Arachis spp." *Plant Pathology* 57, no. 5 (2008): 974–980.

Punja, Zamir K. "The biology, ecology, and control of Sclerotium rolfsii." *Annual Review of Phytopathology* 23, no. 1 (1985): 97–127.

Qi, Feiyan, Ziqi Sun, Hua Liu, Zheng Zheng, Li Qin, Lei Shi, Qingzheng Chen et al. "QTL identification, fine mapping, and marker development for breeding peanut (Arachis hypogaea L.) resistant to bacterial wilt." *Theoretical and Applied Genetics* 135, no. 4 (2022): 1319–1330.

Qin, Hongde, Suping Feng, Charles Chen, Yufang Guo, Steven Knapp, Albert Culbreath, Guohao He et al. "An integrated genetic linkage map of cultivated peanut (Arachis hypogaea L.) constructed from two RIL populations." *Theoretical and Applied Genetics* 124 (2012): 653–664.

Raha, S. K., and D. N. Raychaudhuri. "Studies on the aphids (Homoptera: Aphididae) of Nagaland." Entomon 6 (1981): 317–323.

Rathod, Visha, Rasmieh Hamid, Rukam S. Tomar, Rushika Patel, Shital Padhiyar, Jasminkumar Kheni, P. P. Thirumalaisamy, and Nasreen S. Munshi. "Comparative RNA-Seq profiling of a resistant and susceptible peanut (Arachis hypogaea) genotypes in response to leaf rust infection caused by Puccinia arachidis." *3 Biotech* 10 (2020): 1–15.

Ratnaparkhe, Milind B., Xiyin Wang, Jingping Li, Rosana O. Compton, Lisa K. Rainville, Cornelia Lemke, Changsoo Kim, Haibao Tang, and Andrew H. Paterson. "Comparative analysis of peanut NBS-LRR gene clusters suggests evolutionary innovation among duplicated domains and erosion of gene microsynteny." *New Phytologist* 192, no. 1 (2011): 164–178.

Riley, D. G., A. K. Culbreath, and R. Srinivasan. 2011. "Epidemiology of spotted wilt disease of peanut caused by tomato spotted wilt virus." Virus Research 159, no. 2 (2011): 101–109.

Saito, Kazuki, and Fumio Matsuda. "Metabolomics for functional genomics, systems biology, and biotechnology." *Annual Review of Plant Biology* 61, no. 1 (2010): 463–489.

Salanoubat, Marcel, S. Genin, François Artiguenave, J. Gouzy, S. Mangenot, M. Arlat, A. Billault et al. "Genome sequence of the plant pathogen Ralstonia solanacearum." *Nature* 415, no. 6871 (2002): 497–502.

Sharma, H. C., G. Pampapathy, S. L. Dwivedi, and L. J. Reddy. "Mechanisms and diversity of resistance to insect pests in wild relatives of groundnut." *Journal of Economic Entomology* 96, no. 6 (2003): 1886–1897.

Sharma, H. C., P. C. Stevenson, and CLL Gowda Heliothis. "H eliothis/Helicoverpa Management: Emerging Trends and Prospects for Future Research." *Heliothis/Helicoverpa Management* (2005): 465–474.

Shinozaki, Kazuo, and Hitoshi Sakakibara. "Omics and bioinformatics: an essential toolbox for systems analyses of plant functions beyond 2010." *Plant and Cell Physiology* 50, no. 7 (2009): 1177–1180.

Shirasawa, Kenta, Ramesh S. Bhat, Yogendra P. Khedikar, Venkataswamy Sujay, Rohini M. Kolekar, Sharanabasappa B. Yeri, Mallenahally Sukruth et al. "Sequencing analysis of genetic loci for resistance for late leaf spot and rust in peanut (Arachis hypogaea L.)." *Frontiers in Plant Science* 9 (2018): 1727.

Singh, M. K., V. Chandel, V. Hallan, R. Ram, and A. A. Zaidi. "Occurrence of Peanut stripe virus on patchouli and raising of virus-free patchouli plants by meristem tip culture." *Journal of Plant Diseases and Protection* 116, no. 1 (2009): 2–6.

Song, Hui, Pengfei Wang, Changsheng Li, Suoyi Han, Chuanzhi Zhao, Han Xia, Yuping Bi, Baozhu Guo, Xinyou Zhang, and Xingjun Wang. "Comparative analysis of NBS-LRR genes and their response to Aspergillus flavus in Arachis." *PLoS One* 12, no. 2 (2017): e0171181.

Srinivasa Rao, M., D. Manimanjari, M. Vanaja, C. A. Rama Rao, K. Srinivas, V. U. M. Rao, and B. Venkateswarlu. "Impact of elevated CO2 on tobacco caterpillar, Spodoptera litura on peanut, Arachis hypogea." *Journal of Insect Science* 12, no. 1 (2012): 103.

Stalker, H. T., and W. V. Campbell. "Resistance of wild species of peanut to an insect complex." *Peanut science* 10, no. 1 (1983): 30–33.

Stalker, H. T., M. K. Beute, B. B. Shew, and K. R. Barker. "Registration of two root-knot nematode-resistant peanut germplasm lines.(Registrations of Germplasm)." *Crop Science* 42, no. 1 (2002): 312–314.

Subrahmanyam, P., D. McDonald, L. J. Reddy, S. N. Nigam, and D. H. Smith. "Origin and utilization of rust resistance in groundnut." In *Durability of Disease Resistance*, edited by Th. Jacobs and J. E. Parlevliet, 147–158. Dordrecht: Springer Netherlands, 1993.

Sujay, V., M. V. C. Gowda, M. K. Pandey, R. S. Bhat, Y. P. Khedikar, H. L. Nadaf, B. Gautami et al. "Quantitative trait locus analysis and construction of consensus genetic map for foliar disease resistance based on two recombinant inbred line populations in cultivated groundnut (Arachis hypogaea L.)." *Molecular Breeding* 30 (2012): 773–788.

Tarimo, A. J. P., and A. K. Karel. "Insect pests of groundnut (Arachis hypogaea L.) in Tanzania and their control." Proceedings of Second Regional Groundnut Workshop for Southern Africa, 10–14 February 1986, Bulawayo, Zimbabwe, pp. 95–102. Patancheru, India: ICRISAT, 1987.

Tay, Wee Tek, Miguel F. Soria, Thomas Walsh, Danielle Thomazoni, Pierre Silvie, Gajanan T. Behere, Craig Anderson, and Sharon Downes. "A brave new world for an old world pest: Helicoverpa armigera (Lepidoptera: Noctuidae) in Brazil." *PLoS One* 8, no. 11 (2013): e80134.

Termorshuizen, Aad J. "Fungal and fungus-like pathogens of potato." In *Potato Biology and Biotechnology*, 643–665. Elsevier Science BV, 2007.

Thiessen, Lindsey D., and Jason E. Woodward. "Diseases of peanut caused by soilborne pathogens in the Southwestern United States." *International Scholarly Research Notices* 2012, no. 1 (2012): 517905.

Tirumalaraju, Sivananda V., Mukesh Jain, and Maria Gallo. "Differential gene expression in roots of nematode-resistant and-susceptible peanut (Arachis hypogaea) cultivars in response to early stages of peanut root-knot nematode (Meloidogyne arenaria) parasitization." *Journal of Plant Physiology* 168, no. 5 (2011): 481–492.

Tojo, Sumio, Yoichi Hayakawa, and Piyawan Phaophan. "Strains in the common cutworm, Spodoptera litura (Lepidoptera: Noctuidae) with differing host ranges." *Applied Entomology and Zoology* 43, no. 4 (2008): 491–496.

Torres, Adriana Mabel, Germán Gustavo Barros, Sofia Alejendra Palacios, Sofia Noemi Chulze, and Paola Battilani. "Review on pre-and post-harvest management of peanuts to minimize aflatoxin contamination." *Food Research International* 62 (2014): 11–19.

Trudgill, David L., and Vivian C. Blok. "Apomictic, polyphagous root-knot nematodes: exceptionally successful and damaging biotrophic root pathogens." *Annual Review of Phytopathology* 39, no. 1 (2001): 53–77.

Tshilenge-Lukanda, L., Nkongolo KKC, A. Kalonji-Mbuyi, and Kizungu RV. "Epidemiology of the groundnut (*Arachis hypogaea* L.) leaf spot disease: genetic analysis and developmental cycles." *American Journal of Plant Sciences* (2012).

Vadivel, Anguraj Arun K. "Gel-based proteomics in plants: time to move on from the tradition." *Frontiers in Plant Science* 6 (2015): 369.

Van der Merwe, P. J. A., P. Subrahmanyam, F. M. Kimmins, and J. Willekens. "Mechanisms of resistance to groundnut rosette." *International Arachis Newsletter* 21 (2001): 43–46.

Varshney, Rajeev K., Manish K. Pandey, Pasupuleti Janila, Shyam N. Nigam, Harikishan Sudini, M. V. C. Gowda, Manda Sriswathi, T. Radhakrishnan, Surendra S. Manohar, and Patne Nagesh. "Marker-assisted introgression of a QTL region to improve rust resistance in three elite and popular varieties of peanut (Arachis hypogaea L.)." *Theoretical and Applied Genetics* 127 (2014): 1771–1781.

Wang, Houmiao, Yong Lei, Liyun Wan, Liying Yan, Jianwei Lv, Xiaofeng Dai, Xiaoping Ren, Wei Guo, Huifang Jiang, and Boshou Liao. "Comparative transcript profiling of resistant and susceptible peanut post-harvest seeds in response to aflatoxin production by Aspergillus flavus." *BMC Plant Biology* 16 (2016): 1–16.

Wang, Hui, R. Varma Penmetsa, Mei Yuan, Limin Gong, Yongli Zhao, Baozhu Guo, Andrew D. Farmer et al. "Development and characterization of BAC-end sequence derived SSRs, and their incorporation into a new higher density genetic map for cultivated peanut (Arachis hypogaea L.)." *BMC Plant Biology* 12 (2012): 1–11.

Wang, Tong, Erhua Zhang, Xiaoping Chen, Ling Li, and Xuanqiang Liang. "Identification of seed proteins associated with resistance to pre-harvested aflatoxin contamination in peanut (Arachis hypogaea L)." *BMC Plant Biology* 10 (2010): 1–11.

Wang, Zhihui, Dongxin Huai, Zhaohua Zhang, Ke Cheng, Yanping Kang, Liyun Wan, Liying Yan, Huifang Jiang, Yong Lei, and Boshou Liao. "Development of a high-density genetic map based on specific length amplified fragment sequencing and its application in quantitative trait loci analysis for yield-related traits in cultivated peanut." *Frontiers in Plant Science* 9 (2018): 827.

War, Abdul Rashid, Michael Gabriel Paulraj, Mohd Yousf War, and Savarimuthu Ignacimuthu. "Jasmonic acid-mediated-induced resistance in groundnut (*Arachis hypogaea* L.) against Helicoverpa armigera (Hubner)(Lepidoptera: Noctuidae)." *Journal of Plant Growth Regulation* 30 (2011): 512–523.

War, Abdul Rashid, Suraj Prasad Sharma, and Hari Chand Sharma. "Differential induction of flavonoids in groundnut in response to Helicoverpa armigera and Aphis craccivora infestation." *International Journal of Insect Science* 8 (2016): IJIS–S39619.

Wightman, J. A., and P. W. Amin. "Groundnut pests and their control in the semi-arid tropics." *International Journal of Pest Management* 34, no. 2 (1988): 218–226.

Wightman, J. A., and G. V. Rao. "Groundnut pests." *World Crop* (1994): 395–479. doi: 10.1007/978-94-011-0733-4-11

Xu, Ping, Hui Li, Xiaohua Wang, Ge Zhao, Xiaofei Lu, Shengjie Dai, Xiaoyu Cui, Mei Yuan, and Zhenning Liu. "Integrated analysis of the lncRNA/circRNA-miRNA-mRNA expression profiles reveals novel insights into potential mechanisms in response to root-knot nematodes in peanut." *BMC Genomics* 23, no. 1 (2022): 239.

Xu, Z. "Epidemiology and seedborne transmission of peanut stripe virus." In Proceedings of the 2002 International Congress on Plant Virology, edited by L. Wang, 122–128. Beijing: China Agricultural Press, 2002.

Xu, Z., Z. Zhang, and Z. Chen. "Seed transmission of peanut stripe virus and its prevalence in peanut seeds." Journal of Phytopathology 132, no. 3 (1991): 219–224.

Yan, Liying, Zhihui Wang, Wanduo Song, Pengmin Fan, Yanping Kang, Yong Lei, Liyun Wan et al. "Genome sequencing and comparative genomic analysis of highly and weakly aggressive strains of Sclerotium rolfsii, the causal agent of peanut stem rot." *BMC Genomics* 22 (2021): 1–15.

Yu, Bolun, Dongxin Huai, Li Huang, Yanping Kang, Xiaoping Ren, Yuning Chen, Xiaojing Zhou et al. "Identification of genomic regions and diagnostic markers for resistance to aflatoxin contamination in peanut (Arachis hypogaea L.)." *BMC Genetics* 20 (2019): 1–13.

Yuksel, Bayram, James C. Estill, Stefan R. Schulze, and Andrew H. Paterson. "Organization and evolution of resistance gene analogs in peanut." *Molecular Genetics and Genomics* 274 (2005): 248–263.

Zhou, Xiaojing, Youlin Xia, Junhua Liao, Kede Liu, Qiang Li, Yang Dong, Xiaoping Ren et al. "Quantitative trait locus analysis of late leaf spot resistance and plant-type-related traits in cultivated peanut (Arachis hypogaea L.) under multi-environments." *PLoS One* 11, no. 11 (2016): e0166873.

Zhang, Wang, Jun-Lin Xian, Chao Sun, Chun-Ming Wang, Li Shi, and Wei-Chang Yu. "Preliminary study of genome editing of peanut FAD2 genes by CRISPR/Cas9". *Acta Agronomica Sinica* (2021): 1481–1490.

Zhao, Xiaobo, Chunjuan Li, Caixia Yan, Juan Wang, Cuiling Yuan, Hao Zhang, and Shihua Shan. "Transcriptome and proteome analyses of resistant preharvest peanut seed coat in response to Aspergillus flavus infection." *Electronic Journal of Biotechnology* 39 (2019): 82–90.

Zhao, Chuanzhi, Tingting Li, Yuhan Zhao, Baohong Zhang, Aiqin Li, Shuzhen Zhao, Lei Hou et al. "Integrated small RNA and mRNA expression profiles reveal miRNAs and their target genes in response to Aspergillus flavus growth in peanut seeds." *BMC Plant Biology* 20 (2020): 1–16.

Zhao, Yongli, Chong Zhang, Hua Chen, Mei Yuan, Rick Nipper, C. S. Prakash, Weijian Zhuang, and Guohao He. "QTL mapping for bacterial wilt resistance in peanut (Arachis hypogaea L.)." *Molecular Breeding* 36 (2016): 1–11.

12 Omics Advancement in Improving Abiotic Stress Tolerance in Peanut

Xiaoyuan Chi, Na Chen, Lijuan Pan, Jing Xu, Xiao Jiang, Xiangzhen Yin, and Xuhong Zhao

12.1 INTRODUCTION

Abiotic stresses, like salt, cold, or drought, have adverse impacts on plant growth and yield. To adapt to such stresses, plants evolve complicated signaling cascades that regulate the expression of stress-related genes involved in the direct or indirect improvement of stress resistance (Huang et al., 2012; Ritonga et al., 2021). Omics approaches offer a holistic view of the molecules that make up a cell or organism, allowing the identification of genes (genomics), metabolites (metabolomics), mRNA (transcriptomics), and proteins (proteomics) in a non-biased biological context. Functional genomics has helped to detect stress-related genes in crops (Sharma and Pandey, 2017; Singh et al., 2019). Accessibility to the whole genome sequence of numerous plant species and recent developments in genomic approaches promise to deliver methods for locating stress-responsive genes at the genome-wide level (Razzaq et al., 2021).

Peanut (*Arachis hypogaea* L.), which is the main edible oil source, is an important oil seed crop (Pruthvi et al., 2014). Abiotic stresses have a serious impact on peanut growth and yield. For example, cold stress seriously influences its germination, development, bloom, and yields (Chang et al., 2019; Liu et al., 2017; Sorensen et al., 2009; Upadhyaya et al., 2009; Wang et al., 2003). Salinity decreases peanut germination and dry matter production (Singh et al., 2010; Meena et al., 2016), damaging the photosynthetic apparatus (Qin et al., 2011) and inducing K and Ca deficiencies (Leidi et al., 1992) and yield losses (Singh et al., 2010). Drought can seriously affect the plants during flowering and the quality of peanut kernels during the podding period (Miao et al., 2021; Wang et al., 2019; Jiang and Ren, 2004), leading to a decline in peanut quality and yield (Girdthai et al., 2010; Jongrungklang et al., 2013). In addition, peanut often encounters various metal stresses during its life history, which not only limits yield, but also threatens human health due to the accumulation of toxic metals (Liu et al., 2017; Yu et al., 2019).

Patel et al. (2022) investigated the differential regulation of physio-biochemical and metabolomics parameters in peanut (*Arachis hypogaea* L.) under individual (salt, drought, cold, and heat) and combined stress treatments using multivariate correlation analysis. The results showed that combined heat, salt, and drought stress compounds the stress effect of individual stresses. Combined stresses that included heat had the highest electrolyte leakage and lowest relative water content. Lipid peroxidation and chlorophyll contents did not significantly change under combined stresses. Biochemical parameters, such as free amino acids, polyphenol, starch, and sugars, significantly changed under combined stresses compared to individual stresses. Free amino acids increased under combined stresses that included heat; starch, sugars, and polyphenols increased under combined stresses that included drought; and proline concentration increased under combined stresses that included salt. Researchers can use metabolomics data obtained under different individual and combined stresses to identify molecular phenotypes involved in the acclimation response of plants to changing abiotic stress conditions (Patel, Khandwal et al., 2022).

Development of peanut cultivars with enhanced abiotic stress tolerance is one of the most sustainable approaches for its successful production. Although peanut omics research is in its infancy, it has received more and more attention in recent years. In this regard, researchers are making efforts to understand the stress tolerance mechanism, discover genes, and investigate the interaction of genetic and environmental factors. Researchers have developed several omics approaches, tools, and resources for peanut. Modern sequencing technologies have greatly accelerated genomics and transcriptomics studies in peanut. These advancements facilitate quantitative trait loci (QTL) mapping, genome-wide association studies (GWAS), and genomic selection (GS). Meanwhile, researchers have also made progress in other omics branches, including proteomics, metabolomics, and ionomics (Figure 12.1).

12.2 TRANSCRIPTOMICS FOR PEANUT ABIOTIC STRESS TOLERANCE

Transcriptomics is a discipline that studies gene transcription and transcriptional regulation in cells at the overall level, and can reveal the molecular mechanism of specific biological processes and the pathogenesis of diseases (Wang et al., 2009). Revealing the expression of abiotic stress at the whole genome level from the transcriptional level is crucial for understanding the complex regulatory network related to stress adaptation and tolerance. Researchers

DOI: 10.1201/9781003360988-12

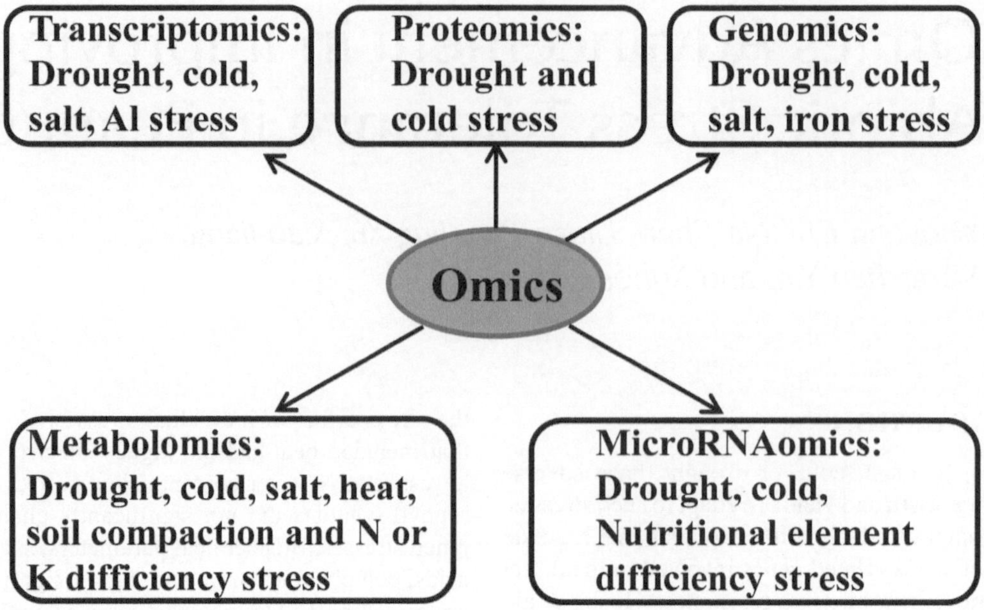

FIGURE 12.1 The role of omics approaches for peanut improvement under abiotic stresses.

construct the transcriptional regulatory network of the genome by analyzing and screening functional genes that are differentially expressed in various tissues and organs of peanut under different abiotic stresses, across different growth and development stages, and in response to different environmental stress factors, to identify the relationship between key functional genes and resistance. This will help to understand the injury mechanism of stress factors and the adaptation mechanism of peanut to stress from the transcriptional level.

12.2.1 Drought Tolerance

Sun et al. (2013) analyzed and compared the transcriptional expression profiles of peanut before and after drought treatment. They identified 935 differentially expressed genes (DEGs), 323 of which are involved in the upregulation of secondary metabolic processes, biomacromolecular metabolism, energy metabolism, and other metabolic pathways under drought stress. Nine genes related to flavonoid metabolism showed significant upregulation under drought stress. Semi-quantitative RT-PCR assays revealed that 15% PEG treatment for 6 hours significantly induced the expression of *AhPAL*. The result showed that flavonoid metabolism might play an important role in peanut responding to drought stress. Researchers conducted a comparative transcriptome analysis of a drought-tolerant and a drought-sensitive peanut variety (Ren et al., 2021). Comparative transcriptomic analysis identified 5,376 DEGs common to both varieties, while 2,993 genes specifically changed in the drought-tolerant variety. This study lays the foundation for the analysis of the molecular mechanism of drought tolerance and promotes the genetic improvement of peanut drought tolerance. Zhao et al. (2018) used whole-transcriptome sequencing to analyze the differential gene expression of the drought-resistant peanut cultivar J11 under drought stress. Researchers identified 51,554 genes, including 49,289 known genes and 2,265 unknown genes. They identified 4648 DEGs by comparing the gene expression profiling differences between the treatment and control groups. A large number of genes were associated with drought stress, including transcription factors and genes involved in photosynthesis–antenna proteins, carbon metabolism, and the citrate cycle. These results provide insights into the diverse mechanisms that underlie the successful establishment of drought resistance in peanut, thereby facilitating the identification of important genes in peanut related to drought management. The molecular mechanism related to drought resistance of peanut still needs to be further improved.

12.2.2 Salt Stress

Chen et al. (2016) screened genes regulated by salt stress in peanut roots using microarray techniques. They identified 4828 upregulated and 3752 downregulated probe sets in peanut roots subjected to 3 h and 48 h of salt stress. The main up-egulated biological processes involved in salt stress responses included transcription regulation, stress response, and metabolism and biosynthetic processes. The main downregulated biological processes included transport processes, photosynthesis, and development. Zhang, Shi et al. (2020) constructed six transcriptome sequencing libraries including 24 h salt treatments and control samples from the young leaves of peanut (Zhang, Shi et al., 2020). A comprehensive analysis between two groups detected 3,425 DEGs, including 2,013 upregulated genes and 1,412 downregulated genes. Of these DEGs, 141 transcription factors (TFs) mainly consisting of MYB, AP2/ERF, WRKY, bHLH, and HSF were identified in response to salinity stress. These identified DEGs' response to salt tolerance of cultivated peanut will provide a solid foundation for improving salt-tolerant peanut genetic manipulation in the future. Many NAC genes are involved in the response to salt stress(Yuan et al., 2020). To explore NAC genes involved in the response to salt stress, researchers analyzed the published transcriptome sequencing results of cultivated peanut under salt treatment (Zhang, Shi et al., 2020). Under salt treatment, the expression level of 28 genes was upregulated by 2-fold, whereas the expression of 15 genes was downregulated more than 2-fold. The expression of 8 genes was significantly upregulated more than 5-fold, the greatest expression reached 17-fold, and the expression of 6 genes was downregulated more than 5-fold.The molecular mechanisms related to peanut response to salt stress mainly come from the study of a few independent genes, and it is difficult to obtain systematic biological genetic information. Advancements in transcriptomic-related techniques enable systematic studies of the molecular mechanisms underlying plant stress responses.

12.2.3 Cold Stress

Identifying cold-tolerant genes and revealing the possible regulation mechanism in peanut represent one of the important ways to create new cold-tolerant varieties and improve the cold tolerance of peanut (Zhang, Jiang et al., 2020). High oleic acid (HOA) peanut is sensitive to cold stress especially during germination, which limits its application in low temperature areas. To understand the molecular mechanism of cold responses in HOA peanuts at the germination stage, Zhang et al. (2019) selected four HOA peanut varieties with different cold tolerances in field experiments and analyzed their genome-wide gene regulation under low temperatures. They identified 139,429 unigenes by high-throughput sequencing and transcriptome analysis. Among these, the detected 3520 common DEGs between two groups of cold-tolerant and cold-sensitive peanuts, and the number of upregulated genes was greater than that of downregulated genes in the cold-tolerant peanuts. A study was conducted to compare the physiology, transcriptome, and lipid structure of the cold-tolerant variety NH5 and the cold-sensitive variety FH18 at different time points. Transcriptome analysis showed that lipid metabolism (including membrane lipid metabolism and fatty acid metabolism) might be an important factor in cold tolerance of peanut. They identified 59 cold tolerance genes involved in lipid metabolism (Zhang et al., 2020). Plants tolerate cold stress by regulating gene networks controlling cellular and physiological traits to modify growth and development. Transcription factor (TF)-directed regulation of transcription within these gene networks is key to eliciting appropriate responses. Identification of traits related to cold tolerance is helpful in the breeding of cold-tolerant crops. Jiang et al. (2020) conducted comparative studies of transcriptome analysis in two peanut varieties with different cold tolerance. They identified 87 TF families including 2328 TF genes. Among them, they found that 445 TF genes exhibited significant differential expression between the two peanut varieties under low temperature stress. These studies contribute to a better understanding of the complex mechanism of peanut cold tolerance factor in response to low temperature stress, and provide valuable resources for the study of the evolutionary history and biological functions of peanut cold tolerance genes. In this experiment, 16 upregulated probe sets had calcium binding activity. Three calmodulin-binding protein genes, one CDPK gene and one calmodulin gene were also upregulated. These proteins may induce the early signaling transduction during cold treatment in peanut (Chen et al., 2014).

12.2.4 Other Stresses

In addition, some metal elements are essential nutrients for the growth of higher plants. When lacking, they will affect the normal growth and development of plants. When excessive, they will also produce toxic effects on plants. Aluminum (Al) toxicity is an important factor in limiting peanut growth on acidic soil. The transcriptome analysis of the root tips (0–1 cm) of peanut cultivar ZH2 (Al-sensitive) and 99-1507 (Al-tolerant) was carried out (Xiao et al., 2021). The study identified 8,587 DEGs. GO and KEGG pathway analysis excavated a group of important Al-responsive genes related to organic acid transport, metal cation transport, transcription regulation and programmed cell death (PCD). These genes were promising targets to modulate Al tolerance in peanuts.

12.3 PROTEOMICS FOR PEANUT ABIOTIC STRESS TOLERANCE

Proteomic analysis is a powerful technique for identifying key regulators involved in various biological processes by providing precise information on molecular alterations within pathways of interest. Peanut plants are prone to different abiotic stresses such as drought, salinity, and extreme temperatures, which can have detrimental effects on plant growth and productivity.

Kottapalli (2009) identified 49 non-redundant proteins involved in a variety of cellular functions (e.g. cell wall strengthening, signal transduction, energy metabolism, cellular detoxification, and gene regulation) that could influence the molecular mechanism of water-deficit stress tolerance in peanut under water-deficit stress. Carmo (2019) analyzed proteins in root tissues of *A. duranensis* subjected to a progressive water deficit in soil and compared their differential abundance to transcript expression profiles obtained by RNA-seq and qRT-PCR. Where 31 proteins were identified, most were associated with stress response and drought perception, including a chitinase-2 (unique to stressed condition), an MLP-like protein, a glycine-rich protein DOT1-like, a maturase K and heat shock-related proteins (HSP70—an isoform unique to the control, and HSP17.3).

Wang et al. (2020) analyzed differentially expressed proteins of peanut buds in response to cold stress using an iTRAQ-based technique, and a major upregulation of a crucial protein such as valosin-containing protein (VCP) and many other important pathway proteins related to cold stress were observed. Li (2022) also found 69 differentially accumulated proteins (DAPs) in drought stress using proteomic analysis, and the overexpression of *AhLEA2* one of the DAPs was shown to enhance drought resistance. Furthermore, Chen (2021) identified 104 DAPs during imbibitional chilling stress, and reduced cytokinin oxidase abundance may be due to downregulated gene expression of corresponding genes, resulting in elevated cytokinin content. In addition, quantitative proteomics analysis by Liu (2021) suggested that ethanol fermentation might be the primary anaerobic fermentation pathway, and waterlogging-tolerant cultivars might rely on other metabolic pathways, such as malic acid metabolism, to generate pyruvate to alleviate energy shortage under waterlogging stress.

12.4 METABOLOMICS FOR PEANUT ABIOTIC STRESS TOLERANCE

Metabolomics is a rapidly evolving field of research that involves the comprehensive analysis of small molecules, or metabolites, in biological systems. This field integrates analytical chemistry, bioinformatics, and systems biology to understand the metabolic pathways involved in various biological processes. Under drought conditions induced by using 20% PEG 6000, sugars, sugar alcohols, organic acids, and fatty acids were found to be the major groups of metabolites altered, and the levels of total metabolites were found to be more pronounced in leaves than roots (Gundaraniya et al., 2020). Similarly, under soil compaction stress and N deficiency stress, the content of some amino acids (lysine) and fatty acids (especially the lipid composition of plasma membrane) increased, which relates to abiotic stress and inhibited the TCA cycle metabolism in peanut roots (Yang et al., 2022). In addition, under soil potassium deficiency stress, K-deficiency increased MDA, O_2, and H_2O_2 levels in both leaf and root of peanut (Patel, Fatnani et al., 2022). Under waterlogging stress, Zeng (2024) revealed that waterlogging stress inhibited the gene expression in starch and sucrose metabolism, using transcriptome sequencing and widely targeted metabolomic analysis, which reduced sucrose input and transformation ability. In addition, waterlogging stress upregulated the expression of the gene encoding tryptophan decarboxylase (AhTDC), and overexpression of *AhTDC* in *Arabidopsis* significantly reduced the seed length, width, and weight. Under salinity stress, Cui (2018) found that leaf relative electrolyte leakage (REC) levels were higher when subjected to salt stress. Patel, Khandwal et al. (2022) investigated the differential regulation of physio-biochemical and metabolomics parameters in peanut under individual (salt, drought, cold, and heat) and combined stress treatments using multivariate correlation analysis. The analysis showed that biochemical parameters such as free amino acids, polyphenols, starch, and sugars significantly changed under combined stresses compared to individual stresses. Gao (2024) investigated the effects of exogenous calcium on the antioxidant defense system in peanut seedling roots that are under salt stress by using indices including the transcriptome and absolute quantitative metabolome of flavonoids. In addition, it was found that under salinity, stress weakened the antioxidant defense capacity of enzymatic systems and effectively inhibited the antioxidant capacity of the linked AsA-GSH cycle. Using transcriptomic and metabolomics analysis, Zhang et al. (2024) revealed that naringenin, pinocembrin, prunetin, JA, and ABA responded to salt stress with different candidate metabolites. In addition, the expression of the JAZ protein-encoding genes (LOC112696383 and LOC112790545) was highly correlated with JA content in the plant hormone signal transduction gene, suggesting that they may be key factors in the response to salt stress.

12.5 MIRNAOMICS FOR PEANUT ABIOTIC STRESS TOLERANCE

MicroRNAs (miRNAs), a class of non-coding small RNAs, cleave transcripts and inhibit translation, making them useful for post-transcriptional regulation (Zhang, 2015; Zhang et al., 2006; Zhang and Unver, 2018). During recent years, more and more miRNAs have been identified in peanuts (Chen et al., 2019; Chi et al., 2011; Figueredo et al., 2020; Gao et al., 2017; Li et al., 2021; Ma et al., 2018; Ren et al., 2021; Tong et al., 2021; Zhang et al., 2017, 2022; Zhao et al., 2010). MiRNAs play crucial roles in not only regulating peanut growth and development (Figueredo et al., 2020; Gao et al., 2017; Ma et al., 2018) but also in peanut adaptations to various abiotic stresses (such as drought, cold, deficiencies of nutritional elements) (Chen et al., 2019; Li et al., 2021; Ren et al., 2021; Tong et al., 2021; Zhang et al., 2022).

Drought-responsive miRNAs in peanuts were identified using transcriptomic analysis in NH5 (drought-tolerant) and FH18 (drought-sensitive) (Ren et al., 2021). Based on comparative transcriptomic analysis, 73 miRNAs were differentially expressed in the drought-tolerant variety specifically under drought stress; of these, two key candidate miRNAs, novel miR_416 and novel miR_73, were identified, and the majority of their target genes were enriched in phenylpropanoid biosynthesis, linoleic acid metabolism, and cutin, suberine, and wax biosynthesis. This study lays the foundation for the analysis of the molecular mechanism of drought tolerance and promotes the genetic improvement of peanut drought tolerance.

Zhang et al. (2022) profiled cold-responsive miRNAs in peanuts using deep sequencing in cold-sensitive cultivar (WQL20) alongside a tolerant variety (WQL30) and identified 407 known miRNAs and 143 novel peanut-specific miRNAs. Researchers validated the expression of selected novel miRNAs using northern blotting and confirmed six known cold-responsive miRNAs (ahy-miR2118, ahy-miR160, ahy-miR482, ahy-miR1511, ahy-miR396, and ahy-miR162). They then identified and validated 12 target genes (*AhARF10*, *AhARF17*, *AhDCL6*, *AhDCL16*, *AhGRF1*, *AhGRF4*, *AhWDRL1*, *AhWDRL2*, *AhSRF*, *AhSPIL1*, *AhDR1*, and *AhDR2*) of these six cold-responsive miRNAs using degradome sequencing and qRT-PCR. Transient expression analysis in *Nicotiana benthamiana* revealed that miR160, miR482, and miR2118 might play positive roles, while miR396, miR162, and miR1511 play negative roles in regulating peanut cold tolerance. This study lays the foundation for understanding miRNA-dependent cold stress responses in peanuts. The identified correlations between miRNAs and their response to cold stress could serve as markers in breeding programs or as tools for improving cold tolerance in peanuts. Li et al. (2021) systematically explored the mechanism of miRNAs related to root growth and development in response to N or K deficiency in peanuts. N deficiency significantly decreased primary root length and lateral root number, which might be associated with the upregulation of miR160, miR167, miR393, and miR396, and the downregulation of AFB3 and GRF. The upregulation of miR156, miR390, NAC4, ARF2, and AFB3, along with the downregulation of miR160, miR164, miR393, and SPL10, likely contributed to the growth of primary and lateral roots under K deficiency. Overall, roots responded differently to the N or K deficiency stresses in peanuts, potentially due to the miRNA-mediated pathway and mechanism. The observation of the expression of the miRNAs and their targeted mRNAs in peanut plants exposed to N and K deficiency could reveal the mechanisms of the plant suffering from nutrient deficiency stress and help find candidate genes to develop important tools for plant breeding.

Chen et al. (2019) revealed a miRNA-mediated regulatory network of embryo abortion under calcium deficiency in peanut. In this study, with the assistance of a recently established platform for genome sequences of wild peanut species, a total of 29 known and 132 potential novel miRNAs were discovered in 12 peanut-specific miRNA families. Among the identified miRNAs, 87 differentially expressed during early embryo development under calcium deficiency and sufficiency conditions, and 117 target genes of the differentially expressed miRNAs. Integrated analysis of miRNAs and transcriptome expression revealed 52 differentially expressed target genes of 20 miRNAs. The expression profiles for some differentially expressed targets by gene chip analysis were consistent with the transcriptome sequencing results. These results show that miRNAs actively modulated seed and embryo development–related genes (*TCP3*, *AP2*, *EMB2750*), cell division and proliferation–related genes (*HsfB4* and *DIVARICATA*), and plant hormone signaling pathway–related genes (*CYP707A1* and *CYP707A3*), among others, during early embryo development. These findings provided for the first time new insights into miRNA-mediated regulatory pathways involved in peanut embryo abortion under calcium deficiency.

Tong et al. (2021) studied the regulatory network of aluminum (Al)-induced programmed cell death (PCD) using transcriptomics, miRNAs, and degradome analysis in the root tips of two peanut cultivars, ZH2 (Al-sensitive, S) and 99-1507 (Al-tolerant, T). Through deep sequencing, they identified 2,284 miRNAs and observed differential expression in 147 miRNAs under Al stress. Additionally, they validated 19,237 target genes of 749 miRNAs through degradome sequencing. GO and KEGG analyses of these differential miRNA targets revealed that the pathways of synthesis and degradation of ketone bodies, the TCA cycle, and peroxisome responded to Al stress. The combined analysis of the degradome datasets uncovered 89 miRNA–mRNA interactions (including 45 miRNAs such as miR10420, miR6300, miR2592, and miR396) that may regulate PCD under Al stress. Tong et al. constructed regulatory networks based on these differentially expressed miRNAs and their targets related to PCD. These findings offer a valuable platform for researching Al-induced PCD and provide new insights into genetic engineering for plant stress responses.

12.6 GENOMICS-ASSISTED BREEDING FOR PEANUT IMPROVEMENT

Genomics-assisted breeding (GAB) represents a crucial frontier in enhancing the balance between sustainable agriculture, environmental preservation, and global food security. Its precision and efficiency hold the promise of developing resilient crops, reducing resource utilization, and safeguarding biodiversity, ultimately fostering a more sustainable and secure food production system (Hafeez et al., 2023). Integrating modern genomics tools into crop breeding is necessary to fast-track breeding efficiency and rapid progress. When combined with speed breeding methods, this integration can substantially accelerate the breeding process, leading to faster access of improved varieties to farmers (Raza et al., 2024). GAB can accelerate the process of developing resistant cultivars by using diagnostic markers (Luo et al., 2019).

Climate change imposes several abiotic stresses, such as drought, nutrient imbalances, temperature fluctuations, and soil salinity. These stresses significantly impact crop productivity, including that of peanuts (Benitez-Alfonso et al., 2023; Raza, Mubarik et al., 2023; Raza, Tabassum et al., 2023; Raza et al. 2023, 2021, 2022; Rivero et al.,2022; Varshney et al., 2021; Zandalinas et al., 2021). For example, researchers used widespread phenotyping data from multiple locations and seasons to conduct QTL analysis of drought and iron deficiency tolerance–related traits in peanuts (Pandey et al., 2020). This genome-wide analysis recognized 19 major QTLs, explaining up to 33.9% of the PVE. Researchers also discovered numerous imperative candidate genes in these QTL regions that may be involved in plant growth, seed and pod development, and photosynthesis under stress conditions. These discoveries could be used to design molecular markers for marker-assisted breeding, aiming to boost peanut yield under drought and iron-deficient soil conditions after validation (Pandey et al., 2020). Previously, 52 QTLs correlated with nine yield-related traits were discovered using a combination of phenotyping and genotyping data under well-watered and water-stress conditions at two locations in West Africa. Nonetheless, these QTLs had low PVE, with values of less than 12% (Faye et al., 2015). In a recent study, researchers treated a RIL population of peanut with drought stress tolerance and discovered three major QTLs for pod yield per plant and nine QTLs for specific dry weight. These outcomes provide potential targets for genetic mutation and the improvement of linked markers to enhance drought stress tolerance in peanuts (Ghosh et al., 2022). In another recent study, researchers identified 11 and 12 main-effect QTLs for visual chlorotic rating and SPAD chlorophyll meter reading, respectively. Three QTLs were major for visual chlorotic rating, and two were major for SPAD chlorophyll meter reading. Researchers identified a SNP marker for visual chlorotic rating located in an influential protein-binding gene (*Arahy.QA0C1*). The study identified new QTLs and validated existing ones for iron deficiency chlorosis tolerance, providing valuable genomic resources for marker-assisted breeding (Tayade et al., 2022). Researchers identified 32 QTLs for iron deficiency chlorosis tolerance in peanut using an RIL population. The QTLs detected had low PVE for yield-related traits, but two major QTLs for visual chlorosis rating at 60 and 90 days. Researchers can use the validated QTLs/markers for marker-assisted breeding in peanuts (Pattanashetti et al., 2020). In another recent study, BSA-seq was used to identify genomic regions and SNPs linked with low temperature tolerance in high oleic acid peanut. Researchers identified a 2.29 Mb genomic interval on chromosome A05 harboring 122 genes implicated in abiotic stress response and plant-pathogen interactions. This discovery elucidates molecular mechanisms underlying low-temperature tolerance and provides foundational insights for future marker-assisted breeding programs (Sun et al. 2022). GWAS analysis using the 58 K SNP Axiom_Arachis array to identify SNPs associated with salinity tolerance and radical root length in peanut found three putative SNPs located on chromosomes Aradu.A03, Araip.B01, and Araip.B05 that were significantly related to salinity tolerance and radical root length (Zou et al., 2020). Future studies will focus on different abiotic stresses in peanuts to design climate-smart peanut cultivars.

12.7 CONCLUSION AND FUTURE DIRECTIONS

In summary, abiotic tolerance of plants is a complex process, involving the regulation of a set of TF families and various genes. Elucidating the specific mechanisms and promoting the genetic improvement of abiotic tolerance in peanuts is critically important for peanut extension. Analyzing and screening functional genes or proteins differentially expressed in various tissues and organs of peanuts under different abiotic stresses, across different growth and development stages, and in response to various environmental stress factors will help to understand the relationship between key functional genes and resistance. This understanding will aid in comprehending the injury mechanisms of stress factors and the adaptation mechanisms of peanuts at the transcriptional or protein level. Global metabolic profiling analysis holds the promise of permitting simultaneous monitoring of precursors, intermediates, and products of metabolic pathways. Additionally, other omics studies can further deepen our understanding of the regulatory mechanisms of abiotic stress in peanuts. Although peanut genome sequencing is available and more and more omics studies have been conducted in recent years, in contrast to the abundant information on abiotic stress in other plant species, there are scarce reports with regards to peanut. The mechanism of abiotic stress in peanut is still in its infancy. For the abiotic stress–related genes obtained through omics analysis, in-depth functional and action mechanism studies will be required in the future. In recent years, omics studies have mainly focused on drought, cold and salt stress, while omics reports about metal stress, heat stress, or nutrient deficiency stress have been scarce. Of all the omics studies, transcriptomics is the most studied in peanut, while metabolomics, proteomics, and other studies are relatively few. In addition, since the peanut genetic transformation system remains incomplete, researchers must continue efforts to improve peanut abiotic stress resistance through genetic engineering.

REFERENCES

Benitez-Alfonso, Yoselin, Beth K. Soanes, Sibongile Zimba, Besiana Sinanaj, Liam German, Vinay Sharma, Abhishek Bohra et al. "Enhancing climate change resilience in agricultural crops." *Current Biology* 33, no. 23 (2023): R1246–R1261, https://doi.org/10.1016/j.cub.2023.10.028.

Carmo, Lílian S. T., Andressa C. Q. Martins, Cinthia C. C. Martins, Mário A. S. Passos, Luciano P. Silva, Ana C. G. Araujo, Ana C. M. Brasileiro, Robert N. G. Miller, Patrícia M. Guimarães, and Angela Mehta. "Comparative proteomics and gene expression analysis in *Arachis duranensis* reveal stress response proteins associated to drought tolerance." *Journal of Proteomics* 192 (2019): 299–310, https://doi.org/10.1016/j.jprot.2018.09.011.

Chang BoWen, Zhong Peng, Liu Jie, Tang ZhongHua, Gao YaBing, Yu HongJiu, and Guo Wei. "Effect of low-temperature stress and gibberellin on seed germination and seedling physiological responses in peanut." [In Chinese] *Acta Agronomica Sinica* 45, no. 1 (2019): 118–130, https://doi.org/10.3724/SP.J.1006.2019.84043.

Chen, H., N. Liu, R. Xu, X. Chen, Y. Zhang, R. Hu, X. Lan, Z. Tang, and G. Lin. "Quantitative proteomics analysis reveals the response mechanism of peanut (*Arachis hypogaea* L.) to imbibitional chilling stress." *Plant Biology* 23, no. 3 (2021): 517–527, https://doi.org/10.1111/plb.13238.

Chen, Hua, Qiang Yang, Kun Chen, Shanshan Zhao, Chong Zhang, Ronglong Pan, Tiecheng Cai et al. "Integrated microRNA and transcriptome profiling reveals a miRNA-mediated regulatory network of embryo abortion under calcium deficiency in peanut (*Arachis hypogaea* L.)." *BMC Genomics* 20 (2019): 392, https://doi.org/10.1186/s12864-019-5770-6.

Chen, Na, Maowen Su, Xiaoyuan Chi, Zhimeng Zhang, Lijuan Pan, Mingna Chen, Tong Wang, Mian Wang, Zhen Yang, and Shanlin Yu. "Transcriptome analysis reveals salt-stress-regulated biological processes and key pathways in roots of peanut (*Arachis hypogaea* L.)." *Genes & Genomics* 38 (2016): 493–507, https://doi.org/10.1007/s13258-016-0395-5.

Chen, Na, Qingli Yang, Dongqing Hu, Lijuan Pan, Xiaoyuan Chi, Mingna Chen, Zhen Yang et al. "Gene expression profiling and identification of resistance genes to low temperature in leaves of peanut (*Arachis hypogaea* L.)." *Scientia Horticulturae* 169 (2014): 214–225, https://doi.org/10.1016/j.scienta.2014.01.043.

Chi, Xiaoyuan, Qingli Yang, Xiaoping Chen, Jinyan Wang, Lijuan Pan, Mingna Chen, Zhen Yang, Yanan He, Xuanqiang Liang, and Shanlin Yu. "Identification and characterization of microRNAs from peanut (*Arachis hypogaea* L.) by high-throughput sequencing." *PLOS ONE* 6, no. 11 (2011): e27530, https://doi.org/10.1371/journal.pone.0027530.

Cui, Feng, Na Sui, Guangyou Duan, Yiyang Liu, Yan Han, Shanshan Liu, Shubo Wan, and Guowei Li. "Identification of metabolites and transcripts involved in salt stress and recovery in peanut." *Frontiers in Plant Science* 9 (2018): 217, https://doi.org/10.3389/fpls.2018.00217.

Faye, Issa, Manish K. Pandey, Falalou Hamidou, Abhishek Rathore, Ousmane Ndoye, Vincent Vadez, and Rajeev K. Varshney. "Identification of quantitative trait loci for yield and yield related traits in groundnut (*Arachis hypogaea* L.) under different water regimes in Niger and Senegal." *Euphytica* 206 (2015): 631–647, https://doi.org/10.1007/s10681-015-1472-6.

Figueredo, María Soledad, Damien Formey, Johan Rodríguez, Fernando Ibáñez, Georgina Hernández, and Adriana Fabra. "Identification of miRNAs linked to peanut nodule functional processes." *Journal of Biosciences* 45 (2020): 62, https://doi.org/10.1007/s12038-020-00034-5.

Gao, Chao, Pengfei Wang, Shuzhen Zhao, Chuanzhi Zhao, Han Xia, Lei Hou, Zheng Ju, Ye Zhang, Changsheng Li, and Xingjun Wang. "Small RNA profiling and degradome analysis reveal regulation of microRNA in peanut embryogenesis and early pod development." *BMC Genomics* 18 (2017): 1–18, https://doi.org/10.1186/s12864-017-3587-8.

Gao, Yan, Xuan Dong, Rongjin Wang, Fei Hao, Hui Zhang, Yongyong Zhang, and Guolin Lin. "Exogenous calcium alleviates oxidative stress caused by salt stress in peanut seedling roots by regulating the antioxidant enzyme system and flavonoid biosynthesis." *Antioxidants* 13, no. 2 (2024): 233, https://doi.org/10.3390/antiox13020233.

Ghosh, Subhasini, Supriya S. Mahadevaiah, S. Anjan Gowda, Sunil S. Gangurde, Mangesh P. Jadhav, Anil A. Hake, P. Latha et al. "Genetic mapping of drought tolerance traits phenotyped under varying drought stress environments in peanut (*Arachis hypogaea* L.)." *Euphytica* 218, no. 12 (2022): 168, https://doi.org/ 10.1007/s10681-022-03120-x.

Girdthai, Teerayoot, Sanun Jogloy, Nimitr Vorasoot, Chutipong Akkasaeng, Sopone Wongkaew, C. Corley Holbrook, and Aran Patanothai. "Heritability of, and genotypic correlations between, aflatoxin traits and physiological traits for drought tolerance under end of season drought in peanut (*Arachis hypogaea* L.)." *Field Crops Research* 118, no. 2 (2010): 169–176, https://doi.org/10.1016/j.fcr.2010.05.007.

Gundaraniya, Srutiben A., Padma S. Ambalam, and Rukam S. Tomar. "Metabolomic profiling of drought-tolerant and susceptible peanut (*Arachis hypogaea* L.) genotypes in response to drought stress." *ACS Omega* 5, no. 48 (2020): 31209–31219, https://doi.org/10.1021/acsomega.0c04601.

Hafeez, Aqsa, Baber Ali, Muhammad Ammar Javed, Aroona Saleem, Mahreen Fatima, Amin Fathi, Muhammad Siddique Afridi, Veysel Aydin, Mükerrem Atalay Oral, and Fathia A. Soudy. "Plant breeding for harmony between sustainable agriculture, the environment, and global food security: an era of genomics-assisted breeding." *Planta* 258, no. 5 (2023): 97, https://doi.org/10.1007/s00425-023-04252-7.

Huang, Guo-Tao, Shi-Liang Ma, Li-Ping Bai, Li Zhang, Hui Ma, Ping Jia, Jun Liu, Ming Zhong, and Zhi-Fu Guo. "Signal transduction during cold, salt, and drought stresses in plants." *Molecular Biology Reports* 39 (2012): 969–987, https://doi.org/10.1007/s11033-011-0823-1.

HuiFang, Jiang, and Ren Xiaoping. "The effect on SOD activity and protein content in groundnut leaves by drought stress." [In Chinese] *Acta Agronomica Sinica* 30, no. 2 (2004): 169–174.

Jiang, Chunji, He Zhang, Jingyao Ren, Jiale Dong, Xinhua Zhao, Xiaoguang Wang, Jing Wang et al. "Comparative transcriptome-based mining and expression profiling of transcription factors related to cold tolerance in peanut." *International Journal of Molecular Sciences* 21, no. 6 (2020): 1921, https://doi.org/10.3390/ijms21061921.

Jongrungklang, N., B. Toomsan, N. Vorasoot, S. Jogloy, K. J. Boote, G. Hoogenboom, and A. Patanothai. "Drought tolerance mechanisms for yield responses to pre-flowering drought stress of peanut genotypes with different drought tolerant levels." *Field Crops Research* 144 (2013): 34–42, https://doi.org/10.1016/j.fcr.2012.12.017.

Kottapalli, Kameswara Rao, Randeep Rakwal, Junko Shibato, Gloria Burow, David Tissue, John Burke, Naveen Puppala, Mark Burow, and Paxton Payton. "Physiology and proteomics of the water-deficit stress response in three contrasting peanut genotypes." *Plant, Cell & Environment* 32, no. 4 (2009): 380–407, https://doi.org/10.1111/j.1365-3040.2009.01933.x.

Leidi, E. O., M. Silberbush, M. I. M. Soares, and S. H. Lips. "Salinity and nitrogen nutrition studies on peanut and cotton plants." *Journal of Plant Nutrition* 15, no. 5 (1992): 591–604, https://doi.org/10.1080/01904169209364343.

Li, C., C. Yan, Q. Sun, J. Wang, C. Yuan, Y. Mou, S. Shan, and X. Zhao. "Proteomic profiling of *Arachis hypogaea* in response to drought stress and overexpression of *AhLEA2* improves drought tolerance." *Plant Biology* 24, no. 1 (2022): 75–84, https://doi.org/10.1111/plb.13351.

Li, Lijie, Qian Li, Kyle E. Davis, Caitlin Patterson, Sando Oo, Wanying Liu, Jia Liu et al. "Response of root growth and development to nitrogen and potassium deficiency as well as microRNA-mediated mechanism in peanut (*Arachis hypogaea* L.)." *Frontiers in Plant Science* 12 (2021): 695234, https://doi.org/10.3389/fpls.2021.695234.

Liu, Dengwang, Jian Zhan, Zinan Luo, Ningbo Zeng, Wei Zhang, Hao Zhang, and Lin Li. "Quantitative proteomics and relative enzymatic activities reveal different mechanisms in two peanut cultivars (*Arachis hypogaea* L.) under waterlogging conditions." *Frontiers in Plant Science* 12 (2021): 716114, https://doi.org/10.3389/fpls.2021.716114.

Liu, H. L., X. S. Chen, F. J. Yang, D. M. Bai, X. P. Sun, Y. C. Lu, X. P. Ren, H. F. Jiang, and H. Y. Gao. "Research of identification method on low temperature resistance of peanut germplasm resources phenotype." [In Chinese] *Journal of Peanut Science* 46, no. 3 (2017): 20–25, https://doi.org/10.14001/j.issn. 1002-4093.2017.03.004.

Luo, Huaiyong, Manish K. Pandey, Aamir W. Khan, Bei Wu, Jianbin Guo, Xiaoping Ren, Xiaojing Zhou et al. "Next-generation sequencing identified genomic region and diagnostic markers for resistance to bacterial wilt on chromosome B02 in peanut (*Arachis hypogaea* L.)." *Plant Biotechnology Journal* 17, no. 12 (2019): 2356–2369, https://doi.org/10.1111/pbi.13153.

Ma, Xingli, Xingguo Zhang, Kunkun Zhao, Fapeng Li, Ke Li, Longlong Ning, Jialin He, Zeyu Xin, and Dongmei Yin. "Small RNA and degradome deep sequencing reveals the roles of microRNAs in seed expansion in peanut (*Arachis hypogaea* L.)." *Frontiers in Plant Science* 9 (2018): 349, https://doi.org/10.3389/fpls.2018.00349.

Meena, H. N., Murlidhar Meena, and R. S. Yadav. "Comparative performance of seed types on yield potential of peanut (*Arachis hypogaea* L.) under saline irrigation." *Field Crops Research* 196 (2016): 305–310, https://doi.org/10.1016/j.fcr.2016.06.006.

Miao, H.C., Li, Q.,Hou, X.F.,Jia, D.H., Shi, B.X., Ding, H., Li, L.M., Zhang, Z.,M. "Effects of drought at different growth stages on growth and yield of peanut. " [In Chinese] *Xinjiang Agricultural Sciences* 58, no. 3 (2021): 441–449, https://doi.org/10. 6048 /j. issn.1001-4330.2021.03.006.

Pandey, M. K., Gangurde, S. S., Sharma, V., Pattanashetti, S K., Naidu, G. K., Faye, I., Hamidou, F., et al. Improved genetic map identified major QTLs for drought tolerance- and iron deficiency tolerance-related traits in groundnut. *Genes (Basel).* 12 (2020): 37, https://doi.org/10.3390/genes12010037.

Patel, Jaykumar, Deepesh Khandwal, Babita Choudhary, Dolly Ardeshana, Rajesh Kumar Jha, Bhakti Tanna, Sonam Yadav, Avinash Mishra, Rajeev K. Varshney, and Kadambot HM Siddique. "Differential physio-biochemical and metabolic responses of peanut (*Arachis hypogaea* L.) under multiple abiotic stress conditions." *International Journal of Molecular Sciences* 23, no. 2 (2022): 660, https://doi.org/10.3390/ijms23020660.

Patel, Monika, Dhara Fatnani, and Asish Kumar Parida. "Potassium deficiency stress tolerance in peanut (*Arachis hypogaea*) through ion homeostasis, activation of antioxidant defense, and metabolic dynamics: Alleviatory role of silicon supplementation." *Plant Physiology and Biochemistry* 182 (2022): 55–75, https://doi.org/10.1016/j.plaphy.2022.04.013.

Pattanashetti, Santosh K., Manish K. Pandey, Gopalakrishna K. Naidu, Manish K. Vishwakarma, Omprakash Kumar Singh, Yaduru Shasidhar, Ishwar H. Boodi et al. "Identification of quantitative trait loci associated with iron deficiency chlorosis resistance in groundnut (*Arachis hypogaea*)." *Plant Breeding* 139, no. 4 (2020): 790–803, https://doi.org/10.1111/pbr.12815.

Pruthvi, Vittal, Rama Narasimhan, and Karaba N. Nataraja. "Simultaneous expression of abiotic stress responsive transcription factors, *AtDREB2A*, *AtHB7* and *AtABF3* improves salinity and drought tolerance in peanut (*Arachis hypogaea* L.)." *PLoS One* 9, no. 12 (2014): e111152, https://doi.org/10.1371/journal.pone.0111152.

Qin, L -Q., L. Li, C. Bi, Y -L. Zhang, S -B. Wan, J -J. Meng, Q -W. Meng, and X -G. Li. "Damaging mechanisms of chilling-and salt stress to *Arachis hypogaea* L. leaves." *Photosynthetica* 49 (2011): 37–42, https://doi.org/10.1007/s11099-011-0005-3.

Raza, Ali, Savita Bhardwaj, Md Atikur Rahman, Pedro García-Caparrós, Madiha Habib, Faisal Saeed, Sidra Charagh, Christine H. Foyer, Kadambot HM Siddique, and Rajeev K. Varshney. "Trehalose: A sugar molecule involved in temperature stress management in plants." *The Crop Journal* 12, no. 1 (2024): 1–16, https://doi.org/10.1016/j.cj.2023.09.010.

Raza, Ali, Sidra Charagh, Pedro García-Caparrós, Md Atikur Rahman, Vincent H. Ogwugwa, Faisal Saeed, and Wanmei Jin. "Melatonin-mediated temperature stress tolerance in plants." *GM Crops & Food* 13, no. 1 (2022): 196–217, https://doi.org/10.1080/21645698.2022.2106111.

Raza, Ali, Hua Chen, Chong Zhang, Yuhui Zhuang, Yasir Sharif, Tiecheng Cai, Qiang Yang et al. "Designing future peanut: the power of genomics-assisted breeding." *Theoretical and Applied Genetics* 137, (2024): 66, https://doi.org/10.1007/s00122-024-04575-3.

Raza, Ali, Muhammad Salman Mubarik, Rahat Sharif, Madiha Habib, Warda Jabeen, Chong Zhang, Hua Chen et al. "Developing drought-smart, ready-to-grow future crops." *The Plant Genome* 16, no. 1 (2023): e20279, https://doi.org/10.1002/tpg2.20279.

Raza, Ali, Javaria Tabassum, Ali Zeeshan Fakhar, Rahat Sharif, Hua Chen, Chong Zhang, Luo Ju et al. "Smart reprogramming of plants against salinity stress using modern biotechnological tools." *Critical Reviews in Biotechnology* 43, no. 7 (2023): 1035–1062, https://doi.org/10.1080/07388551.2022.2093695.

Raza, Ali, Javaria Tabassum, Himabindu Kudapa, and Rajeev K. Varshney. "Can omics deliver temperature resilient ready-to-grow crops?." *Critical Reviews in Biotechnology* 41, no. 8 (2021): 1209–1232, https://doi.org/10.1080/07388551.2021.1898332.

Razzaq, Muhammad Khuram, Muqadas Aleem, Shahid Mansoor, Mueen Alam Khan, Saeed Rauf, Shahid Iqbal, and Kadambot HM Siddique. "Omics and CRISPR-Cas9 approaches for molecular insight, functional gene analysis, and stress tolerance development in crops." *International Journal of Molecular Sciences* 22, no. 3 (2021): 1292, https://doi.org/10.3390/ijms22031292.

Ren, Jingyao, He Zhang, Xiaolong Shi, Xin Ai, Jiale Dong, Xinhua Zhao, Chao Zhong, Chunji Jiang, Jing Wang, and Haiqiu Yu. "Genome-wide identification of key candidate microRNAs and target genes associated with peanut drought tolerance." *DNA and Cell Biology* 40, no. 2 (2021): 373–383, https://doi.org/10.1089/dna.2020.6245.

Ritonga, Faujiah Nurhasanah, Jacob Njaramba Ngatia, Yiran Wang, Muneer Ahmed Khoso, Umar Farooq, and Su Chen. "AP2/ERF, an important cold stress-related transcription factor family in plants: A review." *Physiology and Molecular Biology of Plants* 27 (2021): 1953–1968, https://doi.org/10.1007/s12298-021-01061-8.

Rivero, Rosa M., Ron Mittler, Eduardo Blumwald, and Sara I. Zandalinas. "Developing climate-resilient crops: improving plant tolerance to stress combination." *The Plant Journal* 109, no. 2 (2022): 373–389, https://doi.org/10.1111/tpj.15483.

Sharma, Manisha, and Girdhar K. Pandey. "Genomics and functional genomics of stress-mediated signaling in plants: Volume I." *Current Genomics* 18, no. 6 (2017): 467–468, https://doi.org/10.2174/138920291806170929123912.

Singh, A. L., K. Hariprasanna, V. Chaudhari, H. K. Gor, and B. M. Chikani. "Identification of groundnut (*Arachis hypogaea* L.) cultivars tolerant of soil salinity." *Journal of Plant Nutrition* 33, no. 12 (2010): 1761–1776, https://doi.org/10.1080/01904167.2010.503779.

Singh, Baljeet, Neha Salaria, Kajal Thakur, Sarvjeet Kukreja, Shristy Gautam, and Umesh Goutam. "Functional genomic approaches to improve crop plant heat stress tolerance." *F1000Research* 8, (2019): 1721, https://doi.org/10.12688/f1000research.19840.1.

Sorensen, Ronald B., Russell C. Nuti, and Christopher L. Butts. "Yield and plant growth response of peanut to midseason forage harvest." *Agronomy Journal* 101, no. 5 (2009): 1198–1203, https://doi.org/10.2134/agronj2009.0047.

Sun Ai-Qing, Zhang Jie-Dao, Wan Yong-Shan, Liu Feng-Zhen, Zhang Kun, and Sun Li. "In silico expression profile of genes in response to drought in peanut." [In Chinese] *Acta Agronomica Sinica* 39, no 6 (2013): 1045–1053, https://doi.org/10.3724/SP.J.1006.2013.01045.

Sun, Hongxi, Liang Ren, Feiyan Qi, Haixin Wang, Shutao Yu, Ziqi Sun, Bingyan Huang et al. "BSA-seq approach identified candidate region and diagnostic marker for chilling tolerance of high oleic acid peanut at germination stage." *Agronomy* 13, no. 1 (2022): 18, https://doi.org/10.3390/agronomy13010018.

Tayade, Ankur D., Babu N. Motagi, Mangesh P. Jadhav, Anjum S. Nadaf, Rajshekar V. Koti, Sunil S. Gangurde, Vinay Sharma, Rajeev K. Varshney, Manish K. Pandey, and Ramesh S. Bhat. "Genetic mapping of tolerance to iron deficiency chlorosis in peanut (*Arachis hypogaea* L.)." *Euphytica* 218 (2022): 46, https://doi.org/10.1007/s10681-022-02996-z.

Tong, Bin, Yusun Shi, Aaron Ntambiyukuri, Xia Li, Jie Zhan, Aiqin Wang, Dong Xiao, and Longfei He. "Integration of Small RNA and degradome sequencing reveals the regulatory network of al-induced programmed cell death in peanut." *International Journal of Molecular Sciences* 23, no. 1 (2021): 246, https://doi.org/10.3390/ijms23010246.

Upadhyaya, H. D., L. J. Reddy, S. L. Dwivedi, C. L. L. Gowda, and Sube Singh. "Phenotypic diversity in cold-tolerant peanut (*Arachis hypogaea* L.) germplasm." *Euphytica* 165 (2009): 279–291, https://doi.org/10.1007/s10681-008-9786-2.

Varshney, Rajeev K., Rutwik Barmukh, Manish Roorkiwal, Yiping Qi, Jana Kholova, Roberto Tuberosa, Matthew P. Reynolds, Francois Tardieu, and Kadambot HM Siddique. "Breeding custom-designed crops for improved drought adaptation." *Advanced Genetics* 2, no. 3 (2021): e202100017, https://doi.org/10.1002/ggn2.202100017.

Wang, C., Bo Cheng, Y. Zheng, J. Sha, A. Li, and X. Sun. "Effects of temperature to seed emergence, seedling growth and anthesis of peanut."[In Chinese] *Journal of Peanut Science* 32, no. 4 (2003): 7-11.

Wang, Xiangfeng, Axel A. Elling, Xueyong Li, Ning Li, Zhiyu Peng, Guangming He, Hui Sun, Yijun Qi, X. Shirley Liu, and Xing Wang Deng. "Genome-wide and organ-specific landscapes of epigenetic modifications and their relationships to mRNA and small RNA transcriptomes in maize." *The Plant Cell* 21, no. 4 (2009): 1053–1069, https://doi.org/10.1105/tpc.109.065714.

Wang, X., Y. Li, H. Zhu, X. Y. Chi, L. R. Wu, L. G. Zhao, J. S. Wang, and S. L. Yu. "Directional screening on peanut mutant with high oil content and correlation between oil content and leaf water potential." [In Chinese] *Journal of Qingdao Agricultural University* 36, no. 1 (2019): 30–33, https://doi.org/1674-148X(2019)01-0030-04.

Wang, X. J., Y. Shen, D. L. Sun, N. F. Bian, P. X. Shi, Z. M. Zhang, Z. D. Chen, Y. H. Liu, Y. Shen, and X. Wang. "iTRAQ-based proteomic reveals cell cycle and translation regulation involving in peanut buds cold stress." *Russian Journal of Plant Physiology* 67 (2020): 103–110, https://doi.org/10.1134/S1021443720010239.

Xiao, Dong, Xia Li, Yun-Yi Zhou, Li Wei, Chanthaphoone Keovongkod, Huyi He, Jie Zhan, Ai-Qin Wang, and Long-Fei He. "Transcriptome analysis reveals significant difference in gene expression and pathways between two peanut cultivars under Al stress." *Gene* 781 (2021): 145535, https://doi.org/10.1016/J.GENE.2021.145535.

Yang, Liyu, Qi Wu, Haiyan Liang, Liang Yin, and Pu Shen. "Integrated analyses of transcriptome and metabolome provides new insights into the primary and secondary metabolism in response to nitrogen deficiency and soil compaction stress in peanut roots." *Frontiers in Plant Science* 13 (2022): 948742, https://doi.org/10.3389/fpls.2022.948742.

Yu, Rugang, Qun Jiang, Chen Xv, Lien Li, Sijia Bu, and Gangrong Shi. "Comparative proteomics analysis of peanut roots reveals differential mechanisms of cadmium detoxification and translocation between two cultivars differing in cadmium accumulation." *BMC Plant Biology* 19 (2019): 137, https://doi.org/10.1186/s12870-019-1739-5.

Yuan, Cuiling, Chunjuan Li, Xiaodong Lu, Xiaobo Zhao, Caixia Yan, Juan Wang, Quanxi Sun, and Shihua Shan. "Comprehensive genomic characterization of NAC transcription factor family and their response to salt and drought stress in peanut." *BMC Plant Biology* 20 (2020): 454, https://doi.org/10.1186/s12870-020-02678-9.

Zandalinas, Sara I., Felix B. Fritschi, and Ron Mittler. "Global warming, climate change, and environmental pollution: recipe for a multifactorial stress combination disaster." *Trends in Plant Science* 26, no. 6 (2021): 588–599, https://doi.org/ 10.1016/j.tplants.2021.02.011.

Zeng, Ruier, Tingting Chen, Xi Li, Jing Cao, Jie Li, Xueyu Xu, Lei Zhang, and Yong Chen. "Integrated physiological, transcriptomic and metabolomic analyses reveal the mechanism of peanut kernel weight reduction under waterlogging stress." *Plant, Cell & Environment* 47 (2024): 3198–3214, https://doi.org/10.1111/pce.14936.

Zhang, Baohong. "MicroRNA: a new target for improving plant tolerance to abiotic stress." *Journal of Experimental Botany* 66, no. 7 (2015): 1749–1761, https://doi.org/10.1093/jxb/erv013.

Zhang, Baohong, Xiaoping Pan, George P. Cobb, and Todd A. Anderson. "Plant microRNA: a small regulatory molecule with big impact." *Developmental Biology* 289, no. 1 (2006): 3–16, https://doi.org/10.1016/j.ydbio.2005.10.036.

Zhang, Baohong, and Turgay Unver. "A critical and speculative review on microRNA technology in crop improvement: Current challenges and future directions." *Plant Science* 274 (2018): 193–200, https://doi.org/10.1016/j.plantsci.2018.05.031.

Zhang, Chun, C. C. Tang, J. Y. Wang, L. Guo, L. Wang, and W. Li. "Advances on transcriptome of plants under stresses." [In Chinese] *Journal of Biology* 34, no. 2 (2017): 86–90, https://doi.org/10.3969/j.issn.2095-1736.2017.02.086.

Zhang, Gaohua, Yu Shutao, Wang He, and Wang Xuda. "Transcriptome profiling of high oleic peanut under low temperatureduring germination." [In Chinese] *Hereditas (Beijing)* 41, no. 11 (2019): 1050–1059, https://doi.org/10.16288/j.yczz.19-097.

Zhang, Hao, Xiaobo Zhao, Quanxi Sun, Caixia Yan, Juan Wang, Cuiling Yuan, Chunjuan Li, Shihua Shan, and Fengzhen Liu. "Comparative transcriptome analysis reveals molecular defensive mechanism of *Arachis hypogaea* in response to salt stress." *International Journal of Genomics* 2020, no. 1 (2020): 6524093, https://doi.org/10.1155/2020/6524093.

Zhang, He, Chunji Jiang, Jingyao Ren, Jiale Dong, Xiaolong Shi, Xinhua Zhao, Xiaoguang Wang et al. "An advanced lipid metabolism system revealed by transcriptomic and lipidomic analyses plays a central role in peanut cold tolerance." *Frontiers in Plant Science* 11 (2020): 1110, https://doi.org/10.3389/fpls.2020.01110.

Zhang, He, Xiao-Long Shi, Jing-Yao Ren, Jia-Le Dong, Xin-Hua Zhao, Chun-Ji Hang, and Hai-Qiu Yu. "Transcriptional responses of peanut seedings under cold stress."[In Chinese] *Journal of Shenyang Agricultural University* 51, no. 1 (2020): 96–104, https://doi.org/10.3969/j.issn.1000-1700.2020.01.011.

Zhang, Nan, He Zhang, Zhenghao Lv, Baiyi Bai, Jingyao Ren, Xiaolong Shi, Shuli Kang, Xinhua Zhao, Haiqiu Yu, and Tianhong Zhao. "Integrative multi-omics analysis reveals the crucial biological pathways involved in the adaptive response to NaCl stress in peanut seedlings." *Physiologia Plantarum* 176, no. 2 (2024): e14266, https://doi.org/10.1111/ppl.14266.

Zhang, Tingting, Shuhao Hu, Caixia Yan, Chunjuan Li, Xiaobo Zhao, Shubo Wan, and Shihua Shan. "Mining, identification and function analysis of microRNAs and target genes in peanut (*Arachis hypogaea* L.)." *Plant Physiology and Biochemistry* 111 (2017): 85–96, https://doi.org/10.1016/j.plaphy.2016.11.018.

Zhang, Xin, Chao Ren, Yunyun Xue, Yuexia Tian, Huiqi Zhang, Na Li, Cong Sheng, Huifang Jiang, and Dongmei Bai. "Small RNA and degradome deep sequencing reveals the roles of microRNAs in peanut (*Arachis hypogaea* L.) cold response." *Frontiers in Plant Science* 13 (2022): 920195, https://doi.org/10.3389/fpls.2022.920195.

Zhao, Chuan-Zhi, Han Xia, Taylor Price Frazier, Ying-Yin Yao, Yu-Ping Bi, Ai-Qin Li, Meng-Jun Li, Chang-Sheng Li, Bao-Hong Zhang, and Xing-Jun Wang. "Deep sequencing identifies novel and conserved microRNAs in peanuts (*Arachis hypogaea* L.)." *BMC Plant Biology* 10 (2010): 3, https://doi.org/10.1186/1471-2229-10-3.

Zhao, Xiaobo, Chunjuan Li, Shubo Wan, Tingting Zhang, Caixia Yan, and Shihua Shan. "Transcriptomic analysis and discovery of genes in the response of *Arachis hypogaea* to drought stress." *Molecular Biology Reports* 45 (2018): 119–131, https://doi.org/10.1007/s11033-018-4145-4.

Zou, Kunyan, Dongwoo Kang, Ki-Seung Kim, and Tae-Hwan Jun. "Screening of salinity tolerance and genome-wide association study in 249 peanut accessions (*Arachis hypogaea* L.)." *Plant Breeding and Biotechnology* 8, no. 4 (2020): 434–438, https://doi.org/10.9787/PBB.2020.8.4.434.

13 Potential of Modern Biotechnological Tools in Conferring Abiotic Stress Tolerance in Peanut

Haiqiu Yu, Jing Wang, Xinhua Zhao, He Zhang, and Xiaolong Shi

ABBREVIATIONS

Net photosynthetic rate (Pn)
Stomatal conductance (Gs)
Intercellular CO_2 concentration (Ci)
Maximum fluorescence (Fm)
Variable fluorescence (Fv)
Maximum photochemical efficiency (Fv/Fm)
Actual photochemical efficiency (ΦPSII)
Photochemical quenching coefficient (qP)
Apparent electron transport efficiency (ETR)
Non-photochemical quenching coefficient (NPQ)
Ratio of arachidonic acid to linoleic acid (O/L)
Total fatty acid (TFA)
Saturated (SFAs)
Monounsaturated fatty acids (MUFAs)
Polyunsaturated fatty acids (PUFAs)
Reactive oxygen species (ROS)
Singlet oxygen ($^1O_2$)
Hydroxyl radical ($^{\cdot}OH$)
Superoxide anion (O^{2-})
Hydrogen peroxide (H_2O_2)
Superoxide dismutase (SOD)
Peroxidase (POD)
Catalase (CAT)
Ascorbate peroxidase (APX)
Glutathione peroxidase (GPX)
Monodehydroascorbate reductase (MDHAR)
Dehydroascorbate reductase (DHAR)
Glutathione reductase (GR)
Glutathione S-transferase (GST)
Malondialdehyde (MDA)
Abscisic acid (ABA)
Salicylic acid (SA)
Leucine-rich-repeat receptor-like kinases (LRR-RLK)
Calcium dependent protein kinases (CDPKs)
Phosphatase and mitogen-activated protein kinases (MAPKs)
CBF expression (ICE)-C-repeat binding factor/DRE binding factor 1 (CBF/DREB1)-Cold-responsive (COR)
BRASSINOSTEROID-INSENSITIVE2 (BIN2)
Protein phosphatases of type 2C (PP2C)
Sucrose non-fermenting1 (SNF-1)–related protein kinase 2s (SnRK2s)
Dimethylallyl diphosphate (DMAPP)
Isopentenyl diphosphate (IPP)
Mevalonate (MVA)
K^+ outward-rectifying (KOR)
Guard cell outward rectifying K^+ channel (GORK)
Non-selective cation channels (NSCCs)
Plasma membrane (PM)
Salt overly sensitive (SOS1)
High-affinity K^+ transporter (HKT)
Volatile organic compounds (VOCs)
Arbuscular mycorrhizal fungi (AMF)
Plasma membrane intrinsic proteins (PIPs)
programmed cell death (PCD)
COMPACT ROOT ARCHITECTURE 2 (CRA2)
Nitrate transporter (NRT)
Phosphatidyl choline (PC)
Phosphatidyl choline (PE)
Tricarboxylic acid (TCA)
Marker-assisted selection (MAS)
Marker-assisted backcrossing (MABC)
Whole-genome resequencing (WGRS)
Genome-wide selection (GS)
Recombinant inbred lines (RIL)
Quantitative trait loci (QTL)
Restriction fragment length polymorphisms (RFLPs)
Amplified fragment-length polymorphism (AFLP)
Cleaved amplified polymorphic sequence (CAPS)
Variable number of tandem repeats (VNTR)
Simple sequence repeat (SSR)
Inter-simple sequence repeat (ISSR)
Inverse sequence-tagged repeat (ISTR)
Sequence tag site (STS)
Single nucleotide polymorphisms (SNPs)
Single cell multi-omics (SCM-omics)
Spatial muti-omics (SM-omics)
Weighted gene co-expression network analysis (WGCNA)
Differentially expressed genes (DEGs)
Kyoto Encyclopedia of Genes and Genomes (KEGG)
Transcription factors (TFs)

DOI: 10.1201/9781003360988-13

Transcription activator-like nucleases (TALENs)
Zinc-finger nucleases (ZFNs)
The maximum growth rate (V_m)
The occurrence time of maximum growth rate (T_m)
Vascular bundle (Vb)
Medullary cavity (Mc)
Lignifying cells (Lc)
Structure of protein (Pt)
Liposome (Ls)
Starch grain (Sg)

13.1 INTRODUCTION

Peanut (*Arachis hypogea* L.), which originated in South America, is distributed in more than 100 countries around the world, among which China, India and the United States are the three largest production areas (Zhuang et al., 2019). Increasing global warming causes extreme high temperatures and various unpredictable natural disasters. It is more common for land surface drying to accelerate and evapotranspiration to be enhanced, resulting in drought incidence and severity (IPCC). Due to the uncontrolled characteristics of drought, it damages all developmental stages of peanut directly or indirectly. Drought restricts seed imbibition resulting in late/less emergence and seedling establishment. In addition, it causes abnormal physiological processes like loss of turgidity, a reduced rate of carbon assimilation, less gaseous exchange, and increased oxidative damage, which lead to delayed development of leaf, root, and even the whole plant. The plant suffers various enzymatic activities and metabolic and transcriptional levels change, resulting in a decline in productivity (Ren et al., 2022a; Wang et al., 2023).

In order to avoid drought during spring in northeast China, early sowing is implemented (Zhang et al., 2019). But cold damage has become the key limiting factor because chilling damage initially occurs at the cellular and organ levels. The bio membrane system, including cell membrane, nuclear membrane, and organelle membrane, is the initial site of injury, particularly in terms of its structure, function, stability, and enzyme activity, thereby resulting in substantial metabolic imbalance, especially involving respiration and photosynthesis. These changes in turn affect the plant growth and development and eventually incur damage at the whole-plant level, leading to the occurrence of chilling damage (Zhang et al., 2019).

Extreme climate change and irrational human activities have also caused soil salinization, which has affected approximately 7% of land area (900 million hectares) worldwide (Bijalwan et al., 2021). Approximately 90 million hectares of land area are affected in China, leading to severe land degradation and crop yield reduction (FAO, 2021; Jamil et al., 2011). It causes ion toxicity, osmotic stress, Na^+ accumulation, and photosynthesis inhibition. K^+ has played a compromised role in increased Na^+ so that the K^+/Na^+ ratio becomes an important parameter for assessing salt tolerance in peanut. A higher K^+/Na^+ ratio serves as a promotor to enhance stomatal function and enzyme activation and to regulate cell osmosis and vacuole pressure (Shi et al., 2020).

Around the world, the vast majority of peanut is planted in low-yield fields with poor soil nutrients. Peanut yield reduction caused by nutrient stress (nitrogen, phosphorus, potassium, calcium, iron, etc.) has become a focus of attention. Due to poor soil nutrients and unscientific fertilization measures, the nitrogen fixation ability of peanut root nodules decreased, root development was hindered, and nutrient absorption efficiency was reduced.

Therefore, to cope with abiotic stress in the future, it is crucial to understand the morph-physiological and mole-biological response mechanisms of peanut and to utilize modern biotechnological tools to improve abiotic tolerance.

13.2 DROUGHT

13.2.1 Drought Stress on the Development of Peanut

When peanut is subjected to drought, water content in cells undergoes exosmosis, cell contraction occurs and cell metabolism changes, and changes in growth are inevitable. The plant morphological characteristics of peanut are the most direct manifestation of drought stress, such as root reduction, stem shortening, leaf area and number reduction, leaf structure damage, flower bud differentiation inhibition, pod development rate reduction, and so on (Li et al., 2010).

Drought not only reshapes roots but also inhibits root growth, in which total root length, root weight, root volume, root surface area, and average root diameter are all increased to reach water, while with prolonged drought, it is significantly inhibited and shows a decreasing trend (Wang et al., 2023). Similarly, drought has negative effects on shoot development. Drought mainly prohibits stem and branch growth with significant dwarfed characteristics and at the same time reduces lateral branch length, total branch number, and stem diameter, as well as the quality of flowers (Liu et al., 2015; Miao et al., 2021; Tang et al., 2014). Leaves are the most sensitive to water availability. Leaves usually become smaller and yellow, and gradually wilting curls them into a certain shape and angle, and the ratio of new and green leaves decreases. Moreover, functional leaf area, leaf area coefficient, and stomatal conductance decreases, while specific leaf weight increases (Pruthvi et al., 2014; Liu et al., 2022). When soil moisture is less than 40% of the maximum water-holding capacity, leaf growth stops. In order to maintain water balance under drought stress, peanut gradually close stomata and reduce water evaporation and the stomatal conductivity of leaves, thus reducing intercellular CO_2 concentration. In severe drought, mesophyll cells become damaged and enzymes relating to photosynthesis inactivate, thus affecting the transpiration rate and photosynthetic rate of peanut (Miao et al., 2021). The net photosynthetic rate (Pn), stomatal conductance (Gs), intercellular CO_2 concentration (Ci), maximum photochemical efficiency (Fv/Fm), actual photochemical efficiency (ΦPSII), and photochemical quenching coefficient (qP) of leaves decrease gradually with drought stress (Pilon et al., 2018; Dong et al., 2020). Drought stress reduces photosynthetic efficiency by increasing heat dissipation of excess light energy, which leads to changes in fluorescence parameters (Efeoğlu et al., 2009). Pilon et al. (2018) found that under drought stress, photosynthesis, actual quantum yield, and electron transport chain of photosynthetic system II all decreased significantly, and chlorophyll content continuously decreased with prolonged stress (Sun et al., 2024). The gradual closing of stomata causes the temperature of leaves to rise continuously. Lauriano et al. (1997, 2004) combined in situ gas-exchange measurements with in vitro evaluation of the electron transfer process of isolated chloroplasts and the activity of Rubisco and found that drought limits photosynthesis primarily by inhibiting the efficiency of thylakoid reactions, particularly those that occur in photosystem II (Figure 13.1).

The suitable water content in soil during the flower bud differentiation stage of peanut is 60%; if the water content is too low, the inhibition of flower bud primordium formation is induced, the flower bud differentiation lags, and the flower bud quality decreases, leading to reduced microspore number or pollen abortion (Ding et al., 2013; Wang et al., 2000). Actually, drought induces *rhohB* expression, and increases the probability of reactive oxygen accumulation in anthers and pollen, which inhibit pollen viability, germination rate, and pollen tube growth, resulting in a lower pod setting rate (Luo et al., 2023). If the soil water content is less than 50%, the number of flowers decreases significantly and even causes the termination of the flowering process and yield failure (Banavath et al., 2018). Drought decreases yield at seedling stage, mainly because it inhibits leaf growth, resulting in decreased photosynthetic capacity of the peanut population. Compared with the whole growth stage, seedling stage drought only affects the numbers of branches and double full fruit, but has no obvious effect on yield and other yield components (Miao et al., 2021; Zhang et al., 2013; Cheng et al., 2010). With the intensity of drought stress, the pod number decreases, the pod weight reduces, and the yield eventually decreases (Bhalani et al., 2019). Also, drought significantly decreases the ratio of arachidonic acid to linoleic acid (O/L) at different growth stages by reducing their storage characteristics. The fatty acid content in seeds under drought stress is decreased, but the protein and soluble sugar is increased, which is caused by inhibiting the transformation of sugar to fat. In addition, drought stress induces the production of *Aflatus* and *Aspergillus parasiticus*, which are particularly prone to contaminate peanut (Gao, 2022). Wang et al. (2010) found that under both drought and aflatoxin stress, drought-tolerant varieties expressed more differential proteins, including low molecular weight heat shock protein, oxalate oxidase, ascorbate peroxidase, and so on. These protein spots were mainly involved in molecular functional processes such as defense response, signal transduction,

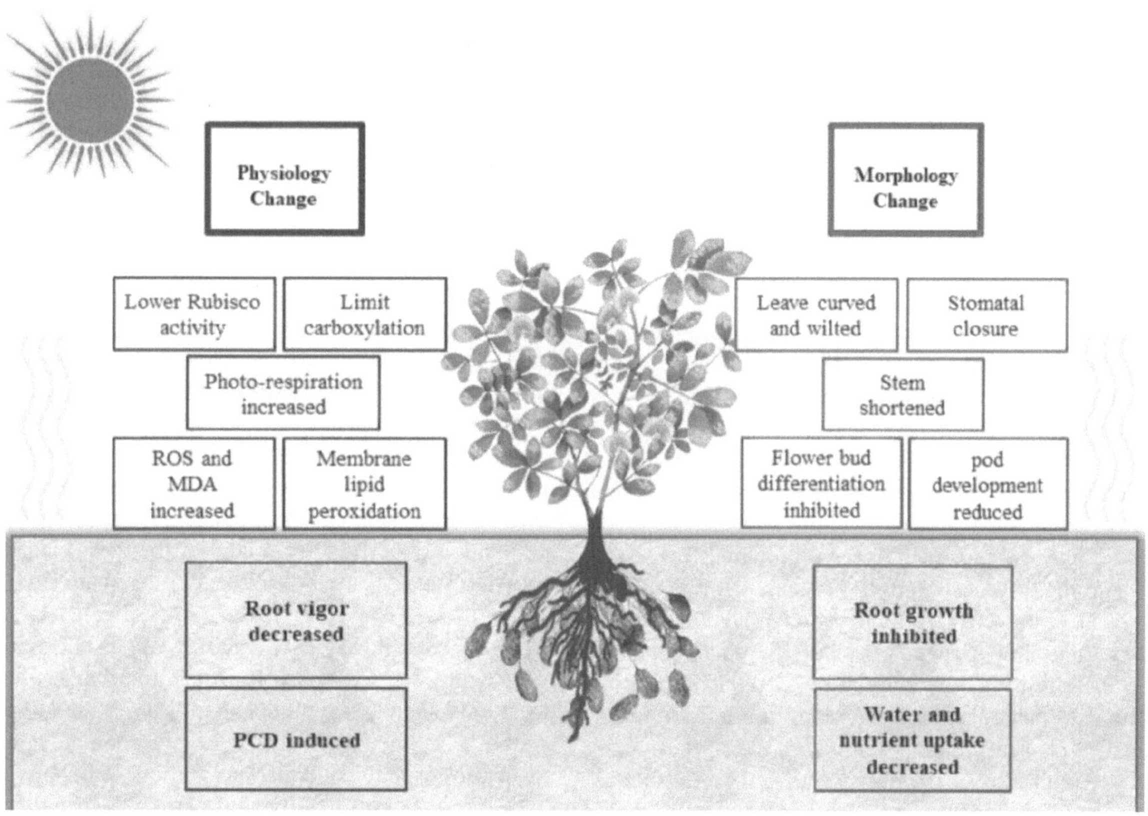

FIGURE 13.1 Effects of drought stress on the morpho-physiological responses of peanut.

transcriptional regulation, and protein storage. Therefore, selecting drought-resistant peanut varieties can not only reduce aflatoxin pollution but also improve peanut yield.

13.2.2 Drought Stress on Physiology of Peanut

Plants under drought stress usually accumulate reactive oxygen species (ROS), including singlet oxygen ($^1O_2$), hydroxyl radical (^-OH), superoxide anion (O^{2-}), and hydrogen peroxide (H_2O_2). ROS can react with proteins, lipids, and DNA to induce oxidative stress, resulting in impaired normal cell function. Generally, ROS production is linearly correlated with the severity of drought stress, and several organelles such as chloroplasts, mitochondria, and peroxisome are the primary sites for ROS production under drought stress (Bhatnagar-Mathur et al., 2009). Under drought stress, superoxide dismutase (SOD), peroxidase (POD) and other protective substances have a synergistic effect in plant, which can efficiently remove ROS produced under drought stress and reduce or avoid the damage of membrane lipid peroxidation to plants (Bhatnagar-Mathur et al., 2009; Bhalani et al., 2019). When drought stress appears, cell water potential decreases significantly, and osmotic potential is reduced by cells actively using organic solutes or absorbing ions, thereby enhancing the ability to absorb and store water to ensure the full expansion of cells and tissues, reducing the damage caused by drought stress, stabilizing their structure and function, and ultimately enhancing drought resistance (Zhu et al., 2016; Banavath et al., 2018). With the aggravation of drought stress, peroxidation occurs in the cell membrane system, malondialdehyde (MDA) gradually accumulates, and membrane permeability increases, causing ion extravazation. Therefore, the content of MDA reflects the degree of damage to the cell membrane (Liu et al., 2015).

13.2.3 Drought Stress on Molecular Biology of Peanut

Drought tolerance is the result of a coordinated action in plants, which activates a range of cellular, molecular, and physiological mechanisms, including inducing or/and inhibiting the expression of various genes, improving the antioxidant system, and reducing transpiration. Some leucine-rich-repeat receptor-like kinases (LRR-RLK) are significantly induced in drought tolerant varieties at early stage. These play crucial roles in the plant stress response and represent the largest subfamily of receptor-like kinases in plants (Liu et al., 2017), which perceive signals, transduce the signal downstream, and finally phosphorylate specific substrates, thereby mediating cellular signaling transduction (Yuan et al., 2010). E3 ubiquitin protein ligase is another key transcriptome module with negative regulation. Under drought stress, it is downregulated by degrading receptor-like protein kinases and ABA receptors (Ren et al., 2022b).

ABA as a key regulator of drought defense hormone has been intensively studied and has been found to be significantly induced under dehydration (Lee and Luan, 2012). The amount of ABA is accumulated via de novo biosynthesis, hydrolysis, and transportation. RT-PCR showed that *AhZEP*, *AhNCED1*, *AhAAO2*, and *AhABA3* relating to ABA biosynthesis increased; *AhBG11* and *AhBG24* of the ß-glucosidase hydrolysis gene were also activated, and the transportation family genes *AhABCG22.1* and *AhABCG22.2* were strongly expressed (Long et al., 2019). All this activity demonstrates that the responding signal mechanisms work interactively and collaboratively. Therefore, increased ABA binds to exclusive receptor of PYRABACTIN RESISTANCE1 and PYR1-LIKE (PYP/PYL) and inactivates protein phosphatases of type 2C (PP2C), thereby automatically activating sucrose non-fermenting1 (SNF-1)-related protein kinase 2s (SnRK2s). Then, the activation of SnRK2s initiates molecular and physiological responses to drought stress (Mahmood et al., 2020). Wang et al. (2023) found the gene encoding PYLs increased significantly and the gene encoding PP2C reduced in drought tolerant varieties, but it was adverse in drought sensitive cultivars, which indicates that smooth ABA signal transduction is crucial to elicit abiotic responses and develop adaptive responses in peanut.

Most of the differentially expressed transcripts are concentrated in the MYB and bHLH gene families, in which the increased ABA transmission pathway and regulated stomata movement saving water (Zhao et al., 2018). During the early stages of drought, drought-resistant cultivars induce the expression of a heat shock protein gene (*AhHSP62*), which increases ABA sensitivity and subsequently decreases the water loss rate in leaves (Zhang et al., 2018). Recently, Yan et al. (2020) cloned one gene with the specific capability of identifying the light regulation signal, FAP1-5, observably increasing the ABA content and the enzymatic activity of SOD, POD, and CAT, which demonstrated that it positively affected the ABA synthesis pathway to enhance the drought resistance of peanut.

Normally, under drought stress, primary metabolites, ABA, and secondary metabolites are emitted to interact with the environment (Tarkowska and Strnad, 2018). There is other name for primary metabolites, like terpenoids, terpenes or isoprenoids, which originated from C5 monomeric precursor's conjugation of dimethylallyl diphosphate (DMAPP) and isopentenyl diphosphate (IPP). DMAPP and IPP is namely derived from acid of the mevalonate (MVA) route and sugar of methylerythritol phosphate (MEP) route located in cytosol and plastid, respectively (Boureau et al., 2013; Alicandri et al., 2020). MEP is the main pathway to get DMAPP and IPP, which are also known as the terpenoids skeleton synthetic pathway. At seedling stage, drought resistance cultivars prefer to upregulate genes relating to the terpenoids skeleton synthetic pathway, but it is adverse in drought-sensitive cultivars, which indicates that drought resistance cultivars adopts fine terpenoids skeleton synthetic pathway to produce enough precursor to both primary and secondary metabolism path way to get those vital metabolites (Wang et al., 2023). Therefore, terpenoids have a vital function in the precursors of ABA, photosynthetic pigments (carotenoids), electron carriers (ubiquinone and plastoquinone), and key components of membrane structures (phytosterols) (Tholl 2015; Abbas et al., 2017). ABA as sesquiterpenes is derived from the cleavage of tetriterpenes (carotenoids) (Buchanan et al., 2020). Under drought stress, increased accumulation of sesquiterpenes simultaneously activates the activity of protective enzymes to resist the attack of reactive oxygen (Zhou et al., 2020).

Phenylpropanoid biosynthesis and flavonoid biosynthesis represent the critical drought response pathway, in which phenylpropanoid biosynthesis is located in the linkage position of primary and secondary metabolism by converting phenylalanine into the coenzyme A (CoA)-activated hydroxycinnamoyl thioester for monolignol and flavonoid biosynthesis (Ferrer et al., 2008; Vogt, 2010). Flavonoid is another kind of material that serves to protect peanut from the invasion of ROS under drought stress. Sugimoto et al. (2022) found that flavonoid deficiency stimulated the upregulation of ROS accumulation, thereby attenuating the activity of the oxidase scavenging system. Consequently, terpenoids play an important role in drought stress in peanut. Research is ongoing to uncover how the terpenoids regulate the synthesis of ABA, phenylpropanoid, and flavonoid.

13.2.4 Application of Different Biotechnological Tools in Improving Drought Tolerance in Peanut

With the rapid development of biotechnology and molecular biology and the completion of plant genome sequencing information, techniques such as molecular marker–assisted selection, functional gene excavation and transgenosis, map cloning, and gene editing have become common means of breeding and germplasm innovation in addition to traditional breeding methods and genetic engineering technologies (Raza et al., 2024).

13.2.4.1 Molecular Markers

Drought resistance is a complex quantitative trait, which is affected by the effects of micropotent polygenes and the environment. Therefore, it is very important to locate peanut drought-resistance related genes and develop peanut drought resistance–related molecular markers on this basis for improving the efficiency of drought resistance breeding via marker-assisted breeding. However, research on drought resistance–related genes and the genetic mechanism of peanut is still in the initial stages, and the foundation is relatively weak. With the completion of the whole genome sequencing of peanut, research into peanut drought resistance has been developed to some extent.

The construction of genetic maps based on molecular markers has become an important activity in the field of quantitative traits research. Genetic linkage maps cover the whole genome at the whole genome level, using SSR, SNP, AFLP and other molecular markers as reference, through whole genome resequencing. Their main function is to detect quantitative trait loci (QTL) for agronomic and quality traits of cultivated peanut.

Boontang et al. (2013) took drought-resistant varieties and control varieties as materials, established two water conditions (normal water supply and drought), analyzed and recorded data on specific leaf weight and relative water content, and used 265 pairs of SSR primers to detect genetic variability, among which 89 pairs of SSR primers showed good polymorphism. There were 20 SSR markers specific to the control varieties. The results showed that SSR markers accurately identify peanut genotypes with different drought-resistance traits. Varshney et al. (2009) constructed the first genetic linkage map of cultivated peanut, which included 135 SSR markers covering 22 linkage groups and identified drought-resistance traits such as transpiration efficiency, specific leaf area, and SPAD value. The results showed that between two and five QTLs could be identified for each trait on average. This provides an important reference for the selection of drought-resistant molecular marker–assisted breeding. Ravi et al. (2011) analyzed nine traits related to the drought resistance of peanut, including specific leaf area, chlorophyll content, canopy conductance, transpiration efficiency, and total dry weight using QTL Cartographer and QTL Network software, respectively. They also identified 105 major QTLs distributed in 54 intervals and 65 major QTLS distributed in 40 intervals, which explained 3.48%–33.36% and 1.30%–15.01% of phenotypic variation. Among them, 12 intervals containing multiple trait QTLs could be detected by the two mapping methods, and a total of 37 colocalized QTLs were obtained. Gautami et al. (2012) constructed a synergistic genetic map based on three isolated populations, and 153 major QTLs and 25 epistatic QTLs were identified as correlated with transpiration, transpiration efficiency, leaf dry weight, and total dry weight, respectively, through joint analysis of drought resistance–related traits in the three isolated populations. Faye et al. (2015) used the RIL population constructed by TAG 24×ICGV 86031 to identify 52 QTLs related to drought resistance traits such as plant height, chlorophyll content, 100 grain weight, and harvest index, among which the QTLs related to chlorophyll content were located on LG_AhVIII.

13.2.4.2 Omics

Omics plays a vital role in bridging the gap between phenotype and genotype by gathering data both on phenotype and alteration by drought stress. Transcriptome sequencing (RNA-seq) is simple to operate, not limited to the known genome sequence information, and can obtain low-abundance expressed genes, offering the advantages of high throughput, high sensitivity, low cost, and wide application fields. Transcriptome studies are the basis and starting point of gene function and structure studies. Using the new generation of high-throughput sequencing, we can quickly and comprehensively obtain all RNA sequence information of target cells of a certain species in a specific state, such as discovering new transcripts, understanding gene expression levels, mining single nucleotide polymorphisms (SNPs), structural variations, and so on.

Recently, more omics have appeared, like lipidomics, proteogenomics, miRNAomics, and primeomics, to help us understand the variations in plant at the genetic, proteomic, and metabolomic levels. Multi-omics plays a leading role in the complex evaluation of drought response of peanut, for drought stress involves modifications ranging from DNA to the whole cell, whole plant, or entire field. Multi-omics can provide a holistic approach to understanding biological processes at the cellular, tissue, or molecular mechanisms and components of cellular biology.

Moreover, with the rapid development of modern molecular biology, single cell multi-omics (SCM-omics) and spatial muti-omics (SM-omics) have emerged to facilitate exploiting the composition and function of cells or sub-cells in tissues and organs (Xie et al., 2023). More importantly, the integration of omics data has been an important problem which requires biologists and mathematicians to work together to collect, analyze, and mine data. In order to efficiently exploit multi-omics data, bioinformatics tools, platforms, packages, online repositories, and algorithms need to be provided (Gupta et al., 2023).

13.2.4.3 Gene Engineering

In recent years, through the application of genetic engineering technology to peanut genetic transformation, a number of high quality transgenic peanut lines have been cultivated. At present, the methods of transforming peanut varieties by transgenic technology mainly include oil synthesis, sensitization, and disease resistance. DREB transcription factors can regulate transcription molecules of drought response genes, thereby enhancing drought resistance of crops. Wang et al. (2018) transformed the *AlDREB2A* gene of swertia salt-tolerant plants into peanut by the transgenic method, which significantly increased the water content and proline content of its leaves and significantly enhanced its drought resistance (Table 13.1).

Genome editing is a technology for targeted DNA modification, and it has gone through many stages, such as ZFNs, TALEN, CRISPR, and so on. In recent years, as more and more plant genomes have been sequenced, nucleases based on synthetic nucleases have gradually developed into a research hotspot for the study of the synergistic regulation of single genes or multiple genes (Zhang et al., 2014). Among them, CRISPR/Cas9 is a new technology that has developed rapidly in recent years and has greatly promoted the improvement of peanut, such as via receptor kinases NFRs that act early in the symbiotic signaling pathway and are involved in recognizing and binding Nod factors (Shu et al., 2020); the *FAD2* gene, a key gene controlling the conversion of oleic acid dehydrogenation to linoleic acid (Li et al., 2019; Yuan et al., 2019); and the FatB acyl-ACP thioesterase gene, which encodes thioesterase of the saturated fatty acid carbon chain (Pan et al., 2020). CRISPR is rarely used in drought resistant peanut breeding, but with the improvement of peanut genome information, it will play an important role in breeding practice and variety improvement.

13.3 CHILLING TEMPERATURE

13.3.1 Chilling Stress on the Development of Peanut

If crops want to accumulate a certain amount of biomass or complete a certain growth period, their life activities must be carried out in a sufficient temperature range. Too low an ambient temperature can inhibit the growth and development of crops or even cause death. Chilling, that is, exposure to low non-freezing temperatures in the range 0–20°C, has been recognized for many years as a stress having a major impact on plant growth. The influence and damage degree of chilling stress on peanut production are closely related to the season, the duration, and the growth period. It causes serious harm to many biological processes of peanut, including seed germination, seedling growth, blossom, and pod development (Table 13.2).

The optimum and minimum temperatures for peanut seed germination are 25–37°C and 12–15°C, respectively. In general, the germination rate of peanut seeds is positively correlated with the ambient temperature when it is between the minimum temperature and the optimum temperature. But if the ambient temperature during peanut germination stage is below 12°C, the seed vitality will be inhibited, and even the phenomena of seed decay and vitality loss will occur, eventually resulting in slow emergence or severe seedling shortage in the field. It is worth noting that the minimum temperature required for seed germination of high-oleic acid peanut is at least 3–5°C higher than that of ordinary peanuts, that is, above 18–19°C.

The optimal and minimum temperatures for the peanut seedling stage are 25–35°C and 14–16°C, respectively (Upadhyaya et al., 2009). If a low temperature below the optimum temperature is encountered in production at this stage, the growth of the peanut seedling will be significantly inhibited, and the lower the temperature, the more obvious the damage symptoms. Zhang et al. (2021) found that the morphology of peanut seedlings changes to different extents after 6°C chilling stress. Most of them suffer dehydration, wilting, yellowing, fading, and other phenomena, some even dried up and died, and only a few could still remain upright and completely green. When the average daily temperature at seedling stage falls from 20.4°C to 19.0°C, the daily increase rates of dry matter and leaf area of peanut seedlings decrease significantly, which are 11.4% and 26.1% lower than the optimal temperature, respectively (Wang et al., 2003). Chilling stress occurring at the seedling stage of peanut also can affect the growth and development of the root system, which shows that the length and number of main roots and lateral roots are decreased, the

TABLE 13.1
A List of Genes under Drought Stress

No.	Gene	Material	Potential Application	Reference
1	*Gdi-15*	Groundnut	*Gdi-15* was homology with flavonol 3-O-glucosyl-transferase (F3OGT) adds glucose molecules at the 3'OH of anthocyanidins after leucoanthocyanidins promoting anthocyanidins synthesis which is discussed in resurrection plants thereby increasing antioxidative activity.	Gopala Krishna et al., 2001
2	*AhLEAs 1-19*	Fleur 11, 73-30; Luhua-14	All 19 LEA transcript accumulation is associated with tolerance acquisition mechanisms; it is critical for further identifying the specific function and role of those LEA genes.	Dramé et al., 2007; Su et al., 2011
3	*Gdi-15*, *Elip*, *SAG12*, VATPase	K-134	Deciphering the critical function of the genes and elucidating the biological mechanism of drought tolerant cultivars.	Govind et al., 2009; Ranganayakulu et al., 2012
4	*AhERF1–6*	Huayu19	Those candidate genes have different expression patterns, and the specific biological function still needs to be identified.	Chen et al., 2012
5	*AhNAC3*	Shanyou52; *Nicotiana tabacum* L. cv. NC90	*AhNAC3* is used as drought-induced transcription factor to improving peanut drought tolerance by improve superoxide scavenge activity and piling up some benefit moleculars.	Liu et al., 2013
6	HSP70, CuZnSOD, RD-22, my-inositol phosphate synthase	C76-16, A-104, Georgia, Green, AP-3	Those candidate genes have different mechanisms in different genotypes, functioned differently. The employment of those genes would facilitate drought-tolerant plant selection.	Dang Phat et al., 2013
7	*AhRabG3b*	Xuzhou68-4, Luhua11	*AhRabG3b* is used as potential candidate gene to improve dehydration and salinity ability of peanut.	Sui et al., 2013
8	*AhNCED1*	Yueyou 7; Shanyou 523	Drought resistant cultivar is of capability to rapidly induce *AhNCED1* expression improving drought tolerance by accumulating more ABA level.	Hu et al., 2013
9	*AhAREB1*	Shanyou 523	*AhAREB1* plays a vital role in drought tolerance by modulating ROS accumulation and ABA levels.	Li et al., 2013
10	*AhAREB1/AhNAC2*	Yueyou 7	Under drought stress, *AhAREB1/AhNAC2* bound in promoter region of *AhNCED1* and performing negative regulator of ABA level induced by drought.	Liu et al., 2016
11	*AhGLK1*	Yueyou 7	As a transcription factor, *AhGLK1* and *AhPORA* helps to promote chlorophyll content and photosynthesis.	Liu et al., 2018
12	*AhDREB1*	Yueyou 7	Utilizing histone deacetylases (HDACs) inhibitor trichostatin A (TSA) promotes *AhDREB1*, expression thereby upregulating *RD29A*, *P5CS1*, *P5CS2*, and *NCED1* expression to increase ABA levels and improving drought tolerance.	Zhang et al., 2018
13	ABA production (*AhZEP*, *AhNCED1* and *AhNCED3*, *AhABA2*, *AhAAO1* and *AhAAO2*, *AhABA3*, *AhBG11* and *AhBG24*), catabolism (*AhCYP707A3*, *AhUGT71K1* and *AhUGT73B4*); transport (*AhABCG22-1* and *AhABCG22-2*)	Yueyou 7	In peanut leaf, ABA sustains homeostasis by producing, catabolizing, and transporting it at the kinetic level.	Long et al., 2019

(*Continued*)

TABLE 13.1 (CONTINUED)
A List of Genes under Drought Stress

No.	Gene	Material	Potential Application	Reference
14	*AhGLK1*, *AhHDA1*	Yueyou 7	Employing *AhHDA1* and *AhGLK1* as an ON/OFF switch for *AhCAB* and *AhPORA* expression during drought stress.	Liu et al., 2020
15	*AhbHLH112*	HY9303	Under drought stress, *AhbHLH112* is capable of activating *AhPOD* and improving ABA response level by co-working with *AtNCED3* and *AtRD29A*.	Li et al., 2021
16	*AhSAUR3*	Tifrunner; *Arabidopsis*	Although *AhSAURs* has different functions in abiotic stress, it is promising to continually extract the function in light of evolutionary divergence.	Liu et al., 2022
17	*AhGLK1*	Yueyou 7	Hair roots of peanut serve as suitable material to study gene function of leaf.	Liu et al., 2023
18	*AhTGA11*	NH5, FH18	The *AhTGA* genes participate in hormone regulation to provide drought tolerance for sustainable growth and development.	Zhong et al., 2023
19	*AhRabG3f*	Xuzhou 68-4	Expression of several stress-responsive genes.	Song et al., 2012
20	*AhKCS1*	K-6	*AhKCS1* could prevent non-stomatal water loss from peanut.	Lokesh et al., 2019
21	*AtNHX1*	Golden and BARI-2000	Transforming *AtNHX1* into peanut could improve drought tolerance with higher proline content.	Asif et al., 2011
22	*AtDREB2A*, *AtHB7*, *AtABF3*	TMV2	Co-expression of stress responsive TFs in one cultivar could activate multiple drought response pathway to improve drought stress tolerance.	Pruthvi et al., 2014
23	*AtHDG11*	JL-24	Transforming *AtHDG11* into peanut is validating for improving drought tolerance by reshaping the plant and up-regulating drought response genes expressed.	Banavath et al., 2018
24	*AtDREB1A*	JL 24; RD2, RD11, and RD33 with carrying the rd29A:DREB1A gene regenerating in T6 to T9 generation; GG20 and its transgenic plants	Since *AtDREB1A* is successfully transformed into two different lines (JL 24 and GG20) and the transgenic lines exhibit brilliant drought tolerance and much better yield than wild lines, it can be used to develop drought-tolerant cultivars with better yield.	Bhatnagar-Mathur et al., 2004, 2007, 2009, 2014; Bhalani et al., 2019
25	*Alfin1*, *PgHSF4*, *PDH45*	GPBD4	It is feasible to gather regulatory genes simultaneous expressed to improve drought-adaptive trait with higher root growth, more relative water content and biomass.	Ramu et al., 2016
26	*MuNAC4*	Narayani	Transforming *MuNAC4* into peanut is validating for improving drought tolerance by increasing lateral root growth, antioxidative enzyme and reduced membrane damage.	Pandurangaiah et al., 2014
27	*PgeIF4A*	JL-24	Transgenic lines of peanut with *PgeIF4A* have great drought tolerance with increased shoot and tap root length, rubust lateral root formation, and photosynthetic and antioxidative enzyme ability and it could be used in drought cultivar improvement.	Rao et al., 2017
28	*MuWRKY3*	K-6	Transforming *MuWRKY3* into peanut could serve as transcript factor switch for drought resistant gene expression, thereby improving drought tolerance.	Kiranmai et al., 2018
29	*SbASR-1*	JL-24, TMV-2, TAG-24, Dh-3-30	Transforming *SbASR-1* into peanut could serve as an LEA protein to improve drought tolerance by up-regulating response genes.	Tiwari et al., 2015
30	*AVP-1*	New Mexico; Valencia A	It is promising that transforming *AVP-1* into peanut will improve drought tolerance by increasing biomass and photosynthetic rate.	Qin et al., 2013

TABLE 13.2
Effects of Chilling Stress on Growth Stages of Peanut (Zhang et al., 2019)

Growth Stages	Optimal Temperature (°C)	Minimum Temperature (°C)	Symptoms of Cold Injury	References
Germination stage	25–37	12–15	Slow germination or even loss of the germination ability; no seedling emergence in a large area in the field	Bell et al., 1994; Yan, 2001; Nakano et al., 2006; Chang et al., 2019
Seedling stage	25–35	14–16	Seedlings grow slowly or even stop growing; leaves become dehydrated, wilted, yellowed, and even withered to death; plant growth weakened	Awal and Ikeda, 2002; Prasad et al., 2006; Upadhyaya et al., 2009
Flower-pegging stage	25–28	22	Pollination hindrance; delayed flowering; decrease in the number of flowers; decrease in the number of needles	Kakani et al., 2002; Wang et al., 2003; Liu et al., 2017
Pod-setting stage	23–25	15	Pods develop slowly or even stop developing; decrease in the number of pods per plant; Increase of empty shell rate	Song and Wang, 1979; Ketring, 1984; Nigam et al., 1998
Pod-filling stage	23–27	4	Pods decay; seed mildew rate increased and inactivated	Ketring, 1984; Upadhyaya et al., 2009; Sorensen et al., 2009; Liu et al., 2017

thickness of lateral roots is increased, and the vigor of seedling roots is declined.

The optimum temperature for peanut plant growth during flower-pegging stage is 25–28°C. Too low or too high temperature could have adverse effects on flowering and fertilization. When the average daily temperature during the flowering stage was 24.9°C, the number of flowers was the highest, up to 77 per plant. When the average daily temperature dropped to 22.6°C, the number of flowers was only 45, with a decrease of 71%. When the average daily temperature rose to 26.4°C, the number of flowers was only 55, with a decrease of 40% (Wang et al., 2003). According to the observation of Shenyang Agricultural University, if the ambient temperature during the peanut flower-pegging stage is lower than 15°C, the morphology of flower organs such as flag valve length, flag valve width, and calyx tube length could be significantly inhibited and reduced by 25.60%, 20.61%, and 21.53%, respectively, compared with the normal temperature. Chilling stress can also lead to the decrease of pollen viability, stigma activity, and style length, resulting in fertilization obstacles and lower seed setting rate.

The optimum temperature for the development of peanut pods is 23–27°C. An average daily temperature below 20°C or above 40°C would have serious impacts on the formation and development of pods. If low temperatures below 20°C are encountered during the pod-filling stage, the phenomena of premature senescence such as leaf shedding and stem wither would appear. This could shorten the functional period of photosynthetic product transfer to pods, significantly decrease the pod fullness and 100-fruit weight, and eventually cause severe production reduction, which may reach 20%–30% in individual years. When the average ground temperature of the fruiting layer is lower than 18°C, pod development could even stop.

13.3.2 Chilling Stress on Physiology of Peanut

Lyons (1973) proposed that chilling damage initially occurs at the cellular and organ levels. The biomembrane system, including the cell membrane, nuclear membrane, and organelle membrane, is the initial site of injury, particularly in terms of its structure, function, stability, and enzyme activity, thereby resulting in substantial metabolic imbalance. The regulatory mechanism of biomembrane fluidity is one of the principal mechanisms by which plants accommodate to changes in temperature conditions (Figure 13.2). It is affected by the distribution ratio of various lipids on the membrane and the unsaturation of the glycerol lipid group (Barrero-Sicilia et al., 2017). When peanut plants are subjected to chilling stress, membrane lipids change from liquid crystal state to the gel state, which can cause the cessation of protoplast flow and an increase in membrane permeability, resulting in electrolyte leakage and loss of balance of intracellular ions (Murata et al., 1982). To maintain turgidity and the original metabolic process, various organic and inorganic substances, such as inorganic salt, proline, betaine, soluble sugars, and soluble proteins, accumulate in plant cells via osmotic regulation, which leads to an increase in the concentration of cell fluid and a decrease in osmotic potential (Kavi Kishor and Sreenivasulu, 2014).

Damage to the plant biomembrane system at low temperatures is also related to membrane lipid peroxidation and protein destruction induced by reactive oxygen species (ROS) (Grant et al., 2014). Membrane lipid peroxidation refers to a series of free radical reactions on double bonds of unsaturated fatty acid on the membrane, which is initiated by oxygen free radicals (O_2^-, H_2O_2, ·OH) on unsaturated fatty acids in lipids (Thomas et al., 2016). ROS *in vivo* would produce abundantly and accumulate rapidly under chilling stress in amounts far beyond the scavenging ability of the antioxidant system, which breaks down the original equilibrium state of ROS. At this point, ROS begin to attack biological macromolecules, such as membrane lipids, nucleic acids, and proteins. The structure of the membrane system is also destroyed, resulting in a decrease in the photosynthetic rate, development of metabolic disorders, and massive accumulation of toxic substances in plants (Choudhury et al., 2017). Chilling-tolerant cultivars can resist external environmental stress by relying on the antioxidant enzyme system, which scavenges ROS and superoxide anion free radicals produced in plant cells and mainly includes superoxide dismutase (SOD), peroxidase (POD), catalase (CAT), ascorbate peroxidase (APX), glutathione peroxidase (GPX), monodehydroascorbate reductase (MDHAR), dehydroascorbate reductase (DHAR), glutathione reductase (GR), and glutathione S-transferase (GST) (Tian et al., 2015).

Plants can modulate the stability and fluidity of membranes by changing the unsaturation of fatty acids in membrane lipids, which is of great significance for organisms in terms of resisting chilling stress (Karabudak et al., 2014). Zhang et al. (2022) detected 22 kinds of free fatty acids from peanut leaves by using gas chromatography-mass spectrometry. The main classes of free fatty acids in various peanut cultivars are similar under chilling or normal temperatures, including palmitic acid (16:0), stearic acid (18:0), oleic acid (18:1), linoleic acid (18:2), linolenic acid (18:3), arachidic acid (20:0), and so on. However, there is great diversity in their contents between the two cultivars under chilling stress. The total fatty acid (TFA) contents in chilling-tolerant and chilling-sensitive genotypes are increased by 4% and 0.7%, respectively. At the same time, the content of saturated (SFAs) and monounsaturated fatty acids (MUFAs), including C16:0, C16:1, C18:0, and C18:1, are obviously decreased in the chilling-tolerant genotype but almost unchanged in the chilling-sensitive genotype. The polyunsaturated fatty acids (PUFAs) C18:2 and C18:3 markedly increase in the chilling-tolerant genotype but increase more slowly in the chilling-sensitive genotype. In general, the chilling-tolerant genotype could accumulate more PUFAs (increased by 23%) than the chilling-sensitive genotype (increased by 2%) and exhibits the genetic characteristic of having more PUFAs during the long-term adaptation to low temperatures.

Photosynthesis is the most sensitive physiological process of plants subjected to temperature stress. Chilling-induced negative effects on any components of the

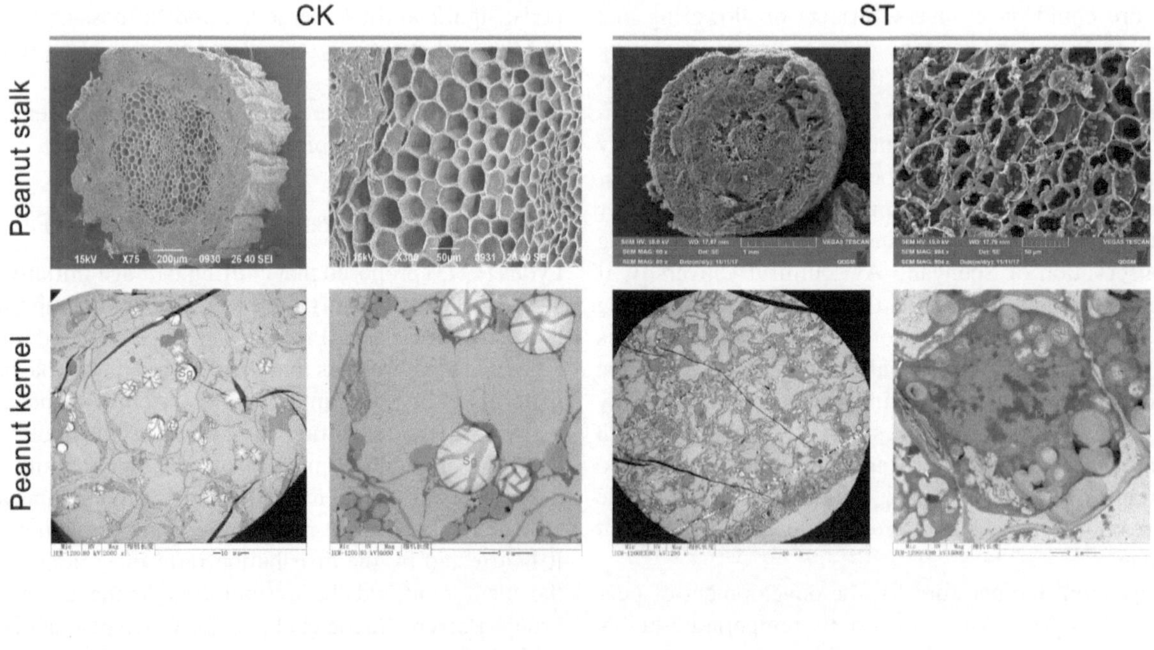

FIGURE 13.2 Effects of salt stress on microstructure of peanut stalk and kernel. Vb: vascular bundle; Mc: medullary cavity; Lc: lignifying cells; Pt: protein; Ls: liposome; Sg: starch grain.

photosynthesis systems may lead to a reduction in overall photosynthetic performance. This includes changing the biological characteristics of thylakoid membrane, reducing stomatal conductance, destroying carbon reduction cycle reaction, limiting photosynthetic electron transport, and reducing the formation of photosynthetic products (Zhang et al., 2016). According to the investigation of Zhang et al. (2022), the stomatal limitation is the major cause of inhibited photosynthesis of peanut seedlings under chilling stress. The inhibition of photosynthesis can also reduce the utilization of absorbed light energy, and excess light energy may result in the generation of toxic ROS (Kojima et al., 2007). The photoinhibition or photooxidation induced by chilling stress may significantly decrease the maximum photochemical efficiency of photosystem II (Fv/Fm), actual photochemical efficiency of photosystem II (ΦPSII), photochemical quenching (qP), and apparent electron transport efficiency (ETR).

13.3.3 Chilling Stress on Molecular Biology of Peanut

The physiological and biochemical changes that occur when plants adapt to constantly changing environmental conditions are determined by extensive reprogramming of gene expression (Calixto et al., 2018). A comprehensive understanding of the changes in the expression profiles of genes associated with specific physiological processes is necessary to reveal the molecular mechanism of chilling tolerance. Genome-scale technologies such as microarrays and RNA-Seq have enabled characterization of the complexity inherent in the transcriptome and have greatly facilitated genetic research involving non-model plants. Based on the weighted gene co-expression network analysis (WGCNA) of physiological indices and transcriptional profiles, Zhang et al. (2022) identified 20 candidate hub genes regulating peanut chilling tolerance. Among them, hub genes encoding CAT (*arahy.RQ5XX8*), MYC4 (*arahy.0RH9QK*), allene oxide synthase (*AOS3*, *arahy.F73TU2*), glutathione S-transferase (*GSTs*, *arahy.41Y8SN* and *arahy.06EJ88*), protein phosphatase 2C (*PP2C*, *arahy.5434G2*), E3 ubiquitin-protein ligase UPL5 (*arahy.U26HEC*), and zinc finger protein 8 (*ZFP8*, *arahy.5ZNX8B*) were likely to positively regulate peanut chilling tolerance, but hub genes encoding polyamine oxidase PAO2 (*arahy.UIG78U* and *arahy.AM4HSV*), NAC2 (*arahy.2FYA1K*), and NAC72 (*arahy.9ZTQ0N*) were likely to negatively regulate peanut chilling tolerance.

Transcription factors (TFs) act as transcriptional activators or repressors and play a central role in the regulation of development, metabolic processes, and biotic and abiotic stresses (Ng et al., 2018). Generally, TFs exist in the upstream of plant growth and the development regulatory network, which can simultaneously regulate the expression of multiple genes. In molecular breeding, improving or enhancing the regulatory capacity of one crucial transcription factor can promote the function of several genes, which can obtain a comprehensive improvement effect compared with changing the expression of a single gene. Jiang et al. (2020) investigated the global expression patterns of entire TFs in two peanut varieties with different chilling-tolerant abilities by using comparative transcriptome analysis. A set of TF genes in response to chilling stress were identified. Among these TF genes, *arahy.0RH9QK* (*MYC2*), *arahy.2KRW4X* (*MYC4*), *arahy.ILFA61* (*PIF3*), *arahy.U3PP50* (*HAL3A*), *arahy.new4888* (*ERF13*), *arahy.E3SE4T* (*WRKY70*), *arahy.EVAW89* (*WRKY53*), and *arahy.B6D6GK* (*WRKY22*) have essential roles in peanut chilling tolerance. Based on the KEGG enrichment analysis, the hub genes of the bHLH TF family are mainly involved in the jasmonic acid (JA) signaling pathway; most hub genes of C2H2 and ERF TF families are participated in abscisic acid (ABA) signaling pathway; and the hub genes of WRKY gene family are mostly enriched in JA signaling, salicylic acid (SA) signaling, MAPK signaling, and plant–pathogen interaction pathways.

The signal transduction of chilling stress is a complex process. Low temperature is initially perceived by the plasma membrane, either due to a change in membrane fluidity or with the help of sensors like Ca^{2+} permeable channels, histidine kinases, receptor kinases, and phospholipases. Subsequently, cytoskeleton reorganization and cytosolic Ca^{2+} influx takes place. An increase in cytosolic Ca^{2+} is sensed by calcium-dependent protein kinases (CDPKs), phosphatase and mitogen-activated protein kinases (MAPKs), which transduce the signals to switch on transcriptional cascades (Gusain et al., 2023). At present, the induction of the CBF expression (ICE)-C-repeat binding factor/DRE binding factor 1 (CBF/DREB1)-Cold-responsive (COR) transcriptional cascade is the most in-depth and conservative mechanism by which plants adapt to chilling stress. ICE1 can directly bind to the MYC element in the promoter region of CBFs to regulate the expression of *CBFs*, and CBF can induce the expression of *CORs* by binding to the promoter element CRT/DRE of CORs, thereby enhancing plant chilling tolerance (Shi et al., 2018). The expression of *CBFs* can also be also finely regulated by some upstream regulatory factors, including TFs and posttranscriptional modification of proteins (Ding et al., 2019). For example, the SUMO E3 ligases SAP and MIZ 1 (SIZ1) and the protein kinase OPEN STOMATA 1 (OST1) can increase the stability of ICE1 through ubiquitination and phosphorylation, respectively, thereby improving plant chilling tolerance. Conversely, the protein stability of ICE1 and CBF can be destroyed by the phosphorylation of MPK3/6 and BRASSINOSTEROID-INSENSITIVE2 (BIN2), resulting in weakened expression of CORs and reduced plant chilling tolerance (Ye et al., 2019). Although the CBF-COR pathway plays a critical role in modulating chilling stress responses, only 10%–20% of CORs in plants are regulated by CBFs (Jia et al., 2016). Numerous TFs and functional genes that regulate plant chilling tolerance through CBF-independent pathways remain to be identified, especially in non-model plants.

13.3.4 Chilling Stress on Application of Different Biotechnological Tools in Improving Chilling Temperature Tolerance in Peanuts

In crop production, the selection and breeding of chilling-tolerant varieties is the fundamental way to solve chilling damage occurring in farmland. In recent years, the advancement of biotechnology to address plant productivity and stress tolerance has been sparked by the emergence of contemporary genetic engineering methodologies and high throughput biological research tools. The identification and isolation of hub genes is the center of biotechnology for achieving genetic improvement of target traits. Its combination with molecular assisted breeding can greatly shorten the breeding process and is of great significance for guiding crop production. Presently, with the application of expression sequence labeling, cDNA microarray, RNA-sequencing, and forward and reverse genetics in peanut research, some progress has been made in the identification and regulation mechanism of chilling-induced genes or QTLs in peanut.

Marker-assisted selection (MAS) technology can be carried out in the early generation and with higher accuracy, which may greatly shorten breeding time, accelerate the breeding process, and provide a reliable basis for innovative germplasm. In the last two decades, using the most successful approach, namely MAS or marker-assisted backcrossing (MABC), some molecular markers have successfully been developed in peanut for tolerance to chilling stress. Bai et al. (2018) evaluated peanut cultivars with different chilling-tolerance levels by using 90 pairs of SSR primers with good polymorphism. Liu et al. (2019) used 187 recombinant inbred lines (RIL), derived from a cross between two peanut cultivars (Xuhua 13 × Zhonghua 6), as mapping population, and successfully constructed a genetic linkage map of cultivated peanut containing 325 markers, covering 20 linkage groups, with a total length of 1493.79cM and an average distance between markers of 4.60cM. In addition, a total of ten QTLs associated with chilling tolerance were detected. Among them, four major QTLs (qLT-TLG1-1, qLTTLG1-2, qLTTLG6-1, and qLT-TLG7-1) were repeatedly detected in two of the chilling environments, which could be used for marker-assisted breeding of chilling tolerance.

The transfer of chilling tolerance genes into plants through genetic engineering technologies can more efficiently enhance plant chilling tolerance and improve the breeding efficiency of chilling tolerance varieties. Presently, genetic engineering techniques such as *Agrobacterium tumefaciens* or DNA-bombardment-mediated transformation are used as powerful tools that complement conventional breeding and expedite rapid progress in peanut improvement. In addition, reverse genetics is also one of the most straightforward strategies to link a candidate gene and its function. Techniques used for the generation of expression data include northern blot analysis, subtractive screening of cDNA, screening of cold-responsive genes using a GUS reporter in T-DNA-tagged plants, quantitative RT-PCR, and cDNA microarray analysis. More than 70% of published chilling tolerance genes were cloned based on transcription data. For example, Wang et al. (2021) identified 3,620 genes related to chilling tolerance from the peanut genome, which were mainly enriched in the phenylpropane biosynthesis pathway. Zhang et al. (2019) obtained 139,429 unigenes by high-throughput transcriptome sequencing of chilling-resistant and chilling-sensitive high-oleic peanut varieties under chilling stress at germination stage, which provided data resources for the in-depth understanding of the molecular mechanism of high-oleic peanut adaptation to chilling stress and the screening of chilling-tolerant genes in peanut.

13.4 SALINITY

Soil salinization has become a global environmental problem that severely affects the sustainability of agriculture. High concentration of soluble salinity ions (Na^+) in salinized soil adversely affects the physical, chemical, and biological properties of soil, making soil unable to provide normal nutrient conditions for plant growth. Meanwhile, excessive Na^+ causes physiological toxicity to plants, affects the geographical distribution and limits their growth, and poses serious threats to food security (Zhu, 2016).

13.4.1 Salinity Stress on the Development of Peanut

Salinity stress remarkably decreases seed viability, delays seed germination, and inhibits the growth of germ and radicle. Plant roots are the main organs of nutrient and water uptake from soil, and are therefore also the first organs to face salinity stress. Gui et al. (2019) found that salinity stress accelerates the corkification of roots and affects the uptake of water and nutrients by roots, thereby preventing normal crop growth and development. Previous data showed that the peanut plant height and dry weight under salt stress (ST) was significantly reduced compared with the control. Logistic equation fitted growth curves showed that the maximum growth rate (V_m) of peanut plant height and dry weight was significantly lower than that of CK, and the occurrence time of maximum growth rate (T_m) was significantly earlier than that of CK (Shi, 2018). Therefore, salt stress causes plant senescence and tends to advance and shorten the period of plant growth, which decreases leaf area index and accelerates leaf withering and yellowing, ultimately affecting the photosynthetic performance of crops.

The decrease in photosynthesis and photosynthetic assimilation capacity results in insufficient material basis for plants to maintain their normal growth, leading to the lengthening of the growing period, the shortening of the vegetative and reproductive periods, and ultimately crop yield reduction. Shi (2018) regularly investigated the pod development of peanut, and compared with control, the V_m and T_m of pod development (including dry weight and volume of pod and kernel) of peanut were significantly reduced under salt stress. These results suggest that salt stress inhibits plant growth and reduces the transport of photosynthetic products to the peanut pod, resulting in a decrease in yield.

The stalk of peanut is an important organ to transport photosynthates and nutrients from shoot to pod, which plays an extremely important role in pod filling. The vascular bundle (Vb) in the stalk is the main channel for water and nutrients, and constitutes a physical support for the plant. Scanning electron microscopy showed the tissue structure of pod stalk was destroyed under salt stress, the thickness of epidermal cells increased, and the diameter ratio increased by 59.97%. The cells surrounding the medullary cavity (Mc) were disrupted, forming many cavities. The vascular bundle also undergoes a structural change during salt stress, and a large number of lignifying cells (Lc) are produced in the inner wall of the vascular bundles, indicating that salt stress accelerates cell lignification and causes plant senescence. Moreover, the effects of salt stress on peanut kernel inclusions were discovered by transmission electron microscopy. The most intuitive reflection is that the osmotic pressure induced by salt stress changes the structure of protein (Pt), liposome (Ls), and starch grain (Sg) in the kernel (Figure 13.2). The liposome is directly related to the oil content of the peanut kernel, which is also the main reason for the low oil content and poor quality of peanuts under salt stress.

In practical peanut production, heterogeneous distribution of soil salinity would lead to uneven seedlings in the peanut field. Especially in the later stages of peanut growth, an increase in soil salinity will lead to serious root rot and stem rot, and then cause a large number of peanut plants to die and dry up, which seriously affects the yield and quality of peanut.

13.4.2 Salinity Stress on Physiology of Peanut

The effect of salinity stress on plants has two main phases. The first phase is osmotic stress (reduction of external water potential). When the soil salt concentration in the root zone is high, the water uptake capacity of the root system is significantly reduced due to the decrease in cell membrane permeability; it can even cause plant cells to lose water, resulting in physiological drought (Roy et al., 2014). Osmotic stress also leads to stomatal closure and therefore precludes the diffusion of CO_2 into the plant cells. Stoma closure can be explained by a rapid decrease in xylem-pressure under salt stress (Wegner et al., 2011). This drop in xylem pressure is known to happen within a few minutes at the onset of salt stress, and the water deficit signals sensed by the root system can be quickly transmitted to the stem, where these signals are decoded and alter stem metabolism (Zhou et al., 2023).

The second phase of the salt stress is ion imbalance (also known as "ionic stress" or "ion toxicity"). When Na^+ or Cl^- is accumulated to a high concentration in plant tissues, it can cause specific ion toxicities (Munns and Tester, 2008); ion toxicity can frustrate the synthesis of protein and decrease the activity of enzymes, which leads to the disorder of the physiological activity of crops, premature leaf senescence, and tissue necrosis (Roy et al., 2014). Generally, Na^+ is a non-essential element in a variety of crops (except for a few C4-crop), and the accumulation of excessive amounts of Na^+ (Na^+ toxicity) can inhibit the activities of many involved in the primary metabolism, Calvin cycle, phenylpropanoid pathway, glycolysis, polyamines, and starch synthase. Excessive accumulation of Na^+ can also induce cytosolic K^+ efflux, resulting in intracellular ion homeostasis imbalance, oxidative stress, interference with Ca^{2+} and K^+ function, nutritional deficiency, and even cell death (Thorne et al., 2020). K^+ is an essential macronutrient that plays an important role in enzyme activation, osmotic regulation, membrane potential regulation, and cytoplasmic pH homeostasis (Zhou et al., 2020). Therefore, maintaining a balanced K^+/Na^+ ratio has emerged as a crucial factor for the regulation of salt stress in plants. Currently, there are relatively few studies regarding the toxicity of Cl^-. Studies have reported that shoot Cl^- exclusion is crucial for salinity tolerance in plants (Geilfus, 2018). In some salt-sensitive crops, higher Cl^- concentrations in shoots are positively correlated with physiological dysfunction. However, this phenomenon is not consistent in halophytes (Moir-Barnetson et al., 2016). Therefore, most recent scholars agree that the effect of Cl^- on plant performance may not be due to its own toxicity, but to the deficiency of macronutrient caused by Cl^-, since the absorption of NO_3^- and SO_4^{2-} is mediated by the same non-selective anion transporter as Cl^- (Bazihizina et al., 2019). It would seem that the adverse effects induced by Na^+ or Cl^- are not due to ionic toxicity per se, but to interference with the absorption or metabolism of other essential ions. Therefore, the term "ion imbalance" is more appropriate to explain this effect of salt stress.

Salt stress can induce osmotic stress, ionic toxicity, and complex secondary effects, thereby resulting in the overproduction of reactive oxygen species (ROS), such as superoxide anions ($O_2^{\cdot -}$), hydrogen peroxide (H_2O_2), singlet oxygen ($O_2^{\cdot}$), hydroxyl radicals (OH^-), and so on (Hasanuzzaman et al., 2020). In plant cells, chloroplasts, mitochondria, and plasma membranes are major sites of ROS generation under salt stress (Ding et al., 2018). Consequently, the accumulation of ROS will destroy the normal physiological functions of cells and lead to cell metabolism disorders.

13.4.3 Salinity Stress on Molecular Biology of Peanut

Regulation of gene expression in salinity stress includes a wide array of mechanisms that are used by plants to upregulate or downregulate the production of specific gene products (protein or RNA). From transcriptional initiation to

RNA processing and to post-translational modification of a protein, various gene regulatory mechanisms have been identified in the central dogma. A large number of genes and transcription factors are upregulated in response to salinity in different plant species, which serve diverse functions. These genes are mainly classified into the following functional categories: ion transport or homeostasis (e.g., *SOS* genes, *AtNHX1*, and *H⁺-ATPase*), senescence-associated genes (e.g., *SAG*), molecular chaperones (e.g., *HSP* genes), and dehydration-related transcription factors (e.g., *DREB*) (Gupta and Huang, 2014).

In the case of ion transport or homeostasis, Na^+ enters root xylem vessels via both apoplastic and symplastic pathways and directly interferes with ion homeostasis in cells. Especially competitive substitution of K^+ at the binding site of biological enzymes directly reduces the enzyme activity, resulting in cell metabolism disorders, even culminating in cell death or apoptosis. In addition, Na^+ enters the cytosol and causes plasma membrane depolarization, leading to K^+ efflux via depolarization-activated K^+ outward-rectifying (KOR) channels, such as guard cell outward rectifying K^+ channel (GORK) and non-selective cation channels (NSCCs). Thus, crop cells prevent membrane depolarization by retaining a negative potential inside the cell to enhance intracellular K^+ retention (inhibiting K^+ efflux) and maintain a high K^+/Na^+. Studies have shown that plasma membrane (PM) H^+-ATPases and tonoplast (TP) H^+-ATPases/H^+-PPases, K^+ transporters, vacuolar Na^+/H^+ exchangers (NHX), and salt overly sensitive (SOS1) Na^+/H^+ antiporters are important participants in maintaining an inside negative membrane potential (Fang et al., 2021). These pathways act in cooperation to mitigate the harmful effects of crop growth caused by salt stress and low K^+. SOS1 has been identified as a Na^+/H^+ antiporter localized in epidermal cells in the root apex, where it actively extrudes Na^+ from the cytosol into the rhizosphere, providing the first line of resistance to prevent the absorption of Na^+ (Shi et al., 2002). Furthermore, salt stress can increase the activity of H^+-ATPase and H^+-ATPase, allowing more H^+ to be pumped into the vacuole, thereby enhancing the ability of transfers Na^+ from the cytoplasm to the vacuole by increasing the transmembrane electrochemical gradient (Ye et al., 2019). Na^+ can also be compartmentalized through ion antiporters such as NHX7 (also named SOS1) within cell membrane and NHX1 within vacuolar membrane to maintain a low Na^+ level, both of whose activity is regulated by calcium-dependent SOS2/SOS3 kinase complexes (Almeida et al., 2017).

13.4.4 Salinity Stress on Application of Different Biotechnological Tools in Improving Salinity Tolerance in Peanuts

13.4.4.1 K^+ Homeostasis is Necessary during Salinity Stress

Stress tolerance is often associated with accumulation of small organic molecules (e.g. proline, free amino acid, soluble protein, glycine betaine, and polyamines), which can act as protective compounds against various stress factors (Zhang et al., 2021). K^+ is also the major inorganic osmolyte for tissue osmotic adjustment under salt stress conditions. The ability to retain K^+/Na^+ homeostasis is critical for plant adaptation to salt stress. In this perspective, exogenous application of K is a feasible method to maintain K^+ homeostasis and osmotic balance and promote crop growth under salt stress (Chakraborty et al., 2016). Our previous study found that the external application of K improves the Na^+, K^+ selective absorption (SA) and selective transport (ST), while reducing the Na^+/K^+ ratio in various organs (Figure 13.3A, B). The increase of K^+ content in plants improved the photosynthetic potential (Figure 13.3C) by increasing the osmotic regulation ability (Figure 13.3D) and reducing the leaf salinity hazard coefficient (LSHC) (Figure 13.3E), thereby ultimately enhancing salt tolerance in peanut (Figure 13.3F) (Shi et al., 2020).

In recent years, multiple transporter genes and proteins encoding K^+ transport channel have been identified, including HKT (high-affinity K^+ transporter) and NHX. The high-affinity K^+ transporter (HKT) is responsible for the transport of Na^+ and K^+, and plays a crucial role in unloading Na^+ from the xylem and maintaining a high intracellular K^+/Na^+ ratio in plant cells (Deinlein et al., 2014). Intracellular NHX proteins are Na^+, K^+/H^+ antiporters involved in K^+ homeostasis, endosomal pH regulation, and salt tolerance (Barragán et al., 2012). Therefore, improving and regulating plant K^+/Na^+ homeostasis through molecular or conventional means could serve as an important way to improve peanut salt tolerance.

13.4.4.2 Rhizosphere Microbes Enhance Plant Salt Tolerance

Beneficial rhizosphere microorganisms that are closely associated with roots are mainly composed of plant growth promoting rhizobacteria (PGPR) and endophytic bacteria. The PGPR use several mechanisms to help plants tolerate abiotic stresses, and this process is termed induced systemic tolerance (IST) (Yang et al., 2009). For most bacteria, secrete phosphate dissolution, 1-aminocyclopropane-1-carboxylate deaminase (ACCD), volatiles, siderophore, indoleacetic acid (IAA), and exopolysaccharides (EPS) are critical for IST. Yuan et al. (2016) reported that PGPR with ACCD activity enhanced salt resistance of plants in saline soil. PGPR can also induce crop resistance to salt stress by producing ACCD and catalyzing the conversion of ACC (1-aminocyclopropane-1-carboxylic acid, the immediate ethylene biosynthesis precursor) to ammonia and α-ketobutyrate (Barnawal et al., 2014). In addition, PGPR can also improve salt tolerance by altering endogenous hormone accumulation and increasing nutrient absorption of plants (Kang et al., 2014). In particular, PGPR can decrease Na^+ uptake in leaves, which

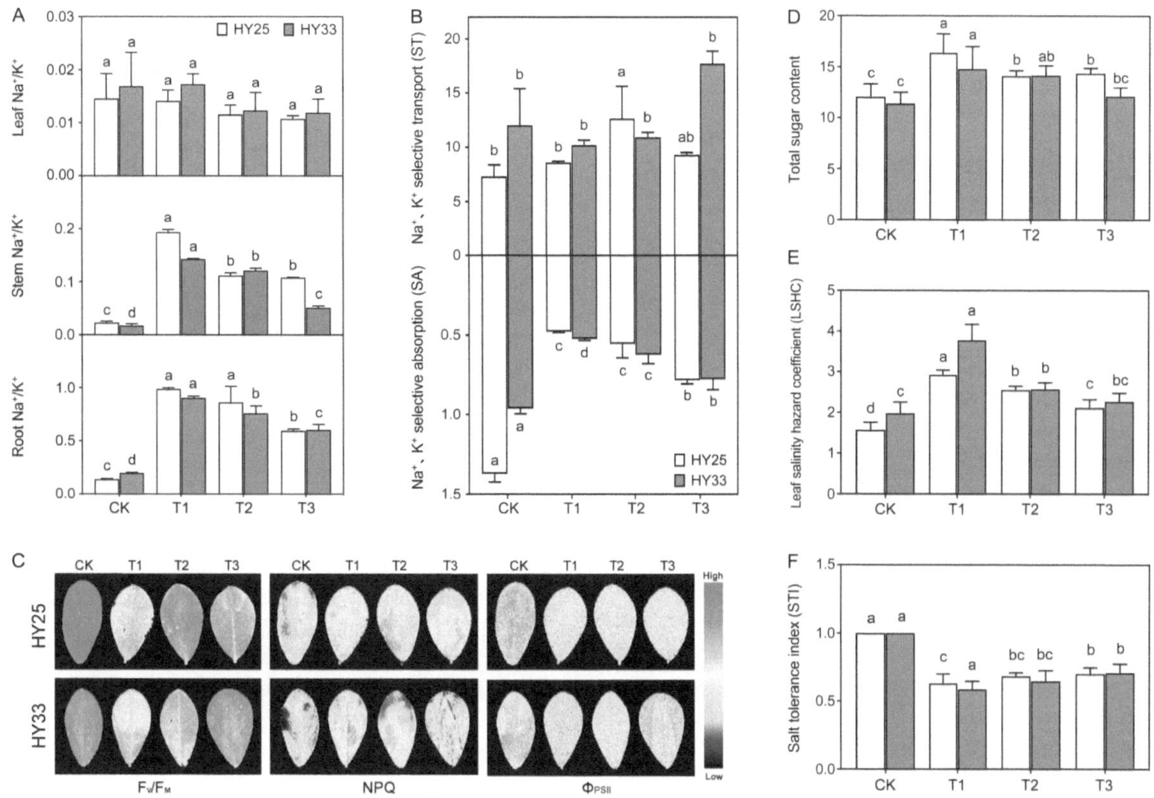

FIGURE 13.3 Effects of potassium fertilizer application on Na^+, K^+ homeostasis and salt tolerance in peanut. CK: 0.00% NaCl + 0 kg ha^{-1} K_2O; T1: 0.25% NaCl + 0 kg ha^{-1} K_2O; T2: 0.25% NaCl + 85 kg ha^{-1} K_2O; T3: 0.25% NaCl + 170 kg ha^{-1} K_2O. (A) Na^+/K^+ in different organs; (B) Na^+, K^+ selective absorption (SA) and selective transport (ST); (C) total sugar content; (D) leaf salinity hazard coefficient (LSHC); (E) chlorophyll fluorescence images; (F) salt tolerance index (STI).

results in increased K^+ uptake and K^+/Na^+ ratio (Ha et al., 2014). One possible reason for this mechanism is that EPS production by certain PGPR may not only bind to cations, but may also promote biofilm formation on the root surface, thereby limiting Na^+ influx into the root system. Therefore, the production of EPS and the formation of biofilm during rhizosphere colonization are conducive to resistance to salt stress or salt-induced osmotic stress in agriculture. Some PGPR strains release volatile organic compounds (VOCs) that can directly and/or indirectly mediate increases in plant biomass, disease resistance, and abiotic stress tolerance (Liu and Zhang, 2015). For example, VOCs produced by *Pseudomonas simiae* can significantly upregulate vegetative storage proteins, glutamine amidohydrolase, and *RuBisCO* large chain protein, which is considered to be a key protein in inducing IST in soybeans under salt stress (Vaishnav et al., 2015). *Alcaligenes faecalis* strain JBCS1294 (β-Proteus) mainly releases adipic acid and butyric acid, and induces salt tolerance in crops through recombinant auxin and gibberellin pathways (Bhattacharyya et al., 2015).

In addition to bacteria, arbuscular mycorrhizal fungi (AMF) can also improve the adaptability of host crops to salt stress. In a symbiotic relationship with a plant, mycelium can form intracellular hyphal structures within living host cells, or extending from the root surface (intraradical mycelium) into the surrounding soil (extraradical mycelium). These structures can improve nutrient and water uptake efficiency in crops and increase the K^+/Na^+ ratio by selectively enhancing K^+ and Ca^{2+} absorption, thereby indirectly avoiding the detrimental effects of soil salinity (Ruiz-Lozano et al., 2012). In response to salt stress, PGPR and AMF can positively regulate the expression of plasma membrane intrinsic proteins (PIPs) coding genes and are involved in aquaporin activity, thus promoting the efficient water absorption of crops under salt stress. Colonization of beneficial microorganisms such as AMF were able to maintain intracellular osmotic balance by increasing the concentration of osmolyte and carbohydrates, and thereby improve the salt tolerance of crops (Evelin et al., 2013). Therefore, improving and maintaining beneficial rhizosphere microbial diversity is of great significance for improving crop environmental tolerance. At present, in addition to improving the microbial community through traditional measures (planting patterns, fertilization, etc.), identification and isolation of key microbial taxa and inoculation of synthetic microbial community are all potential technical means to improve the environmental tolerance of peanut.

13.5 NUTRIENT IMBALANCE

13.5.1 Nutrient Imbalance on the Development of Peanut

13.5.1.1 Nutrient Imbalance on the Root Architecture of Peanut

The most effective adaptive response is to change root architecture to extract more nutrients from the soil with minimal increased metabolic cost in nutrient unbalanced soil environments, which is an important basis for crop root type improvement and stress resistance breeding (Karlova et al., 2021). Under deficiency of N, P, and K, root length and biomass generally decreased, while root length per shoot biomass was enhanced. When N supply is insufficient, long and thin taproots are formed, hindering lateral root growth and reducing lateral root length and density (Tian et al., 2008). However, low N stress limits the growth and development of peanut roots to a great extent, and then affects the formation of above-ground parts. At the same time, the root cortex tissue cells and the length of mature cells increases, more aerenchyma forms, the tissue structure optimizes, and N absorption improves (Mi et al., 2016). In contrast to the easy mobility of N, P and K are of low mobility in the soil and are therefore mainly concentrated in the topsoil layer. In order to adapt to low P or K content in soil, leguminous crops have formed a "surface soil foraging" configuration with relatively increased root distribution, smaller root angle, and weakened root to underground (Lynch and Brown, 2001). In particular, the geotropism trajectories of basal roots, adventitious roots, and lateral roots and the plasticity of these processes have strengthened the exploration and exploitation of shallow soil layers by legumes such as peanuts, where the availability of P and K is greatest in many soils. By remobilizing P from the root cortex, the metabolic cost of soil exploration has been reduced and more P has been obtained (Postma and Lynch, 2011). The cell wall relaxed, expanded, and distorted, and plasmolysis occurred under calcium deficiency in peanut roots. The cell membrane function of lateral root is severely damaged, the development of lateral root is hindered, and the root growth point is necrotic (Zhou and Lin, 1996).

13.5.1.2 Nutrient Imbalance on the Leaf of Peanut

When crops suffer from nutrient imbalance, the most immediate visual symptom is leaf chlorosis, which then induces autophagy and necrosis-type programmed cell death (PCD) and produces necrotic patches (Demidchik et al., 2018). These reactions are accompanied by the generation of reactive oxygen species (ROS) and ion disequilibrium, which is induced by electrolyte/K^+ leakage through ROS-activated ion channels, such as the outwardly-rectifying K^+ channel GORK and non-selective cation channels. Under low K stress, leaf color gradually becomes dim from old to new, from the tip to the edge, and from the outside to the inside, and they eventually shrivel and die. This is because the accumulation of putrescine in the leaves, so when color is lost, many brown spots are produced (Liu et al., 2023). The lignification process of the leaf cell wall is blocked with wilting, and even with signs of chlorosis and necrosis, resulting in decreased leaf photosynthesis and reduction in yield and quality. In addition, N deficiency can also lead to leaf chlorosis. Under medium nitrogen stress, peanut leaves were small and thin, and the lower leaves were light yellow. However, under severe N deficiency stress, the upper leaves of peanut plants lost green coloring, and the lower leaves become yellow and dried up (Jiang et al., 2020). On the side, under low Ca stress, the vacuole membrane of peanut leaves is broken and the compartmentation disappears. Organelles and calcium-containing particles are scattered throughout the cell space, and normal cell metabolism is damaged. The chloroplast structure is obviously destroyed and the envelope is broken. The arrangement of then grain-lamellar structure is disordered and the content of calcium in the matrix is obviously reduced. The chloroplasts are loose and enlarged, and the upper leaves are yellowed and scorched (Yang et al., 2015).

13.5.1.3 Nutrient Imbalance on the Nodule of Peanut

Under the condition of low N supply in soil, the number of root nodules is lower than the normal N supply in peanut. Although increased N application promotes the absorption and accumulation of N in peanut, excessive N application greatly limits the infection of rhizobia and the formation of root nodules (Liu et al., 2011). The N fixation rate of rhizobium is negatively correlated with N application rate. This is mainly reflected in the decrease in nodule N fixation; the decrease in nodule number, size, and dry weight; and the low N utilization rate (Yang et al., 2022). The increase of N application significantly decreases the nodulation ability of peanut. Guo et al. (2022) reported that the highest root nodule number and nodule dry matter of the strong nodulation ability varieties (HH16 and XH11) and the weak nodulation ability varieties (NH5 and TYH1) were gained under low N application (75kg•hm^{-2} and 45kg•hm^{-2}) (Guo et al., 2022). The ^{15}N isotope labeling test shows that the total N uptake of peanut plants showed a parabolic trend with the increase in N application rate, and the ratio of N supply from root nodules, soil, and fertilizer is about 5:3:2 at the podding stage (Figure 13.4). Under high N soil, the residual amount and loss of N increases significantly, while under low and medium N soil, the loss rate of N decreases significantly. The main reason is that high nitrogen application inhibits the recognition and infection of rhizobium on peanuts and other legumes, and inhibits the synthesis of flavonoids, which is a signal substance of root nodules (Dardanelli et al., 2009).

13.5.1.4 Nutrient Imbalance on the Yield and Related Traits of Peanut

Nutrient deficiency leads to excessive accumulation of carbohydrate in peanut leaves, hindering the development of

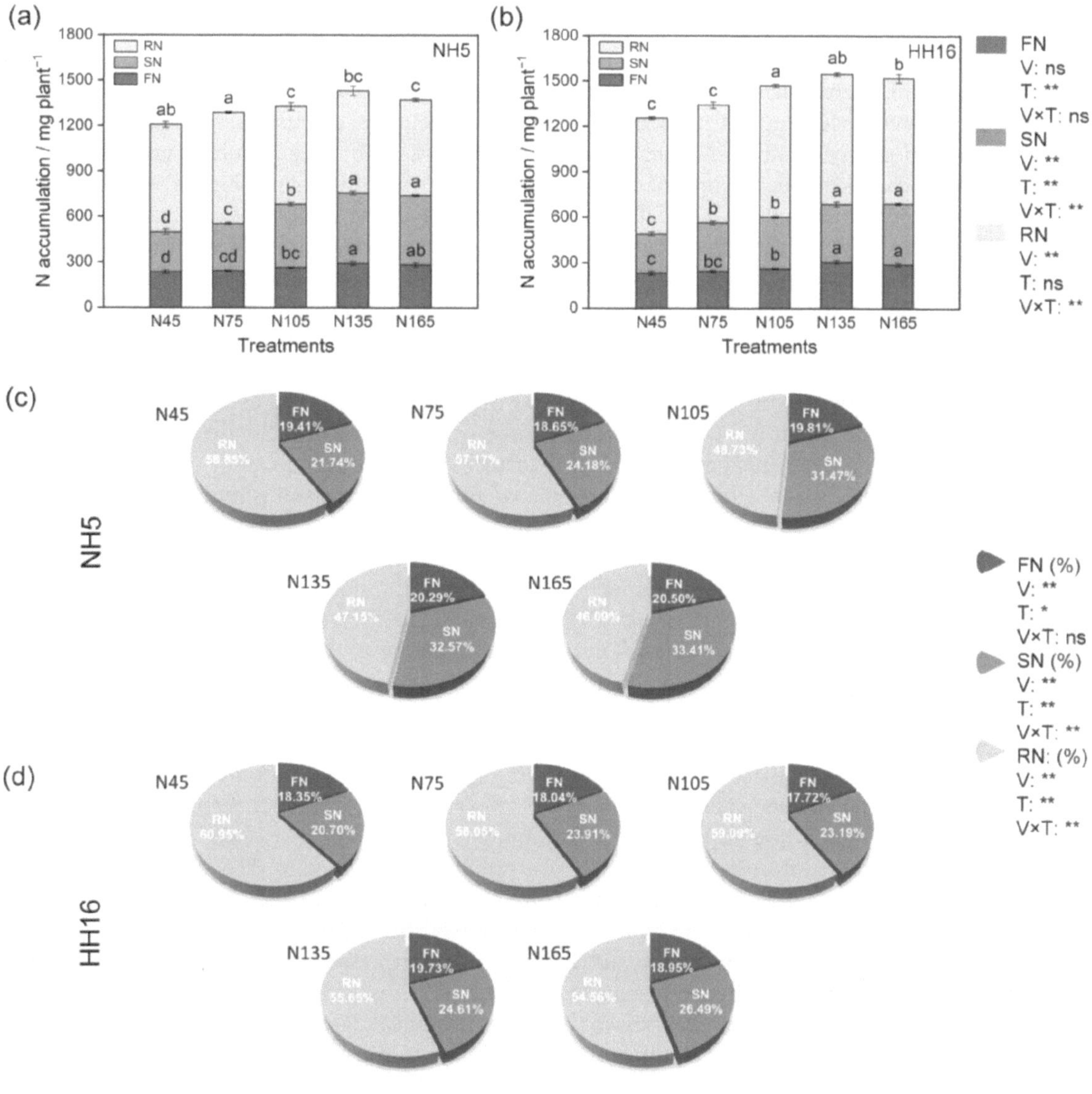

FIGURE 13.4 (a) and (b) are the effects of N application on N supply from soil, fertilizer, and root nodules. (c) and (d) are the effects of N application on N supply proportions from soil, fertilizer, and root nodules. Blue (FN) is N supply from fertilizer, orange (SN) is N supply from soil, and yellow (RN) is N supply from root nodules. Different lowercase letters represent significantly different values ($p< 0.05$) within one variety under five treatments. V represents variety and T represents N fertilizer treatment. ** is $p<0.01$; * is $p<0.05$; ns, not significant.

reproductive organs and the formation of products. This results in premature of leaf senescence, reduced pod yield, and poor grain quality (Patel et al., 2022). Huang (2020) showed that under low potassium stress, pod number, satiated fruit number, and 100 kernel weight per plant significantly decreased. Further investigation shows that under low potassium stress, the year, cultivar, treatment, and cultivar×treatment significantly ($p < 0.05$) affected the peanut 100-pod weight and 100-kernel weight (Liu et al., 2023). The above evidence shows that low potassium stress has significantly reduced peanut yield and related traits, and low potassium tolerance varieties are less affected. The main reason for nutrient deficiency reducing peanut yield is embryo abortion. In the absence of free Ca^{2+} in soil, the number of full fruits, full fruit weight, kernel weight, and yield of peanut are significantly reduced, resulting in yield reduction.

13.5.1.5 Nutrient Imbalance on the Physiology of Peanut

Under the stress of low K, the content of malondialdehyde in peanut leaves is increased, and the activities of superoxide dismutase (SOD) and peroxidase (POD) are higher than those under normal K treatment (Liu et al., 2023). The decrease in osmotic regulation caused by low K stress is compensated for by an increase in the soluble protein content. The net photosynthetic (Pn), transpiration rate (Tr), and stomatal conductance (gs) of peanut leaves are significantly decreased, and the chlorophyll content and carotenoid content in the lower part are significantly decreased compared to those in the upper part, resulting in the increase of minimum fluorescence (F0) under dark adaptation. Maximum fluorescence (Fm), variable fluorescence (Fv), and maximum photochemical efficiency (Fv/Fm) decreased and photochemical quenching coefficient (qP) reduced, but non-photochemical quenching coefficient (NPQ) increased. This shows that the capture, transformation, and transmission of light energy by photosynthetic apparatus are hindered. The increase of heat dissipation leads to more loss of light energy, which leads to a decrease in the photosynthetic performance of leaves.

It is worth noting that the amounts of MDA, hydrogen peroxide (H_2O_2), and superoxide anion (O^{2-}) in leaves are significantly increased. The membrane lipid peroxidation of leaves results in the destruction of the membrane structure, and the stress degree of the lower leaves is significantly higher than that of the middle and upper leaves, resulting in the decline of leaf photosynthesis. Similarly, under N deficiency stress, the contents of chlorophyll a, b, and a+b are significantly decreased in peanut leaves. N deficiency stress has resulted in a continuous decrease in Pn, Tr, and gs, whereas there is a significant increase in Ci in peanut leaves. Photosynthesis-related enzymes, including Rubisco and fructose 1-6 diphosphate aldolase activities, are significantly decreased. Antioxidant enzyme activities such as CAT, SOD, and POD are decreased, while the levels of ROS and MDA are significantly increased (Gao et al., 2023).

Under low K stress, the K content in the roots, stems, leaves, and fruit needles decreases significantly, whereas the sodium content in the root system significantly increases, indicating that the peanut plant actively absorbs more Na^+ to replace the K^+ function. The root length, root surface area, root volume, and root activity are significantly decreased, and root respiration and oxygen consumption rate decline. The main reason is that low K stress has significantly increased MDA, $O^{2\bullet-}$, and H_2O_2 in peanut leaves and roots. In addition, the contents of proline, free amino acids, and soluble sugars are significantly increased in peanut leaf and root under low K stress, while the relative contributions of organic and inorganic penetrants are significantly decreased (Patel et al., 2022). The osmotic potential value is significantly increased, and the contribution of inorganic osmotic substances (Na^+, Ca^{2+} and K^+) to osmotic regulation under low K stress is significantly higher than that of organic osmotic substances. MDA content and ROS (such as $O_{2\cdot-}$ and H_2O_2) in leaves and roots are significantly increased, while SOD, CAT, and APX activities are significantly increased, and GPX enzyme activities are significantly decreased.

13.5.1.1 Nutrient Imbalance on the Molecular Biology of Peanut

A low N level induces C-terminally encoded peptides (CEPs) formation in roots. CEP is transported to the stem and identified by COMPACT ROOT ARCHITECTURE 2 (CRA2), promoting miR2111 expression in the stem. Thus, it plays a positive regulating role in the number of nodules and increases the number of nodules. However, at high N levels, miR169c is expressed and has a direct negative effect on NFYA-C in the regulation of nodules (Xu and Wang, 2023). On the other hand, nitrate metabolism in nodules is also one of the important reasons that nitrate N inhibits nodule fixation. Nitrate N can improve the activities of nitrate reductase and nitrite reductase in the root nodules of leguminous crops. Under the action of nitrate reductase, nitrate is reduced to nitrite and accumulates in the cytoplasm. However, nitrite is an inhibitor of the Nase activity of bacteria. Therefore, excessive accumulation of nitrite directly inhibits the nitrogenase activity of root nodules and further inhibits the N fixation of root nodules. When the level of nitrate in the soil is higher than a certain threshold, the induction of *NIN* by Nod factor is weakened, and *NLP1* and other *NLPs* responded to the nitrate signal and accumulated in the nucleus. *NLP1* inhibits *NIN* activation of downstream *CRE1* and *NF-YA1* genes by directly binding and competing for binding sites on target gene promotors, and ultimately inhibits root nodule formation and N fixation (Lin et al., 2018; Nishida et al., 2021).

N uptake, transport, and storage in terrestrial plants are mainly accomplished by nitrate transporter (NRT). However, when NO^{3-} concentration is lower than 1 $mmol{\cdot}L^{-1}$, the absorption is mainly completed by the high affinity transport system (HATS) composed of high affinity nitrate transporter (NRT2) (Wang et al., 2018). Under N deficiency stress, *AhNRT2.7a* located in the cell membrane was highly expressed in response to low N deficiency stress in peanut stems and leaves (Wang et al., 2022). Moreover, gene ontology (GO) enrichment analysis reveals that "response to nitrate" and "plasma membrane" of peanut roots are significantly enriched GO terms under different N stress conditions (Yang et al., 2022). The differentially expressed genes (DEGs) are mainly enriched in GO terms related to carbohydrate and protein metabolism, such as "carbohydrate derivatives catabolism" and "amino acid transport". According to KEGG's (*Kyoto Encyclopedia of Genes and Genomes*), enrichment analysis of different N deficiency response genes, these genes are mainly involved in pathways of N metabolism, amino acid metabolism, and carbohydrate metabolism.

Low K stress significantly affected the expression of miRNAs and targeted mRNAs in peanut roots (Li et al., 2021). Among them, the upregulation rate of miR167 is 6.32 times, while the upregulation rate of miR156, miR164, miR165, miR166, miR169, miR393, miR396, and miR847 is more than 3 times. Low K stress also has changed the expression of protein-coding genes, among which the expression levels of 7 potential miRNA targets are significantly increased, and the expression level of RF2 is upregulated by 6.47 times. This shows that peanut roots regulate miRNAs and target mRNAs to promote taproot and lateral root growth to adapt to the stress response of low K stress.

The metabolites secreted by crop roots are direct participants in plant physiological and biochemical reactions, and changes in their metabolome under nutrient imbalance can largely reflect the response and defense of plants to stress (Arruda and Barreto, 2020). Under different N deficiency stresses, 262 different metabolites are detected, and these metabolites are closely related to N deficiency stress in peanut roots, so they are defined as N deficiency stress–related metabolites (Yang et al., 2022). A large number of metabolites are allocated to amino acid metabolism and carbohydrate metabolism pathways. Active changes to membrane lipids in plant roots have been observed under nutritional stress (Nakamura, 2013).

One of the main processes by which plants phenotypically regulate abiotic stress is through the regulation of their biofilm fluidity, a mechanism that is influenced by species characteristics and lipid composition on the membrane (Li et al., 2016; Zhanget al., 2020). Many previous studies have shown that the effective measure of crop resistance to cell membrane damage is to maintain the stability of cell membrane by accumulating phosphatidyl choline (PC) and phosphatidyl choline (PE) in cell membrane (Wang et al., 2022; Zhanget al., 2020). These results indicate that many differential metabolites and genes in peanut roots are closely related to amino acid metabolism and the tricarboxylic acid (TCA) cycle under N deficiency stress. The synergistic effects of the TCA cycle and carbohydrate biosynthesis pathway maintain the carbon homeostasis of crops under different nutrient stresses (Li et al., 2021; Zuo et al., 2004). There are significant differences in the expression levels of seven genes related to the TCA cycle under N deficiency stress, which leads to a decrease in the accumulation of key metabolites of the TCA cycle (citrate, oxaloacetate, and malate).

There is a close relationship between the production and accumulation of flavonoids in legume plants and the level of N supply in the soil. Flavonoids, a typical phenyl-C metabolite, play a key role in root nodule formation and the response to environmental stress in leguminous crops (Dong and Song, 2020; Ghitti et al., 2022; Shen et al., 2022). The flavonoids secreted by the roots of leguminous crops activate the nod gene expression of rhizobia, inducing the signaling molecule nod factor, and then the flavonoid-NOD protein complex enables the roots of leguminous crops and rhizobia to recognize each other and promote the formation of rhizobia (Bosse et al., 2021; Hassan and Mathesius, 2012). Crops can capture more iron and aluminum from their respective phosphates by releasing carboxylate, thus making P more soluble (Hodge, 2009). This in turn increases root development in P-rich areas (usually in topsoil) in response to P deficiency (Rogers and Benfey, 2015).

13.5.2 Application of Different Biotechnological Tools in Improving Nutrient Imbalance Tolerance in Peanuts

In recent decades, genomic research on peanuts has made significant progress and seen rapid development. This includes the genome sequence assembly of wild diploid progenitors, wild tetraploids, and two subspecies of cultivated tetraploids. Additionally, researchers have conducted resequencing of multiple germplasm lines, whole genome transcriptome mapping, single-cell sequencing, and both high- and low-cost genotyping analyses (Bertioli et al., 2016; Chen et al., 2016, 2019; Pandey et al., 2020; Zhuang et al., 2019). In addition to natural mutations, researchers use various biotechnology tools such as X-rays, gamma rays, EMS, and sodium azide to create mutations that enhance nutrient uptake and increase nutrient accumulation (Chen et al., 2011). Recently, RNA-guided Cas9 nucleases from the microbial CRISPR (clustered regularly interspaced short palindromic repeat)-Cas system have emerged as powerful, efficient, and accurate tools for developing new crops with improved agronomic and nutritional traits for genome editing in a variety of organisms (Guo et al., 2023; Hsu et al., 2014; Jang and Joung, 2019; Ran et al., 2013). Neelakandan et al. (2022) developed two constructs, harboring the regular gRNA scaffold and containing extended scaffold plus terminator for gene editing in peanut (Neelakandan et al., 2022). In addition, Shu et al. (2020) applied the CRISPR/Cas9 tool to explore the function of NFR genes responding to rhizobia in peanut hairy root transformation system (Shu et al., 2020). These studies indicate that CRISPR-Cas9 can be applied to the peanut transformation system for peanut functional genome research, especially the function of genes in the root.

13.6 MODERN BIOTECHNOLOGICAL TOOLS IN IMPROVING ABIOTIC STRESSES

13.6.1 Genetic Engineering

During environmental adaptation, plants initiate a cascade of responses ranging from cellular to physiological levels. These responses occur in a coordinated spatiotemporal manner and are interconnected, collectively forming an integrated molecular mechanism for stress adaptation. The functions of these genes have also been extensively studied and can be divided into two categories according to their functions. The first category is functional genes, which encode proteins (aquaporin proteins, late-embryogenesis-abundant proteins, osmoregulated proteins, and antioxidant enzymes) that directly against external environmental stress. The second category is regulatory genes; the proteins encoded by these genes are factors that play a regulatory role in signal transduction pathways and stress gene expression (Bovill et al., 2013). Elucidating the role of these genes is important for understanding plant response to stress and improving plant environmental tolerance through genetic engineering.

The tolerance of plants to stress is a quantitative trait controlled by multiple genes. With the rapid development of sequencing technologies that are used for the cognition of the plant genome, an increasing number of plant genomes involved in environmental stress response and tolerance have been cloned and analyzed. Transcription factors (TFs) control a series of downstream genes by binding cis-elements in promoter regions, thus regulating gene expression at the transcriptional level (Ma et al., 2018). TFs involved in stress resistance can regulate the simultaneous expression of multiple stress-resistance genes and genes involved in the transport of stress signals, and can thus enhance stress resistance in plants. Therefore, cloning and isolating the TFs of stress resistance genes and clarifying their functions are of great significance for revealing the signal transduction and molecular mechanism of gene expression regulation in plant response to stress, as well as for improving crop stress resistance through genetic engineering. Use genetic engineering technology to introduce exogenous beneficial genes into crop genomes can improve the stress resistance of plants, and allow these crops to grow on marginal land. However, the expression of resistance genes will greatly affect the yield and quality of crops, so it is still the goal of modern biotechnology efforts to produce crops eith the ability to resist extreme environmental changes and ensure crop yield and quality (Zhu, 2016).

13.6.2 Gene Editing Technology

Genome editing tools enable precise changes in an organism's DNA by introducing targeted mutation, insertion/deletion (InDel), and specific sequence alteration via recruiting specific nucleases. In the past, meganucleases (Puchta et al., 1993), transcription activator-like nucleases (TALENs) (Zhang et al., 2013), zinc-finger nucleases (ZFNs) (Zhang et al., 2010), and, more recently, CRISPR/Cas9 (Jiang et al., 2013) have been developed and used for genome editing. The clustered regularly interspaced short palindromic repeats CRISPR/Cas9 has been demonstrated to be the most successful genome editing system across a wide range of organisms, including plants, given its ability to induce desired mutations in an efficient, cheaper, faster, and accurate manner (Chen et al., 2019). Additionally, this system can detect and cleave complementary DNA sequences in the genome, which may further help enable and promote the commercialization of the resulting edited crop and food products to overcome societal apprehensions about the involvement of "foreign DNAs."

CRISPR/Cas9 can accelerate plant breeding through changing genomes speedily in a precise and expected way. CRISPR/Cas9-mediated genome editing has currently become a prevalent technology and has been extensively used to develop resistance in crop plants (Wolter et al., 2019). It can be used to knock out, replace and insert some related genes, thereby promoting the development of abiotic/biological stress-tolerant crop plants (Li et al., 2022). The application of multiplex genome editing would pave the way for the development of genome-edited crops engineered for tolerance against multiple traits in a single transformation event. Thus, it is predicted that genome editing will become the technology of choice (Jogam et al., 2022). The lack of validated targets for genome editing would be the main obstruction to unlocking the CRISPR potential to develop stress-tolerant crop plants.

13.6.3 Bioinformatics

Plant functioning and responses to abiotic stresses largely involve regulations at the transcriptomic level via complex interactions of signal molecules, signaling cascades, and regulators. Nevertheless, all the signaling networks involved in responses to abiotic stresses have not yet been fully established. In recent years, the plant science and molecular biology community has developed databases about genes, proteins, protein–protein interactions, protein–DNA interactions, and ontologies, which are valuable sources of knowledge for deciphering such regulatory and signaling networks. The use of these data and the development of bioinformatics tools help to make sense of transcriptomic data in specific contexts, such as that of abiotic stress signaling, using functional biological approaches.

Plant biotechnology combined with omics has the potential to solve a number of issues that currently hinder agriculture, such as diseases and pests, pressures from the environment, and climate change. Omics include but are not limited to genomics, transcriptomics, proteomics, epigenetics, metabolomics, miRNAomics, epigenomics, and phenomics (Haas et al., 2017), and all have been used to improve peanut cultivars. Genomics-assisted breeding (GAB) has demonstrated great potential for improving peanut varieties. High-quality genome assembly and well-annotated genomes are very crucial for GAB. The successful assemblies of the genomic sequences of wild diploid progenitors, wild tetraploid, and both the subspecies of cultivated tetraploids provide a cornerstone for the functional genomics research and genetic improvement of peanuts (Bertioli et al., 2016, 2019; Zhuang et al., 2019). Moreover, whole-genome resequencing (WGRS) of mapping populations has facilitated the development of high-density single nucleotide polymorphism (SNP)-based genetic mapping and genome-wide SNP genotyping array, which were developed for fine mapping and candidate gene discovery for stress resistance in peanut (Agarwal et al., 2018).

Although marker-assisted selection approaches have been used to develop superior peanut lines, technological advancements in sequencing and high-throughput genotyping can enhance genetic diversity and advance generation and genomic selection, as well as faster candidate gene discovery, in peanut breeding programs (Varshney et al., 2019). For example, RNA sequencing (RNA-seq), which belongs to the transcriptomic approach, can improve genome annotation and gene discovery, especially for the genes that encode for proteins and non-coding RNAs. Understanding the full metabolic networks involving genes, transcripts, proteins, and metabolites in biological systems is currently crucial because it is extremely difficult to succeed with the strategy of expression of a few single genes in peanut. In this regard, it becomes important to conduct a comprehensive analysis using functional genomics tools such as transcriptomics, proteomics, and metabolomics to unveil the genetic and metabolic responses of specific plant species to abiotic stresses (Pandey et al., 2020).

13.6.4 Molecular Assistant Breeding

The limited success of conventional breeding in improving plant stress resistance can be attributed to the fact that breeders tend to combine genetic material under ideal conditions. In addition, the complex nature of abiotic stresses and plant responses to stress at various stages of development further complicate the selection of resistant varieties. Therefore, the breeding of elite cultivars usually takes years. With the rapid development of molecular biology and genomics, molecular selection breeding has received increasing attention in resistance breeding of crops. Molecular selection breeding occurred through two stages: the first stage is molecular marker-assisted selection (MAS), and the second stage is genome-wide selection (GS) (Xu et al., 2012). MAS is a breeding technique that selects offspring lines with the help of molecular markers linked to the target trait for obtaining excellent single strains containing target genes. Molecular markers and their location in the genome play a vital role in the improvement of crops and the selection of molecular breeding strategies. The sorts of molecular markers can be roughly divided into the following categories: 1) DNA–DNA hybridization–based molecular markers, such as restriction fragment length polymorphisms (RFLPs), cleaved amplified polymorphic sequence (CAPS), and variable number of tandem repeats (VNTR); 2) PCR (polymerase chain reaction)–based molecular markers, including simple sequence repeat (SSR), inter-simple sequence repeat (ISSR), inverse sequence-tagged repeat (ISTR), and sequence tag site (STS); (3) molecular markers based on PCR combined with restrictive enzyme digestion, such as amplified fragment-length polymorphism (AFLP); (4) molecular markers based on single nucleotide polymorphic, such as single nucleotide polymorphisms (SNP) markers (Nadeem et al., 2019, Manimekalai et al., 2020).

Compared with the first generation of RFLP markers based on DNA hybridization, PCR-based SSR markers greatly reduce the time and expense involved in genotyping, thus promoting the rapid development and application of molecular breeding. Compared with SSR markers, SNP markers have a wider coverage in the genome. In recent years, with the rapid development of sequencing technology, a large number of SNP markers in a variety of crops have been developed and high-density genetic maps have been constructed, thus laying a solid foundation for molecular breeding. With the continuous improvement of the reference genome sequences for many legume crops, modern breeding techniques mainly based on molecular marker-assisted selection and genomic aggregation breeding techniques have significantly shortened the breeding cycle (Cazzola et al., 2021). Even if genomes are not yet available, the cheapening of genotyping by sequencing techniques and/or array-based genotyping is facilitating molecular analysis in legume crop species that have not yet been sequenced. Because of this, the number of available markers of all types is rapidly increasing for all species, allowing increasingly precise mapping of QTL/genes of interest and placing us in a position to apply marker-assisted selection or even genomic selection in species where it was unthinkable a few years ago (Pandey et al., 2021, Diego, 2023). Thus, a comprehensive breeding strategy for abiotic stress tolerance should include the following steps and approaches: 1) conventional breeding and germplasm selection, especially of wild relevant species; 2) elucidation of the specific molecular control mechanisms in tolerant and sensitive genotypes; 3) biotechnology-oriented improvement of selection and breeding procedures through functional genomics analysis, the use of molecular probes and markers for selection among natural and bred populations, and transformation with specific genes; and 4) improvement and adaptation of current agricultural practices (Wang et al., 2003).

REFERENCES

Abbas, F., Ke, Y. G., Yu, R. C., Yue, Y. C., Amanullah, S., Jahangir, M. M., & Y. P. Fan. Volatile terpenoids: multiple functions, biosynthesis, modulation, and manipulation by genetic engineering. *Planta* 246 (2017): 803–816. https://doi.org/10.1007/s00425-017-2749-x

Agarwal, G., Clevenger, J., Pandey, M. K., Wang, H., Shasidhar, Y., Chu, Y., & J. C. Fountain, et al. High-density genetic map using whole-genome resequencing for fine mapping and candidate gene discovery for disease resistance in peanut. *Plant Biotechnology Journal* 16 (2018): 1954–1967. https://doi.org/10.1111/pbi.12930

Alicandri, E., Paolacci, R. A., Osadolor, S., Sorgona, A., Badiani, M., & M. Ciaffi. On the evolution and functional diversity of terpene synthases in the Pinus species: a review. *Journal of Molecular Evolution* 88 (2020): 253–283. https://doi.org/10.1007/s00239-020-09930-8

Almeida, D. M., Oliveira, M. M., & N. J. M. Saibo. Regulation of Na+ and K+ homeostasis in plants: towards improved salt stress tolerance in crop plants. *Genetics and Molecular Biology* 40 (Suppl 1) (2017): 326–345. https://doi.org/10.1590/1678-4685-GMB-2016-0106

Arruda, P., & P. Barreto. Lysine catabolism through the saccharopine pathway: Enzymes and intermediates involved in plant responses to abiotic and biotic stress. *Frontiers in Plant Science* 11 (2020): 587. https://doi.org/10.3389/fpls.2020.00587

Asif, M. A., Zafar, Y., Iqbal, J., Iqbal, M. M., Rashid, U., Ali, G. M., Arif, A., & F. Nazir. Enhanced expression of AtNHX1, in transgenic groundnut (*Arachis hypogaea* L.) improves salt and drought tolerance. *Molecular Biotechnology* 49 (2011): 250–256. https://doi.org/10.1007/s12033-011-9399-1

Awal, M. A., & T. Ikeda. Effects of changes in soil temperature on seedling emergence and phenological development in field-grown stands of peanut (*Arachis hypogaea*). *Environmental and Experimental Botany* 47 (2002): 101–113. https://doi.org/10.1016/S0098-8472(01)00113-7

Bai, D. M., Xue, Y. Y., Zhao, J. J., Huang, L., Tian, Y. X., Quan, B. Q., & H. F. Jiang, et al. Identification of cold-tolerance during germination stage and genetic diversity of SSR markers in peanut landraces of Shanxi Province. *Acta Agronomica*

Sinica 44 (2018): 1459–1467. https://doi.org/10.3724/SP.J.1006.2018.01459

Banavath, J. N., Chakradhar, T., Pandit, V., Konduru, S., Guduru, K. K., Akila, C. S., & C. Puli. Stress inducible overexpression of AtHDG11 leads to improved drought and salt stress tolerance in peanut (*Arachis hypogaea* L.). *Frontiers in Chemistry* 6 (2018): 34. https://doi.org/10.3389/fchem.2018.00034

Barrero-Sicilia, C., Silvestre, S., Haslam, R. P., & L. V. Michaelson. Lipid remodelling: unravelling the response to cold stress in Arabidopsis and its extremophile relative *Eutrema salsugineum*. *Plant Science* 263 (2017): 194–200. https://doi.org/10.1016/j.plantsci.2017.07.017

Barnawal, D., Bharti, N., Maji, D., Chanotiya, C. S., & A. Kalra. ACC deaminase-containing Arthrobacter protophormiae induces NaCl stress tolerance through reduced ACC oxidase activity and ethylene production resulting in improved nodulation and mycorrhization in *Pisum sativum*. *Journal of Plant Physiology* 171 (2014): 884–894. https://doi.org/10.1016/j.jplph.2014.03.007

Barragán, V., Leidi, E. O., Andrés, Z., Rubio, L., De Luca, A., Fernández, J. A., Cubero, B., & J. M. Pardo. Ion exchangers NHX1 and NHX2 mediate active potassium uptake into vacuoles to regulate cell turgor and stomatal function in Arabidopsis. *Plant Cell* 24 (2012): 1127–1142. https://doi.org/10.1105/tpc.111.095273

Bazihizina, N., Colmer, T. D., Cuin, T. A., Mancuso, S., & S. Shabala. Friend or foe? Chloride patterning in halophytes. *Trends in Plant Science* 24 (2019): 142–151. https://doi.org/10.1016/j.tplants.2018.11.003

Bell, M. J., Gillespie, T. J., Roy, R. C., Michaels, T. E., & M. Tollenaar. Peanut leaf photosynthetic activity in cool field environments. *Crop Science* 34 (1994): 1023–1029. https://doi.org/10.2135/cropsci1994.0011183X003400040035x

Bertioli, D. J., Cannon, S. B., Froenicke, L., Huang, G., Farmer, A. D., Cannon, E. K., & X. Liu, et al. The genome sequences of *Arachis duranensis* and *Arachis ipaensis*, the diploid ancestors of cultivated peanut. *Nature Genetics* 48 (2016): 438–446. https://doi.org/10.1038/ng.3517

Bertioli, D. J., Jenkins, J., Clevenger, J., Dudchenko, O., Gao, D., Seijo, G., & S. C. M. Leal-Bertioli, et al. The genome sequence of segmental allotetraploid peanut *Arachis hypogaea*. *Nature Genetics* 51 (2019): 877–884. https://doi.org/10.1038/s41588-019-0405-z

Bhalani, H., Thankappan, R., Mishra, G. P., Sarkar, T., Bosamia, T. C., & J. R. Dobaria. Regulation of antioxidant mechanisms by *AtDREB1A* improves soil-moisture deficit stress tolerance in transgenic peanut (*Arachis hypogaea* L.). *PLoS One* 14 (2019): e0216706. https://doi.org/10.1371/journal.pone.0216706

Bhatnagar-Mathur, P., Devi, M.J., Serraj, R., Yamaguchi-Shinozaki, K., Vadez, V., & K.K. Sharma. Evaluation of transgenic groundnut lines under water limited conditions. *International Arachis Newsletter* 24 (2004): 33–34.

Bhatnagar-Mathur, P., Devi, M. J., Reddy, D. S., Lavanya, M., Vadez, V., Serraj, R., Yamaguchi-Shinozaki, K., & K. K. Sharma. Stress-inducible expression of AtDREB1A in transgenic peanut (*Arachis hypogaea* L.) increases transpiration efficiency under water-limiting conditions. *Plant Cell Reports* 26 (2007): 2071–2082.

Bhatnagar-Mathur, P., Rao, J. S., Vadez, V., Dumbala, S. R., Rathore, A., Yamaguchi-Shinozaki, K., & K. K. Sharma. Transgenic peanut overexpressing the DREB1A transcription factor has higher yields under drought stress. *Molecular Breeding* 33 (2014): 327–340. https://doi.org/10.1007/s11032-013-9952-7

Bhatnagar-Mathur, P., Devi, M. J., Vadez, V., & K. K. Sharma. Differential antioxidative responses in transgenic peanut bear no relationship to their superior transpiration efficiency under drought stress. *Journal of Plant Physiology* 166 (2009): 1207–1217. https://doi.org/10.1016/j.jplph.2009.01.001

Bhattacharyya, D., Yu, S. M., & Y. H. Lee. Volatile compounds from Alcaligenes faecalis JBSC1294 confer salt tolerance in *Arabidopsis thaliana* through the auxin and gibberellin pathways and differential modulation of gene expression in root and shoot tissues. *Plant Growth Regulation* 75 (2015): 297–306. https://doi.org/10.1007/s10725-014-9953-5

Bijalwan, P., Jeddi, K., Saini, I., Sharma, M., Kaushik, P., & K. Hessini. Mitigation of saline conditions in watermelon with mycorrhiza and silicon application. *Saudi Journal of Biological Sciences* 28 (2021): 3678–3684. https://doi.org/10.1016/j.sjbs.2021.05.019

Boontang, S., Tantisuwichwong, N., Jogloy, S., Akkasaeng, C., Vorasoot, N., & A. Patanothai. SSR-based genetic variability among peanut genotypes differing in specific leaf weight and relative water content. *African Journal of Biotechnology* 12 (2013): 4053–4064. https://doi.org/10.5897/AJB12.2498

Bovill, W. D., Huang, C. Y., & G. K. McDonald. Genetic approaches to enhancing phosphorus-use efficiency (PUE) in crops: challenges and directions. *Crop & Pasture Science* 64 (2013): 179–198. https://doi.org/10.1071/CP13135

Bosse, M. A., Silva, M. B. D., Oliveira, N., Araujo, M. A., Rodrigues, C., Azevedo, J. P., & A. R. D. Reis. Physiological impact of flavonoids on nodulation and ureide metabolism in legume plants. *Plant Physiology and Biochemistry* 166 (2021): 512–521. https://doi.org/10.1016/j.plaphy.2021.06.007

Boureau, L., Pribat, A., Mortain-Bertrand, A., Bert, L. S., Rolin, D., Teyssier, E., & P. Gallusci. Metabolic engineering of isoprenoid biosynthesis. In *Natural Products*. Berlin Heidelberg: Springer-Verlag, 2013.

Buchanan, B. B., Gruissem, W., & R. L. Jones. *Biochemistry & Molecular Biology of Plants* (2nd ed.). Translated and edited by Qu, L. J., Zhao, J. D., Qin, G. J., Zhong, S., Li, J. G., & Gu, H. Y. Beijing, China: Science Publishing & Media Ltd, 2020.

Calixto, C. P. G., Guo, W., James, A. B., Tzioutziou, N. A., Entizne, J. C., Panter, P. E., & H. Knight, et al. Rapid and dynamic alternative splicing impacts the Arabidopsis cold response transcriptome. *The Plant Cell* 30 (2018): 1424–1444. https://doi.org/10.1105/tpc.18.00177

Cazzola, F., Bermejo, C. J., Gatti, I., & E. Cointry. Speed breeding in pulses: an opportunity to improve the efficiency of breeding programs. *Crop & Pasture Science* 72 (2021): 165–172. https://doi.org/10.1071/cp20462

Chakraborty, K., Debarati, B., Har, N. M., & K. Kuldeepsingh. External potassium (K^+) application improves salinity tolerance by promoting Na^+-exclusion, K^+-accumulation and osmotic adjustment in contrasting peanut cultivars. *Plant Physiology and Biochemistry* 103 (2016): 143–153. https://doi.org/10.1016/j.plaphy.2016.02.039

Chen, K., Wang, Y., Zhang, R., Zhang, H., & C. Gao. CRISPR/Cas genome editing and precision plant breeding in agriculture. *Annual Review of Plant Biology* 70 (2019): 667–697. https://doi.org/10.1146/annurev-arplant-050718-100049

Chen, X. P., Li, H., Pandey, M. K., Yang, Q., Wang, X., Garg, V., & H. Li, et al. Genome of the peanut A-genome progenitor (*Arachis duranensis*) provides insights into geocarpy, oil biosynthesis, and allergens. *Proceedings of the National Academy of Sciences* 113 (2016): 6785–6790. https://doi.org/10.1073/pnas.1600899113

Chen, X., Wang. M. L., Holbrook, C., Culbreath, A., Liang, X. Q., Brenneman, T., & B. Z. Guo. Identification and characterization of a multigene family encoding germin-like proteins in cultivated peanut (*Arachis hypogaea* L.). *Plant Molecular Biology Reporter* 29 (2011): 389–403. https://doi.org/10.1007/s11105-010-0237-6

Cheng, X., Zhao, C. X., Wang, M. L., Wang, Y. F., & G. P. Shan. Effects of drought stress at different growth stages on drought resistance index and yield of peanut. *Journal of Qingdao Agricultural University (Natural Science)* 27, no. 4 (2010): 282–284+288. https://doi.org/10.3969/J.ISSN.1674-148X.2010.04.006

Choudhury, F. K., Rivero, R. M., Blumwald, E., & R. Mittler. Reactive oxygen species, abiotic stress and stress combination. *The Plant Journal* 90 (2017): 856–867. https://doi.org/10.1111/tpj.13299

Dang Phat, M., Chen, C. Y., & C. Holbrook. Evaluation of five peanut (*Arachis hypogaea* L.) genotypes to identify drought responsive mechanisms utilising candidate-gene approach. *Functional Plant Biology* 40 (2013): 1323–1333. https://doi.org/10.1071/FP13116

Dardanelli, M. S., González, P. S., Medeot, D. B., Paulucci, N. S., Bueno, M. Á., & M. B. Garcia. Effects of peanut rhizobia on the growth and symbiotic performance of *Arachis hypogaea* under abiotic stress. *Symbiosis* 47 (2009): 175–180. https://doi.org/10.1007/BF03179977

Deinlein, U., Stephan, A. B., Horie, T., Luo, W., Xu, G., & J. I. Schroeder. Plant salt-tolerance mechanisms. *Trends in Plant Science* 19 (2014): 371–379. https://doi.org/10.1016/j.tplants.2014.02.001

Demidchik, V., Tyutereva, E. V., & O. V. Voitsekhovskaja. The role of ion disequilibrium in induction of root cell death and autophagy by environmental stresses. *Functional Plant Biology* 45 (2018): 28–46. https://doi.org/10.1071/FP16380

Diego, R. Plant breeding is needed to meet agroecological requirements: legume crops as a case study. *Outlook on Agriculture* 52 (2023): 294–302. https://doi.org/10.1177/00307270231195641

Ding, F., Wang, G., & S. Zhang. Exogenous melatonin mitigates methyl viologen-triggered oxidative stress in poplar leaf. *Molecules* 23 (2018): 2852–2865. https://doi.org/10.3390/molecules23112852

Ding, H., Zhang, Z. M., Dai, L. X., Ci, D. W., Qin, F. F., Ma, D. C., Li, M., Song, W. W., & T. Kang. Effects of drought stress on root growth characteristics of peanut during mid-to-late growth stages. *Chinese Journal of Eco-Agriculture* 21, no.12 (2013): 1477–1483. https://doi.org/10.3724/SP.J.1011.2013.30447

Ding, H., Zhang, Z. M., Dai, L. X., Kang, T., Ci, D. W., & Song, W. W. Effects of drought stress on the root growth and growth and development and physiological characteristics of peanut. *Chinese Journal of Applied Ecology* 24 (2013): 1586–1592. https://doi.org/10.13287/j.1001-9332.2013.0335

Ding, Y., Shi, Y., & S. Yang. Advances and challenges in uncovering cold tolerance regulatory mechanisms in plants. *The New Phytologist* 222 (2019): 1690–1704. https://doi.org/10.1111/nph.15696

Dong, Q. Q., Ai, X., Zhang, Y. Z., Zhang, K. C., Zhou, D. Y., Wang, X. G., & C. J. Jiang, et al. Effect of drought stress on physiological characteristics and yield in different tolerant peanut. *Journal of Shenyang Agricultural University* 51, no.1 (2020): 18–26. https://doi.org/10.3969/j.issn.1000-1700.2020.01.003

Dong, W., & Y. Song. The significance of flavonoids in the process of biological nitrogen fixation. *International Journal of Molecular Sciences* 21 (2020): 5926.

Dramé, K. N., Clavel, D., Repellin, A., Passaquet, C., & Y. Z. Fodil. Water deficit induces variation in expression of stress-responsive genes in two peanut (*Arachis hypogaea* L.) cultivars with different tolerance to drought. *Plant Physiology and Biochemistry* 45 (2007): 236–240. https://doi.org/10.1016/j.plaphy.2007.02.002

Efeoğlu, B., Ekmekci, Y., & N. Cicek. Physiological responses of three maize cultivars to drought stress and recovery. *South African Journal of Botany* 75 (2009): 34–42. https://doi.org/10.1016/j.sajb.2008.06.005

Evelin, H., Giri, B., & R. Kapoor. Ultrastructural evidence for AMF mediated salt stress mitigation in *Trigonella foenum-graecum*. *Mycorrhiza* 23 (2013): 71–86. https://doi.org/10.1007/s00572-012-0449-8

Fang, S., Hou, X., & X. Liang. Response mechanisms of plants under saline-alkali stress. *Frontiers in Plant Science* 12 (2021):667458. https://doi.org/10.3389/fpls.2021.667458

FAO. (2021). Soil salinization as a global major challenge. Retrieved from http://www.fao.org/global-soil-partnership/resources/highlights/detail/en/c/1412475/

Faye, I., Pandey, M. K., Hamidou, F., Rathore, A., & R. K. Varshney. Identification of quantitative trait loci for yield and yield related traits in groundnut (*Arachis hypogaea* L.) under different water regimes in Niger and Senegal. *Euphytica* 206 (2015): 631–647. https://doi.org/10.1007/s10681-015-1472-6

Ferrer, J. L., Austin, M. B., Stewart, J. C., & J. P. Nobel. Structure and function of enzymes involved in the biosynthesis of phenylpropanoids. *Plant Physiology and Biochemistry* 46 (2008): 356–370. https://doi.org/10.1016/j.plaphy.2007.12.009

Gao, G. Q. Evaluation of drought resistance and expression analysis of *AhPLP* gene family in Peanut (*Arachis hypogaea* L.). MD diss., Henan Agricultural University, 2022.

Gao, W. J. Effects of different nitrogen levels on growth and development and nitrogen metabolism in drought-stressed peanuts. *Dissertation for Master's Degree*, Shenyang Agricultural University, 2023.

Gautami, B., Pandey, M. K., Vadez, V., Nigam, S. N., Ratnakumar, P., Krishnamurthy, L., & T. Radhakrishnan, et al. Quantitative trait locus analysis and construction of consensus genetic map for drought tolerance traits based on three recombinant inbred line populations in cultivated groundnut (*Arachis hypogaea* L.). *Molecular Breeding* 30 (2012):757–772. https://doi.org/10.1007/s11032-011-9660-0

Geilfus, C. M. Review on the significance of chlorine for crop yield and quality. *Plant Science* 270 (2018): 114–122. https://doi.org/10.1016/j.plantsci.2018.02.014

Ghitti, E., Rolli, E., Crotti, E., & S. Borin. Flavonoids are intra- and inter-kingdom modulator signals. *Microorganisms* 10 (2022): 2479. https://doi.org/10.3390/microorganisms10122479

Gopala Krishna, R., Kumar, G., Krishna Prasad, B. T., Mathew, M. K., & M. U. Kumar. A stress responsive gene from groundnut, Gdi-15, is homologous to flavonol 3-Oglucosyltransferase involved in anthocyanin biosynthesis. *Biochemical and Biophysical Research Communication* 284 (2001): 574–579. https://doi.org/10.1006/bbrc.2001.4992

Govind, G., Harshavardhan, V. T., Patricia, J. K., Dhanalakshmi, R., Senthil, K. M., Sreenivasulu, N., & M. Udayakumar. Identification and functional validation of a unique set of drought-induced genes preferentially expressed in response to gradual water stress in peanut. *Molecular Genetics and Genomics* 281 (2009): 591–605. https://doi.org/10.1007/s00438-009-0441-y

Grant, O. M., Brennan, D., Mellisho, P., Salas, C. D., & C. J. Dix. Impact of enhanced capacity to scavenge reactive oxygen species on cold tolerance of tobacco. *International Journal of Plant Sciences* 175 (2014): 544–554. https://doi.org/10.1086/675976

Gui, G., Lv, C., Stevanato, P., Li, R., Liu, H., Yu, L., & Y. Wang. Transcriptome analysis of salt-sensitive and tolerant genotypes reveals salt-tolerance metabolic pathways in sugar beet. *International Journal of Molecular Sciences* 20 (2019): 5910–5929. https://doi.org/10.3390/ijms20235910

Guo, P., Wang, J. Y., Shi, X. L., Ren, J. Y., Chen, C., Zhang, P., & H. Y. Xiong, et al. Effects of nitrogen application rate on nodule characteristics and nitrogen utilization in different peanut genotypes. *Journal of Shenyang Agricultural University* 53, no.4 (2022): 385–393. https://doi.org/10.3969/j.issn.1000-1700.2022.04.001

Guo, Y., Zhao, G., Gao, X., Zhang, L., Zhang, Y., Cai, X., Yuan, X., & X. Guo. CRISPR/Cas9 gene editing technology: a precise and efficient tool for crop quality improvement. *Planta* 258 (2023): 36. https://doi.org/10.1007/s00425-023-04187-z

Gusain, S., Joshi, S., & R. Joshi. Sensing, signalling, and regulatory mechanism of cold-stress tolerance in plants. *Plant Physiology and Biochemistry* 197 (2023):107646. https://doi.org/10.1016/j.plaphy.2023.107646

Gupta, B., & B. Huang. Mechanism of salinity tolerance in plants: physiological, biochemical, and molecular characterization. *International Journal of Genomics* (2014): 701596. https://doi.org/10.1155/2014/701596

Gupta, S., Kaur, R., Sharma, T., Bhardwaj, A., Sharma, S., Sohal, J. S., & S. V. Singh. Multi-omics approaches for understanding stressor-induced physiological changes in plants: an updated overview. *Physiological and Molecular Plant Pathology* 126 (2023):102047. https://doi.org/10.1016/j.pmpp.2023.102047

Ha, C. V., Leyva-González, M. A., Osakabe, Y., Tran, U. T., Nishiyama, R., Watanabe, Y., & M. Tanaka, et al. Positive regulatory role of strigolactone in plant responses to drought and salt stress. *Proceedings of the National Academy of Sciences* 111 (2014): 851–856. https://doi.org/10.1073/pnas.1322135111

Haas, R., Zelezniak, A., Iacovacci, J., Kamrad, S., Townsend, S., & M. Ralser. Designing and interpreting 'multi-omic' experiments that may change our understanding of biology. *Current Opinion in Systems Biology* 6 (2017): 37–45. https://doi.org/10.1016/j.coisb.2017.08.009

Hasanuzzaman, M., Bhuyan, M. H. M. B., Faisal, Z., Ali, R., Mohammad, M. S., Al Jubayer, M., Masayuki, F., & F. Vasileios. Reactive oxygen species and antioxidant defense in plants under abiotic stress: revisiting the crucial role of a universal defense regulator. *Antioxidants* 9 (2020): 681–733. https://doi.org/10.3390/antiox9080681

Hassan, S., & U. Mathesius. The role of flavonoids in root–rhizosphere signalling: Opportunities and challenges for improving plant–microbe interactions. *Journal of Experimental Botany* 63, no. 9 (2012): 3429–3444.

Hodge, A. Root decisions. *Plant, Cell & Environment* 32 (2009): 628–640.

Hsu, P. D., Lander, E. S., & F. Zhang. Development and applications of CRISPR-Cas9 for genome engineering. *Cell* 157 (2014): 1262–1278.

Hu, B., Hong, L., Liu, X., Xiao, S. N., Lv, Y., & L. Li. The higher expression and distribution of 9-cis-epoxycarotenoid dioxygenase1 (*AhNCED1*) from *Arachis hypogaea* L. contribute to tolerance to water stress in a drought-tolerant cultivar. *Acta Physiologiae Plantarum* 35 (2013): 1667–1674. https://doi.org/10.1007/s11738-012-1209-2

Huang, H. Y. Effects of low potassium stress on peanut growth and nutrient uptake & utilization in peanut. *Dissertation for Master's Degree*, Shenyang Agricultural University, 2020.

Jang, G., & Y. H. Joung. CRISPR/Cas-mediated genome editing for crop improvement: current applications and future prospects. *Plant Biotechnology Reports* 13 (2019): 1–10. https://doi.org/10.1007/s11816-018-0509-4

Jamil, A., Riaz, S., Ashraf, M., & M. R. Foolad. Gene expression profiling of plants under salt stress. *Critical Reviews in Plant Sciences* 30 (2011):435–458. https://doi.org/10.1080/07352689.2011.605739

Jia, Y., Ding, Y., Shi, Y., Zhang, X., Gong, Z., & S. Yang. The CBFs triple mutants reveal the essential functions of CBFs in cold acclimation and allow the definition of CBF regulons in Arabidopsis. *The New Phytologist* 212 (2016): 345–353. https://doi.org/10.1111/nph.14088

Jiang, C. J., Zhang, H., Ren, J. Y., Dong, J. L., Zhao, X. H., Wang, X. G., & J. Wang, et al. Comparative transcriptome-based mining and expression profiling of transcription factors related to cold tolerance in peanut. *International Journal of Molecular Sciences* 21 (2020): 1921. https://doi.org/10.3390/ijms21061921

Jiang, W., Zhou, H., Bi, H., Fromm, M., Yang, B., & D. P. Weeks. Demonstration of CRISPR/Cas9/sgRNA-mediated targeted gene modification in Arabidopsis, tobacco, sorghum, and rice. *Nucleic Acids Research* 41 (2013): e188. https://doi.org/10.1093/nar/gkt780

Jogam, P., Sandhya, D., Alok, A., Peddaboina, V., Allini, V. R., & B. Zhang. A review on CRISPR/Cas-based epigenetic regulation in plants. *International Journal of Biological Macromolecules* 219 (2022): 1261–1271. https://doi.org/10.1016/j.ijbiomac.2022.08.182

Kakani, V. G., Prasad, P. V. V., Craufurd, P. Q., & T. R. Wheeler. Response of in vitro pollen germination and pollen tube growth of groundnut (*Arachis hypogaea* L.) genotypes to temperature. *Plant, Cell and Environment* 25 (2002): 1651–1661. https://doi.org/10.1046/j.1365-3040.2002.00943.x

Karabudak, T., Bor, M., Özdemir, F., & İ. Türkan. Glycine betaine protects tomato (*Solanum lycopersicum*) plants at low temperature by inducing fatty acid desaturase7 and lipoxygenase gene expression. *Molecular Biology Reports* 41 (2014): 1401–1410. https://doi.org/10.1007/s11033-013-2984-6

Karlova, R., Boer, D., Hayes, S., & C. Testerink. Root plasticity under abiotic stress. *Plant Physiology* 187 (2021):1057–1070. https://doi.org/10.1093/plphys/kiab392

Kang, S. M., Radhakrishnan, R., Khan, A. L., Kim, M. J., Park, J. M., Kim, B. R., Shin, D. H., & I. J. Lee. Gibberellin secreting rhizobacterium, *Pseudomonas putida* H-2-3 modulates the hormonal and stress physiology of soybean to improve the plant growth under saline and drought conditions. *Plant Physiology and Biochemistry* 84 (2014): 115–124. https://doi.org/10.1016/j.plaphy.2014.09.001

Kavi Kishor, P. B., & N. Sreenivasulu. Is proline accumulation per se correlated with stress tolerance or is proline homeostasis a more critical issue? *Plant, Cell and Environment* 37 (2014): 300–311. https://doi.org/10.1111/pce.12157

Ketring, D. L. Temperature effects on vegetative and reproductive development of peanut. *Crop Science* 24 (1984): 877–882. https://doi.org/10.2135/cropsci1984.0011183X002400050012x

Kiranmai, K., Rao, G. L., Pandurangaiah, M., Nareshkumar, A., Reddy, V. A., Lokesh, U., Venkatesh, B., Johnson, A. M. A., & Sudhakar, C. A novel WRKY transcription factor, MuWRKY3 (Macrotyloma uniflorum Lam. Verdc.) enhances drought stress tolerance in transgenic groundnut (*Arachis hypogaea* L.) plants. *Frontier in Plant Science* 9 (2018): 346. https://doi.org/10.3389/fpls.2018.00346

Kojima, K., Oshita, M., Nanjo, Y., Kasai, K., Tozawa, Y., Hayashi, H., & Y. Nishiyama. Oxidation of elongation factor G inhibits the synthesis of the D1 protein of photosystem II. *Molecular Microbiology* 65 (2007): 936–947. https://doi.org/10.1111/j.1365-2958.2007.05836.x

Lauriano, J. A., Campos, P. S., Ramalho, J. C., Lidon, F. C., Guedes, M. E., & M. do C. Matos. Partial decline of *Arachis hypogaea* L. photosynthesis triggered by drought stress. *Photosynthetica* 33 (1997): 81–90. https://doi.org/10.1023/A:1022179322721

Lauriano, J. A., Ramalho, J. C., Lidon, F. C., & M. do C. Matos. Peanut photosynthesis under drought and re-watering. *Photosynthetica*, 42 (2004): 37–41.

Lee, S., & Luan, S. (2012). ABA signal transduction at the crossroad of biotic and abiotic stress responses. *Plant, Cell and Environment* 35 (2004): 53–60. https://doi.org/10.1111/j.1365-3040.2011.02426.x

Li, C. J., Yan, C. X., Sun, Q. X., Wang, J., Yuan, C. L., Mou, Y. F., Shan, S. H., & X. B. Zhao. The bHLH transcription factor AhbHLH112 improves the drought tolerance of peanut. *BMC Plant Biology* 21 (2021): 540. https://doi.org/10.1186/s12870-021-03318-6

Li, G. L., Wu, H. X., Wen, L., Shao, K., Li, Z. Y., & S. Y. Zhang. Progress in physiological and molecular mechanism of drought resistant in crop. *Chinese Agricultural Science Bulletin* 26 (2010): 185–191.

Li, K., Zhao, K. K., Ning, L. L., Ma, Q., Li, Z. F., Ma, X. L., Zhang, X. G., & D. M. Yin. Construction of CRISPR/Cas9 multi-target knockout vector of FAD2 gene in peanut. *Shandong Agricultural Sciences* 51 (2019): 56–62, 67.

Li, L., Li, Q., Davis, K. E., Patterson, C., Oo, S., Liu, W., & J. Liu, et al. Response of root growth and development to nitrogen and potassium deficiency as well as microRNA-mediated mechanism in peanut (*Arachis hypogaea* L.). *Frontiers in Plant Science* 12 (2021): 695234. https://doi.org/10.3389/fpls.2021.695234

Li, Q., Shen, W., Zheng, Q., Fowler, D. B., & J. Zou. Adjustments of lipid pathways in plant adaptation to temperature stress. *Plant Signaling & Behavior* 11 (2016): e1058461. https://doi.org/10.1080/15592324.2015.1058461

Li, X. Y., Liu, X., Yao, Y., Li, Y. H., Liu, S., He, C. Y., Li, J. M., Lin, Y. Y., & L. Li. Overexpression of *Arachis hypogaea* AREB1 gene enhances drought tolerance by modulating ROS scavenging and maintaining endogenous ABA content. *International Journal of Molecular Sciences* 14 (2013): 12827–12842. https://doi.org/10.3390/ijms140612827

Li, Y., Wu, X., Zhang, Y., & Q. Zhang. CRISPR/Cas genome editing improves abiotic and biotic stress tolerance of crops. *Frontiers in Genome Editing* 4 (2022): 987817. https://doi.org/10.3389/fgeed.2022.987817

Lin, J. S., Li, X., Luo, Z., Mysore, K. S., Wen, J., & F. Xie. NIN interacts with NLPs to mediate nitrate inhibition of nodulation in *Medicago truncatula. Nature Plants* 4 (2018): 942–952. https://doi.org/10.1038/s41477-018-0319-2

Liu, D. W., Wang, J. G., Li, L., Tan, H. J., Ma, J., & S. Lu. Responses of different peanut cultivars to drought and waterlogging stress and physiological mechanism. *Acta Ecologica Sinica* 35 (2015): 3817–3824. https://doi.org/10.5846/stxb201309272375

Liu, H. L., Bai, D. M., Ning, Q., Guo, J. B., Xu, Z. J., Chen, X. S., Sun, X. P., Jiang, H. F., & H. Y. Gao. Mapping of QTL for low temperature tolerance trait during peanut seed inhibition. *Journal of Northeast Agricultural Sciences* 44 (2019): 5–11+47.

Liu, H. L., Chen, X. S., Yang, F. J., Bai, D. M., Sun, X. P., Lu, Y., Ren, X. P., Jiang, H. F., & H. Y. Gao. Research of identification method on low temperature resistance of peanut germplasm resources phenotype. *Journal of Peanut Science* 46 (2017): 20–25.

Liu, J. L., Zhao, C. X., Wu, N., Wang, Y. F., & M. L. Wang. Effects of drought and rewatering at seedling stage on photosynthetic characteristics and water use efficiency of peanut. *Scientia Agricultura Sinica* 44 (2011): 469–476. https://doi.org/10.3864/j.issn.0578-1752.2011.03.005

Liu, P. L., Du, L., Huang, Y., Gao, S. M., & M. Yu. Origin and diversification of leucine-rich repeat receptor-like protein kinase (LRR-RLK) genes in plants. *BMC Ecology and Evolution* 17, no. 1 (2017): 47. https://doi.org/10.1186/s12862-017-0891-5

Liu, S., Li, M. J., Su, C. L., Ge, K., Li, L. M., Li, X. Y., Liu, X., & L. Li. Negative feedback regulation of ABA biosynthesis in peanut (*Arachis hypogaea*): a transcription factor complex inhibits AhNCED1 expression during water stress. *Scientific Reports* 6 (2016): 37943. https://doi.org/10.1038/srep37943

Liu, X., & H. Zhang. The effects of bacterial volatile emissions on plant abiotic stress tolerance. *Frontiers in Plant Science* 6 (2015): 774–779. https://doi.org/10.3389/fpls.2015.00774

Liu, X., Li, L. M., Li, M. J., Su, L. C., Lian, S. M., Zhang, B. H., Li, X. Y., Ge, K., & L. Li. AhGLK1 affects chlorophyll biosynthesis and photosynthesis in peanut leaves during recovery from drought. *Scientific Reports* 8 (2018): 2250. https://doi.org/10.1038/s41598-018-20542-7

Liu, X., Li, L. M., Zhang, B. H., Zeng, B. H., Zeng, L. D., & L. Li. AhHDA1-mediated AhGLK1 promoted chlorophyll synthesis and photosynthesis regulates recovery growth of peanut leaves after water stress. *Plant Science* 294 (2020): 110461. https://doi.org/10.1016/j.plantsci.2020.110461

Liu, X., Li, S., Wu, J. L., Zhang, B. Y., Li, X. Y., Yan, Y. C., & L. Li. Overexpression of *Arachis hypogaea* NAC3 in tobacco enhances dehydration and drought tolerance by increasing superoxide scavenging. *Plant Physiology and Biochemistry* 70 (2013): 354–359. https://doi.org/10.1016/j.plaphy.2013.05.018

Liu, Y., Song, P., Zhang, Y., Zhou, D., Dong, Q., Jia, P., & H. Yu, et al. Physiological mechanism of photosynthetic, nutrient, and yield responses of peanut cultivars with different tolerances under low K stress. *Agronomy* 13 (2023):185. https://doi.org/10.3390/agronomy13010185

Liu, Y. H., Shen, Y., & Z. D. Chen. Identification of drought tolerance in peanut (*Arachis hypogaea*) germplasm at seedling stage. *Agricultural Science & Technology* 16 (2015): 1973–1876. https://doi.org/10.16175/j.cnki.1009-4229.2015.09.010

Liu, Y. Y., Xiao, L. N., Chi, J. X., Li, R. C., Han, Y., Cui, F., Peng, Z. Y., Wan, S. B., & G. W. Li. Genome-wide identification and expression of SAUR gene family in peanut (*Arachis hypogaea* L.) and functional identification of AhSAUR3 in drought tolerance. *BMC Plant Biology* 22 (2022): 178. https://doi.org/10.1186/s12870-022-03564-2

Lokesh, U., Venkatesh, B., Kiranmai, K., Nareshkumar, A., Amarnathareddy, V., Rao, G. L., Johnson, A. M. A., Pandurangaiah, M., & C. Sudhakar. Overexpression of β-ketoacyl Co-A synthasel gene improves tolerance of drought susceptible groundnut (*Arachis hypogaea* L.) cultivar K-6 by increased leaf epicuticular wax accumulation. *Frontiers in Plant Science* 9 (2019): 1869. https://doi.org/10.3389/fpls.2018.01869

Long, H. T., Zheng, Z., Zhang, Y. J., Xing, P. Z., Wan, X. R., Zheng, Y. X., & L. Li. An abscisic acid (ABA) homeostasis regulated by its production, catabolism, and transport in peanut leaves in response to drought stress. *PLoS ONE*, 14 (2019): e0213963. https://doi.org/10.1371/journal.pone.0213963

Luo, X., Bai, Y. N., Sun, K., Zhang, W., & C. C. Dai. Reduced pollen activity in peanut (*Arachis hypogaea* L.) by long-term monocropping is linked to flower water deficit. *Plant and Soil* 482 (2023): 427–452. https://doi.org/10.1007/s11104-022-05699-5

Lynch, J. P., & K. M. Brown. Topsoil foraging an architectural adaptation of plants to low phosphorus availability. *Plant and Soil*, 237 (2001): 225–237.

Lyons, J. M. Chilling injury in plants. *Annual Review of Plant Physiology* 24 (1973): 445–466.

Ma, X., Zhang, Y., Turečková, V., Xue, G. P., Fernie, A. R., Mueller-Roeber, B., & S. Balazadeh. The NAC transcription factor SlNAP2 regulates leaf senescence and fruit yield in tomato. *Plant Physiology* 177 (2018): 1286–1302. https://doi.org/10.1104/pp.18.00292

Mahmood, T., Khailid, S., Abdullah, M., Ahmed, Z., Shah, K. N. M., Ghafoor, A., & X. M. Du. Insights into drought stress signaling in plants and the molecular genetic basis of cotton drought tolerance. *Cells* 9 (2020:)105. https://doi.org/10.3390/cells9010105

Manimekalai, R., Suresh, G., Kurup, H. G., Athiappan, S., & M. Kandalam. Role of NGS and SNP genotyping methods in sugarcane improvement programs. *Critical Reviews in Biotechnology* 40 (2020): 865–880. https://doi.org/10.1080/07388551.2020.1765730

Mi, G., Chen, F., Yuan, L., & F. Zhang. Ideotype root system architecture for maize to achieve high yield and resource use efficiency in intensive cropping systems. *Advances in Agronomy* 139 (2016): 3–97. https://doi.org/10.1016/bs.agron.2016.05.002

Miao, H. C., Li, Q., Hou, X. F., Jia, D. H., Shi, B. X., Ding, H., Li, L. M., & Z. M. Zhang. Effects of drought at different growth stages on growth and yield of peanut. *Xinjiang Agricultural Sciences* 58 (2021): 441–449. https://doi.org/10.6048/j.issn.1001-4330.2021.03.006

Moir-Barnetson, L., Veneklaas, E. J., & T. D. Colmer. Salinity tolerances of three succulent halophytes (*Tecticornia* spp.) differentially distributed along a salinity gradient. *Functional Plant Biology* 43 (2016): 739–750. https://doi.org/10.1071/FP16025

Munns, R., & M. Tester. Mechanisms of salinity tolerance. *Annual Review of Plant Biology* 59 (2008): 651–681. https://doi.org/10.1146/annurev.arplant.59.032607.092911

Murata, N., Sato, N., Takahashi, N., & Y. Hamazaki. Compositions and positional distribution of fatty acids in phospholipids from leaves of chilling-sensitive and chilling-resistant plants. *Plant and Cell Physiology* 23 (1982): 1071–1079. https://doi.org/10.1093/oxfordjournals.pcp.a076437

Nadeem, M., Li, J., Yahya, M., Wang, M., Ali, A., Cheng, A., Wang, X., & C. Ma. Grain legumes and fear of salt stress: focus on mechanisms and management strategies. *International Journal of Molecular Sciences* 20 (2019): 799. https://doi.org/10.3390/ijms20040799

Nakamura, Y. Phosphate starvation and membrane lipid remodeling in seed plants. *Progress in Lipid Research* 52 (2013): 43–50. https://doi.org/10.1016/j.plipres.2012.07.002

Nakano, T., Suzuki, K., Fujimura, T., & Shinshi, H. Genome-wide analysis of the ERF gene family in Arabidopsis and rice. *Plant Physiology* 140 (2006): 411–432.

Neelakandan, A. K., Wright, D. A., Traore, S. M., Ma, X., Subedi, B., Veeramasu, S., & G. He. Application of CRISPR/Cas9 system for efficient gene editing in peanut. *Plants* 11 (2022): 1361. https://doi.org/10.3390/plants11101361

Ng, D. W., Abeysinghe, J. K., & M. Kamali. Regulating the regulators: the control of transcription factors in plant defense signaling. *International Journal of Molecular Sciences* 19 (2018): 3737. https://doi.org/10.3390/ijms19123737

Nigam, S. N., Rao, R. C. N., & J. C. Wynne. Effects of temperature and photoperiod on vegetative and reproductive growth of groundnut (*Arachis hypogaea* L.). *Journal of Agronomy and Crop Science* 181 (1998): 117–124. https://doi.org/10.1111/j.1439-037X.1998.tb00406.x

Nishida, H., Nosaki, S., Suzuki, T., Ito, M., Miyakawa, T., Nomoto, M., & T. Suzaki, et al. Different DNA-binding specificities of NLP and NIN transcription factors underlie nitrate-induced control of root nodulation. *Plant Cell* 33 (2021): 2340–2359. https://doi.org/10.1093/plcell/koab103

Pan, L. L., Ji, H. C., Huang, J. B., Huai, D. X., Lei, Y., Sui, J. M., Tang, Y. Y., et al. AhFatB gene editing using pollen-tube pathway and Agrobacterium-mediated method in peanut. *Huabei Nongxuebao (Acta Agriculturae Boreali-Sinica)* 35 (2020): 64–70.

Pandey, A. K., Rubiales, D., Wang, Y., Fang, P., Sun, T., Liu, N., & P. Xu. Omics resources and omics-enabled approaches for achieving high productivity and improved quality in pea (*Pisum sativum* L.). *Theoretical and Applied Genetics* 134 (2021): 755–776. https://doi.org/10.1007/s00122-020-03751-5

Pandey, M. K., Pandey, A. K., Kumar, R., Nwosu, C. V., Guo, B., Wright, G. C., Bhat, R. S., et al. Translational genomics for achieving higher genetic gains in groundnut. *Theoretical and Applied Genetics* 133 (2020): 1679–1702. https://doi.org/10.1007/s00122-020-03592-2

Pandurangaiah, M., Rao, G. L., Sudhakarbabu, O., Nareshkumar, A., Kiranmai, K., & U. Lokesh. Overexpression of horsegram (*Macrotyloma uniflorum* (Lam.) Verdc.) NAC transcriptional factor (MuNAC4) in groundnut confers enhanced drought tolerance. *Molecular Biotechnology* 56 (2014): 758–769. https://doi.org/10.1007/s12033-014-9754-0

Patel, M., Fatnani, D., & A. K. Parida. Potassium deficiency stress tolerance in peanut (*Arachis hypogaea*) through ion homeostasis, activation of antioxidant defense, and metabolic dynamics: alleviatory role of silicon supplementation. *Plant Physiology and Biochemistry* 182 (2022): 55–75. https://doi.org/10.1016/j.plaphy.2022.04.013

Pilon, C., Snider, J. L., & V. Sobolev. Assessing stomatal and non-stomatal limitations to carbon assimilation under progressive drought in peanut (*Arachis hypogaea* L.). *Journal of Plant Physiology* 231 (2018): 124–134. https://doi.org/10.1016/j.jplph.2018.09.007

Postma, J. A., & J. P. Lynch. Theoretical evidence for the functional benefit of root cortical aerenchyma in soils with low phosphorus availability. *Annals of Botany* 107 (2011): 829–841. https://doi.org/10.1093/aob/mcq199

Prasad, P. V. V., Boote, K. J., Thomas, J. M. G., Allen, L. H., & D. W. Gorbet. Influence of soil temperature on seedling emergence and early growth of peanut cultivars in field conditions. *Journal of Agronomy and Crop Science* 192 (2006): 168–177. https://doi.org/10.1111/j.1439-037X.2006.00198.x

Pruthvi, A., Narasimhan, R., & Nataraja, K. N. Simultaneous expression of abiotic stress responsive transcription factors, *AtDREB2A*, *AtHB7* and *AtABF3* inmproves salinity and drought tolerance in peanut (*Arachis hypogaea* L.). *PloS ONE* 9 (2014): e111152. https://doi.org/10.1371/journal.pone.0111152

Puchta, H., Dujon, B., & B. Hohn. Homologous recombination in plant cells is enhanced by in vivo induction of double strand breaks into DNA by a site-specific endonuclease. *Nucleic Acids Research* 21 (1993): 5034–5040. https://doi.org/10.1093/nar/21.22.5034

Qin, H., Gu, Q., Kuppu, S., Sun, L., Zhu, X., Mishra, N., Hu, R., et al. Expression of the Arabidopsis vacuolar H^+-pyrophosphatase gene AVP1 in peanut to improve drought and salt tolerance. *Plant Biotechnology Report* 7 (2013): 345–355. https://doi.org/10.1007/s11816-012-0269-5

Ran, F. A., Hsu, P. D., Lin, C. Y., Gootenberg, J. S., Konermann, S., Trevino, A. E., Scott, D. A., et al. Double nicking by RNA-guided CRISPR Cas9 for enhanced genome editing specificity. *Cell* 154 (2013): 479–480. https://doi.org/10.1016/j.cell.2013.09.040

Ranganayakulu, G., Chandraobulreddy, P., Thippeswamy, M., Veeranagamallaiah, G., & C. Sudhakar. Identification of drought stress-responsive genes from drought tolerant groundnut cultivar (*Arachis hypogaea* L. Cv K-134) through analysis of subtracted expressed sequence tags. *Acta Physiologia Plantarum* 34 (2012): 361–377. https://doi.org/10.1007/s11738-011-0835-4

Ramu, V. S., Swetha, T. N., Sheela, S. H., Babitha, C. K., Rohihi, S., Reddy, M. K., Tuteja, N., et al. Simultaneous expression of regulatory genes associated with specific drought-adaptive traits improves drought adaptation in peanut. *Journal of Plant Biotechnology* 14 (2016): 1008–1020. https://doi.org/10.1111/pbi.12461

Rao, T. S. R. B., Naresh, J. V., Reddy, P. S., Reddy, M. K., & Mallikarjuna, G. Expression of *Pennisetum glaucum* eukaryotic translaitonal initiation factor 4A (*PgelF4A*) confers improved drought, salinity, and oxidative stress tolerance in groundnut. *Frontier in Plant Science* 8 (2017): 453. https://doi.org/10.3389/fpls.2017.00453

Ravi, K., Vadez, V., Isobe, S., Mir, R. R., Guo, Y., Nigam, S. N., Gowda, M. V. C., et al. Identification of several small main-effect QTLs and a large number of epistatic QTLs for drought tolerance-related traits in groundnut (*Arachis hypogaea* L.). *Theoretical and Applied Genetics* 122 (2011): 1119–1132. https://doi.org/10.1007/s00122-010-1517-0

Raza, A., Chen, H., Zhang, C., Zhuang, Y. H., Sharif, Y., Cai, T. C., Yang, Q., et al. Designing future peanut: the power of genomics-assisted breeding. *Theoretical and Applied Genetics* 137 (2024): 66. https://doi.org/10.1007/s00122-024-04575-3

Ren, J. Y., Guo, P., Zhang, H., Shi, X. L., Ai, X., Wang, J., Jiang, C. J., et al. Comparative physiological and coexpression network analyses reveal the potential drought tolerance mechanism of peanut. *BMC Plant Biology* 22 (2022a): 460. https://doi.org/10.1186/s12870-022-03848-7

Ren, J. Y., Jiang, C. J., Zhang, H., Shi, X. L., Ai, X., Li, R. Y., & Dong, J. L., et al. LncRNA-mediated ceRNA networks provide novel potential biomarkers for peanut drought tolerance. *Physiologia Plantarum* 174 (2022b): e13610. https://doi.org/10.1111/ppl.13610

Rogers, E. D., & P. N. Benfey. Regulation of plant root system architecture: implications for crop advancement. *Current Opinion in Biotechnology* 32 (2015): 93–98. https://doi.org/10.1016/j.copbio.2014.11.015

Roy, S. J., Negrao, S., & M. Tester. Salt-resistant crop plants. *Current Opinion in Biotechnology* 26 (2014): 115–124. https://doi.org/10.1016/j.copbio.2013.12.004

Ruiz-Lozano, J. M., Porcel, R., Azcón, C., & R. Aroca. Regulation by arbuscular mycorrhizae of the integrated physiological response to salinity in plants: new challenges in physiological and molecular studies. *Journal of Experimental Botany* 63 (2012): 4033–4044. https://doi.org/10.1093/jxb/ers126

Shen, N., Wang, T., Gan, Q., Liu, S., Wang, L., & B. Jin. Plant flavonoids: classification, distribution, biosynthesis, and antioxidant activity. *Food Chemistry* 383 (2022): 132531. https://doi.org/10.1016/j.foodchem.2022.132531

Shi, H., Quintero, F. J., Pardo, J. M., & J. K. Zhu. The putative plasma membrane Na^+/H^+ antiporter Sos1 controls long-distance Na^+ transport in plants. *The Plant Cell* 14 (2002): 465–477. https://doi.org/10.1105/tpc.010371

Shi, X. (2018). Effects of calcium application on growth and development, physiological characteristics, and pod plumpness of peanut under salt stress (Master's thesis, Xinjiang Agricultural University). https://doi.org/10.27431/d.cnki.gxnyu.2018.000174

Shi, X. L., Zhou, D. Y., Guo, P., Zhang, H., Dong, J., Ren, J. Y., Jiang, C. J., et al. External potassium mediates the response and tolerance to salt stress in peanut at the flowering and needling stages. *Photosynthetica* 58 (2020): 1141–1149. https://doi.org/10.32615/ps.2020.070

Shi, Y., Ding, Y., & S. Yang. Molecular regulation of CBF signaling in cold acclimation. *Trends in Plant Science* 23 (2018): 623–637. https://doi.org/10.1016/j.tplants.2018.04.002

Shu, H. M., Luo, Z. L., Peng, Z., & J. P. Wang. The application of CRISPR/Cas9 in hairy roots to explore the functions of AhNFR1 and AhNFR5 genes during peanut nodulation. *BMC Plant Biology* 20 (2020): 417. https://doi.org/10.1186/s12870-020-02614-x

Song, L., Li, R., Xiang, X., Wang, J., Qiao, L., Song, X., Pei, Y., & J. Sui. Overexpression of stress-inducible small GTP-binding protein AhRab7 (*AhRabG3f*) in peanut (*Arachis hypogaea* L.) enhances abiotic stress tolerance. *Journal of Food Agricultural Environment* 10 (2012): 888–894.

Song, Y. K., & D. P. Wang. A preliminary study on the effect of temperature on flowering and pod formation of peanut (*Arachis hypogaea* L.). *Chinese Journal of Oil Crop Science* 1 (1979): 32–37.

Sorensen, R. B., Nuti, R. C., & C. L. Butts. Yield and plant growth response of peanut to midseason forage harvest. *Agronomy Journal* 101 (2009): 1198–1203. https://doi.org/10.2134/agronj2009.0047

Su, L., Zhao, C. Z., Bi, Y. P., Wan, S. B., Xia, H., & X. J. Wang. Isolation and expression analysis of LEA genes in peanut (*Arachis hypogaea* L.). *Journal of Bioscience* 36 (2011): 223. https://doi.org/10.1007/s12038-011-9058-5

Sui, J. M., Li, R., Fan, Q. C., Song, L., Zheng, C. H., Wang, J. S., Qiao, L. X., & S. L. Yu. Isolation and characterization of a stress-responsive small GTP-binding protein AhRabG3b in peanut (*Arachis hypogaea* L.). *Euphytica* 189 (2013): 161–172. https://doi.org/10.1007/s10681-012-0739-4

Sugimoto, K., Zager, J. J., Aubin, B. S., Lange, B. M., & G. A. How. How flavonoid deficiency disrupts redox homeostasis and terpenoid biosynthesis in glandular trichomes of tomato. *Plant Physiology* 188 (2022): 1450–1468. https://doi.org/10.1093/plphys/kiab488

Sun, Z. X., Liu, W., Wang, X. N., Ai, X., Li, Z., Zhou, D. Y., Ma, Q., Li, Y., Wang, J., Ma, X., & X. Wang. Transcriptome-based spatiotemporal analysis of drought response mechanisms in two distinct peanut cultivars. *International Journal of Molecular Sciences* 25, no. 22 (2024): 11895. https://doi.org/10.3390/ijms252211895

Tang, X. M., Chen, Y., Zhong, R. C., Jiang, J., Xiong, F. Q., He, L. Q., Han, Z. Q. et al. Study on drought tolerance in Guangxi peanut varieties (strains) and relative physiological characteristics of seedling stage. *Guangdong Agricultural Science* 14 (2014): 9–13. https://doi.org10.16768/j.issn.1004-874x.2014.14.019.

Tarkowska, D., & M. Strnad. Isoprenoid-derived plant signaling molecules: biosynthesis and biological importance. *Planta* 247 (2018): 1051–1066. https://doi.org/10.1007/s00425-018-2878-x

Tholl, D. Biosynthesis and biological functions of terpenoids in plants. *Advances in Biochemical Engineering/Biotechnology* 148 (2015): 63–106. https://doi.org/10.1007/10_2014_295

Thomas, A. H., Catalá, Á., & M. Vignoni. Soybean phosphatidylcholine liposomes as model membranes to study lipid peroxidation photoinduced by pterin. *Biochimica et Biophysica Acta* 1858 (2016): 139–145. https://doi.org/10.1016/j.bbamem.2015.11.002

Thorne, S. J., Hartley, S. E., & F. J. M. Maathuis. Is silicon a panacea for alleviating drought and salt stress in crops? *Frontiers in Plant Science* 11 (2020): 1221–1236. https://doi.org/10.3389/fpls.2020.01221

Tian, Q., Chen, F., Liu, J., Zhang, F., & G. Mi. Inhibition of maize root growth by high nitrate supply is correlated with reduced IAA levels in roots. *Journal of Plant Physiology* 165 (2008): 942–951. https://doi.org/10.1016/j.jplph.2007.02.011

Tian, X., Liu, Y., Huang, Z., Duan, H., Tong, J., He, X., Gu, W., Ma, H., & L. Xiao. Comparative proteomic analysis of seedling leaves of cold-tolerant and -sensitive spring soybean cultivars. *Molecular Biology Reports* 42 (2015): 581–601. https://doi.org/10.1007/s11033-014-3803-4

Tiwari, V., Chaturvedi, A. K., Mishra, A., & B. Jha. Introgression of the SbASR-1 gene cloned from a halophyte *Salicornia brachiata* enhances salinity and drought endurance in transgenic groundnut (*Arachis hypogaea*) and acts as a transcription factor. *PLoS One* 10 (2015): e0131567. https://doi.org/10.1371/journal.pone.0131567

Upadhyaya, H. D., Reddy, L. J., Dwivedi, S. L., Gowda, C. L. L., & S. Singh. Phenotypic diversity in cold-tolerant peanut (*Arachis hypogaea* L.) germplasm. *Euphytica* 165 (2009): 279–291. https://doi.org/10.1007/s10681-008-9786-2

Vaishnav, A., Kumari, S., Jain, S., Varma, A., & D. K. Choudhary. Putative bacterial volatile-mediated growth in soybean (*Glycine max* L. Merrill) and expression of induced proteins under salt stress. *Journal of Applied Microbiology* 119 (2015): 539–551. https://doi.org/10.1111/jam.12866

Varshney, R. K., Bertioli, D. J., Moretzsohn, M. C., Vadez, V., Krishnamurthy, L., Aruna, R., Nigam, S. N., Moss, B. J., Seetha, K., & K. Ravi. The first SSR-based genetic linkage map for cultivated groundnut (*Arachis hypogaea* L.). *Theoretical and Applied Genetics* 118 (2009): 729–739. https://doi.org/10.1007/s00122-018-3252-x

Varshney, R. K., Pandey, M. K., Bohra, A., Singh, V. K., Thudi, M., & Saxena, R. K. Toward the sequence-based breeding in legumes in the post-genome sequencing era. *Theoretical and Applied Genetics* 132 (2019): 797–816.

Vogt, T. Phenylpropanoid biosynthesis. *Molecular Plant* 1 (2010): 2–20. https://doi.org/10.1093/mp/ssp106

Wang, C., Cheng, B., Zheng, Y., Sha, J., Li, A., & X. Sun. Effects of temperature on seed emergence, seedling growth, and anthesis of peanut. *Journal of Peanut Science* 32 (2003): 7–11.

Wang, F. Q., Wang, M. L., Zheng, Z. R., & Z. H. Song. The effects of drought on peanut flower bud differentiation in the seedling stage. *Chinese Journal of Oil Crop Sciences* 22 (2000): 370–373.

Wang, J., Chen, H. N., Shi, D. C., Yu, T. Y., Yan, C. X., Sun, Q. X., & Yuan, C. L., et al. Functional analysis of *ahnrt2.7a* in response to low-nitrogen in peanut. *Scientia Agricultura Sinica* 55 (2022): 4356–4372. https://doi.org/10.3864/j.issn.0578-1752.2022.22.003

Wang, J., Yu, Y., Jiang, C. J., Sun, Z. X., Wang, X. N., Wang, Z. Y., & Ren, J. Y., et al. Comparative analysis of physiology-anatomy and transcriptome-metabolome involving acute drought stress response of root between two distinct peanut cultivars at seedling stage. *Environmental and Experimental Botany* 24 (2023): 105442. https://doi.org/10.1016/j.envexpbot.2023.105442

Wang, T., Zhang, E., Chen, X., Li, L., & X. Liang. Identification of seed proteins associated with resistance to pre-harvested aflatoxin contamination in peanut (*Arachis hypogaea* L.). *BMC Plant Biology* 10 (2010): 267. https://doi.org/10.1186/1471-2229-10-267

Wang, W., Vinocur, B., & A. Altman. Plant responses to drought, salinity, and extreme temperatures: towards genetic engineering for stress tolerance. *Planta* 218 (2003): 1–14. https://doi.org/10.1007/s00425-003-1105-5

Wang, X., Liu, Y., Han, Z., Chen, Y., Huai, D., Kang, Y., & Z. Wang. Integrated transcriptomics and metabolomics analysis reveal key metabolism pathways contributing to cold tolerance in peanut. *Frontiers in Plant Science* 12 (2021): 752474. https://doi.org/10.3389/fpls.2021.752474

Wang, Y. Y., Cheng, Y. H., Chen, K. E., & Y. F. Tsay. Nitrate transport, signaling, and use efficiency. *Annual Review of Plant Biology* 69 (2018): 85–122. https://doi.org/10.1146/annurev-arplant-042817-040056

Wegner, L. H., Stefano, G., Shabala, L., Rossi, M., Mancuso, S., & S. Shabala. Sequential depolarization of root cortical and stelar cells induced by an acute salt shock: implications for Na^+ and K^+ transport into xylem vessels. *Plant, Cell & Environment* 34 (2011): 859–869. https://doi.org/10.1111/j.1365-3040.2011.02291.x

Wolter, F., Schindele, P., & H. Puchta. Plant breeding at the speed of light: the power of CRISPR/Cas to generate directed genetic diversity at multiple sites. *BMC Plant Biology* 19 (2019): 176. https://doi.org/10.1186/s12870-019-1775-1

Xie, B. C., Gao, D. F., Zhou, B. Q., Chen, S., & L. R. Wang. New discoveries in the field of metabolism by applying single-cell and spatial omics. *Journal of Pharmaceutical Analysis* 13 (2023): 711–725. https://doi.org/10.1016/j.jpha.2023.06.002

Xu, P., & E. Wang. Diversity and regulation of symbiotic nitrogen fixation in plants. *Current Biology* 33 (2023): R543–R559. https://doi.org/10.1016/j.cub.2023.04.053

Xu, Y. B., Lu, Y. L., Xie, C. X., Gao, S. B., Wan, J. M., & B. M. Prasanna. Whole-genome strategies for marker-assisted plant breeding. *Molecular Breeding* 29 (2012): 833–854. https://doi.org/10.1007/s11032-012-9699-6

Yan, C. X., Li, C. J., Sun, Q. X., Zhang, H., Wang, J., Yuan, C. L., Shan, S. H., & X. B. Zhao. Cloning and functional analysis of FAR1-5 transcription factor in peanut. *Journal of Peanut Science* 49 (2020): 16–20. https://doi.org/10.14001/j.issn.1002-4093.2020.02.003

Yan, Q. C. *Seed Science*. Beijing: China Agriculture Press, 2001.

Yang, J., Kloepper, J. W., & C. M. Ryu. Rhizosphere bacteria help plants tolerate abiotic stress. *Trends in Plant Science* 14 (2009): 1–4. https://doi.org/10.1016/j.tplants.2008.10.004

Yang, L., Wu, Q., Liang, H., Yin, L., & P. Shen. Integrated analyses of transcriptome and metabolome provides new insights into the primary and secondary metabolism in response to nitrogen deficiency and soil compaction stress in peanut roots. *Frontiers in Plant Science* 13 (2022): 948742. https://doi.org/10.3389/fpls.2022.948742

Yang, S., Wang, F., Guo, F., Meng, J. J., Li, X. G., & S. B. Wan. Calcium contributes to photoprotection and repair of photosystem II in peanut leaves during heat and high irradiance. *Journal of Integrative Plant Biology* 57 (2015): 486–495. https://doi.org/10.1111/jipb.12249

Ye, X., Wang, H., Cao, X., Jin, X., Cui, F., Bu, Y., & Liu, H., et al. Transcriptome profiling of *Puccinellia tenuiflora* during seed germination under a long-term saline-alkali stress. *BMC Genomics* 20 (2019): 589–605. https://doi.org/10.1186/s12864-019-5860-5

Ye, K., Li, H., Ding, Y., Shi, Y., Song, C., Gong, Z., & S. Yang. BRASSINOSTEROID-INSENSITIVE2 negatively regulates the stability of transcription factor ICE1 in response to cold stress in Arabidopsis. *The Plant Cell* 31 (2019): 2682–2696. https://doi.org/10.1105/tpc.19.00058

Yuan, M., Zhu, J., Gong, L. M., He, L. Q., Lee, C., Han, S. Y., Chen, C., & G. H. He. Mutagenesis of *FAD2* genes in peanut with CRISPR/Cas9-based gene editing. *BMC Biotechnology* 19 (2019): 24. https://doi.org/10.1186/s12896-019-0557-y

Yuan, Q., Druzhinina, I. S., Pan, X., & Z. Yuan. Microbially mediated plant salt tolerance and microbiome-based solutions for saline agriculture. *Biotechnology Advances* 34 (2016): 1245–1259. https://doi.org/10.1016/j.biotechadv.2016.08.005

Yuan, T., Yang, H., He, K., Gou, X. P., Lin, H. H., Clouse, S. D., & J. Li. Genome-wide cloning and sequence analysis of leucine-rich repeat receptor-like protein kinase genes in *Arabidopsis thaliana*. *BMC Genomics* 11 (2010): 19. https://doi.org/10.1186/1471-2164-11-19

Zhang, B. H., Su, L. C., Hu, B., & L. Li. Expression of *AhDREB1*, an AP2/ERF transcription factor gene from peanut, is affected by histone acetylation and increases abscisic acid sensitivity and tolerance to osmotic stress in *Arabidopsis*. *International Journal of Molecular Sciences* 19 (2018): 1441. https://doi.org/10.3390/ijms19051441

Zhang, F., Maeder, M. L., Unger-Wallace, E., Hoshaw, J. P., Reyon, D., Christian, M., & Li, X., et al. High-frequency targeted mutagenesis in *Arabidopsis thaliana* using zinc finger nucleases. *Proceedings of the National Academy of Sciences of the United States of America* 107 (2010): 12028–12033. https://doi.org/10.1073/pnas.0914991107

Zhang, F., Wen, Y., & X. Guo. CRISPR/Cas9 for genome editing: progress, implications, and challenges. *Human Molecular Genetics* 23 (2014): 40–46. https://doi.org/10.1093/hmg/ddu125

Zhang, G., Ahmad, M. Z., Chen, B., Manan, S., Zhang, Y., Jin, H., Wang, X., & J. Zhao. Lipidomic and transcriptomic profiling of developing nodules reveals the essential roles of active glycolysis and fatty acid and membrane lipid biosynthesis in soybean nodulation. *The Plant Journal* 103 (2020): 1351–1371. https://doi.org/10.1111/tpj.14805

Zhang, H., Dong, J., Zhao, X., Zhang, Y., Ren, J., Xing, L., & Jiang, C., et al. Research progress in membrane lipid metabolism and molecular mechanism in peanut cold tolerance. *Frontiers in Plant Science* 10 (2019): 838. https://doi.org/10.3389/fpls.2019.00838

Zhang, H., Jiang, C., Lei, J., Dong, J., Ren, J., Shi, X., & Zhong, C., et al. Comparative physiological and transcriptomic analyses reveal key regulatory networks and potential hub genes controlling peanut chilling tolerance. *Genomics* 114 (2022): 110285. https://doi.org/10.1016/j.ygeno.2022.110285

Zhang, H., Jiang, C., Ren, J., Dong, J., Shi, X., Zhao, X., & Wang, X., et al. An advanced lipid metabolism system revealed by transcriptomic and lipidomic analyses plays a central role in peanut cold tolerance. *Frontiers in Plant Science* 11 (2020): 1110. https://doi.org/10.3389/fpls.2020.01110

Zhang, H., Jiang, C., Yin, D., Dong, J., Ren, J., Zhao, X., & Zhong, C., et al. Establishment of a comprehensive evaluation system for peanut cold tolerance and screening of cold-tolerance germplasm. *Acta Agronomica Sinica* 47 (2021): 1753–1767. https://doi.org/10.3724/SP.J.1006.2021.04182

Zhang, G. H., Yu, S. T., Wang, H., & X. D. Wang. Transcriptome profiling of high oleic peanut under low temperature during germination. *Yi Chuan* 41 (2019): 1050–1059. https://doi.org/10.16288/j.yczz.19-097

Zhang, J., Yuan, H., Yang, Y., Fish, T., Lyi, S. M., Thannhauser, T. W., Zhang, L., & L. Li. Plastid ribosomal protein S5 is involved in photosynthesis, plant development, and cold stress tolerance in *Arabidopsis*. *Journal of Experimental Botany* 67 (2016): 2731–2744. https://doi.org/10.1093/jxb/erw106

Zhang, P., Duo, T., Wang, F., Zhang, X., Yang, Z., &G. Hu. De novo transcriptome in roots of switchgrass (*Panicum virgatum* L.) reveals gene expression dynamics and action network under alkaline salt stress. *BMC Genomics* 22 (2021): 82–96. https://doi.org/10.1186/s12864-021-07368-w

Zhang, Y., Zhang, F., Li, X., Baller, J. A., Qi, Y., Starker, C. G., Bogdanove, A. J., & D. F. Voytas. Transcription activator-like effector nucleases enable efficient plant genome engineering. *Plant Physiology* 161 (2013): 20–27. https://doi.org/10.1104/pp.112.205179

Zhang, Z. M., Dai, L. X., Song, W. W., Ding, H., Ci, D. W., Kang, T., & S. B. Wan. Effect of drought stresses at different growth stages on peanut leaf protective enzyme activities and osmoregulation substances content. *Acta Agronomica Sinica* 39 (2013): 133–141. https://doi.org/10.3724/SP.J.1006.2013.00133

Zhao, X. B., Yan, C. X., Zhang, H., Wang, J., Li, C. J., Xie, H. F., & S. H. Shan. Differential expression of transcription factor families in peanut (*Arachis hypogaea* L.) under drought stress. *Journal of Agricultural Biotechnology* 26 (2018): 1143–1154. https://doi.org/10.3969/j.issn.1674-7968.2018.07.006

Zhong, C., Liu, Y., Li Z., Wang, X. G., Jiang, C. J., Zhao, X. H., Kang, S. L., Liu, X. B., Zhao, S. L., Wang, J., Zhang, H.,

Huang, Y. N., Yu, H. Q., & Xue, R. F. Genome-wide analysis reveals regulatory mechanisms and expression patterns of *TGA* genes in peanut under abiotic stress and hormone treatments. *Fronties in Plant Science* 14 (2023): 1269200. https://doi.org/10.3389/fpls.2023.1269200

Zhou, J., Zhou, H., Chen, P., Zhang, L., Zhu, J., Li, P., & J. Yang. Genome-wide survey and expression analysis of the KT/HAK/KUP family in *Brassica napus* and its potential roles in the response to K^+ deficiency. *International Journal of Molecular Sciences* 21 (2020): 9487–9505. https://doi.org/10.3390/ijms21249487

Zhou, W., & B. Lin. Studies on the symptoms and ultrastructural characteristics of calcium deficiency in peanuts. *Scientia Agricultura Sinica* 4 (1996): 54–58+99–101.

Zhou, X. L., Weng, L. L., Jiang, D. C., & C. P. Xiao. Photosynthetic characteristics and physiological response to stress of *Atractylodes auriculata* under drought stress. *Journal of Chinese Medicinal Materials* 43 (2020): 15–19.

Zhu, J. K. Abiotic stress signaling and responses in plants. *Cell* 167 (2016): 313–324. https://doi.org/10.1016/j.cell.2016.08.029

Zhuang, W., Chen, H., Yang, M., Wang, J., Pandey, M. K., Zhang, C., & Chang, W. C., et al. The genome of cultivated peanut provides insight into legume karyotypes, polyploid evolution, and crop domestication. *Nature Genetics* 1 (2019): 865–876. https://doi.org/10.1038/s41588-019-0402-2

Zuo, Y. M., Chen, Q., & F. S. Zhang. The mechanisms of root exudates of maize in improvement of iron nutrition of peanut in peanut/maize intercropping system by 14C tracer technique. *Acta Agriculturae Nucleatae Sinica* 18 (2004): 43–46. https://doi.org/10.1007/BF02873091

14 Potential of Modern Biotechnological Tools for Enhancing Resistance to Major Biotic Stresses in Peanut

Boshou Liao, Yong Lei, Huaiyong Luo, and Liying Yan

14.1 INTRODUCTION

Peanut or groundnut (*Arachis hypogaea* L.), a legume species that originated in South America, is widely cultivated in more than 100 countries in the world as an important oilseed and cash crop. During 1961–2021, the global peanut harvest area increased from 1.664 million ha to 3.272 ha, yield increased from 849.3 kg/ha to 1648.1 kg/ha, and production increased from 14.134 million tons to 53.927 million tons (FAOSTAT 2022). In 2021, Africa and Asia together accounted for 95.52% of the global peanut harvest area and 89.96% of global production. India, China, Nigeria, Sudan, Senegal, Myanmar, Tanzania, Niger, Guinea, Chad, the United States, and Democratic Republic of Congo were the leading 12 countries for harvest area for the year of 2021, while China, India, Nigeria, the United States, Sudan, Senegal, Myanmar, Argentina, Guinea, Chad, Brazil, and Indonesia were the leading 12 countries for production (FAOSTAT 2022).

Globally, a number of biotic and abiotic factors largely restrict the production and productivity of peanut. Abiotic stresses in peanut, including drought, waterlogging, heat, cold, acid and saline-alkali soils, and poor soil fertility, affect a range of different regions, and they are addressed in Chapter 13 of this book. In terms of biotic stresses affecting the peanut crop, there are many fungal, bacterial, virial, and nematode diseases and insect pests. Late leaf spot, early leaf spot, rust, stem rot, and *Aspergillus* infection and aflatoxin contamination are the most widely distributed biotic stresses (Bera et al. 2022). Regionally, groundnut rosette disease in Africa, bacterial wilt, peanut bud necrosis disease, leaf miner, and *Spodoptera* in South and/or South East Asia; corn earworm, lesser cornstalk borer, southern corn rootworm, nematodes, and tomato spotted wilt virus in North America; and peanut smut in South America are important constraints to peanut production (Bera et al. 2022). Biotic and abiotic stresses together can cause annual crop losses exceeding 3.2 billion US$, and half of these losses could be recovered through genetic enhancement for resistance or tolerance (Dwivedi et al. 2003).

Marked progresses on genetic enhancement in peanut for various stress resistances or tolerances have been achieved through conventional and molecular approaches (Bangaru et al. 2023; Bera et al. 2022; Bhat et al. 2022). Moreover, the newly-developed genomic resources and biotechnological tools are becoming more and more popular and efficient for genetic improvement in peanut. In particular, the accomplishments of genome sequencing and analyzing of the progenitor diploids (*Arachis duranensis* and *Arachis ipaensis*) (Bertioli et al. 2016; Chen et al. 2016; Lu et al. 2018), primitive tetraploid (*Arachis monticola*) (Yin et al. 2019), and the cultivated tetraploids (Tifrunner, Fuhuasheng, and Shitouqi) (Bertioli et al. 2019; Chen et al. 2019; Zhuang et al. 2019) were the milestone events for promoting the peanut improvement using genomic and biotechnological tools. In addition to genomic sequencing, a lot of bioinformatics data including whole-genome re-sequencing (WGRS), transcriptomes, and methylomes are available in various research agencies. Efforts in stress tolerance trait mapping have successfully mapped resistance to early leaf spot, late leaf spot and rust, and other resistances including those to stem rot (Dodia et al. 2019), aflatoxin production (Yu et al. 2020), tomato spotted wilt virus (Agarwal et al. 2018, 2019), bacterial wilt (Luo et al. 2017), aphid, and the vector of groundnut rosette disease (Herselman et al. 2004). Further efforts are underway to map more stress-related traits and integrate multiple resistances.

In this chapter, the biotic stresses affecting peanut and progress on resistance improvement are comprehensively reviewed, and the potential for the utilization of genomic and biotechnological tools in stress tolerance enhancement is discussed. Even through weeds are also important biotic constraints to peanut production, they are not covered in this chapter as they have to date been controlled mainly by spraying herbicides rather than through host-plant resistance. As limited research has been implemented in improving resistance to special herbicides in peanut, the concerned contents are also not included in this chapter.

14.2 MAJOR BIOTIC STRESSES IN PEANUT AND THEIR BIOLOGICAL FEATURES

Peanut is easily affected by a lot of diseases and insects across diverse environments worldwide. To date, more than 30 diseases and 40 insects have been reported to

DOI: 10.1201/9781003360988-14

affect peanut plants and pods/seeds, among which more than 20 diseases and insects can cause substantial yield and economic losses at both global and regional levels. The pathogenic organisms infecting peanut include fungi, bacteria, viruses, and nematodes (McDonald et al. 1998), and the major insects affecting peanut include foliage feeders, sap suckers, and root and pod feeders (Bera et al. 2022). The most popular biotic stresses affecting peanut and their biological features are briefly described as follows.

14.2.1 Fungal Diseases

14.2.1.1 Early Leafspot

Early leafspot of peanut is caused by *Cercospora arachidicola* Hori. This fungal disease is distributed across all the peanut producing regions in the world, being secondary only to the late leafspot (caused by *Phaeoisariopsis personata*) in terms of both the distribution area and the overall economic losses. *C. arachidicola* mostly infect peanut leaves, but infection on stem, branch, petiole, stipule, and even peg is also frequent (Liao 2012; McDonald et al. 1985; Shokes and Culbreath 1997). At about ten days after infection by the pathogen conidia on leaves, the symptoms of early leafspot are first recognizable as small necrotic yellow or brown flecks, and the flecks will then become light to dark brown spots ranging from 1 to 10 mm in diameter. The spots will further develop into mature, sporulating lesions within less than a week. The mature lesions are dark brown on the adaxial (upper) leaflet surface where most sporulation occurs, and a light shade of brown on the abaxial (lower) leaflet surface. Normally, a yellowish halo is often present around the lesions (this may be not obvious in some peanut genotypes). In most cases, defoliation can happen in the infected plants with visible symptoms, and fully defoliation and even death of the whole plant can be seen in heavily diseased fields. Worldwide, peanut yield losses caused by early leafspot range from 10% to over 50% (without chemical control), but vary largely from place to place and from season to season.

The early leafspot pathogen *C. arachidicola* is soilborne. The pathogen can overwinter in the form of sporodochium and mycelium in infected peanut debris, mostly in the soil, under cold environments, and can also survive from season to season on volunteer peanut plants and infected crop debris in warm tropical or subtropical regions. Teleomorphs of the pathogenic fungus are observed, but the ascospores are not the key primary inoculum. When the temperature and moisture are favorable, conidia of *C. arachidicola* are produced directly from mycelium in crop debris in the soil. Under natural conditions, infection by *C. arachidicola* on peanut plant normally occurs earlier than the late leafspot pathogen, and symptoms of both early and late leafspots can be seen within 3–5 weeks after sowing in some cases. Normally, the first infection in the field happens on older leaflets near the soil surface. The conidia produced on the early infected leaves are carried by wind, rain splash, insects, and even artificial operational tools to the later-formed leaves and to adjacent plants. The *C. arachidicola* fungus can grow at temperatures under 10–35°C, but temperatures in the 25–30°C range with high relative humidity favor infection and disease development. When the temperature is below 10°C or over 37°C, disease development is largely restricted. The disease severity can be more serious in upland soils with relatively poor fertility and erratic drought stress. Even though leafspot can be seen in peanut in newly-planted areas far from traditional growing areas, no authentic host species outside the genus *Arachis* have been confirmed, indicating that the conidia could be disseminated by wind over a very long distance. The pathogen may also be disseminated by surface contaminated pods or seeds over long distances.

General control measures against early leafspot have been developed. Planting peanut varieties with desirable resistance wherever available is the most feasible and cost-effective approach, and marked efforts have been made in the last four decades in genetic enhancement for host-plant resistance with substantial progress. Cultural control measures such as rotating peanut with non-host crops, cleaning crop residuals after harvest, eradicating volunteer peanut plants, adjusting sowing time, and controlling weeds are commonly recommended. Spraying of chemical fungicides is useful in controlling early late leafspot. More than ten chemicals have been used in different regions in the world (McDonald et al. 1985), and more chemicals are newly developed and recommended. A few observations in the United States have hinted that application of certain fungicides for controlling leafspots may somehow increase the severity levels of soilborne *Sclerotium rolfsii* (Backman et al. 1997) and *Sclerotinia minor* (Porter 1980). Above all, breeding peanut cultivars with high yield, improved quality, and high level resistance has become the prioritized strategy for comprehensive controlling of early leafspot in peanut.

14.2.1.2 Late Leafspot

Late leafspot of peanut is caused by *Phaeoisariopsis personata* (Berk. & Curt.) von Arx. This fungal disease is distributed across all the peanut producing regions in the world, with the largest distribution area and overall economic losses among the peanut diseases. *P. personata* mostly infects peanut leaves, and also infects stem, branch, petiole, stipule, peg, and even pod in late growth stages (Liao 2012; McDonald et al. 1985; Shokes and Culbreath 1997). Late leafspot can occur solely or together with early leafspot. The invasion process and latent period of late leafspot are similar to early leafspot. Small chlorotic spots appear on leaflets about 10 days after infection and develop into mature and sporulating lesions within another 5 days (McDonald et al. 1985). The diameters of late leafspot lesions (more nearly circular) range from 1 to 5 mm, normally smaller than that of early leafspot, but large lesions can also be observed under some conducive environments or in some special peanut genotypes. The color of late leafspot lesions is darker than that of early leafspot. Only in a few peanut genotypes a slight chlorotic halo can be seen around the lesions. Most sporulation of *P. personata* occurs on the abaxial (lower) surface of leaflets, and the sporodochia are dark and conidia are borne in tight clusters arranged in concentric rings (McDonald et al. 1985). Serious defoliation due to late leafspot can be seen in heavily diseased fields, and yield losses could be as high as 70% on highly susceptible peanut cultivars without chemical control. Yield losses due to both early and late leafspots are highly related to levels of defoliation and harvesting time (McDonald et al. 1985).

The late leafspot pathogen *P. personata* is also soilborne. The pathogen can overwinter or survive from season to season in the form of mycelium and/or sporodochium in infected peanut debris or volunteer peanut plants. Mycelium in lesions on stems, petioles, and pegs is more likely to overseason than that on leaflets. Conidia produced on crop residue in the soil are the main source of initial inoculum (McDonald et al. 1985). Teleomorphs of *P. personata* are observed, but the ascospores are not regarded as a key primary inoculum in the field. When the temperature and moisture are favorable, conidia of *P. personata* are produced directly from mycelium in crop debris in the soil. Unlike the *C. arachidicola*, *P. personata* produces intercellular haustoria during the infection process. Under natural conditions, late leafspot symptoms can be seen 5 weeks after sowing but normally become serious in the late growth stages. Like *C. arachidicola*, the *P. personata* conidia produced on the early infected leaves are carried by wind, rain splash, insects, and even artificial operational tools in the field to the later-formed leaves and to adjacent plants. Temperatures in the 25–30°C range with high relative humidity favor infection and disease development of late leafspot. In most cases, late leafspot can cause more severe damage over a short period because of the capacity of *P. personata* to produce more spores per lesion than *C. arachidicola* (McDonald et al. 1985). No authentic host species outside the genus *Arachis* are confirmed for *P. personata*, but the conidia could be disseminated by wind or other means over very long distances because the disease can be seen in all peanut-producing regions, including remote places where peanut is newly planted.

Control of late leafspot has attracted extensive efforts in peanut production. The major control approaches for late leafspot are similar to that of early leafspot. Planting peanut cultivars with resistance is the most feasible and cost-effective approach. Genetic enhancement for host-plant resistance to late leafspot has been widely stressed, and considerable progress has been achieved. Cultural control measures, including crop rotation, field sanitation after harvest, eradication of volunteer peanut plants, adjustment of sowing time, and weed control, are generally necessary. Chemical control is useful for reducing the severity of late leafspot.

14.2.1.3 Rust

Rust of peanut is caused by *Puccinia arachidis* Speg. This disease has been reported in most peanut-producing regions in the world even though it was largely confined to Central and South America before 1969. Infection of pathogen conidia on peanut leaflets results in small chlorotic spots, which develop into orange-colored pustules within 2 or 3 weeks depending on the environmental factors and susceptibility of peanut cultivars. Spores of *P. arachidis* can infect all aerial parts except for flowers and pegs of peanut plant. Rust pustules appear first on the abaxial surface and may also be formed later on the adaxial surface opposite those on the abaxial surface, with diameters ranging from 0.5 to 1.4 mm. Pustules on the stem are elongate. Most peanut leaves infected with rust become necrotic and dry up but tend to remain attached to the plant, in contrast to the rapid defoliation caused by leafspots (Liao 2012). Yield losses up to 50% from rust damage were reported in heavily diseased regions.

The rust pathogen *P. arachidis* is airborne, and its urediniospores can be disseminated for long distances by air flows and by the movement of infected crop debris or pods/seeds surface-contaminated with urediniospores. The pathogen is thought to survive from season to season mainly on volunteer peanut plants. No authentic host species are known outside the genus *Arachis*. Airborne urediniospores are the primary inoculum in peanut fields. Under field conditions, the spores released from the mature pustules are the source of repeated infections on the same leaflets as well as neighboring plants, giving the plants a fully burned appearance. Temperatures ranging between 20 and 28°C, free water on the leaf surface, and high relative humidity are the favorable environmental conditions for disease development.

Control measures against rust disease in peanut include cultural practices, chemical control, and planting of resistant peanut cultivars. The volunteer peanut plants should be eradicated as far as possible because they can support perpetuation of the pathogen across seasons and locations. Sowing times should be adjusted to avoid infection of the crop from outside spore sources and to avoid the environmental conditions conducive to disease build-up. Several chemicals have been proven to be effective in reducing rust severity in peanut. A lot of rust-resistant germplasm sources in cultivated peanut and wild *Arachis* species have been identified. Marked progress in developing rust-resistant high-yielding cultivars has been made in several countries. Further research efforts are needed for integrating rust resistance with resistance to other major diseases, especially the early and late leafspots.

14.2.1.4 Web Blotch

Web blotch of peanut is caused by *Phoma arachidicola* Marasas, Pauer & Boerema. The teleomorphs of the pathogen is *Didymella arachidicola* (Chock.) Taber, Pettit & Philey. Web blotch is reported in most temperate peanut producing regions in the world, including South Africa,

South America, and North Asia. In China, this disease occurs in the north and northeastern producing zones where the weather in the late growth stage is generally cool and moisture is high (Yu 2011). However, the disease can also occur in central parts of China if the environmental factors are conducive. The first signs of web blotch are small, dark brown, or tan blotches or netlike brown lesions on the adaxial surface of leaves. The small blotches can then become numerous and enlarge to form dark brown, roughly circular lesions. When the lesions mature, they appear darker with dull and rough surfaces, and in heavily diseased fields all leaves can be in dark in appearance and serious defoliation can be seen (Liao 2012). Yield losses of 10% due to web blotch are usual in the temperate producing regions, while yield losses up to 50% have been reported in some extreme cases.

Peanut web blotch is an airborne fungal disease. It is generally severe under cool and moist conditions. Prolonged leaf wetness periods at temperatures below 21°C favor the disease development. The pathogen survives in infected crop debris or on volunteer peanut plants. Pycnidia can develop on fallen leaves and provide the initial inoculum to infect subsequent peanut crops. No other host species has been reported for *P. arachidicola.*

Control measures for peanut web blotch include cultural, chemical approaches and the use of resistant peanut cultivars. Crop rotation and management of infected crop debris and volunteer peanut plants are normally needed. Spraying fungicides is useful in reducing disease severity. In most cases, leafspots usually appear before web blotch, and chemical spraying for controlling leafspots may prevent web blotch to some extent. Peanut germplasm accessions with resistance to web blotch have been identified, and moderately resistant commercial cultivars have been developed (Liao 2014).

14.2.1.5 Stem Rot

Stem rot of peanut is caused by *Sclerotium rolfsii* Sacc. This soilborne fungal disease is also called white mold, southern blight, and southern stem rot. It occurs in all the major peanut growing regions in the world (Deepthi and Reddy 2013). The disease usually occurs in the middle and late growth stages of peanut. The pathogen infects the base of peanut stem or branches near the soil surface. The infected tissues appear dark brown, and necrotic lesions can be seen within a few days. When the temperature and moisture are conducive, quick spreading of *S. rolfsii* mycelium from the base of infected peanut plants and even from detached tissues such as fallen leaves can form wide white sheaths on the soil surface and spread to adjacent plants. Yellowing and wilting of lateral branch, main stem, and entire plant can be seen in infected plants. Infected peanut pods are usually rotted. A large number of spherical sclerotia ranging from 0.5 to 2.0 mm in diameter are produced from the mycelia, first being white and consequently becoming dark brown (Liao 2012). Yield losses due to stem rot range from 10% to 30% in most places, while over 80% in heavily diseased sites has been recorded (Backman and Brenneman 1997).

Peanut stem rot is a typical soilborne fungal disease. The pathogenic fungus has a wide host range of over 200 plant species, making it difficult to eradicate. Considerable diversity of *S. rolfsii* in growth rate, virulence, mycelial compatibility, and sclerotial production has been found, and the isolates with high level pathogenicity can cause severe symptoms in all available peanut genotypes (Yan et al. 2021). The pathogen overwinters with sclerotia or mycelia in soil and crop debris. The sclerotia, capable of surviving for many years near the soil surface, serve as the main sources of primary inoculum in the field. The pathogen can also be transmitted by contaminated peanut pods or kernels, irrigation water, insects, and even field operational tools.

Control measures for peanut stem rot include cultural, chemical approaches and planting of resistant cultivars. Crop rotation, weed control, and deep plowing can somehow reduce the inoculum buildup. Use of certain fungicides for seed treatment and field spraying can reduce disease severity (Akgul et al. 2011; Backman and Brennenman 1997; Grichar 1995). As investigations in the United States indicated that spraying Benomyl for controlling leafspots could increase the severity of stem rot because Benomyl reduced the levels of *Trichoderma* spp, which could suppress the population of *S. rolfsii*, the choice of fungicides for controlling foliar and soilborne diseases should be coordinated. To date, only moderately resistant peanut germplasm lines and cultivars have been reported, and considerable variation in resistance has been found. High level resistance to stem rot exists in the wild *Arachis* species for future transfer into high yielding cultivated genotypes.

14.2.1.6 Pod Rot

Pod rot of peanut is caused by several soilborne pathogenic fungi causing similar symptoms including *Pythium myriotylum* Drechs. *Rhizoctonia solani* Kuhn, *Fusarium solani* (Mart.) Sacc., *Fusarium oxyspourm* Schlechtend, Snyder & Hans, and *Neocosmospora vasinfecta* (Atkinson) Smith. Pod rot is reported in most peanut producing areas in the world, with variable incidences and yield losses. Symptoms of pod rot appear in the late growth stage of peanut. Differences in pathogenesis or pod deteriorating processes in peanut among different rot-causing pathogens have not been well recorded. Normally, tan to brown lesions on pod surfaces can be seen after infection (Liao 2012). Then the lesions enlarge and merge together to cover the whole pod surface, and the shell tissues become deteriorated. The symptoms of the rotted pods can be a dry decay or a greasily wet decay. The color of rotted pods can become deep tan, dark brown, or black depending on pathogen species, soil type, moisture, and rotting process. In most cases, the infected plants do not show obvious above-ground symptoms when infected, except for some deep red or purple pigments on stem/branches, which can be seen in some peanut genotypes. Moreover, the root system of plants infected in pods is also generally not affected. A high rate of pod decay can result in a darker green plant canopy and even prolonged flowering, probably due to the declined transfer of carbohydrates into pods.

Peanut pod rot is a typical soilborne disease. The complicated interactions among the different causing pathogenic fungi are not well investigated. The disease can be caused by single or mixtures of the pathogenic fungi. The pathogens of pod rot are believed to easily survive in soils but their buildup is highly influenced by environmental factors including soil nutrition status. The severity of peanut pod rot is also highly affected by soil type, peanut genotype, temperature, moisture, and management operations such as irrigation. Nematodes are believed to increase pod rot incidence somehow (Csinos and Bell 1997).

Cultural measures including necessary rotation, prevention of water logging, and application of gypsum are useful in reducing pod rot severity. Irrigation of cool underground water during warm weather should be avoided. No cost-effective chemical control option is available against peanut pod rot. Varietal difference in peanut to pod rot is wide, and planting resistant peanut cultivars is prioritized in integrated control. A panel of Spanish type lines including Zhonghua 6 with resistance to bacterial wilt are resistant to pod rot (Liao 2014). No specialization between peanut genotypes and the rot-causing pathogen species is observed.

14.2.1.7 Other Fungal Diseases

14.2.1.7.1 Aspergillus Infection

Peanut pods and seeds are easily infected by *Aspergillus* fungi including *A. flavus* Link ex Fries, *A. parasiticus* Speare, and *A. niger* van Tieghem. Both the infection of *Aspergillus* fungi and the consequent aflatoxin contamination in peanuts and peanut products have already attracted extensive efforts in terms of scientific research and industrial management (see Chapter 16 of this book). Moreover, *Aspergillus* infection of peanut plant and seed can cause seedling diseases. High levels of infection by *A. flavus* in peanut seed may result in rotting of seeds before emergence and the death of seedlings, known as "yellow mold." Under field conditions, brown necrotic lesions with sporulating *A. flavus* can be seen on the cotyledons, radicles, and hypocotyls of ungerminated and germinated seeds. The same symptoms can also be present in emerged seedlings, resulting in stunt and chlorotic plants with weakly developed roots. The incidence of yellow mold can be high in peanut seeds harvested from fields stressed by pre-harvest drought. Varietal resistance in pod shell and seed to infection of *A. flavus* can reduce yellow mold incidence. Another *Aspergillus* infection of peanut seed and young seedling is that by *A. niger* resulting in pre-emergence rot and collar rot in young plant (Liao 2012). The incidence of collar rot is affected by the vigor of seeds, temperature, and susceptibility of peanut cultivars. Peanut seeds sown under dry and cool conditions can result in increased incidences. Varietal difference has been observed for resistance to *A. niger* infection, indicating there is potential for genetic enhancement.

Peanut smut is a special fungal disease confined to South America to date, caused by *Thecaphora frezii* Carranza & Lindquist. This disease causes hypertrophy of pod tissues, and seeds colonized with teliospores have a smutted mass appearance (de Blas et al. 2021). Yield losses due to smut can be as high as 35%. The problem was first observed in some wild *Arachis* materials in Brazil but has also been recently reported in wild species in Bolivia. Peanut smut is difficult to control as both cultural management and chemical treatment have not been effective. In Argentina, the disease is found in all peanut growing areas. Traditional commercial peanut cultivars in South America were all susceptible. However, resistance has been identified in some wild *Arachis* species and has even transferred to elite cultivars (de Blas et al. 2021).

14.2.2 Bacterial Diseases

14.2.2.1 Bacterial Wilt

Bacterial wilt in peanut is caused by *Ralstonia pseudosolanacearum* Safni, a pathogen that has undergone several name changes since its initial identification as *Bacillus solanacearum* by E. F. Smith in 1896. It was later known as *Pseudomonas solanacearum* and *Ralstonia solanacearum* before being renamed in 2014 due to advances in DNA profiling. Despite these changes, *R. solanacearum* is still a commonly used term in the research community. Bacterial wilt is the most significant bacterial disease affecting peanut, with initial reports dating back to Indonesia in 1905. While it has been a minor concern in the US states like North Carolina and Georgia for several decades, it remains prevalent in parts of Asia and Africa, including China, Indonesia, Malaysia, Uganda, and Vietnam (Mehan et al. 1994). The peanut yield loss due to bacterial wilt in some hotspots can be over 80% in susceptible cultivars. Interestingly, this disease does not pose a significant problem in India, even where peanut planting is widespread and local strains of *R. solanacearum* affect many other crops. The disease manifests throughout the peanut growing season, beginning when the bacteria enter through natural openings or wounds in the roots, subsequently invading the water-conducting tissues. This results in bacterial multiplication, blockage, and disruption of water transport, leading to rapid wilting and potentially the death of the plant or its branches. The affected plants typically exhibit dark discoloration in the xylem and pith, and when stems are cut and placed in water, a milky white ooze of bacteria is evident (Liao 2012).

R. pseudosolanacearum is a soilborne organism with a broad host range, affecting over 400 plant species across more than 50 families. Strains from southern China are notably more virulent compared to those from the north. The bacterium thrives at temperatures between 30°C and 35°C, and its survival is aided by high soil moisture and acidified soils that are common in tropical and subtropical regions. While the bacteria can be seed-transmitted at a rate of 4%–15%, the viability of these bacteria significantly decreases when seed moisture content drops below 9%, making seed-borne transmission less likely.

Bacterial wilt of peanut can be effectively controlled by comprehensive approaches including proper crop rotation, crop residual sanitary, soil treatment, biocontrol, and planting resistant cultivars. Planting highly wilt-resistant peanut cultivars is the most practical control method in all diseased regions.

14.2.2.2 Other Bacterial Diseases

Even though some bacterial diseases have been found in peanut, only the bacterial leafspot caused by a *Pseudomonas* species (unidentified) is considered as a key disease in India, Vietnam, and Zimbabwe. After infection by the pathogen, small, circular to irregular and light brown lesions can occur on lower leaves of young plants. The lesions can enlarge and may develop chlorotic halos. Under favorable conditions for disease development, the lesions coalesce, and the leaflets become chlorotic and shed prematurely (Subrahmanyam 1997). The disease is sporadically distributed with limited effect on peanut production.

14.2.3 Viral Diseases

14.2.3.1 Tomato Spotted Wilt Virus (TSWV)

Tomato spotted wilt caused by tomato spotted wilt virus (TSWV) in peanut was first reported in Brazil and has since been reported in the United States, South Africa, India, and Australia (Culbreath et al. 2003). Peanut bud necrosis virus (PBNV) was once considered a strain of TSWV, but was later classified as a distinct virus in the *Tospovirus* genus (Reddy et al. 1992). TSWV causes a wide array of symptoms on peanut, including concentric ringspots, various patterns of chlorosis of leaflets, plant stunting, small pods and kernels, and reddish discoloration and cracking of seed coats. Roots of affected plants typically show varying degrees of necrosis, which can result in whole plant death. The effects of the disease on peanut plants and pods are correlated with the time at which initial symptoms appear. Early infection and disease development with symptoms normally result in higher yield reduction. The disease has been reported to cause more serious yield losses in the southeastern United States (Culbreath et al. 1992a, 1992b). TSWV and its thrip vectors, *Frankliniella fusca* and *F. occidentalis*, present a difficult and complicated challenge in both epidemiological investigation and disease management. Control measures against TSWV in peanut include use of resistant peanut cultivars, chemical spraying, and cultural practices (Brown et al. 1996). Even though the virus is transmitted by thrips, simply controlling the vector typically has not resulted in control of the disease. No single measure can currently provide adequate control against the disease where severe epidemics occur. However, interdisciplinary investigations have resulted in development of integrated management systems which have been successfully deployed in many regions for minimizing losses to this disease in the United States.

14.2.3.2 Peanut Stripe Virus (PStV)

The peanut stripe virus (PStV) is reported in China and Southeast Asia and also in the United States. The symptoms caused by the most widely distributed PStV variants are irregular green blotches on young leaflets that persist as the leaflets age. The dominate PStV isolate in China can cause a mild mottle symptom. PStV is transmitted by many aphid species in a nonpersistent manner. Different variants of PStV have different aphid transmission frequencies. The seed transmission frequency can be as high as 37% in inoculated plants even it is normally less than 5% under natural infection. The most important control measure against PStV is to plant virus-free peanut seeds. Peanut cultivars resistant to aphid may reduce field transmission and disease incidence. There is possibility of transferring viral coat protein genes into the cultivated peanut for generating PStV-resistant genetic stocks (Higgins et al. 2004).

14.2.3.3 Peanut Stunt Virus (PSV)

The peanut stunt virus (PSV) is important in China and some African countries such as Sudan. Yield losses due to PSV infection can be as high as 75% under extreme conditions. Normally, PSV can cause severe dwarfting of entire plants or some branches. Shortening of petioles, reduction of leaflet size, chlorosis, and malformation can be observed. Early infection by PSV can result in reduced pod number and pod size (Liao 2012). PSV is transmitted by three aphid species, *Aphis craccivora*, *Aphid spiraecola*, and *Myzus persicae*, in a nonpersistent manner. The virus could be seed-transmitted but the frequency is very low compared to that in most other viral pathogens. As some forage legumes (such as white clover) are the primary inoculum in field, peanut should not be planted in field close to such legumes. Difference in tolerance to PSV is observed among different peanut cultivars.

14.2.3.4 Groundnut Rosette Virus (GRV)

Rosette disease is a serious problem in peanut in certain African countries south of the Sahara. The disease is caused by a complex of groundnut rosette virus (GRV) and a satellite RNA. Three types of rosette disease (chlorotic, green, and mosaic) are observed in peanut. The chlorotic rosette occurs throughout Africa south of the Sahara. If young plants are infected, they will produce progressively smaller, chlorotic, curled, and distorted leaflets. If older plants are infected, symptoms may be restricted to a few branches or to the apical portion of plant. Early infection causes stunted plants and severe reduction in the number and size of pods. The green rosette occurs in western Africa and Uganda. Young leaflets show mild chlorotic mottling and isolated flecks. Plants infected early are darker green and severely stunted. Mosaic rosette occurs only in eastern and central Africa. Young leaflets show conspicuous mosaic symptoms. The above different forms of rosette are determined by variants of satellite RNA. The viruses associated with all three types of rosette disease are transmitted by aphid (*A. craccivora*) in a persistent or circulative manner. Peanut plants are considered to be the main source of inoculum, but the virus is not transmitted by seeds. Rosette disease is controlled by cultural practices including destroying all volunteer and unharvested infected plants, planting early in season and at high seeding rate, maintaining plant stands, and applying insecticides when aphids are present. Rosette-resistant peanut germplasm lines have been identified, and improved resistant cultivars can be used in disease management.

14.2.3.5 Other Viral Diseases

In addition to the above viral diseases, several other viruses can also affect peanut in different locations, including peanut mottle virus (PMV), cucumber mosaic virus (CMV), peanut clump virus (PCV), and bud necrosis (PBN) (Ghanekar et al. 1979; Reddy et al. 1983). Detailed information on these minor viruses in peanut can be referred to in related books or papers (Demski and Reddy, 1997; Reddy 1991).

14.2.4 Nematode Diseases

The major nematodes affecting peanut are *Meloidogyne arenaria* (Neal) Chitwood, *Meloidogyne hapla* Chitwood, and *Meloidogyne javanica* (Treub.) Chitwood; they are also known as peanut root-knot nematode, northern root-knot nematode, and javanese root-knot nematode, respectively. These nematodes are widely distributed around the world (McSorley et al. 1992), while *M. hapla* mostly exists in areas around 35°N latitude in the northern hemisphere and the remaining two species exist in the vast regions between 35°S and 35°N. Peanut producing areas in the southeastern United States and coastal northern China are heavily affected by nematodes. Peanut plants infected by root-knot nematodes commonly develop galls (with various sizes) which internally enlarge roots. Pegs and pods can also be infected, and develop galls consequently. The galls caused by *M. halpa* are generally smaller than those caused by *M. arenaria*. Infection with *M. halpa* can result in a dense and bushy root system. Development of root systems is significantly reduced in infected peanut plants, with various stunting and chlorosis. Yield losses due to nematode injury are reported to range from 20% to 90% (Kokalis-Burelle and Rodriguez-Kabana 1997). Infection of nematodes is believed to aggravate the severity of other soilborne diseases such as stem rot and pod rot.

Under field conditions, nematodes can overwinter by eggs and larva. Crop debris containing nematode galls may be spread by farming operations or running water, and can produce widely distributed infestation sites. Infected peanut seeds are source of infection and can help dissemination over long distances. Nematodes can be distributed through their own movement but remain mostly localized.

Control measures against nematodes in peanut production include rational rotation with less sensitive crop species, use of nematocides, and planting resistant peanut cultivars. While variability in response to nematodes in the cultivated peanuts has been observed, high-level resistance has been found in some wild *Arachis* species and has been transferred into the cultigens.

14.2.5 Insects

Worldwide, there are several kinds of insect pests affecting peanut, including foliage feeders and defoliators, sap suckers, and root and pod feeders, with wide economic importance. The damage by insect pests in peanut is highly variable across seasons and locations depending on the complex interactions among insects, peanut cultivars, and environmental factors.

14.2.5.1 Foliage Feeders

Several insect species have been reported to affect peanut plants by eating the leaves, including *Spodoptera litura* Fabricius, *Spodoptera exigua* Hübner, *Spodoptera littoralis* Boisduval, and *Helicoverpa armigera* Hübner. The adults of these foliage feeders are light-brown moths and lay eggs mostly on the upper leaf surfaces of peanut. The larvae can eat leaves, young tissues of bulbs, and even pegs. In particular, *S. litura* is problematic during the peg initiation stage, pod development stage, and maturation stages of crop growth (Singh and Sachan 1992). Other key foliage feeders, *Amsacta albistriga* Walker and *Amsacta moori* Butler, are the most common hairy caterpillars that target peanuts. They are voracious feeders, and can cause significant yield reduction in some regions. Peanut leaf miner, *Aproaerema modicella* Deventer is a usual pest of peanuts in South and South-East Asian countries, with the young larvae digging into the leaves of hatcheries, depositing single gleaming white eggs on the underside of the leaflets. In addition, other foliage feeders affecting peanut are *Sympiezomias velatus* Chevrolat, *Epicauta gorhami* Marseul, *Chondracris rosea* De Geer, and *Anomis flava* Fabricius (Xu 2008). Most of the above foliage feeders can eat many other plant species, which favor their survival and reproduction. However, peanut yield losses due to damage by foliage feeders are highly variable in different seasons and locations, and could be over 50% in extreme cases (Islam et al. 1983).

Technical measures for controlling foliage feeders in peanut include agronomic, physical, biological, and chemical approaches. Selecting peanut cultivars relatively resistant to insects, deeply plowing soil before sowing, and intercropping crops with higher plant heights such as corn or sorghum for reducing egg-laying by adult moths on peanut leaves are recommended agronomic measures. Sunflower and castor plants can allure *Spodoptera* spp and thus may be sown to collect egg masses and larval instars as trap crops (Zhou 2009). Removing the alternative hosts and weeds is effective in reducing insect population. Physical methods such as light traps or pheromone traps can be used to collect moths of defoliators. The migration of larvae of red hairy caterpillar can be reduced by digging deep trenches in fields.

Several biological agents are useful with varied efficiency, including egg-larval parasitoid and larval parasitoid species *Apantele ruficrus*, *A. kazak*, *Cotesia marginiventris*, *Campoletes chloridae*, and *Hyposoter didymator* (Braune 1982). *Trichogramma* can parasitize on eggs and young larvae of red hairy caterpillar. While several pesticides are useful in controlling foliage feeders, most defoliators have gained tolerance to most of the available commercialized pesticides (Ramakrishnan et al. 1984; Naeem Abbas et al. 2014), making chemical control increasingly difficult.

14.2.5.2 Sap Sucking Pests

The major sap-sucking insects affecting peanut are peanut aphid (*Aphis craccivora* Koch), thrips (*Scirtothrips dorsalis* Hood, *Megalurothrips distalis* Kanrny, *Thrips palmi* Karny, *Frankliniella schultzei* Trybom), and jassids (*Empoasca vitis* Gothe, *E. flavescens* Fabricius, *E. notate* Melichar). Aphid is one of the most popular and injurious pests of peanut with a worldwide distribution. *A. craccivora* can cause damage directly by injuring the plant growth, or indirectly by transmission of viruses to infested plants. During the seedling, vegetative, and reproductive stages, aphids tend to feed on leaves, flowers, shoots, pegs, and even immature pods. The highest losses in yield due to direct damage by aphid may be up to 10% (Xu 2008). The spread of plant viruses such as peanut stripe (PStV), peanut mottle (PMV), and peanut stunt (PSV) by aphid can cause serious indirect damage. Thrips are small in size (less than 2 mm long) and are slim insects having fringed wings that live in flowers and folded leaflets of peanut plants. Adults and larvae are mobile, and adults have wings of their own (Liao 2012). The sap is sucked from the surface of the leaflets by nymphs and adults, which causes white patches on the upper surface of the leaves and necrotic patches on the lower surface. As the leaflets expand, they split as newly developing leaflets are distorted due to formation of patchy necrotic areas that puncture eventually. Peanut seedlings are mostly injured by thrips. Thrips are also vectors for many viruses like PBNV and TSWV. Jassids (leafhoppers) are another important foliage-sucking pest of peanut and act as limiting factors in the successful cultivation of peanut. Both nymphs and adults suck the sap from the tender leaf and mostly from the lower surface of the leaflet, causing whitening of the veins, yellowing in the form of patches of the leaflets, leaf curling and necrosis (necrosis of leaf tips in V shape), stunted growth, and eventually death of plants. Jassids also act as a vector of TSWV and other viruses (Demski and Reddy 1997; Singh et al. 1992).

Technical measures for controlling sap-sucking pests in peanut are similar to other insects. Early sowing enables peanut plants to initiate flowering before aphids' arrival, and dense sowing provides a barrier to aphids' entry into the field (Mayeux 1984). Field sanitary measures are crucial within crops and between seasons to prevent the transmission of viruses by aphids. Virus-infected plant materials should be eliminated after harvest, and volunteer plants or weeds that harbor viruses should be destroyed. Thrip populations in peanut field can be substantially reduced by several cultural practices such as manipulating sowing dates to avoid peak thrips dispersal and during the susceptible seedling period (McKeown et al. 2001; Culbreath et al. 2010). Several insecticides, including chlorinated hydrocarbons, organophosphates, carbamates, and pyrethroids, have been used for control of aphids. Novel insecticides such as diamides and spinosyns have been effective against thrips (Marasigan et al. 2016, 2018). Seed treatment with pesticides can protect for a month against sap-sucking pests. However, chemical pesticides should not be used indiscriminately but should be used depending on the economic threshold level of insect populations. Release of the reduviid predators such as *Rhynocoris marginatus* (Sahayaraj and Martin 2003), *R. kumraii* (Sahayaraj and Ravi 2007), and *Chrysoperla zastrowi sillemi*, a chrysopid predator (Baskaran and Rajavel 2013), and spraying fungus *Verticillium lecanii* can reduce populations of aphids (Sahayaraj and Namachivayam 2011). Considerable differences in resistance or tolerance to thrips among peanut genotypes have been observed.

14.2.5.3 Root and Pod Feeders

The important root and pod feeders affecting peanut are white grub species (including *Lachnosterna* (= *Holotrichia*) *consanguinea* Blanch, *L. oblita* Faldermann, *F. parallela* Motschulsky, *L. corpulenta* Motschulsky) and termites (mostly *Microtermes* spp. and *Odontotermes* spp.). The widely distributed white grubs are important soil-inhabiting polyphagous pests of peanut. Their adults are normally dark brown in color and emerge out of the soil within a few days after rainfall. The eggs are white and round in shape, while the larvae are whitish yellow in color, fleshy, and C-shaped. In their second, third, and fourth instar larval stages, the young grubs all feed on various organic matter and fine rootlets, while the mature grubs can feed on both peanut roots and pods. In heavily infested peanut fields, wide patches of dead plants can be found, and the remaining plants are often stunted and wilted. Yield losses due to white grub injury in endemic areas can vary from 20% to 80%. In most regions of the world, peanut plants are also harmed by several termites, mostly *Microtermes* spp. and *Odontotermes* spp. The termites burrow within the peanut root and stem and make holes in the pods, and the affected pods become more vulnerable to *Aspergillus* infection in the late growth stage (Xu 2008).

Control of white grubs and other soil-inhabiting insects has proven difficult in peanut production. Agronomic, physical, biological, and chemical approaches are recommended. Ploughing exposes the pupae to scorching solar radiation, and predation by birds is useful. Use of light traps and pheromone traps is practical. Planting or using trap crops such as castor is also useful. Unfortunately, desirable resistant peanut genotypes against white grubs have not been discovered.

14.3 GERMPLASM DISCOVERY, GENETICS, AND BREEDING FOR RESISTANCE TO MAJOR BIOTIC STRESSES IN PEANUT

Host-plant resistance or tolerance in peanut, as in any other crop, is one of the most important components for integrated management of various biotic stresses including diseases and insect pests (Bhat et al. 2022; Kona et al. 2024). The earliest discovery of and breeding for biotic stress resistance in peanut can be traced back to the early 1920s for resistance to bacterial wilt conduced in Indonesia (Buddenhagen and Kelman 1964). Since then, many research efforts have been made in resistance enhancement for the important diseases including leafspots, rust, bacterial wilt, TSWV, GRV, and nematodes with marked progress(Kona et al. 2024), although resistance to insect pests has been relatively less addressed. In general, successful breakthroughs in resistance enhancement to most biotic stresses, like any other traits, largely depends on variation of resistance, access of resistance sources/genetic stocks, genetic natures and mechanisms, screening techniques and facilities, and coordinated integration of resistance with other key traits.

The genetic resources in peanut and their evolutionary structures are well summarized in this book (see Chapter 2 of this book), and its genetic diversity is grouped into different gene pools as suggested by Singh and Simpson (1994). In total, more than 100,000 peanut accessions are conserved in various national and global genebanks, including those in Argentina, Brazil, China, India, the United States, and the International Crops Research Institute for the Semi-Arid Tropics (ICRISAT) (Barkely et al., 2016). Several peanut core and mini-core collections have been selected to represent the genetic diversities and simplify their characterization and utilization (Barkley et al. 2016). The core or mini-core collections have been extensively assessed for stress resistance or tolerance. However, because it derived just from a single hybridization between diploid ancestors, the cultivated peanut possesses a narrow genetic base and genetic variability in response to biotic stresses. Fortunately, in addition to the large cultivated peanut germplasm collections, various genebanks and nurseries around the world also maintain a large number of wild relative accessions in genus *Arachis*, consisting of 80 species belonging to nine sections (Valls and Simpson 2005). The wild *Arachis* accessions are well proven to harbor more diversified resistances to many biotic and abiotic stresses. Through continuous efforts, elite resistances have been successfully transferred from some wild species into cultivated peanuts, showing a great potential in terms of enhancing more resistances in cultivated peanuts, especially for diseases or insect pests against which resistances in the cultigen are absent or weak.

However, several factors such as fertility obstacles due to species incompatibilities and ploidy level differences, the association of stress resistances with traits agronomically unadapted, and difficulty in monitoring introgressed segments have hampered the transfer of genes from wild species. Many methods have been used for introgressing wild *Arachis* genes into cultivated peanut with varied success, in which the hexaploid and tetraploid routes have been most successful. In the hexaploid route, a triploid hybrid derived from a cross between the cultivated allotetraploid species and the diploid wild species is colchicine treated to produce a hexaploid plant, followed by generations of selfing to select tetraploid plants with resistance to multiple diseases (Stalker et al. 1979; Stalker and Beute 1993; Reddy et al. 1996). To overcome the problems of barriers between the cultivated species and the wild species and to get rid of undesirable gene blocks, novel genetic engineering techniques are expected to be used in peanut improvement.

The techniques for resistance screening and characterization have been well established in peanut for a few diseases such as rust, late leafspot, early leafspot, bacterial wilt, TSWV, GRV, nematodes, *Aspergillus* infection and aflatoxin contamination. The recent progress in developing molecular markers and marker-assisted selection approaches has accelerated the genetic enhancement of resistance to biotic stresses in peanut.

14.3.1 Resistance to Fungal Foliar Diseases

14.3.1.1 Early and Late Leafspots

Early leafspot (ELS) and late leafspot (LLS) are the two most widely distributed foliar diseases in peanut across the world, and have attracted more research efforts on resistance enhancement worldwide compared to other diseases. As early as the 1940s, research efforts were initiated in the United States for identifying peanut genotypes with resistance to leafspots (Sun 1998). By the late 1970s, 21 accessions resistant to ELS and 6 resistant to LLS had been identified in the United States (McDonald et al. 1985). Since then, more systematic screening and breeding for leafspot resistance have been implemented. A lot of peanut germplasm accessions were screened against ELS, LLS, and rust at the International Crops Research Institute for the Semi-Arid Tropics (ICRISAT) by using natural conditions and artificial facilities in Hyderabad, India, as well as in other collaborative networks worldwide (McDonald et al. 1985; Barkley et al. 2016). Extensive resistance screening against the foliar diseases has also been conducted in China, India, and African countries (Liao 2014; Yu 2011). Relatively, more screening work has been done for LLS than for ELS because LLS has been more serious, with wider geographic distribution and favorable field testing conditions.

To date, more than 100 cultivated peanut accessions have been identified as resistant to ELS and LLS, and some of them are also resistant to rust. Most of these accessions showed stable resistance across different regions, while some were less stable, probably due to their sensitivity to photoperiod variation (Sun 1998; Yu 2011). It is generally accepted that high-level resistance to leafspot in the cultivated peanut is mostly found in the Valencia type (subspecies *fastigiata* var *fastigiata*) lines collected from South

America (Singh et al. 1997); in particular, the Valencia-type lines (such as PI259747) originating from the region of Tarapoto (Peru) possess a high degree of resistance to rust and moderate levels to LLS (Singh et al.1997). However, their utilization in groundnut breeding has been very limited because of many undesirable attributes, including thick shell, low productivity, poor adaptation, late maturity, and highly reticulated and constricted pods, which are commercially unacceptable Anderson et al. 1993). Later on, screening of the germplasm originating from secondary centers of peanut diversity resulted in the identification of some resistant sources with better agronomic backgrounds. But even then, low productivity is the major constraint for diverse use (Singh et al. 1997; Yu 2011). With the development of mini core in peanut (Upadhyaya et al. 2002), 184 accessions were screened for LLS and rust under artificial epiphytotic conditions (Sudini et al. 2015). The accessions showed significant variation in disease response, and 53 accessions were moderately resistant to LLS in the mini core. Several lines including ICGs 4389, 6993, 11426, 4746, 6022, and 11088 were selected as superior accessions in terms of disease resistance and yield potential.

Meanwhile, many wild *Arachis* species have been identified as highly resistant or even immune to LLS, including *A. diogoi*, *A. cardenasii*, *A. glabrata*, *A. stenosperma*, *A. repens*, *A. appressipila*, *A. paraguariensis*, *A. villosulicarpa*, and *A. hagenbeckii* (Subrahmanyam et al. 1985). Some of these resistant species belong to the *Arachis* section that is cross-compatible with the cultivated peanut, and some belong to other cross-incompatible sections. For ELS, several resistance sources were identified in some wild species including *A. stenosperma* and *A. diogoi* (Foster et al. 1981).

Use of the diploid *A. cardenasii* has resulted in a lot of breeding lines with higher levels of resistance to LLS than is found in cultivated peanuts. Efforts in utilizing the diploid *Arachis* species to transfer LLS resistance to cultivated peanut through developing synthetic allotetraploids at ICRISAT were successful, and some tetraploid accessions obtained were highly resistant (Mallikarjuna et al. 2011). In transferring resistance from the wild species, some amphidiploids were also developed using *A. duranensis* and *A. ipaensis* (Fávero et al. 2006) and *A. gregoryi* and *A. linearifolium* in the United States (Simpson and Starr 2001; Simpson et al. 2003). ICRISAT has also developed many tetraploids and amphidiploids peanuts using wild *Arachis* species, and some synthetic amphidiploids such as ISATGR 278-18 (*A. duranesis* × *A. batizocoi)* and ISATGR 5B *(A. magna* × *A. batizocoi*) developed have been used in backcross breeding programs (Kumari et al. 2014).

Resistance to ELS or LLS is shown to be complex on a genetic basis. The resistance to ELS and LLS were reported as being controlled by recessive multiple genes, but the number of genes/loci were different in different research cases, probably due to difference in the research materials involved. Several studies revealed that the leafspot resistance was quantitative and governed by both additive and non-additive gene effects along with maternal effects (Anderson 1986; Dwivedi et al. 1993). The resistance to LLS was reported to be independent from the ELS resistance (Abdou et al. 1974). Several components including initial infection, latent period, sporulation level, size of lesions, and defoliation rate are related to resistance reaction (Green and Wynne 1986; Chiteka et al. 1988; Anderson et al. 1993; Waliyar et al. 1993, 1995), and can be used in characterizing the resistance especially under laboratory conditions. In the aspect of resistance mechanisms, wild *Arachis* species resistant to LLS in sections Erectoides, Triseminalae, Extranervosae, Rhizomatosae, and Caulorhize have small and nonsporulating lesions, whereas species in section *Arachis* have accessions either with nonsporulating lesions or with variably sporulating lesions. Frequency of infection and defoliation vary greatly within each section and species (Subrahmanyam et al. 1985).

In most breeding programs dedicated to leafspot resistance, the method for scoring resistance levels developed by ICRISAT (McDonald et al. 1985) has been frequently used. The 1–9 scoring protocol can be easily followed in field screening, and can also be used in greenhouse screening with artificial inoculation. Artificial inoculation on detached leaves has also been used in screening resistance and characterizing resistance components. For improving breeding efficiency, extensive research efforts have been made in discovering markers for LLS and rust resistance. Various peanut recombinant inbred lines (RILs), cultivars, backcross lines, and synthetic tetraploids developed from wild diploids were used for validating LLS resistance-linked markers. In a study, GM1954, GM1009, and GM1573 markers showed significant association with LLS resistance, and the RIL 78-1 was resistant to LLS and rust and recorded 30% more pod yield than GPBD 4 (control) (Sukruth et al. 2015).

Some released peanut cultivars tolerant to ELS in India and the United States are BG 3, Bailey, C-99R, CSMG 84-1, DP 1, GG 7, Florida 07, Georganic, ICGS 44, M335, ICGS 76, and M 522 (Bera et al., 2022). LLS resistant or tolerant cultivars released from the United States include C-99R, Florida 07, Florida MDR 98, Southern Runner, TUF Runner, and TM 727 (Gorbet et al. 1987). In India, ICGV 86590, ICGV 86325, ICG (FDRS) 10, Girnar 1, K 134, GBPD 4, BSR 1, R 8808, VRI (Gn) 5, CSMG 84-1, and RG 141 were released as resistant to LLS. In China, several improved peanut cultivars such as Yueyou 7, Yueyou 13, Yueyou 79, and Zhongkaihua 10 with tolerance to leafspots and rust have been developed by using some Valencia type resistant parents. These cultivars are extensively cultivated in southern parts of China. Zhonghua 12 is an LLS and rust tolerant cultivar developed by using ICGV 86699 (an interspecific hybrid derivative) as resistance source (Liao 2014). The elite cultivar named GPBD 4 (Spanish bunch type) with desirable LLS and rust resistance was released in India (Bera et al. 2022; Gowda et al. 2002) and has also been used as resistance source in breeding at least 5 relased cultivars (Bahat et al. 2024). Its resistant parent was

ICGV 86855, consisting of genes from *A. cardenasii* (due to chromosome segments transferring). This could be a good example of achievement from wide hybridization for resistance transferring. Other interspecific derivatives such as ICGVs 99006, 99013, 99004, 99003, and 99001 developed at ICRISAT with LLS resistance could also be desirable parents for use in resistance breeding (Dwivedi et al. 2002).

While marked progress has been made in genetic enhancement of resistance to leafspots in peanut, the available resistant cultivars still need to be further improved for several reasons. Firstly, the levels of leafspot resistance are not ideal, and fungicide is still needed when disease pressures are high. Secondly, most leafspot resistant cultivars, including those developed using resistant genes from wild species, possess a thick shell and low shelling ratio. Thirdly, the leafspot-resistant cultivars normally possess a longer growth period, and the late maturity seems to be correlated with the resistance. Lastly, most of the available cultivars with foliar disease resistances are highly sensitive to soil-borne diseases and aflatoxin contamination. Therefore, further breeding efforts are very much needed to address these problems.

14.3.1.2 Rust

Rust is a wide spreading foliar disease in most tropic and sub-tropic regions in the world. Efforts to identify resistance sources against rust can be traced back to the 1940s. A Valencia type landrace named Tarapoto, collected from Peru, was the first peanut genotype identified as resistant to rust, as well as to LLS. This germplasm line was named aPI259747 after it was introduced into the United States (Hammons 1987). After peanut research was initiated at ICRISAT in 1976, more than 12,000 peanut accessions had been screened against rust by the late 1980s. More than 100 lines were identified as rust-resistant, among which at least 30 lines are also resistant to LLS. In China, screening for rust resistance was extensively conducted in the 1980s, and more than 200 lines (including some introduced interspecific derivatives) were identified as resistant (Sun 1998). Most of the rust-resistant accessions belong to the Valencia type, and PI 298115 is the only resistant line belonging to the Virginia type (Sun 1998). Moderately resistant landraces belonging to the Chinese Dragon type were identified (Liao 2017).

Through continuous observation, many wild *Arachis* species, such as *A. batizocoi*, *A. correntina*, *A. cardenasii*, *A. duranensis*, *A. diogoi*, *A. pusilla*, and *A. villosa*, have been found to be highly resistant to rust (Abdou et al. 1974; Subrahmanyam et al. 1982; Subrahmanyam et al. 1985; Singh et al. 1997). Meanwhile, most of the wild species in sections Erectoides, Triseminalae, Extranervosae, and Rhizomatosae even show immunity to rust with no recognizable symptoms of the disease appearing even after an incubation period of 40 days (Ramanatha Rao 1987).

Genetically, rust resistance in cultivated peanut was regarded as controlled by a few recessive genes with additive effects (Sun 1998; Yu 2011). Resistance was also reported to have either recessive inheritance or dominant with duplicate recessive or partial dominant, or polygenic inheritance (Bera et al. 2022). There is tolerance to rust in peanut which is also considered as "slow rusting" in nature controlled by multiple genes. On the other hand, the rust resistance transferred from wild species was found to be governed by major dominant gene(s), indicating that genetic natures and mechanisms are different between the cultigen and wild species for response to rust. In several studies for characterizing rust resistance in peanuts, variation in resistance components (including latent period, infection frequency, size of lesions, sporulation, and leaf area damaged) was found among genotypes with similar resistance levels, indicating genetic diversity of resistance to rust. Some sources of rust resistance governed by major genes are relatively easy to transfer into agronomically adaptable and desirable genotypes. One example is GPBD 4, a popular rust-resistant cultivar in India, which was successfully developed using ICGV 86855, an interspecific derivative derived from *A. hypogaea* × *A. cardenasii* (Bera et al. 2022).

Much effort has been made toward developing molecular markers for rust resistance using a linkage analyzing approach. An F2 population derived from Yuanza 9102 (a rust-susceptible line) and ICGV86699 (a rust-resistant cultivar) were applied to screen AFLP markers linked to rust resistance (Hou et al. 2007). By analysis of an F2-segregating population derived from the cross of ICGV86699 and Zhonghua 5, three AFLP markers linked to late leaf spot resistance were identified (Xia et al. 2007). The first detailed study for rust resistance QTLs conducted in cultivated peanut was based on a partial genetic map comprising 56 SSR loci for the TAG24 × GPBD4 recombinant inbred line (Khedikar et al. 2010). This study reported 12 QTLs for rust (explaining 1.7%–55.2% phenotypic variation). The SSR marker tightly linked to the major QTL (IPAHM103; QTLrust01) was then validated among a diverse set of genotypes as well as another population (Sarvamangala et al. 2011). This marker has been used for introgressing the major QTL for rust in peanut breeding programs (Varshney et al. 2014). Important efforts have been made on transferring the genomic region on linkage group AhXV governing rust resistance from GPBD 4 to several rust-sensitive varieties used in India including ICGV 91114, JL 24, and TAG 24 (Bhat et al 2022; Varshney et al. 2014).

In breeding for rust resistance, some resistant interspecific hybrids have been selected from progenies of *A. hypogaea* × amphiploid (*A. batizocoi* × *A. duranensis*) and (*A. correntina* ×*A. batizocoi*) at ICRISAT (Singh 1987). Using the backcross method of triploid, hexaploid, and autotetraploid route, interspecific hybrids ICGV 86699 and ICGV 87165 were developed at ICRISAT. ICGV 86699 is an interspecific derivative from the cross (*A. batizocoi* × *A. duranensis*) × *A. hypogaea* (originally developed at North Carolina State University) with high level resistance to rust and leafspots, groundnut bud necrosis (PBN), and peanut mottle virus (PMV) (Reddy et al. 1996). ICGV 87165 is a high-yielding foliar disease resistant interspecific derivative obtained from *A. hypogaea* × *A. cardenasii*. An interspecific derivative, ICGV-SM 86715, was released as cultivar Veronica in Mauritius. ICGV 87853 was released with high-level resistance to rust (Reddy et al. 2000). In China, several rust-resistant or tolerant cultivars have been developed both by using resistance sources in cultivated peanut (Valancia type) and in wild *Arachis* species (Yu 2011; Liao 2014). Zhonghua 4 and Shanyou 27 with moderate resistance to rust have been extensively cultivated in central and southern regions in China where rust is a problem (Yu 2011). In addition to hybrization, artificial mutation has also been used in improving resistance to rust in peanut.

Breeding for rust resistance in peanut is facing similar problems as resistance improvement against leafspots. Even though high level resistance to rust is easy to obtain among the segregating populations, undesirable genetic linkages between resistance and agronomically acceptable traits exist in most breeding populations.

14.3.1.3 Web Blotch

Resistance or tolerance to web blotch in peanut is variable among the germplasm accessions. Among different botanical types of *A. hypogaea*, cultivars belonging to Spanish and Valencia types are more sensitive compared to those belonging to Virginia type (Yu 2011). PI 274190 with a prostrating growth habit was identified resistant to web blotch. The severity of web blotch in peanut cultivars might be also related to their reactions to low temperature (cold tolerance) in the late growth stage. Normally, peanut cultivars with a longer growth period are relatively more tolerant, including Florunner and Southern Runner, developed in the United States. The resistance to web blotch was related to paplillae formation and hypersentitive reaction (Sun et al. 2022). In a linkage analysis of RILs with Zheng 8903 as the resistant parent, several major QTLs related with web blotch resistance were identified, among which qWBRA04 and qWBRA14 were detected in all five involved environments (Liu et al. 2020). Several wild *Arachis* species showed high level resistance, but transferring of resistance to web blotch into tetraploid peanut has been rarely reported. Further breeding efforts are needed to develop web blotch–resistant peanut cultivars for the producing regions, including northern, northeastern, and northwestern parts of China where erratic cool or cold stress occurs during the late growth stage.

14.3.1.4 Resistance to Soilborne Fungal Diseases

Peanut is a special crop as its pods develop under the soil. Both roots and pods of peanut are attacked by a number of soil-enduring fungi. Even extensive research efforts have been made worldwide, and progress in resistance enhancement or breeding against the soilborne diseases including stem rot, root rot, collar rot, and pod rot caused by several fungi complex has been relatively slow.

14.3.1.4.1 Stem Rot

Considerable variation of resistance or tolerance to stem rot caused by *S. rolfsii* is observed among the available peanut germplasm accessions and released cultivars. In most cases, the *S. rolfsii* strains with low, moderate, and high virulence are all present under natural conditions, and the pathogens have complex interactions with environmental factors and other soilborne microorganisms, making the disease pressures often erratic and highly variable. Moderately resistant germplasm accessions such as NC2, NCAc18016, NCAc18416, and several interspecific hybrid derivatives have been reported (Singh et al. 1997). Several released peanut cultivars in the United States, including Southern Runner, Toalson, Pronto, Georgia Browne, Sunbelt Runner, and Tamrun 96, are relatively (partially) resistant (Gorbet et al. 1987; Branch 1994; Smith et al. 1998; Backman and Brenneman 1997). Resistance has also been identified and transfered from wild Aachis species (Bera et al. 2014a, 2016). Cui et al. (2020) reported QTLs and markers for reisistance to stem rot in peanut. Unfortunately, no highly resistant peanut line has been identified against the highly virulent pathogen strains. To date, resistance to stem rot is attributed to polygenic with minor but additive effects, and is thought to be similar to horizontal or field resistance (Bera et al. 2022). More systematic research on mechanisms of stem rot resistance and potential variation among diverse germplasm and extensive cooperative breeding are needed.

14.3.1.4.2 Pod Rot

Pod rot caused by several soilborne fungi occurs frequently but is often erratic in severity due to the complicated interactions among various factors involved. The major pathogens may possess similar mechanisms in causing rotting symptoms in peanut pods. No specialization between the pathogen species and peanut genotypes has been found, although differences in rotting symptoms exist among the pathogenic fungi. Sensitivity to pod rot is highly genetically driven, and only a few peanut genotypes are highly susceptible. Most released peanut cultivars through multi-location selection present some resistance to pod rot, and many moderately resistant cultivars have been identified (Liao 2014). Several genotypes resistant to *Pythium* pod rot also show resistant to *S. rolfsii* (Smith et al. 1989). Interestingly, several bacterial wilt-resistant accessions are highly resistant to pod rot in different environments from continuous observations. Zhonghua 6 is an improved cultivar with resistance to bacterial wilt, pod rot, and aflatoxin production (Liao 2017). Some interspecific hybrid derivatives show resistance to pod rot. Like other soilborne diseases, resistance to pod rot in peanut is likely attributed to polygenic with additive effects (Liao 2014).

14.3.1.4.3 ***Aspergillus*** *Infection and Aflatoxin Contamination*

Infection by *Aspergillus* fungi can cause problems in peanut seed germination and seedling and adult plant wilting, among which the collar rot caused by *A. niger* occurs frequently in some cropping systems such as the summer-sown peanut immediately after wheat harvest in north China (Xu 2008). Large variations in resistance to collar rot in peanut are seen, but the resistance in peanut germplasm has not been systematically characterized. From available investigations, some bacterial wilt-resistant cultivars are more resistant to collar rot (Yu 2011). The peanut lines with tolerance to stem rot seem to have a lower ratio of collar rot. *In vitro* resistance to seed colonization by *A. flavus* might be helpful in reducing the flaroot and yellow mold in the field.

Besides the plant diseases caused by *Aspergillus* infection mentioned above, more research efforts have been made to address aflatoxin contamination due to fungal infection (Liao 2014; Waliyar et al, 1995). Peanut germplasm accessions with different resistance or tolerance mechanisms including resistance to infection in shell, infection in seed, aflatoxin production in kernel, and preharvest contamination have been identified (Jiang et al. 2021). Preharvest resistance, shell infection resistance, seed infection resistance, and aflatoxin formation resistance are all independently inherited, providing future achievement from gene pyramiding (Ding et al. 2022). Progress has been made in breeding aflatoxin-resistant peanut cultivars with high yield in several breeding programs worldwide. For a comprehensive understanding of aflatoxin resistance in peanuts, readers are encouraged to refer to Chapter 16 of this book, which provides detailed insights into this critical area of research.

14.3.1.4.4 Other Fungal Diseases

Sclerotinia blight caused by *S. minor* can be serious in some peanut growing regions. Against that, several peanut cultivars including Virginia 81B, Virginia 93B, Southwest Runner, Tamspan-90, and Tamrun OL07 released in the United States possess some resistance (Akem et al. 1992). Moderately resistant cultivars such as VA-98R, VA93B, and Perry have been utilized commercially (Chappell et al. 1995). QTLs for resistance to *Sclerotinia* blight were identified, and one of the QTLs coincides with a previously reported leaf spot resistance QTL (Liang et al. 2020).

Peanut smut caused by *Thecaphora frezii* Carranza & Lindquist is a problem in South America. Smut resistance has been identified in some wild *Arachis* species and further transferred to peanut cultivars. Two consistent quantitative trait loci (QTLs) related to smut resistance, qSmIA08 and qSmIA02/B02, were identified on chromosome A08 and A02/B02, respectively. The combined genotypic effects of both QTLs reduced smut incidence by 57%, providing new insights into the genetic architecture of smut resistance that may aid in breeding new resistant cultivars (De Blas et al. 2021).

For other minor soilborne fungal diseases, Georgia Browne was found to have partial resistance to *R. solani*. Resistance to both *Pythium* spp. and *R. solani* was found in some Spanish type cultivars (Beute 1997; Brenneman 1997).

14.3.2 Resistance to Bacterial Wilt

Peanut is one of the few crops where bacterial wilt disease has been successfully controlled through the use of wilt-resistant cultivars. High levels of resistance to bacterial wilt in peanuts were reported as early as the 1920s. Schwarz 21, the first resistant cultivar, was developed in Indonesia and has served as a stable resistance source for over a century. Since 1990, ICRISAT has also contributed by screening diverse peanut germplasm accessions in Indonesia, leading to the identification of novel resistant lines. Similarly, resistant germplasm accessions have been identified in Vietnam.

Since the 1950s, several Chinese research institutions have conducted extensive germplasm evaluations, identifying three original resistance sources including Xiekangqing, Taishan Sanlirou and Taishan Zhengzhu (Sun et al. 1981). Over the past decades, China has made significant advancements in bacterial wilt resistance in peanuts. These include identifying resistant germplasm, investigating the pathogenesis of the disease, elucidating the inheritance of resistance, and breeding high-yielding resistant cultivars (Liao et al. 1986, 1992; 2005; Zhang et al. 2023). Since the 1980s, China has released a series of commercially successful resistant cultivars, effectively controlling serious bacterial wilt outbreaks (Liao et al. 2014). To date, over 150 highly resistant germplasm accessions and improved cultivars have been identified and developed globally (Liao 2017).

Additionally, more than 30 wild *Arachis* species and numerous interspecific hybrid derivatives have been identified as resistant to bacterial wilt (Tang and Zhou 2000; Liao and Holbrook 2007). Research has also focused on molecular markers for bacterial wilt resistance in peanuts. Several resistance-related QTLs and candidate loci/genes have been identified through linkage mapping, QTL-seq, and other analyses using recombinant inbred lines (RIL) and F2 populations, particularly by Chinese institutions. These efforts have led to the development of SSR-selective markers, SNP diagnostic molecular markers, and KASP markers for marker-assisted selection (MAS) (Zhao et al. 2016; Luo et al. 2020). Candidate genes identified through these studies are now being targeted for fine mapping and functional validation, paving the way for further advancements in peanut resistance breeding (Luo et al. 2020).

For readers seeking a deeper understanding of the genetics and mechanisms underpinning bacterial wilt resistance in peanuts, Chapter 17 provides a detailed exploration of the complex interplay of these genetic factors.

14.3.3 Resistance to Viral Diseases

14.3.3.1 TSWV

Most of the research on resistance to TSWV in peanut has been conducted in the United States. Variation in TSWV resistance has been observed both in cultivated peanuts and wild *Arachis* species. Some wild *Arachis* species including *A. cardenasii*, *A. diogoi*, *A. correntina*, and *A. pusilla* showed no infection to TSWV under field conditions. In the

United States, the first released peanut cultivar with average resistance to TSWV was Southern Runner (Culbreath et al. 1992a, 1992b, 1994, 1996). Additional cultivars possessing TSWV resistance similar to Southern Runner, including Georgia Browne, Georgia Green, Florida MDR 98, and Tamrun 96, were further released (Branch 1996; Culbreath et al. 1994, 1996). Tifrunner was released in 2005 with significantly higher resistance to TSWV than the moderately resistant cultivar Georgia Green (Holbrook and Culbreath 2007). Florida-EP™113 is a new runner-type variety with superior TSWV resistance. Currently, most peanut cultivars grown in the southeastern United States have higher level resistance to TSWV. Major QTLs for TSWV resistance have been identified (Agarwal et al. 2018, 2019), and can be used for further fine mapping and marker-assisted selection in peanut breeding programs.

14.3.3.2 PStV

Extensive efforts were once made to evaluate peanut germplasm for resistance to PStV, but no resistant source was found from some 9,000 diverse accessions of the cultivated peanut. However, a negative reaction was observed in many wild *Arachis* accessions (Culver et al. 1987; Prasada Rao et al. 1991), and *A. glabrata*, *A. villosa*, and *A. correntina* even showed high level resistance to PStV (Yu 2011). Peanut cultivars belonging to the Virginia type possess lower seed transmission rate (below 1% in some lines) compared to those belonging to the Spanish type, which would be valuable in breeding.

14.3.3.3 PSV

No PSV-resistant peanut genotype has been identified yet, but tolerance under field conditions with high PSV infection pressure is different among peanut cultivars. Some peanut genotypes, such as Zhonghua 3, possess low field incidence, less plant stunt symptom, and reduced yield loss and could be utilized in breeding as well as in production in north China. On the other hand, high-level resistance to PSV has been identified in several wild *Arachis* species based on the research conducted in ICRISAT, India, and in China (Yu 2011). Highly PSV-resistant interspecific hybrid derivatives have been identified and could be used as resources in breeding.

14.3.3.4 Groundnut Rosette Virus

The GRV is mostly confined to certain regions in Africa. GRV resistance sources were first discovered in Senegal in 1952, and subsequently they were used as parents in developing high-yielding, rosette-resistant peanut varieties. Resistance sources to rosette disease are available in several genotypes (Bock et al. 1990; Subrahmanyam et al. 1998; Naidu et al. 1999; Nigam and Rock 1990). Subrahmanyam et al. (2001) identified several wild *Arachis* species resistant to all the three causative agents of rosette, and variations in resistance reaction were observed among interspecific hybrid populations (Bera et al. 2014b). Resistance to rosette virus is controlled by a monogenic dominant or two independent recessive genes, so these resistances are relatively easy to transfer into agronomically desirable types (Nigam and Bock 1990; Olorunju et al. 1992). In Nigeria, UGA2 (Samnut21), M572.80I (Samnut22), and ICGV-IS96894 (Samnut23), medium duration and resistance to GRV were released in 2001, and several early maturing varieties with resistance such as Samnut24, Samnut25, and Samnut26, were released for production (Ajeigbe et al. 2015).

14.3.3.5 Other Viral Diseases

Two wild *Arachis* species, namely *A. diogoi* and *A. pusilla*, also exhibited no infection from peanut mottle virus (Demski and Sowell 1981).

14.3.4 Resistance to Nematode Diseases

Research on resistance to nematodes in peanut has been relatively limited compared to that for foliar diseases. Considerable variation of reaction to nematodes has been observed among the germplasm accessions in the cultivated species. In some germplasm clusters originated from China, the frequency of resistant materials was relatively higher (Yu 2011). In China, nine resistant accessions were identified from field screening of some 6000 entries (Yu 2011). However, only moderately resistant germplasm lines belonging to the cultivated species have been found, while higher levels of resistance have been identified from the wild *Arachis* species.

To date, effective efforts have been made in transferring nematode resistance from wild species to cultigens in the United States (Holbrook and Noe 1990; Nelson et al. 1989). Both reproductive resistance and hypersensitive necrosis to *Meloidogyne* spp. have been reported in tetraploids derived from complex crosses of *A. hypogaea* comprising of *A. batizocoi*, *A. cardenasii*, and *A. diogoi* that are resistant to nematodes. By another route, crossing wild species and doubling chromosome numbers with colchicine, amphidploid tetraploid stocks were developed from *A. batizocoi* × (*A. cardenasii* × *A. diogoi*), among which TxAG-6 with strong resistance to nematodes was registrated and later used as a source in breeding several nematode-resistant cultivars including COAN and NemaTAM. TxAG-7, originated from a backcross of Florunner with TxAG-6, was resistant to *Meloidogyne* (Simpson et al. 1993). A backcross program was also used to introduce root knot nematode resistance from TxAG-7 into Florunner, resulting in the release of COAN, the first peanut cultivar with *M. arenaria* resistance (Simpson and Starr 2001). Subsequently, introgressing genes from TxAG-6 to *A. hypogaea* resulted in the release of NemaTAM (Simpson et al. 2003) and Webb (Simpson et al. 2013). The development of NemaTam was regarded the first successful example of marker-assisted selection (MAS) in breeding for disease resistant peanut variety (Bhat et al. 2022). Another cultivar, Tifguard, was developed based on improved nematode-resistant markers and was resistant to nematode, root-knot nematode, and TSWV (Holbrook et al. 2008). 9.

Garcia et al. (1996) reported that resistance to nematode in *A. cardenasii* was governed by two genes, dominant in nature, where one gene, designated as *Mag*, was responsible for inhibiting root galling and the other one, named *Mae*, was responsible for hindering egg production by *M. arenaria*. In another study, the resistance to *M. arenaria* in COAN originated from wild species was governed by a single gene of dominant nature (Simpson and Starr 2001). QTLs and molecular markers for nematode resistance were reported (Burow et al. 2014; Carpentieri et al. 2014).

Resistance improvement for resistance to nematodes has been less addressed outside the United States, although problematic nematodes are regarded as a constraint to peanut production in some regions in the world. In northern China, the major nematode is *M. halpa*, against which resistance enhancement has not been systematically addressed. Obviously, breeding for nematode resistance is largely influenced by complicated and costly screening procedures. More efficient marker-assisted selection technical approaches are to be developed and employed in breeding.

14.3.5 Resistance to Insects

In peanut, variation in resistance to insect pests among the diverse germplasm accessions has been observed, and a number of resistant or tolerant genotypes have been reported (Nigam et al. 1991, Yu 2011), although research on insect resistance has attracted relatively less effort. Relatively more resistant accessions against aphids and thrips have been identified, and these genotypes normally performed less GRV or PBV infection as these two insects are vectors of virus (Yu 2011). EC36892 is a genotype with stable resistance to aphids. Mechanisms of resistance or tolerance to insects consist of several aspects such as repellence, antibiosis, immunity, physical obstacles, and avoidance that can be used alone or in combination in screening. Reduced damage in peanut leaves due to thrips and jassids is related to the density, distribution, and length of trichome, as well as thickness of cuticles. Reduction in growth and fecundity in aphids is an antibiosis reaction in resistant peanut genotypes (Padgham et al. 1990). Resistance against *A. craccivora* reported in ICG 12991 is governed by a single recessive gene (Minja et al. 1999).

Compared to the cultivated peanut, wider variation in insect resistance exists in some wild *Arachis* species (Sharma et al. 2003; Yu 2011). *A. batizocoi*, *A. diogoi*, *A. correntina*, *A. villosa*, *A. spegazzini*, *A. cardenasii*, *A. stenosperma*, *A. duranensis*, *A. rigonii*, *A. paraguariensis*, *A. pusilla*, *A. glandulifera*, *A. ipaensis*, and *A. repens* are species that possess resistance to thrips (Michelotto et al. 2017; Yang et al. 1993). *A. cardenasii*, *A. duranensis*, *A. kempff-mercadoi*, *A. monticola*, *A. stenosperma*, *A. paraguariensis*, *A. pusilla*, and *A. triseminata* showed multiple resistances to leaf miner and thrips. *A. cardenasii*, *A. appressipila*, *A. ipaensis*, and *A. paraguariensis* showed antibiosis to *Spodoptera* and resistance to leaf feeding pests such as *Helicoverpa armigera* (Sharma et al. 2003). However, success in transferring pest resistance from wild species to high-yielding cultivars is rarely reported.

Due to the difficulty in screening a huge number of germplasm accessions and segregating populations under erratic and variable insect strains, insect resistance breeding in peanut has been rarely conducted. In India, ICGV 87160 (ICG (FDRS), Serenut 10R, SGV0023, SGV 002, SGV 0053, SGV 0084, Samnut 22, and Samnut 23 are released peanut cultivars that were reported to have higher yields in leaf miner–infested fields with field tolerance. Robut 33-1 was identified as resistant to thrips and confirmed to have low infection frequency of bud necrosis virus (BNV) under field conditions (Wightman et al. 1994). A higher tolerance to leaf miner and *Spodoptera* spp. in a breeding line ICGV 86031 was seen as an enhanced ability of the vegetative tissue to regrow after defoliation (Wightman and Rao 1994). Limited efforts have been made in developing insect-resistant peanut genetic stocks via genetic transformation of exotic genes (Bera et al. 2022; Tiwari et al. 2008).

14.4 POTENTIAL OF BIOTECHNOLOGICAL TOOLS FOR EFFICIENTLY ENHANCING RESISTANCE OR TOLERANCE TO IMPORTANT BIOTIC STRESSES IN PEANUT

Peanut production in most regions in the world is expected to further expand in order to meet the increased market need, but production is expected to be affected more frequently and even more seriously by diverse constraints including diseases and insect pests, against which host-plant resistance is fundamental for integrated management (Pandey et al. 2015). Based on the available progress, more intensive and coordinated research on genetic enhancement for developing improved peanut cultivars with multiple resistances to the major diseases and insects is very much needed in most countries or regions (Bera et al. 2022; Isleib et al. 1994; Janila et al. 2013; Liao 2014; Nigam et al. 1991). As the conventional techniques in peanut breeding are not enough in terms of of resistance identification and characterization, gene or trait transferring into high-yielding genetic background, pyramiding of multiple resistant genes, and creation of special novel resistant genotypes, further development and application of modern biotechnological tools are crucial. In order to develop more resistant cultivars, novel biotechnologies are essential for discovering elite resistance genes, dissecting resistance mechanisms, selecting resistance with high efficiency, and integrating resistances with other key traits. Recent advancements in developing novel biotechnological tools in genomics, transcriptomics, proteomics, metabolomics, genetic transformation, and gene editing have huge potential to improve stress tolerances in peanut in several aspects.

14.4.1 Dissection of Mechanisms for Stress Response

An understanding of mechanisms of resistance or tolerance to biotic stresses is very much needed for developing efficient strategies for varietal improvement. The mechanisms of resistance to foliar diseases in peanut such as LLS, ELS, and rust have been well investigated, while those to most soilborne fungal diseases such as stem rot, root rot, and pod rot, remain unknown. Bacterial wilt has been found to be related to pathogen infection and disease development, but the root structural differentiation might be related to the response to the disease in peanut (Liao 2014). The response mechanisms of peanut to insect pests have been less addressed. Mechanisms of resistance or tolerance to biotic stresses are also highly variable among different genetic sources. For example, different resistance mechanisms are seen between cultivated peanut and wild *Arachis* species for foliar diseases including rust, LLS, bacterial wilt (Sing et al. 1997; Yu 2011). Further investigation and understanding of stress response mechanisms in the diversified germplasm pools would be necessary for critically selecting resistance sources and modifying breeding objectives for special stress factors and different ecological regions. The tools of genomic sequencing, RNA sequencing, proteomic profiling, genetic transformation, and gene editing can be used in characterizing the mechanisms of stress responses in peanut.

14.4.2 Discovery of Resistance Genes or Loci

Few genes or loci for stress resistance have been identified in peanut, but further discovery of resistant genes/loci is highly necessary for more effective and precise operation for resistance enhancement. Generally, there are several genes controlling common stress responses in plants, including R genes, LOX genes, and PR 10. In peanut, the above genes with general physiological or biochemical defense functions against stresses have not been well characterized and utilized in breeding. They need to be further characterized by using approaches including genome sequencing and RNA sequencing.

In the cultivated peanut and in *Arachis* species, more specific resistant genes and loci can be further discovered by various means. Resistances to rust, leafspot, bacterial wilt, viruses, and nematodes in some cultivated genotypes or wild species are controlled by major genes which can be easily isolated and transferred in breeding. These genes are expected to be isolated by using combined genomic tools. For peanut diseases with limited geographic distribution, such as TSWV and smut, it is also essential to discover resistant genes whenever possible because these diseases might become more widespread or more serious in the future.

More genes or loci with minor genetic effects for stress tolerance are also expected to be further mined by using combined genomic tools. For example, some differentially expressed candidate R-genes were significantly correlated with resistance to leafspot, among which a majority are receptor like kinases (RLKs) and receptor like proteins (RLPs) with function of sensing the presence of pathogen at the cell surface and initiating protection response. The largest group was receptor-like cytoplasmic kinases (RLCKs) VII, which are involved in pattern-triggered kinase signaling, resulting in the production reactive oxygen species (ROS). When mapped to peanut genomes, it was found that 36 R-genes existed in most chromosomes except for A09 and B09. Identification and association of R-genes involved in leafspot resistance would facilitate genetic selection of LLS resistant lines with desirable agronomic traits (Dang et al. 2021). As major loci have been identified for resistance to infcetion by *Aspergillus flavus* (Jiang et al. 2021), aflatoxin production (Huai et al. 2025), bacterial wilt (Luo et al. 2019; Wang et al. 2018; Zhao et al. 2016), TWSV (Pandey et a. 2017a; Tseng et al. 2016), leafspot (Bera et al. 2022; Pandey et al. 2017a, 2017b; Shoba et al. 2013) and rust (Mondal and Badigannavar 2018), novel genes with resistance function and breeding value are expected to be isolated.

Genomic information is highly crucial for identifying and characterizing novel functional genes and alleles for stress resistance or tolerance. With the progress of whole genome sequencing of several diploid and tetraploid species in genus *Arachis*, the genomic information platforms are largely elevated and highly accessible (Bera et al. 2022; Bertioli et al. 2016, 2019; Zhuang et al. 2019). Historically, genomic research in peanuts woeldwide did not begin until the late 1980s, primarily due to the large genome size (~2.7 Gb). Sequencing of expressed sequence tags (ESTs) was the primary method used in the research on peanut genomic information (Proite et al. 2007; Guo et al. 2008; Bi et al. 2010). Later on, low-density genetic maps were constructed by using inter- and intra-specific crossing combinations to map genes conferring resistance to diseases such as leafspots and nematodes. High-density maps were then constructed with the development of more markers during the 2000s and 2010s, which marked the start of peanut genomics (Paterson et al. 2004; Stalker et al. 2009; Pandey et al. 2012; Ozias-Akins et al. 2017) and the pace picked up in the post-genome sequencing era. As a first step towards characterizing the genome of cultivated peanut, the genomes of the two diploid ancestors (*A. duranensis* V14167 and *A. ipaensis* K30076) of cultivated peanut were sequenced and analyzed (Bertioli et al. 2016). Both these accessions were collected from the most likely geographic region of *A. hypogaea.* Meanwhile, researchers sequenced *A. duranensis* (PI475845 from Bolivia), along with four synthetic tetraploids and their six diploid parents (two A genomes and four B genomes, including the suspected B genome progenitor, *Arachis ipaensis*), to gain insights into peanut evolution. Based on the draft genome of *A. duranensis*, the gene models with 50,324 protein-coding genes were proposed (Chen et al. 2016). In addition, Lu et al. (2018) sequenced *A. ipaensis* and recorded ~1.39 Gb genome with

39,704 predicted protein-encoding genes. Researchers developed the first reference-quality assembly of the *A. monticola* (PI263393) genome, with a size of approximately 2.62 Gb (Yin et al. 2018). Researchers analyzed and demonstrated the efficiency of the current strategy for *de novo* assembly of the highly complex allotetraploid species, using whole-genome shotgun sequencing, single-molecule real-time sequencing, high-throughput chromosome conformation capture technology, and BioNano optical genome maps. Subsequently, Yin et al. (2020) re-sequenced 17 wild *Arachis* diploids from AA, BB, EE, KK and CC groups and 30 tetraploids and compared the previously sequenced genome of *A. monticola*. In 2019, two reference genomes, one for the subsp. *fastigiata* and the other for subsp. *hypogaea* of the cultivated tetraploid, were both reported in *Nature Genetics*. The IPGI-led initiative (Bertioli et al. 2019) completed the sequencing of Tifrunner (PI644011), a runner type belonging to *Arachis hypogaea* subsp. *hypogaea* by deploying several modern sequencing and assembly technologies such as PacBio and Hi-C data/technology. On the basis of this, a genome of ~2.56 Gb with 20 pseudomolecules and 66,469 predicted genes was reported. Similar advanced technologies were deployed by two independent efforts in China, leading to the development of high-quality reference genome assemblies for Shitouqi (Zhuang et al. 2019) and Fuhuasheng (Chen et al. 2019), both belonging to *A. hypogaea* subsp. *fastigiata*. Shitouqi (Zhh0235) is a well-known landrace from south China and also an elite breeding parent (of many improved cultivars) belonging to subspecies *fastigiata* and botanic type *vulgaris* (Spanish type), while Fuhuasheng is another landrace from north China. Researchers reported a genome of approximately 2.54 Gb with 83,709 predicted genes across 20 pseudomolecules for Shitouqi, while Fuhuasheng had a genome of about 2.55 Gb with 83,087 predicted genes across 20 pseudomolecules. The above progress has provided a strong basis for gene discovery in genus *Arachis*.

Meanwhile, numerous efforts have been made in peanut to collect and study the transcriptome through initial efforts with microarrays (Chen et al. 2012; Zhou et al. 2014), real-time polymerase chain reaction (PCR), and RNA sequencing. Luo et al. (2005) attempted to identify resistance genes in response to leafspot using microarray and real-time PCR. They showed similarity with genes encoding lipid transfer protein precursors, pathogenesis-related protein, glucosyl hydrolase family protein, and LRR protein family. In another study, several genes for proteins were identified using transcriptomic and proteomic analysis, and these genes were involved in cell-wall strengthening, hypersensitive cell death, resistance-related proteins, metabolism, and signal transduction in resistant wild species *A. diogoi* against LLS pathogen (Kumar and Kirti 2015). Zhou et al. (2016) confirmed that nucleotide-binding-leucine-rich repeat (NB-LRR)-encoding genes were involved in the LLS resistance mechanism by comparing QTL locations for LLS resistance from genetic linkage map with the physical map. A total of 214 expressed R-genes were identified from cultivated species that are naturally infected by early and late spot pathogens (Dang et al. 2019). Furthermore, efforts were made to identify the association of specific R-genes to leafspot resistance for providing molecular targets for marker-assisted breeding strategies (Dang et al. 2021).

Transcriptomic analysis has been employed to identify differentially expressed genes for resistance to ELS (Gong et al. 2020), LLS (Han et al. 2017) and rust (Rathod et al. 2020a; Rathod et al. 2020b). The results suggested that a few major genes and several factors mediate resistance to ELS disease, showing the characteristics of quantitative traits in defense responses. Most of these studies identified some defense-related genes. Molecular responses of the wild species challenged with the LLS pathogen were investigated using cDNA-AFLP and 2D proteomic study. A total of 233 differentially expressed genes involved in cell wall strengthening, hypersensitive reaction, and resistance-related proteins were identified in the wild species *A. diogoi* (Kumar and Kirti 2015). Transcriptomic analysis in the *A. flavus* resistant peanut genotype J11 led to the detection of 663 differentially expressed genes. Further functional analysis revealed that these genes encoded a wide range of defense or PR proteins (pathogenesis related proteins). Changes in the expression patterns of these genes might contribute to peanut resistance in *A. flavus*. Bosamia et al. (2020) used RNA-Seq to unravel the mechanisms of resistance to stem rot using a resistant (NRCG-CS85) and susceptible (TG37A) genotype and suggested the PAMP-triggered immunity as a probable mechanism of resistance. Differentially expressed genes and translated proteins in wild species indicated its defense mechanism upon interaction with pathogen and provide an initial breakthrough in terms of finding genes possibly involved in sensing or recognizing and early signaling responses to fight infection through the subsequent development of resistance.

Proteomics is constantly advancing to bridge the gap between DNA sequence, transcriptome, and phenotype under the diverse and dynamic stages of growth and development. Since proteins can influence key phenotypic performance and are the products of genes and epigenetic or post-translational mechanisms, population proteomics has the potential to provide insights into functional and meta-population ecology, adaptation, and acclimation processes under various conditions. Proteomics approaches also help identify genetic loci underlying disease resistance for the development of biomarkers. In peanut, the low DNA polymorphism coupled with high morphological variation might involve differences in proteomics. Katam et al. (2014) reviewed the progress made on proteomics in peanut, especially on peanut allergens and adaptive responses to various stresses. Proteomics analyses have been reported for response to high-oleic acid content in seed (Liu et al. 2020), cadmium detoxification and translocation (Yu et al. 2019), allergen production (Mamone et al. 2019), response to water stress (Kottapalli et al. 2009; Katam et al. 2016), gynophore development (Sun et al. 2013; Zhao et al. 2015), response to salinity (Jain et al. 2006), and polyphenol content (Muralidharan et al. 2021). Kumar and Kirti (2015) employed 2D proteomics to confirm a few selected DEGs identified (233) through cDNA-AFLP when *A. diogoi* was challenged with LLS pathogen, from which three candidate genes (*AdLEA*, *AdSGT1* and *AdTLP*) were further characterized and confirmed for their function in imparting disease resistance. *A. diogoi* late embryogenesis abundant (AdLEA) protein-coding gene was characterized and shown to impart resistance to multiple stresses in tobacco (Sharma et al. 2016). *AdSGT1* (suppressor of G2 allele of SKP1), an essential signaling component in R-gene-mediated resistance response against various plant pathogens (Kumar and Kirti 2015), and *AdTLP*, a pathogen-induced thaumatin-like protein gene (Singh et al. 2013), were also characterized. These genes were found to be candidate genes for enhancing stress resistance in crop plants, and might be also valuable in peanut.

14.4.3 Creation of Novel Genetic Stocks

Genetic transformation along with conventional hybridization and artificial mutation are key techniques to create novel genetic stocks, including materials with superior resistance to stresses in plant. As a consequence of ploidy barrier between the cultivated peanut and the wild *Arachis* species, introgression of stress-related genes from the diploid progenitors by conventional breeding is complex. Introgression lines developed by crossing wild species with cultivated peanut normally have undesirable gene blocks. To overcome the problem of lack of beneficial genes within crossable germplasm stocks, genetic engineering/recombinant DNA techniques such as *Agrobacterium tumefaciens* through mediated or direct transfer of desired genes from wild species would be an ideal option to improve resistance against various diseases (Vasavirama and Kirti 2012) and insects. Resistance to several fungal and viral diseases has been achieved through the use of transgenes coding for cell wall components such as chitinase, glucanase, PR proteins, coat proteins, bacterial chloroperoxidase, oxalate oxidase, RNA interference (RNAi), and crystal proteins. Sunkara et al. (2013) reviewed the use of chitinase, glucanase, Rs-AFP2 (*Raphanus sativus* antifungal protein-2), and SniOLP (*Solanum nigrum* osmotin like protein) for LLS and ELS; oxalate, chitinase, and glucanase for *Sclerotinia* blight; chitinase for rust; and anionic peroxidase, glucanase, stilbene synthasesynthetic peptide D4E1, chitinase, mod1, nonheme chloroperoxidase (cpo), LOX 1, and Pn LOX 3 against *A. flavus* infection and aflatoxin production. When compared to the parental genotype, transgenic lines of the Okrun harboring chitinase gene from rice and glucanase genes from alfalfa showed a 43%–100% reduction in *Sclerotinia* blight incidence (Chenault et al. 2005). Two genes, Rchit and CHI, coding for chitinase enzymes against *Fusarium* wilt and leafspot fungi have been evaluated for inheritance in peanut transgenic events (Rohini and Sankara 2001; Iqbal et al. 2011, 2012). Based on the characterization, LLS incidence was decreased in transgenic lines of peanut expressing a defensin gene (BjD) from mustard (Anuradha et al. 2008). Transgenics with cDNA sequence of barley oxalate oxidase conferred enhanced resistance to *Sclerotinia* blight (Livingstone et al. 2005). Transgenics developed using bacterial non-heme chloroperoxidase gene from *Pseudomonas pyrrocinia* and rice chitinase gene showed hyphal growth inhibition of *A. flavus* (Niu et al. 2009; Prasad et al. 2013). The complete nucleotide sequence (4019 nts long) and genome organization (4 ORFs) of GRV are known (Taliansky et al. 1996). Peanut cultivar JL 24 was transformed with the GBNV nucleoprotein gene at ICRISAT, and T2 transgenic events were tested for virus resistance, which can provide reliable GBNV resistance that can be bred into other peanut cultivars through back-cross breeding programs. At ICRISAT, JL24 was also transformed with IPCV-H coat protein and replicase genes to induce pathogen-derived resistance. Genetically modified peanut cultivars that carry viral coat protein gene exhibited high levels of resistance to PStV (Franklin et al. 1993). Further, transgenic peanut plants of Gajah and NC 7 that contained untranslatable full length sequence (*CP2*) and translatable *CP* gene with an N-terminal truncation (*CP4*) of PStV offered resistance to the virus (Higgins et al. 2004). Insertion of viral nucleocapsid protein-coding gene (*tswvnp*) in peanut genome has resulted in resistance to TSWV (Brar et al. 1994). Furthermore, by activating RNA silencing, a natural virus defense mechanism, high-level resistance or immunity can be induced in plants (Waterhouse et al. 2001). RNAi technology such as RNA silencing or homologous gene co-suppression are powerful methods for developing resistance to viruses in peanut (Wang et al. 2000). At ICRISAT, an RNAi-mediated approach is being used to counteract the effect of the PBNV genome's nonstructural silencing suppressor gene (NSs gene). Transformed plants with specific small RNAs, the products of RNA silencing, were highly resistant to PStV infection, and the resistance was stably inherited over five generations (Dietzgen et al. 2004). Resistance derived from pathogens by introducing GRAV or GRV genes/genome sequences, or SatRNA-derived sequences that inhibit/slow down GRV replication, is a possible strategy against GRD via transgenic plant generation (Taliansky et al. 1996). Cry1 EC gene against *S. litura* (Tiwari et al. 2008) and cry1 X gene against *H. armigera* and *S. litura* (Entoori et al. 2008) are two synthetic genes that have shown promise against their respective insect pests. When the trypsin inhibitor gene from cowpea was introduced into peanut, it increased tolerance to insects (Xu et al. 2003). However, the success and achievement of transformation techniques is still poor due to allopolyploidy, genotype specificity, low transformation and regeneration efficiency, and low level of transgene expression. Although many transgenic lines have been developed against biotic stresses, to date no transgenic peanut cultivar has been released commercially. As reported by Sundaresha et al. (2016), resistance against *A. flavus* and *C. arachidicola* was developed in JL 24, ICGV 89104, and ICGV 86031 via overexpressing a tobacco glucanase gene.

With the identification of the candidate genes in peanut and other sources, transgenics have been developed in peanut for improving various traits, including resistance to foliar diseases. Pathogenesis-related (PR) proteins, SniOLP (*Solanum nigrum* osmotin-like protein), and Rs-AFP2 (*Raphanus sativus* antifungal protein-2) were overexpressed in peanut. The transgenic peanut plants showed enhanced disease resistance to LLS based on a reduction in the number and size of lesions on leaves and delay in the onset of disease (Vasavirama and Kirti 2012). ICG 13942 was transformed with *Tcchitinase-I* gene (Marka and Nanna 2021). The T1 transgenic plants when evaluated for ELS, LLS, and rust showed longer incubation, latent period, and lower infection frequencies in comparison to non-transformed plants. A significantly negative correlation was recorded between chitinase activity and the infection frequency by ELS, LLS, and rust pathogens. A chitinase gene from rice (*Rchit*) was introduced into three peanut genotypes through *Agrobacterium*-mediated transformation (Prasad et al. 2013). Evaluation of the transgenic plants for LLS and rust using detached leaf assays showed longer incubation and latent period and lower infection frequencies when compared to their non-transformed counterparts. Overexpression of a fusion defensin gene from radish and fenugreek also improved resistance against leafspots in peanut (Bala et al. 2016).

Ectopic expression of MYB repressor GmMYB3a improved drought tolerance and productivity of transgenic peanuts under water-deficit conditions (He et al. 2020). A novel salt-inducible WRKY transcription factor gene, AhWRKY75, conferred salt tolerance in transgenic peanut (Zhu et al. 2021). Bhalani et al. (2019) reported improved tolerance to soil-moisture-deficit stress among the transgenics expressing *AtDREB1A*, which regulates the antioxidant mechanisms. Expression of *Escherichia coli*-derived mannitol-1-phosphate dehydrogenase (mtlD) in peanut improved photosynthetic, physio-biochemical, and yield parameters under drought stress (Patil et al. 2017). RNA interference (RNAi) could prevent aflatoxin accumulation in transformed peanuts (Arias et al. 2015). Aflatoxin control by exogenous delivery of double-strand RNA (dsRNA) was also demonstrated (Power et al. 2020). These efforts indicated the possibilities for enhancing genetic resistance of peanut in the future.

CRISPRs, a high-throughput genome editing technology focused on the prokaryotic immune system, offers a promising hope for further peanut improvement. Recently, CRISPR/Cas9 technology has become very popular for genome editing, trait discovery, and manipulating genomes in desireable directions. Targeted genome editing technology for functional genes has been an exciting new advancement. This new technology has the potential to be an effective tool in driving disease-fighting varietal development. Targeted genome editing has proven to be effective using zinc finger nucleases (ZFNs) and transcriptional activator-like effector nucleases (TALENs), which involve two DNA binding proteins flanking a sequence of interest (Lloyd et al. 2005; Wright et al. 2005; Cermak et al. 2011; Li et al. 2012; Mahfouz et al. 2011). In peanut, gene editing using CRISPR/Cas9 technology (Yuan et al. 2019) and TALEN-mediated targeted mutagenesis (Wen et al. 2018) has been reported to enhance the oleate content by targeting the *AhFAD2* gene. These successful examples along with advancements in genomics would certainly attract more efforts to improve peanut for various traits including stress resistance or tolerance. Shu et al. (2020) applied the CRISPR/Cas9 tool in peanut hairy root transformation system to explore the function of nod factor receptor (NFR) genes which initiate peanut plant response to *Rhizobia*. While the utilization of CRISPR-based genome modification in peanut is challenging, with advancements in pangenomics and population genomics for exploring the allelic differences underlying the phenotypic variations, this novel technology has ample scope for enhancing the limited resistance available against biotic stresses in peanut.

14.4.4 Enhancment of Selection Efficiency for Tolerances and Resistances

Developing useful markers for MAS breeding is highly necessary for more efficient selection of target traits in stress resistance breeding. Traits like resistance to leafspots and rust are difficult to manage through conventional phenotypic selection because they co-occur, and all lead to defoliation. However, such traits can be handled with genomics-assisted tools like markers. The advantages of genomics-assisted breeding (GAB) have been well demonstrated through translational genomics for breeding LLS- and rust-resistant peanut genotypes (Pandey et al. 2012). Integration of genomic tools with conventional breeding has accelerated peanut improvement with precision in breeding for traits with complex genetic backgrounds. Considerable progress has been made on GAB with the development of several thousands of markers, several genetic maps, dense consensus genetic maps, QTL mapping, and molecular breeding.

The availability of genome-wide genetic markers is essential for trait mapping and marker-assisted breeding. In peanut, the initial efforts on isozyme and seed protein analyses identified only limited variability among the cultivated peanuts (Lu and Pickersgill 1993; Stalker et al. 1994), although substantial diversity really exists within cultivated peanut genotypes in terms of various morphological, physiological and agronomic traits (Stalker 1992). Random amplified polymorphic DNA (RAPD) and restriction fragment length polymorphism (RFLP) approaches failed to detect any DNA variation in cultivated peanut (Halward et al. 1991, 1992; Kochert et al. 1991; Paik-Ro et al. 1992). However, these approaches could identify genetic variability among various wild species (Halward et al. 1991). Later, He and Prakash (1997) reported polymorphic DAF and AFLP markers in cultivated peanut. SSR markers were developed in the cultivated peanut using DNA library, and the polymorphism was detected. SSRs were also developed from *A. pintoi* to identify the variation of this species (Palmieri et al. 2002) and the accessions belonging to the section *Caulorrhizae* (*Arachis*, Fabaceae) (Palmieri et al. 2005). Fifty-six SSRs were developed from the cultivated peanut from SSR-enriched library (He et al. 2003), of which 19 showed polymorphism. STMS markers were also developed in cultivated peanut to detect variation (Ferguson et al. 2004).

Several efforts were made to develop genic and non-genic SSRs (Moretzsohn et al. 2004, 2005; Hong et al. 2008; Zhang et al. 2012; Huang et al. 2016; Peng et al. 2016). Single-locus markers offer an advantage over multi-locus markers in genetic and breeding studies since their alleles can be assigned to specific genomic locus. Zhou et al. (2016) developed 1,790 single-locus SSR markers from the *de novo* assembly of peanut sequence reads. Using the reference genome sequences of *A. duranensis* and *A. ipaensis*, Luo et al. (2017) identified 264,135 and 392,107 SSRs, from which 84,383 and 120,056 SSR markers were developed. CAPS markers were developed for detecting the mutations at *AhFAD2A* and *AhFAD2B* (Chu et al. 2009). Diversity array technology (DArT) (Kilian 2008) and diversity array technology sequencing (DArTSeq) (Shaibu et al. 2020) marker platforms have also been developed for peanut. Shirasawa et al. (2012) developed 504 AhTE markers using *AhMITE1*-enriched libraries. The representative *AhMITE1* exhibited a mean length of 205.5 bp and a GC content of 30.1%, with AT-rich, 9 bp target site duplications and 25 bp terminal inverted repeats. Later, Shirasawaet al. (2012) developed additional 535 AhTE markers using transposon-enriched libraries of other peanut cultivars. Since these AhTE markers were highly polymorphic and user-friendly (Kolekar et al. 2016), they were successfully used to construct linkage maps (Shirasawa et al. 2013) and to identify QTL for resistance to LLS and rust. Moreover, AhTE markers were also used for marker-assisted backcross breeding in peanut (Yeri and Bhat 2016). With the availability of the genome sequences of the diploid progenitors of peanut (Bertioli et al. 2016), efforts were made to identify the genome-wide distribution of *AhMITE1* (Gayathri et al. 2018). For this, a set of diverse genotypes (33), including the genetically unstable peanut mutants which show hyperactivity of *AhMITE1* (Hake et al. 2017), were used to discover the *AhMITE1* insertion polymorphic sites. WGRS reads from these diverse genotypes were analyzed using the computational method polymorphic TEs and their movement detection (PTEMD) (Kang et al. 2016) for the *de novo* discovery polymorphic sites and to develop 2957 *AhMITE1* markers (Gayathri et al. 2018).

The advent of next-generation sequencing and genotyping technologies has enabled the detection of more SNPs in peanut, which have emerged as the marker of choice in mapping (Bertioli et al. 2014), and several studies have reported identifying the SNP markers for mapping and population genomics. Hale et al. (2020) assessed the methods to reduce per-sample costs in high-throughput targeted sequencing projects, minimal equipment, and consumable requirements for targeted sequencing, which would promote utilization of the improved molecular breeding technology.

14.4.5 Integration of Multiple Resistances with Elite Genetic Background

Breeding for stress tolerance in any crop should pay attention to the coordination or balance of the key target traits. A comprehensive approach incorporating traditional and molecular breeding with transgenic techniques would offer solutions to the complex problems presently confronting peanut improvement. The last decade witnessed rapid development of genomic resources such as large-scale molecular markers, genetic maps, and genome sequences and their deployment in genomic-assisted breeding (GAB) in peanut. There are three GAB approaches, namely marker-assisted backcrossing (MABC), marker-assisted recurrent selection (MARS), and genomic selection (GS). MABC and MARS require trait association, while GS does not need such analysis. Realizing the limitations associated with MABC and MARS in terms of capturing small-effect genetic factors, GS has emerged as the most promising, efficient, and cost-effective breeding approach which captures both small- and large-effect genetic factors. GS promises to achieve higher genetic gains to improve complex traits such as yield and oil content in peanut (Pandey et al. 2020). If integrated with rapid generation advancement technology such as speed breeding, GS can make remarkable advances and have a positive impact on breeding programs in peanut (Pandey et al. 2020). The recent progress in developing high-desnity genotyping arrays has accelerated the mapping and selection process for multiple traits in peanut (Bangaru et al. 2023). More efforts are needed in pyramiding multiple resistances into peanut genotypes with desirable genetic backgrounds. Superior resistant peanut cultivars need to possess reasonable resistance to foliar diseases, including LLS and rust, and soilborne diseases such as stem rot and pod rot, and even some resistance to aflatoxin contamination. It is worth noting that most bacterial wilt-resistant peanut accessions possess some resistance to stem rot and aflatoxin, but some accessions resistant to foliar diseases are more sensitive to pod rot and *Aspergillus* infection. As the high oleate trait is becoming more important and popular in peanut breeding, attention should be given to coordinating the oleate content with multiple stress tolerances.

14.5 FUTURE PROSPECTS

The peanut production and supply chain are increasingly important globally as this crop is of high value in enhancing human nutrition, supporting livestock development, and sustaining eco-friendly agricultural systems. With distribution expanding, planting area increasing, cropping system changing, and climate warming, peanut production around the world will continue to be affected by many traditional biotic and abiotic stresses as well as some other newly emerging stresses in the future. Resistance or tolerance to the major biotic stresses would be still among the key aims in genetic enhancement in peanut for maintaining production competency of the crop, protecting the environment, and meeting the increased market demand. Based on the available progress in improving peanut resistance and developing biotechnological tools, there are desirable genomic and genetic resources to speed up the process of resistance improvement. Meanwhile, the following aspects should be emphasized for further improvement of stress resistances in peanut. Firstly, more research is needed in characterizing the biological nature of the key biotic stresses. Special emphasis should be given to soilborne fungal diseases, which are becoming more and more serious in many peanut-producing regions. Understanding the pathogenesis of peanut diseases, especially for those less studied before, is critical for deciding proper strategies of resistance improvement. Secondly, peanut breeding objectives for resistance or tolerance to the different stresses need to be modified based on the stress features, availability of resistance sources, complexity of the resistances/tolerances, and correlation with other key traits. Thirdly, novel resistant sources and even functional genes need to be discovered and characterized by using the biotechnological tools with high throughput. Special emphasis should be given to identifying resistant genes from the wild *Arachis* species, especially from those which have been less utilized because of the traditionally encountered obstacles such as incompatibility and ploidy difference. The sequenced genomic information of the available database and newly generated data in the future would provide useful supports for gene and locus identification for stress resistances. Fourthly, several approaches including interspecific hybridization, mutation breeding, genetic transformation and gene editing will be used to create novel genetic stocks for stress resistance or tolerance. By using genetically modified approaches, it is hoped to develop superior resistance to some of the tough stresses, such as aflatoxin contamination, soilborne disease complex, and insect pests. Fifthly, more efficient biomarkers should be developed for efficient and cheap selection of resistance, especially for those that are difficult in conventional phenotyping. More efforts are required to saturate peanut linkage maps so that MAS can be more practical. Finally, more breeding efforts are needed in pyramiding multiple resistances into elite genetic backgrounds as any single resistance is normally not enough in most natural fields, for which new breeding technologies, such as genomic selection and genome editing, can be employed. This would be highly promising in developing next-generation superior peanut cultivars with high level multiple resistances, high productivity, and improved quality traits.

REFERENCES

Abdou, Y. A. M., W. C. Gregory, and W. E. Cooper. "Sources and Nature of Resistance to *Cercospora arachidicola* Hori and *Cercosporodium personatum* (Beck & Curtis) Deighton in Arachis Species." *Peanut Science* 13 (1974): 6–11. https://doi.org/10.3146/i0095-3679-1-1-3.

Agarwal, G., J. Clevenger, M. K. Pandey, H. Wang, Y. Shasidhar, et al. "High-Density Genetic Map Using Whole-Genome Resequencing for Fine Mapping and Candidate Gene Discovery for Disease Resistance in Peanut." *Plant Biotechnology Journal* 16, no. 11 (2018): 1954–1967. https://doi.org/10.1111/pbi.12930.

Agarwal, G., J. Clevenger, S. M. Kale, H. Wang, M. K. Pandey, et al. "A Recombination Bin-Map Identified a Major QTL for Resistance to Tomato Spotted Wilt Virus in Peanut (*Arachis hypogaea*)." *Scientific Reports* 9 (2019): 18246. https://doi.org/10.1038/s41598-019-54747-1.

Ajeibe, H. A., F. Waliyar, C. A. Echekwu, K. Ayube, B. N. Motagi, et al. *A Famer's Guide to Groundnut Production in Nigeria*. International Crops Research Institue for the Semi-Arid Tropics (ICRISAT), Patancheru, India, 2015.

Ahmad, S., B. Nawade, C. Sangh, G. P. Mishra, T. C. Bosamia, et al. "Identification of Novel QTLs for Late Leaf Spot Resistance and Validation of a Major Rust QTL in Peanut (*Arachis hypogaea* L.)." *Biotechnology* 10 (2020): 1–13. https://doi.org/10.1007/s13205-020-02446-4.

Akem, C. N., H. A. Melouk, and O. D. Smith. "Field Evaluation of Peanut Genotypes for Resistance to Sclerotinia Blight." *Crop Protection* 11 (1992): 345–348. https://doi.org/10.1016/0261-2194(92)90061-9.

Anderson, W. F., J. C. Wynne, and C. C. Green. "Potential for Incorporation of Early and Late Leaf Spot Resistance in Peanut." *Plant Breeding* 97 (1986): 163–170. https://doi.org/10.1111/J.1439-0523.1986.TB01047.X.

Anderson, W. F., C. C. Holbrook, and T. B. Brenneman. "Resistance to *Cercosporidium personatum* within Peanut Germplasm." *Peanut Science* 20 (1993): 53–57. https://doi.org/10.3146/i0095-3679-13-1-4.

Anuradha, T. S., K. Divya, S. K. Jami, and P. B. Kirti. "Transgenic Tobacco and Peanut Plants Expressing a Mustard Defensin Show Resistance to Fungal Pathogens." *Plant Cell Reports* 27 (2008): 1777–1786. https://doi.org/10.1007/s00299-008-0596-8.

Arias, R. N., P. M. Dang, and V. S. Sobolev. "RNAi-mediated Control of Aflatoxins in Peanut: Method to analyze Mycotoxin Production and Transgenic Expression in the Peanut/*Aspergillus* Pathosystem." *Journal of Visualized Experiments* (2015). https://doi.org/10.3791/53398. https://www.jove.com/53398.

Backman, P. A., and T. B. Brenneman. "Stem Rot." In *Compendium of Peanut Diseases*, 2nd ed., edited by N. Kokalis-Burelle, D. M. Porter, R. Rodriguez-Kabana, D. H. Smith, and P. Subrahmanyam, 36–37. St. Paul, MN: APS Press, 1997.

Bala, M., T. Radhakrishnan, A. Kumar, G. P. Mishra, J. R. Dobraia, et al. "Overexpression of a Fusion Defensin Gene from Radish and Fenugreek Improves Resistance Against Leaf Spot Diseases Caused by *Cercospora arachidicola* and *Phaeoisariopsis personata* in Peanut." *Turkish Journal of Biology* 40, no. 1 (2016): 139–149. https://doi.org/10.3906/biy-1412-46.

Bangaru, K. A. Mathew, R. Bagudam, A. P. Wankhade, A. Purohit, E. N. Yohane, S. Abady, D. B. Deshmukh, and J. Pasupuleti. "Next-Generation Crop Breeding Approaches for Improving Disease Resistance in Groundnut *(Arachis hypogaea* L.)." In *Diseases in Legume Crops*, edited by U. C. Jha, H. Nayyar, K. D. Sharma, E. J. B. von Wettberg, P. Singh, and K. H. M. Siddique, 195–232. Springer, 2023.

Barkley, N. A., H. D. Upadhyaya, B. Liao, and C. C. Holbrook. "Global Resources of Genetic Diversity in Peanut." In *Peanuts: Genetics, Processing & Utilization*, edited by H. T. Stalker and R. F. Wilson, 67–109. Academic Press and AOCS Press, 2016. ISBN: 978163067383.

Baskaran, R. K. M., and D. S. Rajavel. "Management of Sucking Insect Pests of Peanut Through Bioagents and Botanicals." *Annals of Plant Protection Sciences* 21, no. 2 (2013): 286–290.

Bera, S. K., S. V. Kasundra, J. H. Kamdar, B. C. Ajay, C. Lal, et al. "Variable Response of Interspecific Breeding Lines of Peanut to *Sclerotium rolfsii* Infection Under Field and Laboratory Conditions." *Electronic Journal of Plant Breeding* 5 (2014a): 22–29.

Bera, S. K., J. H. Kamdar, A. K. Maurya, and P. Dash. "Molecular Diversity and Association of Simple Sequence Repeat Markers with Bud Necrosis Disease in Interspecific Breeding Lines and Cultivars of Peanut (*Arachis hypogaea* L.)." *Australian Journal of Crop Science* 8, no. 5 (2014b): 771–781.

Bera, S. K., J. H. Kamdar, S. V. Kasundra, and P. P. Thirumalaisami. "Identification of Groundnut Genotypes and Wild Species Resistant to Stem Rot Using an Efficient Field Screening Technique." *Electronic Journal of Plant Breeding* 7, no. 1 (2016): 61–70. https://doi.org/10.5958/0975-928X.2016.00009.0.

Bera, S.K., Kirti Rani, J.H. Kamdar, M.K. Pandey, H. Desmae, C.C. Holbrook, M.D. Burow, N. Manivannan, R.S. Bhat, M. D. Jasani, S.S. Bera, A.M. Badigannavar, G. Sunkad, G.C. Wright, P. Janila, and R.K. Varshney. "Genomic Designing for Biotic Stress Resistant Peanut." In *Genomic Designing for Biotic Stress Resistant Oilseed Crops*, edited by C. Kole, 137–214. Springer Nature, AG, Switzerland, 2022.

Bertioli, D. J., P. Ozias-Akins, Y. Chu, K. M. Dantas, S. P. Santos, et al. "The Use of SNP Markers for Linkage Mapping in Diploid and Tetraploid Peanuts." *Genes, Genomes, Genetics* 4 (2014): 89–96. https://doi.org/10.1534/g3.113.007617.

Bertioli, D. J., S. B. Cannon, L. Froenicke, G. Huang, A. D. Farmer, et al. "The Genome Sequences of *Arachis duranensis* and *Arachis ipaensis*, the Diploid Ancestors of Cultivated Peanut." *Nature Genetics* 48 (2016): 438–446. https://doi.org/10.1038/ng.3517.

Bertioli, D. J., J. Jenkins, J. Clevenger, O. Dudchenko, D. Gao, et al. "The Genome Sequence of Segmental Allotetraploid Peanut *Arachis hypogaea*." *Nature Genetics* 51, no. 5 (2019): 877–884. https://doi.org/10.1038/s41588-019-0405-z.

Beute, M.K. "Pythium Diseases." In *Compendium of Peanut Diseases*, 2nd ed., edited by N. Kokalis-Burelle, D. M. Porter, R. Rodriguez-Kabana, D. H. Smith, and P. Subrahmanyam, 27–30. St, Paul, MN: APS Press, 1997.

Bhat, R. S., M. P. Jadhav, P. V. Patil, and K. Shirasawa. "Genomics-Assisted Breeding for Resistance to Leaf Spots and Rust Diseases in Peanut." In *Accelerated Plant Breeding, Volume 4: Oil Crops*, edited by S. S. Gosal and S. H. Wani, 239–278. Springer, 2022.

Bhat, R. S., B. N. Motagi, G. K. Naidu, H. L. Nadaf, V. A. Giritammannavar, and R. V. Hunje. "The Scientific and Economic Impact of the Foliar Disease-Resistant Peanut Variety GPBD 4." *Australasian Plant Pathology* 53 (2024): 285–290.

Bhalani, H., R. Thankappan, G. P. Mishra, T. Sarkar, and T. C. Bosamia. "Regulation of Antioxidant Mechanisms by AtDREB1A Improves Soil-Moisture Deficit Stress Tolerance in Transgenic Peanut (*Arachis hypogaea* L.)." *PLoS One* 14, no. 5 (2019): e0216706. https://doi.org/10.1371/journal.pone.0216706.

Bi, Y., W. Liu, H. Xia, L. Su, C. Zhao, S. Wan, and X. Wang. "EST Sequencing and Gene Expression Profiling of Cultivated Peanut (*Arachis hypogaea* L.)." *Genome* 53, no. 10 (2010): 832–839.

Bock, K. R., A. F. Murant, and R. Rajeshwari. "The Nature of Resistance in Peanut to Rosette Disease." *Annals of Applied Biology* 117 (1990): 379–384. https://doi.org/10.1111/j.1744-7348.1990.tb04224.x.

Bosamia, T. C., S. M. Dodia, G. P. Mishra, S. Ahmad, B. Joshi, P. P. Thirumalaisamy, N. Kumar, A. L. Rathnakumar, C. Sangh, A. Kumar, and R. Thankappan. "Unraveling the Mechanisms of Resistance to Sclerotium tolfsii in Peanut (*Araachis hypogaea* L.) Using Comparactive RNA-Saq Analysis of Resistant and Susceptible Genotypes." *PLoS One* 15, no. 8 (2020): e0236823.

Bosamia, T. C., G. P. Mishra, R. Thankappan, and J. R. Dobaria. "Novel and Stress Relevant EST Derived SSR Markers Developed and Validated in Peanut." *PLoS One* 10 (2015): e0129127. https://doi.org/10.1371/journal.pone.0129127.

Branch, W. D. "Registration of 'Georgia Browne' Peanut." *Crop Science* 34 (1994): 1125–1126.

Branch, W. D. "Registration of 'Georgia Green' Peanut." *Crop Science* 36 (1996): 806. https://doi.org/10.2135/cropsci1994.0011183X003400040058x.

Brar, G. S., B. A. Cohen, C. L. Vick, and G. W. Johnson. "Recovery of Transgenic Peanut (*Arachis hypogaea* L.) Plants from Elite Cultivars Utilizing ACCELL Technology." *Plant Journal* 5 (1994): 745–753. https://doi.org/10.111

Braune, H. J. "Effect of Structure of Host Egg Mass on the Effectiveness of Egg Parasite of *Spodoptera litura* (F.) (Lepidoptera: Noctuidae)." *Drosera* 82, no. 1 (1982): 7–16.

Brown, S., J. Todd, and A. Culbreath. "Effect of Selected Cultural Practices on Incidence of Tomato Spotted Wilt Virus and Populations of Thrips Vectors in Peanuts." *Tospoviruses Thrips Floral Veg Crops* 431 (1996): 491–498. https://doi.org/10.17660/ActaHortic.1996.431.45.

Buddenhagen, I. W., and A. Kelman. "Biological and Physiological Aspects of Bacterial Wilt Caused by *Pseudomonas solanacearum*." *Annual Review of Phytopathology* 2 (1964): 203–230. https://doi.org/10.1146/annurev.py.02.090164.001223.

Burow, M. D., J. L. Starr, C. Park, C. E. Simpson, and A. H. Paterson. "Introgression of Homeologous Quantitative Trait Loci (QTLs) for Resistance to the Root-Knot Nematode [*Meloidogyne arenaria* (Neal) Chitwood] in an Advanced Backcross-QTL Population of Peanut (*Arachis hypogaea* L.)." *Molecular Breeding* 34, no. 2 (2014): 393–406. https://doi.org/10.1007/s11032-014-0042-2.

Carpentieri-Pípolo, V., M. Gallo-Meagher, D. W. Dickson, D. W. Gorbet, M. de Lurdes Mendes, et al. "Molecular Marker Screening of Peanut (*Arachis hypogaea* L.) Germplasm for *Meloidogyne arenaria* Resistance." *African Journal of Biotechnology* 13, no. 26 (2014): 2608–2612. https://doi.org/10.5897/AJB2014.13816.

Cermak, T., E. L. Doyle, M. Christian, L. Wang, Y. Zhang, et al. "Efficient Design and Assembly of Custom TALEN and Other TAL Effector Based Constructs for DNA Targeting." *Nucleic Acids Research* 39 (2011): 82. https://doi.org/10.1093/nar/gkr218.

Chappell, G. F., B. B. Shew, J. M. Ferguson, and M. K. Beute. "Mechanisms of Resistance to *Sclerotinia minor* in Selected Peanut Genotypes." *Crop Science* 35 (1995): 692–696. https://doi.org/10.2135/cropsci1995.0011183X003500030007x.

Chen, X., H. Li, M. K. Pandey, Q. Yang, X. Wang, et al. "Draft Genome of the Peanut A-Genome Progenitor (*Arachis duranensis*) Provides Insights into Geocarpy, Oil Biosynthesis, and Allergens." *Proceedings of the National Academy of Sciences of the United States of America* 113 (2016): 6785–6790. https://doi.org/10.1073/pnas.1600899113.

Chen, X., Q. Lu, H. Liu, J. Zhang, Y. Hong, et al. "Sequencing of Cultivated Peanut, *Arachis hypogaea*, Yields Insights into Genome Evolution and Oil Improvement." *Molecular Plant* 12, no. 7 (2019): 920–934. https://doi.org/10.1016/j.molp.2019.03.005.

Chenault, K. D., H. A. Melouk, and M. E. Payton. "Field Reaction to Sclerotinia Blight among Transgenic Peanut Lines Containing Antifungal Genes." *Crop Science* 45 (2005): 511–515. https://doi.org/10.2135/cropsci2005.0511.

Chu, Y., C. C. Holbrook, and P. Ozias-Akins. "Two Alleles of ahFAD2B Control the High Oleic Acid Trait in Cultivated Peanut." *Crop Science* 49, no. 6 (2009): 2029–2036.

Chu, Y., P. Chee, A. K. Culbreath, T. G. Isleib, C. C. Holbrook, et al. "Major QTLs for Resistance to Early and Late Leaf Spot Diseases Are Identified on Chromosomes 3 and 5 in Peanut (*Arachis hypogaea*)." *Frontiers in Plant Science* 10 (2019): 883. https://doi.org/10.3389/fpls.2019.00883.

Csinos, A. S., and D. K. Bell. "Peanut Pod Rot Complex." In *Compendium of Peanut Diseases*, 2nd ed., edited by N. Kokalis-Burelle, D. M. Porter, R. Rodriguez-Kabana, D. H. Smith, and P. Subrahmanyam, 23–24. St. Paul, MN: APS Press, 1997.

Cui, R., J. Clevenger, Y. Chu, T. Brenneman, T.G. Isleib, et al. "Quantitative Trait Loci Sequencing-Derived Molecular Markers for Selection of Stem Rot Resistance in Peanut." *Crop Science* 60, no. 4 (2020): 2008–2018. https://doi.org/10.1002/csc2.20047.

Culbreath, A. K., J. W. Todd, and J. W. Demski. "Productivity of Florunner Peanut Infected with Tomato Spotted Wilt Virus." *Peanut Science* 19 (1992a): 11–14. https://doi.org/10.3146/I0095-3679-19-1-3.

Culbreath, A. K., J. W. Todd, J. W. Demski, and J. R. Chamberlin. "Disease Progress of Spotted Wilt in Peanut Cultivars Florunner and Southern Runner." *Phytopathology* 82 (1992b): 766–771. https://doi.org/10.1094/PHYTO-82-766.

Culbreath, A. K., J. W. Todd, W. D. Branch, S. L. Brown, J. W. Demski, and J. P. Beasley. "Effect of New Peanut Cultivar Georgia Browne on Epidemics of Spotted Wilt." *Plant Disease* 78 (1994): 1185–1189. https://doi.org/10.1094/PD-78-1185.

Culbreath, A. K., J. W. Todd, D. W. Gorbet, W. D. Branch, R. K. Sprenkel, et al. "Disease Progress of Tomato Spotted Wilt Virus in Selected Peanut Cultivars and Advanced Breeding Lines." *Plant Disease* 80 (1996): 70–73. https://doi.org/10.1094/PD-80-0070.

Culbreath, A. K., J. W. Todd, and S. L. Brown. "Epidemiology and Management of Tomato Spotted Wilt in Peanut." *Annual Review of Phytopathology* 41 (2003): 53–75. https://doi.org/10.1146/annurev.phyto.41.052002.095522.

Culbreath, A. K., B. L. Tillman, R. S. Tubbs, J. P. Beasley, R. C. Kemerait, and T. B. Brenneman. "Interactive Effects of Planting Date and Cultivar on Tomato Spotted Wilt of Peanut." *Plant Disease* 94 (2010): 898–904. https://doi.org/10.1094/PDIS-94-7-0898.

Culver, J. N., J. L. Sherwood, and H. A. Melouk. "Resistance to Peanut Stripe Virus in *Arachis* Germplasm." *Plant Disease* 71 (1987): 1080–1082. https://doi.org/10.1094/PD-71-1080.

Dang, P. M., M. C. Lamb, K. L. Bowen, and C. Y. Chen. "Identification of Expressed R-genes Associated with Leaf Spot Diseases in Cultivated Peanut." *Molecular Biology Reports* 46 (2019): 225–239. https://doi.org/10.1007/s11033-018-4464-5.

Dang, P. M., M. C. Lamb, and C. Y. Chen. "Association of Differentially Expressed R-gene Candidates with Leaf Spot Resistance in Peanut (*Arachis hypogaea* L.)." *Molecular Biology Reports* 48 (2021): 323–334. https://doi.org/10.1007/s11033-020-06049-3.

De Blas, F. J., C. I. Bruno, R. S. Arias, C. Ballén-Taborda, E. Mamani, et al. "Genetic Mapping and QTL Analysis for Peanut Smut Resistance." *BMC Plant Biology* 21 (2021): 312. https://doi.org/10.1186/s12870-021-03023-4.

Deepthi, K. C., and N. P. E. Reddy. "Stem Rot Disease of Peanut (*Arachis hypogaea* L.) Induced by *Sclerotium rolfsii* and Its Management." *International Journal of Life Sciences, Biotechnology, and Pharma Research* 2, no. 3 (2013): 26–38. https://doi.org/10.1007/s13313-020-00748-y.

Demski, J. W., and G. Sowell. "Resistance to Peanut Mottle in *Arachis* spp." *Peanut Science* 8 (1981): 43–44. https://doi.org/10.3146/i0095-3679-8-1-11.

Demski, J. W., and D.R.V. Reddy. "Diseases Caused by Viruses."In *Compendium of Peanut Diseases*, 2nd ed., edited by N. Kokalis-Burelle, D. M. Porter, R. Rodriguez-Kabana, D. H. Smith, and P. Subrahmanyam, 53–59. St. Paul, MN: APS Press, 1997.

Dietzgen, R. G., N. Mitter, C. M. Higgins, R. Hall, P. Y. Teycheney, et al. "Harnessing RNA Silencing to Protect Peanuts from Stripe Disease." *Proceedings of the Fourth International Crop Science Congress*, Brisbane, Australia, 26 September–1 October 2004.

Ding, Y. B., X. K. Qiu, H. Y. Luo, L. Huang, J. B. Guo, et al. "Comprehensive Evaluation of Chinese Peanut Mini-mini Core Collection and QTL Mapping for Aflatoxin Resistance." *BMC Plant Biology* 22 (2022): 207. https://doi.org/10.1186/s12870-022-03582-0.

Dodia, S. M., B. Joshi, S. S. Gangurde, P. P. Thirumalaisamy, G. P. Mishra, et al. "Genotyping-by-Sequencing Based Genetic Mapping Reveals Large Number of Epistatic Interactions for Stem Rot Resistance in Groundnut." *Theoretical and Applied Genetics* 132, no. 4 (2019): 1001–1016. https://doi.org/10.1007/s00122-018-3255-7.

Dwivedi, S. L., D. V. Reddy, S. N. Nigam, G. R. Rao, J. A. Wightman, et al. "Registration of 'ICGV 86031' Peanut Germplasm." *Crop Science* 33, no. 1 (1993): 220. https://doi.org/10.2135/cropsci1993.0011183X003300010069x.

Dwivedi, S. L., J. H. Crouch, S. N. Nigam, M. E. Ferguson, and A. H. Paterson. "Molecular Breeding of Groundnut for Enhanced Productivity and Food Security in the Semi-arid

Tropics: Opportunities and Challenges." *Advances in Agronomy* 80 (2003): 153–221. https://doi.org/10.1016/S0065-2113(03)80004-4.

Dwivedi, S. L., S. Gurtu, and S. N. Nigam. "AFLP Diversity Among Selected Foliar Diseases Resistant Peanut (*Arachis hypogaea* L.) Germplasm." *Indian Journal of Plant Genetic Resources* 15, no. 1 (2002): 46–50.

Entoori, K., R. Sreevathsa, M. K. Arthikala, P. A. Kumar, A. R. V. Kumar, et al. "A Chimeric Cry1X Gene Imparts Resistance to *Spodoptera litura* and *Helicoverpa armigera* in the Transgenic Peanut." *European and Asian Journal of Biological Sciences* 2, no. 1 (2008): 53–65.

FAOSTAT. 2022. Accessed December 6, 2024. http://faostat.fao.org/.

Fávero, A. P., C. E. Simpson, F. M. J. Valls, and N. A. Velo. "Study of Evolution of Cultivated Peanut through Cross Ability Studies Among *Arachis ipaënsis*, *A. duranensis*, and *A. hypogaea*." *Crop Science* 46 (2006): 1546–1552. https://doi.org/10.2135/cropsci2005.09-0331.

Ferguson, M., M. Burow, S. Schulze, P. Bramel, A. Paterson, S. Kresovich, and S. Mitchell. "Microsatellite Identification and Characterization in Peanut (*A. hypogaea* L.)." *Theoretical and Applied Genetics* 108, no. 6 (2004): 1064–1070.

Foster, D. J., H. T. Stalker, J. C. Wynne, and M. K. Beute. "Resistance of *Arachis hypogaea* L. and Wild Relatives to *Cercospora arachidicola* Hori." *Oleagineux* 36 (1981): 139–143.

Franklin, C. I., K. M. Shorrosh, A. N. Trieu, B. G. Cassidy, and R. S. Nelson. "Stable Transformation of Peanut Callus via *Agrobacterium*-Mediated DNA Transfer." *Transgenic Research* 2 (1993): 321–324. https://doi.org/10.1007/BF01976172.

Garcia, G. M., H. T. Stalker, E. Shroeder, and G. Kochert. "Identification of RAPD, SCAR, and RFLP Markers Tightly Linked to Nematode Resistance Genes Introgressed from *Arachis cardenasii* into *Arachis hypogaea*." *Genome* 39, no. 5 (1996): 836–845. https://doi.org/10.1139/g96-106.

Gayathri, M., K. Shirasawa, R. K. Varshney, M. K. Pandey, and R. S. Bhat. "Development of *AhMITE1* Markers through Genome-Wide Analysis in Peanut (*Arachis hypogaea* L.)." *BMC Research Notes* 11, no. 1 (2018): 1–6. https://doi.org/10.1186/s13104-017-3121-8.

Ghanekar, A. M., D. V. R. Reddy, N. Iizuka, P. W. Amin, and R. W. Gibbons. "Bud Necrosis of Peanut (*Arachis hypogaea*) in India Caused by Tomato Spotted Wilt Virus." *Annals of Applied Biology* 93 (1979): 173–179. https://doi.org/10.1111/j.1744-7348.1979.tb06528.x.

Gong, L., S. Han, M. Yuan, X. Ma, A. Hagan, and G. He. "Transcriptomic Analyses Reveal the Expression and Regulation of Genes Associated with Resistance to Early Leaf Spot in Peanut." *BMC Research Notes* 13, no. 1 (2020): 381. https://doi.org/10.1186/s13104-020-05225-9.

Gorbet, D. W., A. J. Norden, F. M. Shokes, and D. A. Knauft. "Registration of 'Southern Runner' Peanut." *Crop Science* 27 (1987): 817.

Gowda, M.V.C., B.N. Motagi, G.K. Naidu, S.B. Diddimimani, and R. Sheshagiri. "GPBD 4: A Spanish Bunch Groundnut Genotype Resistant to Rust and Late Leafspot." *International Arachis Newsletter* 22 (2002): 29–32.

Green, C. C., and J. C. Wynne. "Field Evaluation of the Components of Partial Resistance to Early Leaf Spot in Peanut." *Euphytica* 35 (1986): 561–573. https://doi.org/10.1007/BF00021865.

Grichar, W. J. "Management of Stem Rot of Peanuts (*Arachis hypogaea*) Caused by *Sclerotium rolfsii* with Fungicides." *Crop Protection* 14, no. 2 (1995): 111–115. https://doi.org/10.1016/0261-2194(95)92864-j.

Guo, B, X. Chen, P. Dang, B. T. Scully, X. Liang, C. C. Holbrook, J. Yu, and A. K. Culbreath. "Peanut Gene Expression Profiling in Developing Seeds at Different Reproduction Stages during *Asperillus parasiticus* Infection." *BMC Developmental Biology* 8, no. 12 (2008).

Guo, B., P. Khera, H. Wang, Z. Peng, H. Sudini, et al. "Annotation of Trait Loci on Integrated Genetic Maps of *Arachis* Species." In *Peanuts: Genetics, Processing and Utilization*, edited by H. T. Stalker and R. F. Wilson, 163–207. New York: Academic Press and AOCS Press, 2016. https://doi.org/10.1016/B978-1-63067-038-2.00006-X.

Hake, A. A., K. Shirasawa, A. Yadawad, M. Sukruth, M. Patil, et al. "Mapping of Important Taxonomic and Productivity Traits Using Genic and Non-Genic Transposable Element Markers in Peanut (*Arachis hypogaea* L.)." *PLoS One* 12, no. 10 (2017): e0186113. https://doi.org/10.1371/journal.pone.0186113.

Hale, H., E. M. Gardner, J. Viruel, L. Pokorny, and M. G. Johnson. "Strategies for Reducing Pre-Sample Costs in Target Capture Sequencing for Phylogenomics and Population Genomics in Plants." *Applications in Plant Sciences* 8, no. 4 (2020): e11337.

Halward, T. M., T. H. Stalker, E. A. Larue, and G. Kochert. "Genetic Variation Detectable with Molecular Markers among Unadapted Germplasm Resources of Cultivated Peanut and Related Wild Species." *Genome* 34, no. 6 (1991): 1013–1020.

Halward, T. M., T. H. Stalker, E. A. Larue, and G. Kochert. "Use of Single-primer DNA Amplifications in Genetic Studies of Peanut (*Arachis hypogaea* L.)." *Plant Molecular Biology* 18, no. 2 (1992): 315–325.

Hammons, R. O. "Groundnut Rust Research in the America." In *Groundnut Rust Disease: Proceedings of A Discussion Group Meeting Held at ICRISAT Center*, edited by D. McDonald, P. Subrahmanyam, and J.A. Wightman, 65–71. Patancheru, India, 1987.

Halward, T., H. T. Stalker, and G. Kochert. "Development of an RFLP Linkage Map in Diploid Peanut Species." *Theoretical and Applied Genetics* 87, no. 3 (1993): 379–384. https://doi.org/10.1007/BF01184927.

Han, S., H. Liu, M. Yan, F. Qi, Y. Wang, et al. "Differential Gene Expression in Leaf Tissues Between Mutant and Wild-Type Genotypes Response to Late Leaf Spot in Peanut (*Arachis hypogaea* L.)." *PLoS One* 12, no. 8 (2017): e0183428. https://doi.org/10.1371/journal.pone.0183428.

He, G., and C. S. Prakash. "Identification of Polymorphic DNA Markers in Cultivated Peanut (*Arachis hypogaea* L.)." *Euphytica* 97 (1997): 143–149. https://doi.org/10.1023/A:1002949813052.

He, G., R. Meng, M. Newman, G. Gao, R. N. Pittman, and C. S. Prakash. "Microsatellites as DNA Markers in Cultivated Peanut (*Arachis hypogaea* L.)." *BMC Plant Biology* 3 (2003): 3.

He, Y., S. Mu, Z. He, B. Wang, and Y. Li. "Ectopic Expression of MYB Repressor *GmMYB3a* Improves Drought Tolerance and Productivity of Transgenic Peanuts (*Arachis hypogaea* L.) under Conditions of Water Deficit." *Transgenic Research* 20 (2020): 563–574.

Herselman, L., R. Thwaites, F. M. Kimmins, B. Courtois, P. J. Merwe, et al. "Identification and Mapping of AFLP Markers Linked to Peanut (*Arachis hypogaea* L.) Resistance to the Aphid Vector of Peanut Rosette Disease." *Theoretical and Applied Genetics* 109, no. 7 (2004): 1426–1433. https://doi.org/10.1007/s00122-004-1756-z.

Higgins, C. M., R. M. Hall, N. Mitter, A. Cruickshank, and R. G. Dietzgen. "Peanut Stripe Potyvirus Resistance in Peanut (*Arachis hypogaea* L.) Plants Carrying Viral Coat Protein Gene Sequences." *Transgenic Research* 13 (2004): 59–67. https://doi.org/10.1023/B:TRAG.0000017166.29458.74.

Holbrook, C. C., and J. P. Noe. "Resistance to *Meloidogyne arenaria* in *Arachis* spp. and Its Implication on Development of Resistant Peanut Cultivars." *Peanut Science* 17 (1990): 35–38.

Holbrook, C. C., and A. K. Culbreath. "Registration of 'Tifrunner'Peanut." *Journal of Plant Registration* 1, no. 2 (2007): 124.

Holbrook, C. C., P. Ozias-Akins, Y. Chu, A. K. Culbreath, C. K. Kvien, and T. B. Brenneman. "Registration of 'Tifguard' Peanut." *Journal of Plant Registration* 2, no. 2 (2008): 92–94.

Hong, Y. B., X. Q. Liang, X. P. Chen, H. P. Liu, G. Y. Zhou, et al. "Construction of Genetic Linkage Map Based on SSR Markers in Peanut (*Arachis hypogaea* L.)." *Agricultural Sciences in China* 7 (2008): 915–921. https://doi.org/10.1016/S1671-2927(08)60130-3.

Hou, H.M., B. Liao, Y. Lei, X. Ren, S. Wang, D. Li, H. Jiang, J. Huang, and B. Chen. "Identification of AFLP Markers for Resistance to Peanut Rust." *Chinese Journal of Oil Crop Sciences* 29, no. 2 (2007): 89–92.

Huai, D., L. Huang, X. Xue, B. Yu, Y. Ding, G. Jin et al. "Identification of Candidate Genes Associated with Resistance to Aflatoxin Production in Peanut through Genetic Mapping and Transcriptome Analysis." *Theoretical and Applied Genetics* 138, no. 71 (2025).

Iqbal, M. M., Y. Zafar, F. Nazir, S. Ali, J. Iqbal, et al. "Over Expression of Bacterial Chitinase Gene in Pakistani Peanut (*Arachis hypogaea* L.) Cultivar GOLDEN." *African Journal of Biotechnology* 10 (2011): 5838–5844. https://doi.org/10.5897/AJB11.681.

Iqbal, M. M., F. Nazir, S. Ali, M. A. Asif, Y. Zafar, et al. "Over Expression of Rice Chitinase Gene in Transgenic Peanut (*Arachis hypogaea* L.) Improves Resistance Against Leaf Spot." *Molecular Biotechnology* 50 (2012): 129–136. https://doi.org/10.1007/s12033-011-9426-2.

Islam, W., K. N. Ahmed, A. Nargis, and U. Islam. "Occurrence, Abundance and Extent of Damage Caused by Insect Pests of Peanuts (*Arachis hypogaea*)." *Malaysian Agricultural Journal* 54 (1983): 18–24.

Isleib, T. G., J. C. Wynne, and S. N. Nigam. "Peanut Breeding." In *The Peanut Crop: A Scientific Basis for Improvement*, edited by J. Smartt, 552–623. Londin: Chapman & Hall, 1994.

Jain, S., S. Srivastava, N. B. Sarin, and N. N. Kav. "Proteomics Reveals Elevated Levels of PR 10 Proteins in Saline-Tolerant Peanut (*Arachis hypogaea*) Calli." *Plant Physiology and Biochemistry* 44, no. 4 (2006): 251–259.

Janila, P., S. N. Nigam, M. K. Pandey, P. Nagesh, and R. K. Varshney. "Groundnut Improvement: Use of Genetic and Genomic Tools." *Frontiers in Plant Science* 4 (2013): 23. https://doi.org/10.3389/fpls.2013.00023.

Jiang, Y., H. Luo, B. Yu, Y. Ding, Y. Kang, H. Li, X. Zhou, N. Liu, W. Chen, J. Gup, D. Huai, Y. Lei, H. Jiang, L. Yan, and B. Liao. "High-Density Genetic Linkage Map Construction Using Whole-Genomic Resequencing for Mapping QTLs of Resistance to *Aspergillus flavus* Infection in Peanut." *Frontier Plant Science* 12 (2021). https://doi.org/10.3389/fpis.2021.745408.

Kang, H., D. Zhu, R. Lin, S. O. Opiyo, N. Jiang, S. H. Shiu, and G. L. Wang. "A Novel Method for Identifying Polymorphic Transposable Elements via Scanning of High-throughput Short Reads." *DNA Research* 23, no. 3 (2016): 241–251.

Katam, R., V. Gottschalk, P. Survajahala, G. Brewster, P. Paxton, L. Sakata, C. Lee, C. Williams, B. Guo, and L. M. Latinwo. "Advances in Proteomics Research for Peanut Genetics and Breeding." In *Genetics, Genomics and Breeding of Peanuts*, edited by N. Mallikarjuna and R. K. Varshney, 161–177. CRC Press, Boca, Raton, 2014.

Khedikar, Y. P., M. V. C. Gowda, C. Sarvamangala, K. V. Patgar, H. D. Upadhyaya, et al. "A QTL Study on Late Leaf Spot and Rust Revealed One Major QTL for Molecular Breeding for Rust Resistance in Peanut (*Arachis hypogaea* L.)." *Theoretical and Appllied Genetics* 121 (2010): 971–984.

Kilian, A. "DArt-based Whole Genome Profiling and Novel Information Technologies in Support System of Modern Breeding of Groundnut." In *Proceedings of the Third International Conference for Peanut Genomics and Biotechnology on "Advances in Arachis through Genomics and Biotechnology"(AAGB)-*. ICRISAT, Patacheru, India, 2008.

Kochert, G., T. Halward, W. D. Branch, and C. E. Simpson. "RFLP Variability in Peanut (*Arachis hypogaea* L.) Cultivars and Wild Species." *Theoretical and Applied Genetics* 81, no. 5 (1991): 565–570.

Kokalis-Burelle, N., and R. Rodriguesz Kabana. "Diseases Caused by Nematodes." In *Compendium of Peanut Diseases*, 2nd ed., edited by N. Kokalis-Burelle, D. M. Porter, R. Rodriguez-Kabana, D. H. Smith, and P. Subrahmanyam, 45–53. St. Paul, MN: APS Press, 1997.

Kolekar, R. M., V. Sujay, K. Shirasawa, M. Sukruth, Y. P. Khedikar, M. V. C. Gowda, M. K. Pandey, R. K. Varshney, and R. S. Bhat. "QTL Mapping for Late Leaf Spot and Rust Resistance Using An Improved Genetic Map and Extensive Phenotypic Data on a Recombinnant Inbred Line Population in Peanut (*Arachis hypogaea* L.)." *Eupgytica* 209, no. 1 (2016): 147–156.

Kona, P., C. Sangh, K. T. Ravikiran, B. C. Ajay, and N. Kumar. "Genetic and Genomic Resource to Augment Breeding Strategies for Biotic Stresses in Groundnut." In *Genomics-aided Breeding Strategies for Biotic Stress in Grain Legumes*, edited by A. K. Parthar, A. Bohra, A. Lamichaney, R. K. Mishra, and R. K. Varshney, 359–403. Springer, 2024.

Kottapalli, K. R., R. Rakwal, J. Shibato, G. Burow, D. Tissue, J. Burke, N. Puppala, M. Burow, and P. Payton. "Pgysiology and Proteinmics of the Water-Deficit Stress Response in Three Contrasting Peanut Genotypes." *Plant, Cell & Environment* 32, no.4 (2009): 380–407.

Kumar, D., and P. B. Kirti. "Transcriptomic and Proteomic Analysis of Resistant Host Responses in *Arachis diogoi* Chanllenged with Late Leaf Spot Pathigen, *Phaeoisariopsis personata*." *PLoS One* 10, no. 2 (2015): e0117559.

Kumari, V., M. V. C. Gowda, V. Tasiwal, M. K. Pandey, R. S. Bhat et al. "Diversification of Primary Gene Pool through Introgression of Resistance Allele for Foliar Diseases from Synthetic Amphidiploids to Cultivated Peanut (*Arachis hypogaea* L.)." *The Crop Journal* 2, no. 2–3 (2014): 110–119.

Li, T., B. Liu, M. H. Spalding, D. P. Weeks, and B. Yang. "High Efficiency TALEN-Based Gene Editing Produces Disease-Resistant Rice." *Nature Biotechnology* 30 (2012): 390–392. https://doi.org/10.1038/nbt.2199.

Liang, Y., J. M. Cason, M. R. Baring, and E. M. Septiningsih. "Identification of QTLs Associated with Sclerotinia Blight Resistance in Peanut (*Arachis hypogaea* L.)." *Genetic Resources and Crop Evolution* 68, no. 2 (2020): 629–637. https://doi.org/10.1007/s10722-020-01012-4.

Liao, B. S., W. R. Li, and D. R. Sun. "Genetic Studies on Resistance to Bacterial Wilt in *Arachis hypogaea* L." *Oil Crops of China* 3 (1986): 1–7.

Liao, B. S., R. W. Jiang, Z. J. Tan, R. H. Rao, and G. Y. Tang. "A Study on Characters of Nitrogen-Fixation in Some Groundnut Lines Resistant to Bacterial Wilt." *Oil Crops of China* 4 (1992): 34–37.

Liao, B. S., ed. *A Handbook for Diagnosis of Major Diseases and Insects in Peanut.* Wuhan, China: Hubei Sci-Tech Press, 2012.

Liao, B. S. "Germplasm Characterization and Trait Discovery in Peanut." In *The Peanut Genome*, edited by R. K. Varshney, M. K. Pandey, and N. Puppala, 53–68. Springer, 2017.

Liao, B. S., and C. C. Holbrook. "Groundnut." In *Genetic Resources, Chromosome Engineering and Crop Improvement*, edited by R. J. Singh, 51–87. Boca Raton, FL, USA, CRC Press, 2007.

Liao, B. S., X. Q. Liang, H. F. Jiang, Y. Lei, Z. H. Shan, and X. Y. Zhang. "Progress on Genetic Enhancement for Resistance to Peanut Bacterial Wilt in China." In *Bacterial Wilt Diseases and the Ralstonia solanacearum Species Complex*, edited by C. Allen, P. Prior, and A. C. Hayward, 239–246. St. Paul, MN: The American Phytopathological Society, 2005.

Liao, B. S. "Peanut Breeding." In *Genetics, Genomics and Breeding of Peanuts*, edited by N. Mallikarjuna and R. K. Varshney, 61–78. CRC Press, 2014. https://doi.org/10.1201/b16872.

Liu, N., J. Guo, X. Zhou, B. Wu, L. Huang, et al. "High-Resolution Mapping of a Major and Consensus Quantitative Trait Locus for Oil Content to a 0.8-Mb Region on Chromosome A08 in Peanut (*Arachis hypogaea* L.)." *Theoretical and Applied Genetics* 133, no. 1 (2020): 37–49. https://doi.org/10.1007/s00122-019-03438-6.

Liu, H., Z. Sun, X. Zhang, L. Qin, F. Qi, Z. Wang, P. Du, J. Xu, Z. Zhang, S. Han, S. Li, M. Gao, L. Zhang, Y. Cheng, Z. Zheng, B. Huang, and W. Dong. "QTL Mapping of Web Blotch Resistance in Peanut by High-Throughput Genome-Wide Sequencing." *BMC Plant Biology* 20 (2020): 249. https://doi.org/10.1186/s12870-020-02455-8.

Livingstone, D. M., J. L. Hampton, P. M. Phipps, and E. A. Grabau. "Enhancing Resistance to *Sclerotinia minor* in Peanut by Expressing a Barley Oxalate Oxidase Gene." *Plant Physiology* 137 (2005): 1354–1362. https://doi.org/10.1104/pp.104.057232.

Lloyd, A., C. L. Plaisier, D. Carroll, and G. N. Drews. "Targeted Mutagenesis Using Zinc-finger Nucleases in *Arabidopsis*." *Proceedings of Natl Acad Sci USA* 102, no. 6 (2005): 2232–2237.

Lu, J., and B. Pickersgill. "Isozyme Variation and Species Relationships in Peanut and Its Wild Relatives (*Arcahis* L.-Leguminosae)." *Theoretical and Applied Genetics* 85, no. 5 (1993): 550–560.

Lu, Q., H. Li, Y. Hong, G. Zhang, S. Wen, S. Li, G. Zhou, S. Li, H. Liu, and H. Liu. "Genome Sequencing and Analysis of the Peanut B-Genome Progenitor (*Arachis ipaensis*)." *Frontier Plant Sciences* 9 (2018).

Luo, H., M. K. Pandey, Y. Zhi, H. Zhang, S. Xu, et al. "Discovery of Two Novel and Adjacent QTLs on Chromosome B02 Controlling Resistance Against Bacterial Wilt in Peanut Variety Zhonghua 6." *Theoretical and Applied Genetics* 133, no. 4 (2020): 1133–1148. https://doi.org/10.1007/s00122-020-03537-9.

Luo, M., P. Dang, M. Bausher, C. Holbrook, R. Lee, R. Lynch, and B. Guo. "Identification of Transcripts Involved in Resistance Responeses to Leaf Spot Disease Caused by *Cercosporidium personatum* in Peanut (*Arachis hypogaea*)." *Phytopathology* 95, no. 4 (2005): 381–387.

Luo, H., X. Ren, Z. Li, Z. Xu, X. Li, et al. "Co-localization of Major Quantitative Trait Loci for Pod Size and Weight to a 3.7 cM Interval on Chromosome A05 in Cultivated Peanut (*Arachis hypogaea* L.)." *BMC Genomics* 18 (2017): 58. https://doi.org/10.1186/s12864-016-3456-x.

Mahfouz, M. M., L. Li, M. Shamimuzzaman, A. Wibowo, X. Fang, and J. K. Zhu. "De Novo-Engineered Transcription Activator-Like Effector (TALE) Hybrid Nuclease with Novel DNA Binding Specificity Creates Double-Strand Breaks." *Proceedings of the National Academy of Sciences of the United States of America* 108 (2011): 2623–2628. https://doi.org/10.1073/pnas.1019533108.

Mallikarjuna, N., and D. C. Sastri. "Morphological, Cytological and Disease Resistance Studies of the Intersectional Hybrid between *Arachis hypogaea* L. and *A. glabrata* Benth." *Euphytica* 126, no. 2 (2002): 161–167. https://doi.org/10.1023/A:1016379132324.

Mallikarjuna, N., S. Senthilvels, and D. Hoisington. "Development of New Sources of Tetraploid *Arachis* to Broaden the Genetic Base of Cultivated Groundnut (*Arcahis hypogaea* L.)."*Genetic Resource and Crop Evoluation* 58 (2011): 889–907.

Manone, G., L. D. Stasio, S. D. Caro, G. Picarello, M. A. Nicolai, and P. Ferranti. "Comprehensive Analysis of the Peanut Allergome Combing 2-DE Gel-Based and Gel-free Proteins." *Food Research International* 116 (2019): 1059–1065.

Marasigan, K., M. Toews, J. R. Kemerait, M. R. Abney, A. Culbreath, and R. Srinivasan. "Evaluation of Alternatives to Carbamate and Organophosphate Insecticides against Thrips and Tomato Spotted Wilt Virus in Peanut Production." *Journal of Economic Entomology* 109 (2016): 544–557. https://doi.org/10.1093/jee/tov336.

Marasigan, K., M. Toews, J. R. Kemerait, M. R. Abney, A. Culbreath, and R. Srinivasan. "Evaluation of Alternatives to an Organophosphate Insecticide with Selected Cultural Practices: Effects on Thrips, *Frankliniella fusca*, and Incidence of Spotted Wilt in Peanut Farm Scapes." *Journal of Economic Entomology* 111 (2018): 1030–1041. https://doi.org/10.1093/jee/toy079.

Marka, R., and R. S. Nanna. "Expression of *Tcchitinase-I* Gene in Transgenic Peanut (*Arachis hypogaea* L.) Confers Enhanced Resistance against Leaf Spot and Rust Diseases." *Plant Growth Regulation* 93 (2021): 53–63.

Mayeux, A. "The Peanut Aphid Biology and Control." *Oleagineux* 39, no. 8/9 (1984): 425–434.

McDonald, D., P. Subrahmanyam, R. W. Gibbons, and D. H. Smith. *Early and Late Leafspots of Groundnut*. Information Bulletin No. 21, International Crops Research Institute for the Semi-Arid Tropics, India, 1985.

McDonald, D., D. V. R. Reddy, S. B. Sharma, V. K. Mehan, and P. Subrahmanyam. *Diseases of Peanut*. The Pathology of Food and Pasture Legumes, 63–124, 1998.

McKeown, S., J. Todd, A. Culbreath, D. Gorbet, and J. Weeks. "Planting Date Effects on Tomato Spotted Wilt in Resistant and Susceptible Peanut Cultivars." *Phytopathology* 91 (2001): S60.

McSorley, R., D. W. Dickson, E. M. Candanedo-Lay, T. E. Hewlett, and J. J. Frederick. "Damage Functions for *Meloidogyne arenaria* on Peanut." *Journal of Nematology* 24 (1992): 193–198.

Mehan, V. K., B. S. Liao, Y. J. Tan, A. Robinson-Smith, D. McDonald, and A. C. Hayward. *Bacterial Wilt of Groundnut*. Information Bulletin No. 35, International Crops Research Institute for the Semi-Arid Tropics, India, 1994.

Michelotto, M. D., I. J. de Godoy, M. Z. Pirotta, J. F. dos Santos, E. L. Finoto, and A. Pereira Fávero. "Resistance to Thrips (*Enneothrips flavens*) in Wild and Amphidiploid *Arachis* Species." *PLoS One* 12, no. 5 (2017): e0176811. https://doi.org/10.1371/journal.pone.0176811.

Minja, E. M., P. J. A. van der Merwe, F. M. Kimmins, and P. Subrahmanyam. "Screening Peanut Breeding Lines for Resistance to Aphids, *Aphis craccivora* Koch." *IAN* 19 (1999): 21–23.

Mondal, S., and A. M. Badigannavar. "Mapping of a Dominant Rust Resistance Gene Revealed Two R Genes around the Major Rust_QTL in Cultivated Peanut (*Arachis hypogaea* L.)." *Theoretical and Applied Genetics* 131 (2018): 1671–1681. https://doi.org/10.1007/s00122-018-3106-6.

Moretzsohn, M., M. Hopkins, S. Mitchell, S. Kresovich, J. Valls, and M. Ferreira. "Genetic Diversity of Peanut (Arachis hypogaea L.) and Its Wild Relatives Based on the Analysis of Hypervariable Regions of the Genome." *BMC Plant Biology* 4, no. 1 (2004): 11.

Moretzsohn, M. C., L. Leoi, K. Proite, P. M. Guimarães, S. C. M. Leal-Bertioli, et al. "A Microsatellite Based, Gene-Rich Linkage Map for the AA Genome of *Arachis* (Fabaceae)." *Theoretical and Applied Genetics* 111 (2005): 1060–1071. https://doi.org/10.1007/s00122-005-0028-x.

Moretzsohn, M. C., A. V. Barbosa, D. M. Alves-Freitas, C. Teixeira, and S. C. Leal-Bertioli. "A Linkage Map for the B-Genome of *Arachis* (Fabaceae) and Its Synteny to the A-Genome." *BMC Plant Biology* 9 (2009): 40. https://doi.org/10.1007/s00122-005-0028-x.

Muralidharan, S., Y. Y. Poon, G. C. Wright, P. A. Haynes, and N. A. Lee. "Quantitative Proteomics Analysis of High and Low Polyphenol Expressing Recombinant Inbred Lines (RILs) of Peanut (*Arachis hypogaea* L.)." *Food Chemistry* 334 (2021): 127517.

Naeem, A., S. A. Shad, M. Razaq, A. Waheed, and M. Aslam. "Resistance of *Spodoptera litura* (Lepidoptera: Noctuidae) to Profenofos: Relative Fitness and Cross Resistance." *Crop Protection* 58 (2014): 49–54. https://doi.org/10.1016/j.cropro.2014.01.002.

Naidu, R. A., F. M. Kimmins, C. M. Deom, P. Subrahmanyam, A. J. Chiyembekeza, and P. J. A. Van der Merwe. "Peanut Rosette: A Virus Disease Affecting Peanut Production in Sub-Saharan Africa." *Plant Disease* 83 (1999): 700–709.

Nelson, S. C., C. E. Simpson, and J. L. Starr. "Resistance to *Meloidogyne arenaria* in *Arachis* sp. Germplasm." *Journal of Nematology* 21 (1989): 654–660.

Nigam, S. N., and K. R. Bock. "Inheritance of Resistance to GRV in Peanut." *Annals of Applied Biology* 117 (1990): 553–560.

Nigam, S. N., S. L. Dwivedi, and R. W. Gibbons. "Peanut Breeding: Constraints, Achievements, and Future Possibilities." *Plant Breeding* 61 (1991): 1127–1136.

Niu, C., Y. Akasaka-Kennedy, P. Faustinelli, M. Joshi, K. Rajasekaran, et al. "Antifungal Activity in Transgenic Peanut (*Arachis hypogaea* L.) Conferred by a Nonheme Chloroperoxidase Gene." *Peanut Science* 36 (2009): 126–132. https://doi.org/10.3146/PS08-020.1.

Olorunju, P. E., C. W. Kuhn, and J. W. Demski. "Inheritance of Resistance in Peanut to Mixed Infections of Groundnut Rosette Virus (GRV) and Groundnut Rosette Assistor Virus and a Single Infection of GRV." *Plant Disease* 76 (1992): 95–100. https://doi.org/10.1094/PD-76-0095.

Ozias-Akins, P., E. K. Cannon, and S. B. Cannon. "Genomics Resources for Peanut Improvement." In *The Peanut Genome*, eidted by R. K. Varshney, M. K. Pandey, and N. Puppala, 69–91. Springer, 2017.

Padgham, D. E., F. M. Kimmins, and G. V. Ranga Rao. "Resistance in Peanut (*Arachis hypogaea* L.) to *Aphis craccivora* (Koch.)." *Annals of Applied Biology* 117 (1990): 285–294.

Paik-Ro, O. G., R. L. Smith, and D. A. Knauft. "Restriction Fragment Length Polymorphaism Evaluation of Six Peanut Species within the *Arachis* Section." *Theoretical and Applied Genetics* 84, no. 1 (1992): 201–208.

Palmieri, D. A., A. Hoshino, J. Bravo, C. Lopes, and M. Gimenes. "Isolation and Characterization of Microsatellite Loci from Forage Species *Arachis* pintoi (Genus *Arachis*)." *Molecular Ecology Notes* 2, no. 4 (2002): 551–553.

Palmieri, D. A., M. Bechara, R. Curi, M. Gimenes, and C. Lopes. "Novel Polymorphic Microsatellite Markers in Section Caulorrhizae (*Arachis*, Fabaceae)." *Molecular Ecology Notes* 5, no. 1 (2005): 77–79.

Pandey, M. K., G. Gautami, T. Jayakumar, M. Sriswathi, H. D. Upadhyaya, et al. "Highly Informative Genic and Genomic SSR Markers to Facilitate Molecular Breeding in Cultivated Peanut (*Arachis hypogaea*)." *Plant Breeding* 131, no. 1 (2012): 139–147. https://doi.org/10.1111/j.1439-0523.2011.01911.x.

Pandey, M. K., E. Monyo, P. Ozias-Akins, X. Liang, P. Guimarães, et al. "Advances in *Arachis* Genomics for Peanut Improvement." *Biotechnology Advances* 30 (2012): 639–651. https://doi.org/10.1016/j.biotechadv.2011.11.001.

Pandey, P., V. Ramegowda, and M. Senthil-Kumar. "Shared and Unique Responses of Plants to Multiple Individual Stresses and Stress Combinations: Physiological and Molecular Mechanisms." *Frontiers in Plant Science* 6 (2015): 723. https://doi.org/10.3389/fpls.2015.00723.

Pandey, M. K., H. Wang, P. Khera, M. K. Vishwakarma, S. M. Kale, et al. "Genetic Dissection of Novel QTLs for Resistance to Leaf Spots and Tomato Spotted Wilt Virus in Peanut (*Arachis hypogaea* L.)." *Frontiers in Plant Science* 8 (2017a): 25. https://doi.org/10.3389/fpls.2017.00025.

Pandey, M. K., A. Khan, V. K. Singh, M. K. Vishwakarma, Y. Shasidhar, et al. "QTL-seq Approach Identified Genomic Regions and Diagnostic Markers for Rust and Late Leaf

Spot Resistance in Peanut (*Arachis hypogaea* L.)." *Plant Biotechnology Journal* 15, no. 8 (2017b): 927–941. https://doi.org/10.1111/pbi.12686.

Patil, A. S., R. Thankappan, R. Mehta, R. Yadav, and A. Kumar. "Evaluation of Transgenic Peanut Plants Encoding Coat Protein and Nucleocapsid Protein Genes for Resistance to Tobacco Streak Virus and Peanut Bud Necrosis Virus." *Journal of Environmental Biology* 38 (2017): 187–196. https://doi.org/10.22438/jeb/38/2/MS-191.

Paterson, A. H., H. T. Stalker, M. Gallo-Meagher, S. L. Dwivedi, J. H. Crouch, and E. Mace. "Genomics and Genetic Enhancement of Peanut." *Legume Crop Genomics* (2004): 97–109.

Porter, D. M. "Increased Severity of Sclerotinia Blight in Peanuts Treated with Captafol and Chlorothanonil." *Plant Disease* 64 (1980): 394–395. https://doi.org/10.1094/pd-64-394.

Power, I. L., P. C. Faustinelli, V. A. Orner, V. S. Sobolev, and R. S. Aria. "Analysis of Small RNA Populations Generated in Peanut Leaves after Exogenous Application of dsRNA and ds DNA Targeting Aflatoxin Synthesis Genes." *Scientific Reports* 10, 13820 (2020). https://doi.org/10.1038/s41598-020-70618-6.

Prasad, K., P. Bhatnagar-Mathur, F. Waliyar, and K. K. Sharma. "Overexpression of a Chitinase Gene in Transgenic Peanut Confers Enhanced Resistance to Major Soil-Borne and Foliar Fungal Pathogens." *Journal of Plant Biochemistry and Biotechnology* 22 (2013): 222–233. https://doi.org/10.1007/s13562-012-0155-9.

Prasada Rao, R. D. V. J., A. S. Reddy, S. K. Chakarbarty, D. V. R. Reddy, V. R. Rao, and J. P. Moss. "Identification of Peanut Stripe Virus Resistance in Wild *Arachis* Germplasm." *Peanut Science* 18 (1991): 1–2. https://doi.org/10.3146/I0095-3679-18-1-1.

Proite, K., S. C. Leal-Bertioli, D. J. Bertioli, M. C. Moretzsohn, F. R. da Silva, N. F. Martins, and P. M. Guimaraes. "ESTs from Wild *Arachis* Species for Gene Discovery and Marker Development." *BMC Plant Biology* 7, no. 1 (2007): 7

Qin, H. D., S. P. Feng, C. Chen, Y. F. Guo, S. Knapp, et al. "An Integrated Genetic Linkage Map of Cultivated Peanut (*Arachis hypogaea* L.) Constructed from Two RIL Populations." *Theoretical and Applied Genetics* 124 (2012): 653–664. https://doi.org/10.1007/s00122-011-1737-y.

Ramakrishnan, N., V. S. Saxena, and S. Dhingra. "Insecticide-Resistance in the Population of *Spodoptera litura* (F.) in Andhra Pradesh." *Pesticides* 18, no. 9 (1984): 23–27.

Ramanatha Rao, V. "Origin, Distribution and Taxonomy of *Arachis* and Sources of Resistance to Groundnut Rust (*Puccinia arachidis* Speg.)." In *Groundnut Rust Disease: Proceedings of A Discussion Group Meeting Held at ICRISAT Center*, edited by D. McDonald, P. Subrahmanyam, and J.A. Wightman, 3-15. ICRISAT, Patancheru, India, 1997.

Rathod, V., R. Hamid, R. S. Tomar, S. Padhiyar, J. Kheni, et al. "Peanut (*Arachis hypogaea*) Transcriptome Revealed the Molecular Interactions of the Defense Mechanism in Response to Early Leaf Spot Fungi (*Cercospora arachidicola*)." *Plant Gene* 23 (2020a): 100243. https://doi.org/10.1016/j.plgene.2020.100243.

Rathod, V., R. Hamid, R. S. Tomar, R. Patel, S. Padhiyar, et al. "Comparative RNA-Seq Profiling of a Resistant and Susceptible Peanut (*Arachis hypogaea*) Genotypes in Response to Leaf Rust Infection Caused by *Puccinia arachidis*." *Biotechnology* 10, no. 6 (2020b): 284. https://doi.org/10.1007/s13205-020-02270-w.

Reddy, D. V. R. "Peanut Viruses and Virus Diseases: Distribution, Identification and Control." *Review of Plant Pathology* 70 (1991): 665–678.

Reddy, D. V. R., and K. T. Devi. "Peanuts." In *Viruses and Virus-Like Diseases of Major Crops in Developing Countries*, edited by G. Loebenstein and G. Thottappilly, 397–423. Dordrecht: Kluwer Academic Publishers, 2003.

Reddy, D. V. R., P. W. Amin, D. McDonald, and A. M. Ghanekar. "Epidemiology and Control of Peanut Bud Necrosis and Other Diseases of Legume Crops in India Caused by Tomato Spotted Wilt Virus." In *Plant Virus Disease Epidemiology*, edited by R. T. Plumb and J. M. Thresh, 98–102. Oxford: Blackwell Scientific Publications Ltd., 1983.

Reddy, D. V. R., A. S. Ratna, M. R. Sudarshana, F. Poul, and K. I. Kumar. "Serological Relationships and Purification of Bud Necrosis Virus, a Tospovirus Occurring in Peanut (*Arachis hypogaea* L.) in India." *Annals of Applied Biology* 120 (1992): 279–286. https://doi.org/10.1111/j.1744-7348.1992.tb03425.x.

Reddy, L. J., S. N. Nigam, J. P. Moss, A. K. Singh, and P. Subrahmanyam. "Registration of ICGV 86699 Peanut Germplasm Line with Multiple Disease and Insect Resistance." *Crop Science* 36 (1996): 821. https://doi.org/10.2135/cropsci1996.0011183X003600030072x.

Reddy, L. J., S. N. Nigam, P. Subrahmanyam, F. M. Ismael, N. Govinden, and P. J. A. van der Merwe. "Registration of Groundnut Cultivar Venus (ICGV 87853)." *International Arachis Newsletter* 20 (2000): 29–31.

Rohini, V. K., and R. K. Sankara. "Transformation of Peanut (*Arachis hypogaea* L.) with Tobacco Chitinase Gene: Variable Response of Transformants to Leaf Spot Disease." *Plant Science* 160 (2001): 883–892. https://doi.org/10.1016/s0168-9452(00)00462-3.

Sahayaraj, K., and P. Martin. "Assessment of *Rhynocoris marginatus* (Fab.) (Hemiptera: Reduviidae) as Augmented Control in Peanut Pests." *Journal of Central European Agriculture* 4, no. 2 (2003): 103–110.

Sahayaraj, K., and S. K. R. Namachivayam. "Field Evaluation of Three Entomopathogenic Fungi on Peanut Pests." *Tropicultura* 29, no. 3 (2011): 143–147.

Sahayaraj, K., and C. Ravi. "Evaluation of Reduviid Predators and Plant Products Against Chosen Peanut Pests." *Archives of Phytopathology and Plant Protection* 40, no. 4 (2007): 281–290. https://doi.org/10.1080/03235400600586716.

Sarvamangala, C., M. V. C. Gowda, and R. K. Varshney. "Identification of Quantitative Trait Loci for Protein Content, Oil Content and Oil Quality for Peanut (*Arachis hypogaea* L.)." *Field Crops Research* 122 (2011): 49–59.

Shaibu, A. S., Z. L. Miko, H. A. Ajeigbe, S. G. Mohammed, A. Usman, M. S. Muhammad, M. L. Umar, H. O. Ahmed, A. A. Adnan, and C. Sneller. "Genome-wide Detection of Markers Associated with Early Leaf Spot and Pod Weight in Groundnut Using SNP and SArt Markers." *Journal of Crop Improvement* 1, no. 14 (2020).

Sharma, A., D. Kumar, S. Kumar, S. Rampuria, A. R. Reddy, and P. B. Kirti. "Ectopic Expression of An Atypical Hydrophobic Group 5 LEA Protein from Wild Peanut, Arachis diogoi Confers Abiotica Stress Tolerance in Tobacco." *PLoS One* 11, no. 3 (2016): e0150609.

Sharma, H. C., G. Pampapathy, S. L. Dwivedi, and L. J. Reddy. "Mechanisms and Diversity of Resistance to Insect Pests in Wild Relatives of Peanut." *Journal of Economic Entomology* 96, no. 6 (2003): 1886–1897. https://doi.org/10.1093/jee/96.6.1886.

Shirasawa, K., R. S. Bhat, Y. P. Khedikar, V. Sujay, R. M. Kolekar, S. B. Yeri, M. Sukruth, S. Cholin, B. Asha, M. K. Pandey, R. K. Varshney, and M. V. C. Gowda. "Sequencing Analysis of Genetic Loci for Resistance for Late Leaf Spot and Rust in Peanut (*Arachis hypogaea* L.)." *Frontier Plant Science* 9 (2018): 1727

Shirasawa, K., H. Hirakawa, S. Tabata, M. Hasegawa, H. Kiyoshima, et al. "Characterization of Active Miniature Inverted-Repeat Transposable Elements in the Peanut Genome." *Theoretical and Applied Genetics* 124 (2012): 1429–1438. https://doi.org/10.1007/s00122-012-1798-6.

Shirasawa, K., D. J. Bertioli, R. K. Varshney, M. C. Moretzsohn, S. C. Leal-Bertioli, et al. "Integrated Consensus Map of Cultivated Peanut and Wild Relatives Reveals Structures of the A and B Genomes of *Arachis* and Divergence of the Legume Genomes." *DNA Research* 20 (2013): 173–184. https://doi.org/10.1093/dnares/dss042.

Shoba, D., N. Manivannan, P. Vindhiyavarman, and S. N. Nigam. "Identification of Quantitative Trait Loci (QTL) for Late Leaf Spot Disease Resistance in Groundnut (*Arachis hypogaea* L.)." *Legume Research* 36, no. 5 (2013): 467–472.

Shokes, F. M., and A. K. Culbreath. "Early and Late Leaf Spots." In *Compendium of Peanut Diseases*, 2nd ed., edited by N. Kokalis-Burelle, D. M. Porter, R. Rodriguez-Kabana, D. H. Smith, and P. Subrahmanyam, 160–206. St. Paul: APS Press, 1997.

Shu, H., Z. Luo, Z. Peng, and J. Wang. "The Application of CRISPR/Cas9 in Hairy Roots to Explore the Functions of *AhNFR1* and *AhNFR5* Genes during Peanut Nodulation." *BMC Plant Biology* 20 (2020): 2–15.

Simpson, C. E., and J. L. Starr. "Registration of 'COAN' Peanut." *Crop Science* 41 (2001): 918. https://doi.org/10.2135/cropsci2001.413918x.

Simpson, C. E., J. L. Starr, S. C. Nelson, K. E. Woodard, and O. D. Smith. "Registration of TxAG6 and TxAG7 Peanut Germplasm." *Crop Science* 33 (1993): 1418. https://doi.org/10.2135/CROPSCI1993.0011183X003300060079X.

Simpson, C. E., J. L. Starr, G. T. Church, M. D. Burow, and A. H. Paterson. "Registration of 'NemaTAM' Peanut." *Crop Science* 43 (2003): 1561. https://doi.org/10.2135/cropsci2003.1561.

Simpson, C. E., J. L. Starr, M. R. Baring, M. D. Burow, J. M. Carson, and J. N. Wilson. "Registration of 'Webb' Peanut." *Journal of Plant Registrations* 7 (2013): 265–268.

Singh, A. K. "Hybridization Barriers Among Species of *Arachis* L., Namely of the Section *Arachis* (Including the Peanut) and *Erectoides*." *Genetic Resources and Crop Evolution* 45 (1998): 41–45. https://doi.org/10.1023/A:1008681020851.

Singh, A. K., J. P. Moss, and B. G. Rao. "Incorporation of Rust Resistance from Wild *Arachis* Species into the Cultivated Groundnut." In *Groundnut Rust Disease: Proceedings of A Discussion Group Meeting Held at ICRISAT Center*, edited by D. McDonald, P. Subrahmanyam, and J. A. Wightman, 41–48. Patancheru, India: ICRISAT, 1987.

Singh, K. N., and G. C. Sachan. "Assessment of Yield Loss Due to Insect Pests at Different Growth Stages of Peanut in Pantnagar, Uttar Pradesh India." Crop Protection 11, no. 5 (1992): 414–418. https://doi.org/10.1016/0261-2194(92)90023-X.

Singh, A. K., and C. E. Simpson. "Biosystematic and Genetic Resources." In *The Groundnut Crop: A Scientific Basis for Improvement*, edited by J. Smartt, 96–137. Dordrecht: Springer, 1994.

Sing, A. K., V. K. Mehan, and S. N. Nigam. "Sources of Resistance to Groundnut Fungal and Bacterial Wilt Diseases: An Update and Appraisal." *Information Bulletin* no. 50. Patancheru, AP, India: ICRISAT, 1997.

Singh, N. K., K. R. Kumar, D. Kumar, P. Shukla, and P. B. Kirti. "Characterization of A Pathogen Inducied Thaumation-Like Protein Gene *AdTLP* from *Arachis diogoi*, A Wild Peanut." *PLoS One* 8, no. 12 (2013): e83963.

Smith, O. D., T. E. Boswell, W. J. Grichar, and C. E. Simpson. "Reaction of Select Peanut (*Arachis hypogaea* L.) Lines to Southern Stem Rot and *Pythium* Pod Rot Under Varied Disease Pressure." Peanut Science 16, no. 1 (1989): 9–14. https://doi.org/10.3146/I0095-3679-16-1-3.

Stalker, H. T. "Utilizing Arachis Germplasm Resources." In *Groundnut: A Global Perspective*, edited by S. N. Nigam, 281–296. Patacheru, India: ICRISAT, 1992.

Stalker, H. T. "Utilizing Wild Species for Peanut Improvement." *Crop Sciences* 57 (2017): 1102–1120.

Stalker, H. T., and M. K. Beute. "Registration of Four Interspecific Peanut Germplasm Lines Resistant to *Cercospora arachidicola*." Crop Science 33 (1993): 1117. https://doi.org/10.2135/cropsci2002.3130.

Stalker, H. T., T. Phillips, J. Murphy, and T. Jones. "Variation of Isozyme Patterns among *Arachis* Species." *Theoretical and Applied Genetics* 87, no. 6 (1994): 746–755.

Stalker, H. T., and L. G. Mozingo. "Molecular Markers of *Arachis* and Marker-Assisted Selection." Peanut Science 28 (2001): 117–123. https://doi.org/10.3146/I0095-3679-28-2-13.

Stalker, H. T. A. K. Weissinger, S. Milla-Lewis, and C. C. Holbrook. "Genomics: An Evolving Science in Peanut." Peanut Science 36, no. 1 (2009): 2–10.

Stalker, H. T., J. C. Wynne, and M. Company. "Variation in Progenies of an *Arachis hypogaea* × Diploid Wild Species Hybrid." *Euphytica* 28 (1979): 675–684. https://doi.org/10.1007/BF00038934.

Subrahmanyam, P., D. McDonald, R. W. Gillons. "Variation in *Cercosporidium personatum* Symptoms on Certain Cultivars of *Arachis hypogaea*." *Oleagineux* 37 (1982): 63–67.

Subrahmanyam, P., D. McDonald, R. W. Gibbons, S. N. Nigam, and D. J. Nevill. "Resistance to Rust and Late Leafspot Diseases in Some Genotypes of *Arachis hypogaea*." Peanut Science 9 (1982): 6–10. https://doi.org/10.3146/i0095-3679-9-1-2.

Subrahmanyam, P., D. McDonald, R. J. Reddy. "Peanut Rust: A Major Threat to Peanut Production in the Semiarid Tropics." Plant Disease 69 (1985): 813–819. https://doi.org/10.1094/pd-69-813.

Subrahmanyam, P., P. Moss, D. McDonald, P. V. Subba Rao, and V. R. Rao. "Resistance to Leaf Spot Caused by *Cercosporidium personatum* in Wild *Arachis* Species." Plant Disease 69 (1985): 951–954. https://doi.org/10.1094/PD-69-951.

Subrahmanyam, P., G. L. Hildebrand, R. A. Naidu, L. J. Reddy, and A. K. Singh. "Sources of Resistance to Peanut Rosette Disease in Global Peanut Germplasm." Annals of Applied Biology 132 (1998): 473–485. https://doi.org/10.1111/J.1744-7348.1998.TB05223.X.

Subrahmanyam, P., R. A. Naidu, L. J. Reddy, P. L. Kumar, and M. Ferguson. "Resistance to Peanut Rosette Disease in Wild *Arachis* Species." Annals of Applied Biology 139 (2001): 45–50.

Subrahmanyam, P., and D. McDonald. *Rust Disease of Peanut (Summary in French).* Information Bulletin no. 13. International Crops Research Institute for the Semi-Arid Tropics, Hyderabad, India, 1983.

Subrahmanyam, P. "Bacterial Leafspot." In *Compendium of Peanut Diseases*, 2nd ed., edited by N. Kokalis-Burelle, D. M. Porter, R. Rodriguez-Kabana, D. H. Smith, and P. Subrahmanyam, 43. St. Paul: APS Press, 1997.

Sudini, H. K., H. D. Upadhyaya, S. V. Reddy, U. Naga Mangala, A. Rathore, and K. V. Krishna Kumar. "Resistance to Late Leaf Spot and Rust Diseases in ICRISAT's Mini Core Collection of Peanut (*Arachis hypogaea* L.)." Australasian Plant Pathology 44 (2015): 557–566. https://doi.org/10.1007/s13313-015-0368-1.

Sukruth, M., S. A. Paratwagh, V. Sujay, V. Kumari, M. V. C. Gowda, H. L. Nadaf, B. N. Motagi, S. Lingaraju, M. K. Pandey, R. K. Varshney, and R. S. Bhat. "Validation of Markers Linked to Late Leaf Spot and Rust Resistance, and Selection of Superior Genotypes Among Diverse Recombinant Inbred Lines and Backcross Lines in Peanut (*Arachis hypogaea* L.)." *Euphytica* 204 (2015): 343–351. https://doi.org/10.1007/s10681-014-1339-2.

Sun, D. R., ed. *Breeding of Groundnut*. Beijing: China Agricultural Press, 1998. (in Chinese). ISBN 7-109-05154-4/S.3261.

Sun, D. R., C. R. Chen, and Y. R. Wang. "Resistance Evaluation of Bacterial Wilt (*Pseudomonas solanacearum* E.F. Smith) of Peanut (*Arachis hypogaea* L.) in the People's Republic of China." *Proceedings of the American Peanut Research and Education Society* 13 (1981): 21–28.

Sun, Y., Q. Wang, Z. Li, L. Hou, S. Dai, and W. Liu. "Comparative Proteomics of Peanut Gynophore Development under Dark and Mechanical Stimulation." *Journal of Proteome Research* 12, no. 12 (2013): 5502–5511.

Sun, Z., Y. Cheng, F. Qi, M. Zhang, M. Tian, J. Wang, X. Wu, B. Huang, X. Zhang, and Z. Zheng. "Resistance of Peanut to Web Blotch Caused by *Phoma arachidicola* Is Related to Papillae Formation and the Hypersensitive Response." *Plant Pathology* 71 (2022): 1921–1931.

Sundaresha, S., S. Rohini, V. K. Appanna, et al. "Co-Overexpression of *Brassica juncea* NPR1 (BjNPR1) and *Trigonella foenum-graecum* Defensin (Tfgd) in Transgenic Peanut Provides Comprehensive but Varied Protection Against *Aspergillus flavus* and *Cercospora arachidicola*." Plant Cell Reports 35 (2016): 1189. https://doi.org/10.1007/s00299-016-1945-7.

Sunkara, S., P. Bhatnagar-Mathur, and K. K. Sharma. "Transgenic Interventions in Peanut Crop Improvement: Progress and Prospects." In *Genetics, Genomics and Breeding of Peanuts*, edited by N. Mallikarjuna and R. K. Varshney, 179–216. Boca Raton, FL: CRC Press, Taylor and Francis, 2013. https://doi.org/10.1201/b16872-14.

Taliansky, M. E., D. J. Robinson, and A. F. Murant. "Complete Nucleotide Sequence and Organization of the RNA Genome of Peanut Rosette Umbra Virus." Journal of General Virology 77 (1996): 2335–2345. https://doi.org/10.1099/0022-1317-77-9-2335.

Tang, R. H., and H. Q. Zhou. "Resistance to Bacterial Wilt in Some Interspecific Derivatives of Hybrids Between Cultivated Peanut and Wild Species." Chinese Journal of Oil Crop Sciences 22, no. 3 (2000): 61–65.

Tiwari, S., D. K. Mishra, A. Singh, P. K. Singh, and R. Tuli. "Expression of a Synthetic *cry1EC* Gene for Resistance Against *Spodoptera litura* in Transgenic Peanut (*Arachis hypogaea* L.)." Plant Cell Reports 27 (2008): 1017–1025. https://doi.org/10.1007/s00299-008-0525-x.

Tseng, Y. C., B. L. Tillman, Z. Peng, and J. Wang. "Identification of Major QTLs Underlying Tomato Spotted Wilt Virus Resistance in Peanut Cultivar Florida-EPTM '113.'" BMC Genetics 17 (2016): 128. https://doi.org/10.1186/s12863-016-0435-9.

Upadhyaya, H. D., S. N. Nigam, and R. P. Thakur. "Genetic Enhancement for Resistance to Aflatoxin Contamination in Peanut." In *Summary Proceedings of the Seventh ICRISAT Regional Peanut Meeting for Western and Central Africa*, 6–8 December 2000, Cotonou, Benin, 29–36. Patancheru, India: International Crops Research Institute for the Semi-Arid Tropics, 2002.

Upadhyaya, H.D., P.J. Bramel, R. Ortiz, and S. Singh. "Developing A Mini Core of Peanut for Utilization of Genetic Resources." *Crop Sciences* 42 (2002): 2150–2156.

Valls, J. F. M., and C. E. Simpson. "New Species of Arachis (Leguminosae) from Brazil, Paraguay and Bolivia." *Bomplandia* 14 (2005): 35–63.

Varshney, R. K., M. K. Pandey, P. Janila, S. N. Nigam, H. Sudini, et al. "Marker-Assisted Introgression of a QTL Region to Improve Rust Resistance in Three Elite and Popular Varieties of Peanut (*Arachis hypogaea* L.)." Theoretical and Applied Genetics 127, no. 8 (2014): 1771–1781. https://doi.org/10.1007/s00122-014-2338-3.

Vasavirama, K., and P. B. Kirti. "Increased Resistance to Late Leaf Spot Disease in Transgenic Peanut Using a Combination of PR Genes." Functional & Integrative Genomics 12 (2012): 625–634. https://doi.org/10.1007/s10142-012-0298-8.

Waliyar, F., D. McDonald, P. V. Subba Rao, and P. M. Reddy. "Components of Resistance to an Indian Source of *Cercospora arachidicola* in Selected Peanut Lines." Peanut Science 20 (1993): 93–96. https://doi.org/10.3146/I0095-3679-20-2-7.

Waliyar, F., B. B. Shew, R. Shidahmed, and M. K. Beute. "Effect of Host Resistance on Germination of *Cercospora arachidicola* on Peanut Leaf Surfaces." Peanut Science 22 (1995): 154–157. https://doi.org/10.3146/I0095-3679-22-2-15.

Waliyar, F., K. V. Kumar, M. Diallo, A. Traore, U. N. Mangala, et al. "Resistance to Pre-Harvest Aflatoxin Contamination in ICRISAT's Groundnut Mini Core Collection." European Journal of Plant Pathology 145, no. 4 (2016): 901–913. https://doi.org/10.1007/s10658-016-0879-9.

Wang, M., D. Abbott, and P. M. Waterhouse. "A Single Copy of a Virus-Derived Transgene Encoding Hairpin RNA Gives Immunity to Barley Yellow Dwarf Virus." Molecular Plant Pathology 1 (2000): 401–410. https://doi.org/10.1046/j.1364-3703.2000.00038.x.

Wang, H., Y. M. Shi, Y. Ren, S. L. Li, K. Jiao, M. Yuan, and H. J. Li. "Development of SSR Markers for Root-Knot Nematode Resistance in Peanut." Journal of Peanut Science 37 (2008): 14–17.

Wang, L., X. Zhou, X. Ren, L. Huang, H. Luo, et al. "A Major and Stable QTL for Bacterial Wilt Resistance on Chromosome B02 Identified Using a High-Density SNP-Based Genetic Linkage Map in Cultivated Peanut Yuanza 9102 Derived Population." Frontiers in Genetics 9 (2018): 652. https://doi.org/10.3389/fgene.2018.00652.

Waterhouse, P. M., M. B. Wang, and T. Lough. "Gene Silencing as an Adaptive Defense Against Viruses." *Nature* 411 (2001): 834–842.

Wen, S., H. Liu, X. Li, X. Chen, Y. Hong, H. Li, Q. Lu, and X. Liang."TALEN-mediated Targeted Mutagenesis of Fatty Acid Desaturase 2 (FAD2) in Peanut (*Arachis hypogaea* L.) Promoted the Accumulation of Oleic Acid." *Plant Molecular Biology* 97 (2018): 177–185.

Wightman, J. A., and G. V. R. Rao. "Peanut Pests." In *The Groundnut Crop: A Scientific Basis for Improvement*, edited by J. Smartt, 395–479. World Crop Series. Dordrecht: Springer, 1994.

Wright, D. A., J. A. Townsend, R. J. Jr. Winfrey, P. A. Irwin, J. Rajagopal, P. M. Lonosky, B. D. Hall, M. D. Jondle, and D. F. Voytas. "High-frequency Homologous Recombination in Plants Mediated by Zinc-finger Nucleases." *Plant Journal* 44 (2005): 693–705.

Xia, Y., B. Liao, J. Li, Y. Lei, H. Jiang, F. Cui, and X. Zeng. "Identification of AFLP Markers Linked to Resistance to Late Leaf Spot in Peanut (*Arachis hypogaea* L.)." *Chinese Journal of Oil Crop Sciences* 29, no. 3 (2007): 318–321.

Xu, X. J., ed. *Peanut Diseases, Pests, Weeds and Mice in China*. Beijing: China Agricultural Press, 2008. ISBN 978-7-109-13437-9.

Xu, Z., O. W. Barnett, and P. B. Gibson. "Characterization of Peanut Stunt Virus Strains by Host Reaction, Serology, and RNA Patterns." *Phytopathology* 76 (1986): 390–395. https://doi.org/10.1094/phyto-76-390.

Xu, P. L., L. Shan, Z. J. Liu, F. Wang, B. Zhang, Y. P. Bi, and C. K. Zhang. "Insect Resistant CpTI Gene Transferred into Peanut (*Arachis hypogaea* L.) via *Agrobacterium tumefaciens* and Regeneration of Transgenic Plantlets." Chinese Journal of Oil Crop Sciences 25 (2003): 5–31.

Yan, L. Y., Z. H. Wang, W. D. Song, P. M. Fan, Y. P. Kang, Y. Lei, L. Y. Wan, D. X. Huai, Y. N. Chen, X. Wang, H. K. Sudini, B. S. Liao, et al. "Genome Sequencing and Comparative Genomic Analysis of Highly and Weakly Aggressive Strains of *Sclerotium rolfsii*, the Causal Agent of Peanut Stem Rot." BMC Genomics 22 (2021): 276. https://doi.org/10.1186/S12864-021-07534-0.

Yang, G., K. E. Espelie, J. W. Todd, A. K. Culbreath, R. N. Pittman, and J. W. Demski. "Cuticular Lipids from Wild and Cultivated Peanuts and the Relative Resistance of These Peanut Species to Fall Armyworm and Thrips." Journal of Agricultural and Food Chemistry 41 (1993): 814–818. https://doi.org/10.1021/jf00029a026.

Yeri, S. B., and R. S. Bhat. "Development of Late Leaf Spot and Rust Resistant Backcross Lines in JL 24 Variety of Peanut (*Arachis hypogaea* L.)." Electronic Journal of Plant Breeding 7, no. 1 (2016): 37–41. https://doi.org/10.5958/0975-928X.2016.00005.3.

Yin, D. C. Ji, X. Ma, H. Li, W. Zhang, S. Li, F. Liu, K. Zhao, F. Li, and K. Li. "Genome of An Allotetrapliod Wild Peanut *Arachis monticola*: a *de novo* Assembly." *GigaScience* 7, no. 6 (2018): giy066.

Yin, D., C. Ji, Q. Song, W. Zhang, X. Zhang, C. Chen, C. Wang, G. He, Z. Liang, X. Ma, Z. Li, Y. Tang, Y. Wang, K. Li, L. Ning, H. ZHang, K. Zhao, X. Li, H. Yu, Y. Lei, M. Wang, L. Ma, H. Zheng, Y. ZHang, J. ZHang, W. Hu, and Z.J. Chen. "Comparison of *Arachis monticola* with Diploid and Cultivated Genomes Reveals Asymmetric Subgenome Evolution and Improvement of Peanut." *Advances Sci* 7 (2020): 1901672.

Yu, S. L., ed. *Peanut Genetics and Breeding in China*. Shanghai: Shanghai Scientific and Technology Press, 2011. ISBN 978-7-5478-0610-4.

Yu, B., D. Huai, L. Huang, Y. Kang, X. Ren, Y. Chen, X. Zhou, H. Luo, N. Liu, W. Chen, Y. Lei, R. K. Pandey, H. Sudini, R. K. Varshney, B. S. Liao, H. Jiang. "Identification of Genomic Regions and Diagnostic Markers for Resistance to Aflatoxin Contamination in Peanut (*Arachis hypogaea* L.)." *BMC Genetics* 20 (2019): 32. https://doi.org/10.1186/s12863-019-0734-z.

Yu, B., H. Jiang, M. K. Pandey, L. Huang, D. Huai, X. Zhou, Y. Kang, R. K. Varshney, H. Sudini, X. Ren, H. Luo, N. Liu, W. Chen, J. Guo, W. Li, Y. Ding, Y. Jiang, Y. Lei, B. S. Liao. "Identification of Two Novel Peanut Genotypes Resistant to Aflatoxin Production and Their SNP Markers Associated with Resistance." *Toxins* 12, no. 3 (2020): 156. https://doi.org/10.3390/toxins12030156.

Yu, R., Q. Jiang, C. X, L. L, S. Bu, and G. Shi. "Comparative Proteomics Analysis of Peanut Roots Reveals Differential Mechanisms of Cadmium Detoxification and Translocation between Two Cultivars Differing in Cadmium Accumulation." *BMC Plant Biology* 19, no. 1 (2019): 137.

Yuan, M., J. Zhu, L. Gong, L. He, C. Li, S. Han, C. Chen, and G. He. "Mutagenesisi of FAD2 Genes in Peanut with CRISPR/Cas9 Based Gene Editing." *BMC Biotechnology* 19 (2019): 1–7.

Zhang, J., S. Liang, J. Duan, J. Wang, S. Chen, Z.Cheng, Q. Zhang, X. Liang, and Y. Li. "De novo Assembly and Characterization of the Transcriptome during Seed Development, and Generation of Genic-SSR Markers in Peanut (Arachis hypogaea L.)." *BMC Genomics* 13 (2012): 90.

Zhang, C., H. Chen, T. Cai, Y. Deng, R. Zhuang, N. Zhang, Y. Zeng, Y. Zheng, R. Tang, R. Pan, W. Zhuang. "Overexpression of a Novel Peanut NBS-LRR Gene AhRRS5 Enhances Disease Resistance to *Ralstonia solanacearum* in Tobacco." Plant Biotechnology Journal 15, no. 1 (2017): 39–55. https://doi.org/10.1111/pbi.12589.

Zhang, C., W. Xie, H. Fu, Y. Chen, H. Chen, T. Cai, Q. Yang, Y. Zhuang, X. Zhong, K. Chen, M. Gao, F. Liu, Y. Wan, M. K. Pandey, R. K. Varshney, and W. Zhuang. "Whole Genome Resequencing Identifies Candidate Genes and Allelic Diagnostic Markers for Resistance to *Ralstonia solanacearum* Infection in Cultivated Peanut (*Arachis hypogaea* L.)." Frontiers in Plant Science 13 (2023): 1048168. https://doi.org/10.3389/fpls.2022.1048168.

Zhao, C., S. Zhao, L. Hou, H. Xia, J. Wang, C. Li, A. Li, T. Li, X. Zhang, and X. Wang. "Proteomics Analysis Reveals Differentially Activated Pathways that Operate in Peanut Gynophores at Different Developmental Stages." *BMC Plant Biology* 15, no. 1 (2015): 188

Zhao, Y., C. Zhang, H. Chen, M. Yuan, R. Nipper, et al. "QTL Mapping for Bacterial Wilt Resistance in Peanut (*Arachis hypogaea* L.)." Molecular Breeding 36, no. 2 (2016): 13. https://doi.org/10.1007/s11032-015-0432-0.

Zhou, Z. S. "A Review on Control of Tobacco Caterpillar *Spodoptera litura*." Chinese Bulletin of Entomology 46, no. 3 (2009): 354–361.

Zhou, X., Y. Xia, X. Ren, Y. Chen, L. Huang, et al. "Construction of a SNP-Based Genetic Linkage Map in Cultivated Peanut Based on Large Scale Marker Development Using Next-Generation Double Digest Restriction-Site-Associated DNA Sequencing (ddRADseq)." *BMC Genomics* 15, no. 1 (2014): 351. https://doi.org/10.1186/1471-2164-15-351.

Zhou, X., Y. Xia, J. Liao, K. Liu, Q. Li, et al. "Quantitative Trait Locus Analysis of Late Leaf Spot Resistance and Plant Type-Related Traits in Cultivated Peanut (*Arachis hypogaea* L.) under Multi-Environment." *PLoS One* 11, no. 11 (2016): e0166873. https://doi.org/10.1371/journal.pone.0166873.

Zhu, H., Y. Jiang, Y. Guo, J. Huang, M. Zhou, Y. Tang, J. Sui, J. Wang, and L. Qiao. "A Novel Salt Inducible WRKY Transcription Factor Gene, *AhWRKY75*, Confers Salt Tolerance in Transgenic Peanut." *Plant Physiology and Biochemistry* 160 (2021): 175–183.

Zhuang, W., H. Chen, M. Yang, J. Wang, M. K. Pandey, et al. "The Genome of Cultivated Peanut Provides Insight into Legume Karyotypes, Polyploid Evolution and Crop Domestication." Nature Genetics 51, no. 5 (2019): 865–876. https://doi.org/10.1038/s41588-019-0402-2.

15 The Key Largo Hypothesis

Recent Advances in Understanding Aflatoxin Contamination in Peanut with Omics and Biotechnology

Jake C. Fountain, Sunil S. Gangurde, Manish K. Pandey, Rajeev K. Varshney, and Baozhu Guo

LIST OF ABBREVIATIONS

ROS
: Reactive Oxygen Species

H_2O_2
: Hydrogen Peroxide

YES
: Yeast Extract-Sucrose Medium

YEP
: Yeast Extract-Peptone Medium

CPA
: Cyclopiazonic acid

iTRAQ
: Isobaric tags for relative and absolute quantification

HIGS
: Host-induced gene silencing

SIGS
: Spray-induced gene silencing

dsRNA
: Double-stranded RNA

15.1 BACKGROUND INFORMATION

Aflatoxins are carcinogenic mycotoxins produced by several species of fungi in the order Eurotiales. In peanut, the primary organisms responsible for aflatoxin contamination are *Aspergillus flavus* and *A. parasiticus*, with *A. flavus* being the predominant species (Schroeder and Boller 1973). The early history of aflatoxin, including its discovery and some major milestones, are well reviewed by Pickova et al. (2021). The initial association of aflatoxin contamination, these fungi, and contamination in peanuts dates back to the discovery of aflatoxins themselves. In 1960, an outbreak of aflatoxicosis in poultry termed "Turkey X Disease" occurred, resulting in the deaths of over 100,000 poults (Blount 1961; Klich 2007). This resulted in a concerted research effort with the eventual identification and characterization of fluorescent compounds associated with contaminated peanut meal fed to the affected poults, which was found to be produced by *A. flavus* in 1961 (Nesbitt et al. 1962), eventually resulting in the designation of these as *A. flavus* toxins or aflatoxins. Shortly afterward, in 1963, aflatoxins were also associated with *A. parasiticus* (Austwick and Ayerst 1963). Following the discovery of aflatoxins, their associations with human and animal disease became apparent with aflatoxin exposure in the diet being linked to increased risk of liver cancer, immunosuppression, and negative impacts on child growth and development (Gong et al. 2016; Williams et al. 2004). Instances of large aflatoxin outbreaks resulting in numerous deaths have also occurred (Azziz-Baumgartner et al. 2005; Kamala et al. 2018; Reddy and Raghavender 2007).

In addition to the health and food safety implications, aflatoxin contamination is a serious economic threat facing peanut producers globally. In the United States, average annual losses to the peanut industry due to aflatoxins have been reported as between \$20 and \$60 million (Lamb and Sternitzke 2001; Smith 2023). However, in years with particularly high levels of drought or heat stress, this number can skyrocket. For example, during the 2019 peanut growing season,the US peanut industry lost an estimated \$126 million to aflatoxin issues following severe droughts preceding harvest (Smith 2023). Globally, the costs of aflatoxins in terms of economics and lives are difficult if not impossible to calculate, but recent surveys have implicated aflatoxin exposure in 4.6%–28.2% of global cases of liver cancers (Liu and Wu 2010) and economic losses in the hundreds of millions of dollars annually (Wu 2015). With this level of threat to peanut production and human and animal health, there has been a significant investment by the US and global peanut industries, land-grant universities, non-governmental organizations (NGOs), and public research organizations in recent years into researching the biology and interactions of both peanuts and the *A. flavus* pathogen to develop new methods of combating this issue.

Efforts to prevent aflatoxin contamination can be subdivided into both pre- and post-harvest interventions (reviewed in detail by Torres et al. 2014). Pre-harvest efforts have been focused on the development of resistant germplasm

DOI: 10.1201/9781003360988-15

in breeding programs, prevention of drought stress, harvest timing and maturity estimation, and the use of biological controls including non-aflatoxigenic *A. flavus* strains. Postharvest efforts have focused on moisture management and drying, sanitation and pest control, segregation and sorting of at-risk seed lots, and electronic eye-based sorting in processing lines. These methods have all been incorporated into standard practices by the US peanut industry to mitigate aflatoxin contamination, yet there are still significant economic losses and issues due to this problem every year.

Of the factors which are associated with increased risks of aflatoxin contamination in peanuts, drought stress is likely the most consistent and reoccurring factor. The association of drought stress and aflatoxin contamination in crops like peanut and maize has been known for some time, although the mechanisms underlying this have remained a mystery (Fountain et al. 2014). This observation has been the basis of a number of research questions for our group, including "Why does drought stress lead to exacerbated aflatoxin issues?" and "Why does *A. flavus* produce aflatoxin in the first place?" In this chapter, we review the extensive research efforts that have been made and are ongoing in our collaborative research programs to investigate these questions in the role of drought and drought-associated oxidative stresses in *A. flavus* biology and aflatoxin production regulation as well as applications of genomics and biotechnology for the development of new mitigation strategies in the future.

15.2 OXIDATIVE STRESS TOLERANCE, AFLATOXIN PRODUCTION, AND THE KEY LARGO HYPOTHESIS

The question of "Why does drought stress lead to exacerbated aflatoxin issues?" has been one of the central focuses of our collaborative research programs for the last 15 years, with our original observations from the literature and experimentally being made in maize. It has been long noted that drought stress results in exacerbated preharvest aflatoxin contamination in maize and peanuts (Diener et al. 1987). There may even be carryover activity into postharvest conditions, with maize seeds harvested from drought-stressed plants yielding higher levels of aflatoxin contamination in laboratory bioassays compared to seeds harvested from irrigated plants (Guo et al. 2008). However, an explanation as to why that is the case has remained elusive. Initial observations on drought stress effects on maize pointed to modified expression of known pathogenesis-related protein encoding genes by drought stress (Fountain et al. 2010) including glyoxylase I (Chen et al. 2004), pathogenesis-related protein 10 (PR-10, Chen et al. 2006), and a 14kDa trypsin inhibitor (Chen et al. 1998). Therefore, knowing that drought could affect innate defense gene expression, an upstream regulatory signal needed to be identified.

Examination of the literature on host–*A. flavus* interactions and the influence of drought stress on those interactions consistently seem to point to antioxidant mechanisms as a focal point of defense-related genes differentially expressed in response to *A. flavus* inoculation, drought stress application, or insect damage (reviewed in Fountain et al. 2014; Fountain, Khera et al., 2015). This led us to examine maize drought stress responses in foliar tissues and developing kernels using high-throughput proteomics (Yang et al. 2014), histochemical staining (Yang et al. 2015), and global metabolomics (Yang et al. 2018). Examining differentially expressed proteins in developing maize kernels subjected to drought stress versus irrigated conditions, several biochemical pathways were differentially expressed, including glutathione metabolism, defense-related proteins, and antioxidant scavenger enzymes (Yang et al. 2014). Given the noted connection of drought stress responses with antioxidant enzyme expression, we then investigated the drought stress responses of maize lines with known levels of aflatoxin contamination resistance that positively correlate with their observed tolerance to drought stress. We observed that drought-sensitive, aflatoxin contamination susceptible lines of maize accumulated higher levels of reactive oxygen species (ROS) and reactive nitrogen species (RNS) in their leaf and developing kernel tissues compared to drought-tolerant, more aflatoxin contamination resistant lines (Yang et al. 2015, 2018). Global metabolomics analyses in developing kernel tissues also pointed to a heavy involvement of oxidative stress responses in maize responses to drought stress, with more sensitive lines showing more vigorous responses to stress than more tolerant lines (Yang et al. 2018).

These findings pointed us in the direction of examining the responses of different isolates of *A. flavus* to drought-associated ROS, particularly hydrogen peroxide (H_2O_2), whose scavenging enzymes and the compounds themselves were consistently associated with drought responses in maize. It had already been established that ROS and oxidative spikes are important developmental regulators in *Aspergillus* biology. For example, Jayashree and Subramanyam (2000) showed that ROS spikes are involved in the transition of *A. parasiticus* from idiophasic growth to trophophasic growth and corresponded with the initiation of both conidiation and aflatoxin production. These ROS have also been shown to be associated with the production of aflatoxin and its intermediates, and it was hypothesized that aflatoxin production may provide some form of antioxidant protection as a kind of secondary benefit (Narasaiah et al. 2006).

Based on these studies, we began an initial experiment to compare the levels of oxidative stress tolerance among different isolates of *A. flavus* with differing levels of aflatoxin production, ranging from very high levels of aflatoxin to moderate aflatoxin producers, to commercial atoxigenic biological control isolates, and to atoxigenic field isolates (Fountain et al. 2015). This initial study found that the total amount of oxidative stress that conidia of individual isolates could tolerate and grow in was correlated with their aflatoxin production capability, with highly aflatoxigenic isolates tolerating greater levels of H_2O_2 stress than those with less or no aflatoxin production. This further reinforced the concept that aflatoxin production, if not aflatoxin itself, could be working as a form of antioxidant protection for *A. flavus* and other aflatoxin-producing fungi.

This series of observations and initial studies came to form the basis of a general working hypothesis for the role of environmental stress in aflatoxin contamination we internally termed the "Key Largo Hypothesis" (Figure 15.1). The concept is that under stress free conditions reminiscent of "sitting on the beach in Key Largo," with adequate irrigation and optimum temperatures, aflatoxin contamination will be low due to the lack of physiological stress on host and therefore the infecting pathogen during interactions (Figure 15.1A). In contrast, when under stressful conditions, particularly drought stress and high temperatures, aflatoxin contamination will be higher due to increased physiological stresses in both the host and the fungus in coping with the environmental conditions and in the fungus while responding to stress-related signals from the host (Figure 15.1B). This hypothesis is based mainly on observed responses to stress in sensitive hosts, with elevated ROS in host tissues under drought and heat stress being perceived by infecting *A. flavus*, leading to exacerbated aflatoxin production in response. Alternatively, reduced ROS accumulation in drought-tolerant hosts under stressful conditions or when hosts are responding only to normal growing conditions would lead to normal or reduced aflatoxin contamination. The differences in overall oxidative stress tolerance seen among strains of *A. flavus* with contrasting aflatoxin production capabilities (Fountain et al. 2015) also points to the possible role of aflatoxin production in alleviating oxidative stress in *A. flavus* and aflatoxin production being exacerbated in response to host or environmentally derived oxidative stress (Figure 15.1C).

Testing this hypothesis then became a focus of our research programs from two main angles: *A. flavus* isolate-specific responses to oxidative stress, and the manipulation of antioxidant responses in hosts to examine effects on aflatoxin contamination and host–*A. flavus* interactions. For isolate-specific responses, we employed a series of

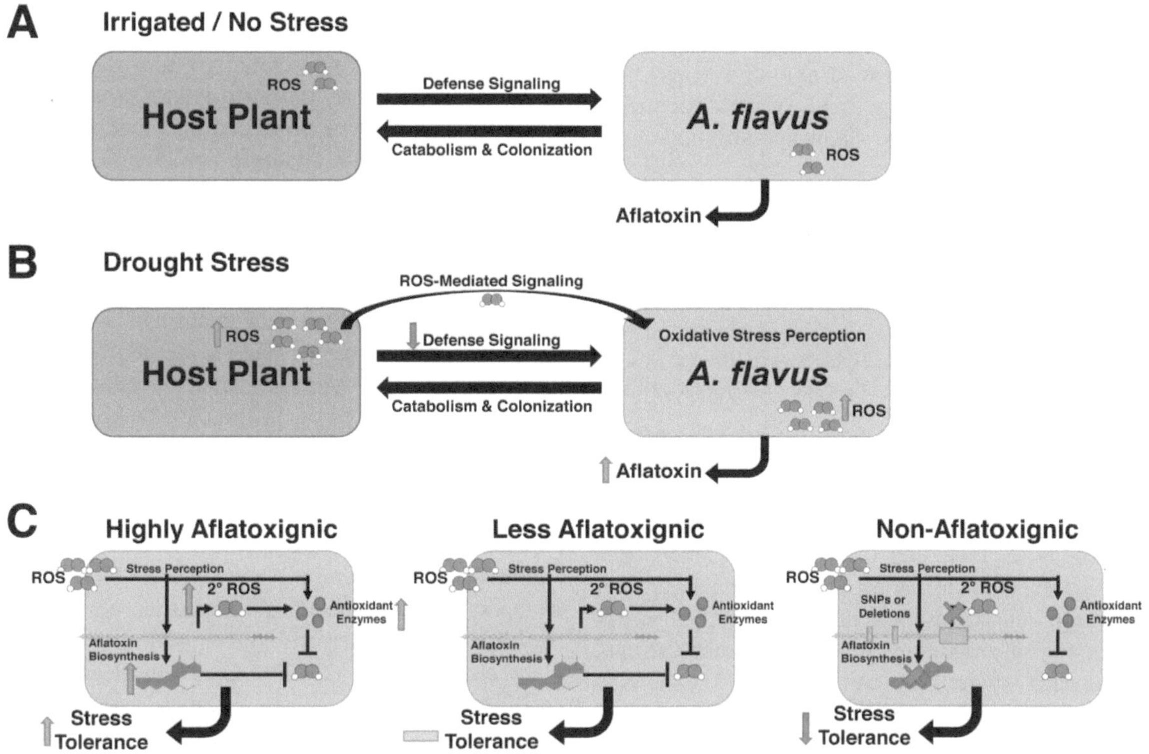

FIGURE 15.1 The "Key Largo" Hypothesis. A. Under irrigated, non-stressed conditions, normal ROS production occurs in both the host plant and in *A. flavus* during their interaction. In this condition, ROS function in both defense signaling in the host and in growth and aflatoxin production regulation in *A. flavus*. B. Under drought stress conditions, ROS production is exacerbated in host plants, resulting in overaccumulation, which can be detrimental to the host. These excess ROS are hypothesized to be perceived by infecting *A. flavus* resulting in the occurrence of oxidative stress to the infecting pathogen and countermeasures in the form of increased antioxidant mechanism expression and aflatoxin production. Under this hypothesis, this perception of excessive ROS production caused by drought stress in the host is the cause of exacerbated aflatoxin production and contamination. C. Differences in isolate aflatoxin production capability may also play a role in overall oxidative stress tolerance. Highly aflatoxigenic isolates exhibit higher levels of oxidative stress tolerance than less or non-aflatoxigenic isolates. This is suspected to be caused by greater levels of aflatoxin biosynthesis-associated oxidase activity, which may cause secondary (2°) ROS bursts and further stimulation of antioxidant mechanisms, and through direct antioxidant reactions between aflatoxins and ROS molecules. The reduction or absence of these mechanisms would, therefore, result in reduced overall oxidative stress tolerance.

experiments focused on the role of aflatoxin production in oxidative stress responses, with the goal of learning more about the practical reasons for aflatoxin production and how the pathogen could respond to ROS perceived during host interactions. In the following sections, we will explore the results of omics-based investigations into *A. flavus* oxidative stress responses, and how they have led to both a new frontier in *A. flavus* comparative genomics and to direct testing of the "Key Largo Hypothesis" using transgenic and genome editing approaches in host plants.

15.3 RESPONSIVE OMICS INVESTIGATIONS INTO DIFFERENT ISOLATE RESPONSES TO OXIDATIVE STRESS

Various omics technologies have become available for studying the responses of organisms to various stimuli at the molecular level. In this program, we examined the responses of various isolates of *A. flavus* with differing levels of oxidative stress tolerance and aflatoxin production capability to H_2O_2-derived oxidative stress *in vitro* using several different approaches, including transcriptomics, proteomics, and metabolomics. These studies have shed light on isolate-specific responses to oxidative stress correlated with their aflatoxin production levels and identified additional secondary metabolite pathways stimulated by oxidative stresses.

15.3.1 Transcriptomics

For the omics studies, we used a common experimental design of growing a set of unique *A. flavus* isolates in stationary liquid cultures in 125mL Erlenmeyer flasks inoculated with conidial suspensions in one of two synthetic culture media: yeast extract-sucrose (YES), which is conducive for aflatoxin production; or yeast extract-peptone (YEP), which is non-conducive for aflatoxin production. The isolates were inoculated into cultures containing different levels of H_2O_2 including 0, 10, 20, or 25mM concentrations. These concentrations were chosen based on our initial study and represent a midpoint level of stress (10mM) and a near maximum level of tolerance minus 5–10 mM for each isolate (Fountain et al. 2015). The isolates were allowed to grow for 7 days with and without oxidative stress in the transcriptome experiment to form mycelial mats before being extracted from the flasks and used for total RNA isolation and sequencing. A total of 64 RNA sequencing libraries were generated, of which two were of poor quality and excluded from subsequent analyses (Fountain, Bajaj, Pandey et al. 2016). This transcriptomics experiment was done in collaboration with Dr. Rajeev Varshney and his Center of Excellence in Genomics and Systems Biology (CEGSB) at the International Crops Research Institute for the Semi-Arid Tropics (ICRISAT) in Hyderabad, India, and the results were presented in three independent manuscripts.

In the first manuscript (Fountain, Bajaj, Pandey et al. 2016), we provided a general overview of the trends and differentially expressed pathways observed in the data in both YES and YEP media. In the second manuscript (Fountain, Bajaj, Nayak et al. 2016). we focused on the responses of aflatoxin-producing isolates in conducive YES medium to oxidative stress. Finally, in the third manuscript (Fountain et al. 2020), we focused on the differences in isolate responses comparing between the conducive and non-conducive culture media, with a focus on signaling mechanisms and additional regulated secondary metabolite pathways.

In this dataset described in all three manuscripts, differential expression was observed in all isolates in response to oxidative stress treatment, with more toxigenic isolates generally exhibiting fewer overall differentially expressed genes than those producing less to no aflatoxin. Aflatoxin biosynthetic genes such as *nor-1*, *verB*, *ver-1*, *omtA*, and *omtB* were upregulated in response to oxidative stress, while some genes were unaffected, such as *aflR*, *VeA*, and members of the sugar cluster. Additional secondary metabolite pathways including those for kojic acid and aflatrem were also regulated by oxidative stress and explored. Kojic acid has been shown to have some antioxidant activity (Chang et al. 2011), while aflatrem has been the subject of fewer investigations. Similar to aflatoxin, aflatrem, an isoprenoid mycotoxin, is produced in a mechanism heavy in oxidase activity. For aflatoxin production, these oxidases have been implicated in the production of secondary ROS bursts that stimulate the enhanced expression of antioxidant mechanisms (Roze et al. 2015). It may, therefore, be conjectured that aflatrem production may provide a similar activity, though it this has yet to be examined. The production of these compounds may provide some degree of antioxidant protection to *A. flavus* in the absence of aflatoxin production in atoxigenic isolates or, in the case of kojic acid, in non-aflatoxin conducive conditions. Other secondary metabolite gene clusters were also differentially expressed in response to oxidative stress. Cyclopiazonic acid (CPA) genes were expressed more in aflatoxigenic isolates and in aflatoxin-conducive YES medium. Also, aflavarin biosynthetic genes were expressed higher in YES medium compared to YEP medium. This is interesting given both the potential role in plant pathogenicity of CPA (Chalivendra et al. 2017), the anti-insect feeding activity of aflavarin (Cary et al. 2015), and the potential toxicity of these two mycotoxins. The connection between medium carbon source and mycotoxin biosynthetic gene cluster expression was also prevalent in the study, with significant effects on aflatoxin, kojic acid, aflatrem, CPA, and aflavarin gene expression. Therefore, oxidative stress and the nutritional composition of the medium the fungus colonizes may significantly affect the mycotoxin repertoire produced during colonization.

In addition to mycotoxin biosynthesis, other mechanisms were significantly regulated by oxidative stress in *A. flavus*. This included several kinase signaling pathways, specifically the SakA/Hogl and MpkA pathways, which have been previously associated with oxidative stress responses (Hagiwara et al. 2016) and cell wall maintenance (Valiante et al. 2008), respectively. In addition, several antioxidant enzyme-coding genes, including catalase, glutathione S-transferase, superoxide dismutase, and thioredoxin, were differentially expressed in response to oxidative stress. Interestingly, these antioxidant mechanisms tended to be more expressed in the aflatoxin non-conducive medium, YEP, pointing to a greater reliance on alternative mechanisms to ameliorate oxidative stress in the absence of aflatoxin production. Together, these observations pointed to a potential role of aflatoxin production in mitigating oxidative stress.

15.3.2 iTRAQ Proteomics

As a follow-up to the transcriptomics experiments, we also examined the overall oxidative stress responses at the protein level of three select isolates of *A. flavus* (Fountain et al. 2018): AF13, a highly aflatoxigenic and stress tolerant isolate which showed the low numbers of differentially expressed genes in the transcriptome studies; NRRL3357, an aflatoxigenic isolate with moderate levels of stress tolerance; and K54A, an atoxigenic isolate which exhibited the lowest degree of oxidative stress tolerance of examined isolates in our studies (Fountain et al. 2015). These three isolates were cultured as described for the transcriptome experiment in the aflatoxin conducive YES medium amended with varying amounts of H_2O_2. They were cultured for 7 days with or without stress as previously described, and the mycelial mats were collected and used for total protein isolation. These proteins were then sequenced and quantified for differential expression analyses using isobaric tags for relative and absolute quantification (iTRAQ) proteomics. Correlative analyses were also performed using the obtained differential protein expression data with the previously mentioned transcriptome results (Fountain, Bajaj, Pandey et al. 2016).

The patterns of protein differential expression observed here corresponded to the transcriptome results, with greater numbers of differentially expressed proteins being observed in the less aflatoxigenic isolate NRRL3357 compared to the more aflatoxigenic isolate AF13 under higher levels of stress. However, at lower levels of stress, more differential expression was seen for AF13 compared to NRRL3357 or the atoxigenic isolate K54A. Overall correlations between the transcriptome and proteome datasets were low based on Pearson correlations between protein and transcript fold changes, though general trends of being up- or downregulated were generally similar between the datasets, with 25.2% of transcripts and proteins showing opposite responses. This low degree of correlation between the datasets may be due to experimental error from the datasets being independently generated, but the possibility of post-transcriptional regulation is also present and may play a role in oxidative stress responses in *A. flavus*. Such regulation has also been hypothesized to occur in *A. flavus* responses to heat stress (Bai et al. 2015).

Of the metabolic pathways differentially expressed at the protein level, antioxidant mechanisms including members of the glutathione metabolic pathway were prominently featured in NRRL3357 under higher levels of stress along with some aflatoxin biosynthetic components, carbohydrate metabolism, and amino acid metabolism. In contrast, in AF13, mainly carbohydrate metabolic pathways including glycolysis, pyruvate metabolism, and N-glycan biosynthesis were differentially expressed. K54A showed responses in glycolysis and amino acid metabolism pathways. In particular, the patterns observed here pointed to potential growth-related expression differences between the isolates, with AF13 showing evidence of gluconeogenesis and carbon starvation at the timepoint observed in the experiment, while NRRL3357 was actively responding to stress and still expressing aflatoxin gene under high stress at the same timepoint. The clear distinction between the isolates here points to a time-dependent response to stress missed in this design, necessitating additional study to observe causal differences in overall oxidative stress responses in these two isolates.

15.3.3 Growth Dynamics and Global Metabolomics

Based on observations in the transcriptome and proteome analyses of the possibility of growth and time-related effects on isolate-specific oxidative stress responses, particularly between NRRL3357 and AF13, we next sought to investigate the effects of oxidative stress on the growth rates of these isolates over time (Fountain et al. 2019). Using an incubating plate reader, NRRL3357 and AF13 were cultured in YES medium amended with increasing levels of H_2O_2 of 0, 10, 15, 20, and 25mM with two different inoculum concentrations of 2.0 x 10^4 conidia/ml or 8.0 x 10^4 conidia/ml with optical density at 405nm being monitored every 15min for 100h. The resulting growth curves revealed stark differences in the growth dynamics between the two isolates. While stress resulted in prolonged lag phases in both isolates, the effects were more pronounced on NRRL3357 than AF13. Lesser concentrations of H_2O_2 were also required to suppress the growth of NRRL3357 than for AF13. Comparisons between inoculum concentrations also showed that increasing inoculum concentration resulted in both earlier initiation of log phase growth and enabled growth at higher levels of H_2O_2 within the time course of the experiment. These observations point to three major conclusions. First, oxidative stress tolerance differences are truly different between these isolates with contrasting levels of aflatoxin production. Second, conidial oxidative stress tolerance and antioxidant capacity may be a determinant of the growth of an isolate exposed to such stress. Third, observed differences in the responses of isolates at higher levels of oxidative stress may not only represent stress responses, but may also be confounded by differences in growth and developmental stages resulting from extended lag phases induced by excessive oxidative stress.

With this in mind, a global metabolomics experiment was designed to examine the differential responses of NRRL3357 and AF13 to a moderate level of oxidative stress (15mM H_2O_2) rather than a concentration close to their maximum levels of tolerance (Fountain et al. 2019). Time would also be considered, with the metabolomics experiment examining responses at both 4 and 7 days in culture. Mycelial tissues from these time points either exposed to or not exposed to oxidative stress were collected and used for global, untargeted metabolomics analysis examining primary metabolism pathways, including carbohydrate (glycolysis and TCA cycle), amino acid (amino acid biosynthesis, amino acid catabolism), and lipid (saturated and unsaturated lipid biosynthesis) metabolic pathways. Secondary metabolic pathways, including glutathione metabolism, polyamine metabolism, and oxylipin formation, were also examined as a part of the metabolite panel.

Comparing between the overall isolate responses at the metabolite level, NRRL3357 exhibited a much more vigorous response to applied oxidative stress compared to AF13, a trend carried over from increased transcriptional and protein-level responses previously observed (Fountain, Bajaj, Pandey et al. 2016, 2018). Comparing between time points, the responses of both isolates was greater and more varied at 4 days compared to 7 days, implying a more active response to ongoing stress at the earlier time point. This confirms the need for examination of earlier responses to oxidative stress in these fungi to better understand the mechanisms they are employing and how aflatoxin production factors into oxidative responses. Specifically, NRRL3357 exhibited metabolic activities consistent with an organism actively working to ameliorate oxidative stress and damage including increased carbohydrate catabolism, glutathione cycle activity, polyamine production, and amino acid and lipid production. However, AF13 showed relatively neutral responses to oxidative stress in the study, implying that this isolate had already dealt with the applied stress and was in a recovery phase at the time points examined. This led to the hypothesis that AF13 possesses a means of more rapidly and completely mitigating oxidative stress than NRRL3357.

15.3.4 Responsive Omics Summary

The overall trends observed in the responsive omics studies examining the responses of both aflatoxin producing and non-producing isolates of *A. flavus* were relatively consistent at the transcript, protein, and metabolite levels. In general, isolates which produced greater levels of aflatoxin tended to show less overall differential expression or accumulation of the target compounds of interest in each study compared to less or non-aflatoxigenic isolates. In particular, the dichotomy between the responses of the isolates AF13 and NRRL3357, which were explored in all three studies, were particularly stark and pointed to the potential for a unique mechanism at work in AF13 which may be absent from other isolates. This has led to additional research questions by our group into why AF13 shows this different and apparently more rapid response to oxidative stress. It also highlighted the potential role of aflatoxin in the mitigation of oxidative stress.

More recent studies performed concurrently and after these omics studies have confirmed the hypothesis that aflatoxin may function as an antioxidant to be, in fact, true. Roze et al. (2015) showed that deletion of aflatoxin pathway intermediate genes resulted in reduced conidial oxidative stress tolerance in *A. parasiticus* through the stimulation of antioxidant expression by secondary ROS bursts generated by oxidase activity during aflatoxin biosynthesis. More definitively, aflatoxin has been recently shown to be able to react with peroxides generated by 2,2'-azobis, 2-amidinopropane (APAB) directly and to provide antioxidant protection to *Escherichia coli* K-12 when provided as a supplement in medium to the bacterium (Finotti et al. 2021; Reverberi et al. 2021). While there are likely other roles of aflatoxins in interactions with other organisms and abiotic conditions in the environment, these findings coupled with our observations point to a definitive role of aflatoxin production and the aflatoxin molecules themselves as antioxidants capable of protecting organisms against environmental oxidative stresses.

In the context of the "Key Largo Hypothesis" (Figure 15.1), these responsive omics results for these select isolates point to an interesting trend. Highly aflatoxigenic isolates like AF13 showed less vigorous responses to stress, while moderate aflatoxin producers like NRRL3357 and atoxigenic isolates like K54A and the biological control strains showed more vigorous responses in all their associated omics studies. While the most straightforward interpretation of this would be that the vigor of these responses is the cause of the difference in oxidative stress tolerance, in light of the time course included in the global metabolomics study (Fountain et al. 2019), the reality is likely more nuanced. These large differences are likely due to growth delays in the less aflatoxigenic isolates resulting from differences in initial conidial stress tolerance, with less aflatoxigenic isolates having less active early stress tolerance leading to more pronounced delays in growth compared to more highly aflatoxigenic isolates. While these responsive differences may represent both oxidative stress responses and differences in the developmental stage, they are likely the result of the initial differences in stress tolerance which are related to aflatoxin production capability and its role in conidial stress tolerance (Roze et al. 2015).

Therefore, the role of aflatoxin production as a stress-alleviating mechanism also factors into the overall "Key Largo Hypothesis" (Figure 15.1C), with more highly aflatoxigenic isolates exhibiting a greater level of tolerance to oxidative stress than less aflatoxigenic or non-aflatoxigenic isolates due to more aflatoxin biosynthesis-associated oxidase activity leading to more secondary ROS bursts and subsequent increases in antioxidant expression (Roze et al. 2015). In addition, the recent implication of aflatoxin compounds in reacting directly with ROS and providing antioxidant protection (Finotti et al. 2021; Reverberi et al. 2021) also points to increased aflatoxin levels in highly aflatoxigenic isolates, resulting in greater stress tolerance overall compared to less or non-aflatoxigenic isolates. Taken together with the observations from host plants and the roles of ROS in drought responses, these data also suggest that aflatoxin production capabilities of the infecting isolates of *A. flavus* may play a role in host interactions, with more toxigenic isolates being less susceptible to host-derived ROS spikes than more moderate producers. However, this will need to be explored experimentally in the future to determine if this is valid.

Together, this interplay of ROS and aflatoxin production seen in these isolates, particularly the contrasting AF13 and NRRL3357 isolates, points to another question: "why are they so different in terms of aflatoxin production and stress tolerance?" While these responsive omics experiments highlighted the existence of these differences and their results, their root causes remain unknown. To determine these causes, we initiated a study in a new direction: comparative genomics.

15.4 GETTING TO THE ROOT: *ASPERGILLUS FLAVUS* COMPARATIVE GENOMICS

Having observed stark differences in the overall oxidative stress responses of *A. flavus* isolates NRRL3357 and AF13 using responsive omics methods, there remained the question of whether there was a novel mechanism at work in AF13 to explain its elevated aflatoxin production and oxidative stress tolerance. In order to explore this and to get to the root of why these isolates exhibit such different responses, we began an effort to produce genome sequences for these isolates. These sequences would then be used for comparative analyses to identify differences in gene content and genome architecture which could explain the isolates' divergent phenotypes.

The field of *A. flavus* genomics began with a collaborative effort by several research groups in the wake of the description of the aflatoxin biosynthetic gene cluster (Yu et al. 2004) as a part of the USDA-ARS Aflatoxin Elimination Workshop working group. In 2003, work began to sequence the genome of the isolate NRRL3357 and was led by Gary Payne at North Carolina State University (NCSU). These efforts resulted in the generation and release of a scaffold-level reference genome for *A. flavus* NRRL3357 in 2005 (Payne et al. 2006, 2007; Yu et al. 2008). This genome was later corrected in 2015 to remove contaminant scaffolds, leaving a reference used by the *Aspergillus* and aflatoxin research communities today (Nierman et al. 2015).

Continuing sequencing efforts have led to the generation of draft assemblies of 213 *A. flavus* genomes from 2005 to 2022 according to the National Center for Biotechnology Information (NCBI 2023) (Figure 15.2). Notably, this has included several comparative studies looking at differences in mating types and sclerotial morphotypes. For example, Gilbert, Mack et al. (2018) released a comparative analysis of the isolate AF70 (a MAT1-2 S-strain) with NRRL3357 (a MAT1-1 L-strain). In addition to comparative analyses, work has also been carried out on generating new reference quality genomes for *A. flavus*. With the advent of newer, more cost-effective sequencing technologies, this kind of effort has become more feasible for individual collaborative groups or even individual labs. Examples of this include a chromosome-level assembly of NRRL3357 led by the University of Wisconsin-Madison and the University of California-Berkeley (Drott et al. 2020; Skerker et al. 2021) and a pair of chromosome arm-level assemblies of AF13 and NRRL3357 produced by our group in collaboration with the University of Georgia, USDA-ARS, and HudsonAlpha (Fountain, Clevenger, Nadon, Wang et al. 2020). Our group has also sequenced several other strains of *A. flavus* for comparative and phylogenomic analyses (Fountain, Clevenger, Nadon, Wang et al. 2020, Fountain, Clevenger, Nadon, Youngblood et al. 2020; Fountain et al. 2021). These and other newly developed reference materials will enable a new generation of investigations into the functional biology of *A. flavus* and its interactions with other organisms.

Our newly generated reference genomes for isolates AF13 and NRRL3357 (Fountain, Clevenger, Nadon, Wang et al. 2020) were also sequenced with a clear biological question in mind: to explore the root causes for the observed phenotypic and responsive omics results observed for these two distinct isolates as previously described to this point. Overall, these genome assemblies were relatively similar, with a total size of 37.439 Mb for AF13 and 36.996 Mb for NRRL3357. The advantages of the paired assemblies, however, came from both a new gene annotation effort and from comparative analyses of gene content and overall genome architecture. Annotation efforts enabled by both previously acquired (Fountain, Bajaj, Pandey et al. 2016) and newly generated transcriptome datasets for both isolates resulted in improved gene annotations over the original NRRL3357 assembly (Nierman et al. 2015). A total of 12,054 and 12,009 genes were predicted for AF13 and NRRL3357, respectively. Comparisons found that there were 45 unique genes in the NRRL3357 assembly, with a majority (35) being presence/absence differences with AF13 along with 10 InDel-associated genes, all of which derived from a 19kb insertion on Chromosome 5.

More unique gene content was found for the AF13 assembly in comparison to NRRL3357 with 153 unique genes identified, 81 of which were presence/absence differences with NRRL3357 and 72 associated with indels. What was interesting about this finding was that 60 of the 72 InDel-associated genes were located in a single, large insertion 310 kb in size located on Chromosome 1. This large insertion was absent from the NRRL3357 assembly, but a similar region was found on Chromosome 8 of the *A. oryzae* genome. Comparisons with other sequenced isolates showed that partial ends of this insertion could be

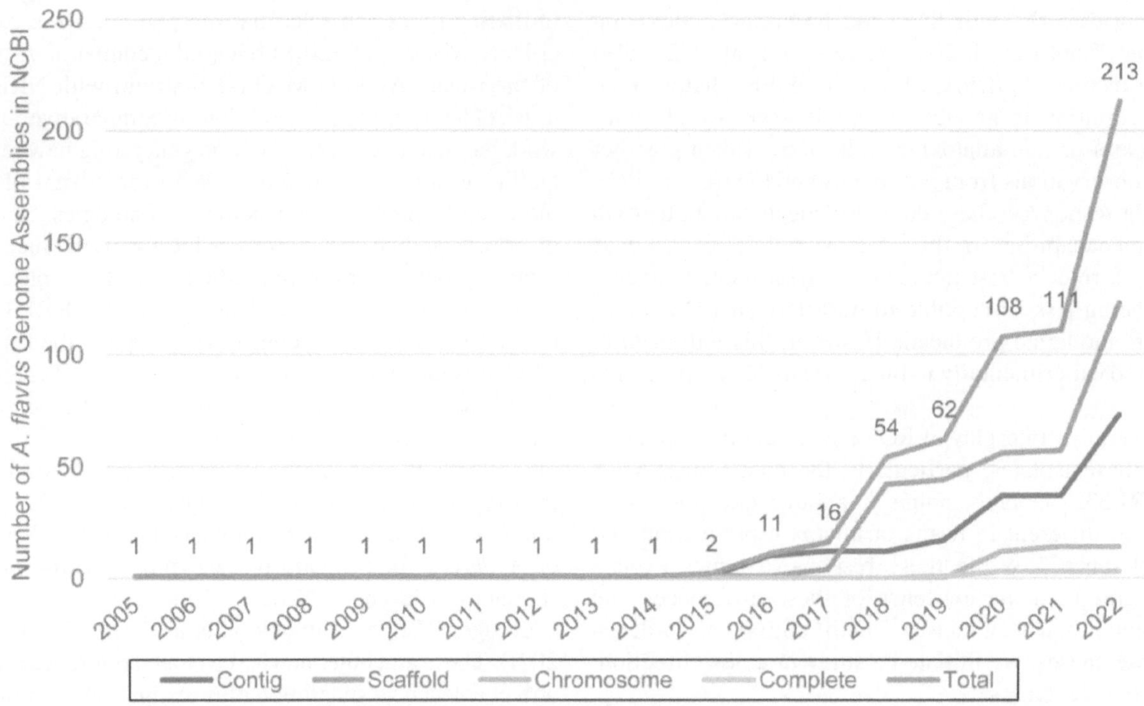

FIGURE 15.2 Number of *Aspergillus flavus* genome assemblies in the NCBI Genebank from 2005 to 2022. Numbers of assemblies were tabulated based on entries returned for search term "*Aspergillus flavus*" in the NCBI Genomes database. A total of 213 assemblies were found. These were composed of three complete (chromosomes and centromeric sequences), 14 chromosome-level, 123 scaffold-level, and 73 contig-level assemblies. Data labels on the plot show the running total number of genomes by year.

found through variant calling, mainly with the first 100kb of the insertion. Thus far, only isolates closely related with AF13, such as Tox4, another isolate sequenced in our study (Fountain, Clevenger, Nadon, Wang et al. 2020), or AF70 (Gilbert, Mack et al. 2018), contained the insertion in its entirety. It is unclear at this point how distributed this insertion is in natural populations of *A. flavus*, but it may be somewhat rare based on current information. Further investigation will be needed to determine its true distribution and how it came to be integrated into the *A. flavus* genome.

Among the genes contained within this interesting insertion, one candidate stood out as a potential regulatory gene that may be involved in aflatoxin production, pathogenicity, and/or oxidative stress tolerance in *A. flavus* and may eventually provide a partial explanation for the increased resilience of AF13 in our previous studies. This gene, termed *atfC* in our study, is a bZIP transcription factor bearing 74.84% homology to the *atfB/atf21* gene. Excitingly, *atfB* has been implicated along with *atfA* to coordinate oxidative stress tolerance and aflatoxin production in *Aspergillus* fungi (Roze et al. 2011, 2013). Based on this, we generated two independent deletion mutant isolates for the *atfC* gene for preliminary characterization. Deletion mutants did show reduced oxidative stress tolerance in culture, though results were variable. No significant effects on aflatoxin production in observed in culture or in inoculated peanut seeds, though one of the mutants did show numerical though non-significant reductions in aflatoxin compared to wild type AF13. While no significant effects on aflatoxin production were observed, the possible effects on oxidative stress tolerance were interesting. pointing to the need for more refined characterization to determine the possible role this gene may have in AF13's response to oxidative stress, aflatoxin production regulation, or peanut pathogenicity.

Overall, this comparative genomics analysis generated both a pair of new reference materials for use by the *Aspergillus* and aflatoxin research communities, and a cadre of putative candidate genes and mechanisms to be explored for their potential roles in regulating stress tolerance and aflatoxin production. Continuing sequencing efforts are currently underway to capitalize on these resources. In the end, the use of these references along with association analyses will pave the way for gene functional discoveries and a better understanding of the mechanisms at play in isolate-specific interactions with host plants and the environment.

15.5 BIOTECHNOLOGY AND THE FUTURE OF AFLATOXIN MITIGATION

Recent advances in our knowledge of aflatoxin production, its regulation, and the biology of the *A. flavus* pathogen provide a basis for the application of new biotechnologies in aflatoxin mitigation. However, in peanut, the application of biotechnology has been limited due to a general lack of acceptance of biotechnology in the peanut industry. This has been driven largely by public perception of transgenic/genetically engineered crops (Hallman et al. 2003; Stewart et al. 2000). Despite this public concern, there has been no reported incidence of food-borne illness due to transgenic methods in other crops commonly occurring in diets such as maize, nor in fibers such as cotton.

Traditional and molecular breeding methods, while powerful and useful for the improvement of a variety of traits in peanut, have been limited in their ability thus far to produce aflatoxin-"resistant" cultivars for commercial use. Therefore, to make significant progress in mitigating aflatoxin issues in peanut in the short term, it may be necessary to revisit transgenic approaches for aflatoxin mitigation, particularly given the successes seen in research applications of these methods. At the least, transgenic approaches are vital to ongoing hypothesis testing and validation of genes identified by biochemical and genetic analyses in peanut.

An example is a transgenic experiment inspired by our previous work linking aflatoxin gene regulation and oxidative stresses associated with drought stress in maize and peanut described in the preceding sections. Given that drought stress stimulates ROS accumulation in host plant tissues and these same ROS can stimulate aflatoxin production and aflatoxin gene expression in culture, we hypothesized (the "Key Largo Hypothesis") that ROS may function in cross-kingdom communication between host plants and *A. flavus*, resulting in exacerbated aflatoxin contamination during drought stress (Figure 15.1). To test this hypothesis, we have begun a series of transgenic experiments to either overexpress antioxidant genes including catalase, superoxide dismutase, or ascorbate peroxidase; or to silence antioxidant gene expression, specifically an isoform of catalase, using CRISPR-Cas9 genome editing (Fountain et al. 2019). This work is currently in progress, but, if successful, will provide a basis for ongoing study into the role of ROS in host–pathogen interactions and aflatoxin regulation, and the development of biomarkers related to antioxidant capacity for use in breeding selection.

While some experiments have been done for transgenic mitigation of aflatoxin issues in peanuts, more have currently been performed in maize to reduce aflatoxin contamination and *A. flavus* growth during infections. For example, experiments to overexpress pathogenesis-related proteins such as PR-10 and a 14-kDa trypsin inhibitor have resulted in significantly reduced aflatoxin and *A. flavus* growth (Chen et al. 2010, 2016). Newer methods have since been developed using applications of RNA interference, specifically the concept of host-induced gene silencing (HIGS). Several recent studies have performed HIGS by using DNA constructs to generate hairpin loops of transcribed RNAs to generate double-stranded RNAs (dsRNAs) in host plants that can be translocated into *A. flavus* for silencing aflatoxin and fungal growth-related genes. Examples of aflatoxin genes targeted using this method in transgenic maize include *aflR* (Masanga et al. 2015), *aflC/pksA* (Thakare et al. 2017), and *aflM/ver-1* (Raruang et al. 2020). Fungal growth genes targeted by HIGS in maize include *amy1* (alpha-amylase) (Gilbert, Majumdar et al. 2018), *alk* (alkaline protease) (Omolehin et al. 2021), and *p2c* (polygalacturonase) (Raruang, 2019). These experiments have shown significant reductions in aflatoxin contamination both greenhouse and field-based screening of these transgenic plants, demonstrating that this HIGS approach may be a valuable method in mitigating aflatoxin issues in the future.

This HIGS approach has also been explored in peanut. One example is Arias et al. (2015), who targeted five aflatoxin-related genes simultaneously including *aflS/aflJ*, *aflR*, *aflC/pksA*, *pes1* (a non-ribosomal peptide synthase gene), and *aflep* (an aflatoxin efflux pump gene). Simultaneous silencing of these genes resulted in significant reductions in aflatoxin contamination. Another example of HIGS for aflatoxin genes in peanut is Sharma et al. (2018), who targeted *aflM/ver-1* and *aflP/omtA* to reduce aflatoxin contamination. This study also examined transgenic overexpression of two antifungal defensin genes, *MsDef1* and *MsDef2*, to reduce aflatoxin contamination.

Together, these studies point to the potential for RNA interference to provide great help in the fight against aflatoxin contamination in maize and peanut. However, in peanut, there remains a reluctance to adopt transgenic varieties in the short to medium term, which hampers the apportion of this technology. A potential work-around to this is a variant of the HIGS concept termed spray-induced gene silencing (SIGS). This concept is similar to HIGS in that it employs dsRNAs or small interfering RNAs (siRNAs) to induce gene silencing in pathogen genes. The difference is the method of delivery. Where in HIGS, the dsRNAs or siRNAs are generated by the host plant and then somehow translocated into the pathogen to initiate gene silencing, SIGS is achieved by the uptake of dsRNAs or siRNAs applied to a host surface by the infecting pathogen to initiate gene silencing. This method has already seen success in several systems, as recently reviewed by de Oliveira Filho et al. (2022). Examples in this review include the confirmed uptake of applied dsRNAs, gene silencing, and reduced lesion sizes for *Aspergillus niger* infections in tomato, apple, and grape by Qiao et al. (2021); and reductions in growth, conidiation, and lesion sizes along with increased fungicide sensitivity of *Fusarium asiaticum*, as well as altered mycelial morphologies and conidiation of other mycotoxigenic species like *F. oxysporum* through dsRNA applications by Gu et al. (2019). Improved efficiencies may also be possible through the use of nanoparticles to protect applied dsRNAs from degradation on surfaces (Gurusamy et al. 2020). These examples highlight the potential of SIGS to silence pathogen genes and limit disease development for surface colonizing fungi. Therefore, it may be possible to use our extensive knowledge of *A. flavus* genomics and the biosynthesis and regulation of aflatoxin production to design dsRNA-based silencing vectors for use in SIGS to mitigate aflatoxin contamination in peanut and maize. It would also have the advantage of providing a non-transgenic method of bringing this type of biotechnological advance to bear on peanut aflatoxin issues, a definite advantage and a possible means of more rapidly deploying this technology to address aflatoxin problems.

15.6 PRACTICAL APPLICATIONS AND FUTURE RESEARCH NEEDS

The preceding findings and future directions provide insights into the role of drought stress and ROS in host–*A. flavus* interactions and aflatoxin production regulation, new genomics resources for researchers, and the potential for applied biotechnology to make progress in aflatoxin mitigation. While these areas are important, they are only a few of the necessary components to combat aflatoxin contamination. Knowledge of the interactions between hosts and *A. flavus* can play a role both pre- and postharvest, though are likely to provide more utility in preharvest infection prevention and reductions in aflatoxin contamination. This will indirectly reduce postharvest contamination issues by reducing inoculum introduced into storage environments. The integration of biotechnology in the development of seed and pod treatments or naturally "resistant" plants may eventually play a role in pre- and postharvest aflatoxin mitigation. However, continuing research into practical aflatoxin prevention strategies is still needed for peanut. Looking toward the near future (5–10 years), there are at least ten major research needs and challenges that must be addressed by the peanut-aflatoxin research community:

1. Developing new varieties of peanut with multi-stress tolerance including drought, heat, and salinity to cope with changing climate and increased risk of aflatoxin contamination;
2. Employing and developing genetic markers, biomarkers, and DNA sequencing technologies in combination with next-generation genetic mapping populations to identify genes and mechanisms consistently associated with reduced aflatoxin contamination and improved drought tolerance;
3. Exploring wild *Arachis* species and exotic germplasm for alleles that may contribute to reduced aflatoxin contamination;
4. Identifying upstream aflatoxin regulatory genes in *A. flavus*, particularly those with potential to co-regulate stress responses and the production of other mycotoxins like cyclopiazonic acid and aflatrem, and the roles of these mechanisms in host interactions;
5. Understanding the genetic diversity in natural populations of *A. flavus* and other aflatoxigenic fungi, and its implications for host interactions;
6. Developing non-GMO, industrially adoptable biotechnological interventions for mitigating aflatoxin contamination in peanut production environments;
7. Developing low cost, non-destructive aflatoxin detection methods for use in breeding screening and industrial detection applications;
8. Improving the formulation and efficacy of biological control organisms and fungicides in mitigating aflatoxin contamination both pre- and postharvest;
9. Developing an accurate risk prediction index for aflatoxin occurrence with an emphasis on loss prevention to farmers and seed lot segregation in storage environments;
10. Encouraging and enabling engagement, collaboration, and knowledge sharing between aflatoxin researchers working on multiple crops, both domestically and globally.

By addressing these needs, significant progress toward mitigating aflatoxin contamination issues in peanut may be made to the benefit of growers, shellers, processers, manufacturers, and consumers around the world.

15.7 CONCLUSIONS

Aflatoxin contamination of peanut is a complex and multifaceted problem that has engaged researchers over the past 63 years. In that time, significant progress has been made in understanding the biology of the *A. flavus* pathogen and its interactions with its environment, and in methods to detect and mitigate aflatoxin contamination in crops. However, there remain significant knowledge gaps in critical areas which prevent adequate prevention to avoid extensive economic losses in severe seasons. The efforts performed by our group in the areas of responsive omics investigations of host plants and *A. flavus* to drought and drought-related oxidative stresses, respectively, have shed additional light on the role ROS may serve in cross-kingdom communication between hosts and *A. flavus* that may contribute to exacerbated aflatoxin issues. Continuing genetic, genomic, and biotechnological studies will enable the harnessing of this information to develop new research directions and, ultimately, solutions to the effects of drought stress on aflatoxin contamination. In addition to this, these same resources and methods will provide novel means of mitigating aflatoxin contamination both pre- and postharvest. Present and future research endeavors, however, must begin bridging the gap between theoretical and applied concepts to develop practical and tangible solutions for growers, producers, and manufacturers.

15.8 ACKNOWLEDGMENTS

We would like to thank the technical staff and collaborators involved in the generation of the data and studies highlighted in this chapter. Work described in this chapter was supported by the Georgia Commodity Commission for Corn, the National Corn Growers Association (NGCA) Aflatoxin Mitigation Center of Excellence (AMCOE), the Georgia Peanut Commission, the Peanut Research Foundation, the National Peanut Board, the International Crops Research Institute for the Semi-Arid Tropics (ICRISAT), the USDA National Institute for Food and Agriculture (NIFA), the US Peanut Federation (USPF), and the USDA Agricultural Research Service (ARS).

REFERENCES

Arias, R. S., Dang, P. M., & Sobolev, V. S. "RNAi-Mediated Control of Aflatoxins in Peanut: Method to Analyze Mycotoxin Production and Transgene Expression in the Peanut/Aspergillus Pathosystem." *Journal of Visualized Experiments* 106 (2015): 53398. https://doi.org/10.3791/53398.

Austwick, P. K. C., & Ayerst, G. "Toxic Products in Groundnuts: Groundnut Microflora and Toxicity." *Chemistry and Industry Review* 2 (1963): 55–61.

Azziz-Baumgartner, E., Lindblade, K., Gieseker, K., Rogers, H. S., Kieszak, S., Njapau, H., Schleicher, R., et al. "Case–Control Study of an Acute Aflatoxicosis Outbreak, Kenya, 2004." *Environmental Health Perspectives* 113, no. 12 (2005): 1779–1783. https://doi.org/10.1289/ehp.8384.

Bai, Y., Wang, S., Zhong, H., Yang, Q., Zhang, F., Zhuang, Z., Yuan, J., Nie, X., & Wang, S. "Integrative Analyses Reveal Transcriptome-Proteome Correlation in Biological Pathways and Secondary Metabolism Clusters in *A. flavus* in Response to Temperature." *Scientific Reports* 5 (2015): 14582. https://doi.org/10.1038/srep14582.

Blount, W. P. "Turkey 'X' Disease." *Journal of the British Turkey Federation* 9 (1961): 52–77. https://doi.org/10.2307/1587768.

Cary, J. W., Han, Z., Yin, Y., Lohmar, J. M., Shantappa, P., Harris-Coward, P. Y., Mack, B., et al. "Transcriptome Analysis of *Aspergillus flavus* Reveals veA-Dependent Regulation of Secondary Metabolite Gene Clusters, Including the Novel Aflavarin Cluster." *Eukaryotic Cell* 14, no. 10 (2015): 983–997. https://doi.org/10.1128/EC.00092-15.

Chalivendra, S. C., DeRobertis, C., Chang, P. K., & Damann, K. E. "Cyclopiazonic Acid Is a Pathogenicity Factor for *Aspergillus flavus* and a Promising Target for Screening Germplasm for Ear Rot Resistance." *Molecular Plant-Microbe Interactions* 30, no. 5 (2017): 361–373. https://doi.org/10.1094/MPMI-02-17-0026-R.

Chang, P. K., Scharfenstein, L. L., Luo, M., Mahoney, N., Molyneux, R. J., Yu, J., Brown, R. L., & Campbell, B. C. "Loss of msnA, a Putative Stress Regulatory Gene, in *Aspergillus parasiticus* and *Aspergillus flavus* Increased Production of Conidia, Aflatoxins, and Kojic Acid." *Toxins* 3, no. 1 (2011): 82–104. https://doi.org/10.3390/toxins3010082.

Chen, Z. Y., Brown, R. L., Damann, K. E., & Cleveland, T. E. "Identification of a Maize Kernel Stress-Related Protein and Its Effect on Aflatoxin Accumulation." *Phytopathology* 94, no. 9 (2004): 938–945. https://doi.org/10.1094/PHYTO.2004.94.9.938.

Chen, Z. Y., Brown, R. L., Damann, K. E., & Cleveland, T. E. "PR10 Expression in Maize and Its Effect on Host Resistance Against *Aspergillus flavus* Infection and Aflatoxin Production." *Molecular Plant Pathology* 11, no. 1 (2010): 69–81. https://doi.org/10.1111/j.1364-3703.2009.00574.x.

Chen, Z. Y., Brown, R. L., Lax, A. R., Guo, B., Cleveland, T. E., & Russin, J. S. "Resistance to *Aspergillus flavus* in Corn Kernels Is Associated with a 14-kDa Protein." *Phytopathology* 88, no. 4 (1998): 276–281. https://doi.org/10.1094/PHYTO.1998.88.4.276.

Chen, Z. Y., Brown, R. L., Rajasekaran, K., Damann, K. E., & Cleveland, T. E. "Identification of a Maize Kernel Pathogenesis-Related Protein and Evidence for Its Involvement in Resistance to *Aspergillus flavus* Infection and Aflatoxin Production." *Phytopathology* 96, no. 1 (2006): 87–95. https://doi.org/10.1094/PHYTO-96-0087.

Chen, Z. Y., Warburton, M. L., Hawkins, L., Wei, Q., Raruang, Y., Brown, R. L., Zhang, L., & Bhatnagar, D. "Production of the 14 kDa Trypsin Inhibitor Protein Is Important for Maize Resistance Against *Aspergillus flavus* Infection/Aflatoxin Accumulation." *World Mycotoxin Journal* 9, no. 2 (2016): 215–228. https://doi.org/10.3920/WMJ2015.1890.

de Oliveira Filho, J. G., da Cruz Silva, G., de Paula Gomes, M., de Sousa, T. L., Ferreira, M. D., & Egea, M. B. "External Application of RNA Interference (RNAi): An Innovative Tool for Controlling Fungi During Food Storage." *Current Opinion in Food Science* 47 (2022): 100872. https://doi.org/10.1016/j.cofs.2022.100872.

Diener, U. L., Cole, R. J., Sanders, T. H., Payne, G. A., Lee, L. S., & Klich, M. A. "Epidemiology of Aflatoxin Formation by *Aspergillus flavus*." *Annual Review of Phytopathology* 25, no. 1 (1987): 249–270. https://doi.org/10.1146/annurev.py.25.090187.001341.

Drott, M. T., Satterlee, T. R., Skerker, J. M., Pfannenstiel, B. T., Glass, N. L., Keller, N. P., & Milgroom, M. G. "The Frequency of Sex: Population Genomics Reveals Differences in Recombination and Population Structure of the Aflatoxin-Producing Fungus *Aspergillus flavus*." *mBio* 11, no. 4 (2020): e00963-20. https://doi.org/10.1128/mBio.00963-20.

Finotti, E., Parroni, A., Zaccaria, M., Domin, M., Momeni, B., Fanelli, C., & Reverberi, M. "Aflatoxins Are Natural Scavengers of Reactive Oxygen Species." *Scientific Reports* 11 (2021): 16024. https://doi.org/10.1038/s41598-021-95325-8.

Fountain, J. C., Bajaj, P., Nayak, S. N., Yang, L., Pandey, M. K., Kumar, V., Jayale, A. S., et al. "Responses of *Aspergillus flavus* to Oxidative Stress Are Related to Fungal Development Regulator, Antioxidant Enzyme, and Secondary Metabolite Biosynthetic Gene Expression." *Frontiers in Microbiology* 7 (2016): 2048. https://doi.org/10.3389/fmicb.2016.02048.

Fountain, J. C., Bajaj, P., Pandey, M. K., Nayak, S. N., Yang, L., Kumar, V., Jayale, A. S., et al. "Oxidative Stress and Carbon Metabolism Influence *Aspergillus flavus* Transcriptome Composition and Secondary Metabolite Production." *Scientific Reports* 6, no. 1 (2016): 38747. https://doi.org/10.1038/srep38747.

Fountain, J. C., Chen, Z. Y., Scully, B. T., Kemerait, R. C., Lee, R. D., & Guo, B. "Pathogenesis-Related Gene Expressions in Different Maize Genotypes Under Drought Stressed Conditions." *African Journal of Plant Science* 4, no. 11 (2010): 433–440. https://doi.org/10.5897/AJPS.9000182.

Fountain, J. C., Chu, Y., Ozias-Akins, P., Chen, Z. Y., Wang, K., Yang, Y., & Guo, B. "Genetic Transformation to Mitigate Drought and Aflatoxin-Related Losses in Peanut." Paper presented at the 51st Annual Meeting of the American Peanut Research and Education Society (APRES), Auburn, AL, July 9–11, 2019.

Fountain, J. C., Clevenger, J. P., Nadon, B., Wang, H., Abbas, H. K., Kemerait, R. C., Scully, B. T., Vaughn, J. N., & Guo, B. "Draft Genome Sequences of One *Aspergillus parasiticus* Isolate and Nine *Aspergillus flavus* Isolates with Varying Stress Tolerance and Aflatoxin Production." *Microbiology Resource Announcements* 9, no. 37 (2020): e00478-20. https://doi.org/10.1128/MRA.00478-20.

Fountain, J. C., Clevenger, J. P., Nadon, B., Youngblood, R. C., Korani, W., Chang, P. K., Starr, D., et al. "Two New *Aspergillus flavus* Reference Genomes Reveal a Large Insertion Potentially Contributing to Isolate Stress Tolerance and Aflatoxin Production." *G3: Genes, Genomes, Genetics* 10, no. 10 (2020): 3515–3531. https://doi.org/10.1534/g3.120.401405.

Fountain, J. C., Clevenger, J. P., Vaughn, J. N., & Guo, B. "Lessons Learned: The Importance of Biological Curation." *Microbiology Resource Announcements* 10, no. 48 (2021): e00473–521. https://doi.org/10.1128/MRA.00473-21.

Fountain, J. C., Khera, P., Yang, L., Nayak, S. N., Scully, B. T., Lee, R. D., Chen, Z. Y., et al. "Resistance to *Aspergillus flavus* in Maize and Peanut: Molecular Biology, Breeding, Environmental Stress, and Future Perspectives." *The Crop Journal* 3, no. 3 (2015): 229–237. https://doi.org/10.1016/j.cj.2015.02.003.

Fountain, J. C., Koh, J., Yang, L., Pandey, M. K., Nayak, S. N., Bajaj, P., Zhuang, W., et al. "Proteome Analysis of *Aspergillus flavus* Isolate-Specific Responses to Oxidative Stress in Relationship to Aflatoxin Production Capability." *Scientific Reports* 8 (2018): 3430. https://doi.org/10.1038/s41598-018-21653-x.

Fountain, J. C., Pandey, A. K., Nayak, S. N., Bajaj, P., Wang, H., Kumar, V., Chitikineni, A., et al. "Transcriptional Responses of Toxigenic and Atoxigenic Isolates of *Aspergillus flavus* to Oxidative Stress in Aflatoxin-Conducive and Non-Conducive Media." *World Mycotoxin Journal* 13, no. 4 (2020): 443–457. https://doi.org/10.3920/WMJ2020.2566.

Fountain, J. C., Scully, B. T., Chen, Z. Y., Gold, S. E., Glenn, A. E., Abbas, H. K., Lee, R. D., Kemerait, R. C., & Guo, B. "Effects of Hydrogen Peroxide on Different Toxigenic and Atoxigenic Isolates of *Aspergillus flavus*." *Toxins* 7, no. 8 (2015): 2985–2999. https://doi.org/10.3390/toxins7082985.

Fountain, J. C., Scully, B. T., Ni, X., Kemerait, R. C., Lee, L. D., Chen, Z. Y., & Guo, B. "Environmental Influences on Maize-*Aspergillus flavus* Interactions and Aflatoxin Production." *Frontiers in Microbiology* 5 (2014): 40. https://doi.org/10.3389/fmicb.2014.00040.

Fountain, J. C., Yang, L., Pandey, M. K., Bajaj, P., Alexander, D., Chen, S., Kemerait, R. C., Varshney, R. K., & Guo, B. "Carbohydrate, Glutathione, and Polyamine Metabolism Are Central to *Aspergillus flavus* Oxidative Stress Responses Over Time." *BMC Microbiology* 19 (2019): 209. https://doi.org/10.1186/s12866-019-1580-x.

Gilbert, M. K., Mack, B. M., Moore, G. G., Downey, D. L., Lebar, M. D., Joardar, V., Losada, L., Yu, J., Nierman, W. C., & Bhatnagar, D. "Whole Genome Comparison of *Aspergillus flavus* L-Morphotype Strain NRRL 3357 (Type) and S-Morphotype Strain AF70." *PLoS One* 13, no. 7 (2018): e0199169. https://doi.org/10.1371/journal.pone.0199169.

Gilbert, M. K., Majumdar, R., Rajasekaran, K., Chen, Z. Y., Wei, Q., Sickler, C. M., Lebar, M. D., Cary, J. W., Frame, B. R., & Wang, K. "RNA Interference-Based Silencing of the Alpha-Amylase (amyl) Gene in *Aspergillus flavus* Decreases Fungal Growth and Aflatoxin Production in Maize Kernels." *Planta* 247 (2018): 1465–1473. https://doi.org/10.1007/s00425-018-2875-0.

Gong, Y. Y., Watson, S., & Routledge, M. N. "Aflatoxin Exposure and Associated Human Health Effects: A Review of Epidemiological Studies." *Food Safety* 4, no. 1 (2016): 14–27. https://doi.org/10.14252/foodsafetyfscj.2015026.

Gu, K. X., Song, X. S., Xiao, X. M., Duan, X. X., Wang, J. X., Duan, Y. B., Hou, Y. P., & Zhou, M. G. "A β2-Tubulin dsRNA Derived from *Fusarium asiaticum* Confers Plant Resistance to Multiple Phytopathogens and Reduces Fungicide Resistance." *Pesticide Biochemistry and Physiology* 153 (2019): 36–46. https://doi.org/10.1016/j.pestbp.2018.10.005.

Guo, B., Chen, Z. Y., Lee, R. D., & Scully, B. T. "Drought Stress and Preharvest Aflatoxin Contamination in Agricultural Commodity: Genetics, Genomics, and Proteomics." *Journal of Integrative Plant Biology* 50, no. 10 (2008): 1281–1291. https://doi.org/10.1111/j.1744-7909.2008.00739.x.

Gurusamy, D., Mogilicherla, K., & Palli, S. R. "Chitosan Nanoparticles Help Double-Stranded RNA Escape from Endosomes and Improve RNA Interference in the Fall Armyworm, *Spodoptera frugiperda*." *Archives of Insect Biochemistry and Physiology* 104, no. 4 (2020): e21677. https://doi.org/10.1002/arch.21677.

Hagiwara, D., Sakamoto, K., Abe, K., & Gomi, K. "Signaling Pathways for Stress Responses and Adaptation in *Aspergillus* Species: Stress Biology in the Post-Genomic Era." *Bioscience, Biotechnology, and Biochemistry* 80, no. 9 (2016): 1667–1680. https://doi.org/10.1080/09168451.2016.1162085.

Hallman, W. K., Hebden, W. C., Aquino, H. L., Cuite, C. L., & Lang, J. T. "Public Perceptions of Genetically Modified Foods: A National Study of American Knowledge and Opinion." Food Policy Institute, Cook College, Rutgers - The State University of New Jersey, 2003. https://ageconsearch.umn.edu/record/18174/. https://doi.org/10.1038/s41538-018-0018-4.

Jayashree, T., & Subramanyam, C. "Oxidative Stress as a Prerequisite for Aflatoxin Production by *Aspergillus parasiticus*." *Free Radical Biology and Medicine* 29, no. 10 (2000): 981–985. https://doi.org/10.1016/S0891-5849(00)00398-1.

Kamala, A., Shirima, C., Jani, B., Bakari, M., Sillo, H., Rusibamayila, N., De Saeger, S., et al. "Outbreak of an Acute Aflatoxicosis in Tanzania During 2016." *World Mycotoxin Journal* 11, no. 3 (2018): 311–320. https://doi.org/10.3920/WMJ2018.2344.

Klich, M. A. "*Aspergillus flavus*: The Major Producer of Aflatoxin." *Molecular Plant Pathology* 8, no. 6 (2007): 713–722. https://doi.org/10.1111/j.1364-3703.2007.00436.x.

Lamb, M. C., & Sternitzke, D. A. "Cost of Aflatoxin to the Farmer, Buying Point, and Sheller Segments of the Southeast United States Peanut Industry." *Peanut Science* 28, no. 2 (2001): 59–63. https://doi.org/10.3146/i0095-3679-28-2-4.

Liu, Y., & Wu, F. "Global Burden of Aflatoxin-Induced Hepatocellular Carcinoma: A Risk Assessment." *Environmental Health Perspectives* 118, no. 6 (2010): 818–824. https://doi.org/10.1289/ehp.0901388.

Masanga, J. O., Matheka, J. M., Omer, R. A., Ommeh, S. C., Monda, E. O., & Alakonya, A. E. "Downregulation of Transcription Factor *aflR* in *Aspergillus flavus* Confers Reduction to Aflatoxin Accumulation in Transgenic Maize with Alteration of Host Plant Architecture." *Plant Cell Reports* 34 (2015): 1379–1387. https://doi.org/10.1007/s00299-015-1794-9.

Narasaiah, K. V., Sashidhar, R. B., & Subramanyam, C. "Biochemical Analysis of Oxidative Stress in the Production of Aflatoxin and Its Precursor Intermediates." *Mycopathologia* 162 (2006): 179–189. https://doi.org/10.1007/s11046-006-0052-7.

National Library of Medicine (US), National Center for Biotechnology Information (NCBI). "Genome, Genome Information by Organism, *Aspergillus flavus*." Accessed February 7, 2023. https://www.ncbi.nlm.nih.gov/genome/browse/#!/eukaryotes/360/.

Nesbitt, B. F., O'Kelly, J., Sargeant, K., & Sheridan, A. N. N. "Aspergillus flavus and Turkey X Disease: Toxic Metabolites of *Aspergillus flavus*." *Nature* 195, no. 4846 (1962): 1062–1063. https://doi.org/10.1038/1951062a0.

Nierman, W. C., Yu, J., Fedorova-Abrams, N. D., Losada, L., Cleveland, T. E., Bhatnagar, D., Bennett, J. W., Dean, R. A., & Payne, G. A. "Genome Sequence of *Aspergillus flavus* NRRL 3357, a Strain That Causes Aflatoxin Contamination of Food and Feed." *Genome Announcements* 3, no. 2 (2015): e00168-15. https://doi.org/10.1128/genomeA.00168-15.

Omolehin, O., Raruang, Y., Hu, D., Han, Z. Q., Wei, Q., Wang, K., Rajasekaran, K., Cary, J. W., & Chen, Z. Y. "Resistance to Aflatoxin Accumulation in Maize Mediated by Host-Induced Silencing of the *Aspergillus flavus* Alkaline Protease (*alk*) Gene." *Journal of Fungi* 7, no. 11 (2021): 904. https://doi.org/10.3390/jof7110904.

Payne, G. A., Nierman, W. C., Wortman, J. R., Pritchard, B. L., Brown, D., Dean, R. A., Bhatnagar, D., Cleveland, T. E., Machida, M., & Yu, J. "Whole Genome Comparison of *Aspergillus flavus* and *A. oryzae*." *Medical Mycology* 44, supplement 1 (2006): S9–S11. https://doi.org/10.1080/13693780600835716.

Payne, G. A., Yu, J., Nierman, W. C., Machida, M., Bhatnagar, D., Cleveland, T. E., & Dean, R. A. "A First Glance into the Genome Sequence of *Aspergillus flavus*." In *The Aspergilli*, edited by G. H. Goldman and S. A. Osmani, 35–44. Boca Raton: CRC Press, 2007. https://doi.org/10.1201/9781420008517-7.

Pickova, D., Ostry, V., Toman, J., & Malir, F. "Aflatoxins: History, Significant Milestones, Recent Data on Their Toxicity, and Ways to Mitigation." *Toxins* 13, no. 6 (2021): 399. https://doi.org/10.3390/toxins13060399.

Qiao, L., Lan, C., Capriotti, L., Ah-Fong, A., Nino Sanchez, J., Hamby, R., Heller, J., et al. "Spray-Induced Gene Silencing for Disease Control Is Dependent on the Efficiency of Pathogen RNA Uptake." *Plant Biotechnology Journal* 19, no. 9 (2021): 1756–1768. https://doi.org/10.1111/pbi.13589.

Raruang, Y., Omolehin, O., Hu, D., Wei, Q., Han, Z. Q., Rajasekaran, K., Cary, J. W., Wang, K., & Chen, Z. Y. "Host-Induced Gene Silencing Targeting *Aspergillus flavus aflM* Reduced Aflatoxin Contamination in Transgenic Maize Under Field Conditions." *Frontiers in Microbiology* 11 (2020): 754. https://doi.org/10.3389/fmicb.2020.00754.

Raruang, Y. "Transgenic Control of Aflatoxin Contamination in Maize Through Host-Induced Gene Silencing Targeting *Aspergillus flavus* Genes Encoding Polygalacturonase (*p2c*) and Versicolorin Dehydrogenase (*aflM*)." PhD diss., Louisiana State University, 2019. https://doi.org/10.1126/sciadv.1602382.

Reddy, B. N., & Raghavender, C. R. "Outbreaks of Aflatoxicoses in India." *African Journal of Food, Agriculture, Nutrition, and Development* 7, no. 5 (2007). https://doi.org/10.18697/ajfand.16.2750.

Reverberi, M., Beccaccioli, M., & Zaccaria, M. "The Role of Aflatoxins in *Aspergillus flavus* Resistance to Stress." In *The Genus Aspergillus: Pathogenicity, Mycotoxin Production and Industrial Applications*, edited by M. Razzaghi-Abyaneh and M. Rai, 1–23. IntechOpen, 2021. https://doi.org/10.5772/intechopen.99003.

Roze, L. V., Chanda, A., Wee, J., Awad, D., & Linz, J. E. "Stress-Related Transcription Factor AtfB Integrates Secondary Metabolism with Oxidative Stress Response in Aspergilli." *Journal of Biological Chemistry* 286, no. 40 (2011): 35137–35148. https://doi.org/10.1074/jbc.M111.253468.

Roze, L. V., Hong, S. Y., & Linz, J. E. "Aflatoxin Biosynthesis: Current Frontiers." *Annual Review of Food Science and Technology* 4 (2013): 293–311. https://doi.org/10.1146/annurev-food-083012-123702.

Roze, L. V., Laivenieks, M., Hong, S. Y., Wee, J., Wong, S. S., Vanos, B., Awad, D., Ehrlich, K. E., & Linz, J. E. "Aflatoxin Biosynthesis Is a Novel Source of Reactive Oxygen Species—A Potential Redox Signal to Initiate Resistance to Oxidative Stress?" *Toxins* 7, no. 5 (2015): 1411–1430. https://doi.org/10.3390/toxins7051411.

Schroeder, H. W., & Boller, R. A. "Aflatoxin Production of Species and Strains of the *Aspergillus flavus* Group Isolated from Field Crops." *Applied Microbiology* 25, no. 6 (1973): 885–889. https://doi.org/10.1128/am.25.6.885-889.1973.

Sharma, K. K., Pothana, A., Prasad, K., Shah, D., Kaur, J., Bhatnagar, D., Chen, Z. Y., et al. "Peanuts That Keep Aflatoxin at Bay: A Threshold That Matters." *Plant Biotechnology Journal* 16, no. 5 (2018): 1024–1033. https://doi.org/10.1111/pbi.12846.

Skerker, J. M., Pianalto, K. M., Mondo, S. J., Yang, K., Arkin, A. P., Keller, N. P., Grigoriev, I. V., & Glass, N. L. "Chromosome Assembled and Annotated Genome Sequence of *Aspergillus flavus* NRRL3357." *G3: Genes, Genomes, Genetics* 11, no. 8 (2021): jkab213. https://doi.org/10.1093/g3journal/jkab213.

Smith, R. "Aflatoxin Costs Peanut Industry Millions Annually." *Southwest FarmPress*, March 10, 2023. https://www.farmprogress.com/peanut/aflatoxin-costs-peanut-industry-millions-annually.

Stewart, C. N., Richards, H. A., & Halfhill, M. D. "Transgenic Plants and Biosafety: Science, Misconceptions and Public Perceptions." *Biotechniques* 29, no. 4 (2000): 832–843. https://doi.org/10.2144/00294bi01.

Thakare, D., Zhang, J., Wing, R. A., Cotty, P. J., & Schmidt, M. A. "Aflatoxin-Free Transgenic Maize Using Host-Induced Gene Silencing." *Science Advances* 3, no. 3 (2017): e1602382. https://doi.org/10.1126/sciadv.160238.

Torres, A. M., Barros, G. G., Palacios, S. A., Chulze, S. N., & Battilani, P. "Review on Pre- and Post-Harvest Management of Peanuts to Minimize Aflatoxin Contamination." *Food Research International* 62 (2014): 11–19. https://doi.org/10.1016/j.foodres.2014.02.023.

Valiante, V., Heinekamp, T., Jain, R., Härtl, A., & Brakhage, A. A. "The Mitogen-Activated Protein Kinase MpkA of *Aspergillus fumigatus* Regulates Cell Wall Signaling and Oxidative Stress Response." *Fungal Genetics and Biology* 45, no. 5 (2008): 618–627. https://doi.org/10.1016/j.fgb.2007.09.006.

Williams, J. H., Phillips, T. D., Jolly, P. E., Stiles, J. K., Jolly, C. M., & Aggarwal, D. "Human Aflatoxicosis in Developing Countries: A Review of Toxicology, Exposure, Potential Health Consequences, and Interventions." *The American Journal of Clinical Nutrition* 80, no. 5 (2004): 1106–1122. https://doi.org/10.1093/ajcn/80.5.1106.

Wu, F. "Global Impacts of Aflatoxin in Maize: Trade and Human Health." *World Mycotoxin Journal* 8, no. 2 (2015): 137–142. https://doi.org/10.3920/WMJ2014.1737.

Yang, L., Fountain, J. C., Ji, P., Ni, X., Chen, S., Lee, R. D., Kemerait, R. C., & Guo, B. "Deciphering Drought-Induced Metabolic Responses and Regulation in Developing Maize Kernels." *Plant Biotechnology Journal* 16, no. 9 (2018): 1616–1628. https://doi.org/10.1111/pbi.12899.

Yang, L., Fountain, J. C., Wang, H., Ni, X., Ji, P., Lee, R. D., Kemerait, R. C., Scully, B. T., & Guo, B. "Stress Sensitivity Is Associated with Differential Accumulation of Reactive Oxygen and Nitrogen Species in Maize Genotypes with Contrasting Levels of Drought Tolerance." *International Journal of Molecular Sciences* 16, no. 10 (2015): 24791–24819. https://doi.org/10.3390/ijms161024791.

Yang, L., Jiang, T., Fountain, J. C., Scully, B. T., Lee, R. D., Kemerait, R. C., Chen, S., & Guo, B. "Protein Profiles Reveal Diverse Responsive Signaling Pathways in Kernels of Two Maize Inbred Lines with Contrasting Drought Sensitivity." *International Journal of Molecular Sciences* 15, no. 10 (2014): 18892–18918. https://doi.org/10.3390/ijms151018892.

Yu, J., Chang, P. K., Ehrlich, K. C., Cary, J. W., Bhatnagar, D., Cleveland, T. E., Payne, G. A., Linz, J. E., Woloshuk, C. P., & Bennett, J. W. "Clustered Pathway Genes in Aflatoxin Biosynthesis." *Applied and Environmental Microbiology* 70, no. 3 (2004): 1253–1262. https://doi.org/10.1128/AEM.70.3.1253-1262.2004.

Yu, J., Payne, G. A., Nierman, W. C., Machida, M., Bennett, J. W., Campbell, B. C., Bhatnagar, D., Dean, R. A., & Cleveland, T. E. "Aspergillus flavus Genomics as a Tool for Studying the Mechanism of Aflatoxin Formation." *Food Additives and Contaminants: Part A* 25, no. 9 (2008): 1152–1157. https://doi.org/10.1080/02652030802213375.

16 Mitigating Peanut Aflatoxin Contamination

Current Advances and Future Perspectives

Yasir Sharif, Shahid Ali Khan, Yuhui Zhuang, Hua Chen, Manish K. Pandey, Pooja Soni, Baozhu Guo, Xinyou Zhang, Rajeev K. Varshney, and Weijian Zhuang

16.1 INTRODUCTION

Peanuts (*Arachis hypogaea* L.) are an economically and nutritionally significant crop, particularly in Asia and Africa, consumed as oil, protein, food, and feed. However, peanuts face a considerable threat of aflatoxin contamination as a major host for both *A. flavus* and *A. parasiticus,* which produce aflatoxins, a potent carcinogen to humans and livestock. Aflatoxins (AFs) can contaminate peanuts during both pre-harvest and post-harvest stages and are resistant to degradation even under extreme conditions such as temperatures above 300°C, complicating removal through conventional processing (Rustom, 1997). The World Health Organization (WHO) estimates that aflatoxin exposure contributes to approximately 25% of liver cancer cases worldwide, with the burden being highest in sub-Saharan Africa and Southeast Asia, where aflatoxin contamination in food staples like peanuts is most prevalent (Liu and Wu, 2010). It is estimated that over 4.5 billion people in developing countries are exposed to aflatoxins (Wild and Gong, 2010). Because *A. flavus* is a facultative parasite, it can infect peanuts in the field and storage, making aflatoxin prevention an international problem (Shabeer et al., 2022). It is well known that the risk of contamination can happen during drying, processing, and transportation (Bediako et al., 2019). To date, no peanut genotype has been found to exhibit stable and high resistance to *A. flavus* under all ecological conditions. Most peanut germplasm lines tested so far remain susceptible to infection under the right environmental stressors, particularly under drought and high temperatures during the pre-harvest stage (Pandey et al., 2019). The search for aflatoxin-resistant varieties continues to be a priority in peanut breeding programs worldwide, driven by the need to safeguard human health and economic livelihoods.

Scientists have been actively researching this issue to address the challenges of aflatoxin contamination in peanuts since the 1960s. Over the decades, there has been significant progress in understanding the dynamics of aflatoxin-producing fungi and their interaction with peanut crops (Soni Gangurde et al., 2020). Early studies focused on identifying resistant peanut varieties, leading to the development of breeding programs aimed at enhancing resistance to *A. flavus* (Yu et al., 2020). With the breakthrough of cultivated peanut genomes in recent years (Zhuang et al., 2019; Bertioli et al., 2019; Chen et al., 2019), rapid advancements in genetic techniques, including candidate gene discovery, marker-assisted selection, transgenics, and genome editing, have been accelerating the creation of more resilient peanut cultivars (Khan et al., 2020; Yu et al., 2023; Neelakandan et al., 2022; Hasan et al., 2021). Research has also expanded into omics technologies, providing deeper insights into peanut–pathogen interactions and aiding in the identification of genetic markers linked to resistance (Wang et al., 2024; Huang et al., 2023). Additionally, biological control methods have shown promise in field trials, such as using non-toxin-producing strains of *A. flavus* to outcompete toxigenic strains (Khan et al., 2021).

This chapter explores the advances in peanut improvement against aflatoxin contamination, focusing on traditional methods and modern biotechnological innovations. It highlights the advancement of omics technologies that enhance our understanding of peanut–pathogen interactions, along with resistance enhancement through MAS breeding, CRISPR-Cas9, and RNA interference (RNAi). Integrating biotechnological strategies with agronomic practices provides a comprehensive approach to addressing aflatoxin challenges and improving food safety.

16.2 AFLATOXIN CONTAMINATION AND ITS DETRIMENTS TO HUMANS

16.2.1 *Aspergillus* Species and Aflatoxin Production

The genus *Aspergillus* encompasses over 200 species (Samson, 1992), including members of the section *Flavi* that significantly impact agriculture (Frisvad et al., 2019). *A. flavus* and *A. parasiticus* are the primary pathogens responsible for aflatoxin contamination, particularly in tropical

DOI: 10.1201/9781003360988-16

and subtropical regions, where warm and humid conditions facilitate their growth (Probst et al., 2014). *A. flavus* predominantly produces aflatoxin B1, the most potent carcinogenic of all aflatoxins, while *A. parasiticus* produces aflatoxins B1, B2, G1, and G2, with aflatoxin G1 being another major contaminant (Amaike and Keller, 2011). *A. flavus* grows well at 28–30 °C and readily colonizes nutrient-rich environments (Lakkireddy et al., 2014). These fungi infect crops like peanuts, maize, and cottonseed both before and after harvest. During the pre-harvest stage, environmental stressors such as drought, high temperatures, and insect damage increase plant vulnerability to *Aspergillus* infection. Postharvest conditions such as improper drying and storage, such as in moisture-rich environments, prommote contamination afterward (Dövényi-Nagy *et al.,* 2020). Due to their resilience to extreme environmental conditions, *Aspergillus* species continue to threaten food safety (Soni Nayak et al., 2020). Geographically *A. flavus* is more prevalent in Asia and Africa, while *A. parasiticus* dominates the Americas. These fungi spread via airborn conidia, which germinate on crops and produce aflatoxins to outcompete other microorganisms (Nji *et al.,* 2023). Key steps and genes involved in the biosynthesis of aflatoxins in *Aspergillus* species are represented in Figure 16.1. Understanding the life cycle and environmental conditions is vital for developing mitigate strategies in vulnerable crops.

16.2.2 Biological Features of Aflatoxins

Aflatoxins (Afs) are highly toxic secondary metabolites with remarkable chemical stability. These polyketide-derived compounds are produced via a complex biosynthetic pathway regulated by gene clusters in *Aspergillus* species (Yu, 2012). Aflatoxin biosynthesis begins with the formation of a polyketide backbone (Figure 16.1), which is subsequently modified through various enzymatic steps, including oxidation, methylation, and cyclization (Ehrlich et al., 2007). Environmental factors such as high temperature, moisture, and nutrients can trigger aflatoxin production by activating the regulatory genes (Loi et al., 2023). The most common aflatoxins are B1, B2, G1, and G2, with AFB1 being the most carcinogenic, particularly to the liver (Saini and Kaur 2012). High chemical stability (like surviving temperatures up to 307 °C) makes it difficult to eliminate from food and can accumulate to contaminate the whole food strain (Marshall et al., 2020). Their lipophilic nature tends to their accumulation in fatty tissues. In livestock, aflatoxin-contaminated feed can lead to the formation of aflatoxin M1, which is excreted in milk, posing further health risks (Alameri et al., 2023)—the resilience and persistence of aflatoxins present a major challenge in both crop production and the food safety.

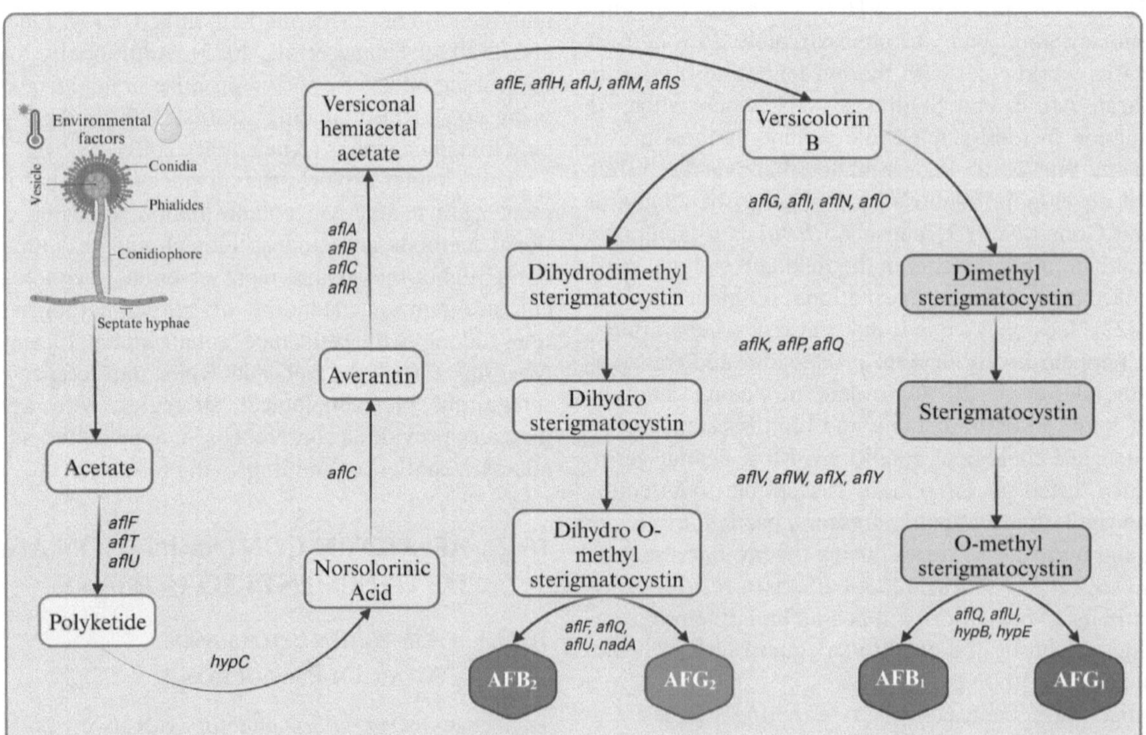

FIGURE 16.1 Schematic representation of the biosynthetic pathway of aflatoxins in *Aspergillus* species, highlighting key metabolic intermediates and the associated genes involved at each step. Environmental factors influence this multistep process, producing toxic aflatoxins (B1, B2, G1, and G2).

16.2.3 Ecological Factors underlying AF Contamination in Peanut

Ecological factors, including temperature, humidity, and rainfall, are key to aflatoxin contamination in peanuts. Aflatoxin production is optimal at 25°C–30°C and with humidity above 85%, conditions common in tropical and subtropical regions (Molnár et al., 2023). Climatic shifts, particularly drought followed by heavy rainfall, increase plant stress and susceptibility to fungal infection (Treseder *et al.,* 2018). Soil moisture extremes, either too little or too much, elevate contamination risks (Bouskill et al., 2016). Pre-harvest drought can injure peanut pods, enabling fungal invasion, especially under high temperatures; prolonged drought leads to aflatoxin production in the field (Pandey et al., 2019). Postharvest contamination is driven by improper drying and storage, as high humidity and temperature promote *Aspergillus* growth (Bradford et al., 2018; Pandey et al., 2019). Effective contamination management will require adapting to shifting climate patterns, particularly temperature and rainfall shifts, which will be vital for managing aflatoxin contamination in peanut production (Nji et al., 2022).

16.2.4 Health Implications

Aflatoxins are potent carcinogens, with aflatoxin B1 classified as a Group 1 carcinogen by the International Agency for Research on Cancer (IARC). AFB1 induces DNA mutations by forming adducts with guanine bases, triggering oncogenes activation and inhibiting tumor suppressor genes (Liu and Wu, 2010). Chronic exposure to aflatoxins, especially in regions with high dietary intake of contaminated foods, is a major cause of hepatocellular carcinoma (HCC) worldwide (Gong et al., 2024). Outbreaks such as that in Kenya in 2004, which caused 200 deaths from aflatoxicosis, underscore the severity of contamination risks (Probst et al., 2007; Williams et al., 2004; Azziz-Baumgartner et al., 2005). The synergistic effect of aflatoxin exposure with hepatitis B virus (HBV) infection further exacerbates the risk of liver cancer (Chu et al., 2018). Beyond carcinogenesis, aflatoxins are immunosuppressive, increasing vulnerability to infections such as malaria and HIV (Gong et al., 2016). In children, chronic exposure leads to growth stunting and developmental delays (Andrews-Trevino et al., 2021). In animals, these toxins not only hinder growth but also act as immunosuppressants (Benkerroum, 2020). The long-term health risks of aflatoxins, including their immunosuppressive effects in humans and animals, contribute significantly to acute illness and chronic disease (Wild and Gong, 2010).

16.2.5 Economic Impact and Global Food Security

Aflatoxin contamination has significant economic consequences, particularly in countries where peanuts and maize are major exports. In Africa, it leads to estimated annual losses of over USD 670 million due to rejected exports (Meijer et al., 2021). Exporters must comply with strict aflatoxin limits set by the European Union (EU) and the United States, where permissible levels are as low as 2 parts per billion (ppb) (Khlangwiset et al., 2011). The economic impact affects the entire supply chain, from the farmer to the exporter (Hussain and Hussein, 2020). In regions with limited food safety infrastructure, such as sub-Saharan Africa and Southeast Asia, aflatoxin contamination exacerbates food insecurity, undermining efforts to meet nutritional needs (Coulibaly et al., 2008). Surveys reveal varying contamination levels in dietary feed, ranging from 22% in Portugal to 82.5% in Iran (Alvarado et al., 2017). Beyond trade, aflatoxin contamination also burdens healthcare systems with costs related to treating aflatoxin-associated diseases, further straining resources in affected regions.

16.2.6 Technologies and Strategies for AF Detection

As aflatoxin is highly detrimental to humans, accurate detection is important. Advances in detection technologies have significantly improved aflatoxin monitoring in crops. Traditional methods such as enzyme-linked immunosorbent assays (ELISA) and high-performance liquid chromatography (HPLC) remain widely used but can be time-consuming and require specialized equipment (Beyene et al., 2019). Newer biosensor technologies, which detect aflatoxins through interactions between biological molecules and toxins, offer rapid, cost-effective, on-site detection suitable for field use (Pisoschi et al., 2024). Polymerase chain reaction (PCR)-based methods target *Aspergillus* DNA, providing an early warning system for potential aflatoxin contamination (Abd El-Aziz et al., 2021). Additionally, artificial intelligence (AI) and machine learning analyze environmental and crop data to predict contamination risks, enabling more targeted interventions (Inglis et al., 2024). These advanced technologies are paving the way for more efficient and proactive approaches to aflatoxin detection and control.

16.3 PEANUT RESISTANCE GENETICS TO AFLATOXIN

16.3.1 Germplasm Evaluation for AF Resistance

Germplasm resources serve as the foundation for genetic studies and assist in discovering important genes and their utilization to improve economically and ecologically important traits (Liu *et al.,* 2020). Mixon and Rogers (1973) were pioneers in proposing the utilization of AFs-resistant peanut cultivars to manage contamination. Several screening programs have been conducted globally to assess the resistance levels of different peanut cultivars and wild species. Germplasm resources from the International Crops Research Institute for the Semi-Arid Tropics (ICRISAT), the US Department of Agriculture (USDA), and the Chinese

Academy of Agricultural Sciences have been instrumental in these efforts (Otyama et al., 2020; Conde et al., 2023; Zhou et al., 2022). The mechanism of AF resistance in peanuts typically involves three aspects: resistance to pod infection, resistance to seed infection, and resistance to AF production (Nigam et al., 2009). In recent decades, several peanut collections have demonstrated resistance to *A. flavus* infection (Dieme et al., 2018). Waliyar et al. (1994) evaluated peanut lines belonging to different agricultural zones of west Africa and India and discovered that ICGV87710, 55–437, J11, and PI337394 F exhibited consistent resistance to seed infection. Holbrook Kvien et al. (2000) reported that drought-tolerant peanut lines PI145681 and Tifton 8 exhibited resistance to AFs. Yong et al. (2004) discovered that two peanut bacterial wilt-resistant cultivars, Zhonghua 6 and Thaishan Zhenzhu, exhibited high resistance to AF contamination. JIANG et al. (2010) evaluated a mini-core population of 155 lines from the ICRISAT and a Chinese core collection of 561 samples. They found eight lines resistant to *A. A. flavus* colonization and eight resistant to AF production. Dieme et al. (2018) inoculated 67 peanut genotypes under a controlled environment and identified AF production-resistant genotype 12CS-104.

Physically injured peanut pods are more prone to *A. flavus* spores than healthy and intact pods (Sudhakar et al., 2007), supporting the view that peanut pods serve as a primary organ of physical defense against fungal invasion. QIU et al. (2019) examined 276 accessions for their resistance to shell infection and found two lines showing a low level of shell infection. The infection index of Yueyou 9 was less than 19%, indicating that it is a highly resistant cultivar (Liao et al., 2009). Njoki and colleagues evaluated six newly improved Kenyan peanut cultivars for their resistance to AF accumulation and found the lowest average disease occurrence of 1.31 ± 1.75% in ICGV-SM 90704 (Njoki et al., 2023). Han et al. (2023) assessed the resistance of 102 Korean peanut accessions to AFB1 to use them in breeding programs. They identified 26 accessions exhibiting lower AFB1 contents compared to the check genotype 55-437 and selected them for developing resistant cultivars. Korani et al. (2017) infected ten peanut accessions with GFP-expressing *A. flavus* strain and observed the GFP signal intensity and fungal infection area after every 8 hours for 3 days. NC 3033 showed the lowest fungal growth, and ICG 1471 showed the lowest AF levels. Two years of field examination also reported lower AF levels in ICG 1471. Under laboratory conditions, seed inoculation of Zhonghua 6 for multiple years proved it to be a highly AFs-resistant cultivar, with an AF content of less than 10% (Liao et al., 2009). Using seeds harvested from different regions and growing seasons and inoculated with multiple *A. flavus* strains, Zhuang et al. (2010) reported an elite cultivar, Minhua 6, characterized by high yield and protein content and stable resistance to seed colonization. Considerable attention has been devoted to examining peanut germplasm for resistance, resulting in the discovery of various robust genotypes such as J11 (Mehan and McDonald, 1980), ICG 12625 (Sujay et al., 2008), AH 7223, and Xinhuixiaoli (SuiLin, 2006), which are being actively utilized in breeding programs or genetic research. Specifically, J11 has become a well-known, highly resistant variety against *A. flavus* infection. Previous studies demonstrated that J11 was resilient to *A. flavus* infection and exhibited resistance to aflatoxin production (Asis et al., 2005). Details of the screening programs and identified *Aspergillus*/aflatoxin resistant varieties/lines are given in Table 16.1.

16.3.2 QTL Mapping for Aflatoxin Resistance

Quantitative trait locus (QTL) mapping has become a powerful tool for understanding the genetic basis of aflatoxin resistance in peanuts. Various studies have identified three distinct host–pathogen interaction levels leading to aflatoxin contamination: in vitro seed colonization (IVSC), pre-harvest aflatoxin contamination (PAC), and aflatoxin production (AP). Despite extensive research, no single peanut genotype has been found to combine all three resistance mechanisms (Pandey et al., 2019). Genome sequencing of both hosts and pathogens has offered valuable insights into their complex genetic architecture and the discovery of novel genes. QTL mapping has discovered an array of QTLs in peanuts that are associated with resistance to *A. flavus* infection by linkage mapping or genome-wide association studies (GWAS). Liang Chen et al. (2009) constructed an SSR linkage map and performed QTL mapping of peanut resistance to *A. flavus* infection. Yu et al. (2019) conducted QTL mapping in a recombinant inbred lines (RILs) population derived from ICG12625 × Zhonghua10, identifying a major QTL on chromosome A10 for resistance to *A. flavus*. Two additional QTLs on chromosomes A03 and A10 were linked to aflatoxin (AF) production. Four QTLs (qAFB1A07, qAFB1B06.1, qAFB2A07, qAFB2B06) showed major and stable effects, explaining 9.32%–21.02% of phenotypic variance (PVE). Khan et al. (2020) used a RIL population of F10 generation from a cross between susceptible YY92 and resistant XHXL cultivars to construct a high-density SNP map and conduct QTL mapping for *A. flavus* resistance. They identified two stable QTLs, qRAF-3-1 and qRAF-14-1, on chromosomes A03 and B04, respectively. QTL qRAF-3-1 was genetically positioned at 1.67 cM and accounted for more than 19% of the PVE, housing a total of 67 genes coding for different TFs, including WRKY, ABC, TGC4, and so on. qRAF-14-1 was localized within a distance of 1.34 cM and contributed to 5.15% of the PVE. The two QTL intervals housed several genes with similar functions, such as RPP13, RKP, PLLS, PPR, TPR, and HMT/DP, which showed differential expressions in response to *A. flavus* infection, and the RPP13, PPR, and lox were screened as candidate genes for validation.

Using a mapping population of 200 RILs from a cross between Zhonghua16 (susceptible) and J11 (resistant), Jiang et al. (2021) discovered six new resistant QTLs on chromosomes A05, A08, B01, B03, and B10, explaining 5.03%–10.87% PVE. Five of the QTLs were associated with J11

TABLE 16.1
Summary of the Screening of Peanut Germplasm for Identification of Aflatoxin-Resistant Genotypes

Resistance Mechanism	Sample Size	*Aspergillus* Species	Screening Jethod	Remarks	Resistant Varieties Identified	Reference
PHAC	31 elite lines	*A. flavus*	Seed colonization test	Bacterial wilt–resistant lines were aflatoxin resistant, but no correlation between two resistances was found	Zhonghua 6, Thaishan Zhenzhu	(Yong et al., 2004)
	13 elite varities	*A. flavus*	mycelial growth on seed surface	Compact testa is more resistant to PHAC	Igola, Serenut 1, Serenut 2, and entry 99527	(Olwari et al., 2013)
	10 elite genotypes	*A. flavus*	Seed infection coverage and intensity analyzer (SICIA)	All genotypes showed PHAC	ICG 1471, NC3033, ICGV 88145, GT-C20	(Korani et al., 2017)
	276 accessions	*A. flavus*	shell infection	Peanut shell is preliminary protection organ	W.h8342 and W.h8333	(QIU et al., 2019)
	6 new Kenyan cultivars	*A. flavus*	Ultrahigh performance liquid chromatography and fluorescence detection (UHPLC-FLD)	ICGV-SM 90704 exhibits better yield characteristics	ICGV-SM 90704	(Njoki et al., 2023)
	102 Korean peanut accessions	*A. flavus*	Ultra-performance liquid chromatography	Percent seed infection index estimation	26 accessions exhibiting lower AFB1 contents	(Han et al., 2023)
PAC	7 breeding lines	*A. flavus*	Conidial heads of *A. A. flavus* (green) and *A. niger* (black)	Linolic acid content did not affect PAC	F1334 and F1344	(Holbrook Wilson et al., 2000)
	20 genotypes	*A. flavus*	Immunoaffinity column fluorometer method	Drought tolerance favors PAC	PI145681, Tifton 8	(Holbrook Wilson et al., 2000)
	11 genotypes	*A. flavus*	Aflatoxin estimation through ELISA	Drought-related traits are associated with PAC	ICGV 98305, ICGV 98330, ICGV 98348, ICGV 98353, Tifton-8	(Girdthai et al., 2010)
IVSC	More than 100 varieites, accessions and breeding lines	*A. flavus*	Visual development of conidial spores	Low seed moisture resists IVSC	P.I. 331326 and P.I. 343419	(Mixon and Rogers, 1973)
	14 genotypes	*A. parasiticus*	Fungal sporulation	Higher moisture reduces infection rate	J-11 and Lampang	(Kisyombe et al., 1985)
	40 elite lines	*A. flavus*	*A. A. flavus* contaminated seed counting	Drought increases IVSC	55–437, PI 337409, PI 337394F, 73–30	(McDonald et al., 1989)
	12 genotypes	*A. flavus, A. paraciticus*	Immunoaffinity column fluorometer method	AF production was significantly affected by (E), (G) and (G × E) interactions	AR-2, GFA-1	(Anderson et al., 1995)

(Continued)

TABLE 16.1 (CONTINUED)
Summary of the Screening of Peanut Germplasm for Identification of Aflatoxin-Resistant Genotypes

Resistance Mechanism	Sample Size	*Aspergillus* Species	Screening Jethod	Remarks	Resistant Varieties Identified	Reference
	35 wild accessions	*A. flavus*	IVSC and AP	*A. pusilla*, *A. chiquitana*, *A. triseminata* species resistant to IVSC and AP	ICG 13212, ICG 11560, ICG 8131, ICG 14875	(Thakur et al., 2000)
	48 elite cultivars and genotypes	*A. flavus*	Infection rate and infection index	IVSC resistance was affected with seed physiologic status resulting from growing environments	Minhua 6, ICGV91284, Yueyou 9	(Zhuang et al., 2007)
IVSC and AP	850 cultivars/ lines	*A. flavus*	Seed colonization and aflatoxin estimation	Some IVSC resistant varieties resist AP	PI 337394F, PI337409, and UF71513	(Mehan et al., 1986)
	25 breeding lines	*A. flavus*, *A. paraciticus*	ELISA for toxin estimation	VAR 27 recorded high *A. A. flavus* infection but low AF level	ICGV87710, 55–437, J11, PI337394 F	(Waliyar et al., 1994)
	155 ICRISAT lines and 561 Chinese core lines	*A. flavus*	Seed infection % and AP	ICRISAT lines are more resistant than Chinese lines	ICG 12625 (resistant to AP) and ICG 4750 (resistant to seed invasion)	(JIANG et al., 2010)
	67 genotypes (58 CSL & 9 NRV)	*A. flavus*	Seed colonization test	Varieties with compact and thicker testa resist IVSC	12CS-104, 73-33	(Dieme et al., 2018)

alleles, significantly enhancing resistance to *A. flavus*. Using 99 Chinese mini-core accessions for GWAS analysis, Ding et al. (2022) identified 16 SNPs/InDels associated with resistance to aflatoxin production. From the RIL population of Zhonghua16 × J11, Jin et al. (2023) identified 11 QTLs for aflatoxin production resistance, with PVE ranging from 4.61% to 11.42%. Three QTLs (qAFTA05.1, qAFTB05.2, and qAFTB06.3) showed major effects, reducing aflatoxin content by 34.77%–47.67% in segregating lines. In a recent study, Yu et al. (2023) identified a major effect QTL, qAFTRA07.1, associated with aflatoxin production resistance (AP), mapped to a 1.98 Mbp interval (Table 16.2). Within this interval, a gene named *AhAftr1* was detected with structural variation (SV) in its leucine-rich repeat (LRR) domain. Transgenic maize plants overexpressing $AhAftr1_{(ZH6)}$ showed a significant decrease in AF production (57.3%) compared to $AhAftr1_{(XH13)}$—furthermore, a molecular diagnostic marker, AFTR.Del.A07, was developed based on this structural variation. Using AFTR.Del.A07, 36 lines exhibiting over 77.67% reduction in aflatoxin content compared to the susceptible control Zhonghua12 (ZH12) were identified from a panel of peanut germplasm accessions and breeding lines (Yu et al., 2023). Detailed information on different high-density fine-mapping programs to identify the resistant QTLs is given in Table 16.2.

16.3.3 Marker Development for AF Tolerance

Marker-assisted breeding (MAB) is a modern breeding technique that uses molecular markers to select desirable traits, such as aflatoxin resistance, at an early stage of plant development. At present, there is a shortage of reliable AF resistance indicators in peanuts that are based on genetic variations such as InDels, SNPs, or SSRs. In the context of aflatoxin resistance, molecular markers linked to resistance genes and QTLs are being developed to facilitate MAB. For example, QTL mapping and GWAS studies have identified markers associated with resistance to *A. flavus* colonization and aflatoxin accumulation. Modern techniques, such as genome-wide association studies (GWAS), offer hope for discovering aflatoxin resistance markers. The development of new peanut varieties could be greatly accelerated with the use of marker-assisted selection (MAS) and marker-assisted backcrossing (MABC). It is necessary to identify markers linked to target traits (Pandey et al., 2017; Varshney et al., 2019). Several successful cases of marker-assisted breeding for aflatoxin resistance have been

TABLE 16.2
List of High-density Genetic Linkage Maps Developed to Identify QTLs for Peanut AF Resistance.

Population	Genotyping Assay	Number of Mapping Lines	Number of Mapped Markers	Mapping Distance (cM)	Number of Genes/ QTLs	PEV %	Reference
Zhonghua10×ICG 12625 (F4-F6 population)	SSRs	140	1175 SSR markers and 42 transposon markers	2038.75	12 QTLs	9.32–21.02	(Yu et al., 2019)
Xinhuixiaoli×Yueyou 92 (F10 population)	SLAF-Seq	314	5022 SNPs	2231.25	2 major QTLs	5.15–19.00	(Khan et al., 2020)
Zhonghua 16×J11 (F8 population)	WGR and SNPs/ InDels	200	233,365 SNPs/InDels	1573.85	6 QTLs	5.03–10.87	(Jiang et al., 2021)
Chinese mini mini core collection	RAD-Seq	99	38,237 SNPs/6207 InDels	N/A	16SNPs/InDels by GWAS	23.33–38.64	(Ding et al., 2022)
Zhonghua 16×J11 (F7-9 population)	SNPs/InDels	200	233365 SNPs/InDels	1573.85	11 QTLs	4.61–11.42	(Jin et al., 2023)
Xuhua13×Zhonghua 6 (F7-9 population)	SNPs/InDels	186	294,330 SNPs/39,530 InDels	N/A	4 QTLs/1 Gene	5.68–13.39	(Yu et al., 2023)

reported. For instance, the peanut variety *Tifguard*, which exhibits moderate resistance to aflatoxin accumulation, was developed using molecular markers linked to resistance QTLs (Holbrook et al., 2024). Lei et al. (2005) discovered two AFLP markers, E45/M53-440 and E44/M53-520, closely associated with the genes responsible for *A. flavus* resistance. To validate the markers, 20 resistant peanut genotypes and seven susceptible genotypes (from F2 progeny of Zhonghua5 × J11) were tested, revealing a strong correlation between the molecular markers and *A. flavus* resistance. A study by Hong et al. (2009) tested 100 SSR markers and found polymorphism for 41 markers with 2-4 alleles per locus across 12 *A. flavus* susceptible and resistant lines. The PIC values ranged from 0.153 to 0.750. Correlation analysis for host resistance and SSR markers identified five markers significantly correlated with *A. flavus* resistance. Notably, the marker pPGSseq19D9 demonstrated the highest correlation, effectively distinguishing between resistant and susceptible varieties, suggesting that pPGSseq19D9 may be linked to a major gene regulating *A. flavus* infection resistance (Hong et al., 2009). A GWAS study on 99 peanut accessions using restriction site-associated DNA sequencing (RAD-Seq) by Yu et al. (2020) identified 60 SNP markers associated with aflatoxin production, explaining 16.87%–31.70% of the PVE. Among these, SNP02686 and SNP19994 exhibited the highest PVE, accounting for 31.70% and 28.91%, respectively. Nadaf et al. (2022) found that three markers, S-21, S-80, and GM-1954, were linked to tolerance against AFs. Based on the discovery of candidates or key genes, more potential markers are being used for breeding applications.

16.4 OMICS STRATEGIES CHARACTERIZING PEANUT RESPONSE TO AFLATOXIN CONTAMINATION

16.4.1 Transcriptomics

A number of transcriptome analyses have provided crucial insights into the molecular responses of peanuts to aflatoxin contamination. Guo et al. (2008) constructed cDNA libraries from resistant (GT-C20) and susceptible (Tifrunner) peanut cultivars, identifying nine and eight upregulated resistance-related genes, respectively, with three shared between the genotypes. Later, Guo et al. (2011) conducted a microarray analysis, revealing that resistant cultivars exhibited upregulation of 62 genes in response to *Aspergillus* infection, with 22 genes consistently expressed across resistant varieties. The research team of Zhuang performed a large scale of 454 sequencing with mixed cDNA from different biotic/abiotic stresses and across tissues of a resistant cultivar Minhua 6, obtaining 110,794 unigenes. Then, a high-density microchip with 101,344 unigene probes was developed and was used to screen genes related to *A. flavus* using RNA samples from pods treated with drought, drought plus *A. flavus* infection, and normal irritation in late growth stage of peanut (Chen et al., 2010). Forty-seven NBS-LRR genes with over three-fold up-/downregulation from Minhua 6 were found in response to *A. flavus* stress (Chen, 2011). Deng et al. (2018) characterized one of the *NB-ARC* genes, *AhRAF4. They* found it showed a distinct response to *A. flavus* infection between resistance and susceptible varieties and played a role in the defense response to the pathogen. Wang et al. (2016) identified 128,725 unigenes from resistant (Zhonghua 6) and susceptible (Zhonghua 12) cultivars, with 30,143 differentially expressed genes (DEGs) and 842 associated with defense, including ADP-ribosylation factors, leucine-rich repeat receptor-like kinases, and pathogenesis-related proteins. Zhao et al. (2019) investigated the correlation between transcriptome and proteome profiles in response to *A. flavus* invasion. They found 663 DEGs out of a total of 13,539 genes. These DEGs included chitinases, pathogenesis-related proteins, and transcription factors. They also found 1,382 proteins, and 314 of them were aberrantly expressed under *A. flavus* infection. However, the correlation between transcriptome and proteome was low. Jayaprakash et al. (2021) used RNA-seq datasets to identify critical genes in peanut immunity, including R genes, pattern recognition receptor (PRR) genes, and thaumatin-like protein 1b. Weighted gene co-expression network analysis (WGCNA) and a comparative transcriptome approach by Cui et al. (2022) identified 18 core genes associated with resistance to *A. flavus*. These genes include MAPK kinase, PRRs, serine/threonine kinase (STK), 1-aminocyclopropane-1-carboxylate oxidase (ACO1), phosphatidylinositol transfer protein, pathogenesis-related proteins (PR10), SNARE protein SYP121, cytochrome P450, pentatricopeptide repeat (PPR) protein, and pectinesterase. Soni et al. (2021) conducted a global transcriptome analysis of resistant and susceptible cultivars, identifying 45,336 transcripts and 57 transcription factor families with key DEGs like resveratrol synthase and ABA-responsive *ABR17.* Another RNA-sequencing study by Soni Nayak et al. (2020) revealed 1,779 unique DEGs in resistant and susceptible lines, implicating pathways related to cell wall metabolism, hormone signaling, flavonoid biosynthesis, and ROS detoxification.

16.4.2 Proteomics

Proteomics, the study of the complete set of proteins within a cell and their structures and functions in physiological pathways, represents a pivotal frontier in understanding biological systems (Wilkins et al., 1998). It has significantly advanced our knowledge of gene regulation and regulatory networks to elucidate the molecular mechanisms underlying host–pathogen interactions and aflatoxin contamination. Employing techniques such as two-dimensional electrophoresis, mass spectrometry, and real-time RT-PCR, Wang et al. (2010) undertook a comprehensive investigation on resistant peanut cultivar YJ-1 and a susceptible cultivar Yueyou 7 under drought, drought with *A. flavus*, and normal irrigation. Their study identified 12 differentially expressed proteins between YJ-1 and Yueyou 7 under varying conditions. Chen (2011) conducted differential expression proteome screening of peanut pericarps by high throughput 2-D gel electrophoresis, and 83 differentially expressed proteins among ~3000 protein spots were analyzed from resistant Minhua 6 infected by *A. flavus* under drought and regular irrigation. Nine proteins were characterized for their defense related to *A. flavus* infection. This provided insight into the physiologic mechanism underlying Aflatoxin contamination. Wang et al. (2012) delved into the intricacies of aflatoxin-responsive proteins in peanut cotyledons infected with aflatoxin-producing *A. flavus* strains. Using a differential proteomics approach, they unveiled a spectrum of proteins crucial in innate immunity and immune signaling, defense induction, cell wall responses, condensed tannin synthesis, penetration resistance, PAMP perception, DNA and RNA stabilization, hypersensitive response, phytoalexin biosynthesis, detoxification, metabolic regulation, and peptidoglycan assembly. Fountain and colleagues used the proteomics approach to gain insight into the regulatory mechanisms of pathogens. They exposed high, moderate, and non-aflatoxigenic strains of *A. flavus* (AF13, NRRL3357 and K54A, respectively) to oxidative stress of H_2O_2 and performed the proteome analysis. They identified 1,173 proteins, 220 differentially expressed proteins controlling the toxic abilities of AF13, NRRL3357, and K54A, indicating that the toxin production is governed by a network of genes that collectively regulate the synthesis of lytic enzymes, oxidative stress resilience, secondary metabolite production, pathogenicity, mycelial development, and carbohydrate metabolism. Bhatnagar-Mathur et al. (2021) conducted a comparative proteomic analysis of susceptible peanut variety JL 24 and its alfalfa defensin gene expressing near-isogenic transgenic derivative. Their findings highlighted the induction of several resistance-associated proteins in the resistant genotypes in secondary metabolic pathways. These included cinnamic acid-4-hydroxylase, chalcone synthase, β-ketoacyl-ACP-reductase, diacylglycerol glycerol-3-phosphate-3-phosphatidyltrans ferase, diacylglycerol acyltransferase, flavanone-3-hydroxylase, lipoxygenase, phenylalanine ammonia-lyase, resveratrol synthase, and monoacylglycerol acyltransferase, suggesting their pivotal roles in resistance.

16.4.3 Metabolomics, Epigenomics, Pangenomics, and Interactomics

Employing untargeted metabolite profiling, researchers demonstrated the significance of Pip in conferring resistance, a finding validated through comprehensive experiments across multiple susceptible and resistant peanut cultivars (Sharma et al., 2021). Expanding the scope, metabolomics has complemented genomics by shedding light on changes in metabolite profiles in response to AF contamination. This integrated approach aids in unraveling the complex network of molecular events involved in peanut–aflatoxin interactions. Kandel et al. (2015) revealed alterations in the levels of specific secondary metabolites, suggesting their potential roles in the intricate defense mechanisms of peanuts.

Epigenomics explores heritable changes in gene expression that do not involve alterations in the DNA sequence. In peanuts, epigenetic modifications such as DNA methylation and histone modifications have been linked to stress responses and disease resistance (Liu et al., 2022). Recent studies suggest that epigenetic regulation may play a role in modulating the expression of defense-related genes in response to *Aspergillus* infection. By understanding the epigenetic mechanisms involved in resistance, breeders can potentially develop peanut varieties with enhanced aflatoxin resistance through targeted epigenetic modifications or by selecting varieties with favorable epigenetic profiles. Pangenomics involves analyzing the complete set of genes present in a species, including those found in different accessions or varieties. In peanuts, pangenomic studies have revealed considerable genetic diversity across different cultivars and wild relatives, some of which harbor unique genes associated with aflatoxin resistance (Bhat et al., 2021; Varshney et al., 2017). By integrating pangenomic data into breeding programs, researchers can access a broader pool of resistance genes. Interactomics studies the interactions between proteins, genes, and other molecules within a biological system. In the context of aflatoxin resistance, interactomic analyses have mapped the complex networks of interactions between peanut and fungal proteins during *Aspergillus* infection (Jayaprakash et al., 2023). These studies have identified key nodes in the interaction network, such as transcription factors and signaling proteins, that play pivotal roles in regulating defense responses. By targeting these nodes through genetic engineering or selective breeding, the overall resistance of peanut plants to aflatoxin contamination can be enhanced.

16.5 PEANUT RESISTANCE MECHANISMS TO AFLATOXINS

16.5.1 Physical Resistance Mechanisms

Physical resistance mechanisms in peanuts play a critical role in preventing *Aspergillus* colonization and subsequent aflatoxin contamination. One of the key structural traits associated with aflatoxin resistance is the seed-coat thickness. A thicker seed coat acts as a barrier, limiting *Aspergillus* spore penetration, thereby reducing infection and toxin production (Commey et al., 2021). Studies have consistently shown that peanut varieties with thicker seed coats, such as the Florunner, exhibit higher resistance to aflatoxin accumulation, particularly under environments conducive to fungal growth (Klich, 2007). Besides thickness, the compactness and permeability of the seed coat also influence resistance. Varieties with more compact seed coats and lower water permeability are less likely to absorb moisture, which is a key factor in promoting fungal growth (Souza, 2023). Lower seed-coat permeability reduces water uptake during post-harvest storage and prevents the onset of fungal infection (Mendu et al., 2022).

Pod structure is another important trait associated with resistance. Peanut varieties with harder and more rigid pods, like *ICGV 91114*, have demonstrated strong resistance during preharvest when pods are in contact with fungal pathogens in the soil (Nadaf et al., 2020). Additional traits, such as a waxy cuticle layer on the seed coat and pod surface, contribute to physical resistance by creating a hydrophobic barrier that reduces water retention and fungal adhesion. Varieties with fewer microcracks or lesions on the seed coat and pod surface tend to be less susceptible to fungal colonization (Liang Chen et al., 2009). Maintaining the structural integrity of the peanut seed coat and pod during growth, harvesting, and storage is essential for reducing aflatoxin contamination.

16.5.2 Biochemical and Molecular Resistance Mechanisms

In addition to physical resistance mechanisms, peanuts possess a variety of biochemical and molecular pathways that contribute to their defense against aflatoxin contamination. One of the primary biochemical defenses involves the production of antifungal proteins, which inhibit the growth and development of *Aspergillus* species. For example, peanuts produce chitinases and glucanases, which degrade the fungal cell wall and prevent the colonization of peanut tissues (Pfliegler et al., 2019). Chitinase activity has been shown to be higher in aflatoxin-resistant peanut varieties, making it a key molecular marker for resistance (Prasad et al., 2013). Secondary metabolites also play a crucial role; compounds, including phytoalexins, flavonoids, and phenolic acids, are synthesized in response to fungal infection and act as chemical barriers to inhibit fungal growth. Phytoalexins, such as resveratrol, are found in higher concentrations in resistant varieties, where they disrupt the metabolic processes of *Aspergillus* species (Sobolev et al., 2018). Additionally, flavonoids and phenolic compounds, like tannins, have antioxidant properties, protecting peanut cells from oxidative damage caused by fungal infection and reducing the accumulation of reactive oxygen species (ROS) that can promote fungal growth (Wang et al., 2023).

Molecular resistance mechanisms in peanuts also involve the activation of stress-related signaling pathways. The jasmonic acid (JA) and salicylic acid (SA) pathways are particularly important in mediating peanut resistance to *Aspergillus* (Jayaprakash et al., 2019). These signaling pathways trigger the expression of defense-related genes, such as those encoding R genes (Deng et al., 2018; Yu et al., 2023) and pathogenesis-related (PR) proteins. Resistant varieties exhibit higher expression levels of PR genes, indicating a more robust defense response to fungal attacks (Liang et al., 2023). Moreover, the role of reactive oxygen species (ROS) in peanut defense mechanisms has gained increasing attention. ROS, generated as a byproduct of normal cellular metabolism, can accumulate in high levels during fungal infection, leading to cellular damage. In resistant peanut varieties, enzymes such as superoxide dismutase (SOD) and catalase (CAT) are upregulated to detoxify ROS, thereby minimizing the oxidative stress caused by *Aspergillus* infection (Sharma et al., 2018). This enzymatic defense mechanism helps maintain cellular integrity and prevents the spread of fungal pathogens within peanut tissues.

16.6 CURRENT AF MANAGEMENT STRATEGIES IN PEANUT

16.6.1 Genetic Methods for Breeding Resistance in Peanut

Developing peanut cultivars resistant to preharvest contamination remains a significant challenge for breeding programs globally (Blankenship et al., 1985). Conventional breeding approaches have had limited success in producing resistant varieties, as many widely cultivated peanut varieties remain susceptible. Given the genetic complexity of this trait, genomics-assisted breeding (GAB) holds promising avenues for the faster development of resistant varieties. Successful application of GAB necessitates the identification of linked and validated genetic markers. In addition, the development of a low-density SNP panel (10–50 SNPs) can facilitate early-generation selection by identifying lines with improved resistance. Despite some progress, pinpointing genomic regions associated with resistance to preharvest contamination has been slow (Liang Zhou et al., 2009), and diagnostic markers for GAB application are still lacking. Nevertheless, several genetic and association mapping studies involving diverse genetic populations are currently underway. In the near future, utilizing multi-parent populations combined with cost-effective sequencing technologies and precise phenotyping is anticipated to facilitate high-resolution mapping for aflatoxin resistance, ultimately leading to a comprehensive panel of diagnostic markers.

Several studies have explored the use of RNA interference (RNAi) to inhibit *A. flavus* growth in peanuts. For instance, an hpRNA construct was employed to suppress the expression of five key genes involved in aflatoxin biosynthesis, including *aflR* (transcriptional activator), *aflS* (co-activator), *aflC* (polyketide synthase), *aflep* (a putative aflatoxin efflux pump), and *pes1* (a nonribosomal peptide synthetase responsible for oxidative stress tolerance) (Arias et al., 2015). Transgenic peanut lines exhibited up to a 100% reduction in AFB1 and AFB2 levels compared to controls. In another study, significant aflatoxin resistance was achieved by overexpressing antifungal defensin genes (*MsDef1* and *MtDef4.2*) while using host-induced gene silencing (HIGS) to target the *aflM* and *aflP* genes involved in aflatoxin biosynthesis (Sharma et al., 2018).

16.6.2 Biotechnological Tools for Solving Aflatoxin Contamination in Peanut

16.6.2.1 Transgenic Plant to Develop Stable Disease-Resistant Peanut

The complex, multigenic nature of AF resistance challenges traditional breeding efforts. Unlike many host–pathogen interactions, *A. flavus* does not conform to typical gene-for-gene resistance mechanisms (Cary et al., 2011). Transgenic technology has emerged as a powerful tool to address the complexities of AF resistance. This approach provides a less time-consuming alternative or complementary method to classical breeding. By introducing genes that confer resistance to AF-producing fungi, transgenic breeding represents a targeted strategy to reinforce the plant's innate defense mechanisms. Transgenic approaches have proven effective in enhancing resistance to *A. flavus* and AF contamination in maize (Thakare et al., 2017; Rajasekaran et al., 2018).

16.6.2.2 RNA Interference-Based Methods

RNAi has emerged as an effective strategy to target key genes involved in *A. flavus* growth and AF production. Researchers have achieved promising results in reducing the impact of *A. flavus* on maize using RNAi-based methods (Majumdar et al., 2017). In the context of AF contamination in peanuts, RNAi presents a promising avenue for developing resistance. By specifically targeting genes involved in the AF biosynthesis pathway or essential for fungal growth, RNAi technology can potentially reduce AF contamination. In a pioneering study, Arias et al. (2015) explored transgenic peanuts expressing hairpin RNA (hpRNA) targeting five AF biosynthesis–related genes—*aflR*, *aflS*, *aflC*, *aflep*, and *pes1*. Remarkably, the RNAi-expressing lines exhibited up to a 100% reduction in AFB1 and AFB2 compared to the control. A subsequent research by Power et al. (2017) employed high-throughput sequencing to identify specific small RNAs in RNAi lines, revealing exclusive matches to the targeted *aflS* and *aflC* genes. Additionally, 39 sRNAs,

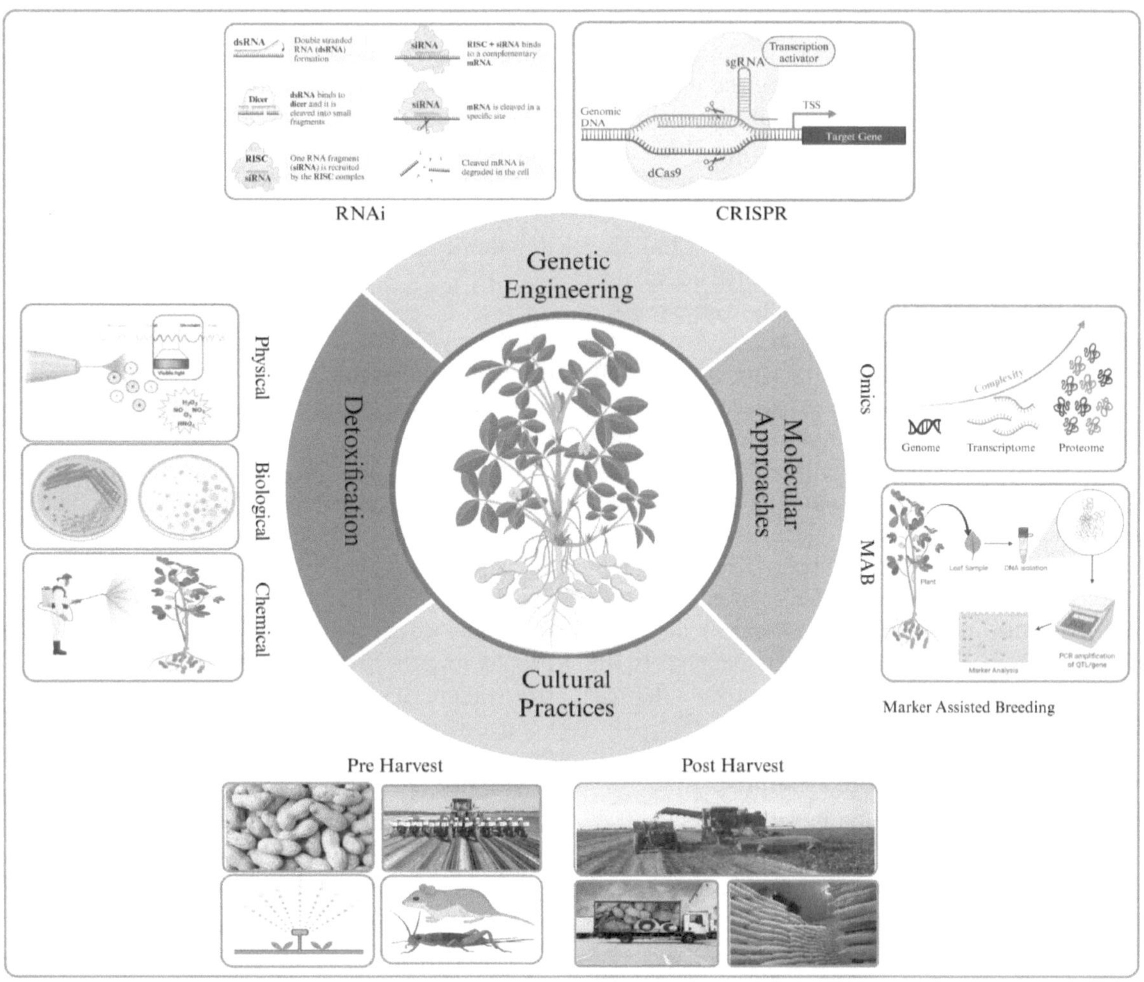

FIGURE 16.2 Integrated approaches to control the aflatoxin contamination in peanuts. These approaches include cultural practices, detoxification (pre- and postharvest methods), genetic engineering and molecular tools such as omics technologies, RNAi, and CRISPR-based gene editing.

mapping without mismatches to the *A. flavus* genome, were found only in the RNAi lines. In another study, Sharma et al. (2018) reported high resistance in transgenic peanuts using RNAi and overexpression techniques. Their study also reported a strong positive correlation between the AF levels and the declined expression of AF biosynthetic genes.

Targeting *A. flavus* genes involved in sporulation and AF production, such as *nsdC*, *veA*, *aflR*, and *aflM*, has proven effective in enhancing resistance and reducing aflatoxin levels in groundnuts to levels below 20 ppb (Prasad et al., 2023). Furthermore, HIGS has downregulated key proteins involved in fungal differentiation and pathogenicity, including calmodulin, transcriptional activator-HacA, kynurenine 3-monooxygenase 2, VeA, and VelC. Strategies for targeted downregulation of AF biosynthetic genes, such as *aflM*, *aflP*, *aflR*, *aflS*, *aflC*, *pes1*, and *aflep* (Sharma et al., 2018; Arias et al., 2015; Power et al., 2020) have shown promising results in developing aflatoxin-resistant genotypes. The summary of management and prevention methods is given in Figure 16.2.

16.6.2.3 Overexpression of Antifungal Genes

Another promising biotechnological approach involves the overexpression of antifungal genes. One of the most commonly studied proteins is chitinase, an enzyme that degrades chitin, a key component of fungal cell walls (Unuofin *et al.*, 2024). Transgenic peanuts with overexpressed chitinase genes have demonstrated significant reductions in fungal colonization and aflatoxin production. Similarly, glucanases—enzymes that break down β-glucans in fungal cell walls—have also been successfully overexpressed in peanuts, resulting in enhanced resistance to *Aspergillus* infection and aflatoxin contamination (Liang et al., 2005). Several antifungal proteins have been characterized as *in vitro* inhibitors of *A. flavus* growth and sporulation and are employed for transgenic applications (Cheng et al., 2008). Researchers are now exploring combining multiple antifungal genes, such as chitinases and glucanases, to develop peanut varieties with broad-spectrum resistance. Successful applications in transgenic maize provide a strong precedent for similar strategies in peanut (Rajasekaran et al., 2018; Schubert et al., 2015). These developments underscore the potential of transgenic approaches to improve aflatoxin resistance in peanut significantly.

16.6.2.4 Specific and Inducible Promoters

Promoters are important tools for molecular research to study the functions of a gene. Tissue-specific and inducible promoters are more important as they can drive genes' expression in a tissue-specific manner or under specific stress. The selection of a suitable promoter is the key step for developing transgenic plants since it influences the cell, tissue, or stage specificity and determines the expression level of the transgene (Koyama et al., 2005). Through cDNA microarray analysis, Zeng (2012) identified peanut pericarp-specific and aflatoxin-induced promoters (*AhLAC1, AhPLEC1, and AhOMT1*) for targeted gene expression. The *AhLAC1* and *AhPLEC1* promoters were identified as pericarp/seed-specific, with motifs linked to pericarp development, while the *AhOMT1* promoter was found to be aflatoxin- and fungal elicitor-specific. Deng et al. (2018) identified an *A. flavus*-induced NBS-LRR gene, *AhRAF4*, derived from *A. duranensis* through recombination. *AhRAF4* exhibits defense-related expression and potential involvement in pericarp development. Its unique evolution, regulatory responses, and role in resistance suggest its potential for genetic improvement in cultivated peanuts to enhance disease resistance and mitigate aflatoxin contamination. NBS-LRR genes are key resistance-related genes. A study identified *AhRAF27*, *AhRAF32*, and *AhRAF41* genes linked to aflatoxin resistance in peanut. These NBS-LRR genes exhibited distinct expression patterns under *A. flavus* infection and drought stress. Differential responses in resistant and susceptible cultivars suggest their roles in defense mechanisms and pericarp development, offering insights for breeding aflatoxin resistance (Bo, 2014). Similarly (Sharif et al., 2022) identified and characterized a pericarp specific expression promoter *AhGLP17-1P*, and *AhAPY2-1p* (Sharif et al., 2023) with specific expression in pericarp tissues. The same authors also identified a pericarp and testa-abunndant expression, driving the strong expression of GUS genes in pericarp and testa of transgenic plants (Sharif et al., 2024). The same team of researchers also worked out the function of *AhOMT1* promoter under *A. flavus* infection (Zhuang et al., 2023). These promoters can provide environment-safe transgenics with reduced aflatoxin contants.

16.6.3 Conventional Management Options

Conventional methods for managing aflatoxin contamination in peanuts include agronomic practices such as crop rotation, soil amendments, and improved irrigation management. Crop rotation with non-host crops effectively reduces *Aspergillus* spores in the soil, limiting the chances of infection in subsequent peanut crops (Pandey et al., 2019). Soil amendments, including lime, gypsum, and organic matter, improve soil health and moisture retention, reducing plant stress and susceptibility to infection (Hendrix et al., 2020). Proper irrigation management is also crucial, as drought stress increases the risk of *Aspergillus* infection (Soni, Gangurde et al., 2020). Ensuring a consistent water supply throughout the growing season is vital, and drip irrigation can effectively maintain soil moisture while reducing aflatoxin levels (Zhu et al., 2004). Together, these conventional methods provide a holistic approach to managing aflatoxin contamination in peanut.

16.6.4 Biological Management for Mitigating AF Contamination

Biological management strategies utilize various microorganisms or biological agents to reduce AF levels in crops. *Trichoderma spp.* compete with *Aspergillus* for space and nutrients while producing enzymes that degrade fungal cell walls, inhibiting fungal growth (Ren et al., 2022). A key biological approach involves introducing naturally occurring non-aflatoxigenic strains of *A. flavus* and *A. parasiticus*. This process entails applying a conidial suspension of these non-aflatoxigenic strains to the soil before planting or directly to the seedlings (Dorner, 2002; Lavkor et al., 2023). Commercial products, such as *Afla-Guard* and *AflaSafe*, have successfully reduced aflatoxin contamination in peanut fields across Africa and the United States (Peles et al., 2021). The application of non-aflatoxigenic *Aspergillus* strains not only manages AF production in the field but also helps prevent contamination during storage. Strains like AF36 have been successfully employed for AF biocontrol (Garcia-Lopez et al., 2021). Atoxigenic strains of *A. flavus* can outcompete toxigenic strains, demonstrating significant potential in reducing AF contamination (Abbas et al., 2017).

16.7 CONCLUSION AND FUTURE PERSPECTIVES

The consequences of aflatoxin contamination for human health and agricultural economies underscore the urgency of effective management. Despite considerable progress, several challenges remain, mainly due to the polygenic nature of aflatoxin resistance and its dependence on environmental conditions. With its increased incidence of drought and heat, climate change further heightens the risk. Moreover, while biotechnological advancements such as transgenic methods and CRISPR-Cas9 gene editing offer promising solutions, their adoption is often constrained by regulatory barriers and public perception. Despite these challenges, there are numerous opportunities for future advancements. The growing availability of whole-genome sequencing data, pangenomics, and epigenomics provides exciting potential for identifying novel resistance genes and regulatory pathways. These insights, combined with molecular genetics tools, are improving our understanding of aflatoxin resistance and susceptibility mechanisms. Moreover, the integration of omics strategies, germplasm evaluation, and marker/genome-assisted breeding contributes to the development of peanut varieties with inherent resistance. The advent of biotechnological tools, particularly transgenics and genome editing, opens new frontiers in achieving stable and precise disease-resistant peanuts. By manipulating specific genes involved in the susceptibility to *A. flavus* infection, researchers aim to mitigate the risk posed by aflatoxins. The integration of precision agriculture technologies, such as remote sensing and artificial intelligence (AI), enables early detection of contamination risks, allowing for timely interventions. The ongoing exploration of functional genomics and other omics tools will deepen our understanding of peanut responses to aflatoxin contamination, facilitating the identification of key genes and pathways for targeted manipulation. Leveraging these insights will enable the development of varieties with enhanced resistance, setting the stage for more robust and effective management practices. Continued research in transgenics may lead to the development of highly specific and tailored genetic modifications that confer durable resistance to aflatoxin-producing pathogens. In conclusion, the multifaceted nature of aflatoxin contamination in peanuts demands a holistic approach that combines conventional and modern biotechnological strategies.

REFERENCES

Abbas, H. K., C. Accinelli, and W. T. Shier. "Biological control of aflatoxin contamination in US crops and the use of bioplastic formulations of *Aspergillus flavus* biocontrol strains to optimize application strategies." *Journal of Agricultural and Food Chemistry* 65 (33) (2017): 7081–7087. doi: 10.1021/acs.jafc.7b01452.

Abd El-Aziz, A. R. M., S. M. Shehata, S. M. Hisham, and A. A. Alobathani. "Molecular profile of aflatoxigenic and non-aflatoxigenic isolates of *Aspergillus flavus* isolated from stored maize." *Saudi Journal of Biological Sciences* 28 (2) (2021): 1383–1391. doi: 10.1016/j.sjbs.2020.11.073.

Alameri, M. M., A. S. Kong, M. N. Aljaafari, H. A. Ali, K. Eid, M. A. Sallagi, W. H. Cheng, A. Abushelaibi, S. E. Lim, and J. Y. Loh et al. "Aflatoxin contamination: an overview on health issues, detection and management strategies." *Toxins* 15 (4) (2023): 246. doi: 10.3390/toxins15040246.

Alvarado, A. M., R. Zamora-Sanabria, and F. Granados-Chinchilla. 2017. "A focus on aflatoxins in feedstuffs: levels of contamination, prevalence, control strategies, and impacts on animal health." In *Aflatoxin-Control, Analysis, Detection Health Risks*, edited by Lukman Bola Abdulra'uf, 116–152. IntechOpen.

Amaike, S., and N. P. Keller. "*Aspergillus flavus*." *Annual Review of Phytopathology* 49 (2011): 107–133. doi: 10.1146/annurev-phyto-072910-095221.

Anderson, W., C. Holbrook, D. Wilson, and M. Matheron. "Evaluation of preharvest aflatoxin contamination in several potentially resistant peanut genotypes." *Peanut Science* 22 (1) (1995): 29–32. doi: 10.3146/pnut.22.1.0007.

Andrews-Trevino, J. Y., P. Webb, G. Shively, A. Kablan, K. Baral, D. Davis, K. Paudel, et al. "Aflatoxin exposure and child nutrition: measuring anthropometric and long-bone growth over time in Nepal." *The American Journal of Clinical Nutrition* 113 (4) (2021): 874–883. doi: 10.1093/ajcn/nqaa397.

Arias, R. S., P. M. Dang, and V. S. Sobolev. "RNAi-mediated control of aflatoxins in peanut: method to analyze mycotoxin production and transgene expression in the Peanut/*Aspergillus* Pathosystem." *JoVE* 106 (2015): e53398. doi: 10.3791/53398.

Asis, R., D. L. Barrionuevo, L. M. Giorda, M. L. Nores, and M. A. Aldao. "Aflatoxin production in six peanut (*Arachis hypogaea* L.) genotypes infected with *Aspergillus flavus* and *Aspergillus parasiticus*, isolated from peanut production areas of Cordoba, Argentina." *Journal of Agricultural Food Chemistry* 53 (23) (2005): 9274–9280. doi: 10.1021/jf051259+

Azziz-Baumgartner, E., K. Lindblade, K. Gieseker, H. S. Rogers, S. Kieszak, H. Njapau, R. Schleicher, et al. "Case–control study of an acute aflatoxicosis outbreak, Kenya, 2004."

Environmental Health Perspectives 113 (12) (2005): 1779–1783. doi: 10.1289/ehp.8384.

Bediako, K. A., K. Ofori, S. K. Offei, D. Dzidzienyo, J. Y. Asibuo, and R. A. Amoah. "Aflatoxin contamination of groundnut (*Arachis hypogaea* L.): Predisposing factors and management interventions." *Food Control* 98 (2019): 61–67. doi: 10.1016/j.foodcont.2018.11.020.

Benkerroum, N. "Chronic and acute toxicities of aflatoxins: mechanisms of action." *International Journal of Environmental Research and Public Health* 17 (2) (2020): 423. doi: 10.3390/ijerph17020423.

Bertioli, D. J., J. Jenkins, J. Clevenger, O. Dudchenko, D. Gao, G. Seijo, S. Leal-Bertioli, et al. "The genome sequence of segmental allotetraploid peanut *Arachis hypogaea*." *Nature Genetics* 51 (5) (2019): 877–884. doi: https://doi:org/10.1038/s41588-019-0405-z.

Beyene, A. M., X. Du, D. E. Schrunk, S. Ensley, and W. K. Rumbeiha. "High-performance liquid chromatography and Enzyme-Linked Immunosorbent Assay techniques for detection and quantification of aflatoxin B(1) in feed samples: a comparative study." *BMC Research Notes* 12 (1) (2019): 492. doi: 10.1186/s13104-019-4538-z.

Bhat, R. S., K. Shirasawa, V. Sharma, S. N. Isobe, H. Hirakawa, C. Kuwata, M. K. Pandey, R. K. Varshney, and M. C. Gowda. 2021. "Population genomics of peanut." In *Population Genomics of Crop Plants*, edited by O. P. Rajora. Springer.

Bhatnagar-Mathur, P., K. Yogendra, S. Parankusam, H. Sanivarapu, K. Prasad, S. B. Lingampali, and K. K. Sharma. "Comparative proteomics provide insights on the basis of resistance to *Aspergillus flavus* infection and aflatoxin production in peanut (*Arachis hypogea* L.)." *Journal of Plant Interactions* 16 (1) (2021): 494–509. doi: 10.1080/17429145.2021.1995058.

Blankenship, P., R. Cole, and T. Sanders. "Comparative susceptibility of four experimental peanut lines and the cultivar Florunner to preharvest aflatoxin contamination." *Peanut Science* 12 (2) (1985): 70–72. doi: 10.3146/pnut.12.2.0006.

Bo, Z. "Resistance identification of *Aspergillus flavus* in RILs and related NBS-LRR gene cloning and analysis of peanut." Master Diss., Fujian Agriculture and Forestry University, 2014.

Bouskill, N. J., T. E. Wood, R. Baran, Z. Ye, B. P. Bowen, H. Lim, J. Zhou, et al. "Belowground response to drought in a tropical forest soil. I. Changes in microbial functional potential and metabolism." *Frontiers in Microbiology* 7 (2016): 525. doi: 10.3389/fmicb.2016.00525.

Bradford, K. J., P. Dahal, J. V. Asbrouck, K. Kunusoth, P. Bello, J. Thompson, and F. Wu. "The dry chain: reducing postharvest losses and improving food safety in humid climates." *Trends in Food Science & Technology* 71 (2018): 84–93. doi: https://doi:org/10.1016/j.tifs.2017.11.002.

Cary, J. W., K. Rajasekaran, R. L. Brown, M. Luo, Z. Y. Chen, and D. Bhatnagar. "Developing resistance to aflatoxin in maize and cottonseed." *Toxins* 3 (6) (2011): 678–696. doi: 10.3390/toxins3060678.

Chen, C. Proteomics Studies of *Aspergilus flavus* stress response in Min hua 6 peanut pod." Masters Diss., Fujian Agriculture and Forestry University, 2011.

Chen, H., B. J. Jiang, J. B. Zheng, D. R. Cook, C. Zhang, G. H. He, C. H. Zhuang, and W. Zhuang "A genetic and genomic resource for mining genes toward improvement peanut yield and quality in cultivated peanut." Paper presented at the Vth International Congress on Legume Genetics and Genomics, California, July 2–8, 2010.

Chen, X., Q. Lu, H. Liu, J. Zhang, Y. Hong, H. Lan, H. Li, et al. "Sequencing of cultivated peanut, *Arachis hypogaea*, yields insights into genome evolution and oil improvement." *Molecular Plant* 12 (7) (2019): 920–934. doi: 10.1016/j.molp.2019.03.005.

Cheng, Z., H. Yin, X. Y. Chen, M. L. Yan, and W. Zhuang. "Cloning and characterization of gene resistant to *Aspergillus flavus* and their transformation into peanut (*Arachis hypogaea* L.)." 3rd International Conference for Peanut Genomics and Biotechnology on Advances in Arachis through Genomics and Biotechnology (AAGB-2008) ICRISAT, India, 4–8 November 2008.

Chu, Y. J., H. I. Yang, H. C. Wu, M. H. Lee, J. Liu, L. Y. Wang, S. N. Lu, *et al.* "Aflatoxin B(1) exposure increases the risk of hepatocellular carcinoma associated with hepatitis C virus infection or alcohol consumption." *European Journal of Cancer* 94 (2018): 37–46. doi: 10.1016/j.ejca.2018.02.010.

Commey, L., T. K. Tengey, C. J. Cobos, L. Dampanaboina, K. K. Dhillon, M. K. Pandey, H. K. Sudini, *et al.* "Peanut seed coat acts as a physical and biochemical barrier against *Aspergillus flavus* infection." *Fungi* 7 (12) (2021): 10.3390/jof7121000.

Conde, S., J. F. Rami, D. K. Okello, A. Sambou, A. Muitia, R. Oteng-Frimpong, L. Makweti, *et al.* "The groundnut improvement network for Africa (GINA) germplasm collection: a unique genetic resource for breeding and gene discovery." *G3 Genes|Genomes|Genetics* 14 (1) (2023). doi: 10.1093/g3journal/jkad244.

Coulibaly, O., K. Hell, R. Bandyopadhyay, S. Hounkponou, and J. F. Leslie. 2008. "Economic impact of aflatoxin contamination in sub-Saharan Africa." In *Mycotoxins: Detection methods, management, public health and agricultural trade*, edited by J. F. Leslie, R. Bandyopadhyay, and A. Visconti, 67–76. CABI Wallingford UK.

Cui, M., S. Han, D. Wang, M. S. Haider, J. Guo, Q. Zhao, P. Du, et al. "Gene co-expression network analysis of the comparative transcriptome identifies hub genes associated with resistance to *Aspergillus flavus* L. in Cultivated Peanut (*Arachis hypogaea* L.)." *Frontiers in Plant Science* 13 (2022): 899177. doi: 10.3389/fpls.2022.899177.

Deng, Y., H. Chen, C. Zhang, T. Cai, B. Zhang, S. Zhou, J. Fountain, R. L. Pan, B. Guo, and W. Zhuang. "Evolution and characterisation of the Ah RAF 4 NB-ARC gene family induced by *Aspergillus flavus* inoculation and abiotic stresses in peanut." *Plant Biology* 20 (4) (2018): 737–750. doi: 10.1111/plb.12726.

Dieme, R. M. A., I. Faye, Y. A. B. Zoclanclounon, D. Fonceka, O. Ndoye, and P. M. Diedhiou. "Identification of sources of resistance for peanut *Aspergillus flavus* Colonization and Aflatoxin contamination." *International Journal of Agronomy* (2018). doi: 10.1155/2018/5468602.

Ding, Y., X. Qiu, H. Luo, L. Huang, J. Guo, B. Yu, H. Sudini, et al. "Comprehensive evaluation of Chinese peanut mini-mini core collection and QTL mapping for aflatoxin resistance." *BMC Plant Biology* 22 (1) (2022): 207. doi: 10.1186/s12870-022-03582-0.

Dorner, J. W. C., J. Richard. "Effect of application of nontoxigenic strains of *Aspergillusflavus* and *A. parasiticus* on subsequent aflatoxin contamination of peanuts in storage." *Journal of Stored Products Research* 38 (4) (2002): 329–339. doi: 10.1016/S0022-474X(01)00035-2.

Dövényi-Nagy, T., C. Rácz, K. Molnár, K. Bakó, Z. Szláma, Á. Jóźwiak, Z. Farkas, I. Pócsi, and A. C. Dobos. "Pre-harvest modelling and mitigation of aflatoxins in maize in a changing climatic environment—A review." *Toxins* 12 (12) (2020): 768. doi: 10.3390/toxins12120768.

Ehrlich, K. C., K. Kobbeman, B. G. Montalbano, and P. J. Cotty. "Aflatoxin-producing *Aspergillus* species from Thailand." *International Journal of Food Microbiology* 114 (2) (2007): 153–159. doi: 10.1016/j.ijfoodmicro.2006.08.007.

Frisvad, J. C., V. Hubka, C. Ezekiel, S. B. Hong, A. Nováková, A. Chen, M. Arzanlou, et al. "Taxonomy of *Aspergillus* section Flavi and their production of aflatoxins, ochratoxins and other mycotoxins." *Studies in Mycology* 93 (2019): 1–63. doi: 10.1016/j.simyco.2018.06.001.

Garcia-Lopez, M. T., Y. Luo, A. Ortega-Beltran, R. Jaime, J. Moral, and T. Michailides. "Quantification of the aflatoxin biocontrol strain *Aspergillus flavus* AF36 in soil and in nuts and leaves of pistachio by real-time PCR." *Plant Disease* 105 (6) (2021): 1657–1665. doi: 10.1094/PDIS-05-20-1097-RE.

Girdthai, T., S. Jogloy, N. Vorasoot, C. Akkasaeng, S. Wongkaew, C. Holbrook, and A. Patanothai. "Associations between physiological traits for drought tolerance and aflatoxin contamination in peanut genotypes under terminal drought." *Plant Breeding* 129 (6) (2010): 693–699. doi: 10.1111/j.1439-0523.2009.01738.x.

Gong, A., M. Song, and J. Zhang. "Current strategies in controlling *Aspergillus flavus* and aflatoxins in grains during storage: a review." *Sustainability* 16 (8) (2024): 3171. doi: 10.3390/su16083171

Gong, Y. Y., S. Watson, and M. N. Routledge. "Aflatoxin exposure and associated human health effects, a review of epidemiological studies." *Food Safety (Tokyo)* 4 (1) (2016): 14–27. doi: 10.14252/foodsafetyfscj.2015026.

Guo, B., X. Chen, P. Dang, B. T. Scully, X. Liang, C. C. Holbrook, J. Yu, and A. K. Culbreath. "Peanut gene expression profiling in developing seeds at different reproduction stages during *Aspergillus parasiticus* infection." *BMC Developmental Biology* 8 (1) (2008): 12. doi: 10.1186/1471-213X-8-12.

Guo, B., N. D. Fedorova, X. Chen, C. H. Wan, W. Wang, W. C. Nierman, D. Bhatnagar, and J. Yu. "Gene expression profiling and identification of resistance genes to *Aspergillus flavus* infection in peanut through EST and microarray strategies." *Toxins* 3 (7) (2011): 737–753. doi: 10.3390/toxins3070737.

Han, S., B. C. Kim, J. Ha, and T. H. Jun. "Selection of resistant varieties to *Aspergillus flavus* by determination of aflatoxin B1 Content in Korean Peanut (*Arachis hypogaea* L.) Accessions." *The Korean Journal of Crop Science* 68 (3) (2023): 175–187. doi: 10.7740/kjcs.2023.68.3.175

Hasan, N., S. Choudhary, N. Naaz, N. Sharma, and R. A. Laskar. "Recent advancements in molecular marker-assisted selection and applications in plant breeding programmes." *Journal of Genetic Engineering and Biotechnology* 19 (1) (2021): 128. doi: 10.1186/s43141-021-00231-1.

Hendrix, M., I. Obed, M. Alice, P. Elijah, Y. Jones, C. Samuel, and B. Rick. "The effects of gypsum on pod-yield and preharvest aflatoxin contamination in selected peanut cultivars of Zambia." *African Journal of Plant Science* 14 (3) (2020): 134–138. doi: 10.5897/AJPS2019.1807.

Holbrook, C. C., J. Clevenger, P. Ozias-Akins, Y. Chu, and T. Brenneman. "Registration of 'TifJumbo'peanut." *Journal of Plant Registrations* 18 (2) (2024): 279–284. doi: 10.1002/plr2.20285.

Holbrook, C. C., C. K. Kvien, K. S. Rucker, D. M. Wilson, J. E. Hook, and M. E. Matheron. "Preharvest aflatoxin contamination in drought-tolerant and drought-intolerant peanut genotypes." *Peanut Science* 27 (2) (2000): 45–48. doi: 10.3146/i0095-3679-27-2-1.

Holbrook, C. C., D. M. Wilson, M. E. Matheron, J. E. Hunter, D. A. Knauft, and D. W. Gorbet. "*Aspergillus* colonization and aflatoxin contamination in peanut genotypes with reduced linoleic acid composition." *Plant Disease* 84 (2) (2000): 148–150. doi: 10.1094/PDIS.2000.84.2.148.

Huang, R., H. Li, C. Gao, W. Yu, and S. Zhang. "Advances in omics research on peanut response to biotic stresses." *Frontiers in Plant Science* 14 (2023): 1101994. doi: 10.3389/fpls.2023.1101994.

Hussain, A. H., and H. Z. Hussein. "Evaluation of *Agaricus* sp. and *Pleurotus* sp. extracts efficiency in *Aspergillus flavus* growth inhibition and aflatoxin B1 reduction." *Systematic Reviews in Pharmacy* 11 (10) (2020): 564–569. doi: 10.31838/srp.2020.10.84.

Inglis, A., A. C. Parnell, N. Subramani, and F. M. Doohan. "Machine learning applied to the detection of mycotoxin in food: a systematic review." *Toxins* 16 (6) (2024): 268. doi: 10.3390/toxins16060268.

Jayaprakash, A., A. Roy, R. R. Thanmalagan, A. Arunachalam, and P. Lakshmi. "Understanding the mechanism of pathogenicity through interactome studies between *Arachis hypogaea* L. and *Aspergillus flavus*." *Journal of Proteomics* 287(2023): 104975. doi: 10.1016/j.jprot.2023.104975.

Jayaprakash, A., A. Roy, R. R. Thanmalagan, A. Arunachalam, and L. Ptv. "Immune response gene coexpression network analysis of *Arachis hypogaea* infected with *Aspergillus flavus*." *Genomics* 113 (5) (2021): 2977–2988. doi: https://doi:org/10.1016/j.ygeno.2021.06.027.

Jayaprakash, A., R. R. Thanmalagan, A. Roy, A. Arunachalam, and L. PTV. "Strategies to understand *Aspergillus flavus* resistance mechanism in *Arachis hypogaea* L." *Current Plant Biology* 20 (2019): 100123. doi: https://doi:org/10.1016/j.cpb.2019.100123.

Jiang, Y., H. Luo, B. Yu, Y. Ding, Y. Kang, L. Huang, X. Zhou, *et al.* "High-density genetic linkage map construction using whole-genome resequencing for mapping QTLs of resistance to *Aspergillus flavus* infection in peanut." *Frontiers in Plant Science* 12 (2021): doi: 10.3389/fpls.2021.745408.

Jiang, H. F., X. P. Ren, S. Y. Wang, X. J. Zhang, J. Q. Huang, B. S. Liao, C. Holbrook, and H. Upadhyaya. "Development and evaluation of peanut germplasm with resistance to *Aspergillus flavus* from core collection." *Acta Agronomica Sinica* 36 (3) (2010): 428–434.

Jin, G., N. Liu, B. Yu, Y. Jiang, H. Luo, L. Huang, X. Zhou, et al. "Identification and pyramiding major QTL loci for simultaneously enhancing aflatoxin resistance and yield components in peanut." *Genes* 14 (3) (2023): 625. doi: 10.3390/genes14030625.

Kandel, Y. R., J. S. Haudenshield, A. Y. Srour, K. T. Islam, A. M. Fakhoury, P. Santos, J. Wang, et al. "Multilaboratory comparison of quantitative PCR assays for detection and quantification of Fusarium virguliforme from soybean roots and soil." *Phytopathology* 105 (12) (2015): 1601–1611. doi: 10.1094/PHYTO-04-15-0096-R.

Khan, R., F. M. Ghazali, N. A. Mahyudin, and N. I. P. Samsudin. "Biocontrol of aflatoxins using non-aflatoxigenic *Aspergillus. flavus*: A literature review." *Journal of Fungi* 7 (5) (2021): 381. doi: 10.3390/jof7050381.

Khan, S. A., H. Chen, Y. Deng, Y. Chen, C. Zhang, T. Cai, N. Ali, et al. "High-density SNP map facilitates fine mapping of QTLs and candidate genes discovery for *Aspergillus flavus* resistance in peanut (*Arachis hypogaea*)." *Theoretical and

Applied Genetics 133 (7) (2020): 2239–2257. doi: 10.1007/s00122-020-03594-0.

Khlangwiset, P., G. S. Shephard, and F. Wu. "Aflatoxins and growth impairment: a review." *Critical Reviews in Toxicology* 41 (9) (2011): 740–755. doi: 10.3109/10408444.2011.575766.

Kisyombe, C. T., M. Beute, and G. Payne. "Field evaluation of peanut genotypes for resistance to infection by *Aspergillus parasiticus*." *Peanut Science* 12 (1) (1985): 12–17. doi: 10.3146/pnut.12.1.0004.

Klich, M. A. "*Aspergillus flavus*: the major producer of aflatoxin." *Molecular Plant Pathology* 8 (6) (2007): 713–722. doi: 10.1111/j.1364-3703.2007.00436.x.

Korani, W. A., Y. Chu, C. Holbrook, J. Clevenger, and P. Ozias-Akins. "Genotypic regulation of aflatoxin accumulation but not *Aspergillus fungal* growth upon post-harvest infection of peanut (*Arachis hypogaea* L.) seeds." *Toxins* 9 (7) (2017): 218. doi: 10.3390/toxins9070218.

Koyama, T., T. Ono, M. Shimizu, T. Jinbo, R. Mizuno, K. Tomita, N. Mitsukawa, et al. "Promoter of *Arabidopsis thaliana* phosphate transporter gene drives root-specific expression of transgene in rice." *Journal of Bioscience and Bioengineering* 99 (1) (2005): 38–42. doi: 10.1263/jbb.99.38.

Lakkireddy, K., K. Kasturi, and K. Rao. "Aflatoxins in food and feed: the science of safe food." *Journal of Food Science and Technology* 3 (2014): 5.

Lavkor, I., T. Ay, S. Sobucovali, I. Var, H. Saghrouchni, A. M. Salamatullah, and A. B. Mekonnen. "Non-aflatoxigenic *Aspergillus flavus*: A promising biological control agent against aflatoxin contamination of corn." *ACS Omega* 8 (19) (2023): 16779–16788. doi: 10.1021/acsomega.3c00303.

Lei, Y., B. Liao, and S. Wang. "Identification of AFLP markers for resistance to seed infection by *Aspergillus flavus* in peanut (*Arachis hypogaea* L.)." *Acta Agronomica Sinica* 31 (10) (2005): 1349.

Liang, B., Y. Bai, C. Zang, X. Pei, J. Xie, Y. Lin, X. Liu, T. Ahsan, and C. Liang. "Overexpression of the first peanut-susceptible gene, AhS5H1 or AhS5H2, enhanced susceptibility to Pst DC3000 in Arabidopsis." *International Journal of Molecular Sciences* 24 (18) (2023): doi: 10.3390/ijms241814210.

Liang, X., X. Chen, Y. Hong, H. Liu, G. Zhou, S. Li, and B. Guo. "Utility of EST-derived SSR in cultivated peanut (*Arachis hypogaea* L.) and Arachis wild species." *BMC Plant Biology* 9(2009): 1–9. doi: 10.1186/1471-2229-9-35.

Liang, X., C. Holbrook, R. Lynch, and B. Guo. "β-1, 3-Glucanase activity in peanut seed (*Arachis hypogaea*) is induced by inoculation with *Aspergillus flavus* and copurifies with a conglutin-like protein." *Phytopathology* 95 (5) (2005): 506–511. doi: 10.1094/PHYTO-95-0506.

Liang, X., G. Zhou, Y. Hong, X. Chen, H. Liu, and S. Li. "Overview of research progress on peanut (*Arachis hypogaea* L.) host resistance to aflatoxin contamination and genomics at the Guangdong Academy of Agricultural Sciences." *Peanut Science* 36 (1) (2009): 29–34. doi: 10.3146/AT07-003.1.

Liao, B., W. Zhuang, R. Tang, X. Zhang, S. Shan, H. Jiang, and J. Huang. "Peanut aflatoxin and genomics research in China: progress and perspectives." *Peanut Science* 36 (1) (2009): 21–28. doi: 10.3146/AT07-004.

Liu, Y., Y. Geng, X. Xie, P. Zhang, J. Hou, and W. Wang. "Core collection construction and evaluation of the genetic structure of Glycyrrhiza in China using markers for genomic simple sequence repeats." *Genetic Resources and Crop Evolution* 67 (2020): 1839–1852. doi: 10.1007/s10722-020-00944-1.

Liu, Y., J. Wang, B. Liu, and Z. Y. Xu. "Dynamic regulation of DNA methylation and histone modifications in response to abiotic stresses in plants." *Journal of Integrative Plant Biology* 64 (12) (2022): 2252–2274. doi: 10.1111/jipb.13368.

Liu, Y., and F. Wu. "Global burden of aflatoxin-induced hepatocellular carcinoma: a risk assessment." *Environmental health perspectives* 118 (2010): 818–24. doi: 10.1289/ehp.0901388.

Loi, M., A. F. Logrieco, T. Pusztahelyi, É. Leiter, L. Hornok, and I. Pócsi. "Advanced mycotoxin control and decontamination techniques in view of an increased aflatoxin risk in Europe due to climate change." *Frontiers in Microbiology* 13 (2023): 1085891. doi: 10.3389/fmicb.2022.1085891.

Majumdar, R., K. Rajasekaran, and J. W. Cary. "RNA Interference (RNAi) as a potential tool for control of mycotoxin contamination in crop plants: concepts and considerations." *Frontiers in Plant Science* 8 (2017): doi: 10.3389/fpls.2017.00200.

Marshall, H., J. P. Meneely, B. Quinn, Y. Zhao, P. Bourke, B. F. Gilmore, G. Zhang, and C. T. Elliott. "Novel decontamination approaches and their potential application for post-harvest aflatoxin control." *Trends in Food Science & Technology* 106 (2020): 489–496. https://doi:org/10.1016/j.tifs.2020.11.001.

McDonald, D., V. Mehan, and S. D. Hall. 1989. Aflatoxin contamination of groundnut: Proceedings of the International Workshop, 6–9 October 1987, ICRISAT Center, India. Patancheru, India: International Crops Research Institute for the Semi-Arid Tropics, 1989.

Mehan, V., and D. McDonald. "Screening for resistance to *Aspergillus flavus* invasion and aflatoxin production in groundnuts." International Crops Research Institute for the Semi Arid Tropoics, Patancheru, Hyderbad, Andhra Pradesh, 1980. https://oar.icrisat.org/7032/1/GIP_2.pdf.

Mehan, V. K., D. McDonald, N. Ramakrishna, and J. H. Williams. "Effects of genotype and date of harvest on infection of peanut seed by *Aspergillus flavus* and subsequent contamination with aflatoxin." *Peanut Science* 13 (2) (1986): 46–50. doi: 10.3146/i0095-3679-13-2-1.

Meijer, N., G. Kleter, M. de Nijs, M. L. Rau, R. Derkx, and H. van der Fels-Klerx. "The aflatoxin situation in Africa: systematic literature review." *Comprehensive Reviews in Food Science and Food Safety* 20 (3) (2021): 2286–2304. doi: 10.1111/1541-4337.12731.

Mendu, L., C. J. Cobos, T. K. Tengey, L. Commey, V. K. Balasubramanian, L. D. Williams, K. K. Dhillon, et al. "Seed coat mediated resistance against *Aspergillus flavus* infection in peanut." *Plant Gene* (2022): 100381. doi: 10.1016/j.plgene.2022.100381.

Mixon, A. C., and K. M. Rogers. "Peanut accessions resistant to seed infection by *Aspergillus flavus*." *Agronomy Journal* 65 (4) (1973): 560–562. doi: 10.2134/agronj1973.00021962006500040011x.

Molnár, K., C. Rácz, T. Dövényi-Nagy, K. Bakó, T. Pusztahelyi, S. Kovács, C. Adácsi, I. Pócsi, and A. Dobos. "The effect of environmental factors on mould counts and AFB1 toxin production by *Aspergillus flavus* in Maize." *Toxins* 15 (3) (2023): doi: 10.3390/toxins15030227.

Nadaf, H., B. Babu, G. Chandrashekhara, D. Savithramma, and M. K. Naik. "Single marker analysis in groundnut for tolerance to in vitro seed colonization by *Aspergillus flavus* and aflatoxin contamination." *Legume Research-An International Journal* 45 (10) (2022): 1301–1308. doi: 10.18805/LR-4224.

Nadaf, H., G. Chandrashekhara, and H. Babu. "Aflatoxin contamination in groundnut under normal moisture and moisture stress field conditions." *Journal of Pharmacognosy and PhytochemistryI* 9 (1) (2020): 1083–1088. doi: 10.47881/264.967x

Neelakandan, A. K., D. A. Wright, S. M. Traore, X. Chen, M. H. Spalding, and G. He. "CRISPR/Cas9 based site-specific modification of FAD2 cis-regulatory motifs in peanut (*Arachis hypogaea* L.)." *Frontiers in Genetics* 13 (2022): 849961. doi: 10.3389/fgene.2022.849961.

Nigam, S., F. Waliyar, R. Aruna, S. Reddy, P. L. Kumar, P. Q. Craufurd, A. Diallo, B. Ntare, and H. D. Upadhyaya. "Breeding peanut for resistance to aflatoxin contamination at ICRISAT." *Peanut Science* 36 (1) (2009): 42–49.

Nji, Q. N., O. O. Babalola, and M. Mwanza. "Aflatoxins in maize: can their occurrence be effectively managed in Africa in the face of climate change and food insecurity?" *Toxins* 14 (8) (2022): 574. doi: 10.3390/toxins14080574.

Nji, Q. N., O. O. Babalola, and M. Mwanza. "Soil *Aspergillus* species, pathogenicity and control perspectives." *Journal of Fungi* 9 (7) (2023): 766. doi: 10.3390/jof9070766.

Njoki, L., S. Okoth, P. Wachira, A. Ouko, J. Mwololo, M. Rizzu, S. Oufensou, and T. Amakhobe. "Evaluation of agronomic characteristics, disease incidence, yield performance, and aflatoxin accumulation among six peanut varieties (*Arachis hypogea* L.) grown in Kenya." *Toxins* 15 (2) (2023): 111. doi: 10.3390/toxins15020111.

Olwari, F., J. Bisikwa, A. Kaaya, and D. Okello. "Tolerance levels of peanut varieties against *Aspergillus flavus* infection." *Journal of Plant Pathology & Microbiology* 4 (2013): 195. doi: 10.4172/2157-7471.1000195.

Otyama, P. I., R. Kulkarni, K. Chamberlin, P. Ozias-Akins, Y. Chu, L. M. Lincoln, G. E. MacDonald, *et al.* "Genotypic characterization of the U.S. Peanut core collection." *G3 Genes|Genomes|Genetics* 10 (11) (2020): 4013–4026. doi: 10.1534/g3.120.401306.

Pandey, M. K., G. Agarwal, S. M. Kale, J. Clevenger, S. N. Nayak, M. Sriswathi, A. Chitikineni, *et al.* "Development and evaluation of a high density genotyping 'Axiom_Arachis' Array with 58 K SNPs for accelerating genetics and breeding in groundnut." *Scientific Reports* 7 (1) (2017): 40577. doi: 10.1038/srep40577.

Pandey, M. K., R. Kumar, A. K. Pandey, P. Soni, S. S. Gangurde, H. K. Sudini, J. C. Fountain, et al. "Mitigating aflatoxin contamination in groundnut through a combination of genetic resistance and post-harvest management practices." *Toxins* 11 (6) (2019): 315. doi: https://doi:org/10.3390/toxins11060315.

Peles, F., P. Sipos, S. Kovács, Z. Győri, I. Pócsi, and T. Pusztahelyi. "Biological control and mitigation of aflatoxin contamination in commodities." *Toxins* 13 (2) (2021): 10.3390/toxins13020104.

Pfliegler, W. P., I. Pócsi, Z. Győri, and T. Pusztahelyi. "The aspergilli and their mycotoxins: metabolic interactions with plants and the soil biota." *Frontiers in Microbiology* 10 (2019): 2921. doi: 10.3389/fmicb.2019.02921.

Pisoschi, A. M., F. Iordache, L. Stanca, E. Mitranescu, L. Bader Stoica, O. I. Geicu, L. Bilteanu, and A. I. Serban. "Biosensors for food mycotoxin determination: a comparative and critical review." *Chemosensors* 12 (6) (2024): 92. doi: 10.3390/chemosensors12060092.

Power, I. L., P. M. Dang, V. S. Sobolev, V. A. Orner, J. L. Powell, M. C. Lamb, and R. S. Arias. "Characterization of small RNA populations in non-transgenic and aflatoxin-reducing-transformed peanut." *Plant Science* 257 (2017): 106–125. doi: 10.1016/j.plantsci.2016.12.013.

Power, I. L., P. C. Faustinelli, V. A. Orner, V. S. Sobolev, and R. S. Arias. "Analysis of small RNA populations generated in peanut leaves after exogenous application of dsRNA and dsDNA targeting aflatoxin synthesis genes." *Scientific Reports* 10 (1) (2020): 1–12. doi: 10.1038/s41598-020-70618-6.

Prasad, K., P. Bhatnagar-Mathur, F. Waliyar, and K. K. Sharma. "Overexpression of a chitinase gene in transgenic peanut confers enhanced resistance to major soil borne and foliar fungal pathogens." *Journal of Plant Biochemistry and Biotechnology* 22 (2013): 222–233. doi: 10.1007/s13562-012-0155-9.

Prasad, K., K. Yogendra, H. Sanivarapu, K. Rajasekaran, J. W. Cary, K. K. Sharma, and P. Bhatnagar-Mathur. "Multiplexed host-induced gene silencing of *Aspergillus flavus* genes confers aflatoxin resistance in groundnut." *Toxins* 15 (5) (2023): 319. doi: 10.3390/toxins15050319.

Probst, C., R. Bandyopadhyay, and P. Cotty. "Diversity of aflatoxin-producing fungi and their impact on food safety in sub-Saharan Africa." *International Journal of Food Microbiology* 174 (2014): 113–122. doi: 10.1016/j.ijfoodmicro.2013.12.010.

Probst, C., H. Njapau, and P. J. Cotty. "Outbreak of an acute aflatoxicosis in Kenya in 2004: identification of the causal agent." *Applied and Environmental Microbiology* 73 (8) (2007): 2762–2764. doi: 10.1128/aem.02370-06.

QIU, X. k., Y. p. KANG, J. b. GUO, B. l. Yu, W. g. CHEN, H. f. JIANG, H. Li, W. t. LI, H. y. LUO, and Y. LEI. "Method for screening resistance of peanut shell to *Aspergillus A. flavus* infection and identification of resistant genotypes." *Chinese Journal of Oil Crop Sciences* 41 (1) (2019): 109. doi: 10.7505/j.issn.1007-9084.2019.01.014.

Rajasekaran, K., R. J. Sayler, C. M. Sickler, R. Majumdar, J. M. Jaynes, and J. W. Cary. "Control of *Aspergillus flavus* growth and aflatoxin production in transgenic maize kernels expressing a tachyplesin-derived synthetic peptide, AGM182." *Plant Science* 270 (2018): 150–156. doi: 10.1016/j.plantsci.2018.02.006.

Ren, X., M. T. Branà, M. Haidukowski, A. Gallo, Q. Zhang, A. F. Logrieco, P. Li, S. Zhao, and C. Altomare. "Potential of Trichoderma spp. for biocontrol of aflatoxin-producing *Aspergillus flavus*." *Toxins* 14 (2) (2022): 86. doi: 10.3390/toxins14020086.

Rustom, I. Y. "Aflatoxin in food and feed: occurrence, legislation and inactivation by physical methods." *Food Chemistry* 59 (1) (1997): 57–67. doi: 10.1016/S0308-8146(96)00096-9.

Saini, S. S., and A. Kaur. "Aflatoxin B1: Toxicity, characteristics and analysis: Mini review." *Global Advanced Research Journal of Chemistry and Material Science* 1 (4) (2012): 063–70.

Samson, R. A. "Current taxonomic schemes of the genus *Aspergillus* and its teleomorphs." *Biotechnology* 23 -(1992): 355–390. doi: 10.1039/b618937g.

Schubert, M., M. Houdelet, K. H. Kogel, R. Fischer, S. Schillberg, and G. Nölke. "Thanatin confers partial resistance against aflatoxigenic fungi in maize (Zea mays)." *Transgenic Research* 24 (5) (2015): 885–895. doi: 10.1007/s11248-015-9888-2.

Shabeer, S., S. Asad, A. Jamal, and A. Ali. "Aflatoxin contamination, its impact and management strategies: an updated review." *Toxins* 14 (5) (2022): 307. doi: 10.3390/toxins14050307.

Sharif, Y., H. Chen, Y. Deng, N. Ali, S. Khan, C. Zhang, W. Xie, et al. "Cloning and functional characterization of a pericarp abundant expression promoter (*AhGLP17-1P*) from peanut (*Arachis hypogaea* L.)." *Frontiers in Genetics* 12 (2022): 821281. doi: 10.3389/fgene. 2021.821281.

Sharif, Y., G. Mamadou, Q. Yang, T. Cai, Y. Zhuang, K. Chen, Y. Deng, et al. "Genome-wide investigation of apyrase (APY) genes in peanut (*Arachis hypogaea* L.) and functional characterization of a pod-abundant expression promoter *AhAPY2-1p*." *International Journal of Molecular Sciences* 24 (5) (2023): 4622. doi: 10.3390/ijms24054622.

Sharif, Y., Y. Zhuang, W. Xie, C. Zhang, K. Chen, Y. Deng, Y. Chen, *et al.* "Molecular cloning and functional identification of a pericarp- and testa-abundant gene's (*AhN8DT-2*) promoter from *Arachis hypogaea*." *International Journal of Molecular Sciences* 25 (14) (2024): doi: 10.3390/ijms25147671.

Sharma, K. K., A. Pothana, K. Prasad, D. Shah, J. Kaur, D. Bhatnagar, Z. Y. Chen, Y. Raruang, J. W. Cary, and K. Rajasekaran. "Peanuts that keep aflatoxin at bay: a threshold that matters." *Plant Biotechnology Journal* 16 (5) (2018): 1024–1033. doi: 10.1111/pbi.12846.

Sharma, S., B. Choudhary, S. Yadav, A. Mishra, V. K. Mishra, R. Chand, C. Chen, and S. P. Pandey. "Metabolite profiling identified pipecolic acid as an important component of peanut seed resistance against *Aspergillus flavus* infection." *Journal of Hazardous Materials* 404 (2021): 124155. doi: 10.1016/j.jhazmat.2020.124155.

Sobolev, V., R. Arias, K. Goodman, T. Walk, V. Orner, P. Faustinelli, and A. Massa. "Suppression of aflatoxin production in Aspergillus species by selected peanut (*Arachis hypogaea*) stilbenoids." *Journal of Agricultural and Food Chemistry* 66 (1) (2018): 118–126. doi: 10.1021/acs.jafc.7b04542.

Soni, P., S. S. Gangurde, A. Ortega-Beltran, R. Kumar, S. Parmar, H. K. Sudini, Y. Lei, et al. "Functional biology and molecular mechanisms of host-pathogen interactions for aflatoxin contamination in groundnut (*Arachis hypogaea* L.) and maize (*Zea mays* L.)." *Frontiers in Microbiology* 11 (2020): 227. doi: 10.3389/fmicb.2020.00227.

Soni, P., S. S. Gangurde, A. Ortega-Beltran, R. Kumar, S. Parmar, H. K. Sudini, Y. Lei, *et al.* "Functional biology and molecular mechanisms of host-pathogen interactions for aflatoxin contamination in groundnut (*Arachis hypogaea* L.) and maize (*Zea mays* L.)." *Frontiers in Microbiology* 11 (2020): 227. doi: 10.3389/fmicb.2020.00227.

Soni, P., S. N. Nayak, R. Kumar, M. K. Pandey, N. Singh, H. K. Sudini, P. Bajaj, et al. "Transcriptome analysis identified coordinated control of key pathways regulating cellular physiology and metabolism upon *Aspergillus A. flavus* infection resulting in reduced aflatoxin production in groundnut." *Journal of Fungi* 6 (4) (2020): 370. doi: 10.3390/jof6040370.

Soni, P., A. K. Pandey, S. N. Nayak, M. K. Pandey, P. Tolani, S. Pandey, H. K. Sudini, *et al.* "Global transcriptome profiling identified transcription factors, biological process, and associated pathways for pre-harvest aflatoxin contamination in groundnut." *Journal of Fungi* 7 (6) (2021): 413. doi: 10.3390/jof7060413.

Souza, T. d. V. e. "A review on peanuts resistance to *Aspergillus flavus* and prospects in susceptibility genes-mediated resistance." Masters Dissertation, Wageningen University and Research, 2023.

Sudhakar, P., P. Latha, M. Babitha, P. Reddy, and P. Naidu. "Relationship of drought tolerance traits with aflatoxin contamination in groundnut." *Indian Journal of Plant Physiology* 12 (3) (2007): 261.

SuiLin, L. "Breeding of peanut varieties with high yield and high resistance to *Aspergillus flavus* Link.ex Fr.in Guangdong." *Guangdong Academy of Agricultural Sciences* 8 (2006): 20–23. doi: 10.3969/j.issn.1004-874X.2006.08.009.

Sujay, V., V. Kusuma, G. Yugandhar, B. Sujatha, M. Gowda, and H. Upadhyaya. "Selection of accessions from minicore to improve disease resistance in groundnut." 3rd International Conference for Peanut Genomics and Biotechnology on Advances in Arachis through Genomics and Biotechnology, 4–8 November 2008.

Thakare, D., J. Zhang, R. A. Wing, P. J. Cotty, and M. A. Schmidt. "Aflatoxin-free transgenic maize using host-induced gene silencing." *Science Advances* 3 (3) (2017): e1602382. doi: 10.1126/sciadv.1602382.

Thakur, R., V. Rao, S. Reddy, and M. Ferguson. "Evaluation of wild *Arachis* germplasm accessions for in vitro seed colonization and anatoxin production by *Aspergillus flavus*." *International Arachis Newsletter* 20 (2000): 44–46.

Treseder, K. K., R. Berlemont, S. D. Allison, and A. C. Martiny. "Drought increases the frequencies of fungal functional genes related to carbon and nitrogen acquisition." *PLoS One* 13 (11) (2018): e0206441. doi: 10.1371/journal.pone.0206441.

Unuofin, J. O., O. A. Odeniyi, O. S. Majengbasan, A. Igwaran, K. M. Moloantoa, Z. P. Khetsha, S. A. Iwarere, and M. O. Daramola. "Chitinases: expanding the boundaries of knowledge beyond routinized chitin degradation." *Environmental Science and Pollution Research International* 31 (26) (2024): 38045–38060. doi: 10.1007/s11356-024-33728-6.

Varshney, R. K., M. K. Pandey, A. Bohra, V. K. Singh, M. Thudi, and R. K. Saxena. "Toward the sequence-based breeding in legumes in the post-genome sequencing era." *Theoretical and Applied Genetics* 132 (3) (2019): 797–816. doi: 10.1007/s00122-018-3252-x.

Varshney, R. K., M. K. Pandey, and N. Puppala. "Future prospects for peanut improvement." In *The Peanut Genome*, 165–169. International Crops Research Institute for the Semi-Arid Tropics (ICRISAT), 2017. https://oar.icrisat.org/10493/1/Future%20Prospects%20for%20Peanut.pdf.

Waliyar, F., H. Hassan, and S. Bonkoungou. "Sources of resistance to *Aspergillus flavus* and aflatoxin contamination in groundnut genotypes in West Africa." *Plant Disease* 78 (7) (1994): 704–708.

Wang, H., Y. Lei, L. Wan, L. Yan, J. Lv, X. Dai, X. Ren, W. Guo, H. Jiang, and B. Liao. "Comparative transcript profiling of resistant and susceptible peanut post-harvest seeds in response to aflatoxin production by *Aspergillus flavus*." *BMC Plant Biology* 16 (1) (2016): 54. doi: 10.1186/s12870-016-0738-z.

Wang, H., R. V. Penmetsa, M. Yuan, L. Gong, Y. Zhao, B. Guo, A. D. Farmer, *et al.* "Development and characterization of BAC-end sequence derived SSRs, and their incorporation into a new higher density genetic map for cultivated peanut (*Arachis hypogaea* L.)." *BMC Plant Biology* 12 (1) (2012): 10. doi: 10.1186/1471-2229-12-10.

Wang, T., E. Zhang, X. Chen, L. Li, and X. Liang. "Identification of seed proteins associated with resistance to pre-harvested aflatoxin contamination in peanut (*Arachis hypogaea* L)." *BMC Plant Biology* 10 (2010): 267. doi: 10.1186/1471-2229-10-267.

Wang, Y., D. Liu, H. Yin, H. Wang, C. Cao, J. Wang, J. Zheng, and J. Liu. "Transcriptomic and metabolomic analyses of the response of resistant peanut seeds to *Aspergillus flavus* infection." *Toxins* 15 (7) (2023): 414. doi: 10.3390/toxins15070414.

Wang, Z., Y. Lei, and B. Liao. "Omics-driven advances in the understanding of regulatory landscape of peanut seed development." *Frontiers in Plant Science* 15 (2024): 1393438. doi: 10.3389/fpls.2024.1393438.

Wild, C. P., and Y. Y. Gong. "Mycotoxins and human disease: a largely ignored global health issue." *Carcinogenesis* 31 (1) (2010): 71–82. doi: 10.1093/carcin/bgp264.

Wilkins, M. R., E. Gasteiger, L. Tonella, K. Ou, M. Tyler, J.-C. Sanchez, A. A. Gooley, *et al.* "Protein identification with N and C-terminal sequence tags in proteome projects11 Edited by R. Huber." *Journal of Molecular Biology* 278 (3) (1998): 599–608. doi: https://doi:org/10.1006/jmbi.1998.1726.

Williams, J. H., T. D. Phillips, P. E. Jolly, J. K. Stiles, C. M. Jolly, and D. Aggarwal. "Human aflatoxicosis in developing countries: a review of toxicology, exposure, potential health consequences, and interventions." *The American Journal of Clinical Nutrition* 80 (5) (2004): 1106–1122. doi: 10.1093/ajcn/80.5.1106.

Hong, Y., L. Shaoxiong, and L. Haiyan. "Correlation analysis of SSR markers and host resistance to *Aspergillus flavus* infection in peanut (*Arachis hypogaea* L.)." *Molecular Plant Breeding* 7 (2009): 360–364.

Yong, L., W. Shengyu, L. Dong, J. Huifang, and L. Boshou. "Evaluation of resistance to aflatoxin production among peanut germplasm with resistance to bacterial wilt." *Chinese Journal of oil Crop Sciences* 26 (1) (2004): 69–71.

Yu, B., D. Huai, L. Huang, Y. Kang, X. Ren, Y. Chen, X. Zhou, et al. "Identification of genomic regions and diagnostic markers for resistance to aflatoxin contamination in peanut (*Arachis hypogaea* L.)." *BMC Genetics* 20 (2019): 1–13. doi: 10.1186/s12863-019-0734-z.

Yu, B., H. Jiang, M. K. Pandey, L. Huang, D. Huai, X. Zhou, Y. Kang, et al. "Identification of two novel peanut genotypes resistant to aflatoxin production and their SNP markers associated with resistance." *Toxins* 12 (3) (2020): 156. doi: 10.3390/toxins12030156.

Yu, B., N. Liu, L. Huang, H. Luo, X. Zhou, Y. Lei, L. Yan, et al. "Identification and application of a candidate gene *AhAftr1* for aflatoxin production resistance in peanut seed (*Arachis hypogaea* L.)." *Journal of Advanced Research* (2023): doi: 10.1016/j.jare.2023.09.014.

Yu, J. "Current understanding on aflatoxin biosynthesis and future perspective in reducing aflatoxin contamination." *Toxins* 4 (11) (2012): 1024–1057. doi: 10.3390/toxins4111024.

Zeng, X. "Isolation and Characterization of pericarp-abundant expression genes and promoters in *Arachis hypogaea*." Masters Dissertation, Fujian Agriculture and Forestry University, 2012.

Zhao, X., C. Li, C. Yan, J. Wang, C. Yuan, H. Zhang, and S. Shan. "Transcriptome and proteome analyses of resistant preharvest peanut seed coat in response to *Aspergillus flavus* infection." *Electronic Journal of Biotechnology* 39 (2019): 82–90. doi: 10.1016/j.ejbt.2019.03.003.

Zhou, X., X. Ren, H. Luo, L. Huang, N. Liu, W. Chen, Y. Lei, B. Liao, and H. Jiang. "Safe conservation and utilization of peanut germplasm resources in the oil crops middle-term genebank of China." *Oil Crop Science* 7 (1) (2022): 9–13. doi: doi:org/10.1016/j.ocsci.2021.12.001.

Zhu, H., M. Lamb, C. Butts, and P. Blankenship. "Improving peanut yield and grade with surface drip irrigation in undulating fields." *Transactions of the ASAE* 47 (1) (2004): 99–106. doi: 10.13031/2013.15875.

Zhuang, W., H. Chen, M. Yang, J. Wang, M. K. Pandey, C. Zhang, W. C. Chang, et al. "The genome of cultivated peanut provides insight into legume karyotypes, polyploid evolution and crop domestication." *Nature Genetics* 51 (5) (2019): 865–876.

Zhuang, W., S.-M. Fang, L. I. Yu, Y.S. Chen, Z. Cheng, and Y.S. Chen. "Screening and identification of resistant peanut varieties and strains to *Aspergillus flavus*." *Fujian Journal of Agricultural Sciences* 22 (3) (2007): 261–265.

Zhuang, Y., Y. Sharif, X. Zeng, S. Chen, H. Chen, C. Zhuang, Y. Deng, M. Ruan, S. Chen, and W. Zhuang. "Molecular cloning and functional characterization of the promoter of a novel *Aspergillus A. flavus* inducible gene (*AhOMT1*) from peanut." *Frontiers in Plant Science* 14 (2023): 1102181. doi: 10.3389/fpls.2023.1102181.

Zhuang, W., S. H. I. Xin-guo, C. Hua, W. Zai-xin, G. de-yi, C. A. I. Lai-long, and L. I. Yu. "Breeding and good characteristics of a new peanut variety, Minhua 6." *Fujian Journal of Agricultural Sciences* 25 (3) (2010): 303–309.

17 Bacterial Wilt in Peanut

Genetic Diversity, Mechanism, Enhancement, and Application in Disease Management

Sheidu Abdullaziz, Boshou Liao, Yuhui Zhuang, Chong Zhang, Yong Lei, Hua Chen, Xinyou Zhang, and Weijian Zhuang

17.1 INTRODUCTION

Bacterial wilt (BW) caused by *Ralstonia solanacearum* is a devastating disease that affects a great number of plant species, including peanut, worldwide. This pathogen invades the plant through its vascular system, causing the blockage of water (Almeida et al. 2015) and essential nutrients to the plants, leading to symptoms such as black or brown discoloration in the stem tissue, stunted growth, and wilting, which leads to the death of the plant (Schell 2000; Wang et al. 2023). This is a soil-borne disease that dwells in the soil over a long period of time, making it difficult to control. Yield losses due to bacterial wilt can range from 20% to 100%, depending on environmental conditions, peanut cultivar susceptibility, and management practices (Gantait et al. 2019). *R. solanacearum*'s adverse effect is more prevalent in the tropics and subtropical regions (Pandey et al. 2020). Peanut bacterial wilt was first reported in Indonesia in 1905, and since then has been reported in many regions. Bacterial wilt in peanut was once of minor importance in the United States, primarily in North Carolina and Georgia, but its prevalence has diminished over several decades. However, in Africa, China, Vietnam, and Malaysia, the disease remains a problem (Obasa & Haynes 2021). In China, bacterial wilt affects over 10% of the total peanut sowing area annually, with varying incidence rates depending on cultivar susceptibility (Bindal & Srivastava 2019; Manda et al. 2020).

The disease severity is caused by many factors, including temperature, soil moisture, and pH levels (Manjeet & Savita 2013). These factors not only affect the spread of the pathogen but also determine the effectiveness of different management strategies. Despite efforts to control the disease, bacterial wilt continues to pose a major challenge to peanut production, especially in regions with conducive environmental conditions for the growth of *R. solanacearum*.

This chapter delves into the latest advancements in the characterization of the bacterial wilt pathogen and the breeding strategies employed to develop resistant peanut varieties. It explores the identification of resistance sources within peanut germplasm, the genetic patterns of resistance, and the application of marker-assisted selection (MAS) and genomic approaches. Additionally, it highlights the progress and challenges in developing bacterial wilt–resistant cultivars. The molecular mechanisms underlying peanut resistance to bacterial wilt are examined, detailing the plant's genetic and biochemical responses to infection. These insights provide a deeper understanding of the interactions between the pathogen and the host plant, revealing key resistance genes and signaling pathways involved in the defense response. Effective management of bacterial wilt also requires an integrated approach.

17.2 ADVANCE IN BACTERIAL WILT PATHOGEN CHARACTERIZATION

Bacterial wilt, caused by the gram-negative soil-borne bacterium *Ralstonia solanacearum*, poses a significant threat to global agriculture due to its broad host range and devastating pathogenicity (Singh et al. 2016; Obasa & Haynes 2021). The pathogenesis of *R. solanacearum* is facilitated by a diverse array of virulence factors, primarily involving the delivery of virulence proteins (AvrA, Pop2, GALA, RipAU, etc.) into the host cytoplasm or apoplast (Abramovitch et al. 2006; Chen et al. 2025). This pathogen utilizes cell wall–degrading enzymes to breach plant defenses and establish infections effectively (Genin & Denny 2012; Malinowski et al. 2024; Ababa, 2024). The pathogen's spread is facilitated by factors like contaminated soil, water, infected plant materials, and agricultural practices. Its prevalence and impact are particularly noted in regions with conducive environmental conditions for bacterial growth (Mansfield et al. 2012; Singh et al. 2024).

The genome of *R. solanacearum* usually comprises a large chromosome and a megaplasmid, encoding a wide array of virulence factors. Genomic sequencing has uncovered its ability to rapidly adapt to diverse environmental niches and hosts (Salanoubat et al. 2002; Coupat et al. 2008). This genomic flexibility underpins its pathogenic success and provides insights into potential targets for disease management (Remenant et al. 2010). Recently, Chen et al. (2021) provided an in-depth analysis of the genome of *R. solanacearum* strains affecting peanuts. The genome of

DOI: 10.1201/9781003360988-17

R. solanacearum is highly complex, comprising approximately 5.8 million base pairs with a G+C content of around 67%. The genomic structure includes one circular chromosome and a large megaplasmid, and a small plasmid. The genome encodes a vast array of genes involved in virulence, environmental adaptability, and metabolic versatility. Key features include the type III secretion system (T3SS), which is a critical pathogenicity determinant allowing the injection of effector proteins into host cells to suppress immune responses and manipulate host cellular processes (Angot et al. 2006; Deslandes & Genin 2014). A knockout mutant library encompassing all 81 type III effectors (T3Es) in the peanut-affecting strain *Rs-P.36220* has recently been developed, informed by genome studies (Chen et al. 2021; Chen et al. 2025). Among these effectors, 13 were identified as having altered pathogenicity based on pathogenicity evaluations of the individual effector mutants in peanut. Some of these effectors have been characterized for their roles in pathogenicity and their interactions with peanut during *Ralstonia solanacearum* infection (Chen et al. 2025). This library serves as a valuable genetic resource for unraveling the mechanisms underlying pathogen–host interactions, providing insights into resistance and susceptibility in peanut. The type II and type IV secretion systems (T2SS and T4SS) are involved in secreting enzymes and transferring DNA, respectively, enhancing the bacterium's ability to infect and cause disease. Additionally, cell wall–degrading enzymes such as cellulases, pectinases, and proteases degrade the plant cell wall, facilitating bacterial invasion and spread (Macho & Zipfel 2014; Preston et al. 2005; Alfano & Collmer 2004; Christie et al. 2005; Abramovitch et al. 2006; Coaker et al. 2005; Shao et al. 2002; Verma et al. 2023; Nizamani et al. 2023).

17.3 IDENTIFICATION OF RESISTANCE SOURCES IN PEANUT GERMPLASM

The global peanut community has made significant strides in collecting and characterizing genetically diverse peanut germplasm to identify sources of resistance to bacterial wilt. Numerous studies have identified sources of bacterial wilt resistance within cultivated peanut germplasm and its wild relatives. Wild *Arachis* species, including the progenitors of *Arachis hypogea*, are valuable sources of genetic diversity and resistance to bacterial wilt (Luo et al. 2019). Interspecific hybridization has facilitated the transfer of resistance alleles into cultivated peanut, and more than 3400 resistant accessions have been identified in wild *Arachis* collections (Stalker 2017). Researchers, most especially from China, have contributed immensely in evaluating diverse peanut accessions for resistance to *R. solanacearum*, focusing on traits such as delayed symptom development, reduced disease severity, and enhanced plant survival. Resistant accessions are characterized by their ability to limit bacterial colonization and spread, maintain vascular integrity, and sustain minimal physiological damage in response to infection. The identification of bacterial wilt–resistant peanut germplasm lines has strongly supported the subsequent breeding of resistant varieties. At the beginning of the twentieth century, Indonesia first discovered a bacterial wilt–resistant peanut variety and bred the world's first bacterial wilt–resistant peanut variety "Schwarz 21" (Schwarz & Hartley 1983). Since the initial selection of resistant peanut varieties, Chinese institutions have extensively conducted germplasm evaluation for BW resistance from which Xiekangqing, Taishansanlirou, and Taishanzhenzhu have been identified and used for breeding programs. In China, more than 150 germplasm accessions and improved cultivars highly resistant to bacterial wilt have been identified and developed (Liao 2014). Resistant germplasm lines have been found in five of the six botanic types of cultivated peanut, mostly existing in the Chinese dragon type and Spanish type. It was interesting to note that the wilt-resistant dragon type lines were only collected from southern China, where bacterial wilt was a problem (Liao et al. 2005).

Nugrahaeni and Rahayu (2017) conducted an evaluation of 150 introduced peanut accessions for resistance to bacterial wilt and identified 23 lines exhibiting resistance. Notably, MLGA 0292 demonstrated the highest level of resistance among these accessions. Similarly, Zhou et al. (2014) evaluated various new peanut cultivars and lines for their resistance to bacterial wilt and identified 8 highly resistant varieties/lines out of 21 tested materials. These included Minhua 821, Minhua 825, Yueyou 256, Yueyou 79, Yueyou 114, Yueyou 9, and Yueyou 202-35, originating from different genetic sources such as Ah1722, Yinduhuapi, and Taishansanlirou.

Furthermore, Zhang et al. (2017) developed a new resistance evaluation method. This method involves culturing toxic bacterium *R. solanacearum* and preparing bacterium suspensions with specific concentrations. Healthy peanut plants are then exposed to these suspensions by leaf-cutting inoculation, and the degree of wilt in their stems and branches is observed after 2–3 weeks (Zhang et al., 2017). Genotypes showing no wilt symptom with concentrations of 10^8 CFU/ml were considered highly resistant, while those exibitng varying degrees of wilt symptoms were classified as moderately resistance to high susceptibility. This approach has proven advantageous in overcoming previous limitations and facilitating practical identification of resistant germplasm in genetic breeding practices, genetic offspring evaluation, and QTL mapping uses (Zhang et al. 2017; Li et al., 2010). While the method of inoculating via wounded root steeping effectively reflects cultivar resistance, it requires a larger inoculation volume and a longer duration of disease presence, which poses significant inconvenience to the experiment. Many new resistant varieties have been found by the leaf cutting method, including Minhua 8, Guihua 1, Guihua 2, Mohua 1 and Mohua 2, Guihua 317, Puhua 1, and so on, which were approved for release to production. This method offers several advantages, including a short evaluation cycle, easy operation, and a smaller inoculation volume. This innovative technique has become an invaluable tool for high throughput, identifying thousands of peanut germplasm with resistance to bacterial wilt caused by *R. solanacearum*. More recently, a novel hydroponic culture inoculation system for bacterial wilt resistance was developed, in which germinated peanut seeds are transplanted onto a plastic plate and cultured in Hoagland's solution. The 4-week old plants are injured in the roots and inoculated with *R. solanacearum* suspension, and disease can be recorded one week after inoculation. This novel method has been proven to be quick, high throughput, precise, and cost-effective (Song et al. 2024).

17.4 RESISTANCE INHERITANCE AND MARKER-ASSISTED SELECTION

17.4.1 Genetic Pattern of Peanut Resistance to Bacterial Wilt

Understanding the genetic basis of resistance to bacterial wilt in peanuts is crucial for developing resistant cultivars. Resistance to bacterial wilt in peanut is controlled by major genes (Zhang et al. 2023). Peanut resistance to bacterial wilt exhibits a heritable pattern, as demonstrated by phenotyping of the F1 and F2 generations in a cross between *Yueyou 92* (var. *vulgaris*, resistant) and *Xinhuixiaoli* (var. *fastigiata*, susceptible). The segregation pattern observed in the F2 generation showed a 1:3 ratio of resistance to susceptibility. This was evident when all *Xinhuixiaoli* plants perished, while all *Yueyou 92* individuals remained alive under identical conditions (Zhao & Zhang 2016). Genetic studies on Chinese dragon varieties revealed that cytoplasmic factors play a role in resistance to bacterial wilt, with dragon type showing partial dominance to bacterial wilt. Different peanut types show varying responses to latent infection, indicating potential genetic differences among subspecies (Jiang et al. 2007). The capacity of latent infection tolerance among different types of peanut is different. Generally, the prostrating and late-maturing resistant lines have a longer incubation period of disease and a delayed disease development. This indicates that the genetic basis of the two subspecies or different types of peanut might be different (Liao 2014).

Before the publication of the two diploid genomes in 2016, less significant progress had been made in understanding the genetic basis of bacterial wilt resistance (BWR) in peanuts. Previous studies using techniques such as SSR (simple sequence repeat) and AFLP (amplified fragment length polymorphism) reported that peanut resistance to bacterial wilt was controlled by multiple loci spread across several chromosomes or linkage groups. These studies provided a foundational understanding of the genetic complexity underlying BWR. The publication of the diploid genomes marked a turning point in BWR research. In 2016, Zhao et al. made a groundbreaking discovery by identifying stable resistance to bacterial wilt located on the B02 chromosome using a mixed genetic map of SNP and SSR markers. The BWR-associated SNPs were discovered through restriction site-associated DNA (RAD) sequencing combined with bulked segregant analysis (BSA), while the SSR markers were derived from a wild B-genome peanut species. This study was pivotal, as it was the first to pinpoint a specific chromosomal location associated with stable BWR using F2 offspring and F8 generation of RIL population. Subsequent research consistently reported the same interval region on the upper end of chromosome 12 (Sub B02), establishing a clearer genetic pattern for BWR. It is now understood that BWR in peanuts is controlled by several closely linked QTLs within a region spanning approximately 1.8Mb–8.7Mb (Zhang et al. 2023).

QTL mapping has proven to be a vital tool in identifying genomic regions associated with disease resistance. Several studies have successfully utilized QTL mapping and marker development for breeding bacterial wilt–resistant peanuts. The use of marker-assisted backcrossing (MABC) to introgress QTLs for bacterial wilt resistance into elite peanut varieties has exhibited lines with enhanced resistance and maintained desirable agronomic traits. Luo et al. (2019) conducted a comprehensive QTL mapping study using a multi-parent advanced generation inter-cross (MAGIC) population. This study identified several novel QTLs on linkage groups A01, A07, and B09. The integration of these QTLs into breeding programs has the potential to enhance the genetic diversity and resistance levels of peanut varieties.

Marker development is essential for the practical application of QTLs in breeding programs. Molecular markers such as SSRs, single nucleotide polymorphisms (SNPs), and AFLPs have been developed and validated for bacterial wilt resistance in peanuts. For instance, Luo et al. (2019) developed SSR markers linked to QTLs for bacterial wilt resistance. These markers were validated in diverse peanut germplasm and showed high accuracy in predicting resistance. However, Wang et al. (2018) developed a set of SNP markers that were highly correlated with resistance traits. These markers were derived from a genome-wide association study (GWAS) conducted on a diverse peanut panel. The markers were successfully used to select resistant lines in a breeding program, demonstrating their practical utility. Similarly, Fang et al. (2022) identified SNP markers associated with resistance QTLs using high-throughput genotyping platforms. These SNP markers have been integrated into breeding programs for marker-assisted selection.

Understanding the genetic pattern of resistance to bacterial wilt in peanuts is critical for effective breeding programs. QTL mapping and marker development have provided valuable insights into the genetic architecture of resistance, enabling the identification and introgression of resistance genes into elite cultivars.

17.4.2 Marker-Assisted Selection and Genomics Approaches

Marker-assisted selection (MAS) is utilized for screening offspring to identify desired genotypes through the application of genetically linked markers. Recent advancements in peanut research have significantly enhanced the mapping of resistance regions against *R. solanacearum*, utilizing high-throughput genomic technologies. Numerous studies have focused on resistance QTL mapping and marker development, enhancing understanding of BWR's genetic basis and facilitating MAS in breeding programs (Jiang et al. 2007b). Valuable markers have been screened for selection and evaluation of bacterial wilt resistance in peanut breeding programs in previous studies (Li et al. 2020; Linh et al. 2016). For instance, Zhao et al. (2016) employed linkage analysis and bulk segregant analysis (BSA) with SSR and SNP markers, identifying significant QTLs on linkage groups LG1 and LG10. These QTLs explained substantial phenotypic variance, ranging from 12% to 21%. LG1 QTL was stably identified in F2, F8, and F9 generation located at the B02 chromosome. Several close-linked SNP markers were developed. Similarly, Wang et al. (2018) mapped a consistent QTL, qBWB02.1, on chromosome B02 using a dense SNP-based genetic linkage map derived from a recombinant inbred line (RIL) population. While major QTL on chromosome B02 was repeatedly detected using different methods and populations, two diagnostic markers were successfully validated in the lines derived from Yuanza 9102, but other susceptible and resistant cultivars could not be distinguished. Further studies by Luo et al. (2019) conducted phenotypic selections and QTL-seq analyses within RIL populations, pinpointing additional QTLs on chromosome B02 and other chromosomes using BSA. Moreover, two major QTLs were detected on chromosome B02 using linkage mapping and QTL-seq based on a RIL population derived from a cross between peanut cultivars Xuhua 13 and Zhonghua 6 (Luo et al. 2020). Notably, Qi et al. (2022) identified a major QTL, qBWA12, for BW resistance, which was fine-mapped to a specific genomic region using high-throughput sequencing and validated with a KASP marker, A12.4097252. Interestingly, Zhang et al. (2023) expanded on these findings by analyzing a population derived from crossing peanut lines Yueyou 92 (resistant) and Xinhuixiaoli (susceptible). The QTL-seq analysis highlighted a candidate genomic region on chromosome 12 associated with BW resistance, encompassing several genes, including those of the nucleotide binding site leucine-rich repeat (NBS-LRR) family known for disease resistance. Diagnostic SNP markers derived from this study showed robust differentiation between resistant and susceptible cultivars, especially those from resistance resource of Xiekangqing. Further identification of robust molecular markers associated with resistance will allow for the rapid screening and selection of phenotypically superior plants, speeding up the breeding cycle and ensuring high-quality, disease-resistant peanut varieties (Fang et al. 2022).

In conclusion, the integration of advanced genomics technologies and MAS has greatly improved the ability to breed peanut cultivars resistant to bacterial wilt. Key QTLs on chromosomes 12/B02 have been identified and mapped, leading to the development of reliable molecular markers. These advancements enhance breeding precision and efficiency, paving the way for high-yielding, disease-resistant peanut varieties.

17.5 DEVELOPING BACTERIAL WILT–RESISTANT PEANUT CULTIVARS

Peanut breeding has achieved significant milestones in developing varieties resistant to bacterial wilt (BW). Fujian Agriculture and Forestry University, China, has successfully developed elite varieties that have proven to be highly resistant. Among these varieties are Minhua 821 and Minhua 825 (Quanhua551 x Sunoleic95), developed using MAS, with high yield and high oleate content. Other notable wilt-resistant cultivars extensively cultivated in China include Minhua 8, Luhua 3, Zhonghua 2, Zhonghua 4, Tianfu 11 (Yueyou 200), Yueyou 92, Yueyou 202-35, Zhonghua 6, Yuanza 9102, Yuhua 17, Guihua 37, Guihua 58, and Zhonghua 21, Jinhua 1012 (a white seed coat variety), and so on. These cultivars are predominantly from regions where bacterial wilt is more prevalent, such as Guangdong, Guangxi, Fujian, Henan, Hubei, and Shandong (Gong et al. 2022).

Zhou et al. (2014) evaluated various new peanut cultivars and lines for their resistance to bacterial wilt and identified 8 highly resistant varieties/lines out of 21 tested materials. These included Yueyou 256, Yueyou 79, Yueyou 114, Yueyou 9, and Yueyou 202-35, originating from different genetic sources such as Ah1722, Yinduhuapi, and Taishansanlirou. Recently, Yin (2020) achieved an advancement by crossing Luhua 10 (EMS mutagenesis) with Baisha 1016, resulting in a new variety with BW resistance and high quality. Similarly, Li et al. (2020) bred Puhua 36 from a cross between Yuanza 9102 and Puhua 9327, producing a variety known for its resistance to BW, high yield, and early maturity.

MAS plays a crucial role in tracking peanut BWR genes and selecting superior progeny (Jiang et al. 2007b). The effectiveness of resistance markers is validated through controlled inoculation of breeding lines with *R. solanacearum*, using methods like soil drenching and stem injection, followed by monitoring for disease symptoms (Chen et al. 2014). The development of these BW-resistant cultivars integrates traditional breeding, MAS, transgenic, genomics, and modern biotechnological tools.

Advancements in genomic technologies, such as genomic selection (GS) and genome-wide association studies (GWAS), have further improved the efficiency of breeding BW-resistant peanuts. GS uses genome-wide markers to predict breeding values, capturing both major and minor QTL effects, while GWAS has identified key QTLs associated with BW resistance, facilitating the rapid development of resistant cultivars

In conclusion, the development of BW-resistant cultivars in peanut demonstrates the effectiveness of combining traditional breeding methods with modern genomic technologies. Continuous research and innovation in this area are critical for addressing the evolving challenges posed by bacterial wilt and ensuring the sustainability of peanut production.

17.6 MOLECULAR MECHANISM UNDERLYING BACTERIAL WILT RESISTANCE

17.6.1 Molecular Responses to Bacterial Wilt Infection

Recent advances in molecular biology have revolutionized understanding of peanut resistance to bacterial wilt by elucidating the genetic and molecular mechanisms underlying this crucial trait. Sophisticated techniques such as transcriptomics, proteomics, and metabolomics have played a pivotal role in these advancements. These approaches have enabled researchers to identify critical genes, proteins, and metabolic pathways activated in response to *R. solanacearum* infection. By comparing resistant and susceptible peanut varieties, scientists pinpoint molecular differences that confer resistance. Key among these technologies is RNA sequencing (RNA-seq), which enables comprehensive gene expression profiling. Studies have revealed specific resistance genes that are upregulated during infection, shedding light on the intricate defense mechanisms of peanut plants (Chen et al. 2014). For instance, downregulation of certain primary metabolite and defense-related genes has been implicated in the resistance disparities observed between peanut genotypes (Chen et al. 2014). Moreover, researchers have identified several genes linked to bacterial wilt resistance, such as qBW-1 and qBW-2, which are quantitative trait loci (QTL) showing promise as candidates for resistance (Zhao et al. 2016). Functional validation of these candidate genes involves advanced techniques like gene silencing through RNA interference and precise gene editing using CRISPR-Cas9 technology. These methods allow researchers to directly observe changes in plant susceptibility when specific genes associated with resistance are manipulated. Among these technologies, RNA sequencing (RNA-seq) stands out, enabling comprehensive profiling of gene expression. Transcriptome analyses have revealed the involvement of salicylic acid (SA) and jasmonic acid (JA) signaling pathways in resistant lines, emphasizing their role in mediating resistance to *R. solanacearum* (Zhao et al. 2014; Guo et al. 2006). Recent studies have identified 150 *AhCN* genes in peanuts, categorized into nine subfamilies with conserved structural features and cis-regulatory elements indicative of their role in hormone signaling and defense (Zhao et al. 2024a). One notable gene, *AhCN34*, was significantly upregulated in the highly resistant peanut variety "H108" compared to the susceptible variety "H107," underscoring its importance in bacterial wilt defense. Epigenetic modifications, such as m6A methylation, further contribute to resistance regulation. For instance, the *AhALKBH15* gene, involved in m6A demethylation, upregulates resistance genes, enhancing defense against pathogens like *R. solanacearum* (Zhao et al. 2024b). These findings suggest that manipulating epigenetic regulation offers promising avenues for improving peanut resistance. The molecular mechanisms behind bacterial wilt resistance involve complex processes where the plant recognizes the pathogen, sends signals within its cells, and activates defense responses to fight off the infection. Integrating these molecular insights into breeding strategies, such as MAS and GS, offers significant promise for developing peanut cultivars with robust resistance. This approach not only improves our understanding of plant–pathogen interactions but also provides sustainable solutions for managing bacterial wilt in peanut cultivation.

In conclusion, advancements in molecular biology have significantly enhanced understanding of peanut resistance to bacterial wilt, offering promising avenues for developing resilient peanut varieties through targeted genetic improvements and informed breeding strategies.

17.6.2 Molecular Mechanisms Underlying Bacterial Wilt Resistance

Understanding the molecular mechanisms of peanut resistance to *R. solanacearum* is crucial for developing effective disease management strategies. Recent studies have highlighted key genetic components involved in peanut's defense against bacterial wilt, providing valuable insights for breeding resistant cultivars (Liao et al. 2005; Jiang et al. 2007; Kona et al. 2024). Peanuts utilize a dual-layered defense system comprising PAMP-triggered immunity (PTI) and effector-triggered immunity (ETI). PTI initiates immune responses through the rapid induction of defense-related genes such as *PR-1*, *PR-2* (β-1,3-glucanase), and *PR-3* (chitinase), triggered by the recognition of pathogen-associated molecular patterns (PAMPs) (Nabi et al. 2024; Kumar et al. 2021; Kawai et al. 2024). These responses are orchestrated by a complex network of signaling pathways involving salicylic acid (SA), jasmonic acid (JA), and ethylene (ET), which mediate systemic acquired resistance (SAR) and induced systemic resistance (ISR). Several genetic components have been identified as key to bacterial wilt resistance, including *qBW-1* and *qBW-2*, which are quantitative trait loci (QTL) with strong potential for breeding applications (Zhao et al. 2016). The *AhRRS5* gene has been shown to confer resistance by triggering a hypersensitive response and activating defense mechanisms via hormone signaling pathways, particularly through the EDS1 pathway (Zhang et al. 2017). Another significant discovery involves the leucine-rich repeat receptor-like kinase gene *AhRLK1*, identified as CLAVATA1. Zhang et al. (2019) demonstrated that *AhRLK1* was upregulated in peanut varieties upon inoculation with *R. solanacearum*. Overexpression of *AhRLK1* in tobacco led to enhanced resistance, suggesting its role in activating defense mechanisms through the hypersensitive response (HR) and hormone defense signaling pathways, particularly via the EDS1 pathway. This pathway is crucial for SA-mediated defense responses, essential for systemic acquired resistance (SAR) and local defense against pathogens.

In addition to *AhRLK1*, transcriptome analyses have identified differentially expressed genes in resistant and susceptible peanut lines. These genes are involved in pathogen recognition, signal transduction, and defense responses, including cell wall reinforcement, production of antimicrobial compounds, and regulation of reactive oxygen species (ROS). The involvement of SA and JA signaling pathways in resistant lines underscores their role in mediating resistance to *R. solanacearum* (Zhao et al. 2014; Guo et al. 2006). Microarray hybridization identified *AhRLK1*, a novel NBS-LRR resistance gene *AhRRS5*, and another gene *AhRRS22*, which were upregulated in resistant cultivars (Zhang et al. 2017). *AhRRS5*, in particular, was shown to participate in defense response through complex regulatory networks involving multiple signaling pathways. Recent QTL fine mapping led to the positional cloning of two CC-NBS-LRR genes (*AhRRS1* and *AhRRS2*), which were patented and shown to significantly enhance resistance when expressed in transgenic plants (CN202210111059.2, CN202210111077.0). Transgenic plants showed significant increases in resistance to bacterial wilt disease. Further investigating the function and signal transduction of these R genes will clarify molecular mechanism underlying peanut resistance to bacterial wilt. Advanced techniques like yeast two-hybrid screening and co-immunoprecipitation have provided insights into protein–protein interactions during immune responses, illuminating signaling cascades for potential resistance improvement (Zhao 2022). These molecular findings not only deepen our understanding of peanut defense mechanisms but also highlight strategic avenues for developing resistant cultivars through marker-assisted selection and genetic engineering, ensuring sustainable disease management in peanut cultivation.

17.7 OTHER MANAGEMENT STRATEGIES FOR BACTERIAL WILT CONTROL

Cultural practices and crop rotation: Breeding cultivars resistant to bacterial wilt is the most cost-effective and environmentally friendly strategy for disease control, particularly in economically important crops like tomato, potato, tobacco, eggplant, pepper, and peanut (Liao 2005). However, studies have revealed that resistant plants may still harbor *R. solanacearum* without displaying wilt symptoms, suggesting the importance of understanding molecular interactions between the pathogen and host plants (Vailleau et al. 2023). Crop rotation improves soil health and manages disease by maintaining soil structure and organic matter, with evidence showing significant reductions in wilt severity and delayed onset when rotating crops like tomato with corn or resistant varieties (Adhikari & Basnyat 1998). Similar benefits have been observed in potato cultivation with rotations involving wheat, sweet potato, and other crops.

Soil amendments and biocontrol agents: Soil amendments, such as increasing calcium concentrations and applying nitrogen, phosphorus, and potassium fertilizers, have been shown to reduce bacterial wilt incidence and enhance crop yields (Kim et al. 2023; Qasim et al. 2023). The use of organic fertilizers and silicon–chitosan combinations has also proven effective in inducing resistance to bacterial wilt (Ye et al. 2022). Biological control agents (BCAs), including bacteria and fungi, offer eco-friendly alternatives to chemical pesticides. Promising BCAs, such as *Pseudomonas* spp. and *Bacillus* spp., suppress bacterial wilt through mechanisms like competition, antibiosis, and induced systemic resistance (Montesinos 2003; Chen et al. 2014).

Chemical treatments and fungicides: Combining these chemical approaches with sustainable practices is essential for minimizing adverse effects. Chemical treatments, including pesticides and non-pesticides, play a role in managing bacterial wilt, although their environmental impact necessitates careful use. Recent studies have shown that zinc oxide nanoparticles and salicylic acid can mitigate salt stress and enhance resistance to bacterial wilt in peanut seedlings (Luthfiana et al. 2023; Ali et al. 2023; Seleiman et al. 2023). Silicon and chitosan also effectively reduce bacterial wilt through induced resistance (Krishnan et al. 2023). Combining these chemical approaches with sustainable practices is essential for minimizing adverse effects.

Integrated pest management (IPM) approaches: IPM strategies against bacterial wilt have shown success through synergistic combinations of methods, including the use of biological control agents (BCAs) alongside their substrates to enhance root colonization (Wamani et al. 2023). Factors contributing to production losses in bacterial wilt management are crucial for developing effective IPM strategies, with cultural practices like crop rotation with resistant cultivars or soil amendments playing a significant role in reducing disease incidence and increasing crop yields (Benti 2023).

17.8 CHALLENGES AND FUTURE PERSPECTIVES

Current breeding strategies for bacterial wilt in peanuts are hindered by incomplete knowledge of the genetic mechanisms of resistance, which complicates the development of universally effective cultivars (Zhang et al. 2023). The pathogen *R. solanacearum* exhibits substantial genetic diversity and adaptability, often overcoming existing resistance (Yang et al. 2022; Zhang et al. 2023). Environmental factors further influence disease expression, complicating management strategies. Additionally, limited genetic resources and challenges associated with integrating resistance traits from wild varieties increased the complexity of breeding. The emergence of virulent pathogen strains and the impacts of climate change necessitate novel breeding strategies and disease-management approaches.

Advances in high-throughput sequencing and genomics technologies are paving the way for the rapid identification of resistance markers, enhancing marker-assisted selection. Genetic engineering and genome editing are emerging as vital tools for introducing targeted resistance traits, potentially revolutionizing peanut breeding (Zhang et al. 2023).

Significant research gaps remain, particularly in understanding the molecular bases of resistance, the mechanisms of host and pathogen interaction, and the pathogen's ecology. Future studies should focus on elucidating these aspects through advanced molecular techniques and exploring the genetic diversity of *R. solanacearum*. There is also a need to develop more sustainable management practices that minimize environmental impacts and enhance the adaptability of peanut farming to changing climatic conditions.

REFERENCES

Ababa, G. "Pathogenic Diversity, Ecology, Epidemiology, and Management Practices of the Potato Bacterial Wilt (*Ralstonia solanacearum*) Disease." *Cogent Food & Agriculture* (2024). https://tandfonline.com.

Abramovitch, R. B., R. Janjusevic, C. E. Stebbins, and G. B. Martin. "Type III Effector AvrPtoB Requires Intrinsic E3 Ubiquitin Ligase Activity to Suppress Plant Cell Death and Immunity." *Proceedings of the National Academy of Sciences of the United States of America* 103, no. 8 (2006): 2851–2856. https://doi.org/10.1073/pnas.0507892103.

Adhikari, T. B., and R. C. Basnyat. "Effect of Crop Rotation and Cultivar Resistance on Bacterial Wilt of Tomato in Nepal." *Canadian Journal of Plant Pathology* 20, no. 3 (1998): 283–287.

Alfano, J. R., and A. Collmer. "Type III Secretion System Effector Proteins: Double Agents in Bacterial Disease and Plant Defense." *Annual Review of Phytopathology* 42, no. 1 (2004): 385–414. https://doi.org/10.1146/annurev.phyto.42.040103.110731.

Ali, Q., M. Ahmad, M. Kamran, S. Ashraf, M. Shabaan, B. H. Babar, U. Zulfiqar, F. U. Haider, M. A.l Ali, and M. S. Elshikh. "Synergistic Effects of Rhizobacteria and Salicylic Acid on Maize Salt-stress Tolerance." *Plants* 12, no. 13 (2023): 2519.

Almeida, I. M. G. de, L. M. R. Rodrigues, and L. O. S. Beriam. "Xanthomonas fuscans Subsp. fuscans Causing Wilt Symptoms in Bean Plants (*Phaseolus vulgaris*) in Brazil." *Arquivos Do Instituto Biológico* 82 (2015): 1–3. https://doi.org/10.1590/1808-1657000042013.

Angot, A., N. Peeters, E. Lechner, F. Vailleau, C. Baud, L. Gentzbittel, E. Sartorel, P. Genschik, C. Boucher, and S. Genin. "Ralstonia solanacearum Requires F-box-like Domain-Containing Type III Effectors to Promote Disease on Several Host Plants." *Proceedings of the National Academy of Sciences of the United States of America* 103, no. 39 (2006): 14620–14625. https://doi.org/10.1073/pnas.0509393103.

Benti, E. A. "Control of Bacterial Wilt (Ralstonia solanacearum) and Reduction of Ginger Yield Loss through Integrated Management Methods in Southwestern Ethiopia." *The Open Agriculture Journal* 17, no. 1 (2023).

Bindal, S., and S. Srivastava. "Bacterial Wilt of Solanaceous Crops: Sign, Symptoms and Management." *Agrica* 8, no. 2 (2019): 134. https://doi.org/10.5958/2394-448x.2019.00019.1.

Chen, K., L. Wang, H. Chen, C. Zhang, S. Wang, P. Chu, S. Li, H. Fu, T. Sun, M. Liu, Q. Yang, H. Zou, and W. Zhuang. "Complete Genome Sequence Analysis of the Peanut Pathogen *Ralstonia solanacearum* Strain Rs-P.362200." *BMC Microbiology* 21, no. 1 (2021): 118. https://doi.org/10.1186/s12866-021-02157-7.

Chen, K., Y. Zhuang, H. Chen, T. Lei, M. Li, S. Wang, L. Wang, et al. "A *Ralstonia* Effector RipAU Impairs Peanut AhSBT1.7 Immunity for Pathogenicity via AhPME-Mediated Cell Wall Degradation." *The Plant Journal* 121, no. 2 (2025): e1721 city via AhPME-mediated cell wall degradation.

Chen, Y., X. Ren, X. Zhou, L. Huang, L. Yan, Y. Lei, B. Liao, J. Huang, S. Huang, W. Wei, and H. Jiang. "Dynamics in the Resistant and Susceptible Peanut (*Arachis hypogaea*) Root Transcriptome on Infection with the *Ralstonia solanacearum*." *BMC Genomics* 15, no. 1 (2014): 1078. https://doi.org/10.1186/1471-2164-15-1078.

Christie, P. J., K. Atmakuri, V. Krishnamoorthy, S. Jakubowski, and E. Cascales. "Biogenesis, Architecture, and Function of Bacterial Type IV Secretion Systems." *Annual Review of Microbiology* 59 (2005): 451–485. https://doi.org/10.1146/annurev.micro.58.030603.123630.

Coaker, G., A. Falick, and B. Staskawicz. "Activation of a Phytopathogenic Bacterial Effector Protein by a Eukaryotic Cyclophilin." *Science* 308, no. 5721 (2005): 548–550. https://doi.org/10.1126/science.1108633.

Coupat, B., F. Chaumeille-Dole, S. Fall, P. Prior, P. Simonet, X. Nesme, and F. Bertolla. "Natural Transformation in the *Ralstonia solanacearum* Species Complex: Number and Size of DNA That Can Be Transferred." *FEMS Microbiology Ecology* 66, no. 1 (2008): 14–24. https://doi.org/10.1111/j.1574-6941.2008.00552.x.

Deslandes, L., and S. Genin. "Opening the *Ralstonia solanacearum* Type III Effector Tool Box: Insights into Host Cell Subversion Mechanisms." *Current Opinion in Plant Biology* 20 (2014): 110–117. https://doi.org/10.1016/j.pbi.2014.05.002.

Fang, Y., H. Liu, Z. Zhang, F. Qi, Q. Chen, H. Liu, L. Wan, X. Wang, M. Tian, D. Lv, Z. Sun, W. Dong, B. Huang, and X. Zhang. "Genomic Identification and Phenotypic Evaluation of Yield Traits and Bacterial Wilt Resistance in High Oleic Peanut Breeding Lines Developed by Marker Assisted Backcrossing." *Euphytica* 218, no. 6 (2022). https://doi.org/10.1007/s10681-022-03032-w.

Gantait, S., J. Panigrahi, I. C. Patel, C. Labrooy, A. L. Rathnakumar, and J. K. Yasin. "Peanut (*Arachis hypogaea* L.) Breeding." In *Advances in Plant Breeding Strategies: Nut and Beverage Crops*, 253–299. Springer International Publishing, 2019. https://doi.org/10.1007/978-3-030-23112-5_8.

Genin, S., and T. P. Denny. "Pathogenomics of the *Ralstonia solanacearum* Species Complex." *Annual Review of Phytopathology* 50, no. 1 (2012): 67–89. https://doi.org/10.1146/annurev-phyto-081211-173000.

Gong, J., S. U. N. Dong-lei, N. Bian, X. Wang, and X. Wang. "Research Progress of Peanut Bacterial Wilt in China." *Chinese Journal of Oil Crop Sciences* 44, no. 6 (2022): 1159–1165. https://doi.org/10.19802/j.issn.1007-9084.2021298.

Guo, B., P. Khera, H. Wang, Z. Peng, H. Sudini, X. Wang, M. Osiru, J. Chen, V. Vadez, M. Yuan, and C. T. Wang. "Annotation of Trait Loci on Integrated Genetic Maps of *Arachis* Species." In *Peanut Science and Technology*, edited by W. R. Stalker and H. T. Stalker. AOCS Press, 2006.

Jiang, H., B. Liao, X. Ren, Y. Lei, E. Mace, T. Fu, and J. H. Crouch. "Comparative Assessment of Genetic Diversity of Peanut (*Arachis hypogaea* L.) Genotypes with Various Levels of Resistance to Bacterial Wilt through SSR and AFLP Analyses." *Journal of Genetics and Genomics* 34, no. 6 (2007): 544–554. https://doi.org/10.1016/s1673-8527(07)60060-5.

Jiang, H., B. Chen, X. Ren, B. Liao, Y. Lei, D. Fu, C. Ma, E. Mace, and J. H. Crouch. "Identification of SSR Markers Linked to Bacterial Wilt Resistance of Peanut with RILs." *Chinese Journal of Oil Crop Sciences* 29, no. 1 (2007b): 26–30.

Kawai, T., M. Ikegawa, D. Ori, and S. Akira. "Decoding Toll-like Receptors: Recent Insights and Perspectives in Innate Immunity." *Immunity* 57, no. 4 (2024): 649–673.

Kim, H. J., S. M. Kim, Y. H. Kim, J. H. Park, D. Kang, J. Yun, R. Shin, and J. Hong. 2023. "Reduced Tomato Bacterial Wilt by Ferrous Chloride Application." Research in Plant Disease 29, no. 1 (2023): 82–87. https://doi.org/10.5423/RPD.2023.29.1.82.

Krishnan, Srividhya, Subramaniyasharma Sivaraman, Sowndarya Jothipandiyan, Ponnusami Venkatachalam, Saravanan Ramiah Shanmugam, and Nithyanand Paramasivam. "Bioprospecting of Aqueous Phase from Pyrolysis of Plant Waste Residues to Disrupt MRSA Biofilms." *Biofouling* 39, no. 2 (2023): 231–243.

Kona, Praveen, Chandramohan Sangh, K. T. Ravikiran, B. C. Ajay, and Narendra Kumar. "Genetic and Genomic Resource to Augment Breeding Strategies for Biotic Stresses in Groundnut." In *Genomics-aided Breeding Strategies for Biotic Stress in Grain Legumes*, 359–403. Singapore: Springer Nature Singapore, 2024.

Kumar, J., A. Ramlal, K. Kumar, A. Rani, and V. Mishra. "Signaling Pathways and Downstream Effectors of Host Innate Immunity in Plants." *International Journal of Molecular Sciences* 22, no. 16 (2021): 9022. https://doi.org/10.3390/ijms22169022.

Li, H., S. Hearne, M. Bänziger, Z. Li, and J. Wang. "Statistical Properties of QTL Linkage Mapping in Biparental Genetic Populations." *Heredity* 105, no. 3 (2010): 257–267. https://doi.org/10.1038/hdy.2010.56.

Li, Z., J. Jing, H. Nie, et al. "Breeding of Puhua 36, a New Peanut Variety with High Resistance to Bacterial Wilt." *Seed Industry Guide* 5 (2020): 31–33.

Liao, B. "A Broad Review and Perspective on Breeding for Resistance to Bacterial Wilt." In *Bacterial Wilt Disease and the Ralstonia Solanacearum Species Complex*, edited by C. Allen, P. Prior, and A. C. Hayward, 225–238. St. Paul, Minnesota: APS Press, 2005.

Liao, B., X. Liang, H. Jiang, Y. Lei, Z. Shan, and X. Zhang. "Progress on Genetic Enhancement for Resistance to Peanut Bacterial Wilt in China." In *Bacterial Wilt Diseases and the Ralstonia solanacearum Species Complex*, edited by C. Allen, P. Prior, and A. C. Hayward, 239–246. The American Phytopathological Society, 2005.

Linh, N. T. T., N. Van Tru, L. T. Bich Thuy, N. Van Thang, N. T. Van, and N. Van Viet. "The Linkage of SSR Markers with Bacterial Wilt Disease Resistance in Peanut." *TAP CHI SINH HOC* 38, no. 2 (2016). https://doi.org/10.15625/0866-7160/v38n2.6195.

Luo, H., M. K. Pandey, A. W. Khan, B. Wu, J. Guo, X. Ren, X. Zhou, Y. Chen, W. Chen, L. Huang, N. Liu, Y. Lei, B. Liao, R. K. Varshney, and H. Jiang. "Next-Generation Sequencing Identified Genomic Region and Diagnostic Markers for Resistance to Bacterial Wilt on Chromosome B02 in Peanut (*Arachis hypogea* L.)." *Plant Biotechnology Journal* 17, no. 12 (2019): 2356–2369. https://doi.org/10.1111/pbi.13153.

Luo, Huaiyong, Manish K. Pandey, Ye Zhi, Huan Zhang, Siliang Xu, Jianbin Guo, Bei Wu et al. "Discovery of Two Novel and Adjacent QTLs on Chromosome B02 Controlling Resistance against Bacterial Wilt in Peanut Variety Zhonghua 6." *Theoretical and Applied Genetics* 133 (2020): 1133–1148.

Luthfiana, Ifa Dwi, Hersanti Hersanti, and Fitri Widiantini. "The Potency of Lysinibacillus Sp. in Carbon Fiber and Zinc Oxide Nanoparticles Mixture to Suppress Ralstonia Solanacearum in Vitro." *International Journal of Environmental, Sustainability, and Social Science* 4, no. 1 (2023): 250–255.

Macho, A. P., and C. Zipfel. "Plant PRRs and the Activation of Innate Immune Signaling." *Molecular Cell* 54, no. 2 (2014): 263–272. https://doi.org/10.1016/j.molcel.2014.03.028.

Malinowski, R., D. Singh, A. Kasprzewska, S. Blicharz, and A. Basińska-Barczak. "Vascular Tissue–Boon or Bane? How Pathogens Usurp Long-Distance Transport in Plants and the Defence Mechanisms Deployed to Counteract Them." *New Phytologist* 243, no. 6 (2024): 2075–2092.

Manda, R. R., V. A. Addanki, and S. Srivastava. "Bacterial Wilt of Solanaceous Crops." *International Journal of Chemical Studies* 8, no. 6 (2020): 1048–1057. https://doi.org/10.22271/chemi.2020.v8.i6o.10903.

Manjeet, A., and P. Savita. "Effect of Soil Moisture and Temperature on the Severity of Macrophomina Charcoal Rot of Sorghum." *Journal of Scientific Research* 4, no. 1 (2013): 155–158.

Mansfield, J., S. Genin, S. Magori, V. Citovsky, M. Sriariyanum, P. Ronald, M. Dow, V. Verdier, S. V. Beer, M. A. Machado, I. Toth, G. Salmond, and G. D. Foster. "Top 10 Plant Pathogenic Bacteria in Molecular Plant Pathology." *Molecular Plant Pathology* 13, no. 6 (2012): 614–629. https://doi.org/10.1111/j.1364-3703.2012.00804.x.

Montesinos, E. "Development, Registration and Commercialization of Microbial Pesticides for Plant Protection." *International Microbiology* 6, no. 4 (2003): 245–252. https://doi.org/10.1007/s10123-003-0144-x.

Nabi, Z., S. Manzoor, S. U. Nabi, T. A. Wani, H. Gulzar, M. Farooq, V. M. Arya, F. S. Baloch, C. Vlădulescu, S. M. Popescu, and S. Mansoor. "Pattern-Triggered Immunity and Effector-Triggered Immunity: Crosstalk and Cooperation of PRR and NLR-Mediated Plant Defense Pathways during Host–Pathogen Interactions." *Physiology and Molecular Biology of Plants* 30, no. 4 (2024): 1–18.

Nizamani, M. M., Q. Zhang, G. Muhae-Ud-Din, and Y. Wang. "High-Throughput Sequencing in Plant Disease Management: A Comprehensive Review of Benefits, Challenges, and Future Perspectives." *Phytopathology Research* 5, no. 1 (2023): 1–17.

Nugrahaeni, N., and M. Rahayu. "Short Communication: Peanut Introduced Germplasm Response Against *Ralstonia* Bacterial Wilt Disease." *Nusantara Bioscience* 9, no. 2 (2017): 138–140. https://doi.org/10.13057/nusbiosci/n090202.

Obasa, K., and L. Haynes. "Two New Bacterial Pathogens of Peanut Causing Early Seedling Decline Disease Identified in the Texas Panhandle." *Plant Disease* (2021). https://doi.org/10.1094/PDIS-07-21-1555-RE.

Pandey, M. K., A. K. Pandey, R. Kumar, C. V. Nwosu, B. Guo, and G. C. Wright. "Translational Genomics for Achieving Higher Genetic Gains in Groundnut." *Theoretical and Applied Genetics* 133 (2020): 1679–1702. https://doi.org/10.1007/s00122-020-03592-2.

Preston, G. M., D. J. Studholme, and I. Caldelari. "Profiling the Secretomes of Plant Pathogenic Proteobacteria." *FEMS Microbiology Reviews* 29, no. 2 (2005): 331–360. https://doi.org/10.1016/j.fmrre.2005.01.001.

Qasim, Muhammad, Jing Ju, Haitao Zhao, Saleem Maseeh Bhatti, Gulnaz Saleem, Saima Parveen Memon, Shahzaib Ali, Muhammad Usama Younas, Nimra Rajput, and Zameer Hussain Jamali. "Morphological and Physiological Response of Tomato to Sole and Combined Application of Vermicompost and Chemical Fertilizers." *Agronomy* 13, no. 6 (2023): 1508.

Qi, F., Z. Sun, H. Liu, Z. Zheng, L. Qin, L. Shi, Q. Chen, H. Liu, X. Lin, L. Miao, M. Tian, X. Wang, B. Huang, W. Dong, and X. Zhang. "QTL Identification, Fine Mapping, and Marker Development for Breeding Peanut (*Arachis hypogaea* L.) Resistant to Bacterial Wilt." *TAG. Theoretical and Applied Genetics* 135, no. 4 (2022): 1319–1330. https://doi.org/10.1007/s00122-022-04033-y.

Remenant, Benoît, Bénédicte Coupat-Goutaland, Alice Guidot, Gilles Cellier, Emmanuel Wicker, Caitilyn Allen, Mark Fegan et al. "Genomes of Three Tomato Pathogens within the Ralstonia Solanacearum Species Complex Reveal Significant Evolutionary Divergence." *BMC Genomics* 11 (2010): 1–16.

Salanoubat, Marcel, S. Genin, François Artiguenave, J. Gouzy, S. Mangenot, M. Arlat, A. Billault et al. "Genome Sequence of the Plant Pathogen Ralstonia Solanacearum." *Nature* 415, no. 6871 (2002): 497–502.

Schell, Mark A. "Control of Virulence and Pathogenicity Genes of Ralstonia Solanacearum by an Elaborate Sensory Network." *Annual Review of Phytopathology* 38, no. 1 (2000): 263–292.

Schwarz, Karl. "Les ménages en République fédérale d'Allemagne 1961-1972-1981." *Population (french edition)* (1983): 565–583.

Seleiman, Mahmoud F., Awais Ahmad, and Thobayet S. Alshahrani. "Integrative Effects of Zinc Nanoparticle and PGRs to Mitigate Salt Stress in Maize." *Agronomy* 13, no. 6 (2023): 1655.

Shao, Feng, Peter M. Merritt, Zhaoqin Bao, Roger W. Innes, and Jack E. Dixon. "A Yersinia Effector and a Pseudomonas Avirulence Protein Define a Family of Cysteine Proteases Functioning in Bacterial Pathogenesis." *Cell* 109, no. 5 (2002): 575–588.

Singh, D., D. Kumar Yadav, G. Chaudhary, V. S. Rana, and R. K. Sharma. "Potential of Bacillus Amyloliquefaciens for Biocontrol of Bacterial Wilt of Tomato Incited by Ralstonia Solanacearum." *Journal of Plant Pathology & Microbiology* 7, no. 327 (2016): 1–6.

Singh, Ravinder, Sehijpreet Kaur, Sukhveer S. Bhullar, Hardeep Singh, and Lakesh K. Sharma. "Bacterial Biostimulants for Climate Smart Agriculture Practices: Mode of Action, Effect on Plant Growth and Roadmap for Commercial Products." *Journal of Sustainable Agriculture and Environment* 3, no. 1 (2024): e12085.

Song, Yuqi, Chen Liu, Keming Yang, Shiqi Sun, Lin Wang, Cansheng Yuan, Yangchun Xu, Wu Xiong, Qirong Shen, and Zhong Wei. "Predatory Protist Promotes Disease Suppression against Bacterial Wilt through Enriching Plant Beneficial Microbes at the Early Stage of Plant Growth." *Plant and Soil* (2024): 1–12.

Stalker, H. Thomas. "Utilizing Wild Species for Peanut Improvement." *Crop Science* 57, no. 3 (2017): 1102–1120.

Vailleau, Fabienne, and Stéphane Genin. "Ralstonia Solanacearum: An Arsenal of Virulence Strategies and Prospects for Resistance." *Annual Review of Phytopathology* 61, no. 1 (2023): 25–47.

Verma, Jay Prakash, Durgesh Kumar Jaiswal, Anand Kumar Gaurav, Arpan Mukherjee, Ram Krishna, and Arthur Prudêncio de Araujo Pereira. "Harnessing Bacterial Strain from Rhizosphere to Develop Indigenous PGPR Consortium for Enhancing lobia (Vigna unguiculata) Production." *Heliyon* 9, no. 3 (2023).

Wamani, Arthur O., James W. Muthomi, Eunice Mutitu, and Wanjohi J. Waceke. "Efficacy of Microbial Antagonists in the Management of Bacterial Wilt of Field-grown Tomato." *Journal of Natural Pesticide Research* 6 (2023): 100051.

Wang, Xiaohua, Ping Xu, Liang Yin, Yan Ren, Shuangling Li, Yanmao Shi, Thomas D. Alcock et al. "Genomic and Transcriptomic Analysis Identified Gene Clusters and Candidate Genes for Oil Content in Peanut (Arachis hypogaea L.)." *Plant Molecular Biology Reporter* 36 (2018): 518–529.

Wang, Xin, Wei Yang, Yu Yang, Min Huang, and Qibing Zhu. "Identification of Tomato Bacterial Wilt Severity Based on Hyperspectral Imaging Technology and Spectrum Transformer Network." *Ecological Informatics* 78 (2023): 102353.

Ye, Meijin, Hong Feng, Jianghong Hu, Qing Yu, and Songqing Liu. "Managingtomato Bacterial Wilt by Suppressing Ralstonia Solanacearum Population in Soil and Enhancing Host Resistance through Fungus-derived Furoic Acid Compound." *Frontiers in Plant Science* 13 (2022): 1064797.

Yang, Yong, Ting Chen, Xiaoqiu Dai, Dong Yang, Yushuang Wu, Huilan Chen, Yixiong Zheng, Qingqing Zhi, Xiaorong Wan, and Xiaodan Tan. "Comparative Transcriptome Analysis Revealed Molecular Mechanisms of Peanut Leaves Responding to Ralstonia solanacearum and Its Type III Secretion System Mutant." Frontiers in Microbiology 13 (2022): 998817.

Yin, D. M. "A New Peanut Variety 'Nongdahua 108'with High Yield and High Resistance to Bacterial Wilt." *China Seed Industry* 11 (2020): 117–118.

Zhang, C., H. Chen, T. Cai, Y. Deng, R. Zhuang, N. Zhang, Y. Zeng, Y. Zheng, R. Tang, R. Pan, and W. Zhuang. "Overexpression of a Novel Peanut NBS-LRR Gene AhRRS5 Enhances Disease Resistance to *Ralstonia solanacearum* in Tobacco." *Plant Biotechnology Journal* 15, no. 1 (2017): 39–55. https://doi.org/10.1111/pbi.12589.

Zhang, C., H. Chen, R. R. Zhuang, Y. T. Chen, Y. Deng, T. C. Cai, S. Y. Wang, Q. Z. Liu, R. H. Tang, S. H. Shan, R. L. Pan, L. S. Chen, K. J. Dietz, and W. J. Zhuang. "Overexpression of the Peanut CLAVATA1-Like Leucine-Rich Repeat

Receptor-Like Kinase AhRLK1 Confers Increased Resistance to Bacterial Wilt in Tobacco." *Journal of Experimental Botany* 70, no. 19 (2019): 5407–5421. https://doi.org/10.1093/jxb/erz274.

Zhang, C., W. Xie, H. Fu, Y. Chen, H. Chen, T. Cai, Q. Yang, Y. Zhuang, X. Zhong, K. Chen, M. Gao, F. Liu, Y. Wan, M. K. Pandey, R. K. Varshney, and W. Zhuang. "Whole Genome Resequencing Identifies Candidate Genes and Allelic Diagnostic Markers for Resistance to *Ralstonia solanacearum* Infection in Cultivated Peanut (*Arachis hypogaea* L.)." *Frontiers in Plant Science* 13 (2023): 1048168. https://doi.org/10.3389/fpls.2022.1048168.

Zhao, C., H. Xia, T. Cao, Y. Yang, S. Zhao, L. Hou, Y. Zhang, C. Li, X. Zhang, and X. Wang. "Small RNA and Degradome Deep Sequencing Reveals Peanut MicroRNA Roles in Response to Pathogen Infection." *Plant Molecular Biology Reporter* 33, no. 4 (2014): 1013–1029. https://doi.org/10.1007/s11105-014-0806-1.

Zhao, K., Y. Li, Z. Li, Z. Cao, X. Ma, R. Ren, K. Wang, L. Meng, Y. Yao, Y. Yang, X. Wang, J. Wang, S. Hu, Y. Li, Q. Ma, D. Cao, K. Zhao, D. Qiu, F. Gong, Z. F. Li, X. Zhang, and D. Yin. "Genome-Wide Analysis of AhCN Genes Reveals the AhCN34 Involved in Bacterial Wilt Resistance in Peanut." *Journal of Integrative Agriculture* (2024a). https://doi.org/10.1016/j.jia.2024.03.006.

Zhao, K., Z. Li, Y. Ke, R. Ren, Z. Cao, Z. Li, K. Wang, et al. "Dynamic N6-Methyladenosine RNA Modification Regulates Peanut Resistance to Bacterial Wilt." *New Phytologist* 242, no. 1 (2024b): 231–246.

Zhao, K., R. Ren, X. Ma, K. Zhao, C. Qu, D. Cao, Q. Ma, Y. Ma, F. Gong, Z. Li, X. Zhang, and D. Yin. "Genome-Wide Investigation of Defensin Genes in Peanut (*Arachis hypogaea* L.) Reveals AhDef2.2 Conferring Resistance to Bacterial Wilt." *The Crop Journal* 10, no. 3 (2022): 809–819. https://doi.org/10.1016/j.cj.2021.11.002.

Zhao, Y., C. Zhang, H. Chen, M. Yuan, R. Nipper, C. S. Prakash, W. Zhuang, and G. He. "QTL Mapping for Bacterial Wilt Resistance in Peanut (*Arachis hypogaea* L.)." *Molecular Breeding* 36 (2016). https://doi.org/10.1007/s11032-015-0432-0.

Zhou, X., Y. Xia, X. Ren, Y. Chen, L. Huang, S. Huang, B. Liao, Y. Lei, L. Yan, and H. Jiang. "Construction of a SNP-Based Genetic Linkage Map in Cultivated Peanut Based on Large Scale Marker Development Using Next-Generation Double-Digest Restriction-Site-Associated DNA Sequencing (ddRADseq)." *BMC Genomics* 15, no. 1 (2014): 351. https://doi.org/10.1186/1471-2164-15-351.

18 Potential of CRISPR/Cas-Based Genome Editing for Peanut Improvement

Guohao He, Anjanasree K. Neelakandan, Sy M. Traore, Mei Yuan, Binita Subedi, Xiangyu Chen, Suman Veeramasu, Xingli Ma, and David A. Wright

18.1 CRISPR/CAS SYSTEM FOR CROP IMPROVEMENT

Historically, crops were created from foraged species when naturally occurring mutants with favorable trait qualities were recognized and retained for seed stocks. Rudimentary breeding programs eventually produced early crop varieties, while yields, along with beneficial agronomic traits, gradually improved. In time, more sophisticated breeding programs were developed, and it was realized that variation was a driving force in crop improvement. During the twentieth century, variation was introduced through mutagenesis programs or introgression of genetic material from wild relatives using conventional breeding approaches. Beginning in the 1980s, transgenic techniques introduced naturally occurring or user-designed recombinant genes randomly into a plant genome using selectable markers to identify transgenic events. The method was later refined to include synthetic genes capable of encoding and expressing sequences for valuable traits, or precisely altering patterns of gene expression at the desired locus through modified transcription factors.

Over the past two decades, the concept of genome editing has been advanced through the idea of creating variation using site specific mutation. Achieving this end required the development of novel tools. Site-specific mutation through precision editing of plant genomes was eventually accomplished using sequence-specific nucleases (SSNs) such as mega nucleases, zinc finger nucleases (ZFNs), and transcription activator–like effector nuclease (TALENs), each type of technology having certain advantages and disadvantages (Baltes and Voytas, 2015; Schiml and Puchta, 2016). Development of the type II clustered regularly interspaced short palindromic repeat (CRISPR) and CRISPR-associated protein 9 (Cas9) system as a genome editing tool greatly simplified the process while lending greater precision to a powerful technique. Awareness of this tool and its utility directly lead to the discovery of a plethora of related CRISPR/Cas systems, a variety of novel tools and techniques, numerous genetic discoveries, and multiple improvements in many crop species.

The type II CRISPR/Cas9 system consists of a single protein (Cas9) that encodes two active nucleases (HNH and RuvC), and a two-part guide RNA (gRNA) made up of trans-activating crispr RNA (tracrRNA) and crispr RNA (crRNA). The protein binds the RNAs to form a complex where the 5' end of the tracrRNA is complementary to the 3' end of the crRNA, while the 5' end of the crRNA is complementary to a specific DNA target. The complex is capable of recognizing and cleaving a specific double-stranded DNA sequence, resulting in a double-stranded break (DSB) in the target sequence, provided that the protospacer adjacent motif (PAM) sequence (NGG) is also correctly positioned adjacent to the target DNA sequence (Louwen et al. 2014; Puchta 2016). Note that in practice, the tracrRNA and crRNA are typically fused to form a single guide RNA (sgRNA) as a means of simplifying system component expression and that the targeting end of the crRNA can be modified to direct the complex to virtually any site in a genome that contains an adjacent PAM sequence.

Cells generally correct induced site-specific DSBs using one of two broadly defined DNA repair mechanisms: non-homologous end-joining (NHEJ) or homology-directed repair (HDR). In eukaryotic cells, NHEJ typically accounts for the majority of repairs used in gene knock-out function. NHEJ is error prone and commonly results in small semi-random insertion or deletion (InDel) mutations at the target site. HDR is utilized to a lesser degree by eukaryotic cells, but it can result in the introduction of template specified mutations at or near the DSB. In this scenario, a relatively large DNA template is designed, which is complementary to the target gene area and carries a single base change or specific set of mutations. For this HDR system, typically many copies of the template are supplied to a cell along with the CRISPR/Cas9 system. After the DSB is formed, the template is utilized by the HDR mechanism, resulting in the introduction of predetermined base changes in the gene of interest.

In an alternative version of genome editing, one Cas9 nuclease domain is inactivated to form a nickase (nCas9) protein that can induce a single strand break (SSB), which is not necessarily mutagenic (Fauser et al., 2014). The partially active nCas9 is fused to a deaminase such as PmCDA1, resulting in targeted deamination of cytosine, and a series of C to T transitions occur in the gene of interest without InDels associated with NHEJ (Eid et al., 2018).

DOI: 10.1201/9781003360988-18

In yet another CRISPR/Cas9 method, single amino acid substitutions are made in both nuclease domains, resulting in a dead (dCas9) protein, which is incapable of cleavage but retains DNA sequence binding specificity (Karlson et al., 2021). If dCas9 is fused to a transcription factor activator or repressor such as VP64 or SRDX respectively, then dCas9 acts as an engineered transcription factor to up- or downregulate gene expression in a targeted manner. In addition, dCas9 can be fused to an enzyme such as a methylase, demethylase, acetylase, or deacetylase to epigenetically modify gene expression (Karlson et al., 2021).

Cas9 isolated from *Streptococcus pyogenes* (SpCas9) is currently the most widely used tool for genome editing in plants. However, more than 30 Cas orthologs have now been applied in plant genome editing projects. Among the orthologs, the second most widely used in plants is the type V CRISPR/Cas12 system; particularly Cas12a (also known as Cpf1) from bacteria such as *Acidaminococcus*, *Lachnospiraceae*, and *Francisella* has been applied in many plant species (Hassan et al., 2021). Compared to Cas9 protein, Cpf1 has several advantages, including using a small guide RNA, targeting T-rich regions of the genome, RNA processing capability that facilitates simpler multiplexing, and generation of DSBs with staggered ends (Table 18.1).

In 2013, three papers reported that Cas9 could be used for genome engineering in plants (Shan et al., 2013; Li et al., 2013; Nekrasov et al., 2013). Since then, the CRISPR/Cas system has become a tremendously successful tool for genome editing, resulting in hundreds of papers. Applications include improvements in agronomic traits such as grain size, fruit color, and oil profile; changes in abiotic and biotic stress tolerance such as salinity, drought, temperature, pathogen, and insect tolerance or resistance; and increased nutritional quality such as low phytic acid or gluten and high lycopene or β-carotene. These improvements result from three broad categories of genome editing: loss of function, gain of function, and modulation of gene expression, as discussed below.

18.2 LOSS OF FUNCTION

In the CRISPR/Cas system, nuclease induced DSBs and repair by NHEJ processes generate semi-random InDel mutations that can cause frameshift of a coding sequence in a targeted gene or structural changes in a promoter or enhancer, leading to functional gene knock-out (Liu et al., 2016). CRISPR/Cas medicated loss of function studies have resulted in improved disease resistance/tolerance in plants, including resistance to viral, fungal, and bacterial pathogens.

18.2.1 Resistance against Viral Pathogens

Plant viruses are parasites that threaten many economically important crops. Viruses form two major genome types: DNA and RNA viruses. DNA viruses consist of double-stranded DNA (dsDNA) and single-stranded DNA (ssDNA), while RNA viruses have four main classes: double-stranded RNA (dsRNA), negative sense single-stranded RNA (-ssRNA), positive sense single-stranded RNA (+ssRNA), and retroviruses (Roossinck et al., 2015). Therefore, plant resistance against viruses can be achieved by either directly targeting the virus genome or targeting host susceptibility (S) genes that are crucial for the replication of viruses in the plant (Zaidi et al., 2018).

Directly targeting a viral genome using CRISPR/Cas9-mediated genome editing to enhance viral resistance was first reported for the beet severe curly top virus (BSCTV) and yellow dwarf virus (BeYDV) DNA viruses in plant models *N. benthamiana* and *A. thaliana* (Baltes et al., 2015; Ji et al., 2015). Both studies employed multiple CRISPR/Cas9 sgRNAs to target coding and non-coding viral genome regions, resulting in reduced virus accumulation and improved resistance. Similar studies showed that sgRNAs targeting the viral capsid protein (CP) coding sequence, the RCRII motif of the replication protein (Rep) coding sequence, and the intergenic region (IR) of tomato yellow leaf curl virus (TYLCV) caused a significant reduction in viral replication and accumulation (Z. Ali et al. 2015; 2016). Ali et al. (2016) observed that viruses with edited coding

TABLE 18.1
The Differences between Cas9 and Cas12a Proteins

Protein	Cas9	Cas12a (Cpf1)
PAM	NGG	TTTN*
Nuclease cut site	Proximal to PAM	Distal to PAM
DSB	Blunt ends	Staggering ends
RNA molecule	crRNA-tracrRNA	crRNA
Nuclease domain	NHN-RuvC	RuvC
Off-target	higher	lower
Editing efficiency	higher	lower

*N can be A, T, C, or G.

regions were still capable of replication, while after IR editing, the viruses were unable to replicate. Unlike SpCas9, which only recognizes dsDNA, FnCas9 from *Francisella novicida* and LwaCas13a from *Leptotrichia wadei* can be used to target RNA viruses. By targeting cucumber mosaic virus (CMV) and tobacco mosaic virus (TMV) in *N. benthamiana* and *Arabidopsis* plants expressing the FnCas9 system, CMV and TMV accumulation reduced by 40%–80% compared with control plants (Zhang et al., 2018). Targeting four different regions of the turnip mosaic virus (TuMV) by Cas13a also showed that target site choice had different effects on viral replication interference (Aman et al., 2018). However, a potential problem with viral escape was highlighted in the reports of Ali et al. (2016) and Mehta et al. (2019), where it was found that 33%–48% of virus genomes carried a conserved single-nucleotide mutation that conferred resistance to CRISPR/Cas9 cleavage. These discoveries demonstrate the need for a CRISPR/Cas9 expression system using multiplexed sgRNAs to generate multiple DSBs in the target viral genome. When multiple sgRNA target sites were used against the chilli leaf curl virus and cotton leaf curl Multan virus genomes, viral accumulation and disease symptoms were significantly reduced (Roy et al., 2019; Yin et al., 2019).

The use of CRISPR/Cas to target host S genes is an alternative approach to conferring resistance against viruses. Viruses use certain host cell factors to maintain their life cycle, and loss of these factors by genome editing results in the development of resistance. The translation initiation factors eIF4E, eIF(iso)4E, and eIF4G encoding cap-binding proteins are required by RNA viruses, in which 5'-terminal protein VPg interacts with eIF4Es to suppress immunity (Langner et al., 2018). Knock-out mutations of these factors reduce the ability of viruses to interact with host proteins and interrupt translation of the potyvirus genomes. Mutated homozygous eif4e showed immunity to cucumber vein yellowing virus (CVYV), zucchini yellowing mosaic virus (ZYMV), and papaya ring spot mosaic virus-W (PRSV-W) (Chandrasekaran et al., 2016). A mutation such as a single base insertion or deletion in eIF(iso)4E conferred complete resistance to the single-stranded RNA potyvirus (+ssRNA) TuMV in Arabidopsis (Pyott et al., 2016). eIF4G mutation alleles in rice improved resistance to rice tungro spherical virus (RTSV) and enhanced grain yield (Macovei et al., 2018). Mutation in an S gene (coilin gene) was induced by delivering CRISPR/Cas9 and sgRNAs into apical meristematic tissues of potato using biolistic bombardment, resulting in resistance to RNA based potyvirus Y (PVY) (Makhotenko et al., 2019). A similar study showed that knocking out the *Va* gene blocked the interaction between Va and VPg proteins and generated PVY resistant tobacco in the T1 generation (Ren et al., 2021).

18.2.2 Resistance against Fungal Pathogens

Diseases caused by fungal pathogens can result in dramatic yield losses in crops and have a detrimental effect on the quality of harvested products. Plants evolve resistance (R) genes to fight fungal pathogens; however, CRISPR/Cas-mediated genome editing of S genes leads to broad spectrum resistance that can confer greater resistance than R gene–conferred resistance (Schie et al., 2014). The mildew resistant locus O (*MLO*) genes encode a membrane-associated protein with seven transmembrane domains and confer susceptibility to fungi causing powdery mildew disease (Acevedo-Garcia et al., 2014). Induced mutation and homozygous *mlo* alleles significantly increase resistance to powdery mildew. Disruptive homozygous mutations of the three homoeoalleles *TaMLO-A1*, *TaMLO-B1*, and *TaMLO-D1* in hexaploid bread wheat confer resistance to powdery mildew disease caused by the pathogen *Blumeria graminis f. sp hordei (Bgh)* (Wang et al., 2014). In tomato, two sites in the *SlMLO1* gene were targeted to create a 48 bp deletion, resulting in resistance to the powdery mildew pathogen *Oidium neolycopersici* without affecting phenotypic traits and yield (Nekrasov et al. 2017). Other S genes controlling plant defense responses, such as the *EDR1* gene (enhanced disease resistance), were mutated using CRISPR/Cas9 in three homologues of *TaERD1* in wheat and showed significant reduction in blast lesions and increased resistance to powdery mildew disease (Zhang et al., 2017). Similarly, the S genes *StDMR6-1* and *StDMR6-2* were edited in potato using CRISPR/Cas9-based technology. The tetra-allelic mutants from *StDMR6-1* gene disruption generated resistance to late blight and did not show any difference in growth phenotype compared with the wild type, whereas mutations in *StDMR6-2* did not show any changes in resistance, but did show pleiotropic effects, indicating that *StDMR6-1* and *StDMR6-2* have different functions (Kieu et al., 2021).

Some transcription factors (TF), acting as negative regulators, are also considered S genes. CRISPR/Cas9 induced mutants in the ethylene responsive factor *OsERF922* gene in rice showed resistance to *Magnaporthe oryzae* without any apparent pleiotropic effects (Wang et al., 2016). Similarly, researchers used CRISPR/Cas9 to mutate the coding region of mitogen-activated protein kinase 5, enhancing resistance against M. grisea in rice (Xie and Yang, 2013). In grape, researchers used multiple CRISPR/Cas gRNAs to target VvWRKY52, a gene that plays a critical role in the plant's response to biotic stress. The resulting plants showed increased resistance to pathogen *Botrytis cinerea* (Wang et al. 2018).

18.2.3 Resistance against Bacterial Pathogens

CRISPR/Cas9 can target certain S genes involved in host–bacterial pathogen interactions to generate resistance to bacterial diseases (Talakayala et al. 2022). Well-known S gene examples of CRISPR-mediated gene editing to enhance resistance to bacterial disease are the *SWEET* genes that encode sugar transporters controlling glucose and sucrose export. Certain bacteria secrete transcription activator-like effectors (TALEs) into plant host cells to trigger *SWEET* gene expression and provide sugar as a nutrient source to enhance infection. Disrupting the interaction between bacterial TALE and plant *SWEET* genes suppresses pathogen growth and increases resistance in the plant. For instance, researchers edited OsSWEET13 and OsSWEET14 in rice using CRISPR/Cas9, resulting in resistance to bacterial blight (Zhou et al., 2015; Xu et al., 2019; Zeng et al., 2020). Likewise, mutations of *SWEET* gene promoters disrupted the host–bacteria interaction and resulted in rice lines with robust, broad-spectrum bacterial blight resistance (Oliva et al., 2019). Moreover, interfering with the linkage between the effector PthA4 and effector-binding element in the promoter of lateral organ boundaries 1 (*CsLOB1*), as well as a mutation in the coding region of *CsLOB1* by CRISPR/Cas9, resulted in resistance to citrus canker in orange and grape (Jia et al., 2017; Peng et al., 2017).

Similarly, *Pseudomonas syringae pv. tomato* secretes coronatine (COR), which is perceived by co-receptor JAZ2, stimulates stomata opening and facilitates bacterial invasion of the plant, resulting in bacterial speck disease. Mutation introduced in the *SlJAZ2* gene by CRISPR/Cas provides resistance against bacterial speck disease in tomato (Ortigosa et al., 2019). Another S gene, downy mildew resistance 6 (*DMR6*), encodes 2-oxoglutarate Fe (II)-dependent oxygenase (2OGO). During pathogen infection, *DMR6* is upregulated and suppresses plant immunity. The mutant *dmr6* conferred broad-spectrum resistance against multiple pathogens including *P. syringae*, *Phytophthora capsici*, and *Xanthomonas spp.* (Thomazella et al., 2021). Knock-out of one *DMR6* orthologue in banana showed enhanced resistance to banana Xanthomonas wilt caused by *Xanthomonas campestris* pv. *musacearum* and no detrimental impact on plant growth (Tripathi et al., 2021). Inducing S gene mutants by genome editing appears to be a promising approach for creating pathogen resistance. Therefore, identification and characterization of more S genes in different plant species will provide targets for CRISPR/Cas-mediate genome editing and a platform for sustainable disease-resistant breeding.

The examples above illustrate the utility of CRISPR/Cas to induce loss of gene function for pathogen resistance; however, loss of function also generates traits for adaptation to drought (Shi et al., 2017), modified ripening profiles (Ito et al., 2015), male sterility, and factors impacting yield (J. Li et al., 2016; Xu et al., 2016).

18.3 GAIN OF FUNCTION

The CRISPR/Cas system can be used to generate gain-of-function mutations by insertion and/or replacement of desired sequences in a gene of interest. Knock-in donor templates achieve this type of genomic modification through either HDR or NHEJ mechanisms of DSB repair. Because knock-in of a donor DNA by HDR is much more challenging to perform in higher plants (Collonnier et al., 2017), there are few reports demonstrating the insertion of genes by HDR. Shi et al. (2017) applied CRISPR/Cas to insert the moderate constitutive maize promoter GOS2 into the 5'-untranslated region of the native *ARGOS8* gene of maize by HDR. Since maize *ARGOS8* is a negative regulator of ethylene responses, increasing its expression can reduce ethylene sensitivity. HDR-generated variants of *ARGOS8* significantly increased grain yield under drought-stress conditions in the field. In another example, using the more efficient NHEJ repair pathway, a DNA fragment was inserted into the intron of the endogenous rice gene *EPSPS*. This insertion altered the amino-acid sequence of the gene product and resulted in herbicide tolerance (M. Li et al., 2016). Again, using NHEJ, a 5.2 kb carotenoid biosynthesis cassette was inserted at two independent genomic target sites, and the obtained rice plants produced seed with high carotenoid content and no yield penalty (Dong et al., 2020). To improve the insertion efficiency by NHEJ, Lu et al. (2020) used chemically modified donor DNA with phosphorylation and phosphorothioate linkages to make insertions into enhancers and promoters of the rice genome. The average efficiency of this technique increased NHEJ by up to 25%. They also reported a method of tandem repeat-HDR (TR-HDR) for precise sequence insertion or replacement by HDR using chemically modified donor DNA and increased HDR efficiency to 6.1%.

Notably, the targeted insertion of a desired sequence into a plant genome may affect the expression of other endogenous genes, especially when the insertion site is within a gene, which may interfere with plant performance (Dong and Ronald, 2021). Insertion of DNA sequences at previously characterized sites, known as genomic safe harbors, can increase the chance of creating desired traits without negative impact (Anand and Jones, 2018). Potential genomic safe harbors can be identified by measuring the expression of CRISPR/Cas-mediated insertion of genes and evaluating plant performance using high-throughput phenomic approaches (Yang et al. 2020). As methods and approaches for targeted gene insertion continue to be improved, effective insertion and replacement of desired genes will potentially become routine practices that benefit plant breeding in the future.

18.4 MODULATION OF GENE EXPRESSION

Gene expression is a multistep process involving cellular controlled production of DNA, RNA, and protein to create a regulatory network. One aspect of this network is transcriptional control of gene expression. The CRISPR/Cas system can be used as a gene regulation tool to modulate transcriptional processes (Karlson et al., 2021). A variant of the Cas9 protein that lacks nuclease activity, dCas9 linked to an effector domain can act as an artificial transcription factor to activate or repress gene expression or modify epigenetic marks. The development of dCas9 transcription factors involved the fusion of dCas9 to multiple activation domains to enhance gene expression. Initially, dCas9 was fused to four copies of the VP16 transcription activator (VP64) or a single copy of the p65 activation domain, resulting in the constructs dCas9-VP64 and dCas9-p65, respectively, which successfully activated GFP reporter gene expression (Gilbert et al., 2013). However, when similar activators like VP64, EDLL (a motif consisting of 24 conserved amino acids), and transcriptional activator–like effector (TALE) were directly fused to dCas9 to activate gene expression in plants, the activation efficiency was found to be lower than expected. In contrast, fusion of the superman repression domain X (SRDX) to dCas9 or dCpf1 effectively suppressed transcription when targeted to specific gene promoters (Piatek et al., 2015; Lowder et al., 2015). To overcome the limitations in activation efficiency, a more potent construct, dCas9-TV, was developed by fusing dCas9 to six copies of the TALE effector and two copies of VP64. This enhanced construct significantly increased target gene expression, achieving upregulation levels from 1.6- to 92-fold compared to the marginal upregulation observed with dCas9-VP64 alone (Li et al., 2017). Comparative studies of dCas9-TV with dCpf1-TV found that dCas9-TV was more suitable for transcriptional activation of target genes than dCpf1-TV. In addition to direct fusion to dCas9, transcriptional effectors, such as VP64, can be fused to the phage coat protein MS2 and then linked to a modified sgRNA containing two RNA hairpin aptamers. Therefore, multiple copies of VP64 can be recruited to a single dCas9 for improved transcriptional activity of a target gene via protein–RNA interaction rather than by direct fusion to dCas9 (Lee et al., 2019).

The success of gene regulation depends on choosing appropriate transcriptional regulators and specific target sites (Moradpour and Abdulah, 2020). To increase the efficiency of CRISPR/dCas9-mediated transcriptional regulation, the distance between the sgRNA target site and transcription start site (TSS) is vital (Karlson et al., 2021). Recruitment of transcriptional modulators close to the TSS in the promoter region effectively regulates the expression of the desired downstream genes. Gong et al. (2020) revealed that sgRNAs within the 350 bp upstream region of the TSS are the most effective for dCas9-TV to activate the transcription of two genes in rice. For activation, sgRNA is often designed to target -400 to -50 bp upstream of the TSS, while -50 to +300 bp at the TSS is commonly used for repression (Gilbert et al., 2013). Generally, dCas-activators are targeted 5' of the TATA box in the promoter while dCas-repressors are targeted 3' of the principal TATA box or a downstream TATA box, thus blocking polymerase access to a specific TATA box.

Manipulation of gene expression to generate a desirable trait in crops depends not only on the transcriptional level but also the translational efficiency of specific mRNAs (Si et al., 2020). Many factors in the 5' untranslated region (5'UTR) can regulate translation (Liang et al., 2016). The upstream open reading frames (uORFs), a class of *cis*-acting elements, with its initiation codon within the 5' UTR of a transcript and termination codon positioned upstream or downstream of the primary ORF (pORF), regulate translation initiation (Silva et al., 2017). Therefore, mutations in 5' UTRs can create or disrupt uORFs and affect pORF translation (Liang et al., 2016). Xu et al. (2017) used two uORFs within the 5' UTR of the *TBF1* promoter to engineer plant disease resistance without fitness costs. TBF1 is an important transcription factor for the growth-to-defense switch upon immune induction, where translation is suppressed by two uORFs. Si et al. (2020) developed CRISPR/Cas9-medicated genome editing tools to manipulate gene translation in plants by designing sgRNAs to destroy the start codon or the sequence around uORF of a target gene promoter. The *uorf* mutants can potentially relieve the inhibition of pORF translation and provide opportunities for fine-tuning a biological trait through altered gene expression.

18.5 APPLICATION OF CRISPR/CAS SYSTEM FOR EFFICIENT GENE EDITING IN PEANUT

CRISPR/Cas-mediated genome editing tools have been applied in a wide range of crops for fundamental research and genetic improvement. However, there are only a few applications of genome editing techniques in peanut. To exploit potential of CRISPR/Cas-mediated genome editing technology for peanut improvement, we selected the *fatty acid desaturase 2* (*FAD2*) genes as targets because of their simplicity, known mutant effects, trait value, and scientific interest. We then designed several constructs for different needs in peanut genome editing applications, including knock-out, base editing, and regulatory editor for manipulation of gene expression.

18.5.1 Construction of Vectors

A total of 15 constructs have been developed with different components of the CRISPR/Cas system that are potentially useful in peanut genome editing in our laboratory. Two constructs (pDW3872 and pDW3877) with an extended sgRNA scaffold or standard sgRNA scaffold, respectively, were generated to compare the effect of different scaffolds on the editing efficiency in peanut. Dang et al. (2015) demonstrated that the efficiency of mutation was enhanced by extending the duplex of the sgRNA scaffold by 5 bp combined with mutating the continuous sequence of Ts at position 4 (T>C) after the guide sequence. The construct pDW3872 begins with the right border sequence and includes a pVs1 replicon with a destroyed BsaI restriction enzyme site, followed by a ColEI replicon. It also contains an NptI kanamycin resistance gene and a spectinomycin resistance gene for bacterial selection. The left border sequence is followed by a NOS terminator and a Bar resistance gene for in planta selection, driven by a 4x 35s promoter. Additionally, the construct includes an Arabidopsis AtU6-26 promoter to drive sgRNA expression, with dual BsaI restriction sites for guide RNA primer pair insertion and an extended sgRNA scaffold ending with a U6 terminator. The modified

Arabidopsis Ubiquitin AtUBI promoter drives the expression of a dicot codon-optimized Cas9 coding sequence tagged with a 3xFLAG and containing an SV40 NLS signal on both the N-terminal and C-terminal ends. The construct includes a BamHI restriction site and ends with a CaMV 35s terminator. pDW3877 has the same components in the same order as pDW3872, except a standard sgRNA scaffold replaces the extended sgRNA scaffold (Figure 18.1). For these constructs, the guide portion of the sgRNA is inserted as phosphorylated and annealed primer pairs designed with single-stranded end sequences that match the *BsaI* digested vector ends.

To design guide RNAs specific to the coding and promoter regions of FAD2 genes, researchers utilized the Cas-Designer tool (Park and Kim, 2022). The design considerations included targeting seed motif regions, which are 8–12 nucleotides proximal to the PAM sequence for Cas9 nucleases and 5–6 nucleotides for Cpf1 nucleases, as these regions are critical for specificity (Liu et al., 2016; Hassan et al., 2021). This strategy was particularly important due to sequence differences between the promoters in the FAD2A and FAD2B genes. To reduce potential off-target effects, the team also designed a 17-bp truncated guide RNA and a longer guide RNA with two additional guanine residues at the 5' end (Fu et al., 2014; Cho et al., 2014).

The cloning strategy for single and dual guide RNA constructs employed a one-step Golden Gate ligation process. For the dual guide RNA constructs targeting both the RY and 2S motifs simultaneously, researchers introduced a sacB gene into the BsaI sites of the Cas9 vectors, generating pDW3936 and pDW3937, respectively. Researchers created supplementary plasmids, designated as pDW3898 and pDW3899, containing a BsaI site, an extended or standard gRNA scaffold, an Arabidopsis U6 promoter, and a second BsaI site. Researchers designed these supplementary plasmids to generate unique overhangs, preventing them from annealing to the Cas9 vector without an intervening oligo pair. The ligation process involved digesting the Cas9 plasmids and supplementary plasmids with BsaI enzyme, followed by annealing the phosphorylated oligo pairs and ligating them into the vectors. The ligation products were then transformed into DH10B E. coli cells. Researchers identified successful dual oligo ligations on LB medium supplemented with kanamycin and sucrose, designating the final dual gRNA constructs as pDW3954 and pDW3955. These constructs were sequence-verified and subsequently mobilized into *Agrobacterium rhizogenes* strain K599 by electroporation for further studies (Neelakandan et al., 2022a, 2022b) (Figure 18.2).

The CRISPR/Cpf1 system possesses unique features such as a single short crRNA that cleaves DNA without an additional trans-activating crRNA (tracrRNA) as used in Cas9, targets T-rich regions of the genome, cleaves distal to the PAM, and generates staggered ends when cleaving target sites. We have generated two constructs (pDW3882 and pDW3886) with two Cpf1 orthologs, *Lachnospiraceae bacterium* ND2006 (LbCpf1) and *Acidaminococcus sp.* BV3L6 (AsCpf1), respectively, to determine the effect of different Cpf1 variants on editing efficiency in peanut.

Moreover, to test potential regulation of gene expression in peanut, the repressor 3xSRDX was fused to dCas9 with the extended scaffold or standard scaffold to generate constructs pDW3874 and pDW3879 or fused to dLbCpf1 resulting in pDW3883. These three constructs target the *FAD2* gene promoters to test repression of gene expression. Similarly, the TV activator was fused to dCas9 with the extended scaffold or standard scaffold to generate pDW3875 and pDW3880 or dLbCpf1 producing pDW3884 and used to test activation of gene expression. To induce point mutations in the coding region or promoter sequence of *FAD2* genes, the sea lamprey CDA (PmCDA1) or rat CDA (rAPOBEC1) coding sequences were fused to nCas9 with different sgRNA scaffolds to assemble pDW3873, pDW3876, pDW3878, and pDW3881, respectively. Rat CDA was also fused to dLbCpf1 to generate construct pDW3885. The construct maps are showed in Figure 18.3.

18.5.2 Introduction of Mutations in the *FAD2* Gene

Studies of natural mutations occurring in the *FAD2* gene have revealed that InDel mutations causing amino acid changes or a frameshift that generates a premature stop codon can reduce the conversion of oleic acid to linoleic acid and thus increase the content of oleic acid in peanut seeds (Jung et al., 2000; Lopez et al., 2000; Bruner at al., 2001). High levels of oleic acid benefits industry and consumers due to the resulting long shelf storage and nutritional values (Zahran and Tawfeuk, 2019). In peanut, we first used the p201B-Cas9 binary vector (Jacobs and Martin, 2016) and a designed gRNA to target the positions of naturally

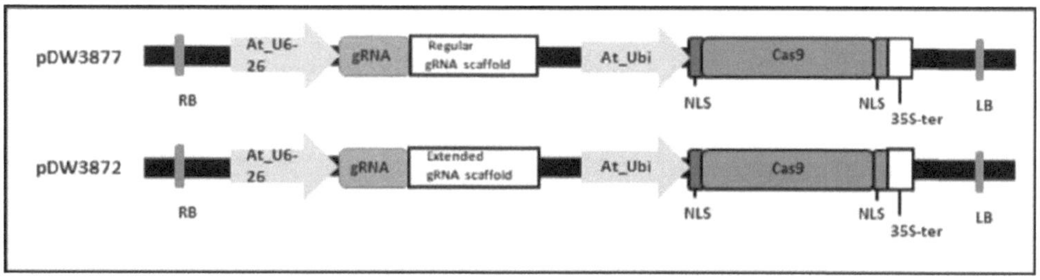

FIGURE 18.1 Schematic representation of the CRISPR/Cas9 binary vector's structures.

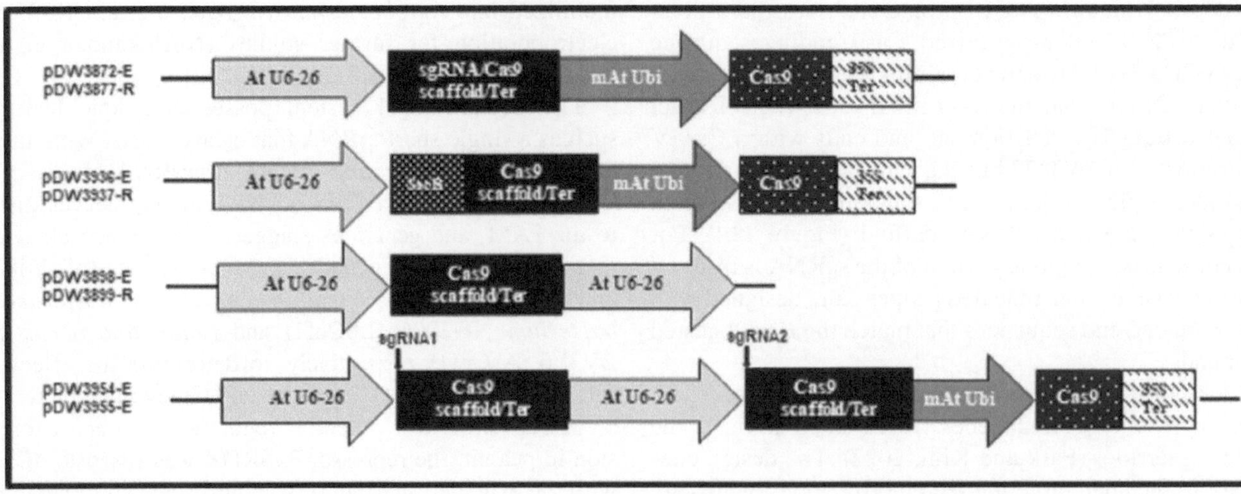

FIGURE 18.2 Schematic representation of construction of vector with dual gRNAs.

occurring mutations of both homoeoalleles of the *FAD2* genes to test if CRISPR/Cas9 would induce known mutations in a variety with wild type *FAD2* genes. The project generated the same mutations as found in natural mutants and one additional point mutation induced in the coding region of the *FAD2* genes (Yuan et al., 2019). However, this construct did not induce deletion mutations as typically generated by CRISPR/Cas-mediated genome editing.

To increase the editing efficiency, we developed optimized constructs as described above and used these constructs to target *FAD2* genes. Using constructs pDW3872 and pDW3877, with different sgRNA scaffolds, more than 90% of detected mutations were deletions ranging from one nucleotide to 57 nucleotides in the coding region of both *FAD2* homoeoalleles (Neelakandan et al., 2022a). Although both constructs are suitable for gene editing in peanut, the editing efficiency for pDW3872 was 32% using the extended scaffold, but only 24% for pDW3877 using the typical scaffold. These constructs were also used to disrupt the combination of *cis*-regulatory elements in the *FAD2* promoters that are thought to be bound by transcription factors. The RY repeat element (CATGCATG) and 2S seed protein motif (CAAACAC) are two crucial *cis*-regulatory elements for regulating expression activity in a number of seed-specific promoters (Reidt et al., 2000; Chatthai et al., 2004). The objective of knocking out the RY and 2S elements in the 5' UTR of *FAD2* genes was for functional characterization of these elements and to potentially generate seeds with increased oleic acid content without affecting the fatty acid composition in other plant tissues. The single sgRNA constructs targeting the distal RY motif resulted in indels from +2 to -33 bp and the standard sgRNA scaffold performed better than the extended gRNA scaffold. The number of mutations detected at the proximal 2S motif was lower than that observed at the RY element and, in contrast, the extended scaffold performed better than the standard sgRNA scaffold. When dual sgRNAs were used to target RY and 2S simultaneously, the use of two gRNAs in the construct had little effect on the editing efficiency. Overall editing efficiency at RY was greater than at 2S regardless of sgRNA order in the constructs, which was similar to the results of the single sgRNA studies. However, the duel sgRNA construct did generate a single instance of a large 1,140 bp deletion between the RY and 2S motifs (Neelakandan et al., 2022b).

Base editor was used in plants to generate loss-of-function mutations by targeting specific motifs or amino acid coding sequences to inactivate or modify enzyme catalytic domains, eliminate start codons (ATG) to abolish translation, alter splicing by disruption of intron/exon junctions, and change codons such as CGA (Arginine), CAA or CAG (glutamine) and TGG (tryptophan), leading to the introduction of premature stop codons (Rees et al., 2018). To establish an efficient base editing system in peanut, two constructs, pDW3873 and pDW3876, with nCas9 fused to different deaminases were tested as tools to induce point mutations in the promoter and the coding sequences of peanut *FAD2* genes. Both constructs harbor the single nuclease null variant nCas9 D10A, to which the PmCDA1 cytosine deaminase was fused to the C-terminal (pDW3873) while rAPOBEC1 deaminase and an uracil glycosylase inhibitor (UGI) were fused to the N-terminal and the C-terminal respectively (pDW3876). Three gRNAs were cloned independently into both constructs, and the functionality and efficiency were tested at three target sites, RY and 2S in the promoter and one site in the coding region of *FAD2* genes. Both constructs displayed base editing activity in which cytosine was replaced by thymine or other bases in the targeted editing window. The construct pDW3873 containing PmCDA1 showed higher efficiency compared to pDW3876 with rAPOBEC1, suggesting that the former is a better base editor in peanut (Neelakandan et al., 2022c).

CRISPR/Cpf1-mediated genome editing has been widely utilized in plants due to its versatility and simplified features (Bandyopadhyay et al., 2020). Among the known 16 different types of Cpf1-family proteins, two Cpf1 orthologs,

A

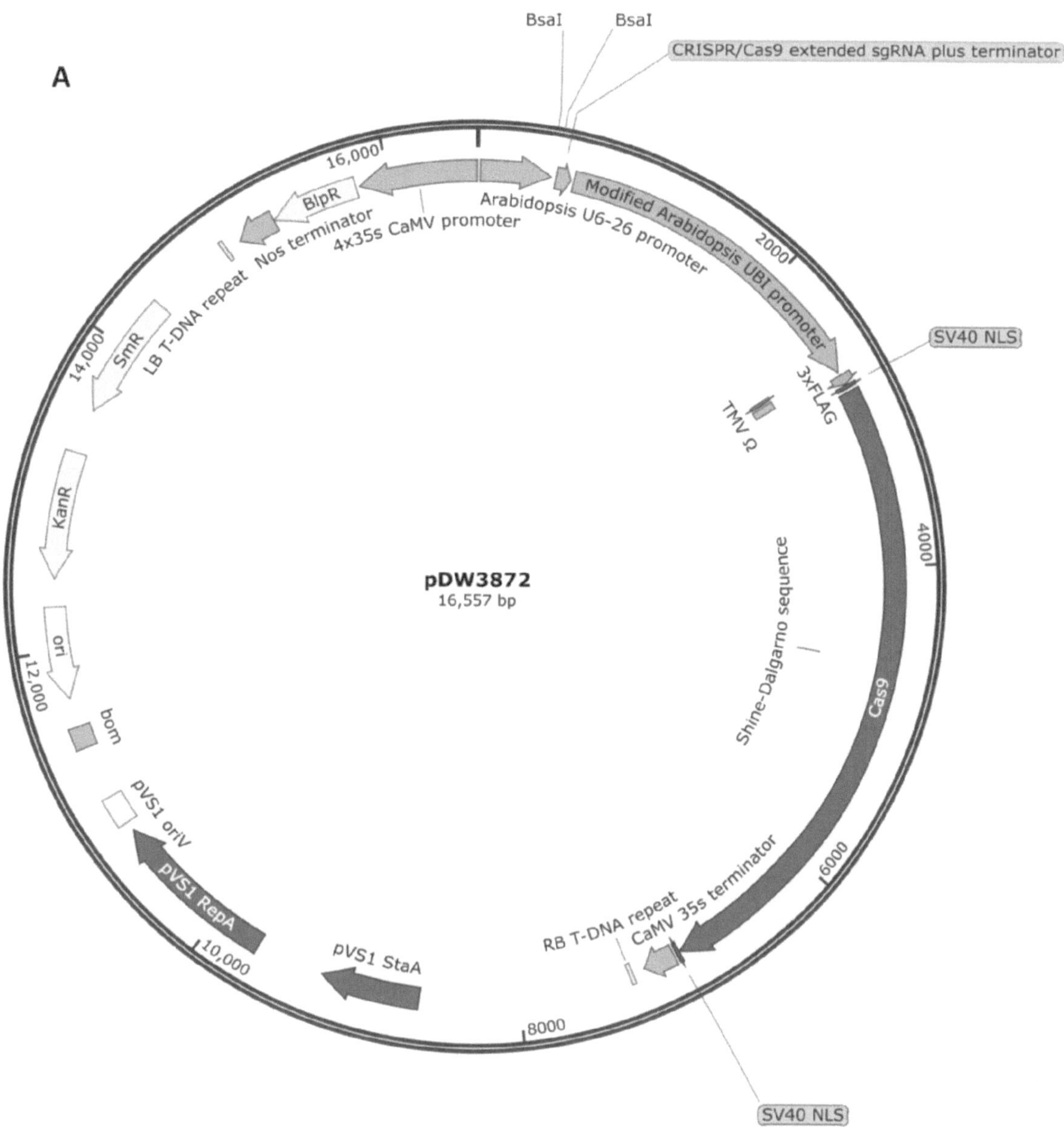

FIGURE 18.3 List of selected construct maps. A. pDW3872-Cas9 with extended scaffold; B. pDW3882-LbCpf1 with modified scaffold; C. pDW3874-dCas9 fused by 3xSRDX; D. pDW3875-dCas9 fused by VP128; E. pDW3873-nCas9 fused by PmCDA1; F. pDW3876-nCas9 fused by rCDA.

AsCpf1 and LbCpf1, demonstrated genome editing activity in plants with varied efficiency (Li et al., 2018). Our preliminary data indicated that the construct pDW3882/LbCpf1 showed higher editing activity compared to the construct pDW3886/AsCpf1 in peanut.

18.5.3 Delivery of CRISPR Components into Peanut Plants

Delivery of exogenous DNAs into plant cells can be achieved by *Agrobacterium*-mediated and biolistic transformation. *Agrobacterium*-medicated transformation is a popular system for delivering genes into peanut cells because of the susceptibility of peanut tissues to infection by a variety of wild type strains of *Agrobacterium tumefaciens* (Lacorte et al., 1991). Transformation in peanut using a disarmed *Agrobacterium* strain was first reported in 1994 (Eapen and George, 1994). However, the success of *Agrobacterium*-medicated transformation in peanut is dependent on optimization of the protocol to include genotypes, explants, and *Agrobacterium* strains used and the conditions of co-cultivation. Despite the fact that significant improvement in transformation protocols has been achieved and genes of interest were delivered into the peanut genome via *Agrobacterium*-medicated transformation (Krishna et al. 2015), only a few peanut genotypes are

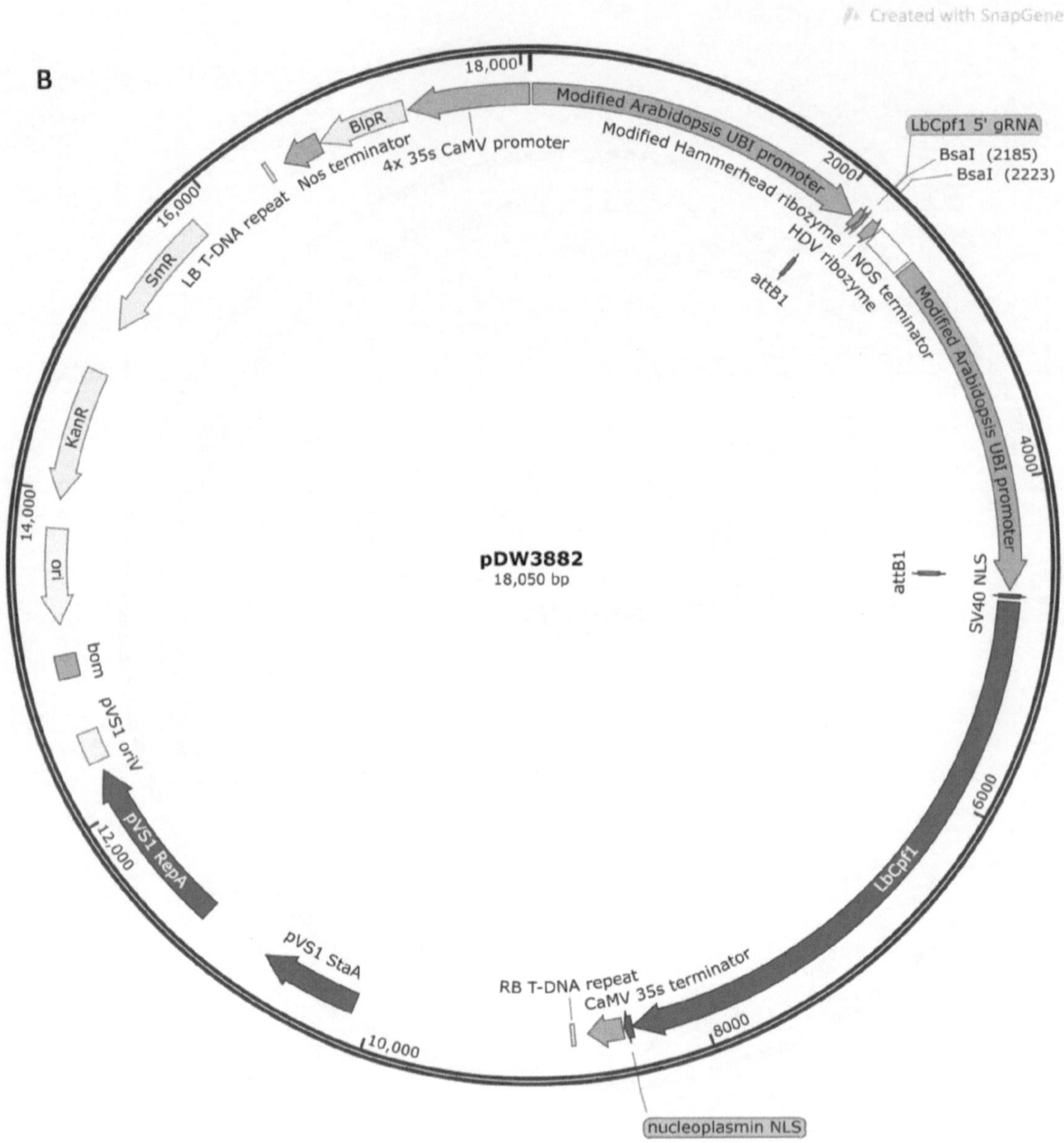

FIGURE 18.3 (Continued)

known to have a high transformation efficiency (Chen et al., 2015). Among these parameters, the major limitation in peanut transformation is the tissue culture procedure for regeneration of adventitious shoots from explants, which is time-consuming, recalcitrant to regeneration, and genotype-dependent (Higgins and Dietzgen, 2000; Umesh and Rao, 2014). Developing an alternative that breaks the tissue culture bottleneck in the transformation system would be considered as advantageous.

In planta transformation methods are alternative approaches that eliminate the need for tissue culture and regeneration steps, and thus are promising techniques (Rohini and Sankara Rao, 2000; Zhang et al., 2005; Kaur and Devi, 2019). An *in planta* transformation procedure using floral parts including floral dip, pollen tube pathway, and direct pollen transformation have been developed depending on the flower structure of specific crop species, all of which are relatively simple, cost-effective, and efficient methods (Clough and Bent, 1998; Wang et al. 2001; Song et al. 2006; Eapen, 2011; Hu et al., 2019). Methods that do not require tissue culture procedures are preferred in crop species that are difficult to regenerate through *in vitro* culture (Song et al., 2006). The advantage of these methods is that transformed seeds can be directly obtained after maturation, though transformation can only be applied in the flower blooming stage of plant development. On the other hand, without tissue culture and regeneration, *in planta* transformation methods do not require skilled workers and relatively expensive laboratory facilities. These methods have been successfully applied to several crop species including maize, melon, onion, papaya, walnut, watermelon, and wheat (Ali et al., 2015; Yang et al., 2017).

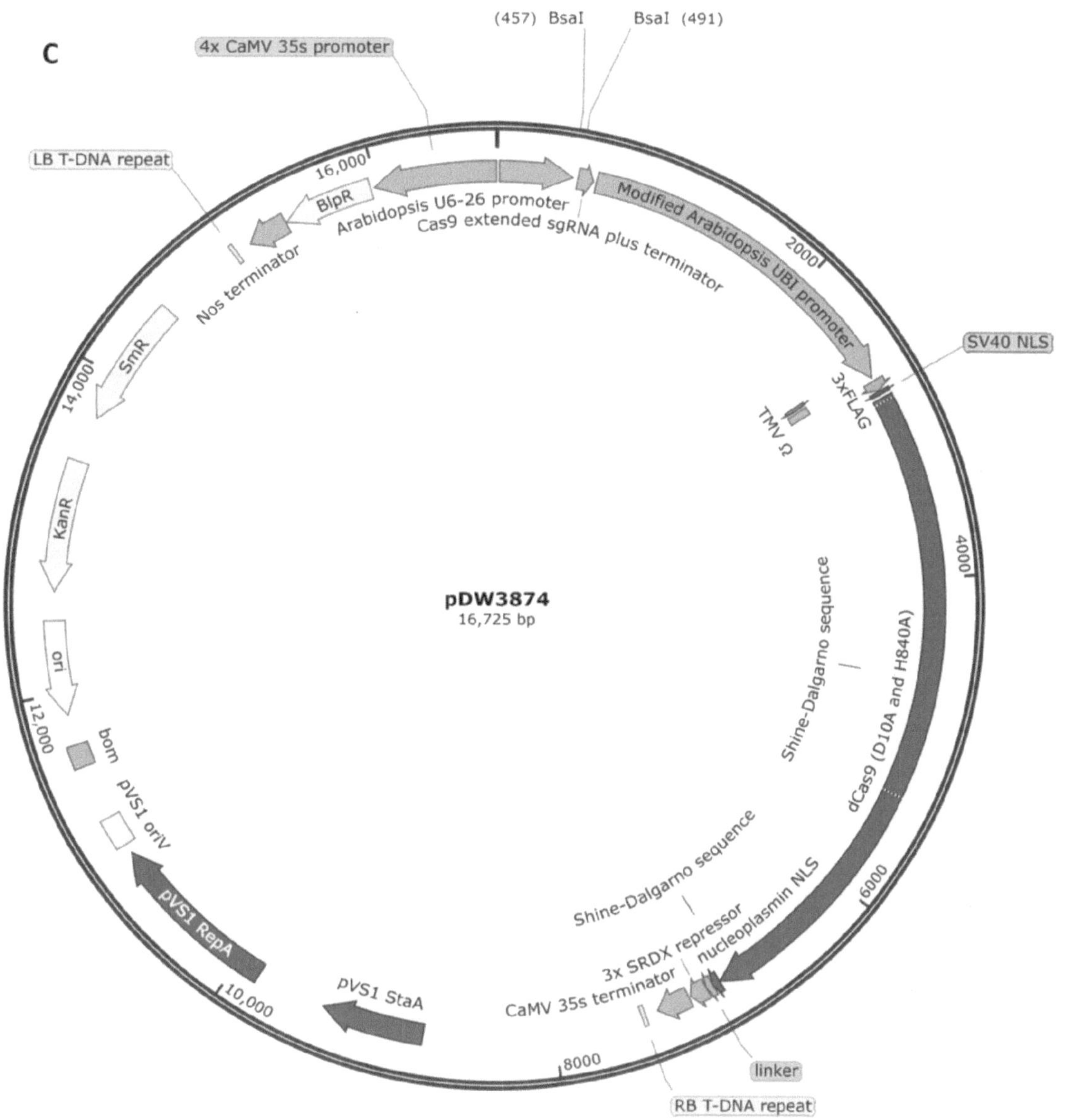

FIGURE 18.3 (Continued)

In peanut, transformation efficiency was improved through a pollen tube approach for gene transfer (Pan et al., 2020).

To exploit the potential of CRISPR/Cas components delivery into peanut cells, we used an *in planta* transformation method called calyx tube (hypanthium) injection-based transformation. Previously validated pDW3872 and pDW3877 constructs were used to target the coding region and RY and 2S motifs in the promoter of *FAD2* genes, and were individually mobilized into *Agrobacterium tumefaciens* strain GV3101. *Agrobacterium* resuspensions (0.6–0.8 of OD_{600}) were injected into the middle of the calyx tube with an insulin syringe between 9 am and 10 am every morning during the flower blossoming season. A total of 2,521 peanut flowers from three genotypes were injected with *Agrobacterium*, and 363 T_0 seeds were harvested. Results showed that 28.9% (105/363) of harvested seeds showed increased oleic acid content, indicating that these 105 seeds contained *FAD2* mutations, although the rate of oleic acid increase in the three genotypes was different. Among edited seeds, 85.7% (90/105) of seeds showed a slight increase in oleic acid content between 55% and 60%, 12.4% (13/105) showed a significant increase of 61%–65%, and 1.9% (2/105) showed a very significant increase in the level of oleic acid at 66%–70%. None of the seeds reached 80% oleic acid levels as seen in naturally occurring mutants. However, amplifications of T_1 plant DNAs showed no mutations in the targets of the *FAD2* genes, and all seeds harvested from T_1 plants derived from these T_0 edited seeds showed the same oleic acid content as wild types. This result might indicate that the CRISPR components were delivered into somatic cells rather than germ cells in T0 seeds using the calyx tube injection method. A further experiment is

D

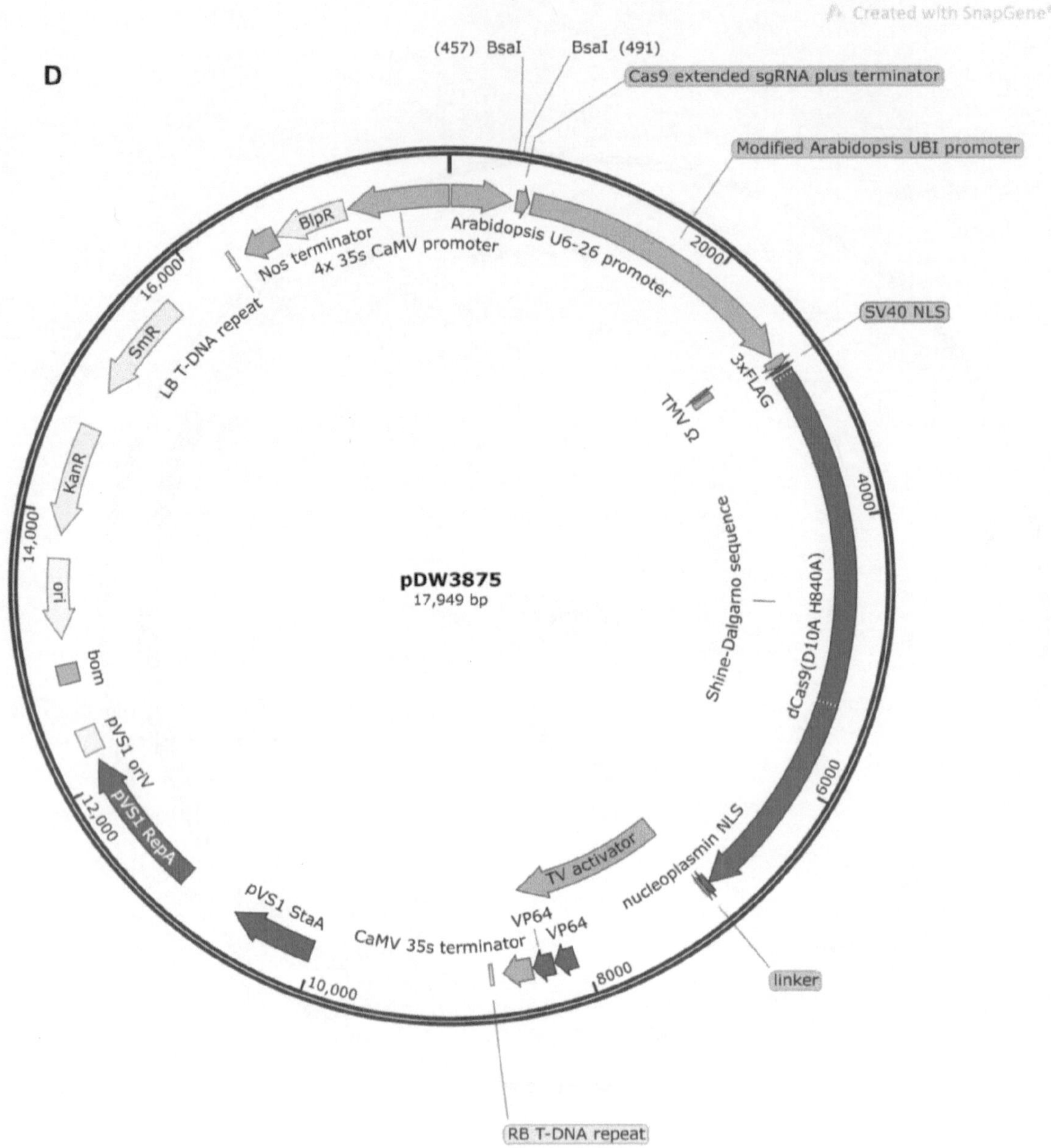

FIGURE 18.3 (Continued)

ongoing to optimize the parameter for inoculation media to determine a suitable time and location of injection to improve delivery of CRISPR/Cas components into reproductive cells with the calyx tube injection method.

Recently, nanoparticle-mediated transformation has shown promise as a tool to deliver CRISPR/Cas components into plant cells. Genetic cargo delivery to plants is a challenge due to the physical barrier of the cell wall; however, the small size of nanoparticles and their highly tunable chemical and physical properties have enabled nanoparticles to bypass biological barriers (Cunningham et al., 2018). Some nanoparticles, such as mesoporous silica nanoparticle (MSN), silicon carbide whiskers (SCW), and carbon nanotube (CNT), have demonstrated high efficiency and species-independent delivery of functional biological cargoes (Chang et al., 2013; Demirer et al., 2019). The benefit from using nanoparticle-mediated transformation is that nanoparticles protect the genetic cargo from cellular enzymatic degradation (Ahmar et al., 2021). However, the uptake of certain nanoparticles by plants can pose risks, including phytotoxic effects caused by vascular blockage, oxidative stress, or structural damage to DNA (Pachapur et al. 2016; Tripathi et al. 2017). In contrast, the use of a carbon nanotube (CNT)-based delivery platform has been shown to avoid such issues, with no toxicity or tissue damage observed in mature plants (Demirer et al., 2019). Since plant nanobiotechnology is still in the early stages of development, efforts are needed to optimize the dose, the delivery method, and the nanoparticle type for efficient nanoparticle-mediated transformation in plants (Ahmar et al., 2021).

E

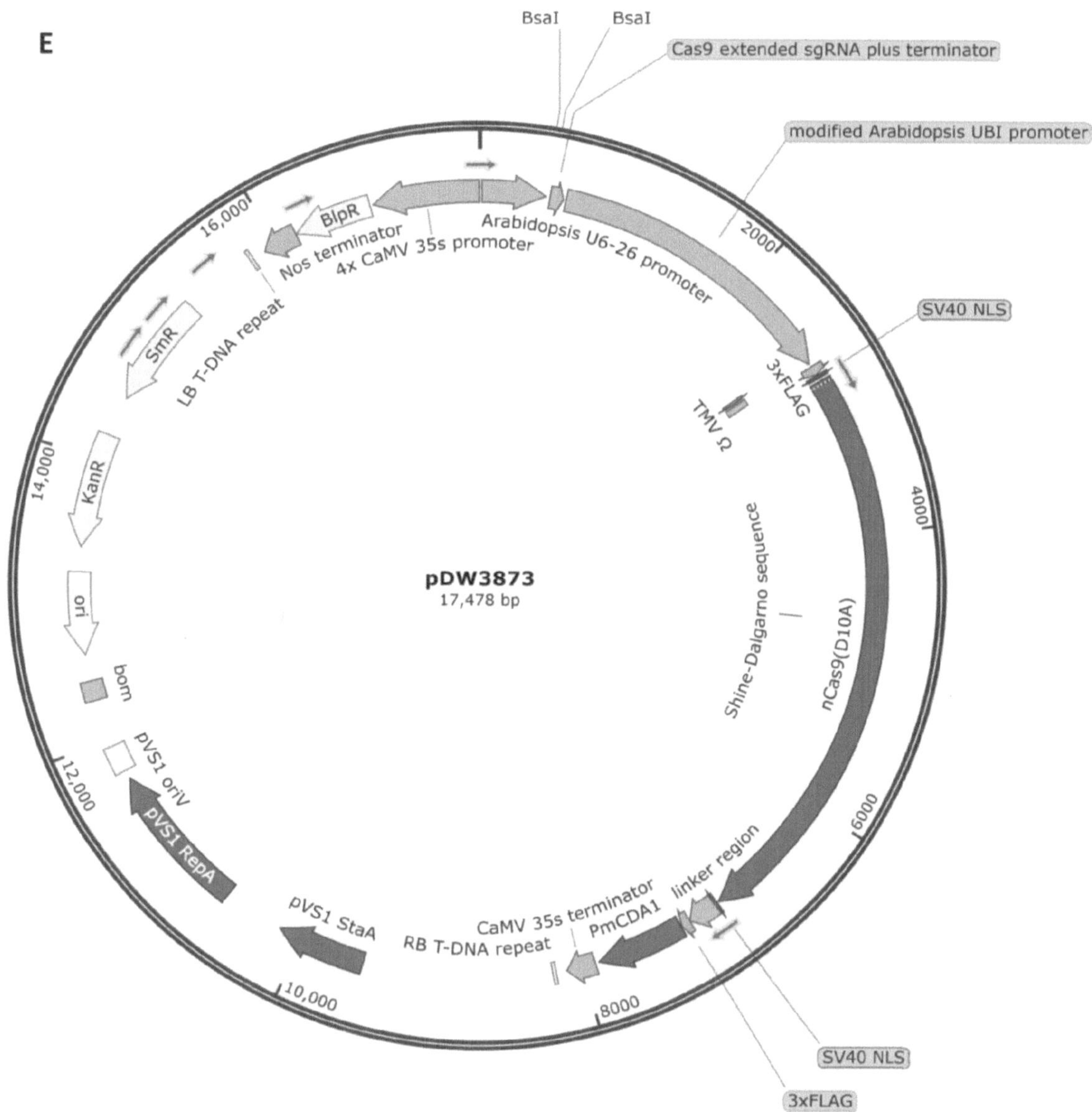

FIGURE 18.3 (Continued)

18.6 CONCLUSIONS AND PROSPECTS

Crop improvement through hybridization and genetic recombination has generated many cultivars with increased yield, higher nutritional value, and enhanced resistance to abiotic and biotic stress. With the application of genomics and genome editing tools, plant breeding becomes more efficient for selection of desirable varieties. CRISPR/Cas-mediated genome editing technology stands out as one of the most important tools due to its simplicity, versatility, and efficiency. This technology is not only effective for fine-tuning gene expression but also has broad applications across various plant species. Despite its potential, the application of genome editing in peanuts lags behind other crops. The challenges include the peanut's allotetraploid genome, the limited number of well-characterized genes, and its relatively large genome. However, the primary obstacle is the species' transformation recalcitrance, which has slowed the adoption of genome editing in peanuts compared to other crops. We have developed CRISPR-based genome editing tools that demonstrate the potential of genome editing for peanut improvement. The toolbox includes different CRISPR/Cas constructs for conventional genome editing and a variety of fusions for base editing and artificial transcription factors designed for knock-out, knock-in, and gene expression manipulation in peanut. Identification of QTLs or genes related to desirable agronomic traits by QTL mapping, RNA-seq, and many other linked techniques could provide genetic resources and more targets for genome editing in peanut. Developing additional novel delivery methods of CRISPR/Cas components into peanut cells is a desirable technology because it is still a major obstacle

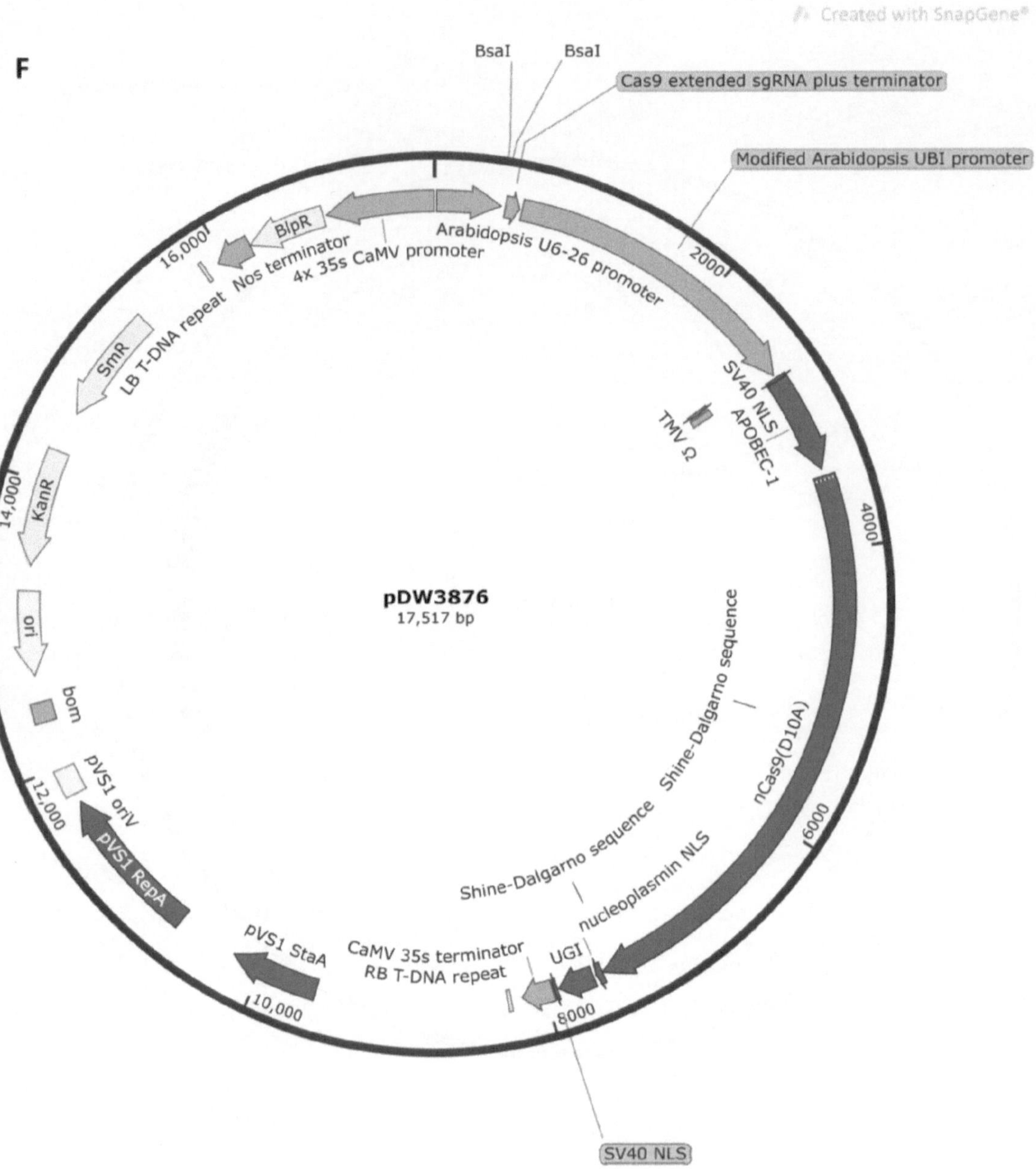

FIGURE 18.3 (Continued)

in peanut genome editing. Once researchers edit the peanut genome using CRISPR/Cas tools, they can easily obtain transgene-free homozygous peanuts through segregation in subsequent generations. Overall, the simplicity and robustness of genome editing approaches provide new avenues for peanut research and breeding.

18.8 ACKNOWLEDGMENTS

The authors are grateful for financial support from USDA/NIFA (Grant No: 2018-67014-27572 and 2018-38821-27758).

REFERENCES

Acevedo-Garcia, J., S. Kusch, and R. Panstruga. Magical mystery tour: MLO proteins in plant immunity and beyond. *The New Phytologist* 204 (2014): 273–281.

Ahmar, S., T. Mahmood, F. Mora-Fiaz, M. S. Shafique, M. S. Chattha, and K. H. Jung. Advantage of nanotechnology-based genome editing system and its application in crop improvement. *Front. Plant Sci.* 12 (2021): 663849.

Cho, S. W., S. Kim, Y. Kim, J. Kweon, H. S. Kim, S. Bae, and J. S. Kim. Analysis of off-target efects of CRISPR/Cas-derived RNA-guided endonucleases and nickases. *Genom Res.* 24 (2014): 132–141.

Ali, Z., A. Abulfaraj, A. Idris, S. Ali, M. Tashkandi, and M. M. Mahfouz. CRISPR/Cas9-mediated viral interference in plants. *Genome Biol.* 16 (2015): 238.

Ali, Z., S. Ali, M. Tashkandi, S. Shan, A. Zaidi, and M. M. Mahfouz. CRISPR/Cas9-mediated immunity to geminiviruses: differential interference and evasion. *Sci Rep.* 6 (2016): 26912.

Aman, R., Z. Ali, H. Butt, A. Mahas, F. Aljedaani, M. Z. Khan, S. Ding, et al. RNA virus interference via CRISPR/Cas13a system in plants. *Genome Biol.* 19 (2018): 1.

Anand, A., and T. J. Jones. Advancing agrobacterium-based crop transformation and genome modification technology for agricultural biotechnology. *Curr Top Microbiol Immunol.* 418 (2018): 489–507.

Baltes, N. J., and D. F. Voytas. Enabling plant synthetic biology through genome engineering. *Trends Biotechnol.* 33 (2015): 120–131.

Baltes, N. J., A. W. Hummel, E. Konecna, R. Cegan, A. N. Bruns, D. M. Bisaro, and D. F. Voytas. Conferring resistance to geminiviruses with the CRISPR– Cas prokaryotic immune system. *Nat Plants.* 1 (2015): 15145.

Bandyopadhyay, A., N. Kancharla, V. S. Javalkote, S. Dasgupta, and T. P. Prutnell. CRISPR-Cas12a (Cpf1): a versatile tool in the plant genome editing tool box for agricultural advancement. *Front Plant Sci.* 11 (2020): 584151.

Bruner, A. C., S. Jung, A. G. Abbott, and G. L. Powell. The naturally occurring high oleate oil character in some peanut varieties results from reduced oleoyl-PC desaturase activity from mutation of Aspartate 150 to Asparagine. *Crop Sci.* 41 (2001): 522–526.

Chandrasekaran, J., M. Brumin, D. Wolf, D. Leibman, C. Klap, M. Pearlsman, A. Sherman, et al. Development of broad virus resistance in non-transgenic cucumber using CRISPR/Cas9 technology. *Mol Plant Pathol.* 17 (2016): 1140–1153.

Chang, F. P., L. Y. Kuang, C. A. Huang, W. N. Jane, Y. Hung, Y. C. Hsing, and C. Y. Mou. A simple plant gene delivery system using mesoporous silica nanoparticles as carriers. *J Mater Chem. B* 39 (2013): 5279–5287.

Chatthai, M., B. S. Forward, D. Yevtushenko, I. Stefanov, L. Osuska, M. Osusky, and S. Misra. 2S storage protein gene of Douglas-fir: characterization and activity of promoter in transgenic tobacco seeds. *Plant Physi Biochem.* 42 (2004): 417–423.

Chen, M. N., Q. L.Yang, T. Wang, N. Chen, L. J. Pan, X. Y. Chi, Z. Yang, et al. *Agrobacterium*-mediated genetic transformation of peanut and the efficient recovery of transgenic plants. *Can J Plant Sci.* 95 (2015): 735–744.

Clough, S. J., and A. F. Bent. Floral dip: a simplified method for *Agrobacterium*-mediated transformation of *Arabidopsis thaliana. Plant J.* 16 (1998): 735–743.

Collonnier, C., A. Guyon-Debast, F. Maclot, K. Mara, F. Charlot, and F. Nogue. Towards mastering CRISPR-induced gene knock-in in plants: Survey of key features and focus on the model Physcomitrella patens. *Methods.* 121–122 (2017): 103–117.

Cunningham, F. J., N. S. Goh, G. S. Demirer, J. L. Matos, and M. P. Landry. Nanoparticle-mediated delivery towards advancing plant genetic engineering. *Trends Biotechnol.* 36 (2018): 882–897.

Dang, Y., G. X. Jia, J. Choi, H. M. Ma, E. Anaya, C. T. Ye, P. Shankar, and H. Q. Wu. Optimizing sgRNA structure to improve CRISPR-Cas9 knockout efficiency. *Genome Biology* 16 (2015): 280.

Demirer, G., H. Zhang, J. L. Matos, N. S. Goh, F. J. Cunningham, Y. Sung, R. Chang, et al. High aspect ratio nanomaterials enable delivery of functional genetic material without DNA integration in mature plants. *Nat Nanotech.* 14 (2019): 456–464.

Dong, O. X., S. Yu, R. Jain, N. Zhang, P. Q. Duong, C. Butler, Y. Li, et al. Marker-free carotenoid-enriched rice generated through targeted gene insertion using CRISPR-Cas9. *Nat Commun.* 11 (2020): 1178.

Dong, O. X., and P. C. Ronald. Targeted DNA insertion in plants. *PNAS.* 118 (2021): e2004834117.

Eapen, S. Pollen grains as a target for introduction of foreign genes into plants: an assessment. *Physiol Mol Biol Plants* 17 (2011): 1–8.

Eapen, S., and L. George. *Agrobacterium tumefaciens* mediated gene transfer in peanut (*Arachis hypogaea* L.). *Plant Cell Rep.* 13 (1994): 582–586.

Eid, A., S. Alshareef, and M. M. Mahfouz. CRISPR base editors: genome editing without double-stranded breaks. *Biochem J.* 475 (2018): 1955–1964.

Fauser, F., S. Schiml, and H. Puchta. Both CRISPR/Cas-based nucleases and nickases can be used efficiently for genome engineering in *Arabidopsis thaliana. Plant J.* 79 (2014): 348–359.

Fu, Y., J. D. Sander, D. Reyon, V. M. Cascio, and J. K. Joung. Improving CRISPR-Cas nuclease specifcity using truncated guide RNAs. *Nat Biotechnol.* 32, no. 3 (2014): 279–284. https://doi.org/10.1038/nbt.2808

Gilbert, L. A., M. H. Larson, L. Morsut, Z. Liu, G. A. Brar, S. E. Torres, N. Stern-Ginossar, et al. CRISPR-mediated modular RNA-guided regulation of transcription in Eukaryotes. *Cell* 154 (2013): 442–451.

Gong, X., T. Zhnag, J. Xing, R. Wang, and Y. Zhao. Positional effects on efficiency of CRISPR/Cas9-based transcriptional activation in rice plants. *aBiotech.* 1 (2020): 1–5.

Hassan, M. M., Y. Zhang, G. Yuan, K. De, J. G. Chen, W. Muchero, G. Tuskan, et al. Construct design for CRISPR/Cas-based genome editing in plants. *Trends in Plant Science* 26 (2021): 1133–1152.

Higgins, C. M., and R. G. Dietzgen. Genetic transformation, regeneration and analysis of transgenic peanut. *ACIAR Technical Report Project* 2000, no. 48 (2000): 1–86.

Hu, D., A. F. Bent, X. L. Hou, and Y. Li. *Agrobacterium*-mediated vacuum infiltration and floral dip transformation of rapid-cycling *Brassica rapa. BMC Plant Biol.* 19 (2019): 246.

Ito, Y., A. Nishizawa-Yokoi, M. Endo, M. Mikami, and S. Toki. CRISPR/Cas9- mediated mutagenesis of the RIN locus that regulates tomato fruit ripening, *Biochem. Biophys Res Commun.* 467 (2015): 76–82.

Jacobs, T. B., and G. B. Martin. High-throughput CRISPR vector construction and characterization of DNA modifications by generation of tomato hairy roots. *J Visualized Experiments* 110 (2016): e53843. https://doi.org/10.3791/53843.

Ji, X., H. Zhang, Y. Zhang, Y. Wang, and C. Gao. Establishing a CRISPR– Cas-like immune system conferring DNA virus resistance in plants. *Nat Plants* 1 (2015): 15144.

Jia, H., Y. Zhang, V. Orbovic, J. Xu, F. F. White, J. B. Jones, and N. Wang. Genome editing of the disease susceptibility gene Cs LOB 1 in citrus confers resistance to citrus canker. *Plant Biotechnol J.* 15 (2017): 817–23.

Jung, S., G. Powell, K. Moore, and A. Abbott. The high oleate trait in the cultivated peanut (*Arachis hypogaea* L.) II. Molecular bais and genetics of the trait. *Mol Gen Genet* 263 (2000): 806–811.

Karlson, C. K. S., S. N. Mohd-Noor, N. Nolte, and B. C. Tan. CRISPR/dCas9-based systems: Mechanisms and applications in plant sciences. *Plants* 10 (2021): 2055.

Kaur, R. P., and S. Devi. In planta transformation in plants: A review. *Agric Rev.* 40 (2019): 159–174.

Kieu, N. P., M. Lenman, E. S. Wang, B. L. Petersen, and E. Andreasson. Mutations introduced in susceptibility genes through CRISPR/Cas9 genome editing confer increased late blight resistance in potatoes. *Scientific Rep.* 11 (2021): 4487.

Krishna, G., B. K. Singh, E. K. Kim, V. K. Morya, and P. W. Ramteke. Progress in genetic engineering of peanut (*Arachis hypogaea* L.) – A review. *Plant Biotech J.* 13 (2015): 147–162.

Lacorte, C. E., E. Mansur, B. Timmerman, and A. R. Cordeiro. Gene transfer into peanut (*Arachis hypogaea* L.) by *Agrobacterium tumefaciens. Plant Cell Rep.* 10 (1991): 354–357.

Langner, T., S. Kamoun, and K. Belhaj. CRISPR crops: plant genome editing toward disease resistance. *Annu Rev Phytopathol.* 56 (2018): 479–512.

Lee, J. E., M. Neumann, D. I. Duro, and M. Schmid. CRISPR-based tools for targeted transcriptional and epigenetic regulation in plants. *PLoS One* 14, no. 9 (2019): e0222778.

Li, J., J. E. Norville, J. Aach, M. McCormack, D. Zhang, J. Bush, G. M. Church, et al. Multiplex and homologous recombination-mediated genome editing in *Arabidopsis* and *Nicotiana benthamiana* using guide RNA and Cas9. *Nat Biotechnol.* 31 (2013): 688–691.

Li, J., X. Meng, Y. Zong, K. Chen, H. Zhang, J. Liu, J. Li, and C. Gao. Gene replacements and insertions in rice by intron targeting using CRISPR-Cas9. *Nat Plants* 2 (2016): 16139.

Li, M., X. Li, Z. Zhou, P. Wu, M. Fang, X. Pan, Q. Lin, et al. Reassessment of the four yield-related genes Gn1a, DEP1, GS3, and IPA1 in rice using a CRISPR/Cas9 system. *Front Plant Sci.* 7 (2016): 377.

Li, Z., D. Zhang, X. Xiong, B. Yan, W. Xie, J. Sheen, and J. F. Li. A potent Cas9-derived gene activator for plant and mammalian cells. *Nat Plants* 3, no. 12 (2017): 930–936. https://doi.org/10.1038/s41477-017-0046-0.

Li, S., X. Zhang, W. Wang, X. Guo, Z. Wu, W. Du, Y. Zhao, and L. Xia. Expanding the scope of CRISPR/Cpf1-mediated genome editing in rice. *Mol Plant* 11 (2018): 995–998. https://doi.org/10.1016/j.molp.2018.03.009.

Liang, X., W. Shen, H. Sun, M. T. Migawa, T.A. Vickers, and S. T. Crooke. Translation efficiency of mRNAs is increased by antisense oligonucleotides targeting upstream open reading frames. *Nat Biotechnol.* 34, no. 8 (2016): 875–880.

Liu, D., R. Hu, K. J. Palla, G. A. Tuskan, and X. Yang. Advances and perspectives on the use of CRISPR/Cas9 systems in plant genomics research. *Current Opinion in Plant Biology* 30 (2016): 70–77.

Lopez, Y., H. L. Nadaf, O. D. Smith, J. P. Connell, A. S. Reddy, and A. K. Fritz. Isolation and characterization of the Delta(12) fatty acid desaturase in peanut (*Arachis hypogaea* L.) and search for polymorphisms for the high oleate trait in Spanish market-type lines. *Theor Appl Genet.* 101 (2000): 1131–1138.

Louwen, R., R. H. Staals, H. P. Endtz, P. van Baarlen, and J. van der Oost. The role of CRISPR-Cas systems in virulence of pathogenic bacteria. *Microbiol. Mol. Rev.* 78 (2014): 74–88.

Lowder, L. G., D. Zhang, N. J. Baltes, J. W. Paul, X. Tang, X. Zheng, D. F. Voytas, et al. A CRISPR/Cas9 toolbox for multiplexed plant genome editing and transcriptional regulation. *Plant Physiol.* 169 (2015): 971–985.

Lu, Y., Y. Tian, R. Shen, Q. Yao, M. Wang, M. Chen, J. Dong, et al. Targeted, efficient sequence insertion and replacement in rice. *Nat. Biotech.* 38 (2020): 1402–1407. https://doi.org/10.1038/s41587-020-0581-5.

Macovei, A., N. R. Sevilla, C. Cantos, G. B. Jonson, I. Slamet-Loedin, T. Cermak. D. F. Voytas, et al. Novel alleles of rice eIF4G generated by CRISPR/Cas9-targeted mutagenesis confer resistance to Rice tungro spherical virus. *Plant Biotechnol. J.* 16 (2018): 1918–1927.

Makhotenko, A. V., A. V. Khromov, E. A. Snigir, S. S. Makarova, V. V. Makarov, T. P. Suprunova, N. O. Kalinina, et al. Functional analysis of coilin in virus resistance and stress tolerance of potato *Solanum tuberosum* using CRISPR-Cas9 editing. *Doklady Biochemistry and Biophysics* 484 (2019): 88–91.

Mehta, D., A. Stürchler, R. B. Anjanappa, S. S. Zaidi, M. Hirsch-Hoffmann, W. Gruissem, and H. Vanderschuren. Linking CRISPR-Cas9 interference in cassava to the evolution of editing-resistant geminiviruses. *Genome Biol.* 20 (2019): 80.

Moradpour, M., and S. N. A. Abdulah. CRISPR/dCas9 platforms in plants: strategies and applications beyond genome editing. *Plant Biotech J.* 18 (2020): 32–44.

Neelakandan, A. K., D. A. Wright, S. M. Traore, X. Chen, M. H. Spalding, and G. He. CRISPR/Cas9 based site-specific modification of *FAD2* cis-regulatory motifs in peanut (*Arachis hypogaea* L). *Front Genet.* 13 (2022b): 849961. https://doi.org/10.3389/fgene.2022.849961.

Neelakandan, A. K., B. Subedi, S. M. Traore, P. Binagwa, D. A. Wright, and G. He. Base editing in peanut using CRISPR/nCas9. *Front Genome Ed.* 4 (2022c): 901444. https://doi.org/10.3389/fgeed.2022.901444.

Neelakandan, A. K., D. A. Wright, S. M. Traore, X. Ma, B. Subedi, S. Veeramasu, M. H. Spalding, et al. Application of CRISPR/Cas9 system for efficient gene editing in peanut. *Plants* 11 (2022a): 1361. https://doi.org/10.3390/plants11101361.

Nekrasov, V., B. Staskawicz, D. J. Weigel, D. G. Jonathan, and S. Kamoun. Targeted mutagenesis in the model plant *Nicotiana benthaminana* using Cas9 RNA-guided endonuclease. *Nat. Biotechnol.* 31 (2013): 691–693.

Nekrasov, V., C. Wang, J. Win, C. Lanz, D. Weigel, and S. Kamoun. Rapid generation of a transgene-free powdery mildew resistant tomato by genome deletion. *Sci. Rep.* 7 (2017): 482.

Oliva, R., C. Ji, G. Atienza-Grande, J. C. Huguet-Tapia, A. Perez-Quintero, T. Li, J. S. Eom, et al. Broad-spectrum resistance to bacterial blight in rice using genome editing. *Nat. Biotechnol.* 37 (2019): 1344–1350.

Ortigosa, A., S. Gimenez-Ibanez, N. Leonhardt, and R. Solano. Design of a bacterial speck resistant tomato by CRISPR/Cas9-mediated editing of SlJAZ2. *Plant Biotechnol J.* 17 (2019): 665–73.

Pachapur, V. L., A. Dalila Larios, M. Cledón, S. K. Brar, M. Verma, R. Y. Surampalli, et al. Interaction of metal oxide nanoparticles with higher terrestrial plants: physiological and biochemical aspects. *Plant Physiol. Biochem.* 110 (2016): 210–225

Pan, L., H. Ji, J. Huang, D. Huai, L. Yong, J. Sui, Y. Tang, et al. *AhFatB* gene editing using pollen-tube pathway and *Agrobacterium* Mediated method in peanut. *ACTA Agric Boreali-SINICA* 35 (2020): 64–70.

Park, M. E., and H. U. Kim. Applications and prospects of genome editing in plant fatty acid and triacylglycerol biosynthesis. *Front. Plant Sci.* 13 (2022): 969844.

Peng, A., S. Chen, T. Lei, L. Xu, Y. He, L. Wu, L. Yao, et al. Engineering canker-resistant plants through CRISPR/Cas9-targeted editing of the susceptibility gene Cs LOB 1 promoter in citrus. *Plant Biotechnol J.* 15 (2017):1509–19.

Piatek, A., Z. Ali, H. Baazim, L. Li, A. Abulfaraj, S. Al-shareef, M. Aouida, et al. RNA-guided transcriptional regulation in planta via synthetic dCas9-based transcription factors. *Plant Biotechnol. J.* 13 (2015): 578–589.

Puchta, H. Using CRISPR/Cas in three dimensions: towards synthetic plant genomes, transcriptomes and epigenomes. *Plant J.* 87 (2016): 5–15.

Pyott, D. E., E. Sheehan, and A. Molnar. Engineering of CRISPR/Cas9- mediated potyvirus resistance in transgene-free Arabidopsis plants. *Mol. Plant Pathol.* 4 (2016): 1–13.

Rees, H. A., and D. R. Liu. Base editing: precision chemistry on the genome and transcriptome of living cells. Nature reviews. *Genetics* 19, no. 12 (2018): 770–788. https://doi.org/10.1038/s41576-018-0059-1

Reidt, W., T. Wohlfarth, M. Ellerström, A. Czihal, A. Tewes, I. Ezcurra, L. Rask, and H. Bäumlein. Gene regulation during late embryogenesis: the RY motif of maturation-specific gene promoters is a direct target of the FUS3 gene product. *The Plant Journal* 21 (2000): 401–408.

Ren, R., Q. Zheng, F. Ni, Q. Jin, X. Wan, and J. Wei. Breeding for PVY resistance in tobacco LJ911 using CRISPR/Cas9 technology. *Crop Breeding and Applied Biotechnology* 21 (2021): e31682116.

Rohini, V. K., and K. Sankara Rao. Transformation of peanut (*Arachis hypogaea* L.): a non-tissue culture based approach for generating transgenic plants. *Plant Sci.* 150 (2000): 41–49.

Roossinck, M. J., D. P. Martin, and P. Roumagnac. Plant virus metagenomics: advances in virus discovery. *Phytopathology* 105 (2015): 716–727.

Roy, A., Y. Zhai, J. Ortiz, M. Neff, B. Mandal, S. K. Mukherjee, and H. R. Pappu. Multiplexed editing of a begomovirus genome restricts escape mutant formation and disease development. *PLos One* 14, no. 10 (2019): e0223765.

Schie, C., and F. L. Takken. Susceptibility genes 101: how to be a good host. *Annu. Rev. Phytopathol.* 52 (2014): 551–581.

Schiml, S., and H. Puchta. Revolutionizing plant biology: multiple ways of genome engineering by CRISPR/Cas. *Plant Methods* 12 (2016): 8.

Shan, Q., Y. Wang, J. Li, Y. Zhang, K. Chen, Z. Liang, K. Zhang, et el. Targeted genome modification of crop plants using a CRISPR-Cas system. *Nat. Biotechnol.* 31 (2013): 686–688.

Shi, J., H. Gao, H. Wang, H. R. Lafitte, R. L. Archibald, M. Yang, S. M. Hakimi, et al. ARGOS8 variants generated by CRISPR-Cas9 improve maize grain yield under field drought stress conditions, *Plant Biotechnol J.* 15 (2017): 207–216.

Si, X., H. Zhang, Y. Wang, K. Chen, and C. Gao. Manipulating gene translation in plants by CRISPR-Cas9-mediated genome editing of upstream open reading frames. *Nat. Protocols* 15, no. 2 (2020): 338–363.

Silva, J., R. Fernandes, and L. Romão. Gene expression regulation by upstream open reading frames in rare diseases. *J. Rare Diseases Research & Treatment* 2, no. 4 (2017): 33–38.

Song, X. M., Y. H. Gu, and G. Y. Qin. Application of a transformation method via the pollen-tube pathway in agriculture molecular breeding. *Life Sci J.* 4, no. 1 (2006): 77–79.

Talakayala, A., S. Ankanagari, and M. Garladinne. CRISPR-Cas genome editing system: A versatile tool for developing disease resistant crops. *Plant Stress* 3 (2022): 100056.

Thomazella, D. P. D. T., K. Seong, R. Mackelprang, D. Dahlbeck, Y. Geng, U. S. Gill, T. Qi, et al. Loss of function of a DMR6 ortholog in tomato confers broad-spectrum disease resistance. *Proc. Natl. Acad. Sci.* 118, no. 27 (2021): e2026152118.

Tripathi, D. K., Shweta, S. Singh, R. Pandey, V. P. Singh, N. C. Sharma, S. M. Prasad, et al. An overview on manufactured nanoparticles in plants: uptake, translocation, accumulation and phytotoxicity. *Plant Physiol. Biochem.* 110 (2017): 2–12.

Tripathi, J. N., V. O. Ntui, T. Shah, L. Tripathi. CRISPR/Cas9-mediated editing of *DMR6* orthologue in banana (*Musa spp.*) confers enhanced resistance to bacterial disease. *Plant Biotechnol J.* 19 (2021): 1291–1293.

Umesh, T. G., and N. N. Rao. An efficient and improved method of in Planta transformation in *Arachis hypogaea* L. *Int J LifeSc Bt & Pharm Res.* 3, no. 4 (2014): ISSN 2250–3137.

Wang, F., C. Wang, P. Liu, C. Lei, Y. Gao, Y. Liu, and K. Zhao. Enhanced rice blast resistance by CRISPR/Cas9-targeted mutagenesis of the ERF transcription factor gene OsERF922. *PLoS One* 11 (2016): e154027.

Wang, J. X., Y. Sun, G. Cui, and J. Hu. Transgenic maize plants obtained by pollen-mediated transformation. *Acta Bot Sin.* 43 (2001): 275–279.

Wang X, M. Tu, D. Wang, J. Liu, Y. Li, Z. Li, Y. Wang, et al. CRISPR/Cas9-mediated efficient targeted mutagenesis in grape in the first generation. *Plant Biotechnol J.* 16 (2018): 844–55.

Wang, Y., X. Cheng, Q. Shan, Y. Zhang, J. Liu, C. Gao, and J. Qiu. Simultaneous editing of three homoeoalleles in hexaploid bread wheat confers heritable resistance to powdery mildew. *Nat. Biotechnol.* 32 (2014): 947–952.

Xie, K., and Y. Yang. RNA-guided genome editing in plants using a CRISPR-Cas system. *Mol. Plant* 6 (2013):1975–1983.

Xu, G., M. Yuan, C. Ai, L. Liu, E. Zhuang, S. Karapetyan, S. Wang, et al. uORF-mediated translation allows engineered plant disease resistance without fitness costs. *Nature* 545, no. 7655 (2017): 491–494.

Xu, R., Y. Yang, R. Qin, H. Li, C. Qiu, L. Li, P. Wei, et al. Rapid improvement of grain weight via highly efficient CRISPR/Cas9-mediated multiplex genome editing in rice. *J Genet. Genomics* 43 (2016): 529–532.

Xu, Z., X. Xu, Q. Gong, Z. Li, Y. Li, S. Wang, Y. Yang, et al. Engineering broad-spectrum bacterial blight resistance by simultaneously disrupting variable TALE-binding elements of multiple susceptibility genes in rice. *Mol. Plant* 12 (2019): 1434–1446.

Yang, L., G. Cui, Y. Wang, Y. Hao, J. Du, H. Zhang, H. M., Wang, et al. Expression of foreign genes demonstrates the effectiveness of pollen-mediated transformation in *Zea mays. Front Plant Sci.* 8 (2017): 383.

Yang, W., H. Feng, X. Zhang, J. Zhang, J. H. Doonan, W. D. Batchelor, L. Xiong, et al. Crop phenomics and high-throughput phenotyping: Past decades, current challenges, and future perspectives. *Mol. Plant* 13 (2020): 187–214.

Yin, K., T. Han, K. Xie, J. Zhao, J. Song, and Y. Liu. Engineer complete resistance to Cotton Leaf Curl Multan virus by the CRISPR/Cas9 system in *Nicotiana benthamiana. Phytopathol Res.* 1 (2019): 9.

Yuan, M., J. Zhu, L. Gong, L. He, C. Lee, S. Han, C. Chen, et al. Mutagenesis of *FAD2* genes in peanut with CRISPR/

Cas9 based gene editing. *BMC Biotechnology* 19 (2019): 24. https://doi.org/10.1186/s12896-019-0516-8.

Zahran, H. A., and H. Z. Tawfeuk. Physicochemical properties of new peanut (*Arachis hypogaea* L.) varieties. *OCL* 26 (2019): 19.

Zaidi, S. S., M. S. Mukhtar, and S. Mansoor. Genome editing: targeting susceptibility genes for plant disease resistance. *Trends Biotechnol.* 36, no. 9 (2018):898–906.

Zeng, X., Y. Luo, N. T. Q. Vu, S. Shen, K. Xia, and M. Zhang. CRISPR/Cas9-mediated mutations of OsSWEET14 in rice cv. Zhonghua11 confers resistance to *Xanthomonas oryzae pv. Oryzae* without yield penalty. *BMC Plant Biol.* 20 (2020): 1–11.

Zhang, T., Q. Zheng, X. Yi, H. An, Y. Zhao, S. Ma, and G. Zhou. Establishing RNA virus resistance in plants by harnessing CRISPR immune system. *Plant Biotechnology J.* 16 (2018): 1415–1423.

Zhang, Y., Y. Bai, G. Wu, S. Zou, Y. Chen, C. Gao, and D. Tang. Simultaneous modification of three homoeologs of TaEDR1 by genome editing enhances powdery mildew resistance in wheat. *Plant J.* 91 (2017): 714–724.

Zhang, Y., X. Yin, A. Yang, G. Li, and J. Zhang. Stability of inheritance of transgenes in maize (*Zea mays* L.) line produced using different transformation methods. *Euphytica* 144 (2005): 11–22.

Zhou, J., Z. Peng, J. Long, D. Sosso, B. Liu, J. S. Eom, S. Huang, et al. Gene targeting by the TAL effector PthXo2 reveals cryptic resistance gene for bacterial blight of rice. *Plant J.* 82 (2015): 632–643.

19 Potentials of Bioinformatics in Peanut Breeding Programs

Ali Razzaq, Sandaru Malaweera Arachchige, and Jianping Wang

19.1 INTRODUCTION

Cultivated peanut (*Arachis hypogaea* L.) belongs to the leguminous or Fabaceae family, *Arachis* genus, a member of Aeschynomeneae tribe (Janila et al., 2016). Peanuts are cultivated in between 40°S and 40°N of arid and semi-arid tropics under diverse ecological and environmental conditions (Holbrook et al., 2016), and they serve as a great food source of edible oil (Griel et al., 2004) and energy snacks.

Breeding for cultivars with superior yield and resilience is the ultimate target of all peanut breeding programs. Although peanut breeding programs have achieved tremendous strides in the past decades, there are still some major gaps such as limited genetic diversity, limited or unavailability of genetic basis of some multi-genic or quantitative traits such as drought tolerance and durable disease resistance, which altogether make it challenging for breeders to develop elite cultivars. To overcome these challenges, there is a need to develop integrated advanced breeding strategies that can take advantage of the informative functionality of peanut genomes, which are holistically and efficiently decoded. Advancements in peanut genomics and transcriptomics have generated a substantial number of datasets, demanding sophisticated bioinformatics tools to analyze and extract any useful information. The current available bioinformatic tools and any new tools under development or to be developed enable us to fully annotate the peanut genome and transcriptome and prevent valuable datasets from being unmined or lost due to software limitations.

Bioinformatics is an interdisciplinary field that can be defined based on two core disciplines, computer programing and molecular biology, and its integrative approach. It focuses on developing and applying algorithms to evaluate and investigate biological datasets to examine the functions and structure of biological polymers and their relation to living organisms (Tapprich et al., 2021). Bioinformatics is becoming central to biological research and has emerged as an outstanding discipline (Xu et al., 2023). During the last decade, great progress has been made in upgrading bioinformatic tools to maximize our ability to gather, store, quantify, and characterize large amounts of datasets. Bioinformatic tools are becoming vital to study biomolecules such as DNA, RNA, protein, and metabolites, thus offering an excellent platform to access the unexplored phenomena of life (Xu et al., 2023).

Recently advanced technologies in high-throughput next generation sequencing (NGS) and profiling have generated huge datasets, including genomes, transcriptomes, proteomes, and metabolomes. As the size and complexity of these omics datasets increases with the higher organisms (Uesaka et al., 2022; Xu et al., 2023), highly sophisticated tools and resources are required to successfully analyze these omics datasets. Therefore, the bioinformatics field needs constant technological upgrades to deal with the new challenges by improving the data analysis toolkits and algorithms (Yang et al., 2021). Selection of a software and its accuracy to represent true results according to specified requirements is a common concern of researchers. However, nowadays, hundreds of software tools and various analysis platforms with detailed descriptions and tutorials are freely available and accessible to all (Tan et al., 2022).

Likewise, bioinformatics plays a crucial role in exploring the genomic, transcriptomic, and proteomic landscape of peanut by designing highly innovative computational programs, databases, software tools, and web platforms to fully annotate peanut genomes, such as by identifying repetitive sequences within the genome, detecting regulatory elements, and annotating functional genes (Bertioli et al., 2019). Furthermore, several software tools are available to define the genomic regions, identify the coding and non-coding RNAs (Ma et al., 2020), and predict gene function through incorporating RNA-seq technologies (Ahmad et al., 2023). Bioinformatics tools allow for sequence quality checks, trimming reads, sequence alignment, SNP calling, QTL identification, gene prediction, gene expression quantification, pathway analysis, phylogenetic studies, and so on (Gomez-Casati et al., 2018).

Bioinformatics helps breeders to select the desirable traits assisted by molecular markers to develop elite crop varieties through quantitative trait loci (QTL) mapping, genome-wide association studies (GWAS), or genome selection (GS) (Mochida et al., 2010). Thus, bioinformatics allows breeders to fully understand the peanut genome, translational genomics, traits genetics, and their regulation mechanisms. It also facilitates breeders to select a superior trait based on data-driven prediction and to fast-track the variety development process. In this chapter, we discuss applications of bioinformatics in peanut research, including genomes, functional genomics, and genetic analysis like

DOI: 10.1201/9781003360988-19

SNP calling, QTL-seq, and RNA-seq analysis in breeding applications.

19.2 BIOINFORMATICS IN PEANUT GENETICS AND GENOMICS

Cultivated peanut is an allotetraploid (2n = 4x = 40, AABB) with two closely related sub-genomes (A and B). Developing a well assembled and annotated reference genome sequence is pivotal for carrying out advanced genetic studies for any plant species, including peanut, during this genomic era. However, there are challenges ahead in dealing with the complex plant genomes having repetitive sequences, which makes it a difficult task to decode the complex plant genomes like peanut, because peanut contains two closely related sub-genomes, A and B, and large amount of repetitive sequence (Kong et al., 2023). Specifically, the complexity of polyploid genomes makes it more difficult to use the same sequencing standard and bioinformatics pipelines, as these pipelines may not be reproducible in all cases. For example, trio binning (Koren et al., 2018) and falcon-phase (Chin et al., 2016) have been developed to assemble the haplotypes of diploid crop, but these are not effective for polyploid crops. In eukaryotic cells, NHEJ typically dominates the repair process and knocks out gene function. Different tools are required to do sequence alignment and assembly, as many tools developed for short reads perform poorly when aligning long reads (Sedlazeck et al., 2018). For variant calling, variant callers differ in their ability to call SNPs or InDels and in their biases towards calling heterozygous and reference variants (Hwang et al., 2015). Applying the right bioinformatics tool for the right application is crucial. Therefore, continuous efforts are needed to improve both sequencing technologies and bioinformatic algorithms/tools and select the appropriate pipelines to assemble and analyze the complex genomic data (Kong et al., 2023).

19.2.1 Bioinformatics in Peanut Genomics

With the most updated NGS platforms and bioinformatic tools, reference genomes of wild and cultivated peanut varieties were generated with great accuracy and cost-effectiveness (Bertioli et al., 2016, 2019). The "International Peanut Genome Initiative" (IPGI) has provided the complete reference genome of cultivated peanut, cv. Tifrunner of 2.7Gb using shotgun sequencing (Bertioli et al., 2016). Parallelly, a Chinese peanut cv. Shitouqi (ssp. *A. h. fastigiata* var. *vulgaris*) was also sequenced providing an assembly genome size of 2.54 Gb. Resequencing of 4 synthetic tetraploids, 18 wild species, and 30 allotetraploid genotypes offers greater knowledge about peanut domestication, evolution, trait biology, and genomic landscape.

Assembling reference genome facilitated comparing and decoding the genomes of various peanut cultivars and was utilized to detect genomic variations linked with important agronomic traits. It has been used for gene functional analysis, marker development, mapping, identification of unique genes regulating biotic/abiotic resistance (Bertioli et al., 2019; Zhuang et al., 2019).

This has been made possible by employing more sophisticated and improved bioinformatics tools. For example, in the sequencing of wild ancestors (*A. ipaensis* and *A. duranensis*) of cultivated peanuts, short reads were generated and aligned by SOAPdenovo (Li, R et al., 2010), which is an excellent de novo genome assembly bioinformatic tool for generating reference genomes. Furthermore, SSPACE (Boetzer et al., 2011) was used to link the pre-assembled scaffolds and linkage map was constructed using Mapmaker Macintosh 2.0 (Lander et al., 1987). MadMapper and MAKER-P pipeline were used for linkage group identification and gene prediction respectively (Campbell et al., 2014). On the other hand, a different pipeline was used to construct the reference genome of cultivated peanut, where long reads and shorts read were created and long reads were aligned using MECAT (Xiao et al., 2017). For gene prediction analysis, SNAP (Korf 2004) and AUGUSTUS (Stanke et al., 2006) were used. Similarly, another study on cultivated peanut provides deep insights into polyploid evolution and domestication (Zhuang et al., 2019), where long reads were aligned and scaffolded using FALCON (https://github.com/PacificBiosciences/FALCON) and LACHESIS for chromosome level scaffolding (Simão et al., 2015). EuGene 4.2 was used for gene prediction and annotation (Foissac et al., 2008). Therefore, bioinformatics plays a pivotal role in generating peanut reference genome, which paves the way for improving peanut cultivars for desired traits and stress resilience.

Not only does developing reference genomes and annotating genes involve genome assembly and annotation, but bioinformatic tools and databases also assist in comparative genomics, the identification and application of genetic markers, genome-wide association studies (GWAS), transcriptomics, gene expression analysis, and studies in functional genomics and genome editing.

Comparative genomics is a field of biological research that compares the genome sequences of different species. The primary goal is to identify similarities and differences in the structure, function, and evolutionary processes of genomes (Hardison, 2003). This comparison helps elucidate species evolution, gene and genetic element functions, and genome organization. By comparing the peanut genome with others, bioinformatics identifies conserved genes and genetic elements, reveals evolutionary relationships, and discovers unique peanut genetic features. Comparative genomics also transfers functional annotations from model organisms to predict unknown gene functions and understand trait genetics. Bioinformatics provides tools and methodologies for analyzing genomic data across species and cultivars. Tools like BLAST (Altschul et al., 1990), MAFFT (Katoh and Standley, 2013), and Clustal Omega (Sievers and Higgins, 2014) identify homologous regions and genes through sequence alignment. Software such as MUMmer (Marçais et al., 2018) and LASTZ (Ghosh, 2015) facilitate whole-genome comparisons. Bioinformatics constructs phylogenetic trees using tools like MEGA (Tamura et al., 2021), PhyML (Guindon et al., 2010), and RAxML (Stamatakis, 2014), illustrating evolutionary relationships. Visualization tools like Circos (Krzywinski et al., 2009) and IGV (Integrative Genomics Viewer) (Robinson et al., 2011) display comparative genomic data, including synteny maps, structural variations, and phylogenetic relationships.

Advanced bioinformatics tools enable comprehensive data analysis, from quality control to variant calling, facilitating marker-assisted selection, genomic selection, and other breeding strategies. By integrating genomic data with bioinformatics, researchers can accelerate the development of peanut varieties with enhanced yield, disease resistance, and environmental adaptability.

19.2.2 Bioinformatics for Peanut Genetic Variant Analysis

19.2.2.1 Genome Aligner Tools

Aligning the sequence reads to the reference genome is one of the most important steps in sequencing data analysis. Fast, accurate, and efficient algorithms are needed to perfectly align the sequence to the reference genome. Currently, the main software tools used in peanut genomics for aligning short sequences to a reference genome are Burrows-Wheeler Transform (BWA) and Bowtie2. Both BWA and Bowtie2 are faster and need less computer memory compared to older methods like MAQ and Novoalign (Peng et al., 2020). BWA is more accurate, while Bowtie2 is faster because of its algorithm. BWA has a lower error rate with higher percentages of successfully mapped reads and higher accuracy as compared to Bowtie. BWA uses breadth-first search, carefully looking through all nearby nodes on the suffix tree. Bowtie uses depth-first search, skipping nodes more aggressively and going back only when needed. BWA-mem is proven to be more accurate than Bowtie 2, especially for longer reads. Bowtie 2, on the other hand, is faster and handles InDels better (Langmead and Salzberg, 2012). In terms of sensitivity and accuracy, BWA is likely the better choice for mapping reads in plant datasets, especially when dealing with samples that are not closely related. On the other hand, Bowtie2 can be a good choice when high accuracy is needed over sensitivity (Yao et al., 2020). Table 19.1 presents a compilation of some freely available aligner tools. For example, Bowtie2 is utilized for diverse purposes, including gene identification for late leaf spot and rust resistance (Shirasawa et al., 2018), nodulation (Peng et al., 2020), branching habit traits (Kayam et al., 2017), high oleic acid content (Shirasawa et al., 2016), leaf spot resistance (Liang et al., 2017), tetrasomic recombination (Clevenger et al., 2017), and genetic diversity analysis (Pandey et al., 2017). BWA is employed for investigating (Agarwal et al., 2018) bacterial wilt (Qi et al., 2022; Luo et al., 2019), web blotch resistance (Liu et al., 2020), major agronomic traits (Zhang et al., 2017), shelling percentage (Luo et al., 2019), seed weight (Wang et al., 2022), purple testa color (Zhao et al., 2020), and production, quality, and yield-related traits (Jadhav et al., 2021; Wang et al., 2019). It is also used to study diversity among wild and cultivated lines (Chopra et al., 2016) and marker development (Kim et al., 2017).

19.2.2.2 SNP Calling

SNP calling is a very important genomics approach to detect single nucleotide changes within DNA. It refers to the single base change at a specific position in the genome that can be used as a very important marker system for genetic studies. There are numerous tools designed to call the SNP across a wide range of organisms including plants (Table 19.2).

In peanut, GATK, Samtools, and SWEEP have been widely used to call SNPs to investigate a number of important traits as depicted in Table 19.2. For example, SWEEP was employed to call variants related to nodulating and non-nodulating (Peng et al., 2020), early and late leaf spot resistance (Agarwal et al., 2018; Clevenger et al., 2018; (Han et al., 2018), tetrasomic recombination and signature of selection (Clevenger et al., 2017), and genetic diversity (Pandey et al., 2017). Likewise, Samtools have been utilized to detect the SNPs in marker development (Kim et al., 2017), high oleic acid (Shirasawa et al., 2016), seed weight (Wang et al., 2022), shelling percentage (Luo et al., 2019), bacterial wilt resistance (Qi et al., 2022; Luo et al., 2019), and late leaf spot and rust resistance (Shirasawa et al., 2018). GATK identified the SNPs linked to aflatoxin production (Yu et al., 2020), web blotch resistance (Liu et al., 2020), sucrose content in kernels (Guo et al., 2023), quality traits (Sun et al., 2022), growth habit related traits (Li et al., 2022), branching habit trait (Kayam et al., 2017), purple testa colour (Zhao et al., 2020), yield-related traits (Wang et al., 2019), and genetic diversity analysis (Nabi et al., 2021; Chopra et al., 2016).

19.2.3 Current Marker Platforms and Genetic Maps

Genetic maps and molecular markers provide a great platform to determine the markers linked with specific traits including abiotic stress tolerance, biotic stress resistance, and agronomic traits in peanuts. Thus, high resolution deep genetic maps encompassing complete genome coverage increase the opportunity to identify closely associated genetic markers linked to important traits via association mapping. There are vast applications of various molecular markers, such as SNP, restriction fragment length polymorphism (RFLP), random amplified polymorphic DNA (RAPD), amplified fragment length polymorphism (AFLP), and simple sequence repeats (SSR) in peanut molecular breeding, trait mapping, and germplasm characterization (Pandey et al., 2012). Evaluating the degree of genetic diversity in population using different DNA markers gives

TABLE 19.1
A List of Mostly Used Freely Available Genome Aligners

Genome Aligners and Mappers	Function and Feature	Source/Availability	Reference
BWA (Burrows-Wheeler Aligner)	Short-read aligner	https://bio-bwa.sourceforge.net/	Li and Durbin, 2009
Bowtie	Short-read aligner	https://bowtie-bio.sourceforge.net/index.shtml	Langmead et al., 2009
Bowtie 2	Short-read aligner	https://bowtie-bio.sourceforge.net/bowtie2/index.shtml	Langmead and Salzberg, 2012
Minimap2	Short-read aligner	https://github.com/lh3/minimap2	Li, 2018
SOAP	Short-read aligner	http://soap.genomics.org.cn	Li et al., 2008
TMAP (Torrent Mapping Alignment Program)	Short-read aligner	https://github.com/nh13/TMAP	
Subread	Short-read aligner	https://subread.sourceforge.net/	Liao, et al., 2013
IGNUMAP	Short-read aligner	https://dna.cs.byu.edu/gnumap/	Clement et al., 2010
NextGenMap	Highly polumorphic Short-read aligner	https://cibiv.github.io/NextGenMap/	Sedlazec et al., 2013
HISAT (Hierarchical Indexing for Spliced Alignment of Transcripts)	Spliced read aligner	http://www.ccb.jhu.edu/software/hisat/index.shtml	Kim et al., 2015
HISAT2	Spliced read aligner for both DNA and RNA	https://daehwankimlab.github.io/hisat2/	Kim et al., 2019
TopHat	Splice junction aligner	https://ccb.jhu.edu/software/tophat/index.shtml	Trapnell et al., 2009
STAR	Spliced read aligner	https://code.google.com/archive/p/rna-star/	Dobin et al., 2013
SOAPsplice	Spliced read aligner	http://soap.genomics.org.cn/soapsplice.html.	Huang et al., 2011
HISAT-3N	Nucleotide conversion aligner	https://daehwankimlab.github.io/hisat2/hisat-3n/	Zhang et al., 2021
GMAP/GSNAP	Plant-based genomic alignment program	http://research-pub.gene.com/gmap/	Wu et al., 2016
PALMapper	RNA short read aligner	https://bioinformaticshome.com/tools/rna-seq/descriptions/PALMapper.html#gsc.tab=0	Jean et al., 2010
STAR-Fusion	Fusion gene aligner	https://hpc.nih.gov/apps/STAR.html	Dobin et al., 2013

TABLE 19.2
List of Software Tools Used for SNP Calling

Software	Feature	Source/Availability
GATK (Genome Analysis Toolkit)	Variant filtering and SNP calling	https://gatk.broadinstitute.org/hc/en-us
Samtools	Small InDels and SNP identification	https://www.htslib.org/
ANGSD (Analysis of Next Generation Sequencing Data)	SNP and genotype calling	http://www.popgen.dk/angsd/index.php/ANGSD
Strelka2	Small-variant calling	https://github.com/Illumina/strelka
SNVer	Unique variant calling	https://snver.sourceforge.net/
Atlas 2	SNP/InDel calling	https://www.hgsc.bcm.edu/software/atlas-2
SOAPsnp	SNP/InDel calling	http://soap.genomics.org.cn/soapsnp.html
SNPTools	SNP calling	https://www.hgsc.bcm.edu/software/snptools
Thunder	InDel calling	https://genome.sph.umich.edu/wiki/Thunder
INDELseek	InDel calling	https://github.com/tommyau/indelseek/
FreeBayes	Variant identification	https://www.geneious.com/plugins/freebayes/#citing
VarDict	Variant calling	https://github.com/AstraZeneca-NGS/VarDict
DeepVariant	Variant calling	https://github.com/google/deepvariant
Bcftools	Variant calling	https://samtools.github.io/bcftools/howtos/variant-calling.html
Platypus	Variant calling	https://github.com/andyrimmer/Platypus
LoFreq	Variant calling	https://csb5.github.io/lofreq/
VarScan	Variant detection	https://varscan.sourceforge.net/
BS-SNPer	SNP calling in bisulfite data	https://github.com/hellbelly/BS-Snper
MethGo	SNP calling in bisulfite data	https://paoyangchen-laboratory.github.io/methgo/
SMAPdigger	SNP calling in bisulfite data	https://github.com/gaosjlucky/SMAPdigger
Bis-SNP	SNP calling in bisulfite data	http://epigenome.usc.edu/publicationdata/bissnp2011
MethylCoder	SNP calling in bisulfite data	https://github.com/brentp/methylcode

knowledge about hybrid origin, population structure, misidentified accessions, genetic variations, genetic redundancy, and relatedness within the group of cultivars (Wang et al., 2011). Molecular markers can also help to manage and utilize genetic resources more efficiently and effectively. DNA markers have been widely used to study the intraspecific and interspecific variations within wild and cultivated peanut genotypes (Pandey et al., 2012). Bioinformatics tools are very critical to generate linkage maps and their visualization and interpretation. These tools help scientists to develop marker platforms and genetic maps before the peanut reference genome is sequenced. For example, Burow et al. (2001) created the first map of tetraploid using RFLP markers by making the cross between synthetic amphidiploid and Florunner. MAPMAKER/EXP v 3.0 was used for linkage analysis to detect the double crossover events. In another study, a cultivated peanut cultivar Fleur11, hybridized with *A. duranensis* x *A. ipaënsis* and a population of 88 individuals were used to create the SSR genetic map. CarthaGene software version 1.0 and Mapdisto software version 1.7.2.4 were used to construct the genetic map. Introgression analysis was performed using CSSL Finder software version 0.8b4 (Fávero et al., 2006). This linkage map comprised 21 groups having 298 markers with 1843.7cM map distance (Foncéka et al., 2009). These linkage maps are very beneficial for peanut trait improvement as they introduce unique QTLs related to disease resistance and other agronomically important traits (Nagy et al., 2010). Approximately, more than 9,000 SSRs markers have been developed by various research groups for peanut that has allowed the development of dense and more saturated genetic maps for *Arachis hypogea* and also facilitated phylogenetic assessment of Arachis genus by employing various bioinformatic tools (Guo et al., 2016; Moretzsohn et al., 2013). A co-dominant SSR marker was developed, and a genetic map was generated using the MAPMAKER tool for nematode resistance via high-resolution mapping that can be employed to screen resistant heterozygous individuals (Nagy et al., 2010) and then analyze markers for accuracy for resistance against peanut root-knock nematode under field tests.

With the advancement of SNP-based genotyping and designing of new bioinformatic tools, several high-density genetic maps of peanut were developed. For example, Zhou et al. (2014) constructed the first SNP-based high-density map of two cultivated peanut genotypes and 166 recombinant inbreed lines (RILs) via the integrated NGS based ddRADseq technique. About 175GB datasets were obtained using pair-end sequencing and aligned via SOAP, which detected 53,257 SNPs among parental lines while 14,663 SNPs were obtained from the RIL progenies using Custom Perl scripts. Approximately 1,765 polymorphic markers were identified that have the ability for genetic map construction. Genetic distance and marker order were calculated using program Joinmap 4.0. This high-density map served as the reference map for *A. hypogaea* and facilitated the marker-assisted breeding of peanut. In another experiment, about 384 accessions of diploid and tetraploid were utilized for SNP genotyping array via whole genome resequencing. Paired end sequences were aligned using Bowtie2, and SNP calling was performed with SWEEP Prime version program. The results identified 54,564 markers which resulted in polymorphic groups among diploid accession, 15,897 markers in cultivated lines, and 47,116 markers within interspecific hybrids and cultivated germplasm (Clevenger et al., 2017).

Genotyping by sequencing (GBS) was employed for the construction of a high-resolution SNP map within 20 linkage groups having 2,753 SNP markers with a genetic distance of 1.34 cM. Sequencing reads were aligned via BWA, and SNP calling was performed via SWEEP software. For linkage map construction and QTL analysis, JoinMap 4.0 version and WinQTLCartgrapher software 2.5. were used, respectively. The result showed that two major QTLs were linked with early leaf spot resistance present on chromosome B04 and A03 (Han et al., 2018). Yu and colleagues tested the 99 Chinese mini-core accessions against the aflatoxin production and identified two resistant accessions. Genotyping was done using RAD-seq, and SNP calling was carried out using GATK and SAMtools; 60 SNP markers were detected strongly linked with aflatoxin having phenotypic variation ranging from 16.87% to 31.70% (Yu et al., 2020). A high-density SNP genetic map using JoinMap v4.1 was constructed on RILs developed from a cross between Tifrunner and NC 3033. The QTL analysis was performed with WinQTLCart v2.5, QTLNetwork v2.0, and QTL IciMapping v4.2 to detect the QTL × environment interactions, epistasis QTLs, and additive QTLs. The result showed 73 SSR markers and 1,451 SNPs along with 33 QTLs associated with stem rot resistance (Luo et al., 2020).

Khan et al. (2020) used the Join Map 4.0 program to construct two high-density SNP-based maps consisting of 5,022 and 1,975 SNP loci respectively. Two QTLs *qRAF-14-1* and qRAF-3-1 were identified that associated with *A. flavus* resistance in cultivated peanut. Gene ontology, UniProt annotations, and other bioinformatic tools were employed to identify candidate genes from which 6 genes having similar function were linked with QTLs. Moreover, some putative genes including cytochrome P450, WRKY, lipoxygenase, and RPP13-like protein 1 were also detected. In another experiment, two SNP-based high-resolution genetic maps with JoinMap (v4.0) were developed with 2,668 and 3,341 loci for NAM_Florida-07 and NAM_Tifrunner, respectively. GWAS analysis using a mixed linear model implemented in R package revealed 17 and 11 SNP trait associations in NAM_Florida-07 and 28 and 19 SNP in NAM_Tifrunner for seed weight and pod weight, respectively. These co-localized SNP trait associations were found to be co-regulated by many genes present on four different chromosomes (Gangurde et al., 2020).

19.2.4 Bioinformatics in Major Trait QTLs Identification

Advancement in genomic tools has led to characterization of most valuable traits and exploration of genetic variations using molecular breeding approaches. Quantitative trait loci (QTL) allow peanut breeders to dissect any specific trait of interest that can be used for crop improvement. Bioinformatics play a critical role in identifying QTLs in peanuts related to biotic/abiotic stress, oil quality, and agronomic morphological and yield-related traits. Several bioinformatics tools and packages have been widely used for QTL analysis, which include QTL Cartographer 2.5 software, R/qtl2 package, QTL-seq_framework1.4.4, MapQTL v6.0, and QTL IciMapping V4.1.

For example, Jin et al. developed an RIL population to identify aflatoxin production–related QTLs in peanut. QTL Cartographer 2.5 software was used in this study. Three major QTLs, *qAFTB06.3*, *qAFTB05.2*, and *qAFTA05.1*, were identified strongly associated with *A. flavus* resistance (Jin et al., 2023). Ding et al. (2022) evaluated the Chinese mini-core collection and identified 14 resistant lines against *A. flavus*, including 5 with aflatoxin production resistance and 7 with seed infection resistance and having shell infection resistance. GATK software used for SNP calling and mixed linear models and general linear models were conducted with the TASSEL 5.0 tool. 16 InDels/SNPs linked with these resistance traits were identified. Zh.h1312 accession possessed QTLs strongly related to aflatoxin resistance. RILs with introgressed genes from *A. batizocoi*, *A. correntina*, and *A. cardenasii* have been evaluated and two QTLs (*qSbIB04* and *qSbIA04*) were found to be associated with sclerotinia blight resistance using the R/qtl2 package. The *qSbIB04* QTL showed 22% phenotypic variation at genetic distance of 13.38 cM, while *qSbIA04* explained 29% phenotypic variation and mapped at 56.39 cM distance (Rosso et al., 2023). Luo et al. (2019) identified a major QTL associated with bacterial wilt using QTL-seq_framework1.4.4 and SnpEff v3.0 software. Similarly, Luo et al. (2020) developed RIL population by crossing Xuzhou68-4 with YZ9012 and constructing an SNP-based genetic map. Windows QTL Cartographer 2.5 software was used for QTL analysis, which showed a stable and major

QTL *qBWB02.1* linked to bacterial wilt resistance. SnpEff v3.0 software was used to predict the function of detected SNPs, and InterPro online server, eggNOG-mapper, and Blast2GO were employed for candidate gene analysis. SNP and SSR markers have been used to construct genetic linkage maps with JoinMap 4 software, and variant calling was performed using SAMtools. QTLs analysis was conducted with WinQTLCart 2.5 and R/qtl package and detect two major QTLs for bacterial wilt resistance having 12%–21% phenotypic variation (Zhao et al., 2016). MapQTL v6.0 package was used to identify a major bacterial wild resistance QTL *qBWA12* using a high-density genetic map constructed by 521 RILs (Qi et al., 2022). Chu et al. (2019) used RILs (GP-NC WS16 x Florida-07) to develop an SNP-based genetic map. QTL cartographer 2.5 program was used to identify 3 QTLs, *qLLSB05*, *qLLSB03*, and *qLLSA05*, having 5%–41% phenotypic variation for early and late leaf spot resistance.

RIL population was used for QTL analysis using QTL IciMapping V4.1 and detected two QTLs, *QTLPBND-02* and *QTLPBND-01*, with 12.38%–16.88% phenotypic variation strongly linked with peanut bud necrosis resistance (Jasani et al., 2021). Zhang et al. (2020) evaluated 120 accessions of a US mini-core collection using SVS software package and GAPIT and detected 46 QTLs against early and late leaf spot resistance, explaining 10.19%–24.11% phenotypic variation. Among these QTLs, 28 were found to be strongly linked with late leaf spot, while 18 QTLs were associated with early leaf spot resistance.

RIL population with 181 lines has been tested under three diverse environmental conditions showing transgressive segregations and continuous distribution across all determined traits. QTL analysis was performed using Cartographer 2.5 Software and IciMapping V4.1 Software and detected 26 novel QTLs having a wide range of heritability (0.87–0.95) with phenotypic variation of 3.79%–52.37%. A major, stable novel QTL region for pod shape has been identified present on chromosome 2 having 543 candidate genes (Zhang et al., 2023b). In another experiment, Zhang et al. studied the pod number per plant by using RIL population. QTL mapping with Cartographer 2.5 Software and IciMapping V4.1 Software identified 11 different QTLs with 5.12%–24.91% phenotypic variations. Two major QTLs were closely linked with pod numbers. Additionally, epistatic interaction involving 26 pairs controlling 43 loci has been observed (Zhang et al., 2023a).

WinQTL Cartographer v2.5_011 software was used to identify 49 QTLs including 7 major QTLs for seed weight, pod weight, and pod area (Chavarro et al., 2020). Recently, Wang et al. (2022) used Cartographer 2.5 software to identify a major QTL (*qSWB06.3*) of peanut seed weight. Whole genome resequencing was used to study the 318 RILs by constructing high-density genetic maps and detecting QTLs related to quality traits. IciMapping results showed 110 QTLs present on 18 chromosomes and 5,461 bin markers located at 2,032 cM genetic distance. A major QTL qA05.1 demonstrated large phenotypic variation corresponding to 1.5 Mb physical interval in four different environments for six fatty acid, protein and oil content (Sun et al., 2022).

Genome-wide association studies (GWAS) are powerful tools for identifying genetic variants associated with specific traits across populations. By scanning the genomes of diverse species, GWAS enable researchers to pinpoint single nucleotide polymorphisms (SNPs) correlated with traits, including agronomically significant characteristics (Pandey et al., 2014). These identified SNPs can be utilized to develop molecular markers for marker-assisted selection (MAS) in breeding programs and to establish mapping populations for specific traits or QTLs in crops (Pandey et al., 2016). Bioinformatics plays a crucial role in GWAS by providing computational tools and methodologies essential for data analysis, interpretation, and visualization. Tools like PLINK are highly valued for their speed and efficiency in handling large datasets, offering extensive functionalities for data manipulation, quality control, and association analysis (Purcell et al., 2007). Trait analysis by association, evolution and linkage (TASSEL), known for its versatility, supports association mapping, linkage disequilibrium analysis, and diversity studies, making it widely used in plant genomics (Bradbury et al., 2007). The genome association and prediction integrated tool (GAPIT) integrates various statistical models and algorithms for genome-wide association and genomic prediction, accommodating high-dimensional data and complex traits (Lipka et al., 2015). EMMAX (efficient mixed-model association eXpedited) and FaST-LMM improve association mapping accuracy by effectively correcting for population structure and kinship using mixed models, crucial in diverse plant populations (Kang et al., 2010; Lippert et al., 2011). FarmCPU (fixed and random model circulating probability unification) enhances GWAS power and computational efficiency by iteratively employing fixed and random effect models, thereby reducing false positives (Liu et al., 2016). Haploview is widely used for analyzing and visualizing linkage disequilibrium and haplotype maps, aiding in understanding genetic structure within plant populations (Barrett et al., 2005).

19.2.5 Bioinformatics in Peanut Transcriptomes

Transcriptome sequencing could be used for gene identification, expression analysis, and SNP discovery. Gene expression atlas of cultivated peanut was developed through RNA-seq and reads assembly of 22 different tissues of Tifrunner were used to capture the maximum genes expression during different growth stages. The sequence read counts were assessed against the genomes of diploid wild ancestors *A. duranensis* and *A. ipaensis*. Several bioinformatic tools have been used for transcriptome assembly and identification of differentially expressed genes. Bowtie and Evidential Gene pipeline were used for alignment and redundancy analysis. Furthermore, gene annotation was performed using Trinotate pipeline and Self-Organizing Maps in R package was used for gene network identification

(Clevenger et al., 2016). This valuable peanut gene atlas is freely available online at peanutbase.org (https://peanutbase.org/gene_expression/atlas_combined). Transcriptome sequencing of 22 *Arachis* accessions, including the cultivated and wild species, were used to study the genetic diversity among these accessions and generated 27,045 and 29,234 unique transcripts of diploid and tetraploid species, respectively. The assembly was aligned, and SNP calling was performed using BWA and GATK tools. A R package SNPRelat was used to identify the unique transcripts (Chopra et al., 2016).

Studying the transcriptomes of different peanut genotypes using various bioinformatic tools has opened up opportunities for the discovery of agronomically important genes. This type of transcriptomic analysis is primarily supported by a bioinformatic pipeline involved in data analysis. Following RNA extraction and cDNA library construction, large volumes of sequencing data are generated, which then undergo analysis through the data analysis pipeline. The transcriptomic data analysis pipeline includes quality checking, adapter trimming, reference genome alignment and assembly, generation of read count matrices, normalization, quantification, and visualization of differentially expressed genes (DEGs), as well as functional annotation using various bioinformatic tools (Stark et al., 2019). Yin et al. (2013) performed transcriptome profiling where Reads Per kb per Million reads methods was used to calculate the expression level of unigenes, and functional annotation was done by using gene ontology (GO) and Kyoto Encyclopedia of Genes and Genomes (KEGG) pathway. 59,236 unigenes were found, from which 18% and 15% were differentially expressed in low-oil and high-oil content varieties, respectively. More than 1,500 unigenes were detected which regulate the fatty acid biosynthesis pathways.

As mentioned earlier, RNA-seq analysis requires a series of bioinformatic tools to analyze, interpret, and formulate the results. For example, HISAT2 was used to align and StringTie to assemble the transcripts. Later, GATK2 or SAMtools were used for SNP calling. EBSeq R package was employed for identification of DEGs, while GO and KEGG analyses were performed using the GOseq R package. KOBAS was used to identify candidate genes related to seed weight/size (Li et al., 2021). In another study, Tophat2 was used to align the reads, and then DEGs were identified using the Cufflinks and Cuffdiff programs. FatiGO software was used for GO enrichment analysis of DEGs (Peng et al., 2017), Similarly, Karmakar et al. (2019) used Tophat2 for alignment and StringTie v1.3.3 for gene expression estimation. DEGs were identified by EdgeR 3.6.7 and WEGO software used for GO analysis. KEGG pathway analysis was used to identify the DEGs related to crack entry mechanisms. RNA-seq data related to pod development was analyzed by Illumina instrument software, while expectation maximization (RSEM) was used to quantify the gene expression. GO was performed through WEGO software for gene annotation and KEGG pathway for pathway analysis (Yang et al., 2017). Bioinformatic analysis of transcriptomic data for oil biosynthesis pathway was conducted using HISAT2 for aligning, RSEM v1.2.8 for gene expression calculation, StringTie v1.0.4 for transcript reconstruction, CPC v0.9-r2 for new transcript prediction, and Phyper for GO and KEGG enrichment analysis (Wang et al., 2021).

Researchers recognize wild relatives of peanut, including *A. duranensis* and *A. stenosperma*, as important sources of genetic diversity for abiotic stress tolerance. Vinson et al. (2018) studied the root transcriptomic response of these wild peanut relatives to dehydration treatments to identify DEGs between treated and control samples. GMAP/GSNAP package was used for alignment, HTSeqCount/0.9.1 for read counting, edgeR package in R for DEG analysis, and Circa software for determining DEG locations on *A. duranensis* chromosomes. Functional annotation was performed using the OrthoFinder algorithm and GO enrichment analysis. Similarly, Jiang et al. (2021) conducted a transcriptomic analysis using Hisat (2.0.4) as the assembler and DESeq (1.10.1) to identify DEGs related to drought stress between two peanut cultivars. Zhao et al. (2021) utilized DESeq2 for differential expression analysis of drought resistance in peanuts, employing Hisat2 (2.2.4) aligner and StringTie (1.3.1) assembler.

Peanut transcriptomic studies are used not only for gene identification, but also for many other studies utilizing various bioinformatic tools and databases, for example, spatial transcriptomics analyzing (Liu et al., 2022), mRNA degradation, microRNAs (Zhang et al., 2022), and long non-coding RNAs identification (Tian et al., 2020). They can also be used for weighted gene co-expression network analysis (WGCNA) (Cui et al., 2022; Zhang et al., 2022).

19.3 PEANUT WEB-BASED BIOINFORMATIC TOOLS

Rapid advances in low-cost sequencing technologies have led to the generation of large-scale peanut genomic data. This data is vital to fast-forward peanut breeding and therefore needs a central hub to store all sorts of omics data. To make these datasets easily accessible, numerous web-based online databases have been developed, which play a significant role in peanut genetic and genomic studies. Peanut online databases provide a central stage to store, access, and analyze omics data and ultimately improve our knowledge about peanut genomics, genetics, transcriptomics, and breeding (Table 19.3). In addition, there are some other online information resources which provide valuable information to peanut research and farmer communities around the globe (Table 19.4).

Among peanut online data resources, PeanutBase is an excellent web database which is specifically designed to help professional researchers and breeders to access all the peanut datasets under one umbrella. This online resource is pivotal to collect, access, and download all types of datasets that are crucial for peanut breeding. The major resource

TABLE 19.3
List of Peanut Online Databases

Database	Link
Peanut Genome Resource (PGR)	http://peanutgr.fafu.edu.cn/
NCBI GenBank	https://www.ncbi.nlm.nih.gov/genbank/
PeanutBase	https://www.peanutbase.org/
Legumepedia	https://cegresources.icrisat.org/legumepedia/index.php
Legume Information System (LIS)	https://www.legumeinfo.org/
TropGenDB—Groundnut	https://tropgenedb.cirad.fr/tropgene/JSP/interface.jsp?module=peanut
Phytozome	https://phytozome-next.jgi.doe.gov/
Legume Integrative Platform	https://www.zhaolab.org/LegumeIP/gdp/
Plaza	https://bioinformatics.psb.ugent.be/plaza/

of PeanutBase is the genomic data of the *Archis* species, including the cultivated peanut. Recently, pangene sets using pandagma tools (https://github.com/legumeinfo/pandagma) have been established and included in PeanutBase database. These pangene sets include one set each from *A. ipaensis* and *A. duranensis* and five sets from *A. hypogaea.*

Another important feature of the PeanutBase database is the collection of genetic maps, markers, and QTLs for *Archis* species, including cultivated peanut. Currently, about 29 genetic maps and 798 markers for different traits are curated from the literature. Similarly, about 489 QTLs including 276 QTLs from cultivated peanut related to numerous agronomic traits have been gathered and stored with publication details (Dash et al., 2016). There are several important bioinformatic tools that have been integrated to the PeanutBase, including gene expression atlases of high-throughput sequencing microarrays for *Arachis* species, which are available at Seq Read Archive, Array Express, and NCBI GEO.

Jbrowse is the most crucial tool which allows researchers to navigate through the genomic regions and visualize the specific regions within the genome via a user-friendly interface. It also provides a customizable display to find and search the target genes or genomic regions within the peanut genomes, which is useful for functional genomics and data integration across diverse Arachis genomes datasets and provides a full annotated landscape of the genomes. Likewise, ZZBrowse provides a platform to visualize the QTL and GWAS datasets to detect associations among the species and conserved regions within the genome of a species. A new interesting artificial intelligence–based tool named LegumeBot has been recently added to PeanutBase, which provides a unique search option to access any information stored in online databases.

Likewise, the legume information system (LIS) database is particularly designed to gather all the genomic and genetic data sets of leguminous crops under one platform (Gonzales et al., 2005). The LIS database is integrated in other freely accessible databases to enable comparative genetic and genomic analysis (Berendzen et al., 2021). LIS has been developed to merge all datasets across all the leguminous species and maintains the new genomic data portals for important legume crops. For this, LIS provides synteny mapping, map viewers, and genome viewers across all sequenced legumes (Dash S et al., 2016).

The main features of LIS include LIS InterMines, which integrate genomic and gene expression data across the ten mines. Other features include a genome browser, gene search tool, LIS sequence server BLAST, phylogenetic tree viewer, genome context viewer, genetic map

TABLE 19.4
List of Online Peanut Information Resources for the Research Community

Peanut Information Resources	Website
The Peanut Institute	https://peanut-institute.com/
The American Peanut Council	https://www.peanutsusa.com/
The Peanut Research Foundation	https://peanutresearchfoundation.org/
The American Peanut Research and Education Society	https://apresinc.com/
The Peanut Information Network System	https://site.caes.uga.edu/pins/
The International Crops Research Institute for the Semi-Arid Tropics	http://www.icrisat.org/
The Integrated Breeding Program	https://www.integratedbreeding.net/
Crop Ontology—Groundnut	https://cropontology.org/search?q=groundnut

viewer, and SNP comparison tool. LIS also offers a specialized tool called Funnotate to annotate nucleotide or protein sequences within the LIS database to detect gene families and homologous gene sequences. The latest online tool developed and integrated into LIS in 2018 is called CoNekT@LIS, and it helps users to analyze and visualize the co-function pathways and co-expression in leguminous crops (Proost and Mutwil, 2018). To date, LIS contains the whole genome sequencing data of 9 legume crops within the SoyBase (http://soybase.org) and reference genome sequence of 4 wild and 1 cultivated peanut species with the collaboration of the PeanutBase database (Dash et al., 2016). There are many important features of tools/browsers at Soybase, PeanutBase, and LIS, including transcriptomes and gene model prediction, detection of centromeric and pericentromeric boundaries, genetic markers availability, and informative gene annotations.

19.4 SUMMARY AND FUTURE PROSPECTS

Progress in genomics and bioinformatics tools has allowed scientists to decode the peanut genome and capture the unique genes, identify QTL hotspots, and detect various SNP associated with different agronomic traits. Bioinformatics tools enable aligning the sequence to read more accurately and thereby helps breeders to develop various molecular markers for marker-assisted selection for biotic and abiotic stresses. There is a need to develop more sophisticated bioinformatics tools that can be integrated with genomics and transcriptomics to investigate the hidden and untouched molecular pathways underlying important regulatory mechanisms in peanut. Ultimately, this integrated approach will lead to the development of elite peanut cultivars having climate resilience and resistance to various diseases.

REFERENCES

Agarwal, G., Clevenger, J., Pandey, M.K., Wang, H., Shasidhar, Y., Chu, Y., Fountain, J.C., Choudhary, D., Culbreath, A.K., Liu, X. and Huang, G., 2018. High-density genetic map using whole-genome resequencing for fine mapping and candidate gene discovery for disease resistance in peanut. *Plant Biotechnology Journal*, *16*(11), pp.1954–1967.

Ahmad, N., Zhang, K., Ma, J., Yuan, M., Zhao, S., Wang, M., Deng, L., Ren, L., Gangurde, S.S., Pan, J. and Ma, C., 2023. Transcriptional networks orchestrating red and pink testa color in peanut. *BMC Plant Biology*, *23*(1), p.44.

Altschul, S.F., Gish, W., Miller, W., Myers, E.W. and Lipman, D.J., 1990. Basic local alignment search tool. *Journal of Molecular Biology*, *215*(3), pp.403–410.

Barrett, J.C., Fry, B., Maller, J.D.M.J. and Daly, M.J., 2005. Haploview: Analysis and visualization of LD and haplotype maps. *Bioinformatics*, *21*(2), pp.263–265.

Berendzen, J., Brown, A.V., Cameron, C.T., Campbell, J.D., Cleary, A.M., Dash, S., Hokin, S., Huang, W., Kalberer, S.R., Nelson, R.T. and Redsun, S., 2021. The legume information system and associated online genomic resources. *Legume Science*, *3*(3), p.e74.

Bertioli, D.J., Cannon, S.B., Froenicke, L., Huang, G., Farmer, A.D., Cannon, E.K., Liu, X., Gao, D., Clevenger, J., Dash, S. and Ren, L., 2016. The genome sequences of Arachis duranensis and Arachis ipaensis, the diploid ancestors of cultivated peanut. *Nature Genetics*, *48*(4), pp.438–446.

Bertioli, D.J., Jenkins, J., Clevenger, J., Dudchenko, O., Gao, D., Seijo, G., Leal-Bertioli, S.C., Ren, L., Farmer, A.D., Pandey, M.K. and Samoluk, S.S., 2019. The genome sequence of segmental allotetraploid peanut Arachis hypogaea. *Nature Genetics*, *51*(5), pp.877–884.

Boetzer, M., Henkel, C.V., Jansen, H.J., Butler, D. and Pirovano, W., 2011. Scaffolding pre-assembled contigs using SSPACE. *Bioinformatics*, *27*(4), pp.578–579.

Bradbury, P.J., Zhang, Z., Kroon, D.E., Casstevens, T.M., Ramdoss, Y. and Buckler, E.S., 2007. TASSEL: Software for association mapping of complex traits in diverse samples. *Bioinformatics*, *23*(19), pp.2633–2635.

Burow, M.D., Simpson, C.E., Starr, J.L. and Paterson, A.H., 2001. Transmission genetics of chromatin from a synthetic amphidiploid to cultivated peanut (*Arachis hypogaea* L.): broadening the gene pool of a monophyletic polyploid species. *Genetics*, *159*(2), pp.823–837.

Campbell, M.S., Law, M., Holt, C., Stein, J.C., Moghe, G.D., Hufnagel, D.E., Lei, J., Achawanantakun, R., Jiao, D., Lawrence, C.J. and Ware, D., 2014. MAKER-P: a tool kit for the rapid creation, management, and quality control of plant genome annotations. *Plant Physiology*, *164*(2), pp.513–524.

Chavarro, C., Chu, Y., Holbrook, C., Isleib, T., Bertioli, D., Hovav, R., Butts, C., Lamb, M., Sorensen, R., A. Jackson, S. and Ozias-Akins, P., 2020. Pod and seed trait QTL identification to assist breeding for peanut market preferences. *G3: Genes, Genomes, Genetics*, *10*(7), pp.2297–2315.

Chin, C.S., Peluso, P., Sedlazeck, F.J., Nattestad, M., Concepcion, G.T., Clum, A., Dunn, C., O'Malley, R., Figueroa-Balderas, R., Morales-Cruz, A. and Cramer, G.R., 2016. Phased diploid genome assembly with single-molecule real-time sequencing. *Nature Methods*, *13*(12), pp.1050–1054.

Chopra, R., Burow, G., Simpson, C.E., Chagoya, J., Mudge, J. and Burow, M.D., 2016. Transcriptome sequencing of diverse peanut (Arachis) wild species and the cultivated species reveals a wealth of untapped genetic variability. *G3: Genes, Genomes, Genetics*, *6*(12), pp.3825–3836.

Chu, Y., Chee, P., Culbreath, A., Isleib, T.G., Holbrook, C.C. and Ozias-Akins, P., 2019. Major QTLs for resistance to early and late leaf spot diseases are identified on chromosomes 3 and 5 in peanut (*Arachis hypogaea*). *Frontiers in Plant Science*, *10*, p.883.

Clement, N.L., Snell, Q., Clement, M.J., Hollenhorst, P.C., Purwar, J., Graves, B.J., Cairns, B.R. and Johnson, W.E., 2010. The GNUMAP algorithm: unbiased probabilistic mapping of oligonucleotides from next-generation sequencing. *Bioinformatics*, *26*(1), pp.38–45.

Clevenger, J., Chu, Y., Chavarro, C., Agarwal, G., Bertioli, D.J., Leal-Bertioli, S.C., Pandey, M.K., Vaughn, J., Abernathy, B., Barkley, N.A. and Hovav, R., 2017. Genome-wide SNP genotyping resolves signatures of selection and tetrasomic recombination in peanut. *Molecular Plant*, *10*(2), pp.309–322.

Clevenger, J., Chu, Y., Chavarro, C., Botton, S., Culbreath, A., Isleib, T.G., Holbrook, C.C. and Ozias-Akins, P., 2018. Mapping late leaf spot resistance in peanut (Arachis hypogaea) using QTL-seq reveals markers for marker-assisted selection. *Frontiers in Plant Science*, *9*, p.83.

Clevenger, J., Chu, Y., Scheffler, B. and Ozias-Akins, P., 2016. A developmental transcriptome map for allotetraploid *Arachis hypogaea*. *Frontiers in Plant Science*, *7*, p.1446.

Cui, Y., Bian, J., Lv, Y., Li, J., Deng, X.W. and Liu, X., 2022. Analysis of the transcriptional dynamics of regulatory genes during peanut pod development caused by darkness and mechanical stress. *Frontiers in Plant Science*, *13*, p.904162.

Dash, S., Campbell, J.D., Cannon, E.K., Cleary, A.M., Huang, W., Kalberer, S.R., Karingula, V., Rice, A.G., Singh, J., Umale, P.E. and Weeks, N.T., 2016. Legume information system (LegumeInfo. org): A key component of a set of federated data resources for the legume family. *Nucleic Acids Research*, *44*(D1), pp.D1181–D1188.

Dash, S., Cannon, E.K., Kalberer, S.R., Farmer, A.D. and Cannon, S.B., 2016. PeanutBase and other bioinformatic resources for peanut. In *Peanuts* (pp. 241–252). AOCS Press.

Ding, Y., Qiu, X., Luo, H., Huang, L., Guo, J., Yu, B., Sudini, H., Pandey, M., Kang, Y., Liu, N. and Zhou, X., 2022. Comprehensive evaluation of Chinese peanut mini-mini core collection and QTL mapping for aflatoxin resistance. *BMC Plant Biology*, *22*(1), p.207.

Dobin, A., Davis, C.A., Schlesinger, F., Drenkow, J., Zaleski, C., Jha, S., Batut, P., Chaisson, M. and Gingeras, T.R., 2013. STAR: ultrafast universal RNA-seq aligner. *Bioinformatics*, *29*(1), pp.15–21.

Fávero, A.P., Simpson, C.E., Valls, J.F. and Vello, N.A., 2006. Study of the evolution of cultivated peanut through crossability studies among Arachis ipaensis, A. duranensis, and *A. hypogaea*. *Crop Science*, *46*(4), pp.1546–1552.

Foissac, S., Gouzy, J., Rombauts, S., Mathé, C., Amselem, J., Sterck, L., de Peer, Y.V., Rouzé, P. and Schiex, T., 2008. Genome annotation in plants and fungi: EuGene as a model platform. *Current Bioinformatics*, *3*(2), pp.87–97.

Foncéka, D., Hodo-Abalo, T., Rivallan, R., Faye, I., Sall, M.N., Ndoye, O., Fávero, A.P., Bertioli, D.J., Glaszmann, J.C., Courtois, B. and Rami, J.F., 2009. Genetic mapping of wild introgressions into cultivated peanut: a way toward enlarging the genetic basis of a recent allotetraploid. *BMC Plant Biology*, *9*(1), pp.1–13.

Gangurde, S.S., Wang, H., Yaduru, S., Pandey, M.K., Fountain, J.C., Chu, Y., Isleib, T., Holbrook, C.C., Xavier, A., Culbreath, A.K. and Ozias-Akins, P., 2020. Nested-association mapping (NAM)-based genetic dissection uncovers candidate genes for seed and pod weights in peanut (*Arachis hypogaea*). *Plant Biotechnology Journal*, *18*(6), pp.1457–1471.

Ghosh, Nandini, 2015. Exact genome alignment. *Theses*, 232.

Gomez-Casati, D.F., Busi, M.V., Barchiesi, J., Peralta, D.A., Hedin, N. and Bhadauria, V., 2018. Applications of bioinformatics to plant biotechnology. *Current Issues in Molecular Biology*, *27*(1), pp.89–104.

Gonzales, M.D., Archuleta, E., Farmer, A., Gajendran, K., Grant, D., Shoemaker, R., Beavis, W.D. and Waugh, M.E., 2005. The Legume Information System (LIS): An integrated information resource for comparative legume biology. *Nucleic Acids Research*, *33*(suppl_1), pp.D660–D665.

Griel, A.E., Eissenstat, B., Juturu, V., Hsieh, G. and Kris-Etherton, P.M., 2004. Improved diet quality with peanut consumption. *Journal of the American College of Nutrition*, *23*(6), pp.660–668.

Guindon, S., Dufayard, J.F., Lefort, V., Anisimova, M., Hordijk, W. and Gascuel, O., 2010. New algorithms and methods to estimate maximum-likelihood phylogenies: Assessing the performance of PhyML 3.0. *Systematic Biology*, *59*(3), pp.307–321.

Guo, B., Khera, P., Wang, H., Peng, Z., Sudini, H., Wang, X., Osiru, M., Chen, J., Vadez, V., Yuan, M. and Wang, C.T., 2016. Annotation of trait loci on integrated genetic maps of Arachis species. In *Peanuts* (pp. 163–207). AOCS Press.

Guo, J., Qi, F., Qin, L., Zhang, M., Sun, Z., Li, H., Cui, M., Zhang, M., Li, C., Li, X. and Zhao, Q., 2023. Mapping of a QTL associated with sucrose content in peanut kernels using BSA-seq. *Frontiers in Genetics*, *13*, p.1089389.

Han, S., Yuan, M., Clevenger, J.P., Li, C., Hagan, A., Zhang, X., Chen, C. and He, G., 2018. A SNP-based linkage map revealed QTLs for resistance to early and late leaf spot diseases in peanut (*Arachis hypogaea* L.). *Frontiers in Plant Science*, *9*, p.1012.

Hardison, R.C., 2003. Comparative genomics. *PLoS Biology*, *1*(2), p.e58.

Holbrook, C.C., Burow, M.D., Chen, C.Y., Pandey, M.K., Liu, L., Chagoya, J.C., Chu, Y. and Ozias-Akins, P., 2016. Recent advances in peanut breeding and genetics. In Stalker, H. T., Wilson R. F. (Eds), *Peanuts Genetics, Processing, and Utilization* (pp.111–145). AOCS Press.

Huang, S., Zhang, J., Li, R., Zhang, W., He, Z., Lam, T.W., Peng, Z. and Yiu, S.M., 2011. SOAPsplice: genome-wide ab initio detection of splice junctions from RNA-Seq data. *Frontiers in Genetics*, *2*, p.46.

Hwang, S., Kim, E., Lee, I. and Marcotte, E.M., 2015. Systematic comparison of variant calling pipelines using gold standard personal exome variants. *Scientific Reports*, *5*(1), p.17875.

Jadhav, M.P., Gangurde, S.S., Hake, A.A., Yadawad, A., Mahadevaiah, S.S., Pattanashetti, S.K., Gowda, M.V., Shirasawa, K., Varshney, R.K., Pandey, M.K. and Bhat, R.S., 2021. Genotyping-by-sequencing based genetic mapping identified major and consistent genomic regions for productivity and quality traits in peanut. *Frontiers in Plant Science*, *12*, p.668020.

Janila, P., Pandey, M.K., Shasidhar, Y., Variath, M.T., Sriswathi, M., Khera, P., Manohar, S.S., Nagesh, P., Vishwakarma, M.K., Mishra, G.P. and Radhakrishnan, T., 2016. Molecular breeding for introgression of fatty acid desaturase mutant alleles (*ahFAD2A* and *ahFAD2B*) enhances oil quality in high and low oil containing peanut genotypes. *Plant Science*, *242*, pp.203–213.

Jasani, M.D., Kamdar, J.H., Bera, S., Sunkad, G. and Bera, S.K., 2021. Novel and stable major QTLs conferring resistance to peanut bud necrosis disease and identification of resistant high yielding peanut breeding lines. *Euphytica*, *217*, pp.1–13.

Jean, G., Kahles, A., Sreedharan, V.T., Bona, F.D. and Rätsch, G., 2010. RNA-Seq read alignments with PALMapper. *Current Protocols in Bioinformatics*, *32*(1), pp.11–6.

Jiang, C., Zhang, H., Ren, J., Dong, J., Zhao, X., Wang, X., Wang, J., Zhong, C., Zhao, S., Liu, X. and Gao, S., 2020. Comparative transcriptome-based mining and expression profiling of transcription factors related to cold tolerance in peanut. *International Journal of Molecular Sciences*, *21*(6), p.1921.

Jiang, C., Li, X., Zou, J., Ren, J., Jin, C., Zhang, H., Yu, H. and Jin, H., 2021. Comparative transcriptome analysis of genes involved in the drought stress response of two peanut (Arachis hypogaea L.) varieties. *BMC Plant Biology*, *21*, pp.1–14.

Jin, G., Liu, N., Yu, B., Jiang, Y., Luo, H., Huang, L., Zhou, X., Yan, L., Kang, Y., Huai, D. and Ding, Y., 2023. Identification and Pyramiding Major QTL Loci for Simultaneously Enhancing Aflatoxin Resistance and Yield Components in Peanut. *Genes*, *14*(3), p.625.

Kang, H.M., Sul, J.H., Service, S.K., Zaitlen, N.A., Kong, S.Y., Freimer, N.B., Sabatti, C. and Eskin, E., 2010. Variance component model to account for sample structure in genome-wide association studies. *Nature Genetics*, *42*(4), pp.348–354.

Karmakar, K., Kundu, A., Rizvi, A.Z., Dubois, E., Severac, D., Czernic, P., Cartieaux, F. and DasGupta, M., 2019. Transcriptomic analysis with the progress of symbiosis in 'crack-entry'legume Arachis hypogaea highlights its contrast with 'infection thread'adapted legumes. *Molecular Plant-Microbe Interactions*, *32*(3), pp.271–285.

Katoh, K. and Standley, D.M., 2013. MAFFT multiple sequence alignment software version 7: Improvements in performance and usability. *Molecular Biology and Evolution*, *30*(4), pp.772–780.

Kayam, G., Brand, Y., Faigenboim-Doron, A., Patil, A., Hedvat, I. and Hovav, R., 2017. Fine-mapping the branching habit trait in cultivated peanut by combining bulked segregant analysis and high-throughput sequencing. *Frontiers in Plant Science*, *8*, p.467.

Khan, S.A., Chen, H., Deng, Y., Chen, Y., Zhang, C., Cai, T., Ali, N., Mamadou, G., Xie, D., Guo, B. and Varshney, R.K., 2020. High-density SNP map facilitates fine mapping of QTLs and candidate genes discovery for Aspergillus flavus resistance in peanut (Arachis hypogaea). *Theoretical and Applied Genetics*, *133*, pp.2239–2257.

Kim, D., Langmead, B. and Salzberg, S.L., 2015. HISAT: a fast spliced aligner with low memory requirements. *Nature Methods*, *12*(4), pp.357–360.

Kim, D., Paggi, J.M., Park, C., Bennett, C. and Salzberg, S.L., 2019. Graph-based genome alignment and genotyping with HISAT2 and HISAT-genotype. *Nature Biotechnology*, *37*(8), pp.907–915.

Kim, K.S., Lee, D., Bae, S.B., Kim, Y.C., Choi, I.S., Kim, S.T., Lee, T.H. and Jun, T.H., 2017. Development of SNP-based molecular markers by Re-sequencing strategy in peanut. *Plant Breeding and Biotechnology*, *5*(4), pp.325–333.

Kong, W., Wang, Y., Zhang, S., Yu, J. and Zhang, X., 2023. Recent advances in assembly of plant complex genomes. *Genomics, Proteomics & Bioinformatics, 21*(3), pp.427–439.

Koren, S., Rhie, A., Walenz, B.P., Dilthey, A.T., Bickhart, D.M., Kingan, S.B., Hiendleder, S., Williams, J.L., Smith, T.P. and Phillippy, A.M., 2018. De novo assembly of haplotype-resolved genomes with trio binning. *Nature Biotechnology*, *36*(12), pp.1174–1182.

Korf, I., 2004. Gene finding in novel genomes. *BMC Bioinformatics*, *5*, pp.1–9.

Krzywinski, M., Schein, J., Birol, I., Connors, J., Gascoyne, R., Horsman, D., Jones, S.J. and Marra, M.A., 2009. Circos: An information aesthetic for comparative genomics. *Genome Research*, *19*(9), pp.1639–1645.

Lander, E.S., Green, P., Abrahamson, J., Barlow, A., Daly, M.J., Lincoln, S.E. and Newburg, L., 1987. MAPMAKER: an interactive computer package for constructing primary genetic linkage maps of experimental and natural populations. *Genomics*, *1*(2), pp.174–181.

Langmead, B. and Salzberg, S., 2012. Fast gapped-read alignment with Bowtie 2. *Nature Methods*, *9*, pp.357–359.

Langmead, B., Trapnell, C., Pop, M. and Salzberg, S.L., 2009. Ultrafast and memory-efficient alignment of short DNA sequences to the human genome. *Genome Biology*, *10*, pp.1–10.

Li, H., 2018. Minimap2: pairwise alignment for nucleotide sequences. *Bioinformatics*, *34*(18), pp.3094–3100.

Li H. and Durbin, R., 2009. Fast and accurate short read alignment with Burrows-Wheeler transform. *Bioinformatics*, *25*, pp.1754–1760.

Li, L., Cui, S., Dang, P., Yang, X., Wei, X., Chen, K., Liu, L. and Chen, C.Y., 2022. GWAS and bulked segregant analysis reveal the Loci controlling growth habit-related traits in cultivated Peanut (*Arachis hypogaea* L.). *BMC Genomics*, *23*(1), p.403.

Li, R., Li, Y., Kristiansen, K. and Wang, J., 2008. SOAP: short oligonucleotide alignment program. *Bioinformatics*, *24*(5), pp.713–714.

Li, R., Zhu, H., Ruan, J., Qian, W., Fang, X., Shi, Z., Li, Y., Li, S., Shan, G., Kristiansen, K. and Li, S., 2010. De novo assembly of human genomes with massively parallel short read sequencing. *Genome Research*, *20*(2), pp.265–272.

Li, Z., Zhang, X., Zhao, K., Zhao, K., Qu, C., Gao, G., Gong, F., Ma, X. and Yin, D., 2021. Comprehensive transcriptome analyses reveal candidate genes for variation in seed size/weight during peanut (*Arachis hypogaea* L.) domestication. *Frontiers in Plant Science*, *12*, p.666483.

Liang, Y., Baring, M., Wang, S. and Septiningsih, E.M., 2017. Mapping QTLs for leafspot resistance in peanut using SNP-based next-generation sequencing markers. *Plant Breeding and Biotechnology*, *5*(2), pp.115–122.

Liao, Y., Smyth, G.K. and Shi, W., 2013. The Subread aligner: fast, accurate and scalable read mapping by seed-and-vote. *Nucleic Acids Research*, *41*(10), pp.e108–e108.

Lipka, A.E., Kandianis, C.B., Hudson, M.E., Yu, J., Drnevich, J., Bradbury, P.J. and Gore, M.A., 2015. From association to prediction: statistical methods for the dissection and selection of complex traits in plants. *Current Opinion in Plant Biology*, 24, pp.110–118.

Lippert, C., Listgarten, J., Liu, Y., Kadie, C.M., Davidson, R.I. and Heckerman, D., 2011. FaST linear mixed models for genome-wide association studies. *Nature Methods*, *8*(10), pp.833–835.

Liu, B., Li, Y. and Zhang, L., 2022. Analysis and visualization of spatial transcriptomic data. *Frontiers in Genetics*, *12*, p.785290.

Liu, H., Sun, Z., Zhang, X., Qin, L., Qi, F., Wang, Z., Du, P., Xu, J., Zhang, Z., Han, S. and Li, S., 2020. QTL mapping of web blotch resistance in peanut by high-throughput genome-wide sequencing. *BMC Plant Biology*, *20*, pp.1–11.

Liu, X., Huang, M., Fan, B., Buckler, E.S. and Zhang, Z., 2016. Iterative usage of fixed and random effect models for powerful and efficient genome-wide association studies. *PLoS Genetics*, *12*(2), p.e1005767.

Luo, H., Pandey, M.K., Khan, A.W., Guo, J., Wu, B., Cai, Y., Huang, L., Zhou, X., Chen, Y., Chen, W. and Liu, N., 2019. Discovery of genomic regions and candidate genes controlling shelling percentage using QTL-seq approach in cultivated peanut (*Arachis hypogaea* L.). *Plant Biotechnology Journal*, *17*(7), pp.1248–1260.

Luo, H., Pandey, M.K., Zhi, Y., Zhang, H., Xu, S., Guo, J., Wu, B., Chen, H., Ren, X., Zhou, X. and Chen, Y., 2020. Discovery of two novel and adjacent QTLs on chromosome B02 controlling resistance against bacterial wilt in peanut variety Zhonghua 6. *Theoretical and Applied Genetics*, *133*, pp.1133–1148.

Luo, Z., Cui, R., Chavarro, C., Tseng, Y.C., Zhou, H., Peng, Z., Chu, Y., Yang, X., Lopez, Y., Tillman, B. and Dufault, N., 2020. Mapping quantitative trait loci (QTLs) and estimating the epistasis controlling stem rot resistance in cultivated peanut (*Arachis hypogaea*). *Theoretical and Applied Genetics*, *133*, pp.1201–1212.

Ma, X., Zhang, X., Traore, S.M., Xin, Z., Ning, L., Li, K., Zhao, K., Li, Z., He, G. and Yin, D., 2020. Genome-wide identification and analysis of long noncoding RNAs (lncRNAs) during seed development in peanut (*Arachis hypogaea* L.). *BMC Plant Biology*, *20*, pp.1–14.

Marçais, G., Delcher, A.L., Phillippy, A.M., Coston, R., Salzberg, S.L. and Zimin, A., 2018. MUMmer4: A fast and versatile genome alignment system. *PLoS Computational Biology*, *14*(1), p.e1005944.

Mochida, K. and Shinozaki, K., 2010. Genomics and bioinformatics resources for crop improvement. *Plant and Cell Physiology*, *51*(4), pp.497–523.

Moretzsohn, M.C., Gouvea, E.G., Inglis, P.W., Leal-Bertioli, S.C., Valls, J.F. and Bertioli, D.J., 2013. A study of the relationships of cultivated peanut (*Arachis hypogaea*) and its most closely related wild species using intron sequences and microsatellite markers. *Annals of Botany*, *111*(1), pp.113–126.

Nabi, R.B.S., Cho, K.S., Tayade, R., Oh, K.W., Lee, M.H., Kim, J.I., Kim, S., Pae, S.B. and Oh, E., 2021. Genetic diversity analysis of Korean peanut germplasm using 48 K SNPs 'Axiom_ Arachis' Array and its application for cultivar differentiation. *Scientific Reports*, *11*(1), p.16630.

Nagy, E.D., Chu, Y., Guo, Y., Khanal, S., Tang, S., Li, Y., Dong, W.B., Timper, P., Taylor, C., Ozias-Akins, P. and Corley Holbrook, C., 2010. Recombination is suppressed in an alien introgression in peanut harboring Rma, a dominant root-knot nematode resistance gene. *Molecular Breeding*, *26*, pp.357–370.

Pandey, M.K., Agarwal, G., Kale, S.M., Clevenger, J., Nayak, S.N., Sriswathi, M., Chitikineni, A., Chavarro, C., Chen, X., Upadhyaya, H.D. and Vishwakarma, M.K., 2017. Development and evaluation of a high density genotyping 'Axiom_Arachis' array with 58 K SNPs for accelerating genetics and breeding in groundnut. *Scientific Reports*, *7*(1), pp.1–10.

Pandey, M.K., Monyo, E., Ozias-Akins, P., Liang, X., Guimarães, P., Nigam, S.N., Upadhyaya, H.D., Janila, P., Zhang, X., Guo, B. and Cook, D.R., 2012. Advances in Arachis genomics for peanut improvement. *Biotechnology Advances*, *30*(3), pp.639–651.

Pandey, M.K., Roorkiwal, M., Singh, V.K., Ramalingam, A., Kudapa, H., Thudi, M., Chitikineni, A., Rathore, A. and Varshney, R.K., 2016. Emerging genomic tools for legume breeding: current status and future prospects. *Frontiers in Plant Science*, *7*, p.455.

Pandey, M.K., Upadhyaya, H.D., Rathore, A., Vadez, V., Sheshshayee, M.S., Sriswathi, M., Govil, M., Kumar, A., Gowda, M.V.C., Sharma, S. and Hamidou, F., 2014. Genomewide association studies for 50 agronomic traits in peanut using the 'reference set'comprising 300 genotypes from 48 countries of the semi-arid tropics of the world. *PLoS One*, *9*(8), p.e105228.

Peng, Z., Liu, F., Wang, L., Zhou, H., Paudel, D., Tan, L., Maku, J., Gallo, M. and Wang, J., 2017. Transcriptome profiles reveal gene regulation of peanut (*Arachis hypogaea* L.) nodulation. *Scientific Reports*, *7*(1), p.40066.

Peng, Z., Zhao, Z., Clevenger, J.P., Chu, Y., Paudel, D., Ozias-Akins, P. and Wang, J., 2020. Comparison of SNP calling pipelines and NGS platforms to predict the genomic regions harboring candidate genes for nodulation in cultivated peanut. *Frontiers in Genetics*, *11*, p.222.

Proost, S. and Mutwil, M., 2018. CoNekT: An open-source framework for comparative genomic and transcriptomic network analyses. Nucleic acids research, *46*(W1), pp.W133–W140.

Purcell, S., Daly, M.J. and Sham, P.C., 2007. WHAP: Haplotype-based association analysis. Bioinformatics, *23*(2), pp.255–256.

Qi, F., Sun, Z., Liu, H., Zheng, Z., Qin, L., Shi, L., Chen, Q., Liu, H., Lin, X., Miao, L. and Tian, M., 2022. QTL identification, fine mapping, and marker development for breeding peanut (*Arachis hypogaea* L.) resistant to bacterial wilt. *Theoretical and Applied Genetics*, *135*(4), pp.1319–1330.

Robinson, J.T., Thorvaldsdóttir, H., Winckler, W., Guttman, M., Lander, E.S., Getz, G. and Mesirov, J.P., 2011. Integrative genomics viewer. *Nature Biotechnology*, *29*(1), pp.24–26.

Rosso, M.H., de Blas, F.J., Massa, A.N., Oddino, C., Giordano, D.F., Seijo, J.G., Arias, R.S., Soave, J.H., Soave, S.J., Buteler, M.I. and Bressano, M., 2023. Two QTLs govern the resistance to Sclerotinia minor in an interspecific peanut RIL population. *Crop Science*, *63*(2), pp.613–621.

Sedlazeck, F.J., Lee, H., Darby, C.A. and Schatz, M.C., 2018. Piercing the dark matter: bioinformatics of long-range sequencing and mapping. *Nature Reviews Genetics*, *19*(6), pp.329–346.

Shirasawa, K., Bhat, R.S., Khedikar, Y.P., Sujay, V., Kolekar, R.M., Yeri, S.B., Sukruth, M., Cholin, S., Asha, B., Pandey, M.K. and Varshney, R.K., 2018. Sequencing analysis of genetic loci for resistance for late leaf spot and rust in peanut (*Arachis hypogaea* L.). *Frontiers in Plant Science*, *9*, p.1727.

Shirasawa, K., Kuwata, C., Watanabe, M., Fukami, M., Hirakawa, H. and Isobe, S., 2016. Target amplicon sequencing for genotyping genome-wide single nucleotide polymorphisms identified by whole-genome resequencing in peanut. *The Plant Genome*, *9*(3), pp. plantgenome2016-06.

Sievers, F. and Higgins, D.G., 2014. Clustal omega. *Current Protocols in Bioinformatics*, *48*(1), pp.3–13.

Simão, F.A., Waterhouse, R.M., Ioannidis, P., Kriventseva, E.V. and Zdobnov, E.M., 2015. BUSCO: assessing genome assembly and annotation completeness with single-copy orthologs. *Bioinformatics*, *31*(19), pp.3210–3212.

Stamatakis, A., 2014. RAxML version 8: a tool for phylogenetic analysis and post-analysis of large phylogenies. *Bioinformatics*, *30*(9), pp.1312–1313.

Stanke, M., Keller, O., Gunduz, I., Hayes, A., Waack, S. and Morgenstern, B., 2006. AUGUSTUS: ab initio prediction of alternative transcripts. *Nucleic Acids Research*, *34*(suppl_2), pp. W435–W439.

Stark, R., Grzelak, M. and Hadfield, J., 2019. RNA sequencing: the teenage years. *Nature Reviews Genetics*, *20*(11), pp.631–656.

Sun, Y., Shang, L., Zhu, Q.H., Fan, L. and Guo, L., 2022. Twenty years of plant genome sequencing: achievements and challenges. *Trends in Plant Science*, *27*(4), pp.391–401.

Sun, Z., Qi, F., Liu, H., Qin, L., Xu, J., Shi, L., Zhang, Z., Miao, L., Huang, B., Dong, W. and Wang, X., 2022. QTL mapping of quality traits in peanut using whole-genome resequencing. *The Crop Journal*, *10*(1), pp.177–184.

Tamura, K., Stecher, G. and Kumar, S., 2021. MEGA11: Molecular evolutionary genetics analysis version 11. *Molecular Biology and Evolution*, *38*(7), pp.3022–3027.

Tan, Y.C., Kumar, A.U., Wong, Y.P. and Ling, A.P.K., 2022. Bioinformatics approaches and applications in plant biotechnology. *Journal of Genetic Engineering and Biotechnology*, *20*(1), p.106.

Tapprich, W.E., Reichart, L., Simon, D.M., Duncan, G., McClung, W., Grandgenett, N. and Pauley, M.A., 2021. An instructional definition and assessment rubric for bioinformatics instruction. *Biochemistry and Molecular Biology Education*, *49*(1), pp.38–45.

Tian, H., Guo, F., Zhang, Z., Ding, H., Meng, J., Li, X., Peng, Z. and Wan, S., 2020. Discovery, identification, and functional characterization of long noncoding RNAs in Arachis hypogaea L. *BMC Plant Biology*, *20*, pp.1–16.

Trapnell, C., Pachter, L. and Salzberg, S.L., 2009. TopHat: discovering splice junctions with RNA-Seq. *Bioinformatics*, *25*(9), pp.1105–1111.Uesaka, K., Oka, H., Kato, R., Kanie, K., Kojima, T., Tsugawa, H., Toda, Y. and Horinouchi, T., 2022. Bioinformatics in bioscience and bioengineering: recent advances, applications, and perspectives. *Journal of Bioscience and Bioengineering*, *134*(5), pp.363–373.

Uesaka, K., Oka, H., Kato, R., Kanie, K., Kojima, T., Tsugawa, H., Toda, Y. and Horinouchi, T., 2022. Bioinformatics in bioscience and bioengineering: recent advances, applications, and perspectives. *Journal of Bioscience and Bioengineering*, *134*(5), pp.363–373.

Vinson, C.C., Mota, A.P.Z., Oliveira, T.N., Guimaraes, L.A., Leal-Bertioli, S.C.M., Williams, T.C.R., Nepomuceno, A.L., Saraiva, M.A.P., Araujo, A.C.G., Guimaraes, P.M. and Brasileiro, A.C., 2018. Early responses to dehydration in contrasting wild Arachis species. *PLoS One*, *13*(5), p.e0198191.

Wang, J., Qi, F., Zheng, Z., Sun, Z., Tian, M., Wang, X., Huang, B., Dong, W. and Zhang, X., 2021. Global transcriptome analyses provide into several fatty acid biosynthesis-related genes in peanut (Arachis hypogaea L.). *Tropical Plant Biology*, *14*(3), pp.267–282.

Wang, J., Yan, C., Li, Y., Li, C., Zhao, X., Yuan, C., Sun, Q. and Shan, S., 2019. GWAS discovery of candidate genes for yield-related traits in peanut and support from earlier QTL mapping studies. *Genes*, *10*(10), p.803.

Wang, M.L., Sukumaran, S., Barkley, N.A., Chen, Z., Chen, C.Y., Guo, B., Pittman, R.N., Stalker, H.T., Holbrook, C.C., Pederson, G.A. and Yu, J., 2011. Population structure and marker–trait association analysis of the US peanut (Arachis hypogaea L.) mini-core collection. *Theoretical and Applied Genetics*, *123*, pp.1307–1317.

Wang, Z., Yan, L., Chen, Y., Wang, X., Huai, D., Kang, Y., Jiang, H., Liu, K., Lei, Y. and Liao, B., 2022. Detection of a major QTL and development of KASP markers for seed weight by combining QTL-seq, QTL-mapping and RNA-seq in peanut. *Theoretical and Applied Genetics*, 135(5), pp.1779–1795.

Wu, T.D., Reeder, J., Lawrence, M., Becker, G. and Brauer, M.J., 2016. GMAP and GSNAP for genomic sequence alignment: enhancements to speed, accuracy, and functionality. *Statistical Genomics: Methods and Protocols*, pp.283–334.

Xiao, C.L., Chen, Y., Xie, S.Q., Chen, K.N., Wang, Y., Han, Y., Luo, F. and Xie, Z., 2017. MECAT: fast mapping, error correction, and de novo assembly for single-molecule sequencing reads. *Nature Methods*, 14(11), pp.1072–1074.

Xu, M., Xu, C., Chen, M., Xiao, Z., Wang, Y., Xu, Y. and Xu, D., 2023. Comparative analysis of commonly used bioinformatics software based on omics. *Gene Reports*, 32, p.101800.

Yang, S., Li, L., Zhang, J., Geng, Y., Guo, F., Wang, J., Meng, J., Sui, N., Wan, S. and Li, X., 2017. Transcriptome and differential expression profiling analysis of the mechanism of Ca2+ regulation in peanut (*Arachis hypogaea*) pod development. *Frontiers in Plant Science*, *8*, p.1609.

Yang, Y., Saand, M.A., Huang, L., Abdelaal, W.B. and Sirohi, M.H., 2021. Applications of multi-omics technologies for crop improvement. *Frontiers in Plant Science*, 12, p.563953.

Yao, Z., You, F.M., N'Diaye, A., Knox, R.E., McCartney, C., Hiebert, C.W., Pozniak, C. and Xu, W., 2020. Evaluation of variant calling tools for large plant genome re-sequencing. *BMC Bioinformatics*, *21*(1), pp.1–16.

Yin, D., Wang, Y., Zhang, X., Li, H., Lu, X., Zhang, J., Zhang, W. and Chen, S., 2013. De novo assembly of the peanut (Arachis hypogaea L.) seed transcriptome revealed candidate unigenes for oil accumulation pathways. *PLos One*, *8*(9), p.e73767.

Yu, B., Jiang, H., Pandey, M.K., Huang, L., Huai, D., Zhou, X., Kang, Y., Varshney, R.K., Sudini, H.K., Ren, X. and Luo, H., 2020. Identification of two novel peanut genotypes resistant to aflatoxin production and their SNP markers associated with resistance. *Toxins*, *12*(3), p.156.

Zhang, H., Chu, Y., Dang, P., Tang, Y., Jiang, T., Clevenger, J.P., Ozias-Akins, P., Holbrook, C., Wang, M.L., Campbell, H. and Hagan, A., 2020. Identification of QTLs for resistance to leaf spots in cultivated peanut (*Arachis hypogaea* L.) through GWAS analysis. *Theoretical and Applied Genetics*, *133*, pp.2051–2061.

Zhang, F., Yang, J., Zhang, N., Wu, J. and Si, H., 2022. Roles of microRNAs in abiotic stress response and characteristics regulation of plant. *Frontiers in Plant Science*, *13*, p.919243.

Zhang, S., Hu, X., Miao, H., Chu, Y., Cui, F., Yang, W., Wang, C., Shen, Y., Xu, T., Zhao, L. and Zhang, J., 2019. QTL identification for seed weight and size based on a high-density SLAF-seq genetic map in peanut (*Arachis hypogaea* L.). *BMC Plant Biology*, *19*, pp.1–15.

Zhang, S., Hu, X., Wang, F., Miao, H., Chu, Y., Yang, W., Cui, F., Xu, S., Guo, J., Yu, H. and Zhou, K., 2023a. Genetic dissection of additive and epistatic quantitative trait loci controlling pod number per plant in peanut (*Arachis hypogaea* L.). *Euphytica*, *219*(3), p.31.

Zhang, S.Z., Hu, X.H., Wang, F.F., Ye, C.H.U., Yang, W.Q., Sheng, X.U., Song, W.A.N.G., Wu, L.R., Yu, H.L., Miao, H.R. and Chun, F.U., 2023b. A stable and major QTL region on chromosome 2 conditions pod shape in cultivated peanut (*Arachis hyopgaea* L.). *Journal of Integrative Agriculture*, *22*(8), pp.2323–2334.

Zhang, Y., Park, C., Bennett, C., Thornton, M. and Kim, D., 2021. Rapid and accurate alignment of nucleotide conversion sequencing reads with HISAT-3N. *Genome Research*, *31*(7), pp.1290–1295.

Zhang, X., Zhang, J., He, X., Wang, Y., Ma, X. and Yin, D., 2017. Genome-wide association study of major agronomic traits related to domestication in peanut. *Frontiers in Plant Science*, *8*, p.1611.

Zhang, Y., Yuan, X., Zhang, Y., Luo, Y., Zhao, K., Zu, F., Tian, Z., Li, J., Zhang, L., He, X. and Gao, J., 2025. GWAS and WGCNA analysis uncover candidate genes associated with drought in Brassica juncea L. *Frontiers in Plant Science*, *16*, p.1551804.

Zhao, N., Cui, S., Li, X., Liu, B., Deng, H., Liu, Y., Hou, M., Yang, X., Mu, G. and Liu, L., 2021. Transcriptome and co-expression network analyses reveal differential gene expression and pathways in response to severe drought stress in peanut (Arachis hypogaea L.). *Frontiers in Genetics*, *12*, p.672884.

Zhao, Y., Ma, J., Li, M., Deng, L., Li, G., Xia, H., Zhao, S., Hou, L., Li, P., Ma, C. and Yuan, M., 2020. Whole-genome resequencing-based QTL-seq identified AhTc1 gene encoding a R2R3-MYB transcription factor controlling peanut purple testa colour. *Plant Biotechnology Journal*, *18*(1), pp.96–105.

Zhao, Y., Zhang, C., Chen, H., Yuan, M., Nipper, R., Prakash, C.S., Zhuang, W. and He, G., 2016. QTL mapping for bacterial wilt resistance in peanut (*Arachis hypogaea* L.). *Molecular Breeding*, *36*, pp.1–11.

Zhou, X., Xia, Y., Ren, X., Chen, Y., Huang, L., Huang, S., Liao, B., Lei, Y., Yan, L. and Jiang, H., 2014. Construction of a SNP-based genetic linkage map in cultivated peanut based on large scale marker development using next-generation double-digest restriction-site-associated DNA sequencing (ddRADseq). *BMC Genomics*, *15*, pp.1–14.

Zhuang, W., Chen, H., Yang, M., Wang, J., Pandey, M.K., Zhang, C., Chang, W.C., Zhang, L., Zhang, X., Tang, R. and Garg, V., 2019. The genome of cultivated peanut provides insight into legume karyotypes, polyploid evolution and crop domestication. *Nature Genetics*, *51*(5), pp.865–876.

20 Peanut Databases
Source and Possible Applications

Chi-Nga Chow, Chien-Wen Yang, Nai-Yun Wu, Yu-hsuan Chiu, and Wen-Chi Chang

20.1 INTRODUCTION

In the era of Big Data, bio-databases have become increasingly important for life-sciences research. The explosion of high-throughput technologies, such as DNA sequencing and proteomics, has generated vast amounts of biological data that need to be organized, stored, and analyzed. Bio-databases provide researchers with a centralized platform to access and share biological data, facilitating collaborations and accelerating scientific discoveries. In addition, the increasing availability of open-access data and the development of new analytical tools have further expanded the utility of bio-databases in biological research. As a result, bio-databases have become indispensable for many areas of life-sciences research, including plant research.

With the advent of high-throughput DNA sequencing technologies, plant genomes can be sequenced rapidly and efficiently, and their information can be stored and analyzed in the genome database. This has revolutionized the field of plant research by providing researchers with an extensive source of genetic data that can be used to study plant biology, evolution, and biotechnology. One of the primary uses of genome databases in plant research is to identify and annotate genes. Genome databases provide researchers access to a wealth of genetic information, including the location and function of individual genes, their regulatory regions, and their interactions with other genes and proteins. This information can be used to study plant development, physiology, and responses to biotic and abiotic stresses. For example, genomic data have been used to identify genes involved in plant responses to drought, salt, and heat stress, which are important for improving plant tolerance to these stresses. Another important application of genome databases in plant research is the identification of genetic markers for plant breeding. Genetic markers are specific DNA pieces that are linked to certain traits or characteristics, such as disease resistance or yield. Genome databases not only provide genetic information for identifying and developing genetic markers for crop improvement but also enable the study of plant evolution and phylogenetics. Plant genomes contain evolutionary information that can help researchers understand the relationships and divergence times of different plant species. Genome databases offer access to the genetic information of a wide range of plant species, which can be used to study plant evolution, construct phylogenetic trees, and investigate important events, such as the origins of photosynthesis and the evolution of flowering plants. Recently, Marchant et al. have provided insights into seed plant evolution by indicating that genes directing seed development were co-opted from those controlling the development of fern species. Additionally, genome databases are crucial for plant biotechnology, which involves genetic engineering to improve plant traits such as yield, quality, and resistance to pests and diseases. For instance, Cai et al. applied a high-quality genome assembly of Spinach to investigate insights into the biology and evolution of Spinash and provided valuable genetic information that can be used to improve Spinach breeding programs. This chapter not only presents peanut-specific genome databases (PeanutBase, PGR, and LIS-ArachisMine) but also introduces several large databases that contain peanut gene information, including NCBI and Phytozome. On the other hand, there are also several useful databases and analysis platforms for peanut gene expression analysis, such as BioStudies ArrayExpress collection and eFP Browser for *Arachis*. The utilities and applications of these peanut databases will be described in detail in the chapter.

20.2 DATABASES RELATED TO PEANUT FUNCTIONAL GENOMICS

In recent years, there has been a growing interest in peanut genome research within the scientific community due to the numerous challenges such as biotic and abiotic stress faced by peanut cultivation, food security, and agricultural sustainability. Recently, genomics sequencing provided tools for understanding the genetic basis of various traits and developing improved varieties with enhanced stress tolerance and higher yield potential. The peanut genome has a complex structure, with tetraploid and a large genome size of approximately 2.6 Gb. Genome sequencing efforts have been made to understand the genomic structure and diversity of peanuts. Several studies have reported the whole-genome sequencing and assembly of different peanut cultivars, including Tifrunner, Shitouqi, and Fuhuasheng. These genomic studies have led to the identification of

 DOI: 10.1201/9781003360988-20

thousands of genetic markers that are useful for breeding programs to improve yield and disease resistance. In addition, genomic studies have facilitated the identification of genes responsible for oil biosynthesis and allergen production, which can aid in developing improved peanut varieties with desirable traits. With the advancements in genome sequencing technology, several peanut genome databases have been developed, providing an invaluable resource for researchers and breeders. These databases provide access to comprehensive genomic information and tools for analysis, aiding in the understanding of the molecular mechanisms underlying various traits in peanuts.

One of the major peanut genome databases is PeanutBase (https://www.peanutbase.org/), a central repository of peanut genomic information. The database contains information on the reference genome of the peanut cultivar Tifrunner, as well as additional genomic data from other peanut cultivars. PeanutBase provides tools for genome browsing, gene expression analysis, and comparative genomics, allowing researchers to study gene function and evolution across different peanut cultivars. Another significant peanut database is Peanut Genome Resource (PGR) (http://peanutgr.fafu.edu.cn/), which contains information on the whole-genome sequencing and annotation of several peanut cultivars. PGR also provides access to transcriptome data and tools for functional annotation and comparative genomics analysis. ArachisMine (https://mines.legumeinfo.org/arachismine/begin.do) is another peanut database that sources data from the Legume Information System (LIS) datastore. It contains the gene annotations, DNA and protein sequences, and genetic markers for *Arachis* species. Moreover, PeanutDB (http://bioinfolab.muohio.edu/txid3818v1) provides a web portal with integrated search, filtering, navigation, and visualization tools for easy access to the transcriptome assembly and annotations for peanuts. Unfortunately, PeanutDB is no longer being maintained, and the current webpage link is not working. In addition, PeanutMap is a web-based platform that allows users to view specific linkage groups of a map set and compare multiple maps based on marker or trait correspondences. It can compare linkage groups among multiple map sets and identify corresponding linkage groups from the results of different research projects. It contains data from two published peanut genome map sets and three map sets of phenotypic traits. PeanutMap's data has been integrated into the LIS website for cross-species comparisons. The current webpage link for PeanutMap is also not functional.

Overall, the availability of these peanut genomic databases has greatly facilitated research in peanut biology and breeding programs. By providing access to genomic data and tools, researchers can better understand the genetic basis of important traits such as stress tolerance and yield potential, and develop improved varieties with desirable traits.

20.3 PEANUT GENOMIC AND EXPRESSION DATABASES

20.3.1 PeanutBase

PeanutBase (https://www.peanutbase.org/) serves as a collective resource by providing the genomes, genes, traits, and other information related to crop improvement of peanut species. As a genome resource, PeanutBase offers the genome sequences of *A. duranensis*, *A. ipaensis*, and *A. hypogaea* cv. Tifrunner. Users can visualize genome sequence and gene structures (i.e., exons and introns) of these three *Arachis* species by using "Genome (GBrowse/JBrowse)" and searching the input sequence against the peanut genomes and genes by using the "BLAST sequence search" (protein and nucleotide sequences as the input). In addition, the "Genome Context Viewer" enables users to conduct comparative analyses between various peanut species. Within this viewer, the "Micro-synteny Visualization" helps users compare chromosome segments with similar gene orders, while the "Dot Plots" and "Macro-synteny Visualization" offer additional visualizations.

PeanutBase collected the gene expression from six resources and used DESeq2 or edgeR to identify the genes that express differentially between two tissues or between treated and untreated samples (e.g., drought and control). Users can download a list of differentially expressed genes from "Gene Expression Resources." The "Gene Expression Resources" integrates various platforms for mining expression data. For instance, "ArchisMine" can provide users with expression data that has been annotated by the new genome annotation (Tifrunner.gnm2.ann2). "MultiQC" offers an interactive web-based report that allows users to review the RNA-seq analysis results. "Comparative Gene Expression" allows users to compare expressions of homologous genes in legumes. (Figure 20.1)

Besides gene expression, PeanutBase has constructed several searchable functions ("Marker-Assisted Selection," "Trait Search Tool," "ArachisPheno," "Germplasm," and "ZZBrowse") to help users access genomic variants, features, and traits which are potentially or practically used to improve breeding and develop molecular markers. In "Germplasm," PeanutBase also provides the images, informational files, and geographic distribution of peanut germplasm.

20.3.2 PGR Databases

Peanut Genome Resource (PGR, http://peanutgr.fafu.edu.cn/) is developed and hosted by Fujian Agriculture and Forestry University and National Cheng Kung University with the help of many other researchers from the international peanut community and other related communities. PGR provides various functions and visualization tools, that is, genome browse, search functions, germplasm/variety, maps/traits, download, and survey gene expressions.

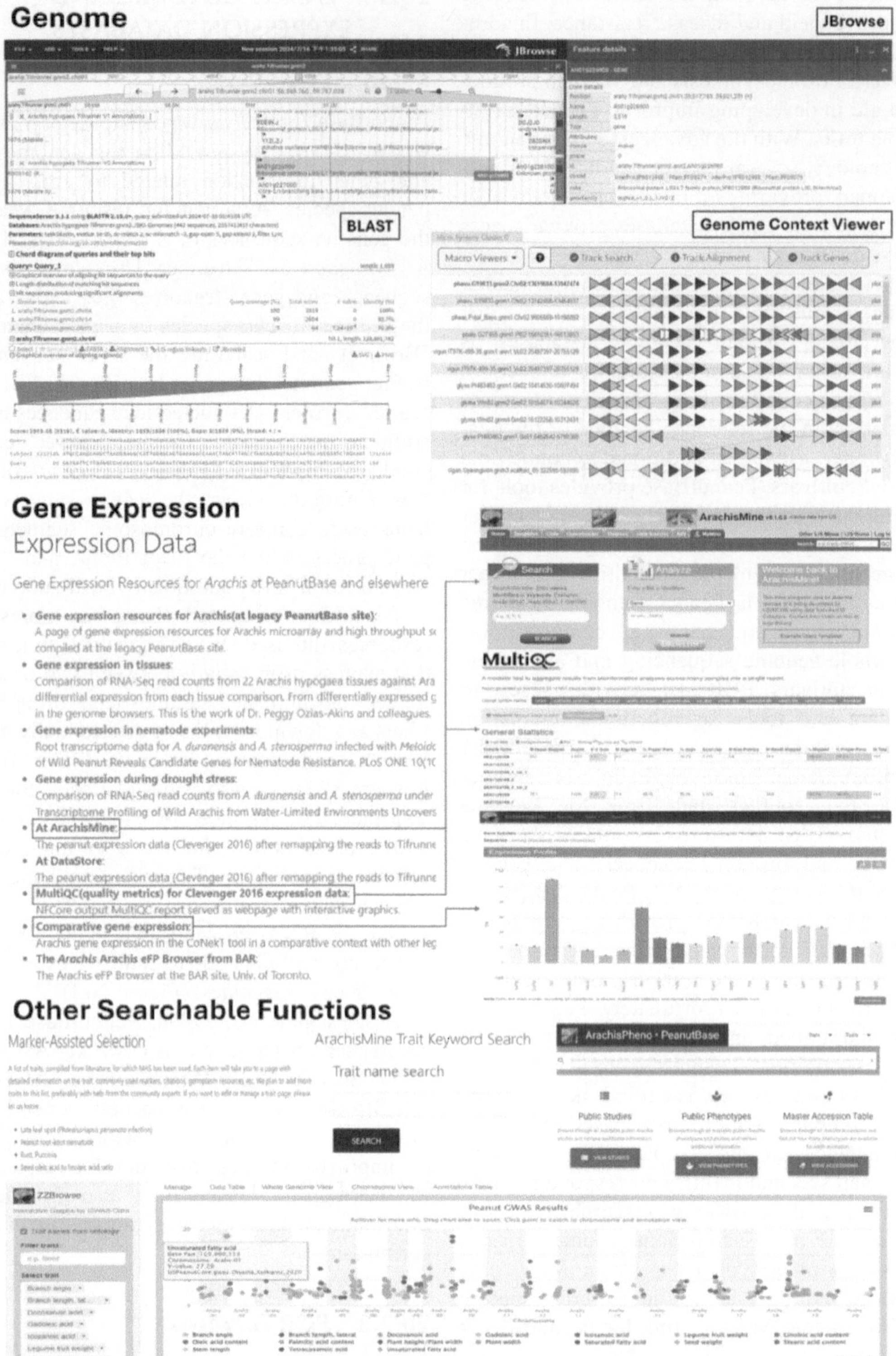

FIGURE 20.1 Screenshot of PeanutBase functions. PeanutBase integrates various genetic and genomic analysis tools and platforms for illustrating peanut data. For instance, it offers "JBrowse," "BLAST," and "Genome Context Viewer" for genomic data; "ArchisMine," "MultiQC," and "Comparative Gene Expression" for expression data; and "Marker-Assisted Selection," "Trait Search Tool," "ArachisPheno," "Germplasm," and "ZZBrowse" for genomic variants, features, and trait data searches.

PGR provides three searchable functions ("Genome Browse," "Search," and "Expression") for *A. hypogaea* var. Shitouqi to facilitate users to rapidly retrieve the information of interest. In "Genome Browse," users can browse peanut genome and view genetic structures (i.e., exons, introns, and five/three prime untranslated regions). Moreover, "Genome Browse" integrates gene expression of 29 tissues (the density of read counts from RNA sequencing data). To retrieve information about a single gene, the "Search" function can be used by inputting gene ID, name, KEGG ortholog, gene ontology, and InterPro ID and executing BLAST tools to align an input sequence (a protein or nucleotide sequence) against peanut genome, mRNA, or protein sequences. In the "Expression" function, users can survey gene expressions in 29 and 32 experimental conditions from transcriptome or microarray data, respectively. The result pages show the information, FPKM (transcriptome), or normalized expression levels (microarray) of the user's selected conditions.

PGR also provides the descriptions, pictures, and information for cultivated peanut species, 46 cultivars, 59 mutants, and 6 typical varieties in China in "Germplasm & Variety," and offers information for SNP and SSR markers, genetic and QTLs of traits in the "Maps & Traits" function. Besides, users also can download genomic data, map data, QTL mapping data, sequences, and a gene list by giving a keyword or a genome region (Figure 20.2).

20.3.3 NCBI Peanut Genome Resources

The NCBI Genome Database (https://www.ncbi.nlm.nih.gov/data-hub/genome/) is a public repository containing a comprehensive collection of genomic data for various organisms. In this section, we introduce the NCBI Genome Database interfaces and two powerful genome analysis tools, "Comparative Genome Viewer" and "NCBI Genome Workbench." The NCBI Genome Database collects 19 genome assemblies of six different *Arachis* species (Table 20.1). With the data package download function (Figure 20.3), users can easily download several genomic data including genome sequence, transcript sequence, protein sequence, annotation files, and a genome assembly data report which records assembly information, for example, accession number, method, and release date (Figure 20.3). The NCBI Genome Database also provides several types of statistical reports, including assembly statistics, annotation reports, quality analysis statistics, and chromosome statistics (Figure 20.4).

One of the powerful genomic analysis tools from NCBI is the Comparative Genome Viewer (https://www.ncbi.nlm.nih.gov/genome/cgv/), which enables users to compare different peanut genomes based on whole-genome assembly–assembly alignments. Users can zoom into a specific alignment by selecting the region of interest and clicking the alignment band for the alignment details panel (Figure 20.5). The Comparative Genome Viewer also provides additional options for users to investigate the genome alignments, such as "Show alignment sequence", and "Download alignment FASTA" (Figure 20.5). With the graphical representation of the sequencing similarity between the selected genomes, the Comparative Genome Viewer can help users investigate genome evolution, domestication, and improvement of the cultivated peanut by comparing its genetic sequence with the ancestor genome. Another powerful tool, the NCBI Genome Workbench (https://www.ncbi.nlm.nih.gov/tools/gbench/), is free desktop software that offers a rich set of integrated tools including graphic view tools, multiple sequence alignment algorithms, and the phylogenetic tree builder. The NCBI Genome Workbench supports several bioinformatics data formats (e.g., FASTA, BED, and GFF) and sources (NCBI databases and local databases) to import data. Users can analyze NCBI data and their in-house genomic data without requiring complicated bioinformatics pipelines or programming. The two major function categories provided by the NCBI Genome Workbench are "Open Viewer" and "Run Tool." For instance, users can view the alignment result file using the "Multiple Sequence Alignment Viewer," which provides several color methods to reveal the alignment results. In addition to viewing the genome sequence, users can select the tool on the "Run Tool" panel to analyze data (Figure 20.6).

20.3.4 Legume Information System (LIS)

The Legume Information System (LIS, https://www.legumeinfo.org/) is a comprehensive database that focuses on 16 genera of legumes, including 11 significant crops and special-focus genera. In addition, LIS offers a download function that includes data on repetitive elements, genome-wide association studies, markers, and genetic maps. As a legume-related database, LIS has developed many tools to help users explore the evolutionary relationships between legume species. Users can retrieve gene families and gene expression data across different legume species and identify potentially conserved regions using the ZZBrowse GWAS/QTL tool. Additionally, LIS offers a SequenceServer BLAST and a map viewer to facilitate genomic sequence comparisons and genetic variant analysis, respectively. LIS provides genome assembly and gene annotation data for three *Arachis* species (Table 20.1), which are integrated into ArachisMine. ArachisMine is an integrated platform for peanut genomics research, providing access to genome annotation, gene expression, genetic variation, and functional annotation data for the peanut genome. This database enables users to find genetic markers associated with specific traits in QTL studies or GWAS, providing genome positions for further analysis.

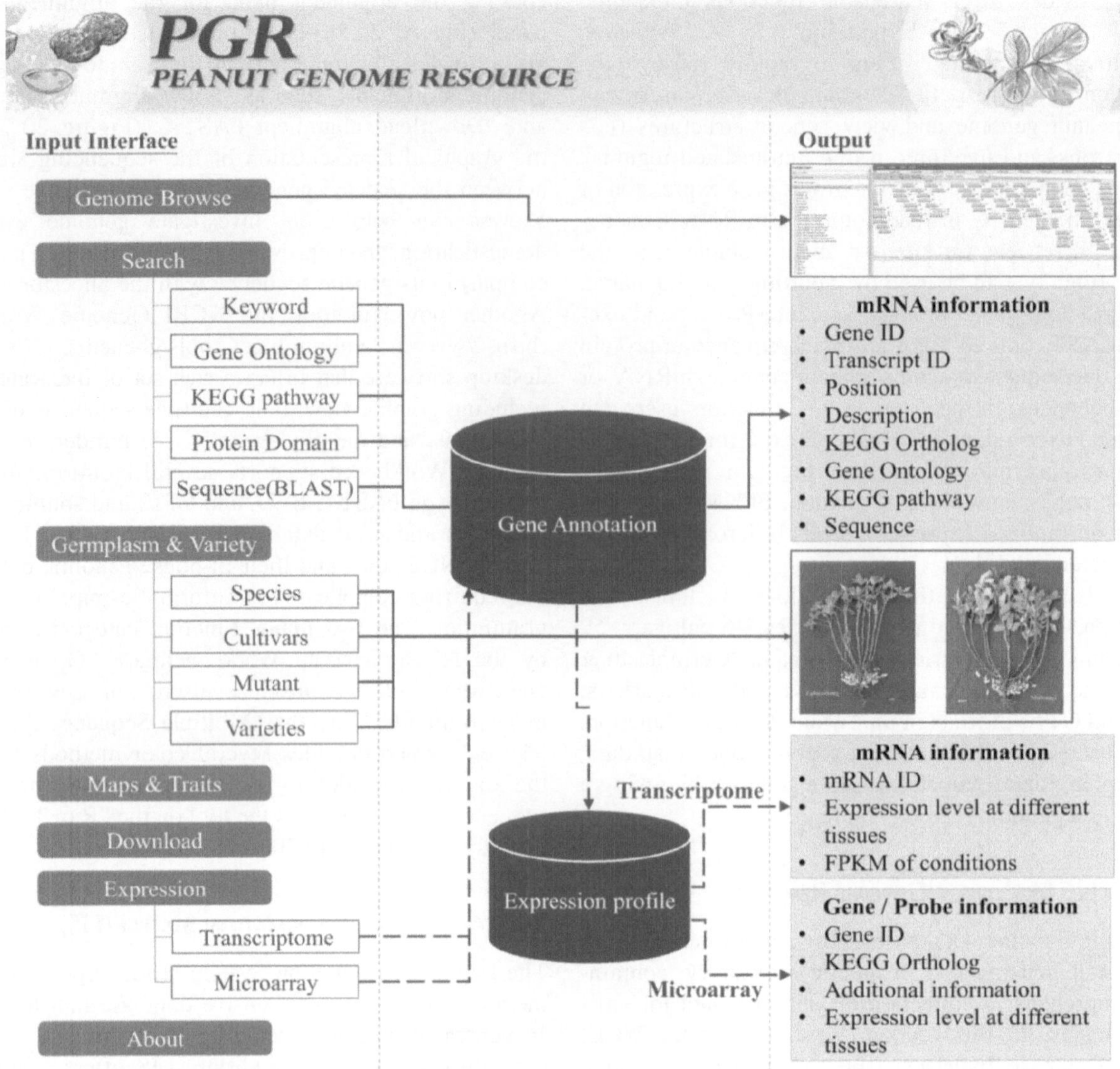

FIGURE 20.2 Overview of the Peanut Genome Resource (PGR) website. PGR provides three searchable functions ("Genome Browse," "Search," and "Expression") for *A. hypogaea* var. Shitouqi and the picture/description information for other cultivated peanut species in China.

20.3.5 Phytozome

Phytozome (https://phytozome-next.jgi.doe.gov/) provides a view of every plant gene's evolutionary history at the sequence level, gene structure, gene family, and genome organization. Phytozome hosts data from various plant species, including model organisms such as *Arabidopsis thaliana*, rice, and *A. hypogaea*. By comparing the sequences, gene structure, and genome organization of genes across different plant species, researchers can gain insights into the evolution of plant genomes and the functions of specific genes. Using JBrowse in Phytozome, researchers can visualize the genomic features of plant species of interest, such as the positions of genes, exons, introns, and regulatory regions.

20.3.6 *Arachis* eFP Browser

The eFP Browser is a database that provides gene expression data for various plant species, including *Arabidopsis*, maize, rice, and *Arachis*, among others. It allows researchers to visualize absolute and relative gene expression patterns in different tissues and developmental stages of the plant, as well as providing expression data from various conditions. The *Arachis* eFP Browser (https://bar.utoronto.ca/efp_arachis/cgi-bin/efpWeb.cgi) is a valuable resource for peanut researchers because it offers a collection of gene expression data for peanut in 22 tissues and developmental stages, and allows for easy comparison of gene expression patterns across different tissues.

TABLE 20.1
The *Arachis* Genome Assemblies in Peanut-Related Databases

Database	Species	Number of Modifiers	Modifier (Cultivar, Ecotype, Isolate)
NCBI	*A. hypogaea*	7	Tifrunner, Shitouqi, GDAAS-fuhuasheng2018, Haihua1, KAUST, Tennessee Red, Bailey_II
	A. paensis	1	K30076
	A. duranensis	4	V14167, PI475845, K30065, K30060
	A. stenosperma	1	V10309
	A. cardenasii	1	K10017
	A. monticola	1	PI 263393
PGR	*A. hypogaea*	1	Shitouqi
PeanutBase	*A. hypogaea*	4	Tifrunner, Shitouqi, Fuhuasheng, Bailey_II
	A. ipaensis	1	K30076
	A. duranensis	1	V14167
	A. stenosperma	1	V10309
	A. cardenasii	1	K10017
LIS	*A. hypogaea*	3	Fuhuasheng, Shitouqui, Tifrunner
	A. duranensis	1	V14167
	A. ipaensis	1	K30076
	A. cardenasii	1	K10017
Phytozome	*A. hypogaea*	1	Tifrunner
Arachis eFP Browser	*A. hypogaea*	1	Tifrunner

20.3.7 ArraryExpress

ArrayExpress is a publicly accessible database that contains high-throughput functional genomics data, including microarray and various sequencing data. The ArrayExpress Data Warehouse is a database of gene expression profiles that have been selected from the repository and consistently re-annotated. Although ArrayExpress was initially published in 2007, as of September 30, 2022, the ArrayExpress interface has been retired. As a result, all datasets have now been migrated into the BioStudies ArrayExpress collection (https://www.ebi.ac.uk/biostudies/arrayexpress). Currently, there are ten transcriptome datasets of *Arachis* in the BioStudies database, including six microarray datasets and four high-throughput sequencing datasets. All the raw data in each dataset are available to download.

20.3.8 Peanut Non-Coding RNA Database

miRNAs are endogenous small ncRNAs that regulate gene expression by targeting messenger RNAs (mRNAs) for degradation or translational repression. miRBase is a vital database for miRNA research, providing an extensive collection of miRNA sequences and annotations. It includes numerous miRNA sequences from various species, such as plants, animals, and viruses. However, there are only 23 miRNA records for *Arachis hypogaea* in miRBase. A miRNA prediction tool named miRDeep-P2 is a fast and precise tool for analyzing miRNA from small RNA sequencing data, making it invaluable for identifying novel miRNAs in plant genomes. These resources are crucial for miRNA annotation in the peanut genome and offer valuable insights into the regulation of gene expression at the post-transcriptional level.

After identifying miRNAs in the peanut genome, determining the miRNA target genes is important for understanding the regulatory mechanisms of miRNAs. The psRNATarget server, developed for high-throughput analysis of next-generation sequencing data, features an efficient distributed computing back-end pipeline that operates on a Linux cluster. Its user-friendly front end incorporates three simplified interfaces for submitting user-generated or preloaded small RNAs and transcript sequences. The server then generates a detailed list of small RNA/target pairs, along with online tools for batch downloading, keyword searching, and result sorting. Users can upload sequencing data from the peanut to reconstruct the relationships among miRNAs and their target genes.

LncRNAs are RNA molecules longer than 200 nucleotides that do not encode proteins. They can regulate gene expression through various mechanisms, including interactions with DNA, RNA, and proteins. Several bioinformatic resources have been developed for predicting plant lncRNAs, such as PLncPRO, CPC2, and PLncDB V2.0. PLncPRO is a valuable tool for predicting lncRNAs in plants using transcriptome data. It employs a machine-learning approach based on the random forest algorithm to distinguish between coding and non-coding transcripts. CPC2 (Coding Potential Calculator 2) is a tool developed

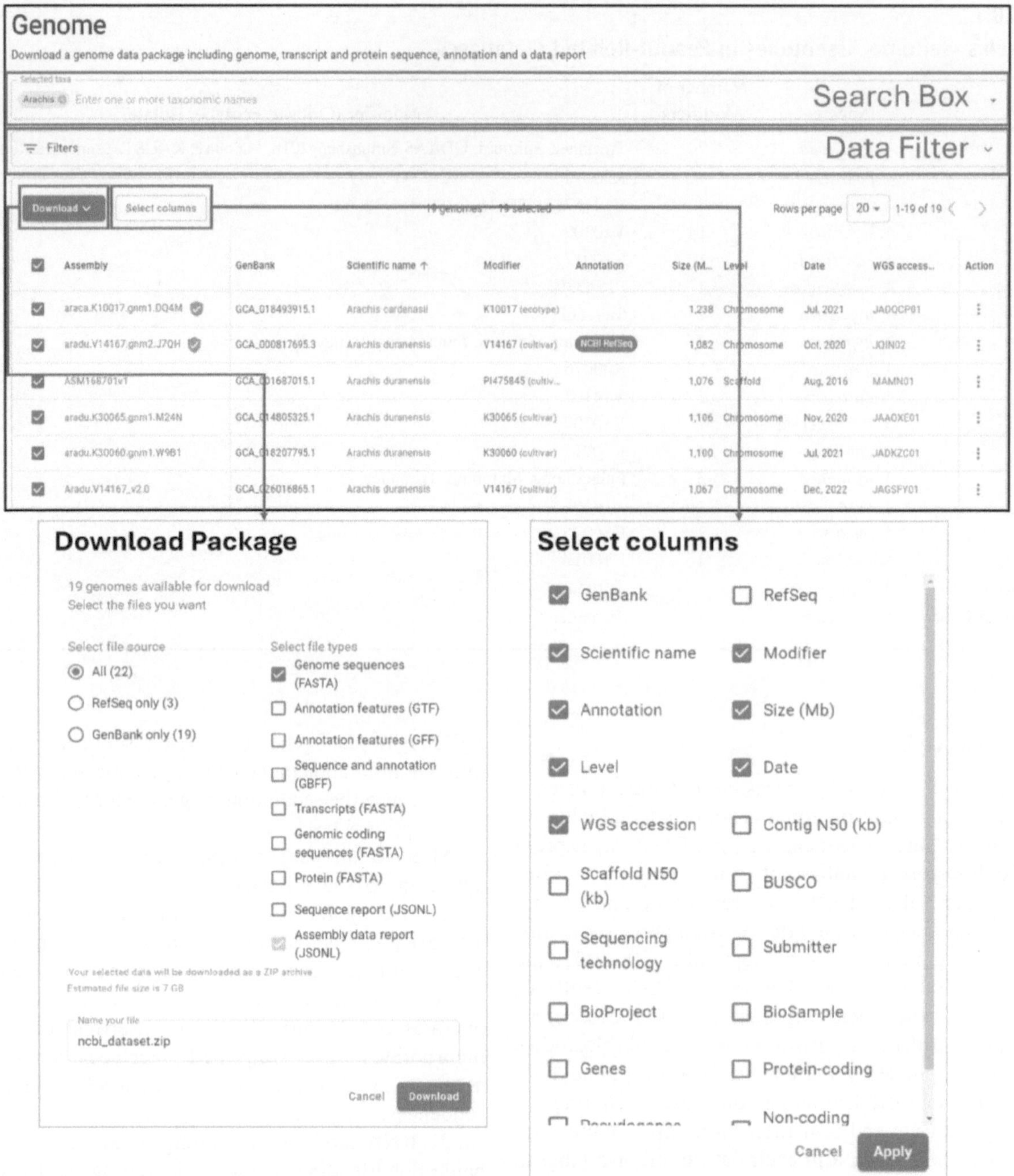

FIGURE 20.3 The NCBI Genome Database search result. The data table provides several functions: data filters, package download, and table column customization.

to predict the coding potential of transcripts, assisting in the identification of potential lncRNAs from transcriptome sequencing data. PLncDB V2.0 is a comprehensive database for plant lncRNAs containing over 1.2 million entries for 80 plant species, including 17,088 predicted lncRNAs from *Arachis ipaensis*.

It is believed that more non-coding RNAs annotated in the peanut genome will advance functional genomic research in peanuts.

20.4 PERSPECTIVES IN PEANUT DATABASE DEVELOPMENT

Over the last decade, peanut genomics research has made significant progress due to the availability of high-throughput sequencing technologies and the development of numerous peanut-specific databases. These databases provide a wealth of information about peanut genomes, transcriptomes, and proteomes, allowing researchers to explore

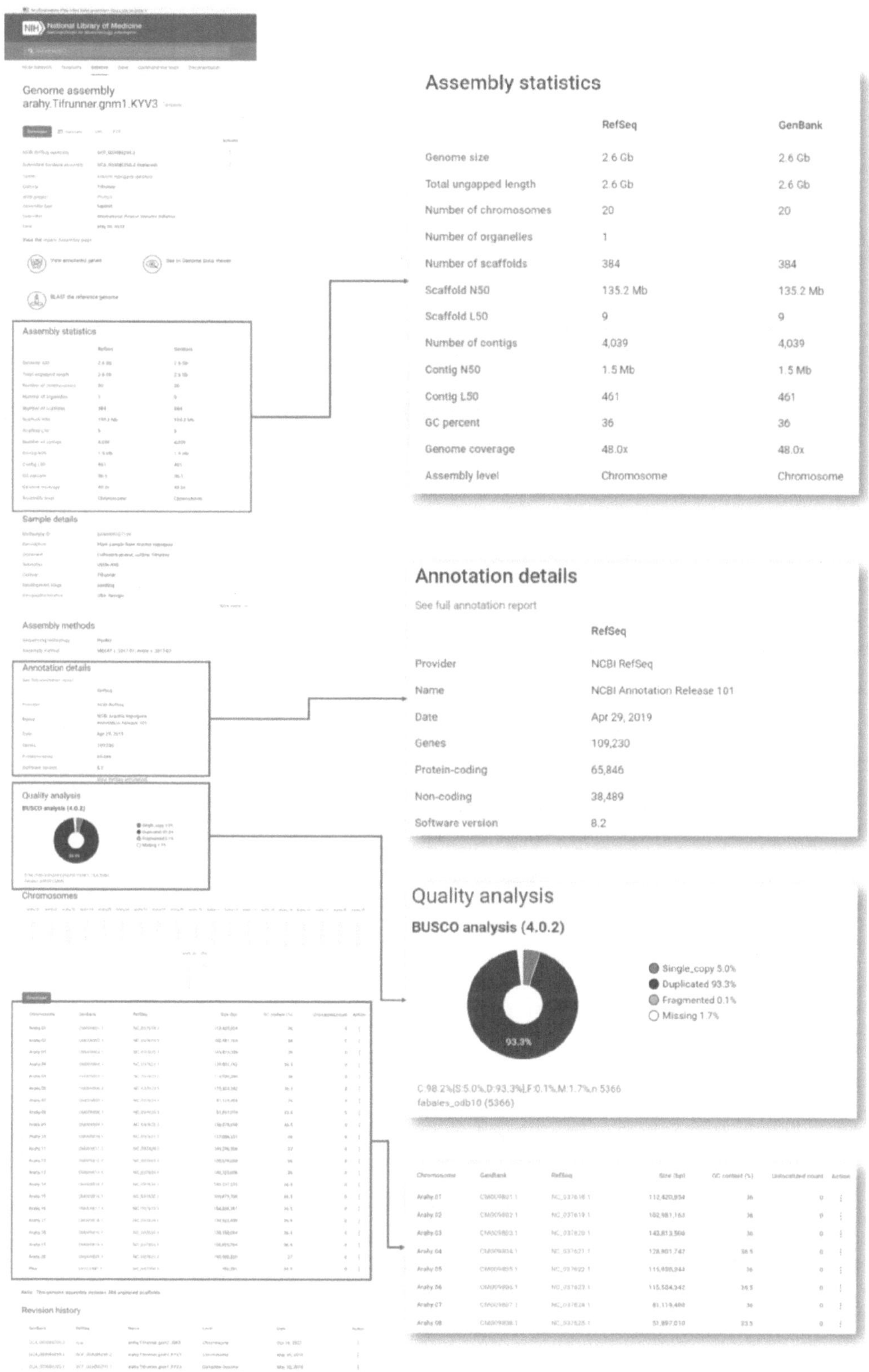

FIGURE 20.4 An overview of the genomic data information page. The NCBI Genome Database provides several statistical reports such as assembly statistics, annotation details, quality analysis, and chromosome statistics.

various aspects of peanut biology, including gene expression, genetic diversity, and stress response.

One trend in database development toward peanut genomics research is the integration of multi-omics data. Multi-omics data integration involves combining data from various sources, such as genomics, transcriptomics, proteomics, and metabolomics, to gain a more comprehensive understanding of biological systems. Several peanut-specific

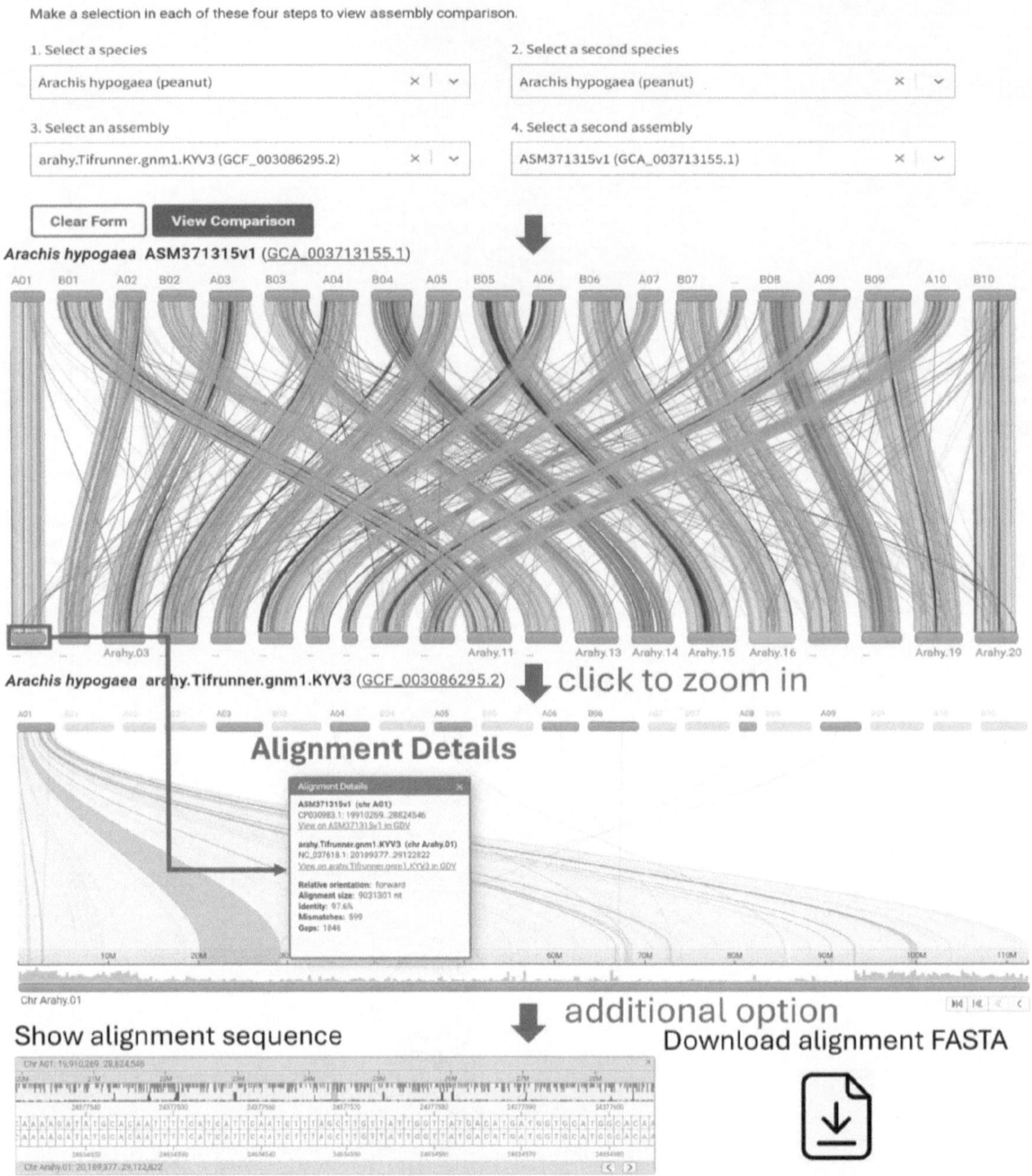

FIGURE 20.5 The print screen of the Comparative Genome Viewer interface. The forward alignments are colored green, while the reverse alignments are colored purple with a twisted shape. Users can click a chromosome to zoom in to see alignments at the chromosome level or click an alignment region to access the additional information, e.g., "Alignment details," "Show alignment sequence," and "Download alignment FASTA".

databases, such as PeanutBase and PGR, have integrated multiple omics data sets to provide a more comprehensive view of peanut biology. For example, PeanutBase includes transcriptomic data from different peanut tissues, developmental stages, and stress treatments, which enables researchers to identify differentially expressed genes involved in peanut-specific traits, such as seed development, oil biosynthesis, and abiotic stress response. The peanut genomics community will be essential in expanding and generating new data in recent databases.

Another trend in database development toward peanut genomics research is the development of user-friendly interfaces and visualization tools. The sheer volume and complexity of data generated by high-throughput sequencing technologies can be overwhelming for researchers without a bioinformatics background. To address this challenge,

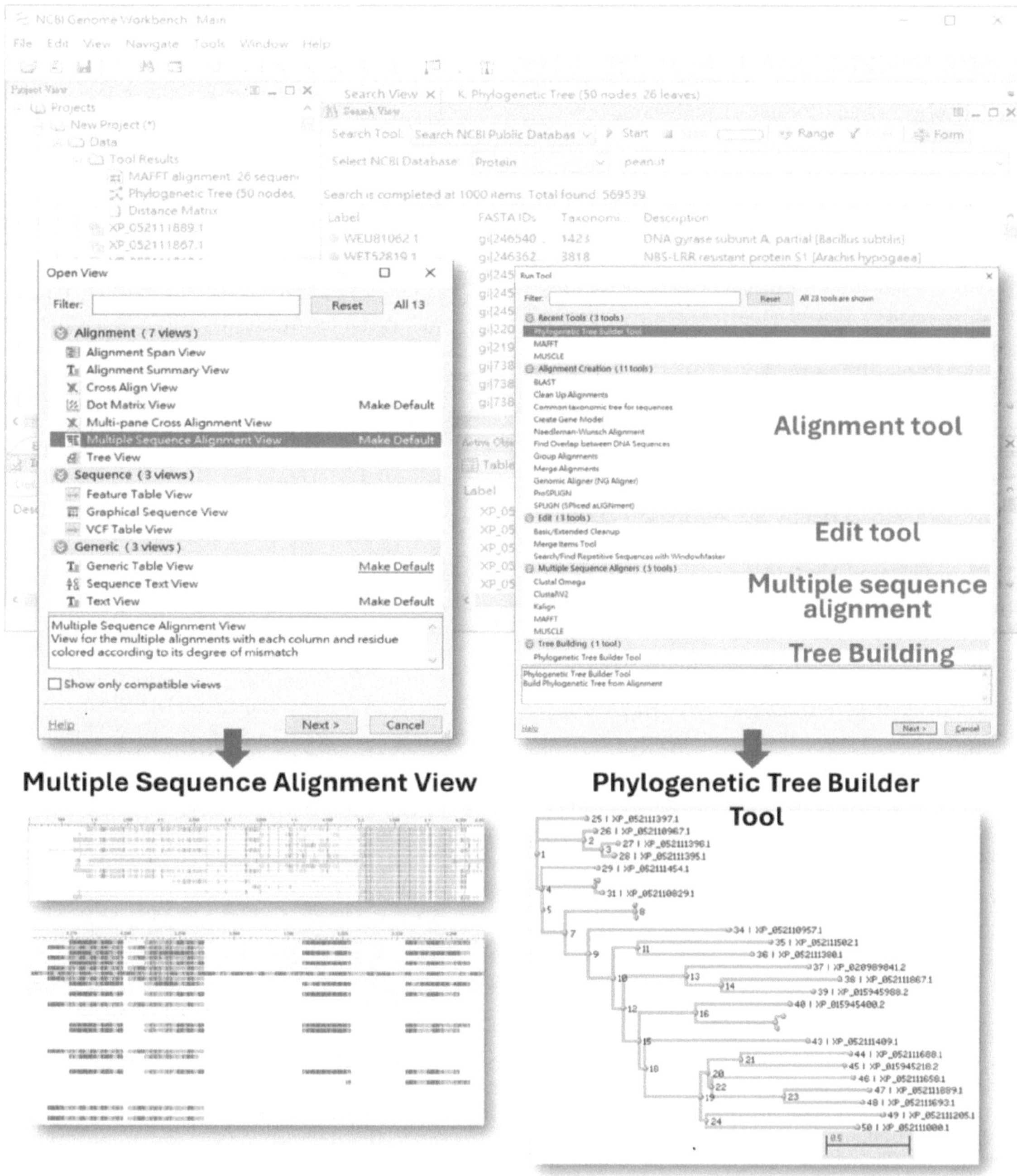

FIGURE 20.6 The print screen of multiple sequence alignment view and phylogenetic tree tool interfaces from NCBI Genome Workbench. NCBI Genome Workbench supports various data formats for data import. The two major functions "Open Viewer" and "Run Tool" allow users to visualize genomic data or perform several downstream analysis functions.

PGR has developed user-friendly interfaces and visualization tools that enable researchers to explore and analyze data more easily. Additionally, several databases provide genetic maps collected manually from literature. In these studies, researchers typically plotted markers on chromosomes to illustrate the genetic linkage between them. However, this visualization method has not been incorporated into existing databases. Further enhancements in visualization are needed in these databases.

A third trend in database development toward peanut genomics research will be the use of machine learning and artificial intelligence (AI) techniques. Machine learning algorithms and AI techniques enable researchers to analyze large datasets more efficiently and accurately than traditional statistical methods. These techniques have not yet been applied to peanut-specific databases. How AI methods are used to identify gene regulatory networks, predict gene function, and classify peanut cultivars based on their agronomic traits will be critical in the future. The integration of machine learning and AI techniques into peanut-specific databases provides an exciting opportunity for peanut genomics research and is likely to accelerate the discovery of novel genes and regulatory pathways.

A peanut-related database for predicting cis-regulatory elements and constructing gene regulatory networks is also very important. Transcription factors (TFs), which are DNA-binding proteins, play crucial roles in regulating gene expression. The availability of genome sequences and genomic variations is helpful for researchers to obtain precise promoter sequences that usually contain TF binding sites (TFBSs). Although existing peanut-related databases do not have the capability to predict TFBSs on promoter sequences, there are several open-access databases of TF binding preferences, such as JASPAR, CIS-BP, and TFBS scanners like FIMO and PoSSuMsearch that enable the prediction of cis-regulatory elements on promoter sequences. Additionally, there are currently no databases designed to provide regulatory information, such as DNA-binding domain, TF families, and binding preferences, of TFs in peanut genomes. The availability of predicted tools and TF information would benefit the understanding of gene regulation and the function of non-coding regions in *Arachis* genomes.

REFERENCES

Barrett, T., S. E. Wilhite, P. Ledoux, C. Evangelista, I. F. Kim, M. Tomashevsky, K. A. Marshall, *et al.* "Ncbi Geo: Archive for Functional Genomics Data Sets--Update." *Nucleic Acids Res* 41, no. Database issue (Jan 2013): D991–5.

Beckstette, M., R. Homann, R. Giegerich, and S. Kurtz. "Fast Index Based Algorithms and Software for Matching Position Specific Scoring Matrices." *BMC Bioinformatics* 7 (Aug 24 2006): 389.

Berendzen, Joel, Anne V. Brown, Connor T. Cameron, Jacqueline D. Campbell, Alan M. Cleary, Sudhansu Dash, Samuel Hokin, *et al.* "The Legume Information System and Associated Online Genomic Resources." *Legume Science* 3, no. 3 (2021): e74.

Bertioli, D. J., S. B. Cannon, L. Froenicke, G. Huang, A. D. Farmer, E. K. Cannon, X. Liu, *et al.* "The Genome Sequences of Arachis Duranensis and Arachis Ipaensis, the Diploid Ancestors of Cultivated Peanut." *Nat Genet* 48, no. 4 (Apr 2016): 438–46.

Cai, X., X. Sun, C. Xu, H. Sun, X. Wang, C. Ge, Z. Zhang, *et al.* "Genomic Analyses Provide Insights into Spinach Domestication and the Genetic Basis of Agronomic Traits." *Nat Commun* 12, no. 1 (Dec 13 2021): 7246.

Chen, N., L. Pan, Z. Yang, M. Su, J. Xu, X. Jiang, X. Yin, *et al.* "A Myb-Related Transcription Factor from Peanut, Ahmyb30, Improves Freezing and Salt Stress Tolerance in Transgenic Arabidopsis through Both Dreb/Cbf and Aba-Signaling Pathways." *Front Plant Sci* 14 (2023): 1136626.

Chen, X., Q. Lu, H. Liu, J. Zhang, Y. Hong, H. Lan, H. Li, *et al.* "Sequencing of Cultivated Peanut, Arachis Hypogaea, Yields Insights into Genome Evolution and Oil Improvement." *Mol Plant* 12, no. 7 (Jul 1 2019): 920–34.

Clevenger, J., Y. Chu, B. Scheffler, and P. Ozias-Akins. "A Developmental Transcriptome Map for Allotetraploid Arachis Hypogaea." *Front Plant Sci* 7 (2016): 1446.

Dai, X., Z. Zhuang, and P. X. Zhao. "Psrnatarget: A Plant Small Rna Target Analysis Server (2017 Release)." *Nucleic Acids Res* 46, no. W1 (Jul 2 2018): W49–W54.

Dash, Sudhansu, Ethalinda K. S. Cannon, Scott R. Kalberer, Andrew D. Farmer, and Steven B. Cannon. "Chapter 8 - Peanutbase and Other Bioinformatic Resources for Peanut." In *Peanuts*, edited by H. Thomas Stalker and Richard F. Wilson, 241–52: AOCS Press, 2016.

Duan, X., E. Schmidt, P. Li, D. Lenox, L. Liu, C. Shu, J. Zhang, and C. Liang. "Peanutdb: An Integrated Bioinformatics Web Portal for Arachis Hypogaea Transcriptomics." *BMC Plant Biol* 12 (Jun 19 2012): 94.

Fornes, O., J. A. Castro-Mondragon, A. Khan, R. van der Lee, X. Zhang, P. A. Richmond, B. P. Modi, *et al.* "Jaspar 2020: Update of the Open-Access Database of Transcription Factor Binding Profiles." *Nucleic Acids Res* 48, no. D1 (Jan 8 2020): D87–D92.

Gaudet, P., A. Bairoch, D. Field, S. A. Sansone, C. Taylor, T. K. Attwood, A. Bateman, *et al.* "Towards Biodbcore: A Community-Defined Information Specification for Biological Databases." *Nucleic Acids Res* 39, no. Database issue (Jan 2011): D7–10.

Goodstein, D. M., S. Shu, R. Howson, R. Neupane, R. D. Hayes, J. Fazo, T. Mitros, *et al.* "Phytozome: A Comparative Platform for Green Plant Genomics." *Nucleic Acids Res* 40, no. Database issue (Jan 2012): D1178–86.

Grant, C. E., T. L. Bailey, and W. S. Noble. "Fimo: Scanning for Occurrences of a Given Motif." *Bioinformatics* 27, no. 7 (Apr 1 2011): 1017–8.

Jesubatham, A. M., and M. D. Burow. "Peanutmap: An Online Genome Database for Comparative Molecular Maps of Peanut." *BMC Bioinformatics* 7 (Aug 11 2006): 375.

Jin, J., P. Lu, Y. Xu, Z. Li, S. Yu, J. Liu, H. Wang, N. H. Chua, and P. Cao. "Plncdb V2.0: A Comprehensive Encyclopedia of Plant Long Noncoding Rnas." *Nucleic Acids Res* 49, no. D1 (Jan 8 2021): D1489–D95.

Kang, Y. J., D. C. Yang, L. Kong, M. Hou, Y. Q. Meng, L. Wei, and G. Gao. "Cpc2: A Fast and Accurate Coding Potential Calculator Based on Sequence Intrinsic Features." *Nucleic Acids Res* 45, no. W1 (Jul 3 2017): W12–W16.

Karsch-Mizrachi, I., T. Takagi, G. Cochrane, and Collaboration International Nucleotide Sequence Database. "The International Nucleotide Sequence Database Collaboration." *Nucleic Acids Res* 46, no. D1 (Jan 4 2018): D48–D51.

Kidokoro, S., K. Watanabe, T. Ohori, T. Moriwaki, K. Maruyama, J. Mizoi, N. Myint Phyu Sin Htwe, *et al.* "Soybean Dreb1/Cbf-Type Transcription Factors Function in Heat and Drought as Well as Cold Stress-Responsive Gene Expression." *Plant J* 81, no. 3 (Feb 2015): 505–18.

Kozomara, A., M. Birgaoanu, and S. Griffiths-Jones. "Mirbase: From Microrna Sequences to Function." *Nucleic Acids Res* 47, no. D1 (Jan 8 2019): D155–D62.

Kozomara, A., and S. Griffiths-Jones. "Mirbase: Annotating High Confidence Micrornas Using Deep Sequencing Data." *Nucleic Acids Res* 42, no. Database issue (Jan 2014): D68–73.

Kuang, Z., Y. Wang, L. Li, and X. Yang. "Mirdeep-P2: Accurate and Fast Analysis of the Microrna Transcriptome in Plants." *Bioinformatics* 35, no. 14 (Jul 15 2019): 2521–22.

Kuznetsov, A., and C. J. Bollin. "Ncbi Genome Workbench: Desktop Software for Comparative Genomics, Visualization, and Genbank Data Submission." *Methods Mol Biol* 2231 (2021): 261–95.

Lee, S. C., M. H. Lim, J. G. Yu, B. S. Park, and T. J. Yang. "Genome-Wide Characterization of the Cbf/Drebl Gene Family in Brassica Rapa." *Plant Physiol Biochem* 61 (Dec 2012): 142–52.

Luo, H., X. Wang, G. Zhan, G. Wei, X. Zhou, J. Zhao, L. Huang, and Z. Kang. "Genome-Wide Analysis of Simple Sequence Repeats and Efficient Development of Polymorphic Ssr Markers Based on Whole Genome Re-Sequencing of Multiple Isolates of the Wheat Stripe Rust Fungus." *PLoS One* 10, no. 6 (2015): e0130362.

Marchant, D. B., G. Chen, S. Cai, F. Chen, P. Schafran, J. Jenkins, S. Shu, *et al.* "Dynamic Genome Evolution in a Model Fern." *Nat Plants* 8, no. 9 (Sep 2022): 1038–51.

Mou, Y., C. Yuan, Q. Sun, C. Yan, X. Zhao, J. Wang, Q. Wang, S. Shan, and C. Li. "Mikc-Type Mads-Box Transcription Factor Gene Family in Peanut: Genome-Wide Characterization and Expression Analysis under Abiotic Stress." *Front Plant Sci* 13 (2022): 980933.

One Thousand Plant Transcriptomes, Initiative. "One Thousand Plant Transcriptomes and the Phylogenomics of Green Plants." *Nature* 574, no. 7780 (Oct 2019): 679–85.

Parkinson, H., M. Kapushesky, M. Shojatalab, N. Abeygunawardena, R. Coulson, A. Farne, E. Holloway, *et al.* "Arrayexpress--a Public Database of Microarray Experiments and Gene Expression Profiles." *Nucleic Acids Res* 35, no. Database issue (Jan 2007): D747–50.

Redsun, S., S. Hokin, C. T. Cameron, A. M. Cleary, J. Berendzen, S. Dash, A. V. Brown, *et al.* "Doing Genetic and Genomic Biology Using the Legume Information System and Associated Resources." *Methods Mol Biol* 2443 (2022): 81–100.

Singh, U., N. Khemka, M. S. Rajkumar, R. Garg, and M. Jain. "Plncpro for Prediction of Long Non-Coding Rnas (Lncrnas) in Plants and Its Application for Discovery of Abiotic Stress-Responsive Lncrnas in Rice and Chickpea." *Nucleic Acids Res* 45, no. 22 (Dec 15 2017): e183.

Sudhansu Dash, Ethalinda K.S. Cannon, Scott R. Kalberer, Andrew D. Farmer, and Steven B. Cannon. "Peanutbase and Other Bioinformatic Resources for Peanut (Chapter 8)." In *In Peanuts Genetics, Processing, and Utilization*, edited by H. Thomas Stalker and Richard F. Wilson, 241–52: AOCS Press, 2016.

Sullivan, A., P. K. Purohit, N. H. Freese, A. Pasha, E. Esteban, J. Waese, A. Wu, *et al.* "An "Efp-Seq Browser" for Visualizing and Exploring Rna Sequencing Data." *Plant J* 100, no. 3 (Nov 2019): 641–54.

Wang, X., M. H. Jia, P. Ghai, F. N. Lee, and Y. Jia. "Genome-Wide Association of Rice Blast Disease Resistance and Yield-Related Components of Rice." *Mol Plant Microbe Interact* 28, no. 12 (Dec 2015): 1383–92.

Weirauch, M. T., A. Yang, M. Albu, A. G. Cote, A. Montenegro-Montero, P. Drewe, H. S. Najafabadi, *et al.* "Determination and Inference of Eukaryotic Transcription Factor Sequence Specificity." *Cell* 158, no. 6 (Sep 11 2014): 1431–43.

Winter, D., B. Vinegar, H. Nahal, R. Ammar, G. V. Wilson, and N. J. Provart. "An "Electronic Fluorescent Pictograph" Browser for Exploring and Analyzing Large-Scale Biological Data Sets." *PLoS One* 2, no. 8 (Aug 8 2007): e718.

Yang, Q., Y. Sharif, Y. Zhuang, H. Chen, C. Zhang, H. Fu, S. Wang, *et al.* "Genome-Wide Identification of Germin-Like Proteins in Peanut (Arachis Hypogea L.) and Expression Analysis under Different Abiotic Stresses." *Front Plant Sci* 13 (2022): 1044144.

Zhang, Q., J. Geng, Y. Du, Q. Zhao, W. Zhang, Q. Fang, Z. Yin, *et al.* "Heat Shock Transcription Factor (Hsf) Gene Family in Common Bean (Phaseolus Vulgaris): Genome-Wide Identification, Phylogeny, Evolutionary Expansion and Expression Analyses at the Sprout Stage under Abiotic Stress." *BMC Plant Biol* 22, no. 1 (Jan 14 2022): 33.

Zhuang, W., H. Chen, M. Yang, J. Wang, M. K. Pandey, C. Zhang, W. C. Chang, *et al.* "The Genome of Cultivated Peanut Provides Insight into Legume Karyotypes, Polyploid Evolution and Crop Domestication." *Nat Genet* 51, no. 5 (May 2019): 865–76.

21 Moving Beyond Conventional to Modern Breeding
An Overview of Fast Breeding Concepts in Peanut

Lixian Qiao, Xiaona Yu, Yanyan Tang, and Hong Zhu

21.1 HISTORY AND TECHNOLOGY OF CULTIVATED PEANUT DOMESTICATION

Peanut (*Arachis hypogaea* L.) is among the most important oil and food legumes and is widely cultivated in tropical and subtropical regions (Bertioli et al., 2016; Bertioli et al., 2019; Zhuang et al., 2019). Most wild peanut species are diploid, while cultivars are allotetraploid (AABB-type genome; 2n = 4x = 40), and the origin of tetraploid cultivated peanut has always been a great concern. However, the wide species that serves as donor to both A and B subgenomes remains unclear. Although some suggest that peanut polyploidy may result from the fusion of unreduced gametes, the prevailing belief is that tetraploid peanut formed through the natural doubling of chromosomes following the cross of wild diploid peanuts. Currently, researchers widely accept that cultivated peanut likely originated from a single, recent hybridization event between two diploid species, followed by polyploidization (Kochert et al., 1996; Cuc et al., 2008; Moretzsohn et al., 2013; Kochert et al., 1991; Ramos et al., 2006; Seijo et al., 2007; Robledo et al., 2010).

Furthermore, findings from geographic distribution, morphology, karyotype, cross-compatibility, and chromosome association in hybrid cells of different wild species suggest that *Arachis batizocoi* may serve as the subgenome B donor species of cultivated peanut.

Later studies found that *A. batizocoi* crossed with other wild species showed a decrease in bivalents and an increase in univalents during chromosome association, and a significant decrease in pollen viability. These results indicated that the genome of *A. batizocoi* differed significantly from that of other wild species (Singh and Moss, 1982, 1984; Smartt et al., 1978). Morphologically, *A. batizocoi* is more similar to *A. fastigiata*, and the chromosomes of F_1 hybrids are more susceptible to trivalent formation after crossing *A. batizocoi* with *A. fastigiata* (a tetraploid subspecies) (Singh and Moss, 1984). Research shows that *A. batizocoi* is more closely related to the subspecies *A. fastigiata* than to *A. hypogea* and likely acts as a subgenome B donor species for *A. fastigiata*. The subgenome B donor species of *A. hypogea* could be the other wild species, which suggests that tetraploid peanut may not be of single origin (Singh and Moss 1984). Most authors support the hypothesis of a single origin for *A. hypogaea* because of the very limited genetic variability among landraces and commercial cultivars (Kochert et al., 1996; Raina et al., 2001; Milla et al., 2005).

However, different authors, using the morphological characteristics and cross-compatibility data, have proposed *A. correntina*, *A. duranensis*, *A. cardenasii* (all with A genome), and *A. batizocoi* and *A. williamsi* (B genome) as probable parents of *A. hypogaea* (Fernandez and Krapovickas, 1994; Lavia, 2011). Kochert et al. (1996) proposed that peanut is single origin. RFLP analysis and cytogenetics studies of 40 materials belonging to 17 species concluded that cultivated peanuts originated from *A. duranensis*. Chromosome doubling of the hybrid or fusion of unreduced gametes in the F_1 (AB) generation formed the heterotetraploid. Subsequently, researchers grew this heterotetraploid in different environments and, through long-term selection, domesticated it into the present different subspecies and botanical varieties (Kochert et al., 1996).

Researchers inferred the origin of tetraploid peanut by analyzing tetraploid cultivated species and diploid and tetraploid wild species using fluorescence in situ hybridization (FISH) and genomic in situ hybridization (GISH). Several studies demonstrated that *A. duranensis* and *A. ipaensis* might be the donor of the A and B subgenomes, respectively. Researchers used FISH to analyze peanut varieties, wild tetraploids, and seven wild diploid peanuts, which they presumed to be the ancestors of tetraploid peanut. The size, number, position on chromosomes, and heterochromatin bands of FISH signals were very similar in peanut cultivar and moticola, while there were great differences in wild diploid peanuts. The result suggests that cultivated peanuts were domesticated from *A. moticola*, based on the combination of FISH results and geographical distribution. The wild ancestors of *A. moticola* are *A. duranensis* and *A. ipaensis* (Seijo et al., 2004).

DOI: 10.1201/9781003360988-21

Subsequently, the same group constructed detailed cytogenetic maps by detecting rDNA loci using FISH and analyzing heterochromatin distribution using DAPI$^+$ staining in 11 species currently assigned to the B or non-A genome, and the results showed that *A. ipaensis* is most likely the B-subgenome donor of tetraploid peanut (Robledo and Seijo, 2010). Seijo et al. (2007) conducted a comparative analysis of these wild diploid with tetraploid species using GISH technique, and the results were consistent with FISH analysis, suggesting that *A. duranensis* and *A. ipaensis* are most likely the A- and B-subgenome donors of tetraploid peanut, respectively. The highly similarity of genomic hybridization results between the cultivated and wild tetraploid peanut *A. monticola* provides evidence for a single origin of tetraploid peanut.

A variety of analyses using molecular markers have been carried out aiming to establish the genetic relationships among species of *Arachis* (Kochert et al., 1991; Raina et al., 2001; Moretzsohn et al., 2013; Milla et al., 2005; Burow et al., 2009). These analyses showed that species with the B genome are usually clearly separated from those with the A genome, and most supported that the ancestors of tetraploid peanut were *A. duranensis* and *A. ipaensis*. In addition, the genome size of *A. hypogaea* is close to the sum of those for *A. duranensis* and *A. ipaensis* (1.25 and 1.56 Gb, respectively), indicating that there has been no large change in genome size since polyploidy (Samoluk et al., 2015).

However, with the development of next-generation sequencing technology, the genetic evolution information of cultivated peanut has also made a breakthrough. On the basis of remarkably high DNA identity of the *A. ipaensis* genome and the B subgenome of cultivated peanut and biogeographic evidence, Bertioli et al. (2016) concluded that *A. ipaensis* may be a direct descendant of the same population that contributed the B subgenome to cultivated peanut (Bertioli et al., 2016). Bertioli et al. (2019) reported the complete chromosome-scale genome sequence of *A. hypogaea* cv. Tifrunner, a runner-type peanut, and demonstrated that the cultivated peanut is of hybrid origin and has a polyploid genome that contains essentially complete sets of chromosomes from two ancestral species. Zhuang et al. (2019) reported that he diploid A genome and tetraploid A subgenome shared 34,266 co-linear genes, whereas the diploid B genome and tetraploid B subgenome shared 38,417, also indicating better preservation of gene co-linearity. Comparison with the sequenced A and B genomes supports *A. ipaensis* or a close relative as the peanut B subgenome progenitor with average >99.5% identity, but much greater divergence (~97% identity) from *A. duranensis*. Therefore, with the completion of the cultivated (*A. hypogaea*) and two wild peanut (*A. duranensis* and *A. ipaensis*) genome sequencing, more systematic and comprehensive data are available to explain the domestication of tetraploid peanut (Bertioli et al., 2016, 2019; Zhuang et al., 2019).

21.2 CROSSBREEDING

21.2.1 Hybridization Breeding

A hybrid is the result of genetic recombination through hybridization, resulting in different variant types that provide an extremely abundant material for selection.

Theoretical basis:

1) Genetic recombination combines the excellent properties of both parents.
2) Gene interactions produce new traits.
3) Gene accumulation produces super-parental traits.

21.2.2 Parental Selection

Breeders select parents based on breeding objectives and variety performance. To select parents with good traits, it is necessary to carry out detailed observation and research on the original breeding material. Generally, breeders should follow these principles:

First, they should select traits with comprehensive excellent performance or outstanding advantages, allowing the variety to complement important traits that have both strengths and weaknesses.

Many important traits of peanut are quantitative traits, and the mean value of all traits in the hybrid offspring is mostly between the two parents. Therefore, the average value of parents can generally determine the performance trend of the hybrid offspring. If the parents have more advantages, their offspring will exhibit good tendency of performance and the chances of good recombination types will be increased. The advantages and disadvantages of the two parents complement each other. For example, researchers successfully crossed Luhua3, a high-resistance and high-yield variety, with Xuzhou 68-4, a high yielding but non-cyanobacteria resistant variety, using Xuzhou 68-4 as the male parent. Luhua3 was able to effectively overcome the shortcomings of Xuzhou 68-4 and generate cyanobacteria-resistant offspring. When selecting parents for hybridization, both parents should have common advantages, but they cannot have mutual disadvantages.

Second, researchers should select parents with significant differences in ecotype and distant kinship. Varieties or materials with different ecological types, geographical origins, and genetic relationships possess distinct genetic material bases, leading to more genetic variation and diverse segregation in their hybrid offspring. In addition to the obvious complementary effects of traits, hybrid offspring often exhibit transgressive inheritance. Therefore, we may have more opportunity to obtain lines with wide adaptability and high yield if we use parents with different ecological, geographical, and genetic relationships.

Third, parents with high general combining ability should be selected as breeding material. Additive genes of the parental varieties determine general compatibility. Varieties with high general combining ability performed well in all combinations crossed with them. Therefore, it is necessary to select varieties with high combining ability as parents to obtain progeny with excellent performance traits.

21.2.3 Backcrossing

A backcross refers to a hybrid F_1 that is crossed with one of the two parents one or more times, such as (A×B) F_1×A or (A×B) F_1×B. The parent used as repeated crosses is called the "recurrent parent," and the other parent is called the "non-recurrent parent." The main purpose of backcross breeding is to integrate one or more advantageous traits from the non-recurrent parent to the recurrent parents, while maintaining most of the excellent traits of the recurrent parents (Yu et al., 2017). Researchers mostly use this approach to improve or overcome the shortcomings of commercial varieties in production. For traits with a simple genetic basis, backcrossing will achieve a good effect.

21.3 ARTIFICIAL MUTAGENESIS BREEDING

Naturally, organisms have a low mutation rate, which artificial mutagenesis can increase. Applying artificial mutagenesis in crop breeding can help develop new traits, and successive artificial selection can inherit those advantageous traits. This process is known as mutagenesis breeding. Several methods can be used to accelerate the mutation process and improve mutation frequency, such as physical irradiation, chemical mutagen and transposon insertion mutation, and so on. In peanut, researchers primarily bred Luhua6 by irradiating mature dry seeds of Baisha1016 with ^{60}Co γ-rays, resulting in new advantages such as 10-day earlier maturity and a 13.6% increase in yield (Qiu et al., 1988). Branch et al. (2003) generated mutants with high oleic acid content by irradiating Georgia Runner with γ-rays, and from these mutants they developed the high oleic acid cultivar "Georgia-02C" through further crossing and selection. Huayu22, a popular peanut cultivar in North China, was developed by ^{60}Co γ-ray irradiation and further hybridization, with characteristics of high yield, good quality, early maturity, disease tolerance, and wide adaptability (Wu et al., 2006). Shanhua8 comes from the irradiation of ^{60}Co γ-ray on hybrids F_1, with 12.2% and 12.7% increased pod and kernel yield respectively compared to control varieties Luhua12 (Bai and Wang, 2008). Yu et al. (2008) obtained the high oleic acid mutant "SPI098" through γ-ray irradiation of a germplasm line "79266," and from this, the high oleic cultivar "Huayu32" was successfully developed. More so, Chu et al. (2009) obtained high oleic acid mutant GA-T2636M by γ-ray irradiation on Georgia Runner. "Hanghua2" is a peanut variety with high and stable yield, disease resistance, and excellent quality. In addition, Guan et al. (2014) developed it from a peanut mutant derived from the cultivar Yueyou13, which had traveled to outer space. Wang, Sui et al. (2015) investigated the effects of fast neutron irradiation on somatic embryogenesis with embryonic leaflet as explants, and new peanut germplasms were developed. Researchers used DNA polymorphisms to detect possible differences between most of the mutants and the wild-type plants (Luhua11), successfully obtaining the new variety Yuhua7 (Wang et al. 2019).

Researchers have also used chemical mutagenesis breeding in peanuts, leading to the development of some new varieties. Sivaram et al. (1985) acquired the C458 type high oleic mutant, with 80% oleic and 5% linoleic, by imposing ethane methyl sulfonate (EMS) on Florunner. Wang et al. (2002) obtained remarkable variations in pod length from L7-1 by sodium azide mutagenesis. Using EMS soaking on mature dry seeds, mutants with increased oil and protein contents were developed (Wang et al., 2008). Huayu16 and Luhua11 were injected with EMS through the pollen tube; mutants with obviously increased pod number and kernel weight per plant were obtained successfully, and Huayu40 was bred and registered (Wang et al., 2010). EMS mutagenesis made Zhonghua5 and Yuanza9102 show different sensitivity on EMS treatment, and the suitable EMS concentration and treatment time for each variety were confirmed respectively (Yin et al., 2009). Zhao et al. (2012) conducted in-vitro mutagenesis with 4 mg/L pingyangmycin (PYM) as the mutagen, and mutants were screened with medium containing 15~20 g/L NaCl. In a germination test with a 0.7% NaCl solution, the germination rate of M3-seeds from 18 NaCl-tolerant regenerated plants was substantially higher than that of the control parent (Huayu22). Using these salt-tolerant mutants as parents, new salt-tolerant peanut varieties Yuhua2 and Yuhua6 were then bred and registered in 2018 (Wang, Qiao et al., 2015). Wang et al. (2020) established a practical and reliable method for creating new peanut varieties with high oil content using in-vitro mutagenesis and directional OP-based selection. As a result, three new peanut varieties—Yuhua4, Yuhua9, and Yuhua14—were bred and registered in 2018. These varieties have oil content levels of 57.7%, 61.1%, and 59.3%, respectively.

Establishment of a mutant pool is essential for new variety breeding and gene function identification. Approximately 2,000 mutants were obtained through in-vitro mutagenesis and tissue culture using embryonic leaflets from peanut varieties Huayu20 and Huayu22 as explants. These were treated with ^{60}Co γ-rays irradiation, fast neutron irradiation, and PYM mutagenesis. As a result, 160 high-oil germplasms and 140 salt-tolerant germplasms were acquired, leading to the breeding and registration of six new high-oil and salt-tolerant varieties (Wang, Sui et al. 2015, Zhao et al., 2012). In another study, mature dry seeds of Luhua11, Huayu36, and Weihua16 were treated with ^{60}Co γ-rays irradiation, fast neutron irradiation, and EMS mutagenesis, and 30,000 M1 single plants were obtained sucessfully, 10,000 of which were reproduced to yield M2 seeds. From these, 421 mutants with significantly altered traits were acquired, including oil content over 58%, oleic acid content over 78%, protein content over 30%, and other traits such as plant height, branch angle, branch number, leaf color, pod type, pod number, and kernel type (Wang and Zhang, 2015).

Another mutant pool came from two lines injected by N^+ and EMS mutagenesis respectively, and a large number of mutants were obtained, including plant height, growth cycle, fertility, pod shape, leaf shape, and kernel quality (Deng et al., 2013; Wang, Zhang et al., 2015). Researchers can use all these mutant pools as essential resources for further identification of functional genes and development of molecular markers, which will benefit efficient molecular breeding.

21.4 MOLECULAR BREEDING

Conventional breeding methods have received due attention by peanut breeders around the world, and at present, most peanut varieties are developed through these methods. However, conventional breeding methods have a series of limitations, such as a long breeding cycle, heavy workload, insufficient special breeding material, and a narrow genetic base. In order to overcome the weaknesses of conventional breeding methods, molecular breeding methods have gained more and more attention. In peanuts, the deployment of molecular technology breeding tools has been successful in bolstering productivity, enhancing nutritional quality, and developing resistance to multiple stresses.

21.4.1 Genetically Modified Organism (GMO)

Genetic modification is a method of breeding new varieties with specific target traits by genetic transformation. Firstly, genes controlling certain traits should be isolated from donor organisms according to the breeding objectives, and then introduced into recipient organisms through genetic engineering. After screening and field experiments, a new transgenic variety or germplasm is developed. In peanut transgenic breeding, researchers can derive the target gene from peanut or other species. They primarily focus on target genes related to disease resistance, insect resistance, herbicide resistance, quality, and abiotic stress resistance. Currently, researchers have made some progress in transgenic peanut research. For example, they transformed the tobacco chitinase gene into peanut cultivars, significantly increasing chitinase activity and developing transgenic plants resistant to peanut leaf spot disease (Rohini and Sankara, 2001). Expression of *Tcchitinase-I* gene in transgenic peanut confers resistance to leaf spot and rust diseases (Marka and Nanna, 2021). Han et al. (2022) found a non-expresser of pathogenesis-related gene from peanut (*AhNPR3*) that could mediate the immune response through regulating the expression of WRKY and PR genes. Zhao, Tian et al. (2022) found that defensin gene *AhDef2.2* enhanced resistance to bacterial wilt in peanuts. Researchers introduced a synthetic cry1EC gene into peanut through *Agrobacterium* transformation, resulting in transgenic plants resistant to the larvae of *Spodoptera litura* (Tiwari et al., 2008). For abiotic stress, Patel et al. (2017) found that overexpression of the mtlD gene in transgenic peanut enhanced salinity stress tolerance and suggested field trials of these salt-tolerant lines for release. Zhu et al. (2021) identified AhWRKY75 as a positive regulator of salt stress resistance in peanut. Liu et al. (2020) proposed that AhHDA1-mediated AhGLK1 regulates the recovery growth of peanut leaves after water stress by promoting chlorophyll synthesis and photosynthesis. The heterologous expression of *GmMYB3a* significantly increased drought tolerance and productivity of transgenic peanuts under conditions of water deficit (He et al., 2020). Ectopic expression of three stress-responsive regulatory genes, *MuNAC4*, *MuWRKY3*, and *MuMYB96*, from horse gram enhanced resistance to drought stress in transgenic peanuts (Venkatesh et al., 2022). Sarkar et al. (2016) demonstrated that the overexpression of *AtDREB1A* from *Arabidopsis* conferred tolerance to soil-moisture deficit stress in transgenic peanuts. Dodo et al. (2008) constructed RNAi vector and transformed peanut hypocotyl explants to obtain transgenic plants lacking immunodominant allergen Arah2. Even though researchers have tried a variety of transformation methods to obtain transgenic peanuts, there are still some problems in the current research, such as low transformation and regeneration efficiency, chimera generation, transgenic plant infertility, and so on. Compared to other crops, the application of transgenic technology in peanut breeding began later and progressed more slowly, and it has not yet reached the stage of industrial production. Therefore, in order to accelerate the research and application of peanut genetic engineering, it is necessary to optimize the genetic transformation system and combine it with traditional breeding methods. Peanut transgenic breeding will certainly make a greater contribution to improving peanut yield and quality in the future.

21.4.2 Marker-Assisted Selection (MAS)

Conventional breeding methods, which select genotypes through phenotype profiles, have disadvantages such as long cycles and low efficiency. Each trait expresses itself during a specific period of plant growth and development; therefore, phenotypic selection is influenced by time. For a quantitative trait, its susceptibility to the environment and the relationship between phenotype and genotype is not clear. Meanwhile, introgression of recessive genes and pyramiding of multiple genes is very difficult using conventional breeding methods. The most effective way to improve genotypes is through direct selection. With the rapid development of modern molecular biology technology, marker-assisted selection (MAS) is a new technology used to analyze the genetic composition of individuals at the molecular level and to realize the direct selection of genotypes, and it can obtain historical recombination episodes and allelic variety not exhibited in structured populations. Since the sequencing of peanut, researchers have developed thousands of markers and QTL mapping for significant agronomic traits and stress resistance. It has also been proved that MAS could be used to introgress multiple recessive genes in less time than conventional breeding. Burow et al. (1996) constructed resistant segregating populations by crossing nematode resistant wild and cultivated species and obtained three RAPD markers related to nematode resistance. Liu et al. (2015) identified five InDels which significantly correlated to tomato spotted wilt virus infection and leaf spot. By BSA-RAD sequencing, Zhao et al. (2016) identified SNP and SSR markers located within two adjacent QTLs controlling bacterial wilt resistance. These markers can provide worthy genetic information for MAS programs and germplasm investigation to design improved peanut cultivars (Kim et al., 2017). Kolekar et al. (2017) used four molecular markers related to disease resistance and successfully obtained foliar disease–resistant and productive lines through molecular marker–assisted backcrossing. Clevenger et al. (2018) identified several markers linked to QTLs for leaf spot resistance. The high oleic acid trait in peanut is due to the simultaneous mutations in the *FAD2A* and *FAD2B* genes (Jung et al., 2000). Chen et al. (2010) developed functional molecular markers to detect allelic variations in *FAD2A* and *FAD2B* alleles. There are variations in phenotypic traits and polymorphism at the genomic DNA level in cultivated peanut varieties. Studying the DNA polymorphism of peanut germplasm resources by molecular marker technology could reveal the genetic variation of peanut germplasm resources at the molecular level. These results lay the foundation for the relationship study of peanut germplasm resources, the construction of genetic maps, and the mining of key genes controlling quality and resistance, which can eventually be used for peanut improvement. In recent years, MAS has also contributed to the design and release of improved varieties. For instance, Varshney et al. (2014) used marker-assisted backcrossing to introgress a rust resistance major QTL into three peanut cultivars to produce 200 introgression lines with improved rust resistance and superior yield. Bera et al. (2018) introgressed high oleic acid alleles from SunOleic 95R into ICGV 05141 using MAS to produced recombinant lines with improved oleic acid content, oil quality, pod yield, and necessary pod/seed traits.

21.4.3 Genome Selection (GS)

Genomic selection is a new breeding method using high-density molecular markers covering the whole genome. It needs to estimate the individual breeding values in combination with phenotype records or pedigree records, assuming that at least one of these markers is in linkage disequilibrium with all QTLs controlling the trait. In GS, precise multi-location phenotyping data and optimum genotyping density are essential to designing training and testing populations for adjusting genomic prediction by different GS models. By constructing prediction models, early prediction and selection of individuals according to the estimated breeding value of the genome are made possible, shortening the selection time, accelerating the breeding process, and saving costs. Genomic selection divides populations into reference and candidate populations, and the reference population for modeling, to estimate the breeding value of the candidate population. In recent years, there has been progress in peanut genome selection. Chaudhari et al. (2019) identified 13,355 genome-wide polymorphic SNPs for performing GS analysis in peanuts using 58K Axiom_Arachis SNP array, and also phenotyped 340 elite peanut lines for 11 agronomic, 11 quality, and 6 foliar disease resistance traits, which could be applied in designing GS estimate models. Chu et al. (2019) constructed a SNP-based linkage map using a recombinant inbred line population and identified major QTLs and molecular markers related to early and late leaf spot resistance. Post-hoc analysis of the RIL subpopulation, not utilized in the QTL mapping, revealed that the flanking markers for these QTLs could successfully be selected for resistant and susceptible lines. Pandey et al. (2020) discovered that naïve interaction models, along with naïve and informed interaction genomic selection (GS) models, effectively improved prediction accuracy by accounting for the complex genotype × environment interactions in peanuts. This finding offers valuable insights into the potential application of GS breeding, particularly for untested genotypes and previously overlooked environments. Zhao, Ren et al. (2022) revealed key candidate loci and 13,125 polymorphic SNP markers associated with important agronomic traits through GWAS analysis, which provided valuable information for further fine mapping of QTLs and improving the quality and yield of peanut using GS. Cason et al. (2023) discovered numerous SNPs connected with sting nematode resistance using 13,306 filtered SNPs and phenotyping, which provided important information for molecular breeding. In the future, breeders should take advantage of GS with conventional breeding to accelerate and design peanut breeding programs.

21.5 SMART BREEDING

Smart breeding of peanuts is the practice of using genetic engineering and data-driven technologies to breed high-yielding, fast-growing, good-quality varieties that can withstand the challenges of climate change or disease. The practice has been around for many years in many other crops (Lavarenne et al., 2018; Mou and Zhao, 2022). What makes it "smart" is the use of data-driven technologies to improve the accuracy and speed of selection, production, and other steps of the breeding process. This saves many resources in terms of labor and materials and lowers costs. The core of using data-driven technology in breeding consists of the following four important aspects;

21.5.1 Genome Assembly, Available in Wild and Cultivated Peanuts

There is a large genomic and phenotypic variation in wild and cultivated peanuts. The genome of Shitouqi was sequenced and assembled using a combination of single-molecule real-time sequencing (SMRT), PacBio, and optical map and chromosome conformation capture (Hi-C) technologies, resulting in a final high-quality peanut genome sequence comprising 2.54 Gb with 20 pseudomolecules (Zhuang et al., 2019). The genome of a landrace peanut Fuhuasheng was assembled using Illumina HiSeq and PacBio technologies in combination with BioNano genome mapping, yielding a high-quality 2.53 Gb of assembled sequence (Chen et al., 2019). *A. monticola* is an allotetraploid wild peanut species and is considered the direct ancestor of cultivated peanut. The genome size of *A. monticola* is ~2.62 Gb (Yin et al., 2018). Once a sufficient number of genome sequences from wild and cultivated peanuts have been obtained, researchers can integrate complementary data sets such as transcriptome sequencing (RNA-Seq) (Ahmad et al., 2022; Rathod et al., 2020), single-cell RNA-seq (scRNA-seq) (Liu et al., 2021), chloroplast genome sequences (Wang et al., 2018), and phenotypic data. This integration will facilitate the identification of alleles or genes associated with peanut agronomic traits. Additionally, it will support comprehensive germplasm characterization at both genomic and agronomic levels. Germplasm resources are the most valuable assets in agricultural production; they are also the basis for variety selection and genetic improvement, and the material basis for breeding new varieties. Identification and evaluation of germplasm resources is an important part of breeding. In recent years, the development of genomics technology has provided new ideas for the identification of peanut germplasm resources. Researchers have used SNP microarrays and SNP typing systems to perform whole-genome resequencing on major peanut cultivars, enabling the identification and evaluation of peanut germplasm resources through whole-genome resequencing methods. Otyama et al. (2019) genotyped 107 peanut mini-core collections from the United States using a 58-K SNP array, revealing an average of 676 polymorphic SNPs per chromosome. These lines were employed for a genome-wide association study (GWAS) on traits related to arachidic and behenic fatty acid composition. In another study, 100 Korean peanut germplasms and 284 USDA core collections were analyzed for population relationships and GWAS, leading to the identification of five SNPs associated with seed ratio on different chromosomes (Zou et al., 2021). Additionally, researchers re-sequenced 203 peanut accessions from around the world for genetic diversity analysis. This study combined GWAS with transgenic experiments and RNA-seq to identify candidate genes, including FAX1 for seed weight and AhDPB2 for seed length (Liu et al., 2022). The application of genomic techniques in germplasm characterization and trait innovation has accelerated the uniform characterization and evaluation of peanut germplasm. As a result, the discovery and selection of suitable accessions or germplasms that correlate with specific traits can be leveraged in future peanut breeding programs to develop improved varieties (Raza et al., 2024).

21.5.2 Joint Analysis of Phenomics and Gene Functional Characterization

Phenomics is of great importance in genetic research. In recent years, phenomics research in crops has made great strides in generating multidimensional phenotypes at the cell, organ, plant, and population levels (Lobos et al., 2017). This includes measuring precise and automated phenotypic information on plant growth, structure, and various proportional components at high throughput. Sensing techniques allow detailed documentation of plant survival and their response to the environment in which they are located, which in turn responds to changes in their environment. These include temperature, humidity, intensity of sunlight, diseases, insects and weeds, soil properties and microorganisms, and so on (Steeneken et al., 2023). Those sensing environmental parameters are the basis for efficient improvement breeding and cultivation management. The natural variations, induced mutations, and quantitative trait loci with good alleles are suitable not only for discovering genetic causes, but also for improving mechanistic knowledge of the relationships between genotype and phenotype. Only a few studies have addressed peanut phenotype detection using spectral discrimination. The study reports that using leaf-level remote sensing and unmanned aerial vehicles (UAV) images can detect bacterial wilt (BW) disease at early disease stages and discriminate the BW severity levels in peanut (Chen et al. 2020).

Phenotyping efficiency is a major constraint to genetic development in breeding programs (Araus et al., 2018), especially in the area of plant high-throughput phenotyping, which is a bottleneck in conventional breeding, MAB, or GS (Crossa et al., 2017). High-quality phenotyping of plants is also necessary to evaluate the results of genetic modifications. This observation has prompted national, regional, and international programs to improve plant phenotyping resources. The development of more reliable, accurate, and cost-effective plant high-quality phenotyping will facilitate the prediction of peanut genotype performance under complex field conditions and can guide future peanut breeding. Currently, research on peanut functional genes remains in the developmental stage. Genomics-based functional genomics offers a novel approach for exploring peanut functional gene research. Research has focused on QTL identification, marker-assisted breeding, and genome-wide gene family identification. Over the past decade, researchers have identified many QTLs and candidate genes in peanut based on genetic mapping, QTL-seq, and GWAS, and have cloned many genes related to important agronomic traits. Mapping and cloning important genes for agronomic traits is the key to molecular breeding. To accurately conduct breeding work and greatly improve breeding efficiency, researchers must analyze the molecular mechanism of plant trait formation at the gene level. Despite the abundance of genetic, phenotypic, and functional genomics information, researchers need to fully exploit and integrate these methods to create a complete and automated genotyping and phenotyping platform. This platform will provide rich genetic information and data for molecular breeding in peanuts.

21.5.3 Genomics-Assisted Breeding (GAB) in Peanuts

Genomic breeding offers numerous possibilities for integrating various genomic tools and technologies. Varshney et al. (2005) proposed genomics-assisted breeding (GAB) by using genomic tools in breeding. This approach has produced several high-yielding, stress-tolerant, and nutritionally improved varieties (Varshney et al., 2021). In peanut, there are now several examples of the use of GAB that have resulted in molecular breeding products. It is therefore important to investigate the possibilities of genomic breeding, especially GAB, in peanut. In recent years, the use of GAB has enabled peanut researchers to increase the effectiveness of target trait selection for future breeding efforts (Pandey et al., 2020; Raza et al., 2024). The integration of genomic methods with HTP via molecular markers enables GAB to guide breeding experiments in predicting phenotype from genotype. With GAB, it is now possible to integrate numerous important traits into a single specific variety. Most importantly, breeders can effectively combine and integrate desirable traits, previously considered independent entities, into a single genotype. In addition, GAB is helping to exploit modern genomic resources and characterize allelic variation to drive the development of germplasm and cultivars over the last decade.

21.5.4 Genome Editing and Transformation System

In recent years, continuous research into genome editing technology and optimization of the gene transformation system has led to increases in crop improvement. These sequencing technologies use whole-genome insights to identify the genetic components that affect desired traits. This enables breeders to develop new varieties more quickly. Researchers worldwide have reported genetic transformation in peanuts. In most studies, researchers used GUS or GFP as the reporter gene base with the Agrobacterium-mediated transformation method. The selection of transformed explants was diverse and included the cotyledon, cotyledon nodules, apical meristem, hypocotyl, young leaves, immature embryos and cotyledons, and embryogenic callus (Krishna et al., 2015; Zhang et al., 2018).

Researchers have developed a rapid and stable genetic transformation system by injecting Agrobacterium tumefaciens into the second stem of peanuts to produce transgenic plants. They transplanted the positive transgenic peanut and covered the injection site with tamping compound. More than 40% of the transgenic plants obtained showed chimeras in the T0 generation, but about 9% of the T1 generation were hybrids rather than chimeras (Zhai et al., 2022). They achieved transgenic peanut resistance to peanut stem necrosis caused by TSV by transferring the TSV coat protein gene through Agrobacterium-mediated transformation. Integration of the transgene in generations T1, T2, and T3 was confirmed by PCR with gene-specific primers (Mehta et al., 2013). The Arabidopsis homeodomain leucine zipper transcription factor (AtHDG11) expressed in transgenic peanut plants with inducible rd29A promoter. Stress-inducible expression of AtHDG11 in three independent homozygous transgenic peanut lines led to improvements in drought and salt tolerance by upregulating known stress-responsive genes (Banavath et al., 2018). The transgenic peanut lines expressing AtDREB1A exhibited enhanced tolerance to soil moisture deficit stress, as evidenced by their significantly lower accumulation of malondialdehyde and hydrogen peroxide compared to the wild type (Bhalani et al., 2019). A fatty acid elongase gene, KCS1, was isolated from the drought-tolerant cultivar K-9 and overexpressed in the drought-sensitive peanut cultivar K-6 control by constitutive CaMV35S promoter. Transgenic peanut overexpressing AhKCS1 showed normal growth and water deficiency conditions (Lokesh et al., 2018). There are also reports on peanut gene function that rely on the hairy root transformation system (Liu et al., 2018; Su et al., 2021). This has limited the observation of many important genes for agronomic traits. It is clear that genetic transformation in peanut is still very difficult and lags far behind crops such as soybean, rice, and maize. The complex genome of tetraploid varieties is one of the most important factors complicating genetic transformation in peanuts. Researchers anticipate overcoming these bottlenecks through continued research and the development of new methods and tools to realize the potential of CRISPR/Cas9 for genome editing in peanuts. The CRISPR system could also introduce beneficial traits from wild peanut relatives into cultivated peanuts to develop future smart varieties.

Smart breeding is a new opportunity for the development of science and technology through Big Data, artificial intelligence, and integrated genomic–enviromic prediction (Xu et al., 2022). It is foreseeable that the development pace of smart breeding will become the core competitive aspect of the peanut breeding industry in the next 10 years. However, in the transition from traditional breeding to molecular breeding and then to smart breeding, increasingly intelligent technologies are being used in the genotypic analysis of individuals in the laboratory and fewer individuals in field trials. As a result, breeding is more predictable, accurate, efficient, economical, and socially and environmentally friendly.

21.6 CONCLUSION AND FUTURE DIRECTIONS

With the widespread application of next-generation sequencing technology and the development of high-throughput and high-resolution genome-wide association analysis technologies, genomics research has improved dramatically. In the big data era, more and more scientists are focusing on using these massive data to explore fundamental biological problems such as genetic composition, evolution mechanisms, and environmental adaptation mechanisms. Future breeding will gradually evolve from traditional population-based breeding to smart breeding with the implementation of genome editing technology, big data, artificial intelligence, and integrated genomic–enviromic prediction (not only genome editing). At the same time, researchers will closely integrate genome-editing technology with other molecular design breeding techniques to improve peanut yield, quality, and disease resistance. Consequently, modern breeding techniques that utilize genome-wide marker data are increasingly suitable for continuously improving populations and enhancing the rate of genetic gain. Furthermore, advancements in genome editing have significantly enhanced the ability to modify peanut genomes.

REFERENCES

Ahmad, N., L. Hou, J. J. Ma, et al. "Bulk RNA-Seq Analysis Reveals Differentially Expressed Genes Associated with Lateral Branch Angle in Peanut." *Genes* 13 (2022). https://doi.org/10.3390/genes13050855.

Araus, J. L., S. C. Kefauver, M. Zaman-Allah, et al. "Translating High-Throughput Phenotyping into Genetic Gain." *Trends in Plant Science* 23 (2018): 451–466.

Bai, L. X., and S. W. Wang. "Characteristics and High Yield Cultivation Techniques of New Peanut Variety Shanhua8." *Shandong Agricultural Sciences* 40, no. 3 (2008): 117–118.

Banavath, J. N., T. Chakradhar, V. Pandit, et al. "Stress-Inducible Overexpression of AtHDG11 Leads to Improved Drought and Salt Stress Tolerance in Peanut (*Arachis hypogaea* L.)." *Frontiers in Chemistry* 6 (2018): 34. https://doi.org/10.3389/fchem.2018.00034.

Bera, S. K., J. H. Kamdar, S. V. Kasundra, et al. "Improving Oil Quality by Altering Levels of Fatty Acids through Marker-assisted Selection of ahfad2 alleles in Peanut (Arachis hypogaea L.)." *Euphytica* 214 (2018): 162.

Bertioli, D. J., S. B. Cannon, L. Froenicke, et al. "The Genome Sequences of *Arachis duranensis* and *Arachis ipaensis*, the Diploid Ancestors of Cultivated Peanut." *Nature Genetics* 48, no. 4 (2016): 438–446. https://doi.org/10.1038/ng.3517.

Bertioli, D. J. Jenkins, J. Clevenger, et al. "The Genome Sequence of Segmental Allotetraploid Peanut *Arachis hypogaea*." *Nature Genetics* 51, no. 5 (2019): 877–884. https://doi.org/10.1038/s41588-019-0405-z.

Bhalani, H., R. Thankappan, G. P. Mishra, et al. "Regulation of Antioxidant Mechanisms by AtDREB1A Improves Soil-Moisture Deficit Stress Tolerance in Transgenic Peanut (*Arachis hypogaea* L.)." *PLoS One* 14, no. 5 (2019): e0216706. https://doi.org/10.1371/journal.pone.0216706.

Branch, W. D. "Registration of 'Georgia-02C' Peanut." *Crop Science* 43, no. 5 (2003): 1883–1884. https://doi.org/10.2135/cropsci2003.1883.

Burow, M. D., C. E. Simpson, M. W. Faries, et al. "Molecular Biogeographic Study of Recently Described B- and A-Genome *Arachis* Species, Also Providing New Insights into the Origins of Cultivated Peanut." *Genome* 52, no. 2 (2009): 107–119. https://doi.org/10.1139/G08-110.

Burow, M. D., C. E. Simpson, A. H. Paterson, et al. "Identification of Peanut (*Arachis hypogaea* L.) RAPD Markers Diagnostic of Root-Knot Nematode (*Meloidogyme arenaria* (Neal) Chitwood) Resistance." *Molecular Breeding* 2, no. 4 (1996): 369–379. https://doi.org/10.1007/BF00564204.

Cason, J. M., C. E. Simpson, M. D. Burow, et al. "Use of Wild and Exotic Germplasm for Resistance in Peanut." *Journal of Plant Registrations* 17, no. 1 (2023): 1–25.

Chaudhari, S., D. Khare, S. C. Patil, et al. "Genotype × Environment Studies on Resistance to Late Leaf Spot and Rust in Genomic Selection Training Population of Peanut (*Arachis hypogaea* L.)." *Frontiers in Plant Science* 10 (2019): 1338.

Chen, T., W. Yang, H. Zhang, et al. "Early Detection of Bacterial Wilt in Peanut Plants through Leaf-Level Hyperspectral and Unmanned Aerial Vehicle Data." *Computers and Electronics in Agriculture* 177 (2020): 105708. https://doi.org/10.1016/j.compag.2020.105708.

Chen, Z., M. L. Wang, N. A. Barkley, et al. "A Simple Allele-Specific PCR Assay for Detecting AFAD2 Alleles in Both A and B Genomes of the Cultivated Peanut for High-Oleate Trait Selection." *Plant Molecular Biology Reporter* 28, no. 4 (2010): 542–548. https://doi.org/10.1007/s11105-010-0172-1.

Chu, Y., P. Chee, A. Culbreath, et al. "Major QTLs for Resistance to Early and Late Leaf Spot Diseases Are Identified on Chromosomes 3 and 5 in Peanut (*Arachis hypogaea*)." *Frontiers in Plant Science* 10 (2019): 883. https://doi.org/10.3389/fpls.2019.00883.

Chu, Y., C. C. Holbrook, and P. Ozias-Akins. "Two Alleles of ahFAD2B Control the High Oleic Acid Trait in Cultivated Peanut." *Crop Science* 49, no. 6 (2009): 2029–2036. https://doi.org/10.2135/cropsci2009.04.0193.

Clevenger, J., Y. Chu, C. Chavarro, et al. "Mapping Late Leaf Spot Resistance in Peanut (*Arachis hypogaea* L.) Using QTL-Seq Reveals Markers for Marker-Assisted Selection." *Frontiers in Plant Science* 9 (2018): 83. https://doi.org/10.3389/fpls.2018.00083.

Crossa, J., P. Pérez-Rodríguez, J. Cuevas, et al. "Genomic Selection in Plant Breeding: Methods, Models, and Perspectives." *Trends in Plant Science* 22, no. 11 (2017): 961–975.

Cuc, L. M., E. S. Mace, J. H. Crouch, et al. "Isolation and Characterization of Novel Microsatellite Markers and Their Application for Diversity Assessment in Cultivated Groundnut (*Arachis hypogaea*)." *BMC Plant Biology* 8, no. 1 (2008): 55. https://doi.org/10.1186/1471-2229-8-55.

Deng, Z. L. Y., J. J. Song, Y. Wang, et al. "Progress of Development and Creating Routes of Peanut Mutants." *Chinese Agricultural Science Bulletin* 29, no. 21 (2013): 12–15.

Dodo, H. W., K. N. Konan, F. C. Chen, et al. "Alleviating Peanut Allergy Using Genetic Engineering: The Silencing of the Immunodominant Allergen Arah2 Leads to Its Significant Reduction and a Decrease in Peanut Allergenicity." *Plant Biotechnology Journal* 6, no. 2 (2008): 135–145. https://doi.org/10.1111/j.1467-7652.2007.00305.x.

Fernandez, A., and A. Krapovickas. "Cromosomas y Evolución en *Arachis* (Leguminosae)." *Bonplandia* 8, no. 1 (1994): 187–220.

Gangurde, S. S., H. Wang, S. Yaduru, et al. "Nested-Association Mapping (NAM)-Based Genetic Dissection Uncovers Candidate Genes for Seed and Pod Weights in Peanut (*Arachis hypogaea*)." *Plant Biotechnology Journal* 18, no. 7 (2020): 1457–1471. https://doi.org/10.1111/pbi.13307.

Guan, J., X. Y. Jiang, X. L. Huang, et al. "Demonstration and Cultivation Technology of the New Peanut Variety 'Hang Hua2'." *Anhui Agricultural Science Bulletin* 20, no. 11 (2014): 48–49.

Han, S. Y., X. M. Zhou, L. Shi, et al. "AhNPR3 Regulates the Expression of WRKY and PR Genes and Mediates the Immune Response of the Peanut (*Arachis hypogaea* L.)." *The Plant Journal* 110 (2022): 735–747.

He, Y., S. Mu, Z. He, et al. "Ectopic Expression of MYB Repressor GmMYB3a Improves Drought Tolerance and Productivity of Transgenic Peanuts (*Arachis hypogaea* L.) under Conditions of Water Deficit." *Transgenic Research* 29 (2020): 563–574.

Jung, S., D. Swift, E. Sengoku, et al. "The High Oleate Trait in the Cultivated Peanut (Arachis hypogaea L.). Isolation and Characterization of Two Genes Encoding Microsomal Oleoyl-PC Desaturases." *Molecular and General Genetics* 263 (2000): 796–805.

Kim, K.-S., D. Lee, S. B. Bae, et al. "Development of SNP-Based Molecular Markers by Re-Sequencing Strategy in Peanut." *Plant Breeding and Biotechnology* 5, no. 4 (2017): 325–333.

Kochert, G., T. Halward, W. D. Branch, et al. "RFLP Variability in Peanut (*Arachis hypogaea* L.) Cultivars and Wild Species." *Theoretical and Applied Genetics* 81, no. 5 (1991): 565–570. https://doi.org/10.1007/BF00226726.

Kochert, G., H. T. Stalker, M. Gimenes, et al. "RFLP and Cytogenetic Evidence on the Origin and Evolution of Allotetraploid Domesticated Peanut, *Arachis hypogaea* (Leguminosae)." *American Journal of Botany* 83, no. 10 (1996): 1282–1291. https://doi.org/10.1002/j.1537-2197.1996.tb13926.x.

Kolekar, R. M., M. Sukruth, K. Shirasawa, et al. "Marker-Assisted Backcrossing to Develop Foliar Disease-Resistant Genotypes in TMV 2 Variety of Peanut (*Arachis hypogaea* L.)." *Plant Breeding* 136, no. 6 (2017): 948–953. https://doi.org/10.1111/pbr.12541.

Krishna, G., B. K. Singh, E. K. Kim, et al. "Progress in Genetic Engineering of Peanut (*Arachis hypogaea* L.)—A Review." *Plant Biotechnology Journal* 13, no. 2 (2015): 147–162. https://doi.org/10.1111/pbi.12251.

Lavarenne, J., S. Guyomarc'h, C. Sallaud, et al. "The Spring of Systems Biology-Driven Breeding." *Trends in Plant Science* 23, no. 9 (2018): 706–720. https://doi.org/10.1016/j.tplants.2018.06.003.

Lavia, G. I., A. M. Ortiz, G. Robledo, et al. "Origin of Triploid *Arachis pintoi* (Leguminosae) by Autopolyploidy Evidenced by FISH and Meiotic Behaviour." *Annals of Botany* 108, no. 1 (2011): 103–111. https://doi.org/10.1093/aob/mcr075.

Liu, H., D. X. Hu, P. X. Du, et al. "Single-Cell RNA-Seq Describes the Transcriptome Landscape and Identifies Critical Transcription Factors in the Leaf Blade of the Allotetraploid Peanut (*Arachis hypogaea* L.)." *Plant Biotechnology Journal* 19, no. 11 (2021): 2261–2276. https://doi.org/10.1111/pbi.13618.

Liu, L. F., P. M. Dang, and C. Y. Chen. "Development and Utilization of InDel Markers to Identify Peanut (*Arachis hypogaea*) Disease Resistance." *Frontiers in Plant Science* 6 (2015): 988. https://doi.org/10.3389/fpls.2015.00988.

Liu, X., L. Li, M. Li, et al. "AhGLK1 Affects Chlorophyll Biosynthesis and Photosynthesis in Peanut Leaves during Recovery from Drought." *Scientific Reports* 8 (2018): 2250. https://doi.org/10.1038/s41598-018-20750-0.

Liu, X., L. M. Li, B. H. Zhang, et al. "AhHDA1-Mediated AhGLK1 Promoted Chlorophyll Synthesis and Photosynthesis Regulates Recovery Growth of Peanut Leaves after Water Stress." *Plant Science* 394 (2020): 110461. https://doi.org/10.1016/j.plantsci.2020.110461.

Liu, Y. Y., L. B. Shao, J. Zhou, et al. "Genomic Insights into the Genetic Signatures of Selection and Seed Trait Loci in Cultivated Peanut." *Journal of Advanced Research* 42 (2022): 237–248. https://doi.org/10.1016/j.jare.2021.12.013.

Lobos, G. A., A. V. Camargo, A. del Pozo, et al. "Editorial: Plant Phenotyping and Phenomics for Plant Breeding." *Frontiers in Plant Science* 8 (2017): 706. https://doi.org/10.3389/fpls.2017.00706.

Lokesh, U., B. Venkatesh, K. Kiranmai, et al. "Overexpression of ss-Ketoacyl Co-A Synthasel Gene Improves Tolerance of Drought Susceptible Groundnut (*Arachis hypogaea* L.) Cultivar K-6 by Increased Leaf Epicuticular Wax Accumulation." *Frontiers in Plant Science* 9 (2018): 1869. https://doi.org/10.3389/fpls.2018.01869.

Marka, R., and R. S. Nanna. "Expression of Tcchitinase-I gene in Transgenic Peanut (*Arachis hypogaea* L.) Confers Enhanced Resistance Against Leaf Spot and Rust Diseases." *Plant Growth Regulation* 93 (2021): 53–63.

Mehta, R., T. Radhakrishnan, A. Kumar, et al. "Coat Protein-Mediated Transgenic Resistance of Peanut (*Arachis hypogaea* L.) to Peanut Stem Necrosis Disease through Agrobacterium-Mediated Genetic Transformation." *Indian Journal of Virology* 24, no. 2 (2013): 205–213. https://doi.org/10.1007/s13337-013-0131-6.

Milla, S. R., T. G. Isleib, and H. T. Stalker. "Taxonomic Relationships among *Arachis* Sect. *Arachis* Species as Revealed by AFLP Markers." *Genome* 48, no. 1 (2005): 1–11. https://doi.org/10.1139/g04-098.

Moretzsohn, M. C., E. G. Gouvea, P. W. Inglis, et al. "A Study of the Relationships of Cultivated Peanut (*Arachis hypogaea*) and Its Most Closely Related Wild Species Using Intron Sequences and Microsatellite Markers." *Annals of Botany* 111, no. 1 (2013): 113–126. https://doi.org/10.1093/aob/mcs245.

Mou, Z. M., and D. K. Zhao. "Gene Rational Design: The Dawn of Crop Breeding." *Trends in Plant Science* 27, no. 7 (2022): 633–636. https://doi.org/10.1016/j.tplants.2022.03.009.

Otyama, P. I., A. Wilkey, R. Kulkarni, et al. "Evaluation of Linkage Disequilibrium, Population Structure, and Genetic Diversity in the U.S. Peanut Mini Core Collection." *BMC Genomics* 20, no. 1 (2019): 481. https://doi.org/10.1186/s12864-019-5872-4.

Pandey, M. K., A. K. Pandey, R. Kumar, et al. "Translational Genomics for Achieving Higher Genetic Gains in Groundnut." *Theoretical and Applied Genetics* 133 (2020): 1679–1702.

Pandey, M. K., S. Chaudhari, D. Jarquin, et al. "Genome-based Trait Prediction in Multi- environment Breeding Trials in Groundnut." *Theoretical and Applied Genetics* 133 (2020): 3101–3117.

Patel, K. G., R. Thankappan, G. P. Mishra, et al. "Transgenic Peanut (*Arachis hypogaea* L.) Overexpressing mtlD Gene Showed Improved Photosynthetic, Physio-Biochemical, and Yield-Parameters under Soil-Moisture Deficit Stress in Lysimeter System." *Frontiers in Plant Science* 8 (2017): 1881.

Qiu, Q. S., R. R. Lu, and D. Z. Zhang. "Breeding and Application of Luhua6, a New Early Maturity Peanut Variety." *Nuclear Agronomy Bulletin* 1988, no. 05 (1988): 208–210.

Raina, S. N., V. Rani, T. Kojima, et al. "RAPD and ISSR Fingerprints as Useful Genetic Markers for Analysis of Genetic Diversity, Varietal Identification, and Phylogenetic Relationships in Peanut (*Arachis hypogaea*) Cultivars and Wild Species." *Genome* 44, no. 5 (2001): 763–772. https://doi.org/10.1139/gen-44-5-763.

Ramos, M. L., G. Fleming, Y. Chu, et al. "Chromosomal and Phylogenetic Context for Conglutin Genes in *Arachis* Based on Genomic Sequence." *Molecular Genetics and Genomics* 275, no. 6 (2006): 578–592. https://doi.org/10.1007/s00438-006-0108-3.

Rathod, V., R. Hamid, R. S. Tomar, et al. "Comparative RNA-Seq Profiling of a Resistant and Susceptible Peanut (*Arachis hypogaea*) Genotype in Response to Leaf Rust Infection Caused by *Puccinia arachidis*." *3 Biotech* 10 (2020): 486. https://doi.org/10.1007/s13205-020-02462-z.

Raza, A., H. Chen, C. Zhang, et al. "Designing Future Peanut: The Power of Genomics-assisted Breeding." *Theoretical and Applied Genetics* 137, no. 3 (2024): 66.

Robledo, G., and G. Seijo. "Species Relationships among the Wild B Genome of *Arachis* Species (Section *Arachis*) Based on FISH Mapping of rDNA Loci and Heterochromatin Detection: A New Proposal for Genome Arrangement." *Theoretical and Applied Genetics* 121, no. 6 (2010): 1033–1046. https://doi.org/10.1007/s00122-010-1372-6.

Rohini, V. K., and R. K. Sankara. "Transformation of Peanut (*Arachis hypogaea* L.) with Tobacco Chitinase Gene: Variable Response of Transformants to Leaf Spot Disease." *Plant Science* 160, no. 5 (2001): 889–898. https://doi.org/10.1016/s0168-9452(01)00332-8.

Samoluk, S. S., G. Robledo, M. Podio, et al. "First Insight into Divergence, Representation, and Chromosome Distribution of Reverse Transcriptase Fragments from L1 Retrotransposons in Peanut and Wild Relative Species." *Genetica* 143, no. 1 (2015): 113–125. https://doi.org/10.1007/s10709-015-9820-5.

Sarkar, T., R. Thankappan, A. Kumar, et al. "Stress Inducible Expression of AtDREB1A Transcription Factor in Transgenic Peanut (*Arachis hypogaea* L.) Conferred Tolerance to Soil-Moisture Deficit Stress." *Frontiers in Plant Science* 7 (2016): 935.

Seijo, G., G. I. Lavia, A. Fernández, et al. "Genomic Relationships between the Cultivated Peanut (*Arachis hypogaea*, Leguminosae) and Its Close Relatives Revealed by Double GISH." *American Journal of Botany* 94, no. 12 (2007): 1963–1971. https://doi.org/10.3732/ajb.94.12.1963.

Seijo, G. J., G. I. Lavia, A. Fernández, et al. "Physical Mapping of the 5S and 18S-25S rRNA Genes by FISH as Evidence That *Arachis duranensis* and *A. ipaensis* Are the Wild Diploid Progenitors of *A. hypogaea* (Leguminosae)." *American Journal of Botany* 91, no. 9 (2004): 1294–1303. https://doi.org/10.3732/ajb.91.9.1294.

Singh, A. K., and J. P. Moss. "Utilization of Wild Relatives in Genetic Improvement of *Arachis hypogaea* L. 2. Chromosome Complements of Species Section *Arachis*." *Theoretical and Applied Genetics* 61, no. 4 (1982): 305–314. https://doi.org/10.1007/BF00264697.

Singh, A. K., and J. P. Moss. "Utilization of Wild Relatives in the Genetic Improvement of *Arachis hypogaea* L.: 5. Genome Analysis in Section *Arachis* and Its Implications in Gene Transfer." *Theoretical and Applied Genetics* 68, no. 4 (1984): 355–364. https://doi.org/10.1007/BF00251185.

Sivaram, M. R. Co. "A New High Yielding Mutant Variety of Groundnut." *Mutation Breeding Newsletter* 25 (1985): 5–6.

Smartt, J. "The Genomes of *Arachis hypogaea* 2. The Implications in Interspecific Breeding." *Euphytica* 27 (1978): 677–680. https://doi.org/10.1007/BF00043157.

Steeneken, P. G., E. Kaiser, G. J. Verbiest, et al. "Sensors in Agriculture: Towards an Internet of Plants." *Nature Reviews Methods Primers* 3 (2023): 60.

Su, L., S. Liu, X. Liu, et al. "Transcriptome Profiling Reveals Histone Deacetylase 1 Gene Overexpression Improves Flavonoid, Isoflavonoid, and Phenylpropanoid Metabolism in *Arachis hypogaea* Hairy Roots." *PeerJ* 9 (2021): e10976. https://doi.org/10.7717/peerj.10976.

Tiwari, S., D. K. Mishra, A. Singh, et al. "Expression of a Synthetic cry1EC Gene for Resistance against *Spodoptera litura* in Transgenic Peanut (*Arachis hypogaea* L.)." *Plant Cell Reports* 27, no. 6 (2008): 1017–1025. https://doi.org/10.1007/s00299-008-0512-7.

Varshney, R. K., A. Bohra, J. Yu, et al. "Designing Future Crops: Genomics-Assisted Breeding Comes of Age." *Trends in Plant Science* 26, no. 6 (2021): 631–649.

Varshney, R. K., A. Graner, and M. E. Sorrells. "Genomics-assisted Breeding for Crop Improvement." *Trends Plant Science* 10, no. 12 (2005): 621–630.

Varshney, R. K., M. K. Pandey, P. Janila, et al. "Marker-assisted Introgression of a QTL Region to Improve Rust Resistance in Three Elite and Popular Varieties of Peanut (Arachis hypogaea L.)." *Theoretical and Applied Genetics* 127 (2014): 1771–1781.

Venkatesh, B., A. R. Vennapusa, N. J. Kumar, et al. "Co-expression of Stress-responsive Regulatory Genes, MuNAC4, MuWRKY3 and MuMYB96 Associated with Resistant-traits Improves Drought Adaptation in Transgenic Groundnut (*Arachis hypogaea* L.) Plants." *Frontiers in Plant Science* 13 (2022): 1055851.

Wang, C. T., Y. Y. Tang, J. Z. Xu, et al. "EMS-Induced Variations in Pod Characters of Peanut." *Nuclear Agronomy Bulletin* 145, no. 2 (2008): 5–9.

Wang, C. T., X. Z. Wang, and Y. Y. Tang. "High Yielding Mutants Achieved by Injecting EMS into Peanut Flower Organs." *Journal of Nuclear Agricultural Sciences* 24, no. 2 (2010): 239–242.

Wang, C. T., X. D. Yang, and D. X. Chen. "Production of Extra-Large Podded and Small Podded Peanut." *Journal of Peanut Science* 31, no. 4 (2002): 5–8.

Wang, J., C. J. Li, C. X. Yan, et al. "A Comparative Analysis of the Complete Chloroplast Genome Sequences of Four Peanut Botanical Varieties." *PeerJ* 6 (2018): e5671. https://doi.org/10.7717/peerj.5671.

Wang, J., L. Shi, Y. Liu, et al. "Development of Peanut Varieties with High Oil Content by In Vitro Mutagenesis and Screening." *Journal of Integrative Agriculture* 19, no. 12 (2020): 2974–2982. https://doi.org/10.1016/S2095-3119(19)62879-6.

Wang, J., J. Sui, Y. Xie, et al. "Generation of Peanut Mutants by Fast Neutron Irradiation Combined with In Vitro Culture." *Journal of Radiation Research* 56, no. 3 (2015): 437–445. https://doi.org/10.1093/jrr/rrv002.

Wang, X., L. Liu, L. Qiao, et al. "Breeding Peanut Variety Yuhua 7 by Fast Neutron Irradiation and Tissue Culture." *Chinese Journal of Biotechnology* 35, no. 2 (2019): 270–280. https://doi.org/10.13345/j.cjb.190020.

Wang, X. J., and X. Y. Zhang, eds. "Study on Peanut Biotechnology." In *Induction and Application of Peanut Mutants*, 222. Chapter 7, Science Press, 2015.

Wang, Y., L. X. Qiao, X. L. Wu, et al. "Effect of In Vitro Mutagenesis with Pingyangmycin and NaCl-Directed Screening on Production and Quality of Offspring in Peanut." *Acta Agriculturae Boreali-Sinica* 30, no. 1 (2015): 202–206.

Wang, Y., X. G. Zhang, H. M. Li, et al. "Agronomic Traits and Quality Analysis of Peanut EMS Mutation Progeny." *Journal of Plant Genetic Resources* 16, no. 4 (2015): 914–919+926.

Wu, L. R., J. Chen, Y. Q. Shi, et al. "Breeding of Huayu 22 by 60Co γ-Rays Mutagenesis Combined with Hybridization." *Journal of Nuclear Agricultural Sciences* 20 (2006): 309–311.

Xu, Y., X. Zhang, H. Li, et al. "Smart Breeding Driven by Big Data, Artificial Intelligence, and Integrated Genomic-enviromic Prediction." *Molecular Plant* 15 (2022): 1664–1695.

Yin, D. M., C. M. Ji, X. L. Ma, et al. "Genome of an Allotetraploid Wild Peanut *Arachis monticola*: A De Novo Assembly." *GigaScience* 7 (2018): 1–9. https://doi.org/10.1093/gigascience/giy127.

Yin, D. M., G. Q. Tai, H. T. Yang, et al. "MS Mutagenesis and Molecular Identification of Peanut Mutant." *Chinese Agricultural Science Bulletin* 25, no. 5 (2009): 53–56.

Yu, M., M. Sun, Y. Guo, et al. "Breeding New Peanut Line with High Oleic Acid Content Using Backcross Method." *Acta Agronomica Sinica* 43, no. 6 (2017): 855–861.

Yu, S. L., C. T. Wang, J. M. Zhang, et al. *Analysis of Peanut Varieties and Their Pedigree in China*. Shanghai Science and Technology Press, 2008.

Zhai, Q., R. Q. Chen, X. H. Liang, et al. "Establishment and Application of a Rapid Genetic Transformation Method for Peanut." *Chinese Bulletin of Botany* 57 (2022): 327–339. https://doi.org/10.3969/j.issn.1674-3466.2022.03.005.

Zhang, J., H. Zhang, J. R. Botella, et al. "Generation of New Glutinous Rice by CRISPR/Cas9-Targeted Mutagenesis of the Waxy Gene in Elite Rice Varieties." *Journal of Integrative Plant Biology* 60, no. 5 (2018): 369–375. https://doi.org/10.1111/jipb.12620.

Zhao, H. L., R. Z. Tian, H. Xia, et al. "High-Density Genetic Variation Map Reveals Key Candidate Loci and Genes Associated with Important Agronomic Traits in Peanut." *Frontiers in Genetics* 13 (2022): 845602. https://doi.org/10.3389/fgene.2022.845602.

Zhao, K., R. Ren, X. L. Ma, et al. "Genome-Wide Investigation of Defensin Genes in Peanut (*Arachis hypogaea* L.) Reveals AhDef2.2 Conferring Resistance to Bacterial Wilt." *The Crop Journal* 10 (2022): 809–819. https://doi.org/10.1016/j.cj.2021.11.001.

Zhao, M., J. Sui, L. Qiao, et al. "An Efficient Regeneration System for Peanut: Somatic Embryogenesis from Embryonic Leaflets." *Journal of Food, Agriculture & Environment* 10 (2012): 527–531.

Zhao, Y., C. Zhang, H. Chen, et al. "QTL Mapping for Bacterial Wilt Resistance in Peanut (Arachis hypogaea L.)." *Molecular Breeding* 36 (2016): 13.

Zhao, Y., J. Ma, M. Li, et al. "Whole-Genome Resequencing-Based QTL-Seq Identified AhTc1 Gene Encoding a R2R3-MYB Transcription Factor Controlling Peanut Purple Testa Colour." *Plant Biotechnology Journal* 18, no. 1 (2020): 96–105. https://doi.org/10.1111/pbi.13178.

Zhu, H., Y. N. Jiang, Y. Guo, et al. "A Novel Salt Inducible WRKY Transcription Factor Gene, AhWRKY75, Confers Salt Tolerance in Transgenic Peanut." *Plant Physiology and Biochemistry* 160 (2021): 175–183. https://doi.org/10.1016/j.plaphy.2021.01.021.

Zhuang, W., H. Chen, M. Yang, et al. "The Genome of Cultivated Peanut Provides Insight into Legume Karyotypes, Polyploid Evolution, and Crop Domestication." *Nature Genetics* 51, no. 5 (2019): 865–876. https://doi.org/10.1038/s41588-019-0402-2.

Zou, K. Y., K. S. Kim, K. Kim, et al. "Genetic Diversity and Genome-Wide Association Study of Seed Aspect Ratio Using a High-Density SNP Array in Peanut (*Arachis hypogaea* L.)." *Genes* 12, no. 8 (2021): 1181. https://doi.org/10.3390/genes12081181.

Index